Amorphous and Nano Alloys Electroless Depositions

Technology, Composition, Structure and Theory

Amorphous and Nano Alloys Electroless Depositions

Technology, Composition, Structure and Theory

Bangwei Zhang
College of Physics
Hunan University
Changsha, China

ELSEVIER
AMSTERDAM • BOSTON • HEIDELBERG
LONDON • NEW YORK • OXFORD
PARIS • SAN DIEGO • SAN FRANCISCO
SINGAPORE • SYDNEY • TOKYO
Chemical Industry Press

Elsevier
Radarweg 29, PO Box 211, 1000 AE Amsterdam, Netherlands
The Boulevard, Langford Lane, Kidlington, Oxford OX5 1GB, UK
225 Wyman Street, Waltham, MA 02451, USA

British Library Cataloguing-in-Publication Data
A catalogue record for this book is available from the British Library

Library of Congress Cataloging-in-Publication Data
A catalog record for this book is available from the Library of Congress

ISBN: 978-0-12-802685-4

For information on all Elsevier publications
visit our website at http://store.elsevier.com/

Publisher: Jonathan Simpson
Acquisition Editor: Simon Tian
Editorial Project Manager: Naomi Robertson
Production Project Manager: Julie-Ann Stansfield
Designer: Mark Rogers

Typeset by Thomson Digital

This book is dedicated to my parents
Zhang Zhongshu and Liu Yuanxiu.
An ordinary Chinese couple both of whom
used their lifetime to foster their three sons
and three daughters.

Contents

Part II
Technology of Electroless Plating – Plating Bath, Critical Parameters, Deposition Rate, and Stability of Plating Bath

2. Electroless Plating Baths of Metals, Binary Alloys, and Multicomponent Alloys

3. Electroless Composite Plating

Preface

The electroless plating (EP) covered in this book is somewhat different from the original usage of the term by Brenner. Brenner's use of the term electroless plating originally meant autocatalytic plating [1]. The term electroless plating, as used in this book, is not limited to the autocatalytic deposition but includes all chemical plating. So, the definition of electroless plating, as used in this book, involves immersing a substrate/workpiece into a bath solution, a metal/alloy or compound layer is then deposited on it via the transfer of electrons between reacting chemical species during the chemical reactions (redox reaction) without any applied current. It should be pointed out that displacement deposition is included in the electroless plating [2–4]. Displacement deposition is also known as galvanic or immersion plating. A more noble metal, that is the redox potential of a metal ion in solution, because it is more positive always deposits on the surface of a less noble metal in the galvanic displacement deposition. The less noble metal acts as a reducing agent. The difference between electroless plating and galvanic displacement deposition is that a reducing agent is necessary in former case, but is not required in latter case. That is why galvanic displacement deposition is simpler and faster [5–8].

There are a lot of advantages of EP compared to traditional electrolytic deposition, such as:

- No external electric source, including rectifiers, batteries or anodes is needed so that the equipment needed for electroless plating is rather simple.
- Excellent uniformity and less porous deposits are produced, even on complex components. On the other hand, there exists the phenomenon of current density concentration at peaks and protrusions in the electrolytic plating process. So, the films produced by electrolytic plating are not always uniform, there are nearly no deposits at protrusions or in the recesses. In addition, the films always have pinholes, which make the deposits susceptible to rust.
- The electroless deposition process is easy to use in volume production.
- Deposits can be produced directly on non-conductive materials such as ceramics and plastics and other so-called hard plating substrates.
- Deposits usually have unique properties, which have applications in many industries.

Compared to the traditional electrolytic deposition process, there are also some drawbacks to electroless plating:

- The EP baths are usually rather more complex than electrolytic deposition baths, so they are more expensive than electrolytic deposition.
- Compared with the electrolytic plating process, EP has a limited life, and therefore the composition of consumption in the bath should be often added, especially during the process of the bulk production.
- Electroless processes result in a higher waste treatment burden than that of electrolytic processes, which causes more troubles of waste solution treatment.

Since electroless plating was discovered by Brenner and Riddell in 1946, nearly seventy years have passed. EP is highly developed today, and is used in most parts of industry. Many basic subjects of EP, such as the improvement of bath solutions and new bath solutions, new alloy deposition, the extension of plating substrates, the mechanisms, especially the theory of electroless plating, nano electroless deposition, properties of in-depth study, and so forth have been studied and gratifying progress has been made. Considerable advances in such issues have been made in recent years, especially since the beginning of the new century. This can be corroborated from the fact that many papers were published in recent years. If you put "electroless plating" into the Elsevier data base, as of November 1, 2014, 10,565 papers can be found. The distribution of these articles is that 309 was published up to 1979; 3,323 from 1980 to 1999; 4,230 from 2000 to 2009; and 2,823 from 2009 to November 1, 2014. If you search within the results using "nano plating", then 1,610 articles can be found up to January 23, 2013, of which only 1 article was published up to 1979; another one was published between 1980 and 1989; 55 from 1990 to 1999; 885 from 2000 to 2009; and 660 from 2010 to January 23, 2013. That is to say, only a few papers on nano electroless plating were published in various journals included in the Elsevier data base before 2000, and most of them were been published after 2001.

Few books on electroless plating have been published, for example, <Electroless plating: fundamentals and applications> by G. O. Mallory and J. B. Hajdu in 1990, and < Electroless nickel plating> by Wolfgang Riedel in 1991. It is obvious that these books can't include recent developments in the materials used in electroless plating because they were published twenty years ago. Some books in electroless plating have been published in Chinese in recent years, in which nearly no nano electroless plating knowledge is included and not too many English papers in electroless plating were cited in any of them. One can see that there is an obvious gap in the existing literature that needs to be filled. It hoped that this book will fill the gap in knowledge left by existing electroless books.

Materials scientists, after studying the subject carefully enough, can streamline a thick metallurgy book into only few words: constituents, structures and properties; how to make steel or alloy material and their applications are not usually included in the contents of a metallurgy book. Similarly, a book on electroless plating can be streamlined into only few words as well: technology (manufacture process), constituents, structures and properties. Usually, how to manufacture electroless plating alloys is closely related to the contents of a book in electroless plating. Most electroless deposition books also include applications of electroless plating alloys. However in order to concentrate the contents of this book in a similar way to a common metallurgy book, and to avoid a lengthy of book, the properties and applications of EP will be not included in this book. In this way, this book will cover manufacturing technology plus the metallurgy of electroless plating alloys. In addition, in order to concentrate the fundamental content and theory of electroless plating, the subjects of pretreatment for electroless plating, post-treatment of electroless deposits, quality control of electroless deposits, stripping of electroless coating, treatment of waste bath solution, and equipment for electroless plating, which are usually included in other electroless plating books will not be included in this book.

There are four parts included in the book: 1, the history of electroless plating; 2, the technology and manufacturing principles of electroless plating: plating bath, affecting parameters, deposition rate and stability of plating bath; 3, the composition, microscopic structure and surface morphology of electroless deposits; and 4, the principles of the formation, mechanism and theory of electroless plating.

Regarding the contents of the four parts of this book, we would like to say more as follows.

The history of electroless plating is narrated in detail in Part 1. It can be divided into several periods: the discovery of electroless plating, the early preparation stage (1946–1959), the slow stage of development (1960–1979), the rapid development and expansion stage (1980–1999), and the stage of nano electroless plating and the deep investigation of the mechanisms and theory of electroless plating (2000–). No electroless plating books have narrated and discussed the whole history of electroless plating in detail, and this is the first time it has been done. The aim is to understand the development process of electroless plating more clearly and properly.

The part on technology includes the subjects of plating baths, affecting parameters, deposition rate, and the stability of plating baths. The aim in this part is not only to describe how to get EP alloy deposits, but also to sort and analyze the huge amount of experimental data and try to find out what is a unified or a regular kernel from the plentiful data, as well as seeking possible explanations. In addition, this part tries to identify the existence of possible conflicting phenomena in the experimental facts, and point out what questions need to be solved. There are three chapters on new subjects (Nano EP, EP Fe based alloy deposits, and Green EP) which appear in this part, and which will be illustrated below.

Some scholars describe electroless nickel plating (EN) as electroless plating. The reason for this is that EN is undoubtedly of the great importance, and has been widely employed in various industries due to its unique physical and chemical properties. However, we think that electroless plating of Fe based alloys should be studied as well, even just from the point of view of pure basic investigation. That is why our group studied the electroless plating of Fe based alloys early on, from the middle of 1980s, and the study subjects of three graduate students in our group were all based on electroless plating in Fe based alloys. Therefore, a chapter on electroless plating in Fe based alloys is included in this part.

As mentioned above, there are nearly no papers on nano EP published before the new century and this area has not been included in nearly all existing electroless plating books. A chapter on nano electroless plating has therefore been included in this part as well.

People have realized that conventional electroless plating is not exactly a "green" technology because many chemicals are added into electroless bath solutions which may be toxic, for example, formaldehyde which is a known carcinogen. Therefore, how to make the technology of electroless plating green technology and to make the process more environmentally friendly are very important. Fortunately, scientists have begun to study the green technology of electroless plating and to make appropriate advances. So, a chapter on the green technology of electroless plating is included in this part, which represents another feature of the book.

In order to narrate more clearly, the baths of electroless plating has been classified according to metals, binary, ternary, quaternary, quinary and multi alloys, but not to the various based alloys as is customary. However, the classification of the bath of electroless plating in binary alloys is still according to various based alloys as usual. Composite electroless deposits are discussed in a subsequent chapter in order to detail their nature. Many papers have reported the results of electroless plating on various substrates including on the so-called hard-coated substrates in order to enhance the applications of electroless plating and to better master the skills of fabrication. However, this subject will be not included in this book because it mainly represents the technical aspects of the subject (which of course is not to say that they are not important).

The plating rate and various affecting parameters are one of the main subjects in EP technology. Many EP papers concentrate on or refer to this subject, so the experimental data in the field are very plentiful. It should be said that nearly no books or review papers concentrate on the description and analysis of these, especially no one has compared them in order to find universal phenomena and laws, to find problems such as the existence of conflict phenomenon, and to point out what needs to be done in the future. A special Chapter is included, and all of these points have been discussed.

The compositions and microstructures of electroless deposits are very important. Clarifying them not only helps in understanding the nature of EP deeply, but also closely connects with the properties and applications of EP deposits. Collecting the SEM micrographs of electroless alloy depositions together and concentrating discussion of them, people may understand the nature of electroless plating deposition more clearly and deeply. The contents of these two aspects are dispersed in many literatures, and no one has collected them together, analyzing and discussing them. That is why we use a part to describe and discuss them.

The part of kinetics, theory and mechanism of electroless plating includes the formation process of electroless deposits, their formation mechanism and the theory describing the formation of deposition. There has been new progress especially in the theory of describing the formation, such as the formation theory of EP amorphous alloys and EP microscopic theory, and these areas are generally no well-covered in existing electroless plating books. Here we pay particular attention to them.

The development of electroless plating in China was not started until the early 1980s because of the delay caused by the well-known Cultural Revolution. Our group was one of the first institutions to study electroless plating in China. Except for the above-mentioned electroless plating of Fe based alloys, we studied some new electroless plating alloy systems, e.g. Cu-Ni-Sn-P quaternary alloy system, and the systematical study of some properties, especially the welding property of such alloy systems. Although the data and results of our research are included in the book, however they form only a small part of the whole book. Most of the results and data cited in the book are from scientific original papers published by various groups, companies and other institutions all over the world. More than 2,100 papers, reports and patents are cited in the book. Thereby, three points must be pointed out.

First, there is a huge amount of literature on electroless plating, only those closely related to the contents of the book were cited, others may not be referred to, which of course does not mean that these works are not important or not significant.

Second, the author, of course, wishes to seek proper use of the information cited. However, I don't know if this is the case, and only the readers can make the right judgments.

Third, the author usually analyzes and evaluates the data and results cited, and may say some words about them. Of course, I don't know if it is real correct and appropriate, so may I ask the readers to make their evaluation and comment.

At the beginning in designing this book, and during the whole writing process, I set a rule for myself that this book must not only well organize the cited results and data, providing a complete knowledge of EP to the readers, but also to analyze them and review them in as detail as possible. For the existed data, besides the achievements and progress, the possible problems and shortcomings must be boldly affirmed and pointed out, and the context and possible development for every subject discussed in the book also must be mentioned. In this way, the author hopes that readers can have an up-to-date understanding of the development process of EP, the basic academic subjects and the nature of EP from the book, thereby they not only can learn and master the existing knowledge of EP, but also can get some inspiration and enlightenment, and begin to think, consider and propose some new problems in the development of EP. This is my highest aim and greatest wish in writing this book. Now, the book is finished, I really do not know whether my rule has been executed, and my hope has been achieved. Only the readers can give right judgment about it.

I have pointed out many times in the book that after the discovery of EP nearly seventy year's ago, great and plentiful achievements have been presented, which have come from the hard investigation of scientists all over the world. Until today, research in this field has continued. There is no doubt to be expected that more and more new and deeper results will appear.

The materials contained within this book from our group were the crystallization of the collective efforts of the whole group in the field of electroless plating research over many years. I wish to express my sincere thanks to all those who participated in the research, including my colleagues, my former graduate students and even many undergraduate students of mine. I'd like to list some but not all of their names here: Dr. Zhang Heng, Dr. Tang Zhaosheng, Dr. Yi Ge, Professor Dr. Hu Wangyu, Professor Dr. Liao Shuzhi, Professor Dr. Xie Haowen, Professor Dr. Shu Xiaolin, Professor Dr. Wang Lingling, and Professor Dr. Ouyang Yifang. I appreciate very much the Max-Plank-Institut für Plasmaphysik (IPP), Germany where I worked for a long period of time, for granting me the Max-Plank Scholarship, which enabled me to cooperate with Dr. E. Taglauer, and Dr. S. Lindig from IPP, in the study of the surface morphology of electroless deposits on Ni-P based alloy systems.

My friend Professor Dr. W. A. Jesser, University of Virginia, spent a lot of time in critically reading part of the manuscript, I am very grateful to him for his selfless help.

The writing of the book took more than four years. My wife has accompanied me day and night, so that I could concentrate on the writing and never feel alone. I really appreciate her help.

I express my thanks to the network center and library in Hunan University for their helping me in searching the relevant data. Particular thanks are due to Mrs. Xiao Wei and Mrs. Huang Xiaoling for helping me to reprint many papers that were difficult to get hold of. Also thanks are due to Mr. Hu Hao for helping me to use the network more smoothly. Many friends and relatives gave me both tangible and intangible help for many years, I express sincerely my gratitude to them.

Zhang Bangwei

Yuelu Mountain, Hunan University, New Year's Day, 2015

REFERENCES

[1] Brenner A. P&SF 1984;71:24.

[2] Mallory O, Hajdu JB. Electroless Plating: Fundamentals and Applications. AESFS: Orlando, FL; 1990.

[3] Carraro C, Maboudian R, Magagnin L. Surf Sci Rep 2007;62:499.

[4] Djokić SS, Cavallotti PL. In: Djokić SS, editor. Electroless Deposition: Theory and Applications, in Modern Aspects of Electrochemistry, 48. Springer; 2010. p. 251–89.

[5] Yae S, Nasu N, Matsumoto K, Hagihara T, Fukumuro N, Matsuda H. Electrochim Acta 2007;53:35.

[6] Lu Xianmao, Chen Jingyi, Skrabalak S E, Xia Younan, Proc. IMechE Vol. 221 Part N: J. Nanoengineering and Nanosystems 2008:1.

[7] Qiu T, Wu XL, Mei YF, Chu PK, Siu. GG. Appl Phys 2005;A 81:669.

[8] Chong Hyun, Rheema Youngwoo, ChoaYong-Ho, Park Deok-Yong, Myung NV. Electrochim Acta 2010;55:1072.

Part I

History of Electroless Plating

Chapter 1

History–From the Discovery of Electroless Plating to the Present

Chapter Outline

Amorphous and Nano Alloys Electroless Depositions. http://dx.doi.org/10.1016/B978-0-12-802685-4.00001-7

It is very important to consider the history of electroless plating over time, so that its full developmental process, from the very beginning of its discovery to the present, can be understood. Studying and understanding its history of plating will not only promote a deeper understanding of electroless plating, but also allow predictions to be made about the direction that research and technological development of electroless plating is taking. To date, although some discussion of this history can be found in a few papers, no paper addresses this topic specifically. That is the motivation for this chapter, which discusses the history of electroless plating. Such a discussion is rarely seen in existing books on this topic; hence, this is an attempt to fill this gap.

The development of electroless plating can be divided into five stages: its discovery, the early stages of development, a period of slow growth, a period of rapid development, and finally a period of deeper development, including nanoelectroless plating. Each period will be discussed in the following sections.

Figure 1.1a shows the number of published articles related to electroless plating from various databases, and also the number of US Patents related to electroless plating from 1947 to November, 2014. Figure 1.1b shows the number of articles related to nanoelectroless plating from various databases from 1947 to November, 2013. The meaning of the variations shown in the figure will be explained and described in the following sections of this chapter.

1.1 DISCOVERY OF ELECTROLESS PLATING

1.1.1 Early Works

The study of electroless plating can be traced back to the middle of the nineteenth century. Early in 1848, Wurth noted that nickel could be deposited by reduction with hypophosphite in a nickel salt solution; he only obtained a black powder [1]. In addition, the bath decomposed spontaneously.

Almost 70 years later, Breteau obtained a rather bright deposit of Ni–P alloys using a similar bath [2] to that used by Wurth. Like the earlier experiments, the bath was not stable but decomposed spontaneously.

On December 5, 1916, Roux was granted a US Patent for a "Process of producing metallic deposits," which was the first patent in electroless plating [3]. The bath he used was the hypophosphite solution: water 1000 g, citrate of nickel or other salt of nickel 100 g, ammonia, alkalis, or alkaline salts 100 g, ammoniacal or alkaline hypophosphite 100 g. He stated in the patent that "the metallic salt may be added in proportion, or, on the contrary, the hypophosphite or phosphate may be added little by little to the said salt." It is obvious that the process of deposition was neither convenient nor suitable in practice. In addition, high concentrations of hypophosphite were used in these baths, which decomposed spontaneously and led to the formation of deposits on anything which was in contact with the solution. So the drawbacks of these baths were very obvious.

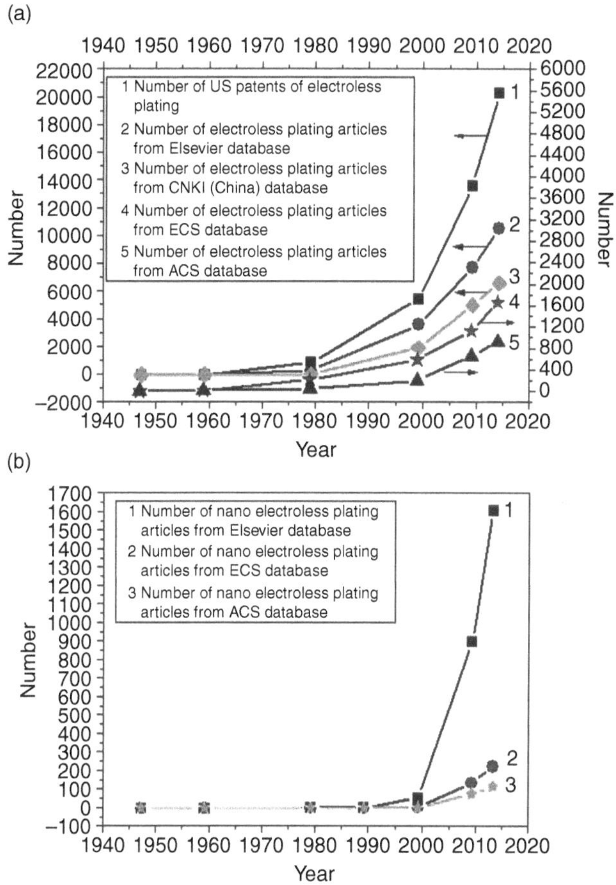

FIGURE 1.1 (a) Number of articles from various databases and of US Patents in electroless plating, and (b) number of articles in nanoelectroless plating from various databases from 1947 to November 2014.

More than 10 years later, in 1931, Pall and Friederici [4], Scholder and Heckel [5], and Scholder and Haken [6] continued the research into electroless plating. They were solely interested in studying the chemical reduction reaction. Pall and Friederici [4] used sodium hypophosphite to rapidly reduce palladium salts to elementary palladium powder, and the bath they used decomposed at the same time. Scholder and Heckel [5] observed that the phosphorus is present in the precipitate as a phosphide, and that the phosphorus content of the deposit increases with increasing the acidity of the plating medium. Scholder and Haken [6] studied the powders obtained by reduction of nickel salts with hypophosphite. They found that the precipitate from acid solutions was essentially a mixture of various nickel phosphides. The precipitate from alkaline solutions contained elementary nickel and a mixture of nickel phosphides. Similarly, they reduced also cobalt salts to cobalt metal powders.

It should be pointed out that the chemical reaction in all of these early works was spontaneous, and the bath was not stable but decomposed, thereby the process was not controlled. In other words, a new plating process has not been established yet, but these early works laid the foundation for the establishment of a new process.

1.1.2 Brenner and Riddell's Work

It was Brenner and Riddell who really first discovered electroless plating while working at the American National Bureau of Standard (NBS) in February 1944. They not only obtained a good quality Ni–P alloy deposit, but the bath they used was also stable. Therefore, Brenner is known throughout the surface finishing industry as the "father of electroless plating." Figure 1.2 shows the first page of the pioneering paper of Brenner and Riddell published in July, 1946.

It is interesting to note that electroless plating was discovered by accident. This point was mentioned by Brenner and Riddell themselves several times. A year after the discovery, Brenner and Riddell published an article: "Nickel plating on steel by chemical reduction" in 1946 [7], which was reprinted by <P&SF> in August, 1998 [8] to commemorate "one of the most significant discoveries in the history of surface finishing as well as Brenner on his 90th birthday." The first sentence in the paper is: "In the course of an investigation on a nickel plating bath, an unusual chemical reaction was accidentally encountered, by means of which nickel deposits of good quality can be produced on a steel or nickel surface without the use of an electrical current" [7]. Brenner wrote in a paper in 1959: "Electroless plating was discovered by accident by Brenner and Riddell in February 1944" [9].

In 1984, in a paper titled "Reminiscences of early electroless plating" [10], Brenner described the process by which electroless plating was discovered in detail. During World War II, ANBS worked on improving firearms. At that time, Brenner and Ridell were investigating the electro-deposition of nickel–tungsten and cobalt–tungsten alloys; therefore, it was logical for them to use these to coat the liners of machine-gun barrels. Using an ammoniacal citrate nickel–tungsten plating bath, they attempted to electroplate a coating onto the interior of the liner. However, they found that the deposits were always highly stressed and cracked, because of the oxidation of the organic components of the bath. It was obvious that the problem of oxidation at the anode must be solved. In order to prevent oxidation they introduced various reducing agents to the bath. No success was achieved until sodium hypophosphite was added to the plating solution. However, the experiment introduced the sodium hypophosphite to the bath caused considerable confusion. They found that there had been substantial gassing during electrolysis, and the cathode current efficiency was about 130%. In addition, the outside of the liner was also plated though only the inside was

U. S. Department of Commerce
National Bureau of Standards

Research Paper RP1725
Volume 37, July 1946

Part of Journal of Research of the National Bureau of Standards

Nickel Plating on Steel by Chemical Reduction

By *Abner Brenner and Grace E. Riddell*

A process has been developed for the production of adherent nickel deposits of good quality on steel without the use of an electric current. The deposition of nickel is brought about by chemical reduction of a nickel salt with hypophosphites in a hot ammoniacal solution. The reaction is catalytic and, under the prescribed conditions of concentration and pH, no reduction occurs in the solution unless certain metals, such as steel or nickel, are introduced into the bath. The reduction then occurs only at the surface of the immersed metal with the production of a coating of nickel of 96 to 97 percent purity.

I. Introduction

In the course of an investigation on a nickel plating bath an unusual chemical reaction was accidentally encountered, by means of which nickel deposits of good quality can be produced on a steel or nickel surface without the use of an electric current. The reaction involves reduction by hypophosphites in a heated nickel solution. In the nickel bath being investigated, trouble was had with the oxidation of some constituents of the bath at an insoluble anode, and to obviate this difficulty typical reducing agents were added. In one of the experiments the surprising result of an apparent cathode current efficiency of 130 percent was obtained, although during the plating there was considerable gassing at the cathode, which usually indicates a low current efficiency. Furthermore, the object being plated, which was a tube with an inside anode, was found to have been completely plated on the outside, although no external anodes had been used. Although this process will not replace electrodeposition of nickel, it may prove useful for special applications.

Contents

FIGURE 1.2 Groundbreaking paper of A. Brenner and G.E. Riddell published in July, 1946.

acting as the anode. This result made them quickly realize that chemical deposition was taking place concurrently with the conventional electro-deposition. So, as Brenner said: "Thus, autocatalytic plating was born." He also said: "Electroless plating was a spinoff of the war effort" [10]. Over the following year, Brenner and Riddell studied the deposition process in both alkaline and acid baths [11]. However, they did not study the application of electroless plating nickel. Brenner had said: "Grace and I never had commercial connections with electroless plating" [10].

It is worth noting several points regarding the discovery of electroless plating and its discoverers. The first is related to the patent problems of electroless plating. When Brenner and Riddell applied for a patent of the electroless plating process, it referred to the prior knowledge that sodium hypophosphite would reduce nickel and cobalt salts to black metallic powders. So, the patent examiner claimed that since the same compounds were used in the powder process and in the electroless plating process, the latter had no novelty. Brenner did not agree with that because a spontaneous chemical reaction that produced a black powder was not the same as a controlled autocatalytic process that produced a smooth, coherent metal coating. They then asked the examiner in their laboratory to see two comparison experiments. The examiner found that the electroless plating process was very different from a spontaneous chemical reaction process. Thus, Brenner said: "The patent examiner was impressed, and we had no further trouble in obtaining a patent" [10]. Three years after their application, two patents were granted in 1950 (US 2,532,283 and 2,532,284). Because the patents that Brenner and Riddell received were "public patents," anyone could use them without payment of a royalty, and they did not receive any money from them [10].

The second point is that no government funds supported the electroless plating project because "it never had the formal status of an approved project at NBS" [10]. Brenner said this with a little indignation when he used the word "orphan" to describe the situation. He said: "Electroless plating remained an orphan and never received a direct government grant. I did submit proposals for further research to a government agency, but they were not accepted" [10]. In addition, Brenner admitted that he was a little annoyed because he had been unsuccessful, whereas a private organization obtained a contract from government to carry out further research into electroless plating. However, Brenner obtained a lot of funding from various government agencies for his academic research, even though it had no obvious practical application.

The third point is the source of the name of electroless plating. The term of "electroless" was not coined by Brenner and Riddell, but by Blum. Brenner said: "Dr. William Blum, NBS section chief, coined the word 'electroless' to indicate that the process was similar to electroplating, in that a metal coating was produced, but that it was dissimilar because no electric current was involved" [10].

The last point is that people even Brenner himself did not predict the commercial use of electroless plating in the early days. The reason was the cost of sodium hypophosphite at that time. Brenner described this situation clearly [10], stating that: "Dr. Blum thought that the process would have little commercial use." As for himself, he said: "I was a little more optimistic, but did not consider that its use would be widespread because of the cost of sodium hypophosphite." Today, the commercial applications of electroless plating are so widespread that they extend to the majority of industries. It is very interesting to look at the circumstances when electroless plating was invented.

It is obvious that the research carried out before 1946 by various authors provided a very good foundation for Brenner and Riddell's discovery of electroless plating, especially because they then knew about the reduction reaction of sodium hypophosphite. These authors completed the discovery of electroless plating, so their work should be widely recognized.

Of course, there were still drawbacks in the Brenner method, especially if it was used as a large-scale industrial production process. The deposition rate was slow, the coating quality was not so good, and the bath solution was not stable. In a paper written by Brenner himself in 1954 [12], he described the problem in detail. He said: "Occasionally the life of an electroless plating bath comes to an untimely and sudden end through spontaneous decomposition. This phenomenon begins first with a mild, uniform evolution of gas throughout the volume of the bath instead of only on the surface of metal parts. The gassing becomes more vigorous, the solution begins to froth somewhat, and a black precipitate appears in the bath. Finally, the gas evolution becomes so vigorous that the entire contents of the vessel foam over."

1.2 EARLY STAGE OF DEVELOPMENT (1940s–1959)

From Fig. 1.1a, one can see that the number of articles (from Elsevier, ECS, or ACS) and patents (from US Patents) related to electroless plating were rather few in this period. Therefore, the feature of the period is that research into electroless plating had just started, and the process was in the early stages of development. A few research studies were continued, a few patents in electroless plating were issued and some companies started to use the process.

1.2.1 Research Works

West noted in his paper published in 1954 [13] that the pH, the temperature, and the lining used in the plating tank were very important when carrying out electroless nickel (EN) barrel plating. He obtained a 0.004 mm thickness deposit in a 30 min deposition. In another paper written in 1954 [14], West reported EN of nonferrous metals such as Al, Cu, Ti, and brass. The bath he used was of the alkaline type with the pH held at approximately 10. West published another paper in 1955 [15], in which he reported that the corrosion resistance as determined by the salt spray (fog) test on a thin film (0.005 mm thickness) of chromium on steel deposited by the electroless process was equal to the same coating thickness of electroplating chromium on steel.

In 1954, McLean and Karten [16] found that small fuse parts for Navy Ordnance could be successfully nickel plated by chemical reduction. In 1954, Brenner [12] recognized that up to that time, only nickel, cobalt, or nickel–cobalt alloys had been deposited by the electroless process in a manner suitable for practical application.

In 1956, Gutzeit and Mapp [17] studied the properties of electroless plating nickel. In 1957, Goldenstein et al. [18] studied the structure of chemically deposited nickel, which is in the amorphous phase.

De Minjer and Brenner measured the deposition rate of EN plating in 1957 [19]. In addition, they examined the effect of the addition of thiourea and found that a dark precipitate appeared in the bath solution.

Gutzeit proposed the "atomic hydrogen" mechanism for EN plating in 1959 [20].

Knanishu wrote a technical report on the deposition of alloy by chemical reduction in 1961 [21]. In this first review, a literature survey of electroless plating was made. He discussed the deposition of alloys of nickel, cobalt, and alloy combinations of Ni, Cr, and Fe, the base metal being steel, Al, or Mg. He found that Co and Ni could be successfully plated onto steel and Al, as was the plating of Ni on Mg.

During this earlier period, the main problems in the plating by chemical reduction were the plating rates, the need for a smooth lustrous surface, and a more stable plating solution [22]. In addition, researchers recognized that when the basis metal does not possess catalytic properties, the reaction must be initiated by using a catalytic metal in contact with the base metal. Once this is done, plating begins, the deposited metal takes over the role of a catalyst, and plating continues as long as the solution is replenished with the salts of the plating metal, and with a sufficient amount of reducing agent such as hypophosphite.

1.2.2 Patents Issued

During this early period, several patents were issued. The electroless plating of nickel was the subject of US Patent 2,532,283 issued on December 5, 1950, to Brenner and Riddell and assigned to the United States government [23]. This patent discloses an electroless plating process which differed from Roux's deposition of nickel in that the Roux reaction was spontaneous and complete, whereas the Brenner et al. process was a catalytic process so controlled that deposition occurred only on catalytic surfaces immersed in the bath. This electroless plating process produces a substantially uniform, smooth coating on the pin surfaces defining the gap.

"A process of chemical nickel plating" was patented on November 10, 1953 (US Patent 2,658,839) [24]. It described the processes of chemical nickel plating and more particularly an improved way of utilizing the bath solution of nickel plating processes.

Also on November 10, 1953, a patent of "a process of chemical nickel plating and bath therefor" (US Patent 2,658,841) [25] was issued. The object of this invention is to provide an improved chemical nickel plating process with many advantages in commercial use over electroplating methods, including low cost of equipment, good adhesion of the coating in the finished product, high corrosion resistance of the coating, and a coating that is uniform, continuous, and much harder. In particular, the invention presents measurements of the relative amounts of plating deposited, how the weight of the plating deposit changes with changes in temperature, in the ratio of nickel ions to hypophosphite ions, and of the pH value of the plating bath. That is to say, researchers started to study the factors affecting the plating bath solution in the very beginning in the 1950s.

In 1954, Gutzeit, in his US Patent 2,694,019 "processes of chemical nickel plating and baths therefor," [26] provides improved processes of chemical nickel plating of catalytic materials employing baths of the nickel cation-hypophosphite anion type, and improved baths to be used in such processes. To increase the stability of the bath (preventing formation of the "black precipitate"), and to increase the normal plating rate of the bath, various baths of the present type have been suggested employing different additives that serve either as buffers or as exaltants.

In 1955, in a US Patent 2,717,218 "chemical nickel plating methods and apparatus" [27], Barrington and Crehan described chemical nickel plating methods and apparatus. The general object of the present invention is to provide a system and apparatus and method for carrying out the chemical nickel plating process on a large-scale industrial basis rendering it feasible to produce railway tank cars and other large containers having interior linings formed essentially of nickel. Also, in 1955 US Patent 2,721,814 "nickel plating by chemical reduction" [28], Jendrzynski and Stapleton discovered that, contrary to prior ideas careering chemical reduction pealing, superior results are obtained employing a bath containing only ions of the metal to be deposited, a buffer and a reducing agent, hence, in the case of nickel plating, nickel ions and preferably a hypophosphite reducing agent.

In 1956, in a US Patent 2,766,138 "processes of chemical nickel plating" [29], Barrington invents improved processes of chemical nickel plating of catalytic materials employing baths of the nickel cation-hypophosphite anion type, and more particularly to continuous processes of chemical nickel plating of the general character.

In 1958, in US Patent 2,828,227 "electroless deposition of vanadium alloys" [30], Eisenberg and Raleigh described a process in which vanadium together with at least one metal selected from the group composed of iron, nickel, cobalt, and chromium can be electrolessly codeposited on a suitable metallic substrate and to electroless plating solutions used in the said process.

On February 17, 1959, a US Patent 2,874,072 "autocatalytic copper plating process and solution" was issued to Cahill and McConnell [31]. This invention relates to a copper plating process and solution, and more particularly, to an autocatalytic chemical reduction process and bath for continuously plating desired thicknesses of copper on subjects having catalytic metallic surfaces. That is to say that the chemical plating of copper started at very soon after the chemical plating of nickel. The reason is the wide application of the electroless copper plating in the electronics industry, and which will be discussed more in detail later.

1.2.3 Preliminary Applications

The Brenner method was not suitable for large-scale industrial production because of the drawbacks mentioned above. However, it could be used on a relatively small scale of production, mostly for batch plating. Therefore, after its discovery, electroless plating stared to be used by a number of manufacturing companies for commercial plating. At this early stage of development, some companies of chemical plating have been established, and some electroless products have been developed.

The Leonhardt Plating Company, founded in 1950, is a family-owned and operated electroplating job-shop in Cincinnati. The company provides EN plating services, except electroplating products. Leonhardt Plating employed 27 people and serviced over 100 customers at that time.

After recognizing the practical application of Kanigen planting (EN plating), General Transport Company (GATC Co.) started to sell licenses for Kanigen plating in the United States, Europe, Australia, and Japan. Onoda Cement (currently Taiheiyo Cement Co., Ltd.) acquired the license in Japan, and established Japan Kanigen Co. Ltd. in 1955. This company began the first electroless plating processing operation at its Tokyo plant in January 1958. Some authors mentioned this event in 1955 as the commercial inception of EN [32].

On August 23, 1955, the Wall Street Journal reported that the Los Angeles General American Transportation Corp. of Chicago was building a new West Coast manufacturing and warehouse building between Compton and Long Beach Calif. The 44,000-square-foot plant would provide facilities for the company's electroless process of nickel plating and other uses.

When Shipley patented a new catalyst for electroless plating, he and his wife, Lucia, founded the Shipley Company in 1957. Revolutionizing the emerging printed circuit industry, Shipley Company introduced catalysts to the market, resulting in it becoming the leading producer of reliable chemical deposition processes.

1.3 SLOW GROWTH OF PERIOD (1960–1979)

During these two decades, one can see from Fig. 1.1a that the number of articles (from Elsevier, ECS, or ACS) and patents (from US Patents) of electroless plating were 309, 214, 37, and 904. Compared with its early stage of development, electroless plating was developing but still not rapidly, hence that we call it the slow growth period. Research studies have spread to various alloy systems and fields, many patents in electroless plating have been issued, and more industries have used the process during this period. The main feature of this period is that either the research or the patents were concentrated on the improvements in the plating bath such as enhancing the stability of bath, speeding up the plating rate, etc. However, the structure and properties of the deposits were not investigated extensively or deeply. The application of electroless plating was not widespread geographically. For example, in China research and development of electroless plating only started around the end of the 1970s and the beginning of the 1980s.

1.3.1 Improvement of the Plating Bath

During these 20 years, people have studied the plating bath more broadly and deeply and from various aspects so as to understand it better. As the plating baths improved, electroless plating, more particularly EN plating, was used more widely in large-scale industrial applications. Although the main publications in this time period concentrated on improvements in the plating baths, some papers and patents began to turn to studying deeper subjects, such as the structure and properties of the deposits, the electrochemical behavior of electroless plating, mechanism of deposition, etc. We will describe them chronologically in the remainder of this section.

In 1960, Metal Hydrides INC was granted US Patent 2,942,990 "metal plating by chemical reduction with borohydrides" [33]. This invention relates to the chemical deposition of nickel, cobalt, and mixtures thereof upon metals, glass, ceramics, and plastics by the reducing action of a borohydride in an aqueous solution of a salt of nickel and/or cobalt.

Also in 1960, in US Patent 2,965,551 "metal plating process" [34], Henri provides a process for the preparation of a metal surface prior to the application thereto of a metallic coating, which process permits obtaining a better adherence and improved decorative effects of the coating to be subsequently applied. The obtained coatings will adhere better than the conventionally known ones, and can be deposited from baths free from undesirable components liable to contaminate the ultimate metallic deposits.

In US Patent 3,011,920 "method of electroless deposition on a substrate and catalyst solution therefor," issued on December 5, 1961, to Shipley [35], an alternative process is described for preparing the substrate prior to electroless coating, in which the noble metal, for example, palladium, is applied directly to the dielectric substrate in the form of a colloidal solution containing the metal in the zero valence state.

In 1962, in a US Patent 3,024,134 "nickel chemical reduction plating bath and method of using same" [36], Nixon and Rowe stated that the principal object of their invention is the provision of an improved chemical reduction plating bath and plating process in which an improved metal deposit is obtained and the formation of undesirable reaction products is minimized. Another two objectives are to provide an improved chemical reduction plating solution with improved stability, and to provide an electroless plating solution adapted to regulate the availability of metallic ions therein.

In 1965, in a US Patent 3,178,311 "electroless plating process" [37], Leopold described plating processes, and more particularly a process for electroless plating of a metallic coating on a base member. The objects of the invention are to provide an improved electroless plating process, to provide an improved process for the plating of a metallic phosphate coating on a base member, to provide an improved electroless plating process that can be carried out at room temperature, and to provide room temperature electroless plating process that results in coating having high conductivity coupled with good characteristics of adherence and hardness.

In 1965, Graham, Lindsay, and Read [38] studied the structure and mechanical properties of electroless nickel, and found that the phosphorus content of electroless nickel from 4.6 to 9.4 w/o affects the structure, strength, ductility, and hardness of the deposits.

In 1967, Nacus [39] studied chemical nickel plating with sodium borohydride or amine-boranes as reducing agents. He found that although experience with these baths is much more limited than with the hypophosphite baths, they appear to have several advantages and may become economically attractive when the price of boron reagents goes down.

Also in 1967, Randin et al. [40] studied electroless nickel by DTA and X-ray. They assumed from their results that electroless nickel in the as-plated state is a supersaturated solid solution of phosphorus in nickel as a metastable intermediate state between that of a mixture of nickel plus phosphorus and the equilibrium system of nickel plus Ni_3P.

Also in 1967, Pont in his US Patent 3,338,726 "chemical reduction plating process and bath" [41] declared that one objective of his invention is to provide a novel chemical plating process for the deposition of nickel and cobalt in the form of their boron-containing alloys from an aqueous solution on the surface of an object. This surface catalyzes the plating or deposition of the nickel or cobalt alloy from the solution. The new aqueous plating solutions are stable, easily formed, and may be readily controlled to plate Ni–B and Co–B alloys on a catalytic surface at a fast plating rate.

In 1969, Elmore and Davis in their paper "mechanism of bonding electroless metal to organic substrates" [42] examined the interfacial surfaces of substrates subsequent to chemical processing and electroless and electrolytic plating by using a scanning electron microscope. Their examination revealed that interlocking surfaces were present which permitted mechanical interaction at the copper–organic interface. When these surfaces were replicated on an epoxy resin, bonding between this surface and electroless copper was obtained.

Also in 1969, Marti and Lanza performed a more complete discussion of suitable plating techniques to obtain the correct balance of hardness and ductility [43]. A suitable bath temperature is in the range of about 110°F to about 120°F, again depending on the chemistry of the bath, with light to medium agitation and no air injection.

In 1970, Rondin and Hintermann [44] performed a calorimetric study of the electroless deposition of nickel, and measured the heat of reaction of chemically reduced nickel by hypophosphite. Muski in his 1970 paper [45] recognized that (1) there is no accepted definition of the functional aspects of EN, and (2) there is little understanding of practical operation of EN, so he wanted to clarify the situation.

In 1971, Feldstein and Lancsek [46] proposed a new technique for investigating the electrochemical behavior of electroless plating baths and the mechanism of EN plating. They found that given a suitable reference point, the plating rate and metal-to-phosphorus ratio can be predicted from the measured potential. They also proposed a model for the electroless deposition of nickel–phosphorus based on a modified hydride mechanism. With the help of this model, the hydrogen evolved during deposition can be quantitatively accounted for.

In 1971, Gabrielli and Raulin investigated EN plating from an acid bath strongly buffered by acetate with a rotating disk electrode [47]. They found that the polarization curves have an anodic peak which is typical of EN plating. They also found that the polarization resistance measurements allow a direct determination of the plating rate. Their experiments show the important inhibitor effect of the dissolved oxygen on the deposition reaction.

Mallory, in his 1971 paper, "the electroless nickel-boron plating bath; effects of variables on deposit properties" [48], found that the composition of the alloy remains constant over long periods of solution operation, and the hardness of Ni–B alloys deposited from solutions operated in the acid pH range lies in the range of 700–800 Vickers hardness number, for the as-plated deposit.

Also in 1971, D'Amico et al. [49] studied "selective electroless metal deposition using patterned photo-oxidation of Sn(II) sensitized substrates." They found that surface deposits containing hydrated Sn(II) species are generated when substrates are immersed in stannous chloride solutions, and then rinsed in water. These deposits, after a later immersion in an aqueous solution of $PdCl_2$, are capable of activating electroless plating baths to produce a metal deposit on the substrate. The Sn(II) species, however, can readily be photo-oxidized by exposure to ultraviolet light in air, yielding Sn(IV) and thereby destroying the latent catalytic surface. Thus, selective electroless metal deposits on a variety of substrates may be generated by exposure through an appropriate mask.

On January 28, 1972, a US Patent 3,635,758 "electroless metal deposition" [50] was issued to Schneble, Jr., et al. An improved method for enhancing the ductility of electroless metal deposits is provided in which an electroless metal bath contains extraneous ions which have a preferential capacity for being coulombically adsorbed at the outer layer of an electric double layer. This layer is present on an interface in contact with such bath on which the said metal is electrolessly depositing, such ions being present in an amount sufficient to reduce the inclusion of hydrogen in the electrolessly deposited metal.

In 1973, Gobunora et al. studied the "electroless deposition of nickel–boron alloys" [51]. They investigated the conditions for electroless Ni–B plating, the process mechanism, and the structure of the alloys as-plated and after annealing at 150–700°C, and established a correlation of the mechanical and magnetic properties with the changes of phase structure.

Also in 1973, Lelental [52] studied dimethylamine borane (DMAB) as a reducing agent in electroless plating systems. They found that the process of nickel deposition was shown to require a larger metal aggregate for catalysis than that necessary for DMAB decomposition.

On May 25, 1976, US Patent 3,959,531 "improvements in electroless metal plating" was issued to Schneble, Jr., et al. [53]. Electroless metal deposition solutions are provided which comprise a combination of an ion of a metal whose electroless metal deposition is desired; a complexing agent for said ion; a reducing agent for said ion; a pH regulator; and less than about 25 parts per million of metal ions that have an oxidation potential greater than the oxidation potential of the ion of the metal whose electroless deposition is desired.

In 1976, Watanabe and Tanabe [54] studied the formation and morphology of Ni–B amorphous alloys deposited by electroless plating. They found that the grain size of the microcrystals in the plated film decreased with increases in concentration of the reducing agent ($NaBH_4$). Amorphous film can be obtained when more than 400 mg/L reducing agent is added. The increase in the amount of the added reducing agent increases the boron content of the deposited film which is saturated at 11.4 wt% B (41 at%). At this saturation level the Ni–B film becomes amorphous.

In 1977, Büchner and Niederprünt [55] studied the sodium borohydride and amine-boranes, commercially important reducing agents. They carried out a survey of the application of sodiumthorohydride $NaBH_4$ (I) and amine-boranes $R_2NH\text{-}BH_3$(II) in the selective reduction of various organic groups. The field of metallization by chemical means, using I and II, especially the chemical nickel plating of metallic and nonmetallic substrates was attracting more and more scientific as well as industrial interest.

Also in 1977, Scannell in his US Patent 4,038,085 "method of treating electroless nickel plating bath" [56] declared that his invention resides in the addition of large quantities of sodium fluoride to a chemical plating solution for depositing nickel on a catalytic substrate. The purpose of the addition is to increase the rate of deposition of an EN plating bath. In this way, heavy deposits may be made on substrates without substantial surface roughness, and the control and integrity of the properties of the bath can be better maintained, concentrations of additives needed to maintain control of the bath's properties can be reduced, and the life of the plating bath can be extended.

On May 1, 1979, Gulla et al. were granted a US Patent 4,152,164 "electroless nickel plating" [57]. This patent relates to electroless metal deposition and more specifically, to a process in which a plating solution is brought to equilibrium and thereafter operated with the concentration of plating reactants and by-products maintained substantially constant. The plating solution treated in accordance with the invention is one having evaporative losses of at least 1% per plating cycle. Following the process, a plating solution can be operated indefinitely and yields a metal plate of uniform quality and predictable properties at any time during use of the solution. The invention avoids the known problems of by-product build-up and variable concentration of reactants typically associated with the use of such solutions.

On September 25, 1979, a US Patent 4,169,171 "bright electroless plating process and plated articles produced thereby" [58] was issued to Narcus. In this patent, an adherent bright metallic coating that may be any metal or alloy capable of being deposited electrolessly is applied to the surface of a plastic or other dielectric part (or to a metal part) by an electroless process comprising suitable pretreatment of said surface, deposition of a primary electroless copper to produce an underlayer, then depositing a bright secondary electroless coating of the desired metal (the overlayer) with complete elimination of any subsequent electroplating process (usually required to achieve this brightness). The required overlayer may also be limited to selected surface areas of a coated part by mechanically removing portions of the primary electroless copper prior to immersion of the part in the secondary electroless bath to produce the desired overlayer metal.

Also in 1979, Smith [59] studied the mechanics of EN deposition.

1.3.2 Various Electroless Plating Metals

Electroless plating of many metals other than Ni such as Au, Ag, Co, Pt, Sn, etc., and some alloys have been studied during this period.

In 1963, in a US Patent 3,077,285 "tin–nickel–phosphorus alloy coatings" [60], Pranas described Sn–Ni–P alloy coatings, and more particularly to articles of manufacture carrying such coatings. The aim of the invention is to improve the protective properties thereof, and to provide an improved article of manufacture comprising a body carrying an improved protective coating of Ni–P alloy intimately bonded thereto, the outer skin of the coating having tin diffused therein. The inventor declared that a Sn–Ni–P alloy coating is first produced by chemical deposition from a plating bath of nickel cation-hypophosphite anion type and then modified by diffusion tin plating.

On March 21, 1961, a US Patent 2,976,181 "method of gold plating by chemical reduction" was issued to Brookshire [61]. He provides an impinged method for plating suitable surfaces with Au by a controllable chemical reduction process. Since an insoluble gold compound is included in the plating solution, the plating reaction is not adversely affected by light and greater control of the plating process is thus obtained. During the reduction of the insoluble compound, a chelating or complexing soluble compound forms a soluble complex with the insoluble compound. This complex is then reducible upon a suitable surface by the action of a reducing agent.

Luce in his US Patent 3,300,328 "electroless plating of gold" issued on January 24, 1967 [62] declared that his invention relates to coating of base materials with relatively thin layer of gold by means of electroless deposition. On August 6, 1968, Robert in his US Patent 3,396,042 "chemical gold plating composition" [63] invented an entirely different system that will produce gold deposits at the rate of 2–2.5 μm/h at a temperature substantially below that ordinarily used for either immersion or electroless gold. In addition, the baths are stable for a long period of time under production condition. Thus, it has been found that gold deposits from a solution thereof in the presence of palladous ions. Haddad et al. in 1970 performed a process of electroless gold plating [64], which is characterized by rapid production of quality deposits, stability, and controllability. The best results are obtained when the substrate to be plated is presensitized with a very thin layer of a catalytic metal such as palladium, rhodium, or platinum. Photocircuits Corp. in their US Patent 3,589,916 "autocatalytic gold plating solutions" issued on June 29, 1972 [65] declared that improved electroless gold plating baths are provided by formulating in aqueous alkaline media, a water-soluble gold salt; a complexing agent for gold, a water-soluble borohydride or amine borane reducing agent for gold, and a small effective stabilizing amount of a cyanide compound in an amount of between about 5 μm and 500 mg. In the US Patent 3,917,885 "electroless gold plating process" issued on November 4, 1972 [66], Engelhard Minerals & Chemicals Corporation declared that their invention relates to improved electroless plating baths for the autocatalytic deposition of Group IB metals upon substrates, which baths provide improved plating characteristics, including improved plating rates. The electroless plating bath disclosed includes an aqueous solution of an imide complex of the Group IB metal to be plated, an alkali metal cyanide, and a reducing agent, and is maintained at a pH from about 11–14 by the addition of alkali metal hydroxides. "A typical electroless procedure" is described in US Patent 3,700,469, issued to Okinaka on October 24, 1972 [67]. In the patent, conditions and bath compositions are given for electroless plating of pure gold and gold alloys on a variety of substances including ones with gold surfaces. The plating bath contains a gold cyanide complex, and free cyanide is added to stabilize this complex. A borohydride or aliphatic-substituted amine borane is used as the reducing agent and the pH is adjusted by the addition of an alkaline agent. The resulting bath is stable, and yields plating rates of typically 0.5–3 μm/h at about 75°C, and plating thickness up to a micrometer or greater. In 1973, Okinaka published a paper "an electrochemical study of electroless gold deposition reaction" [68]. Their results support the assumptions made previously that (1) the species that reduces aurocyanide ions is BH_3OH^-, an intermediate formed during the hydrolysis of BH_4^-, and (2) the deposition reaction is preceded by competitive adsorption of reacting species. In 1975, Rapson and Groenewald in their paper "the use of gold in autocatalytic plating processes" [69] described the scientific basis of autocatalytic plating, and reviewed electroless Au plating procedures. The effectiveness of Au as an activator in the autocatalytic plating of plastics was also briefly surveyed in their paper.

On February 22, 1977, US Patent 4,009,297 "gold deposition procedures and substrates upon which gold has been deposited" was issued to Redmond et al. [70]. Their invention presents a method for electroless gold plating in which gold ligand complexes are used; a photolytic method for depositing catalytic amounts of gold on a substrate suitable for an electroless metal deposition method whereby the gold complexes used are trivalent gold complex; a method for immersion plating or vaporization plating of gold from complexes, and complexes for the above methods. US Patent 4,091,128 "electroless gold plating bath" was issued to Franz and Vanek on May 23, 1978 [71]. In this patent, a method for preparing an electroless gold plating bath comprising a gold salt and a complexing agent is improved by adding the gold salt as a concentrated solution to a concentrated solution of the complexing agent either maintained at or subsequently heated to the above ambient temperature to eliminate the step of aging the plating bath. The plating efficiency of an electroless gold plating bath comprising a gold salt and an alkali carbonate complexing agent and the quality of the resulting gold film are improved by buffering the bath with an alkali metal bicarbonate, thus maintaining a pH between about 8.5 and about 9.5.

In 1964, at IBM, Dunlap [72] discovered that the coercivity of magnetic deposits from electroless cobalt baths containing hypophosphite ion can be controlled over a broad range of values. The control is accomplished by the addition of small amounts of ferrous ions to the plating bath. In 1965, Kivel and Sallo [73] studied the effect of thiourea on alkaline electroless deposition, and found that the rate of deposition is enhanced for cobalt but not for nickel. A Raney nickel type of mechanism and an electron transfer mechanism are proposed to explain the results. In 1966, Fisher [74] studied the preparation and hard magnetic properties of Co–Zn–P films for high-density recording applications. They found that incorporating Zn increases the coercive force of Co–P film on Mylar substrates by a factor of 2–60. Co–Zn–P films exhibited coercivities ranging from 300 Oe to 1200 Oe, depending on the preparation conditions. In 1969, Morton and Fisher [75] studied the effect of electroless nickel substrate on coercive force and microstructure of certain Co–P films. They found that the increase in coercive force may be due to a decrease in the amorphous characteristics of cobalt films prepared on electroless nickel in comparison to such films on Mylar-adhesive substrate. In 1974, Pearlstein and Weightman [76] developed an acid electroless cobalt plating bath, which utilizes DMAB as the reducing agent and which provides bright deposits at over 13 μm/h at 70°C. The bath, slightly modified, is also usable at room temperature for deposition on palladium-activated nonconductors. Additions of even small quantities of hypophosphite to the DMAB bath resulted in failure to achieve deposition. They studied also effects of bath parameters such as DMAB concentration, bath pH, bath temperature, succinate concentration, and hypophosphite additions on electroless Co plating rate.

In 1969, Pearlstein in his US Patent 3,485,597 "electroless deposition of nickel-phosphorus based alloys" [77] declared that articles of manufacture made from a base metal having an electrolessly deposited nickel–phosphorus based alloy coaling thereon, the alloy containing from more than 1% to about 45% of an element which is codeposited with the alloy, the element being W, Re, Be, Rh, Pd, Pt, Sn, Zn, Mo, or Au. The nickel predominates the phosphorus in quantity.

In 1972, Calligaris et al. were granted a US Patent 3,650,747 "electroless deposition of nickel, cobalt, copper or iron metal, and a bismuth, antimony, lead, tin, titanium, tungsten or chromium salt on a gold, platinum or palladium latent image" [78]. This invention relates to a process that salts of bismuth, antimony, lead, tin, gold, silver, titanium, tungsten, and chromium are used to impart continuous-tone characteristics to image reproductions produced by the electroless deposition of a nonnoble metal onto a recording medium.

In 1974, Gulla was granted a US Patent 3,832,168 "metal finishing alloy of nickel–copper–phosphorus" [79]. He invented a coating of an electroless Ni–Cu–P alloy substantially free of codeposited sulfur, which coating is deposited from an EN solution characterized by the presence of a small amount of copper ions. The copper ions provide an extremely smooth deposit resulting in a lustrous surface appearance. In addition, copper ions can stabilize the plating solution and substantially improve corrosion resistance of the deposit.

In 1977, a US Patent 4,019,910 "electroless nickel polyalloy plating baths" was issued to The Richardson Chemical Company [80]. This invention relates to certain novel, nickel-containing polymetallic alloys produced by electroless plating techniques which contain boron or phosphorus and one or more metals selected from the group consisting of Sn, Cu, Mo, or W as well as to the electroless plating baths for producing such alloys containing an ester complex of polyhydric acid or alcohol such as the diboron ester of glucoheptonic acid.

Hubbell [81] described the "chemically deposited composites: a new generation of electroless coatings" in 1978. He discussed the development of chemically deposited composite coatings (e.g., SiC particles in a Ni–P matrix) and indicated a typical bath composition. He also compared the friction, wear, corrosion behavior, and hardness of the composite coatings with those of electroless Ni.

1.3.3 Electroless Plating Cu

Because of the important application of Cu in the electronic industry, electroless plating Cu was paid particular attention, as discussed further in the following paragraphs.

Early in 1960, Agens was granted a US Patent 2,938,805 "process of stabilizing autocatalytic copper plating solutions" [82]. This prevention relates to a process for stabilizing copper plating solutions used to autocatalytically plate copper onto metals or onto nonmetallic surfaces.

In 1962, at IBM, Marshall developed a process for producing a copper layer with superior properties from the point of view of adhesion, hardness, tarnish resistance, and electrical conductivity [83]. The layer is obtained by electroless deposition of copper from a reducible copper solution on a nonconductive body such as paper. Also, in 1962, Brookshire in his US Patent 3,046,159 "method of copper plating by chemical reduction" [84] used an insoluble compound of copper in the reduction plating solution. The insoluble compound is reduced by including in the planting solution a soluble compound that forms a soluble complex with the insoluble compound. This complex is then reducible upon a suitable surface by the action of a reducing agent.

In 1963, a US Patent 3,075,855 "copper plating process and solutions" was issued to Agens Jr. [85]. His invention relates to an autocatalytic chemical reduction process and bath for plating the desired thickness of copper on objects having a catalytic metallic surface without simultaneously causing the plating of undesired surfaces and/or the precipitation of cuprous oxide from the solution. Also in 1963, John was granted a US Patent 3,095,309 "electroless copper plating" [86]. The invention relates to a new and improved process for the electroless plating of copper, and to a novel and improved electroless copper plating bath. This process can obtain an electroless plating or deposition of a ductile, bright layer of copper on various surfaces such as insulating members, metal, ceramic, and other supporting surfaces. The novel and improved electroless plating bath is simple to use, operates with certainty, and is economical in use. In addition, the invention provides an improved electroless copper depositing of better color than had heretofore been realized, and which may exhibit substantially the same ductility as good electrodeposited copper.

In 1966, Schneble Jr., et al. in their US Patent 3,259,559 "method for electroless copper plating" [87] provided improved chemical methods for imposing a uniform, ragged, firmly adherent, nonporous bright, and ductile deposit of copper on an insulating surface acting as a primary object of their invention. Another object of this invention is to provide methods for making improved printed circuits in which the support for the circuits has improved insulating properties.

Atkinson was granted a US Patent 3,310,430 "material and method for electroless deposition of metal" in 1967 [88]. His invention relates generally to the coating of nonmetallic objects, such as objects made of glass, plastic, etc., with a metal, and more particularly to the coating of such objects with an electrically conducting film or layer of a metal having a low resistance.

In 1968, at IBM, Hall and Koretzky developed a process for electroless plating Cu [89], in which they used DMAB as the reducing agent. Several advantages accrue from the use of DMAB. The copper is deposited from the monovalent state. Its electrical conductivity is high, making it suitable for use as an electrical conductor. The electrolyte is stable up to 90°C. The bath was with the following composition: CuCN 10.7 g/L 0.12 mol/L, KCN 15.6 g/L 0.24 mol/L, NaOH 42.9 g/L 1.07 mol/L, $(CH_3)_2NH\ BH_3$ 28.6 g/L 0.55 M. The boric acid orborates formed by the oxidation of the DMAB consumed in depositing copper act as a buffer and help to stabilize the pH of the solution.

In 1970, Hirohata et al. were granted a US Patent 3,532,519 "electroless copper plating process" [90]. Their patent provides the improvement that comprises measuring the pH of a bath consisting of an aqueous solution of formaldehyde, a copper salt, a complexing agent, and an alkali hydroxide; removing the hydrogen gas included; measuring the copper ion concentration; holding the pH at a predetermined value by feeding into the bath a concentrated aqueous solution of an alkali hydroxide; and supplying a mixture of formaldehyde and copper salt in a predetermined mole ratio to said aqueous solution so as to hold the concentrations of formaldehyde and copper salt in a predetermined value.

In 1971, a US Patent 3,615,736 "electroless copper plating bath" was issued to Enthone, Incorporated [91]. This invention relates to high stability electroless copper plating baths containing as a stabilizer a small amount of *o*-phenanthroline and of iodide ions. Additionally, this invention relates to new concentrate compositions and additive compositions, especially well suited for use in forming the high stability electroless copper plating solutions herein and for addition thereto as a stabilizer, and to a method for the electroless plating of copper involving the use of the high stability electroless copper.

In 1972, Schneble Jr., et al. were granted US Patent 3,650,777 "electroless copper plating" [92]. This patent provides an improved aqueous autocatalytic copper deposition solution that comprises maintaining in a solution, containing complexing and reducing agents for the copper ion and a pH adjuster, a small effective amount of a compound providing a metal value selected from the group consisting of Mo, Nb, W, Rh, rare earths of the actinide series, rare earths of the lanthanide series, and mixtures of the foregoing. Additionally, an improved method of depositing electroless copper is provided which comprises employing the solution herein previously defined.

Dutkewych and Hofman were granted a US Patent 3,770,464 "dry replenishment of electroless copper solutions" in 1973 [93]. Their invention relates a process that comprises addition of replenishers in essentially dry form in a manner that avoids triggering the bath. Typical replenishers that may be added in dry form include one or more of a source of cupric ions such as cupric sulphate, a reducing agent for cupric ions such as paraformaldehyde, a chelating agent for cupric ions, bath stabilizers, brighteners, surfactants, and the like. These ingredients may be added singularly or admixed with each other. Essentially, dry materials are used for replenishment rather than a solution as in the prior art to prevent volume growth of the plating solution.

In 1974, Underkofler and Zucconi were granted a US Patent 3,844,799 "Electroless copper plating" [94]. By means of constant control of the temperature between 70°C and about 80°C, and control of the cyanide ion concentration between 0.0002 and 0.0004 molar, high-quality electroless copper plating processes are carried out from solutions containing cupric ions. The extremely high reliability in printed circuit applications is absolutely essential, so very high-quality plating is required. In addition, the products obtained by the present process have fine grained uniaxial metallurgical structure and can withstand the thermal shock of wave soldering without the cracking of surface lands or plated through holes.

In 1976, Ohno reported in his paper: "Electroless copper plating from an iminodiacetate bath" [95] that the mechanism of electroless copper deposition from a new iminodiacetate (IDA) bath is found to be fully explained by the mixed potential theory. The bath is an aqueous solution of cupric sulphate, iminodiacetic acid, formaldehyde, and sodium hydroxide. As the pH increases, both the oxidation of formaldehyde and the reduction of the Cu-complex are promoted, and hence the deposition rate increases. Under optimum conditions, the rate of deposition from the IDA bath was comparable with that from the EDTA bath.

Arisato and Koriyama were granted a US Patent 4,138,267 "compositions for chemical copper plating" in 1979 [96]. Compositions for chemical copper plating heretofore have the serious drawback that they are harmful to the human body due to the use of formaldehyde serving as the reducing agent and tend to dissipate during plating, deteriorating the environment. Moreover, the disproportionating reaction of formaldehyde inevitably causes the decomposition of the compositions, reducing the stability thereof. In order to eliminate these drawbacks, a boron hydride compound was used as a reducing agent. The compositions consist basically of a water-soluble copper salt, a copper complexing agent, and a boron hydride compound. However, the boron hydride compound still has the drawback that the compositions have very poor stability and are very disadvantageous for use in industrial operations. Furthermore, the compositions give platings which are not satisfactory in gloss. The object of this invention is to provide compositions having outstanding stability, and capable of giving plated coatings having an excellent gloss.

1.3.4 Deposition Substrate

People also studied electroless plating on various substrates to understand the electroless plating better during these 20 years.

In 1961, Shipley Jr. in his US Patent 3,011,920 "method of electroless deposition on a substrate and catalyst solution therefor" [97] provided electroless metal deposit coatings in the manufacture of printed electrical circuits, as linings for nave guide cavities, as an initial coating in electroforming, and for decoration. The principal object of this invention is to provide improved methods and materials for depositing electroless meter coatings.

At the 1961 American Electroplaters Society Convention in Boston, MA, DeLong presented a paper: "plating on magnesium by electrodeposition and chemical reduction methods" [98]. In his paper, DeLong stated that not only does plating on magnesium produce good protection, but also good grounding, conductivity, and solderability in the manufacture and use of electronic equipment in the aircraft and missile fields. In 1965, Gutzeit was granted

US Patent 3,122,578 "chemical nickel plating of magnesium and its alloys" [99]. This patent provides a process and a plating bath upon a Mg or Mg alloy surface of a workpiece that are productive of a continuous smooth coating of Ni–P alloy that is free from blisters and other defects characterized by lack of adhesion.

US Patent 3,357,854 "nickel plating process" was issued to Hays on December 12, 1967 [100]. He has performed an electroless process and bath for plating metal article with nickel containing a water-soluble nickel salt, a hypophosphite and a trace amount of inorganic alkali metabisulfite in an aqueous solution. While the process is primarily intended for the plating on aluminum metal and alloys, it is also suitable for other base metals, for instance iron, steel, copper, or brass.

In 1968, Iwasa, Yokozawa, and Teramoto [101] studied electroless nickel plating on silicon wafers with p–n junctions using conventional solutions, and a pronounced difference in plating rate between p- and n-type surfaces was observed. Further experiments show that the rate difference probably should not only be attributed to the photovoltaic effect generated at the p–n junctions but also to the electronegativity difference between p- and n-type Si.

At IBM TDB in 1970, McBride, and Vlasak proposed "activating a nonconductive substrate with sodium borohydride $NaBH_4$, prior to electroless plating" [102]. This is an improved process for activating a nonconductive substrate, such as glass or ceramic, prior to electroless copper or nickel plating. After the substrate is ultrasonically cleaned in an alkaline solution, and rinsed in deionized water and permitted to dry, it is then placed it into a 0.1–1.0 M $NaBH_4$ water solution for approximately 1 min. Next, the substrate is removed from the $NaBH_4$ solution and the excess is drained off. Finally the substrate is placed directly in an electroless Ni plating solution at 160–200°F or an electroless Cu solution at room temperature (75 °F). Also in 1970, Parker and Shah [103] studied the stress of electroless nickel deposits on beryllium. They concluded from their data that with close control of the plating parameters, reproducible, essentially "stress-free" electroless nickel coatings can be plated on critical optical beryllium surfaces such as aerospace mirrors.

US Patents 3,772,056 "sensitized substrates for chemical metallization" [104] and 3,772,078 "process for the formation of real images and products produced thereby" [105], both issued on November 13, 1973, to Polichette et al. describe processes in which previously-etched, nonmetallic substrates are coated with a solution containing nonprecious metal ions, that is, ions of copper, nickel, cobalt or iron, and dried to form an adherent coating of the metal ions. Thereafter, the metal ions are reduced to the metallic state and the substrate is plated with a compatible electroless plating bath.

IBM TDB, Haddad and Schmeckenbecher, published a report in 1976 "activation of catalyst for electroless plating on molybdenum" [106]. Conventionally, Mo surfaces are activated for electroless nickel plating by a dip in a palladium chloride solution. Mo metal goes into solution and a corresponding amount of Pd metal is deposited on the surface of the Mo. Only relatively small amounts of Pd are deposited, since the Mo surface is readily oxidized in air and the molybdenum oxide does not undergo the exchange reaction with palladium.

In US Patent 3,993,808 "method for electroless plating gold directly on tungsten or molybdenum" [107], issued on November 23, 1976, Inaba stated that gold can be directly plated onto W or Mo by an electroless plating process using a plating solution including a gold compound and having pH not less than 8, and desirably not less than 9. The composition of the plating solution is as follows: gold potassium cyanide $KAu(CN)_2$ 5 g/L, nickel chloride $NiCl_2{\cdot}6H_2O$ 5 g/L, EDTA·2Na 50 g/L, ammonium chloride NH_4Cl 30 g/L.

On January 25, 1977, Miller and Cavitt were granted a US Patent 4,005,229 "novel method for the rapid deposition of gold films onto nonmetallic substrates at ambient temperatures" [108]. They state that Au films are deposited on nonmetallic substrates very rapidly at ambient temperatures by sensitizing and activating a surface of a nonmetallic substrate to be coated and subsequently contacting the activated surface with a solution containing a gold salt which is reduced to a metallic gold film in the presence of divalent mercury, cadmium or lead ions.

Haddad and Suierveld reported "depositing crack free heavy electroless nickel coating" [109] on January 1, 1978. In semiconductor packaging as well as other applications it is frequently desirable to deposit electroless nickel layers. However, when the thickness of the nickel layers so deposited exceeds 125 microinches, stresses within the layer cause microcracking. In this process Ni layers can be electrolessly deposited at thicknesses greater than 125 microinches, and in which the stresses are significantly lower and without microcracks. In the process the underlying layer or substrate is activated in the conventional manner. The underlying layer, for example a Mo layer, is exposed to an aqueous solution of palladium chloride. A layer of Ni having a thickness not to exceeding 100 microinches is then deposited on the sensitized area using the conventional electroless nickel solution. Also in 1978, Vratny in his US Patent 4,122,215 "electroless deposition of nickel on a masked aluminum surface" [110] reported that the method is particularly useful for fabricating bonding pads on Al metallized semiconductor devices and for creating beam leads. The described method deposits a thick Ni layer directly on Al without the use of intermediate layers or surface activation as required in the prior art. The method basically comprises immersion in a stop-etchant that simultaneously removes aluminum oxide and activates the surface; immersion in a solution which activates the Al with nickel ions and deactivates mask material; and immersion in a novel electroless nickel bath.

In 1979, Bandrand in his paper "autocatalytic (electroless) plating on aluminum" [111] reported that autocatalytic (electroless) Ni plating on Al is practiced extensively, even though there are numerous problems in preparing Al alloys and selecting a compatible solution.

1.3.5 Application

The application of electroless nickel deposits during this stage has made more progress than those metals in the early stage of development. For the sake of convenience, the specific contents will be discussed together in the Section 1.4.3.

1.4 RAPID DEVELOPMENT OF PERIOD (1980–1999)

After its period of slow development, electroless plating went to the period of rapid development from the very beginning of the 1980s for the following 20 years. From Fig. 1.1a, one can see that development during this period was rapid, and the number of articles (from Elsevier, ECS, or ACS) and patents (from US Patents) of electroless plating approach 3632, 590, 203, and 5504 in 1999, which are 11.7 and 2.8, 6, and 6 times of those by the end in the period of slow growth. Therefore, we call it a period of rapid growth.

The main features in this period of time are as follows: (1) the nature of electroless plating such as its electrochemical behavior and mechanism of formation was studied in greater depth; (2) as well as studying the electroless bath itself, the properties of the deposits have been studied in greater depth; (3) large scale of application of electroless plating was extended to many industrial fields; (4) ternary and multicomponent alloys and composites have been investigated extensively; (5) China finally started research and development of electroless plating after lagging behind for 20 years. Since then, electroless plating has developed rapidly in China; and (6) electroless plating Fe–B based alloys have been proposed and developed as a new direction and field in the electroless plating. These features will be described briefly in the following sections.

1.4.1 Studying the Nature of Electroless Plating

In order to understand the kinetics and nature of electroless plating, researchers have proposed reaction mechanisms for the plating process. Because EN has been studied in greater depth than other materials, this section will concentrate on EN studies. In 1990, Mallory summarized the principal reaction mechanisms that had been proposed [112] to explain electroless nickel deposition since the publication in 1946 of the paper by Brenner and Riddell [7]. There are four principal mechanisms that were known before the rapid development period. In fact, when Brenner and Riddell [7] published their first discovery of electroless nickel plating, they proposed the so-called atomic hydrogen mechanism. The second mechanism is called the hydride transfer mechanism, which was first suggested by Hersch [113] in 1955. Electrochemical mechanism is the third one [114,115], and the fourth is the metal hydroxide mechanism, which was first proposed by Cavallotti and Salvago in 1968 [116].

Most of the features of EN plating can be explained by the proposed mechanisms. However, none of them can explain all the experimental phenomena satisfactorily. The atomic hydrogen mechanism cannot explain why the stoichiometric utilization of hypophosphite is always less than 50%, and does not account for the simultaneous reduction of nickel and hydrogen [112]. According to the hydride transfer mechanism, hydrogen evolution should be at least equivalent to metal deposition; however, it was contrary to some experimental observations [117]. The electrochemical mechanism predicts the nickel ion concentration must have significant effect on the rate of nickel deposition; however, the experimental fact is just the opposite [112]. As for the metal hydroxide mechanism, it cannot explain the experiments in which the anodic and cathodic reactions take place in a cell with two compartments separated by a glass frit [112, 118].

Because of these factors, the mechanism of electroless plating is still the subject of much research. First, scientists further modified and extended the existing mechanisms. Then, researchers proposed other mechanisms to describe the nature of kinetics in the EN plating process. In the middle of the slow development period (1968), Paunovic proposed a mixed potential mechanism [119]. Many scientists used this mechanism to explain their experimental data or to modify the theory during this stage [120–124]. But, Tzeng argued that the mixed potential theory cannot fully explain the EN plating process [125]. Also, Abrantes and Correia pointed out that the mixed potential theory is unable to describe their results [126]. They proposed their own mechanism but no name was given.

During this period, van den Meeraker proposed the so-called universal mechanism [117]. In fact, his theory is just a summary of the reactions in the existing mechanisms, and did not essentially improve the theoretical situation of electroless plating.

It should be pointed out that the mechanisms of electroless plating vary a great deal, because the process is very complex. The total reaction is composed of an oxidation reaction of a reducing agent, deposition reaction of metal ion, and catalytic reaction. These are affected by various factors such as metal salt(s), complexing agent, additives, pH value, temperature, and stability agent. These reactions and factors affect each other; therefore, it is difficult for the whole mechanism to be elucidated by experimental approaches. Based on such considerations, some scientists thought that finding the generalized correct mechanism may not be a clever idea [127].

By the end of this period, the nature of kinetics of electroless deposition process was fairly well understood [128, 129].

1.4.2 Studying the Properties of Electroless Plating Deposits

There were not many works referring to the properties of electroless plating deposits (EPD) during the slow development period. Many studies of the properties of EPD have appeared since the start of the rapid development period. Almost all aspects of performance have been studied, including physical, mechanical, chemical, electric, magnetic, and welding properties. Here, we will just describe each one of them briefly.

1.4.2.1 Physical Properties

In 1992, Yarkosky and Cacciatore developed a method of preparing Al memory disks with a smooth metal plated finish [130], used for producing extremely smooth metal coatings on zincated aluminum substrates using a special double zincating procedure and/or a specially formulated electroless metal plating bath employing a unique plating process. Ernest et al. in Sandia National Laboratories measured the pore density of the nickel deposits on the copper coupons in 1998 [131]. Their results showed the expected trend of decreasing density with increasing deposit thickness. For deposits thicker than about 50 μm, little further improvement in pore density was demonstrated with increasing thickness. Kerr [132–134] studied the porosity of electroless nickel coatings on mild steel substrates from 1996 to 1998. Two scholars from Utah State University [135] measured the thermal expansion coefficient (CTE) of an EN deposit. They found that the CTE for 11% phosphorus content is from 9 to $12 \times 10^{-6}\ K^{-1}$ for the 0.001 inch plating thickness sample from 100 K to room temperature.

1.4.2.2 Mechanical Properties

Mai et al. measured the hardness of the electroless Ni–P coatings, and found that it increased with temperature to a maximum of 1000 HK 200 at about 350°C [136]. Marshall et al. found that the hardness of deposits increased following heat treatment at 400°C and deposits obtained with superimposed current were generally harder than those produced by electroless deposition only [137]. Parker studied the effects of heat treatment on the properties of EN deposits [138]. They found that EN deposits (heating >200°C) produce structural changes that increase hardness, adhesion and abrasion, and wear resistance. The corrosion resistance, ductility, tensile strength, fatigue life, stress, and magnetic and electrical properties also are affected. The author's group found that codeposition of tungsten in binary electroless Ni–P–W deposit improves the deposit's wear resistance [139]. Duncan and Amey reported in 1984 [140] that the tensile strength and ductility of electroless Ni–P and Ni–B deposits, which are 1100 MPa and >700 MPa, and 0.2 and 1.0% elongation, respectively. In 1993, Yamada et al. [141] studied the strength evaluation and effect of graphite on strength of electroless nickel plating on cast iron. It has been found that the maximum bending strength, about 750 MPa, was obtained with the spheroidal graphite cast iron specimens. And the bending strength of plating film on flaky cast iron gave the minimum value, about 400 MPa. From 1996 to 1999, several papers studied the influence of electroless Ni–P deposits on the fatigue and corrosion-fatigue properties of plain carbon steels [142–146].

1.4.2.3 Tribological and Wear Performances

One of the unique characteristics of EN deposition is the superior wear resistance of the coatings. Many researchers have studied [147–154] the tribological and wear performances of Ni–P electroless deposits during this period of time. Duncan and Arney [155] and Baurand [156] studied also the superior wear resistance of electroless Ni–B coatings. Studied have also found that codeposition of W in binary electroless Ni–P deposit improves the deposit's wear resistance [157]. Ni–P matrix composites with dispersion of micrometric particles have been introduced since the early 1980s in order to improve the tribological behavior of Ni–P. For example, in 1999 Bozzini et al. found that Ni–P coatings showed a lower wear resistance than the corresponding composite (Ni–P(9%)/B_4C) coatings; the wear resistance increased as the heat treatment time increased for both types of coating [158].

1.4.2.4 Electrical and Magnetic Properties

Even in this period of time there were not very many papers published on the electric and magnetic properties of electroless deposits. In 1986, Fernández et al. determined the electrical properties of electroless Ni–P thin films. They found that the phosphorous content of the films increases when the pH of the solution is decreased giving rise to a diminution of the temperature coefficient of the resistance (TCR) [159]. Low TCR films (TCR $\leq$ 100 ppm/°C) are obtained for pH = 3.4. Osaka et al. measured the magnetic and electric properties of electroless Ni–P films after heat treatment as a function of the P content in the deposit in 1988 [160]. They found that the residual Ni determines the film properties of resistivity and saturation magnetization. Moreover, they depend only on the P content and are independent of the preparation conditions. The saturation magnetization (M_s), and the reciprocal of resistivity (ρ^{-1}), are proportional to the amount of the residual Ni after sufficient heat treatment. Khan and Nicholson investigated the effects of solution pH on magnetic properties of chemically deposited Co–Ni–P and Co–P films in 1986 [161]. They found that the microstructure of these films dominates their magnetic properties. The Co–Ni–P film, deposited at optimum pH, supports higher recording densities than the Co–P films. In 1991, Hur et al. measured the magnetic properties of electroless Ni–Cu–P alloy deposits [162]. They found that deposits containing more than about 28.5% Cu showed stable nonmagnetic characteristics even after heating at 400°C, whereas lower Cu content deposits showed ferromagnetism even after annealing at 300°C. Liu et al. found in their experiment in 1991 that electroless Co–Ni–Mn–P alloy films had a good perpendicular property [163].

1.4.2.5 Corrosion Properties

Corrosion resistance is another unique property of EPD. More attention was paid to studying the corrosion properties of EPDs during this period; hence, more studies of the corrosion property of EPD have performed in these 20 years.

The reasons for this are twofold. One is that EPDs have few defects such as porosity meaning that their corrosion properties are superior to those of electrolytic coatings. Another is that the equipment for manufacturing EPD is not complex.

In 1986, Duncan measured the corrosion resistance of high phosphorus EN coatings more in detail, and found that these coatings have exceptional resistance to corrosion and erosion in many environments [164]. In addition, he characterizes the performance of these coatings in a broad range of industrial environments.

Levina et al. in 1989 performed an investigation of properties of a nickel–phosphorous coating in a hydrogen sulfide-containing medium [165]. Their experimental results and the field test data showed that under the action of the hydrogen sulfide-containing formation fluid the nickel–phosphorous coating possesses the necessary corrosion resistance and makes it possible to increase the service life of the working parts of the pumps. Submersible centrifugal electric pumps are one of the basic means of mechanized recovery of petroleum from wells.

In 1990, Tomlinson and Carroll studied the relationship between the substrate roughness, deposit thickness, and the corrosion of electroless nickel coatings [166].

In 1993, Doong et al. measured the corrosion behavior of the electroless Ni–P plating interlayer in the modification of the TiN coating [167]. They found that for a thicker Ni–P interlayer (greater than 5 μm), no appreciable corrosion was observed up to hundreds of hours after salt spray testing.

The preparation, formation, and corrosion properties of electroless Ni–W–P amorphous alloys were studied systematically in 1996 by the author's group [168]. It was confirmed that amorphous deposits with higher phosphorus content showed better corrosion resistance than those with lower phosphorus content in acid or alkaline solutions; also an amorphous deposit showed better corrosion resistance in alkaline solutions than in acid solutions. Compared with the corrosion resistance of electroless amorphous Ni–P deposits, the addition of W into Ni–P deposits did not usually improve the corrosion resistance.

In the last several years of the rapid growth period, some authors investigated the corrosion performances of electroless Ni–P composite coatings, such as Ni–P–Si_3N_4 [169], Ni–P–Polymer (PVC, PVA, PAm) [170], Ni–P–MoS_2 [171], and Ni–P–TiO_2 [172]. Some groups [173,174] concluded that the corrosion resistance of the electroless Ni–P composite coating is significantly less than that of the electroless Ni–P coatings. This situation caused the ASM handbook [173] to believe that electroless Ni–P composite coatings are not normally recommended for applications requiring good corrosion resistance. However, other groups concluded from their experimental data that the corrosion performance of Ni–P composite coatings was found to be satisfactory [170,175]. Therefore, thorough research into the corrosion resistance of electroless Ni–P composite coatings is needed.

1.4.2.6 Soldering Properties

It is not easy to find any regular research papers reporting the soldering properties of electroless nickel deposits before the 1980s, which means that there were not many studies in this field during the early and slow developing period of time. As electroless nickel plating has been increasingly used in the semiconductor microcircuit industry since the 1980s, people have paid more attention to the solderability properties of EN plating. In 1991, Baudrand studied how to use EN to reduce the Au requirements in semiconductor microelectronic industry [176]. He measured quantitatively the wetting force and ultrasonic bond strength of Al wire for Ni–B and Ni–P deposits. Riedel summarized the solderability properties of electroless Ni–B and Ni–P deposits published mainly in the 1980s in his book "Electroless nickel plating" in 1991 [118]. However, there no experimental data was cited or listed in that book.

The author's group measured the solderability of several electroless plating alloy system deposits systematically during the 1990s, including Ni–B, Ni–P, and Ni–P–B [177], Ni–B and Ni–P [178], Ni–Cu–B [179], Ni–W–P [139], Ni–Sn–P, Ni–Cu–P, and Ni–Sn–Cu–P deposits [180].

Other authors also studied the solderability properties of electroless plating deposits. Yi measured the solderability of electroless Au plated onto electroless Ni–B over Kovar substrate in 1991 [181]. Lee and Lin measured the solderability of electroless nickel alloys using a wetting balance technique in 1993 [182]. They also studied the wetting behavior between solder and electroless nickel deposits in 1994 [183]. Gaudiello in 1995 [184] reported an autocatalytic electroless Au process suitable for electronic packaging applications, particularly wirebonding needs. In 1998, Watanabe et al. [185] investigated Au wire bondability of electroless gold plating using a disulfiteaurate complex.

These studies show that the solderability of electroless alloy deposits has made significant progress over this 20-year period.

1.4.3 Large-Scale Application in Many Industries

As development of science and engineering and progress of industry, the application of EN deposits has made some progress in the slow stage of development. However, the large-scale application in many industries started in this rapid development of stage.

There were no complete statistical data for such development of EN deposits during these two periods of stage. However, from the papers published by Baudrand [186], Miller [187], Mayer [188], Hajdu [189], Durkin [190], and Hajdu and Zabrocky [191], some main features can be caught.

Table 1.1 shows the percentage of EN use by industry [186].

Table 1.1 shows that EN use was widespread across almost all industries from the last years of the slow development stage and by the beginning of the

TABLE 1.1 Percentage of EN Use by Industry [186]

Industry	Percentage		Industry	Percentage	
	1979	1982		1979	1982
Plating on a Metal Substrate					
Electronics	20	25	Chemical industry	45	5
Oil and gas production	15	20	Mechanical industry		10
Aerospace	10	5	Food industry		5
Automotive industry	5	5	Textile industry		15
Energy industry	5	5	Printing industry		
			Medical industry		
			Others		
Plating on a Plastic Substrate					
Automotive industry	–	80	Pipe fittings	–	5
Various appliances	–	10	Navigation, entertainment, furniture, etc.	–	5

1980s. From the statistics of Miller [187], the world EN market totaled 1460 short tons in 1987, of which the United States held 540, Europe 280, the Far East (includes Japan, Korea, Chinese Taiwan, and Australia) 300, the Eastern bloc 180, and other countries (South Africa, South America, etc.) 160 (the units are all short tons). The United States accounted for the largest share of world EN market, at 37%. In 1993, the US EN market was estimated to be 1600 tons [188]. Hajdu [189] estimated the annual growth of the market for aluminum memory disks fabricated by EN was 288 million disks (1996), 1997 354 million disks (1997), and 435 million disks (1998). It is a really huge amount.

As for the price of EN, it has gradually decreased during these 40 years. According to Durkin [190], back in the 1970s, the average cost for conventional Ni–P deposits was estimated to be over US $3.50 per mil square foot. In the early 1990s, the estimated average cost was somewhere between $2.00 and $3.00 per mil square foot. The reason for this decrease was as the EN deposits have gotten better and more efficient by having greater stability and longer life, the operating costs have decreased somewhat.

It is interesting to note that the EN market size by product is mainly mid phosphors deposits (65%). High phosphors, low phosphors, and other EN products are only 28%, 5%, and 2%, respectively [191]. So, it is obvious that more studies of high and low-phosphors EN are now needed.

1.4.4 Investigation of Ternary and Multicomponent Alloys and Composites

Before 1980, there were only a few publications on electroless poly alloys and composite coatings. Poly alloys are Ni–P alloys containing ternary, quaternary, or more elements. Electroless composite plating is a process in which Ni–P alloys are codeposited with hard or lubricating particles. In order to enhance the properties of EN coatings, researchers investigated more electroless poly alloys and composite coatings during this stage than in the two early stages.

1.4.4.1 Electroless Multicomponent Alloys

Electroless nickel ternary alloy (Ni–Me–P, Me = Cr, Mo, W, Fe, Cu, Zn, Sn, and Sb), the ternary alloy (Ni–Me–B Me = Mo, Fe, Co, Cu, Sn), and the ternary alloy (Co–Me–P Me = Ni, W) were investigated during this 20-year period. For these electroless nickel ternary alloy deposits, most authors studied Ni–Mo–P [192–195], Ni–W–P [196–200], and Ni–Cu–P [201–206]. Only one or two studies have been reported for other electroless Ni–Me–P alloy deposits, namely: Ni–Fe–P [207], Ni–Cr–P [207], Ni–Re–P [208], Ni–Zn–P [209–211], Ni–Sn–P [212,213], Ni–Sb–P [213]. The situation is the same for Ni–Me–B: Ni–Mo–B [214], Ni–Fe–B [215], Ni–Co–B [216], Ni–Cu–B [217], Ni–Sn–B [212]. Co–Ni–P and Co–W–P was studied by Georgiou et al. [218].

In 1990, Gonzalez et al. reviewed most of the ternary Ni–Me–P (Me = Cr, Mo, W, Fe, Cu, and Sn) alloy deposits [219]. They discussed the electrochemical and kinetic parameters, surface analysis and postplating operations as a link between Ni–P and Ni–Me–P autocatalytic alloys.

Regarding electroless quaternary alloy deposits, only a few reports were published even in the 20 years where development was rapid. Hayashi [220] summarized knowledge of electroless quaternary alloy deposits to 1991. In 1983 and 1989, Osaka et al. [221, 222] prepared electroless Co–Ni–Mn–P deposits, which are used for perpendicular recording media. In 1987, Osaka et al. [223] studied electroless Co–Ni–Re–P deposits. In 1998 and 1999, the author's group systematically investigated the preparation and properties of electroless plating Ni–Cu–Sn–P alloys [224, 225].

Only one paper on electroless quinary alloy deposits was published during this period, on Co–Ni–Mn–Re–P [226]. The aim is to improve the performance of perpendicular magnetic media.

Hajdu and Zabrocky predicted in 2000 that poly alloys would become the next frontier of electroless nickel technology [227].

1.4.4.2 Electroless Composite Deposits

Research into the incorporation of hard or soft particles into electroless Ni–P coatings during this stage included such systems as Ni–P + Cr_2C_3 [228], Ni–P + TiO_2 [175], Ni–P + SiC [229, 230], Ni–P + Al_2O_3 [231], Ni–P + diamond [232], Ni–P + Si_3N_4 [233], Ni–P + PTFE [234, 235], Ni–P + MoS_2

[236, 237], and Ni–P + CF_x [238]. Straffelini et al. studied the electroless composite coatings: Ni–P + BN, Ni–P + PTFE, Ni–P + MoS_2, Ni–P + SiC, and Ni–P + SiC + PTFE [239] in 1999.

During this stage, researchers also investigated another kind of electroless composite coating, namely Ni–P + metallic powder. In 1989, Osaka et al. [240] studied the electroless composite films Ni–P + Zr and Ni–P + Nb. In 1999, Hu et al. [241] investigated the electroless composite coating of Ni–W–P + Ni–Ti(particles)–Re(rare earth) deposited on the surface of diamond. In 1999, Chang et al. [242] studied several kinds of composite powders obtained through electroless copper and silver plating, including SiC_w + Cu; SiC_w + Ag; Al_2O_{3p} + Ag; $Carbon_{sf}$ + Ag; and $Saffil_{sf}$ + Ag. Here, w is whisker, p is particle, and sf is short fiber. The composite powders were successfully hot pressed to produce composite specimens containing various amounts of reinforcements.

1.4.5 Electroless Plating Began and Developed Rapidly in China

China began her EN research and application around the end of the 1970s and at the beginning of 1980s. Therefore, EN research in China started about 20 years later than that in the rest of the world. The main reason for the delay is the well-known Cultural Revolution. However, the development of EN in China has been very rapid. In general terms, by the end of this stage, that is, by around 1999, China had caught up the rest of the world in the main aspects of EN development. It was obvious that in addition to the researchers' hard work, commitment from industry and government are indispensable factors in the development of EN. Because of this powerful commitment, China's EN has been promoted rapidly and has achieved remarkable things. The following facts can sketch out the situation [243–245]. Figure 1.1a, curve 3, shows the number of electroless plating articles from the CNKI(China) database, which indicates that before 1979 only a very few EP papers were published in China, but thereafter China really caught up with EN development in the rest of the world.

After electroless plating had been in development for more than 10 years, The first China Electroless Plating Conference was held in Nanking in 1992. Two hundred people attended this meeting. Eighty seven papers were presented from 22 universities, 13 research departments, and 12 industrial companies for this conference. There were published in the Conference Proceedings edited by the Chinese Corrosion and Protection Society. After that, a conference was held every 2 years in China, hence four conferences were held by the end of this stage.

Besides holding this conference, the Society also opened electroless plating technology training courses in order to train junior scholars and workers.

In 2002, the Society organized a survey for the development of EN in China. The data from the survey can also be used for illustrating the development of EN during this stage.

By 1998, there were 90 EN companies in China compared with only one factory in 1992. These used more than 3000 tons of sodium hypophosphite from the biggest factory, Rhodia and the other five or six companies producing sodium hypophosphite. That means that electroless plating in China advanced quite rapidly in these 20 years.

The application of electroless plating in China has been extended to almost all industries, including large-scale petrochemical facilities, oil pipeline used in exploitation of oil fields, electronics, telecommunications, home appliances, aerospace, petrochemical, military, textile, paper printing, motorcycles, and machinery products. The number of EN products is very large, for example, 4 million computer and printer parts, and 70 million per year for mobile phone buzzers are produced in the Dalian Legend Surface Treatment Co., Ltd.

In addition, China has more than 90 research units in which many scientific researchers and engineers and technical personnel were working on the electroless plating at that time.

1.4.6 Electroless Plating Fe–B Based Alloys Have Been Proposed and Developed

The author's group was one of the units that carried out early research into electroless plating in China. After studying EN for several years, we wondered that why there were no reports on Fe-based electroless plating coatings. Fe, Co, and Ni all belong to the iron group in the periodic table. Electroless plated Ni–P (B) and Co–P (B) alloys had been developed and applied to many industries for nearly 40 years at that time. If Fe-based alloys can be plated by the electroless method, how can we prepare them, and if we can, what are the differences between the technologies of electroless plating Ni-based alloys and Fe-based alloys? Under the guidance of the author, the group started the journey of exploring the possibility of electroless plated Fe-based alloys through a project "the electroless plating Fe–B amorphous alloys" that was a graduate student research thesis topic.

Through searching literature more extensively, only one paper of electroless Fe–P deposit was found, namely, "chemical iron-phosphorus film" written by Ruscior and Croiala in 1971 [246]. The paper pointed out that electroless Fe–P films can only be obtained by using a combination of an electrochemical (galvanic) process caused by the Cu–Al couple and chemical deposition resulted from reduction. However, the authors did not analyze the plating mechanism in detail and gave no details of the electroless deposition rate. In addition, their preparation technology of Fe–P is rather simple, and no structure and properties were measured, except the very simple magnetic property that has been reported. It should be pointed out that this paper was published more than 20 years after the discoveries of Brenner and Riddel [7]. It is surprising that no subsequent papers were published by these authors, and no similar papers referred to electroless plating Fe–P alloys from other researchers were published until the middle of the 1980s, when we had just decided to do the electroless plating Fe–B deposits. Of course, up to the middle of the 1980s, no reports from electroless Fe–B deposits could be found in the literature. Considering the above factors, we had actually chosen a very challenging topic to research.

From the very beginning of our experiments, a Cu–Al couple (Cu plate is the substrate, and the Al plate is suspended in the bath solution with a wire) similar to that of Ruscior and Croiala was used. No Fe–B deposits were obtained even after doing the experiment for several months. The research team discussed and consulted over and over. By changing every parameter in the bath solution one at a time, and by paying attention to every detail in the experiment, finally Fe–B deposits were obtained after more than 1 year. Since then, the deposition parameters, the structure and morphology, the crystallization process, and the mechanical and magnetic properties of amorphous Fe–B alloy deposits have been obtained and investigated systematically. An excellent masters thesis was produced from this project, and several papers were published in national and international journals [247–251].

However even the plated amorphous Fe–B deposits easily got rusty after deposition had been carried out for a significant period of time, which is not like electroless Ni or Co-based alloy deposits. To overcome this difficulty, it seems necessary to add extra elements. After adding W or Mo into Fe–B alloys, the Fe–W–B and Fe–Mo–B deposits do not rust so easily [252, 253]. Following this routine, a graduate student in author's group chose "electroless Fe-based multicomponent alloy deposits" as her MS and PhD project. Multiple alloy systems such as Fe–Mo–B, Fe–W–B, Fe–Sn–B, Fe–Mo–W–B, and Fe–Ni–P–B were successfully deposited, and various properties of the deposited coatings were systemically studied. Thus, more papers of electroless Fe-based alloy deposits from the author's group were published [254–259].

Until 1999, the last year of this stage, we have investigated electroless Fe–B alloy deposits for more than 10 years, the papers published in international journals in English more than 10, which did not include the papers published in Chinese. It is shown that our works have been recognized by the international peers.

1.5 IN-DEPTH DEVELOPMENT AND NANOELECTROLESS PLATING STAGE (2000–PRESENT)

After entering into the new century, electroless plating just like a flying horse ran into new fields. These new fields are the in-depth investigation of the mechanism and theory of electroless plating, the rapid rise and development of nanoelectroless plating, in-depth study of poly alloy systems and composite deposition, and electroless plating on difficult substrates. The first two are most obvious, which is why we call this stage the in-depth development and nanoelectroless plating stage, and it is these two topics that will be discussed in this section.

Before discussing the two points, it is appropriate to glance at recent global production.

In today's developed countries, one can say that it is almost impossible to find an industry that does not use chemical nickel plating technology. In the late 1990s, the EN market in the United States exceeded 1 billion US dollars, Japan, Germany and Europe had a plating annual output of about 5 billion US dollars; hence electroless nickel plating has become a very mature technology. Today, there are less than 1000 EN plating facilities in North America. About two-thirds are job shops, and one-third are captive facilities of manufacturing plants. The largest users were automotive applications at 27%, then the oil/chemicals industry with 19% of the total, and then the electronics industry consuming about 14% of the EN produced in the United States [260]. From various data, it can be estimated that the global electroless plating is about 15 billion US dollars in 2010 [260]. It is a really huge amount.

It is indicated from statistical data that the output value of the electroless nickel market in China has exceeded 40 billion Chinese Yuan (≈ 6 billion US dollars) in 2010. That means that EN in China accounted for 40% of global yield in 2010. Figure 1.3 shows the output value and growth rate from 2006 to 2010 in China [261].

The global rapid development of electroless plating since 2000 is evident from the numbers of articles in various databases and of US Patents in electroless plating (Fig. 1.1a). In the 14 years from 2000 to 2014 (November), the number of articles (from Elsevier, ECS, ACS, or CNKI) and patents (from US Patents) of electroless plating are 10,565; 1,658; 918; 6,625; and 20,345 in 2014, which shows the really rapid growth of the subject after the period of rapid growth in 20 years.

A point should be noticed that there is some repetition of contents in some papers, and even in some MS or PhD theses. The reasons are complicated, and it may in part be due to the huge number of papers published. This phenomenon should not occur. However, discussing such phenomena is not the task of this book; it will not be discussed further. It is hoped that some scholars will analyze and discuss the problem in detail someday.

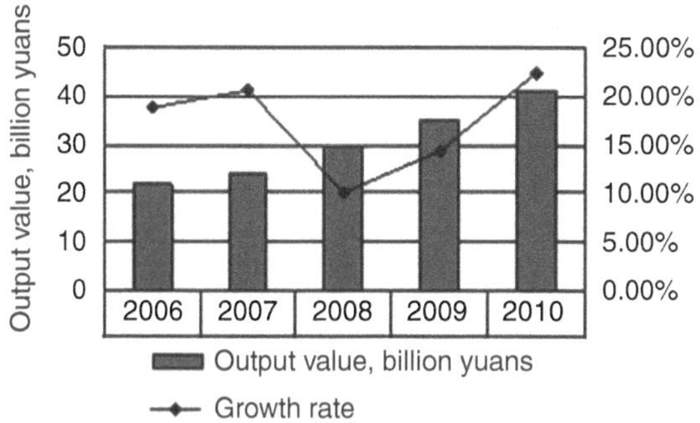

FIGURE 1.3 Development of electroless Ni–P industry from 2006 to 2010 in China [261].

1.5.1 In-Depth Investigation of the Mechanism and Theory in Electroless Plating

The mechanism of electroless plating has been studied since the discovery of EN. However, the experimental data cannot be explained satisfactorily, so work remains to be done in this area. One theory can explain one set of data, and another one can explain a different set, as mentioned in Section 1.4.1. Therefore, attention is still paid to the subject.

There are three main features of the in-depth investigation of the mechanism of electroless plating as detailed in the following paragraphs.

First, the deposition process was investigated, sometimes even using *in situ* monitoring of the deposition to give a deep understanding of the mechanism. For example, evaluation of the initial deposition rate of electroless Ni–P layers by quartz crystal microbalance (QCM) [262], impedance measurements for an EN plating process [263], and observation of the nucleation and growth of electrolessly plated nickel deposited from different bath pHs by transmission electron microscopy (TEM) and QCM [264].

Second, people proposed models to study the mechanism, for example, models of Ni–P electroless deposition in ammoniacal solutions [265], for which a microscopic model of the EN reaction was proposed [266].

Third, people began to study the mechanism of electroless plating using the microscope theory such as *ab initio* molecular orbital approach, density function theory, etc. [267–273]. Because the microscope theory has a very profound physical meaning, we can understand the more fundamental and deeper levels of this problem. This was never done during the former stages.

1.5.2 Rapid Development of Nanoelectroless Plating

The development of nanoelectroless plating basically happened in this period of time (i.e., 2000–2014), as can be shown by the number of published papers on this topic. From the Elsevier, ECS, and ACS Databases, one can find that there were only 57, 5, and 3 papers of nanoelectroless plating published before 1999, but there were 902, 138, and 77 papers published by the end of 2009, respectively. The number of published articles on nanoelectroless plating from the Elsevier, ECS, and ACS Databases were 1610, 226, and 118, respectively, till January 2013. That is to say, there were only a very few nanoelectroless plating studies during the former half the century of electroless plating. But, development has been very fast. This is somewhat accompanied by the development of nanoscience and technology, which has developed rapidly since the new millenium [274].

Nanoelectroless plating can be divided into three categories. The first is electroless nanocomposite plating, which is like electroless composite plating, but the codeposited composite particles are nanosized. For example, EN + Ni-cordierite and EN + diamond nanoparticles [275], EN + diamond and EN + CNT nanocomposite films [276], EN + $A1_2O_3$ nanocoating [277, 278], electroless Ni–B + diamond nanocomposite coatings [279], EN + polymer nanocomposite coatings [280], EN + nano SiC coatings [281], EN + nano-SiO_2 particles [282], electroless Ni–Zn–P + TiO_2 nanocomposite coatings [283, 284], nanocomposite coating of EN + Ag [285], and nanocomposite coating of EN + Au [286].

The second type is electroless plating deposits including metals and alloys onto carbon nanotubes (CNT) and other nanomaterials such as nanopowders. However, work has concentrated on electroless plating deposits onto CNT in recent years. Many reports have been published; here, we just illustrate some examples, such as metal-decorated CNT [287, 288], multiwalled carbon nanotubes (MWCNTs) electroless plated with Co–Fe alloy [289], EN-coated carbon nanofibers [290], multiwalled carbon nanotubes (MWCNT) with EN coating [291], and EN layer on single-walled carbon nanotubes (SWCNT) [292]. It should be pointed out that although this second category is sometimes also called nanocomposite electroless plating, it is very different from the first category because this category deals with its electroless plating onto CNT or other individual nanoobjects, whereas the first one is electroless plating codeposited with nanoparticles.

The third type is the fabrication of large nanofilms and structures or even devices by electroless deposition. Examples of this approach include the following: formation of metal nano- and micropatterns on self-assembled monolayers by pulsed laser deposition through nanostencils and electroless deposition [293], large area 50 nm period grating by multiple nanoimprint lithography and spatial frequency doubling [294], Cu nanotube arrays were fabricated by electroless deposition using porous anodic aluminum oxide membranes as templates [295], nickel monosilicide contact formation in electrolessly etched large arrays of silicon nanowires (SiNWs) deposited onto nickel electrodes [296], and nanocrystalline metal and alloys (Au, Ag, Pd, Ni, and Pd–Ag, Pd–Ni, Pd–Au, Pd–Pb, and Pd–Ni–Au) films coated by electroless deposition onto metal substrates (Cu, Ni) [297].

1.6 SUMMARY AND PROSPECT

After more than 60 years of development, electroless plating has approached a pretty high level. Its applications extend through almost all industrial sectors; it is almost impossible to find an industry operating without the use of EN plating technology in all developed and many developing countries. In addition, the electroless plating technology has emerged as one of the leading growth areas in surface engineering and metal finishing. According to the estimation of Agarwala and Agarwala in 2003 [298], electroless plating technology is growing at a rate of over 15% per annum; certainly no other chemical technique is growing at this rate.

What, then, are the future prospects for electroless plating technology? We would suggest the following trends and directions of development:

Further research is required to obtain lower cost, higher performance nickel and other electroless deposits.

The development of environmentally friendly electroless plating technology is a priority to comply with present and future legislation. The aim is to create a real green electroless plating system.

Significant work is required to elucidate the kinetics and mechanisms of the reactions involved in electroless plating; in particular, *ab initio* approaches are felt to offer potential for improved theoretical description of the process.

Nanoelectroless plating is in the ascendancy and there are many reasons to pursue further developments.

Fe–B based electroless plating alloys have been developed and valuable research carried out for nearly 20 years; however, research relating to applications for these deposits, and their industrial application, is limited. Significant further work is required to identify the problems inherent in these processes and develop solutions for them.

REFERENCES

[1] Wurtz A. Comp Rendus Acad Sci Paris 1844;18:702. 1845;21:149.
[2] Breteau P. Bull Soc Chim 1911;9(4):515.
[3] Roux FA. US Patent 1,207,218 Process of producing metallic deposits, December 5, 1916.
[4] Pall C, Friederici L. Bet Deut Chem Ges 1931;64(B):1766.
[5] Scholder R, Heckel H. Z Unorg Allgem Chem 1931;198:329.
[6] Scholder R, Haken HL. Ber Deut Chem Ges 1931;64(13):2870.
[7] Brenner A, Riddell GE. J Res Nat Bur Stand 1946;37:31. Proc Am Electropla Soc 1946;33:16.
[8] Brenner A, Riddell GE. P&SF 1998;85:54.
[9] Brenner A, History of the electroless plating process, In: Brenner A, editor Symposium on electroless nickel plating. Philadelphia, PA: ASTM Special Technical Publication, No. 265; 1959, p. 1.
[10] Brenner A. P&SF 1984;71:24.
[11] Brenner A, Riddell GE. Plating 1947;39:385.
[12] Brenner A. Metal Finish 1954;52(11):68. 1954;52(12):61.
[13] West HJ. Metal Finish 1954;52(3):64.
[14] West HJ. Metal Finish 1954;52(7):72.
[15] West HJ. Metal Finish 1955;53(1):62.
[16] Maclean JD, Karten SM. A practical application of electroless nickel plating. In: 41st annual proceedings A.Z.S.; 1954, p. 226.
[17] Gutzeit G, Mapp ET. Corr Technol 1956;5:331. Trans Inst Metal Finish 1956;33:383.
[18] Goldenstein A, Rostoker W, Schossberger F. J Electrochem Soc 1957;104:104.
[19] de Minjer CH, Brenner A. Plating 1957;44:1297.
[20] Gutzeit G. Plating 1959;46:1158.
[21] Knanishu J, Technical report—deposition of alloy by chemical reduction, the office of technical services US Department of Commerce, December 4, 1961. http://handle.dtic.mil/100.2/AD273648.
[22] Gutzeit G, Landon RW, A large scale electroless nickel custom plating shop. In: 41st Annual proceedings AZS; 1954, p. 256.

[23] Brenner A, Riddell GE, US Patent 2,532,283 Nickel plating by chemical reduction, December 5, 1950.
[24] Talmey P, Crehan W J, US Patent 2,658,839 Process of chemical nickel plating, November 10, 1953.
[25] Gutzeit G, Krieg A, US Patent 2,658,841 Process of chemical nickel plating and bath therefor, November 10, 1953.
[26] Gutzeit G, US Patent 2,694,019 Processes of chemical nickel plating and baths therefor, November 9, 1954.
[27] Barrington TP, Crehan WJ, US Patent 2,717,218 Chemical nickel plating methods and apparatus, September 6, 1955.
[28] Jendrzynski HJ, Stapleton TF, US Patent 2,721,814 Nickel plating by chemical reduction, October 25,1955.
[29] Brrington TP, US Patent 2,766,138 Processes of chemical nickel plating, October 9, 1956.
[30] Eisenberg PH, Raleigh DO, US Patent 2,828,227 Electroless deposition of vanadium alloys, March 25, 1958.
[31] Cahill AE, McConnell VP, US Patent 2,874,072 Autocatalytic copper plating process and solution, February 17, 1959.
[32] Miller RE. P&SF 1987;74(12):52.
[33] Metal Hydrides Inc, US Patent 2,942,990 Metal plating by chemical reduction with borohydrides, June 28, 1960.
[34] Henri R, US Patent 2,965,551 Metal plating process, December 20, 1960.
[35] Shipley CR, US Patent 3,011,920 Method of electroless deposition on a substrate and catalyst solution therefor, December 5, 1961.
[36] Nixon CF, Rowe RJ, US Patent 3,024,134 Nickel chemical reduction plating bath and method of using same, March 6, 1962.
[37] Leopold C, US Patent 3,178,311 Electroless plating process, April 13, 1965.
[38] Graham AH, Lindsay RW, Read HJ. J Electrochem Soc 1965;112(4):401.
[39] Nacus H. Plating 1967;54:380.
[40] Randin J-P, Maire PA, Saurer E, Hintermann HE. J Electrochem Soc 1967;114:442.
[41] Pont DU, US Patent 3,338,726 Chemical reduction plating process and bath, August 29, 1967. 2000.
[42] Elmore V, Davis KC. J Electrochem Soc 1969;116(10):1455.
[43] Marti JL, Lanza GP. Plating 1969;56(4):377.
[44] Rondin J-P, Hintermann HE. J Electrochem Soc 1970;117:160.
[45] Muski K. Metal Finish 1970;68(12):36.
[46] Feldstein N, Lancsek TS. J Electrochem Soc 1971;118:869.
[47] Gabrielli C, Raulin F. J Appl Electrochem 1971;1:167.
[48] Mallory GO. Plating 1971;58:319.
[49] D'Amico JF, De Angelo MA, Henrickson JF, Kenney JT, Sharp DJ. J Electrochem Soc 1971;118(10):1695.
[50] Schneble, Jr FW, Mccormack JF, Zeblisky RJ, US Patent 3,635,758 Electroless metal deposition, January 18, 1972.
[51] Gobunova KM, Ivanov MV, Moiseevet VP. J Electrochem Soc 1973;120(5):613.
[52] Lelental M. J Elctrochem Soc 1973;120(12):1650.
[53] Schneble, Jr FW, McCormack JF, Zeblisky RJ, US Patent 3,959,531 Improvements in electroless metal plating, May 25, 1976.
[54] Watanabe T, Tanabe Y. Mater Sci Eng 1976;23:97.
[55] Büchner W, Niederprünt H. Pure Appl Chem 1977;49:733.

[56] Scannell WG, US Patent 4,038,085 Method of treating electroless nickel plating bath, July 26, 1977, 2000.

[57] Gulla M, Shipley Jr CR, MacKay HA, US Patent 4,152,164 Electroless nickel plating, May 1, 1979.

[58] Narcus H, US Patent 4,169,171 Bright electroless plating process and plated articles produced thereby, September 5, 1979.

[59] Smith SF. Metal Finish 1979;77(5):60.

[60] Pranas B, US Patent 3,077,285 Tin-nickel-phosphorus alloy coatings, February 12, 1963.

[61] Brookshire RR, US Patent 2,976,181 Method of gold plating by chemical reduction, March 21, 1961.

[62] Luce BM, US Patent 3,300,328 Electroless plating of gold, January 24, 1967.

[63] Robert D, US Patent 3,396,042 Chemical gold plating composition, August 6, 1968.

[64] Haddad MM, Rich DW, Smith MC, IP.com number: IPCOM000073538D, Electroless gold plating, December 1, 1970.

[65] Photocircuits Corp., US Patent 3,589,916 Autocatalytic gold plating solutions, June 29, 1972.

[66] Engelhard Minerals & Chemicals Corporation, US Patent 3,917,885 Electroless gold plating process, November 4, 1972.

[67] Okinaka Y, US Patent 3,700,469 A typical electroless procedure, October 24, 1972.

[68] Okinaka Y. J Electrochem Soc 1973;120(6):739.

[69] Rapson TS, Groenewald T. Gold Bull 1975;8(4):119.

[70] Redmond JP, Andrews DM, Guyler KE, US Patent 4,009,297 Gold deposition procedures and substrates upon which gold has been deposited, February 22, 1977.

[71] Franz H, Vanek JC, US Patent 4,091,128 Electroless gold plating bath, May 23, 1978.

[72] Dunlap WN, IP.com number: IPCOM000095589D, Control of magnetic properties of deposits from an electroless cobalt solution, March 1, 1964.

[73] Kivel J, Sallo JS. J Electrochem Soc 1965;112:1201.

[74] Fisher RD. IEEE Trans Magn 1966;2(4):681.

[75] Morton V, Fisher RD. J Electrochem Soc 1969;116(2):188.

[76] Pearlstein F, Weightman RF. J Electrochem Soc 1974;121:1023.

[77] Pearlstein F, US Patent 3,485,597 Electroless deposition of nickel-phosphorus based alloys, December 23, 1969.

[78] Calligaris E, Callaby R, Rossello M, US Patent 3,650,747 Electroless deposition of nickel, cobalt, copper or iron metal and a bismuth, antimony, lead, tin, titanium, tungsten or chromium salt on a gold, platinum or palladium latent image, March 21, 1972.

[79] Gulla M, US Patent 3,832,168 Metal finishing alloy of nickel-copper-phosphorus, August 27, 1974.

[80] Richardson Chemical Company, US Patent 4,019,910 Electroless nickel polyalloy plating baths, April 26, 1977.

[81] Hubbell FN. P&SF 1978;65(12):58.

[82] Agens MC, US Patent 2,938,805 Process of stabilizing autocatalytic copper plating solutions, May 31, 1960.

[83] Marshall JH, IP.com number: IPCOM000097434D, Producing metallic deposits, December 1, 1962.

[84] Brookshire RR, US Patent 3,046,159 Method of copper plating by chemical reduction, July 24, 1962.

[85] Agens MC Jr, US Patent 3,075,855 Copper plating process and solutions, January 29, 1963.

[86] John ZR, US Patent 3,095,309 Electroless copper plating, June 25, 1963.

[87] Schneble FW Jr, McCormack JF, Zeblisky RJ, Williamson JD, Polichette J, US Patent 3,259,559 Method for electroless copper plating, July 5, 1966.
[88] Atkinson RB, US Patent 3,310,430 Material and method for electroless deposition of metal, March 21, 1967.
[89] Hall DW, Koretzky H, IP.com number: IPCOM000089920D, Electroless deposition of copper, December 1, 1968.
[90] Hirohata H, Oita M, Honjo K, US Patent 3,532,519 Electroless copper plating process, October 6, 1970.
[91] Enthone, Incorporated, US Patent 3,615,736 Electroless copper plating bath, October 26, 1971.
[92] Schneble FW Jr , Zeblisky RJ, McCormack JF, Williamson JD, US Patent 3,650,777 Electroless copper plating, March 21, 1972.
[93] Dutkewych OB, Hofman LA, US Patent 3,770,464 Dry replenishment of electroless copper solutions, November 6, 1973.
[94] Underkofler WL, Zucconi TD, US Patent 3,844,799 Electroless copper plating, June 28, 1974.
[95] Ohno I. Surf Technol 1976;4(6):515.
[96] Arisato Y, Koriyama H, US Patent 4,138,267 Compositions for chemical copper plating, February 16, 1979.
[97] Shipley CR. Jr, US Patent 3,011,920 Method of electroless deposition on a substrate and catalyst solution therefor, December 5, 1961.
[98] DeLong HK. Plating on magnesium by electrodeposition and chemical reduction methods. Presented at the 1961 American Electroplaters Society Convention in Boston, MA, p. 16.
[99] Gutzeit G, US Patent 3,122,578 Chemical nickel plating of magnesium and its alloys, October 12, 1965.
[100] Hays DD, US Patent 3,357,854 Nickel plating process, December 12, 1967.
[101] Iwasa H, Yokozawa M, Teramoto I. J Electrochem Soc 1968;115:485.
[102] IBM TDB, McBride D G, and Vlasak G P, IPcom number: IPCOM000072570D, Activating a nonconductive substrate with sodium borohydride NaBH(4), prior to electroless plating, August 1, 1970.
[103] Parker K, Shah H. J Electrochem Soc 1970;117:1091.
[104] Polichette J, Leech EJ, Nuzzi FJ, US Patents 3,772,056 Sensitized substrates for chemical metallization, November 13, 1973.
[105] Polichette J, Leech EJ, US Patent 3,772,078 Process for the formation of real images and products produced thereby, November 13, 1973.
[106] Haddad MM, Schmeckenbecher AF, IP.com number: IPCOM000086270D, Activation of catalyst for electroless plating on molybdenum, August 1, 1976.
[107] Inaba Y, US Patent 3,993,808 Method for electroless plating gold directly on tungsten or molybdenum, November 23, 1976.
[108] Miller RG, Cavitt RL, US Patent 4,005,229 Novel method for the rapid deposition of gold films onto non-metallic substrates at ambient temperatures, January 25, 1977.
[109] IBM TDB, Haddad MM, Suierveld J, IP.com number: IPCOM000068520D, Depositing crack free heavy electroless nickel coating, January 1, 1978.
[110] Vratny F, US Patent 4,122,215 Electroless deposition of nickel on a masked aluminum surface, October 24, 1978.
[111] Bandrand DW. P&SF 1979;66(12):14.
[112] Mallory GO. The Fundamental aspects of electroless nickel plating. In: Electroless plating: fundamentals and applications. Orlando, FL: American Electroplaters and Surface Finishers Society; 1990. p.1–55.

[113] Hersch P. Trans Inst Metal Finish 1955–56;33:417.
[114] Machu W, El-Gendi S. Metalloberfläche 1959;13:97.
[115] de Minjer CH. Electrodepos Surf Treat 1975;3:261.
[116] Cavallotti P, Salvago G. Electrochim Metall 1968;3:239.
[117] van der Meeraker JEAM. J Appl Electrochem 1981;11:395.
[118] Riedel W. Electroless nickel plating. Stevenage, UK: ASM International Finishing Publications; 1991.
[119] Paunovic M. Plating 1968;51:1161.
[120] Bindra P, Roldan J. J Appl Electrochem 1987;17:1254.
[121] Hessami S, Tobias JL. J Electrochem Soc 1989;136:3611.
[122] Gafin AH, Orchard SW. J Electrochem Soc 1993;140:3458.
[123] Kim YS, Sohn HJ. J Electrochem Soc 1996;143:505.
[124] Stevanovic R, Stevanovic J, Despic A. J Appl Electrochem 1999;29:747.
[125] Tzeng GS. J Appl Electrochem 1996;26:969.
[126] Abrantes LM, Correia JP. J Electrochem Soc 1994;141:2356.
[127] Djokic SS. Electroless deposition of metals and alloys. In: Conway BE, White RE, editors. Modem aspects of electrochemistry, vol. 35. New York: Kluwer Academic/Plenum; 2002. p. 51–133.
[128] Okinaka Y, Osaka T. Electroless deposition processes: fundamentals and applications, vol. 3. Weinheim, Germany: VCH; 1994. p. 55.
[129] Matsubara H, Yonekawa T, Ishino Y, Nishiyama H, Saito N, Inoue Y. Electrochim Acta 2002;47:4011.
[130] Yarkosky EF, Cacciatore PA, US Patent 5,141,778 Method of preparing aluminum memory disks having a smooth metal plated finish, August 25, 1992.
[131] Ernest T, Sorensen R, and Guilinger T. Effectiveness of nickel plating in inhibiting atmospheric corrosion of copper alloy contacts. Albuquerque, NM: Sandia National Laboratories; 1998. www.osti.gov/bridge/servlets/purl/674975-LX14EM/webviewable/.
[132] Kerr C, Barker D, Walsh FC. Trans Inst Metal Finish 1996;74(6):214.
[133] Kerr C. Porosity of electroless nickel coatings on mild steel substrates. PhD Thesis. University of Portsmouth, Portsmouth: 1997.
[134] Reade GW, Kerr C, Barker DB, Walsh FC. Trans Inst Metal Finish 1998;76(4):149.
[135] Folkman SL, Stevens M. Characterization of electroless nickel plating on aluminum mirrors. In: Hatheway AE, editor. Proceedings optomechanical design and engineering, vol. 4771. September 9, 2002. p. 254–64.
[136] Mai QX, Daniels RD, Harpalani HB. Thin Solid Films 1988;166:235.
[137] Marshall GW, Lewis DB, Dodds BE. Surf Coat Technol 1992;53(3):223.
[138] Parker K. P&SF 1981;68(12):71.
[139] Zhang B, Hu W, Zhang Q, Qu X. Mater Charact 1996;37:119.
[140] Duncan RN, Arney TL. P&SF 1984;71(12):49.
[141] Yamada T, Yamamoto A, Fujiwara M, kunugi Y. J. Mater Sci 1993;28:3513.
[142] Puchi ES, Staia MH, Hintermann H, Pertuz A, Chitty JA. Thin Solid Films 1996;290–291:370.
[143] Chitty JA, Staia MH, Pertuz A, Hintermann H, Puchi ES. Thin Solid Films 1997;308–309:430.
[144] Berríos JA, Staia MH, Hernández EC, Hintermann H, Puchi ES. Surf Coat Technol 1998;108–109:466.
[145] Chitty JA, Pertuz A, Hintermann H, Puchi ES. J Mater Eng Perform 1999;81:83.
[146] Contreras G, Fajardo C, Berríos JA, Pertuz A, Chitty J, Hintermann H, Puchi ES. Thin Solid Films 1999;355–356:480.

[147] Gawne DT, Ma U. Wear 1987;120:125.
[148] Parkinson R. Nickel Development Institute Technical Series no. 10081, 1995, 1.
[149] Hur K-H, Jeong J-H, Lee DN. J Mater Sci 1990;25:2573.
[150] Barker D. Trans Inst Metal Finish 1993;71(3):121.
[151] Baudrand D, Bengston J. Metal Finish 1995;93(9):55.
[152] Berkh O, Eskin S, Zahavi J. Metal Finish 1996;94(3):35.
[153] Staia MH, Castillo EJ, Puchi ES, Lewis B, Hintermann HE. Surf Coat Technol 1996;86–87:598.
[154] Apachitei I, Duszczyk J, Katgerman L, Overkamp PJB. Scripta Metall 1998;38(9):1347.
[155] Duncan RN, Arney TL. P&SF 1984;71:49.
[156] Baurand DW. Electroless plating, ASM handbook, Surface engineering 5. Washington, DC: ASM; 1994. p. 290.
[157] Li J, Hu X, Wang D. P&SF 1996;83:62.
[158] Bozzini B, Martini C, Cavallotti PL, Lanzoni E. Wear 1999;225–229:806.
[159] Fernández M, Martínez-Duart JM, Albell JM. Electrochimi Acta 1986;31(1):55.
[160] Osaka T, Usuda M, Koiwa I, Sawai H. Jpn J Appl Phys 1988;27:1885.
[161] Khan MR, Nicholson EL. JMMM 1986;54–57(3):1654.
[162] Hur K-h, Jeong J-h, Lee D-n. J Mater Sci 1991;26:2037.
[163] Liu Z, Chen H, Yang C, He Y. J Phys: Condens Matter 1991;3:9007.
[164] Duncan RN. P&SF 1986;73(7):52.
[165] Levina LA, Zhogov AA, Grigor'ev VS, D'yakov VG, Vashin GZ. Chem Petrol Eng 1989;25(4):215. Translated from Khimicheskoe i Neftyanoe Mashinostroenie, No 4, (1989) 29.
[166] Tomlinson WJ, Carroll MW. J Mater Sci 1990;25:4972.
[167] Doong JC, Duh JG, Tsai SY, Wang JH, Chiou BS. Surf Coat Technol 1993;58(3):157.
[168] Zhang B, Hu W, Qu X, Zhang Q, Zhang H, Tan Z. Trans Inst Metal Finish 1996;74(2):69.
[169] Balaraju JN, Seshadri SK. Trans IMF 1999;77(2):84.
[170] Shoeib MA, Mokhtar SM, Abd El-Ghaffar MA. Metal Finish 1998;96(11):68.
[171] Moonir-Vaghefi SM, Saatchi A. J Hejazi, Metal Finish 1998;95(11):46.
[172] Gierlotka D, Rowinski E, Budniok A, Lagiewka E. J Appl Electrochem 1997;27:1349.
[173] ASM. ASM handbook, vol 5. Materials Park, OH: American Society for Materials; 1991.
[174] Henry J. Metal Finish 1990;88(10):15.
[175] Hussain MS, Such TE. Surf Technol 1981;13:119.
[176] Baudrand DW. P&SF 1981;68(12):57.
[177] Zhang B, Zhang H, Tan Z. Comparison of the solderability properties of electroless plating Ni-B, Ni–P and Ni–P-B alloys. In: Proc. 1th China electroless plating conference. Nanking: The Chinese Corrosion and Protection Society; April 1992. p. 127–132.
[178] Zhang B, Zhang H, Tan Z, Zhang H, Shanghai Met 1992.7.;14(4):44.
[179] Hu W, Wang L, Wu L, Zhang B, Guan H. Trans Inst Metal Finish 1994;72:141.
[180] Xie H, Zhang B. Metal Finish 2001;99(1):40.
[181] Yi X. Electroless gold plating for high density packages. In: 41st proceedings electronic components and technology conference. Atlanta, GA: 1991. p. 26–29.
[182] Lee C-Y, Lin K-L. Jpn J Appl Phys 1994;33:4708.
[183] Lin K-L, Jang J-M. Mater Chem Phys 1994;38(1):33.
[184] Gaudiello JG. In: 45th proceedings electronic components and technology conference. IEEE; 1995. p. 534–7.
[185] Watanabe H, Abe S, Honma H. J Appl Electrochem 1998;28:525.
[186] Baudrand DW. P&SF. 1983;70(12):24.
[187] Miller RD. P&SF 1987;74(2):52.

[188] Mayer LJ. Prod Finish June 1993;48.
[189] Hajdu J. P&SF 1996;83(9):29.
[190] Durkin B. Metal Finish 2004;102(1):21.
[191] Hajdu J, Zabrocky S. Metal Finish 2000;98(5):42.
[192] Koiwa I, Yamada K, Usuda M, Osaka T. Denki Kagagu 1986;54:514.
[193] Koiwa I, Yamada K, Usuda M, Osaka T. J Metal Surf Soc Jpn 1987;38(10):488.
[194] Koiwa I, Usuda M, Yamada K, Osaka T. J Electrochem Soc 1988;135(3):718.
[195] Lee SL, Liang HH. P&SF 1991;78(9):82.
[196] Koiwa I, Osaka T, Sawai H. Japan Metal Surf Technol 1983;34(12):600.
[197] Aoki K, Takano O. P&SF 1990;77(3):48.
[198] Sawai H, Kanamari T, Koiwa I, Shibata S, Nibei K. J Electrochem Soc 1990;137(11):3653.
[199] Guo ZC. Electroplat Pollu Cont 1990;13(3):11.
[200] Liu Z. Corr Prot 1999;20(3):111.
[201] Hur KH, Jeong JH, Lee DN. J Mater Sci 1991;26:2037.
[202] Wang YW, Xiao CG, Deng ZG. P&SF 1992;79(3):82.
[203] Srhiri CA, Chassaing E. P&SF 1992;79(11):68.
[204] Wu Y, Zao Y, Yao M, Hu X. Mater Prot 1995;28(9):6.
[205] Cui Z, Yang X, Wang H, Deng C. Mater Prot 1998;31(1):15.
[206] Nawafune H, Uegaki T, Mizumoto S, Ishikawa M, Nakamura T. Trans Inst Metal Finish 1998;76(6):231.
[207] Iwamatau K. Metal Finish 1989;87(5):25.
[208] Genutien I, Lenkaitieni J, Jusys Z, Luneckas A. J Appl Electrochem 1996;26:118.
[209] Schlesinger M, Meng X. J Electrochem Soc 1990;137(6):1858.
[210] Koyano H, Kato M, Uchida M. P&SF 1991;78(7):68.
[211] Schlesinger M, Meng X, Snyder DD. J Electrochem Soc 1991;138(2):406.
[212] Shimauchi H, Ozawa S, Tamura K, Osaka T. J Electrochem Soc 1994;141(6):1471.
[213] Krasteva N, Armyanov S, Georgieva J, Avramova N, Fotty V. J Electr Mater 1995;24(8):941.
[214] Osaka T, Koiwa I, Yamada K, Nishikawa M, Usuda M. Japan Bull Chem Soc 1987;60:3117.
[215] Kim D, Matsuda H, Aoki K, Takano O. P&SF 1996;83(2):78.
[216] Saito T, Sato E, Matsuoka M, Iwakura C. J Appl Electrochem 1998;28:559.
[217] Hu W, Wu L, Wang L, Zhang B, Guan H. Physica B 1995;212:195.
[218] Georgiou GE, Baiocchi FA, Luftnan HS, Sheng TT, Vasile MJ, Knoell RV. J Electrochem Soc 1991;138(7):2061.
[219] Gonzalez OM, White RE, Cocke DL. P&SF 1990;77(11):63.
[220] Hyashi T. P&SF 1991;78(9):30.
[221] Osaka T, Kasai N, Koiwa I. J Electrochem Soc 1983;130(4):790.
[222] Matsubara H, Toda M, Sakuma T, Homma T, Osaka T, Yamazaki Y, Namikawa T. J Electrochem Soc 1989;136(3):753.
[223] Osaka T, Matsubara H, Yamanishi L, Mizutani H, Goto F. IEEE Trans Magn 1987;MAG-23(5):2356.
[224] Xie H, Zhang B. J Changs Commun Uni 1998;14(3):24.
[225] Zhang B, Xie H, Xu X. Metal Finish 1999;97(10):35.
[226] Takano O, Matsuda H. Metal Finish 1986;84(6):60.
[227] Hajdu J, Zabrocky S. Metal Finish 2000;98(5):42.
[228] Dennis JK, Sheikh ST, Silverstone EC. Trans IMF 1981;59:118.
[229] Onda T, Okayasu N, Yoshihara S, Sato E. Jpn J Surf Finish 1991;42(1):79.
[230] Changcheng X, Xinmin H, Zonggang D, Yanwen W. P&SF 1989;76(6):90.
[231] Zhang H, Zhang B, Tan Z, Tang H. Mech Eng Mater 1990;(4):35.

[232] Feldstein N, Lancsek TS. J Electrochem Soc 1984;131(12):3026.
[233] Balaraju JN, Seshadri SK. J Mater Sci Lett 1998;17(15):1297.
[234] Tulsi SS. Trans Inst Metal Finish 1983;61:147.
[235] Pena-Munoz E, Berçot P, Grosjean A, Rezrazi M, Pagetti J. Surf Coat Technol 1998;107(2-3):85.
[236] Moonir-Vaghefi SM, Saatchi A, Hejazi J. Metal Finish 1997;95(6):102.
[237] Moonir-Vaghefi SM, Saatchi A. Metal Finish 1997;95(11):46.
[238] Island R. Metal Finish 1997;95(3):29.
[239] Straffelini G, Colombo D, Molinari A. Wear 1999;236(1–2):179.
[240] Osaka T, Koiwa I, Usudo M, Aria K, Saito I. J Electrochem Soc 1989;136(4):1124.
[241] Hu G, Liu Y, Yang J, Yang K, Tang F, Jin J. J Cent South Univ Technol 1999;6(1):13.
[242] Chang S-y, Lin J-h, Lin S-j, Theo Z. Kattamis Metall Mater Trans A 1999;30A:1119.
[243] Yang D, Sun Y, Dong B. Electroless plating technology reminiscences and outlook. In: Proceedings of the 1th China electroless plating. Nanking: The Chinese Corrosion and Protection Society; April 1992. p. 1–3.
[244] Zhang Q, Yang D. Surf Eng Inform 2005;5(2):5.
[245] Zhang QJ, Yang D, 12-Year Review – China's technical exchange and development of electroless plating. In: Proceedings of the 7th China electroless plating. Shenzhen: August 10–12, 2004. p. 36–38.
[246] Ruscior C, Croiala E. J Electrochem Soc 1971;18(5):696.
[247] Hu W, Zhang B. Physica B 1991;175:396.
[248] Zhang B, Hu W. Rare Met 1991;10(4):301.
[249] Zhang B, Hu W. Physica B 1993;183:396.
[250] Hu W, Zhang B. Trans Inst Metal Finish 1993;71(1):30.
[251] Hu W, Zhang B. Electropla Protec 1993;13(1):13.
[252] Yi G, Wang L, Ouyang Y, Mao L, Guo Y, Zhang B, Xie H. Metal Finish 1996;94(11):31.
[253] Wang L, Zhang B, Yi K, Ouyang Y, Hu W. J Alloys Comp 1997;255(12):231.
[254] Wang L, Zhao L, Zhang B, Liao S, Ouyang Y, Hu W. Z Metallkd 1997;88(12):945.
[255] Wang L, Zhao L, Zhang B, Liao S, Hu W, Ouyang Y. P&SF 1998;85(12):96.
[256] Wang L, Zhao L, Zhang B, Hu W, Shu X, Sheng X. Z Metallkd 1999;90(5):338.
[257] Wang L, Zhao L, Zhang B, Liao S, Ouyang Y, Hu W, Shu X, Yuan X. P&SF 1999;86(12):84.
[258] Wang L, Zhao L, Zhang B, Hu W, Shu X, Sheng X, Fang Z. J Alloys Comp 1999;287:234.
[259] Wang L, Zhao L, Hu W, Shu X, Yuan X, Zhang B. Acta Metallu Sinica 1999;12(5):733.
[260] Fields B. CEO from Palm International, Products Finishing, www.processcleaning.com/articles/2020-vision-electroless-nickel. Pennington T, editor 2020 Vision: The Future of Coatings, Products Finishing, http://www.processcleaning.com/articles/2020-vision-the-future-of-coatings.
[261] Beijing Qianxun Information Consulting Co, Ltd., China Report Library, Electroless Ni–P industry analysis report, 2011. http://www.baogaoku.com/hyfx/hyfx_397384.html.
[262] Ashassi-Sorkhabi H, Mirmohseni A, Harrafi H. Electrochim Acta 2005;50(28):5526.
[263] Karthikeyan S, Srinivasan KN, Vasudevan T, John S. Portug Electrochim Acta 2006;24:405.
[264] Hiroshi M, Takehide Y, Yoshihiro I, Nobuo S, Hiroshi N, Yasunobu I. Electrochim Acta 2006;52(2):402.
[265] Ebn Touhami M, Chassaing E, Cherkaoui M. Electrochim Acta 2003;48(24):3651.
[266] Malecki A, Micek-Ilnicka A. Surf Coat Technol 2000;123(1):72.
[267] Nakai H, Homma T, Komatsu I, Osaka T. J Phys Chem B 2001;105(9):1701.
[268] Homma T, Komatsu I, Tamaki A, Nakai H, Osaka T. Electrochim Acta 2001;47(1–2):47.
[269] Homma T, Tamaki A, Nakai H, Osaka T. J Electroanal Chem 2003;559:131.

[270] Shimada T, Sakata K, Homma T, Nakai H, Osaka T. Electrochim Acta 2005;51(5):906.
[271] Zeng Y, Liu S, Lihui O, Lihui YJ, Yu S, Wang H, Xiao X. Appl Surf Sci 2006;252(8):2692.
[272] Shimada T. Ph.D. thesis. Theoretical analysis on the reaction mechanisms of reducing agents for electroless deposition processes. Waseda University, Japan, February 2007.
[273] Shimada T, Nakai H, Homma T. J Electrochem Soc 2007;154(4):273.
[274] Zhang B. Physical fundamentals of nanomaterials". Beijing: Chemical Industry Press 2009.
[275] Teh1 K-S, Cheng Y-T, Lin L. J Micromech Microeng 2005;15:2205.
[276] Chao T-Y, Shen G-R, Cheng YT. J Electrochem Soc 2006;153(1):G98.
[277] Fang W, Yu H-y, Sun D-b, Wang X-d, Meng H-m, Yang D. Electropl Finish 2007;26(3):1.
[278] Allahkaram SR, Zoughi M, Rabizadeh T. Micro & Nano Lett 2010;5(5):262.
[279] Kaya B, Gulmez T, Demirkol M. Preparation and properties of electroless Ni-B and Ni-B nanocomposite coatings. In: Proceedings of the world congress on engineering and computer science 2008, WCECS 2008. San Francisco, CA: October 22–24, 2008.
[280] Khalifa ORM, Sakr E. Open Corro J 2009;2:211.
[281] Bigdelia F, Allahkaram SR. Mater Design 2009;30(10):4450.
[282] Rabizadeha T, Allahkaram SR. Mater Design 2011;32(1):133.
[283] Ranganatha S, Venkatesha TV, Vathsala K. Appl Surf Sci 2010;256(24):7377.
[284] Chen W, Gao W, He Y. Surf Coat Technol 2010;204(15):2493.
[285] Ma H, Tian F, Li D, Guo Q. J Alloys Comp 2009;474(1–2):264.
[286] Ma H, Liu Z, Tian F, Sun H. J Alloys Comp 2008;450(1–2):348.
[287] Ang L-M, Andy Hor TS, Xu G-Q, Tung C-H, Zhao S, Wang John LS. Chem Mater 1999;11:2115.
[288] Li X, Ma X, Li S, Wen S, Lun N. J Mater Sci Eng 2004;22(1):48.
[289] Zhao Y, Zheng H, Xue Y, Duan Y. Chin J Aerona 2010;23:377.
[290] Araia S, Endob M, Hashizumec S, Shimojima Y. Electrochem Commun 2004;6(10):1029.
[291] Wang F, Arai S, End M. Carbon 2005;43(8):1716.
[292] Li W, Jin H, Hao Y, Chen T, Dai J, Wang Q. J Nanometer, volume 2011, article ID 348958, 5 pages.
[293] Speets EA, te Riele P, van den Boogaart MAF, Doeswijk LM, Ravoo BJ, Rijnders G, et al. Adv Funct Mater 2006;16(10):1337.
[294] Cui B, Yu ZN, Ge HX, Chou SY. Appl Phys Lett 2007;90:043118.
[295] Li N, Li XT, Yin XJ, Wang W, Qiu SL. Sol State Commun 2004;132(12):841.
[296] Xu WL, Palshin V, Flake JC. J Electrochem Soc 2009;156(7):H544.
[297] Strukova GK, Strukov GV, Batov IE, Sakharov MK, Kudrenk EA, Mazilkin AA. Mater Chem Phys 2010;119(3):377.
[298] Agarwala RC, Agarwala V. Sādhanā 2003;28(3 & 4):475.

Part II

Technology of Electroless Plating – Plating Bath, Critical Parameters, Deposition Rate, and Stability of Plating Bath

This part concentrates on the technology involved in producing electroless plating (EP). This includes the contents of plating bath, critical parameters, deposition rate, and the stability of the plating bath.

Some scholars just refer to electroless nickel (EN) plating as electroless plating. EN is undoubtedly the most important process, being widely employed in various industries due to its unique physical and chemical properties. However, EN is not the same as EP, because besides nickel other metals (Cu, Ag, Au, etc.) and other alloys (e.g., EP Co-based and Fe-based) are used. So, we use the term EP in this book, since this is not limited to EN.

In order to be clear, the bath for electroless plating is classified according to the metal, and whether a binary, ternary, quaternary, quinary, or multialloy is to be produced, but not by various based alloys as usual. However, the classification of the bath of electroless plating in binary alloys is still according to various based alloys as usual.

Composite electroless plating is very important, not only because of its deposits have outstanding performance but also because the study of composite EP allows us to understand its nature better. Composite electroless deposits are therefore discussed in a separate chapter to explore their nature in greater detail.

It was mentioned in Chapter 1 that most of the papers on nano electroless plating were published after 2001, and their contents were not included in any books on this topic. To fill this gap, a chapter on nano electroless plating has been included in this part.

We thought that the electroless plating of Fe-based alloys should be studied as well, from the viewpoint of pure basic investigation. That is why the author's group has studied the electroless plating of Fe-based alloys since the middle of the 1980s, and the research subjects of three graduate students in our group all referred to electroless plating of Fe-based alloys. Therefore, a chapter of electroless plating in Fe-based alloys is included in this part.

Many factors affect EP due to the many parameters that can be altered in a plating bath. A special chapter is included to describe the various parameters that affect the EP process. The importance of plating rate is well known; therefore, the discussion focuses on the effect of various parameters on the plating rate.

People have realized that conventional electroless plating is not exactly a "green" technology because many of the components of the electroless bath solutions may be toxic, for example, formaldehyde is a known carcinogen. Therefore, ways of making electroless plating a "green" technology and making the process more environmentally friendly are very important. Fortunately, scientists have begun to study this area and have made significant advances. A chapter covering green technology for electroless plating is also included in this part.

Chapter 2

Electroless Plating Baths of Metals, Binary Alloys, and Multicomponent Alloys

Chapter Outline

2.1 GENERAL CONSIDERATION FOR ELECTROLESS PLATING BATH SOLUTION

Many people believe that electroless plating is just an autocatalytic plating process. However, EP process involves putting a substrate(s) (workpiece(s)) into plating bath to get a desired deposit through complex chemical reactions.

Amorphous and Nano Alloys Electroless Depositions. http://dx.doi.org/10.1016/B978-0-12-802685-4.00002-9

The basic electroless deposition process without considering all the side reactions can be represented as

$$\text{Reductant} \xrightarrow{\text{Catalytic surface}} \text{Oxidation product} + \text{Me}^- \quad (2.1)$$

$$\text{M}^{z+} + z\text{e} \xrightarrow{\text{S surface}} \text{M}\downarrow \quad (2.2)$$

And the overall electroless reaction can be written as:

$$\text{Reductant} + \text{M}^{z+} \xrightarrow{\text{Catalytic surface}} \text{Oxidation product} + \text{M}\downarrow \quad (2.3)$$

The composition of a modern EP bath is usually very complex, and comprises a metal salt, a reduction agent, a complexing agent, stabilizer, accelerator, buffer, brightening agent, pH adjustment agent, and wetting agent [1–4].

The function of each component of the electroless bath is as follows.

The metal salt provides metal ions, which are the source of metal in the alloy deposits. It is obvious that the metal salt is essential in the bath. The number of metal salts to use depends on what alloy deposits you want, for example, two metal salts are necessary for plating Ni–W–P alloys: $NiSO_4$ and Na_2WO_4 [5], and four metal salts are necessary for plating Co–Ni–Mn–Re–P alloys: $CoSO_4$, $NiSO_4$, $MnSO_4$, and NH_4ReO_4 [6].

The reducing agent is an electron donor, and this reduces the metal ions into metals. In addition, the reducing agent usually provides the nonmetal elements in the alloy deposits, for example, sodium hypophosphite (NaH_2PO_2) provides P, and sodium borohydride ($NaBH_4$) provides B. Of course, the reducing agent is essential in the bath as well. If it is required that the alloy deposit has two nonmetal elements, for example, Ni–P–B alloy, then two reducing agents must be included in the bath: NaH_2PO_2 and $NaBH_4$ [7].

The complexing agent is also known as a chelator, and its role is to complex the Ni ions, preventing excess free Ni ion concentration appearing, and so preventing the precipitation of metal salts and stabilizing the bath. Experiments have indicated that the deposition rate after adding an appropriate complexing agent is greater than that without complexing agent. In addition, complexing agent can also act as pH buffers, preventing the pH of the bath decreasing rapidly. So, modern plating baths almost always have a complexing agent.

The stabilizer can shield catalytically active nuclei, and so it can retard potential spontaneous decomposition of the electroless bath. The stabilizer is also termed a catalytic inhibitor. Excessive use of stabilizer can result in decrease in the plating rate, and can even cause the plating reaction to be completely inhibited. Therefore, trace amounts of stabilizer are always used, and few are used in excess of 10 ppm.

The accelerator can activate the reducing agent and accelerate deposition, so as to enhance the plating rate. The accelerator is also termed the exultant, and acts in opposition to the stabilizer.

The buffer aims to maintain the pH value of bath, so as to stabilize the deposition rate and the quality of deposits.

The wetting agent can increase the wettability of the substrate (or workpiece) surface to be coated. Use of a wetting agent in the plating bath can decrease the contact angle between the solution and the surface, so aiding the escape of H_2 from the substrate surface, which decreases the porosity of the deposits. The wetting agent is in fact a surface active agent, usually as known as a surfactant.

A brightener can make the electroless plating deposits brighter. Usually, the EP deposits have a semiglossy surface, but adding some brightener into the bath causes the surface of the deposit to be more uniform, so enhancing its luster. This means that the brightener is not a necessary component of the electroless bath.

pH regulators maintain the pH value of the bath within a given range during the process of plating deposition. Usually, some hydrogen ions (H^+) will be generated by the chemical reaction in the bath. These will change the pH value of the bath, so a pH regulator is needed. Most typical pH regulators (e.g., H_2SO_4, HCl, NaOH, $NH_3{\cdot}H_2O$, etc.) are compatible with the EN bath and thus can be employed to adjust its pH.

Sections 2.2–2.12 illustrate representative electroless plating bath formulations for various metals, and binary to multialloys. It should be pointed out that in the literature, several different bath solutions may be found for the same electroless metals or alloy plating. Hence, not all possible bath solutions are given here, just some representative bath solutions for every plating bath of metal or alloy.

2.2 PLATING BATH OF ELECTROLESS PURE NICKEL AND NICKEL-BASED BINARY ALLOYS

2.2.1 Pure Ni and Co Metals

In Table 2.1, six bath formulations for electroless pure Ni are listed, which are cited from different papers published in 1967, 1987, 2000, 2003, 2006, and 2010 together with three for pure Co from 1967, 2003, and 2010, respectively.

2.2.2 Ni–P

In Table 2.2, eight bath formulations for electroless Ni–P alloy are listed. The first is just the first bath formulation used by Brenner and Riddell when they discovered electroless nickel, and the remainder are cited from different author's papers published in 1946, 1960, 1980, 1989, 2000, 2005, 2007, and 2010, respectively.

2.2.3 Ni–B

In Table 2.3, six bath formulations for electroless Ni–B alloy are listed, which are cited from different papers published in 1971, 1983, 1988, 1997, 2005, and 2011, respectively.

TABLE 2.1 The Representative Electroless Plating Bath Formulation of Ni and Co Metals From Various Authors

Bath compositions and operating conditions	Ni					Co			
Nickel acetate, mol/L*	60 g/L	0.16			0.004–0.2				
Nickel chloride, mol/L			0.05						
Nickel sulfate, mol/L				0.04		0.005			
Cobalt chloride, mol/L							0.05–0.08	0.08	
Cobalt sulfate, mol/L									0.1
Monoethanolamine (MEA), mol/L*		3.27							
Hydrazine, mol/L*	100 mL/L	2.00	0.10		0.2–1	0.25	1.0	0.8	60 mL/L
Glycolic acid, g/L	60								
Glycine, mol/L			0.30						
Titanium chloride ($TiCl_3$ and $TiCl_4$), mol/L				0.08					
Nitrilotriacetic acid (NTA), mol/L				0.04					
Sodium citrate, mol/L				0.24				0.3	
Sodium hydroxide, g/L					7.5				
Sodium tartrate, mol/L							0.4		
Potassium sodium tartrate, g/L									50
Tetrasodium ethylenediaminetetraacetate acid or editic acid (EDTA), g/L	25								
Na_2EDTA, mol/L					0.016				

Lactic acid, mol/L					0.15				
Stability-containing compound, ppm				10–100					
Amino acid, ppm				100–500					
Thiourea, ppm							2–4		0.5 mg/L
Ammonium chloride, mol/L								0.1	
pH	11	10–13	12	8–9	9.6	11.5 ± 0.1	12	12.4–12.8	11
Temp., °C	90	93–97	85–90	50	74–90	55–75	90	30	85
References	[8]	[9]	[10]	[11]	[12]	[13]	[14]	[15]	[16]

**The original unit was M, which is an old unit, now all M are changed to mol/L, and mM is changed to mmol/L.*

TABLE 2.2 Representative Electroless Plating Bath Formulations for Ni–P Alloys From Various Authors

Bath compositions and operating conditions	Ni–P											
	1			**2**	**3**			**4**	**5**	**6**	**7**	**8**
Nickel sulfate, g/L				25	35	35	30	14.9		30	21	35
Nickel chloride, g/L	30	30	30						55			
Sodium hypophosphite, g/L	10	10	10	25	10	10	10	16.9	18	25	24	20
Sodium hydroxyacetate, g/L					10							
Potassium sodium tartrate, g/L								32.5				
Sodium acetate, g/L						10						5
Sodium citrate, g/L	100		100				100		80	20		15
Sodium pyrophosphate, g/L				50								
Ammonium chloride, g/L	50	100					50					
Lactic acid, g/L											~25	
Propionic acid, g/L											3	
Sulfuric acid, g/L								30.0				
Ammonium hydroxide (28%), mL/L								50–75				
Glycine, g/L										20		
Lead acetate, ppm											3	
Lead(II) nitrate, g/L										0.002		
pH	8–9	8–9	6–8	10–10.5	4.5–5.5	4.5–5.5	9.0–9.5	8.8 ± 0.2	9	4–12	4.6 ± 0.2	4–5
Temp., °C	90	90	90	25–75	90–95	90–95	90–95	90	90	50	85 ± 2	80–90
P in deposit, w/o				4–6	7–9	7–9	5–7	2–3	4.6–12.5 (at%)	13.7–4.1	10.6	8.57
Appearance of deposit	Bright	Dull	Semibright									
References	[17]			[18]	[19]			[20]	[21]	[22]	[23]	[24]

TABLE 2.3 Representative Electroless Plating Bath Formulation for Ni–B Alloys From Various Authors

Bath compositions and operating conditions	Ni–B						
Nickel sulfate, g/L		30					
Nickel chloride, g/L	30		30	20	24	20	10–40
Dimethylamine borane (DMAB), g/L	3.5				10		
Potassium borohydride, g/L			1.5				
Sodium borohydride, g/L		1		0.67		0.48	0.2–1.2
Sodium citrate, g/L		40					
Sodium acetate, g/L					22		
EDTA (ethylenediamine-tetraacetate acid or editic acid), g/L		100					
Ethylene diamine, mL/L			55	44		59 g/L	90
Malonic acid, g/L	34						
Sodium hydroxide, g/L		40	40	40		39	90
Sodium lauryl sulfate, g/L					0.1		
Lead nitrate, mg/L			20			0.0145	0.0145
Lead acetate, g/L		0.05					
pH	5.5	14	14	12	5.5		13.5
Temp., °C	77	90	95	−10 to −5	65–70	95	80, 88, 95
B in deposit, w/o		4		Alkaline bath	Acidic bath		
References	[25]	[26]	[27]	[28]		[29]	[30]

2.3 COBALT-BASED BINARY ALLOYS

2.3.1 Co–P

In Table 2.4, six bath formulations for electroless Co–P alloy are listed, which are cited from different papers published in 1964, 1974, 1987, 1994, 1997, and 2007, respectively.

2.3.2 Co–B

In Table 2.5, five bath formulations for electroless Co–B alloy are listed, which are cited from different papers published in 1974, 1986, 1998, 2002, and 2011, respectively.

TABLE 2.4 The Representative Electroless Plating Bath Formulation of Co–P Alloys From Various Authors

Bath compositions and operating conditions	Co–P					
Cobalt sulfate, g/L	24	30	10	20	0.8 mol/L	0.1 mol/L
Sodium hypophosphite, g/L	20	20	10	35	0.35 mol/L	0.1 mol/L
Sodium citrate, g/L	70		90			0.2 mol/L
Ammonium citrate, g/L		80		30	0.1 mol/L	
Sodium tetraborate, g/L			30			
Glycine, g/L			25			
Ammonium sulfate, g/L	40					
Ammonium chloride, g/L		50				
Ammonium hydroxide, g/L		60				
Boric acid, mol/L						0.3
Sodium laurylsulfate, g/L	0.1					
Maleic acid, mol/L						0–0.01
Succinic acid, mol/L						7.5×10^{-3}
pH	8.5	9.0	9.8	10	10–11	9.2
Temp., °C	92	80	80	60	85–90	80
P in deposit, w/o (without additives)						7.7 ± 0.1
References	[31]	[32]	[33]	[34]	[35]	[36]

2.4 Cu AND COPPER-BASED BINARY ALLOYS

In Table 2.6, five bath formulations for electroless Cu are listed, which are cited from different papers published in 1959, 1967, 1988, 1997, and 2009; also included are one for Cu–Ni from 1998, one for Cu–Zn from 2008, and one for Cu–P from 2011, respectively.

2.5 Au

In Table 2.7, six bath formulations for electroless Au are listed, which are cited from different papers published in 1961, 1970, 1984, 1994, 2002, and 2011, respectively.

2.6 Ag

In Table 2.8, seven bath formulations for electroless Ag are listed, which are cited from different papers published in 1950, 1967, 1971, 1986, 1994, 2000, and 2011, respectively.

TABLE 2.5 Representative Electroless Plating Bath Formulations for Co–B Alloys From Various Authors

Bath compositions and operating conditions	Co–B				
Cobalt sulfate, mol/L	25 g/L		0.1		0.16
Cobalt chloride, mol/L		0.0.126–0.0315		9 g/L	
Dimethylamine borane (DMAB), mol/L	4 g/L		0.05		0.05
Sodium succinate, g/L	25				
Sodium borohydride, mol/L		0.0031–0.0142			
Potassium borohydride, g/L				0.6	
Sodium citrate, mol/L		0.0126			
Sodium tartrate, g/L			0.2 mol/L	60	
Sodium pyroborate, g/L				4	
Citric acid, mol/L					0.62
Sodium sulfate, g/L	15				
pH	5.0	10.2–11.3	6.0	13.5	9.5
Temp., °C	70	≤30	60	45	70
B in deposit, w/o	1.7		2.6	5.1	
References	[32]	[37]	[38]	[39]	[40]

2.7 Pd AND PALLADIUM-BASED BINARY ALLOYS

In Table 2.9, four bath formulations for electroless pure Pd are listed, which are cited from different papers published in 1959, 1969, 2001, and 2011; also included are one for Pd–Cu from 2003; one for Pd–Ru from 2009, three for Pd–P from 1973, 1996, and 2011; and one for Pd–B from 1996, respectively.

2.8 Pt AND PLATINUM-BASED BINARY ALLOYS

In Table 2.10, four bath formulations for electroless Pt are listed, which are cited from different papers published in 1969, 1990, 2002, and 2007. A Pt–Pd alloy from 2009; a Pt–Rh alloy from 2004, and a Pt–Ru alloy from 2010 are also listed, respectively.

2.9 Ru, Rh, Os, AND Cr–P BINARY ALLOYS

In Table 2.11, four bath formulations for electroless Ru are listed, which are cited from different papers published in 1990, 1991, 2011, and 2011. The table also includes Rh from 2002, Os from 1998, and Cr–P from 2004, respectively.

TABLE 2.6 Representative Electroless Plating Bath Formulations for Cu and Copper-Based Binary Alloys From Various Authors

Bath compositions and operating conditions	Cu					Cu–Ni	Cu–Zn	Cu–P
Copper sulfate, mol/L	29 g/L	15 g/L	0.024	0.03	0.047	0.002–0.007		8 g/L
Nickel sulfate, mol/L						0.15		
Zinc oxide, mol/L							0.38	
Copper cyanide, mol/L							0.24	
Tetrasodium ethylenediamine tetraacetic acid, g/L		40						
Sodium potassium tartrate, mol/L	142 g/L						0–0.106	
Sodium EDTA (Versene T), g/L	17							
Sodium hydroxide, g/L	42						0.71 mol/L	
Sodium carbonate, g/L	25							
Formaldehyde (37%), mL/L	167	6						
Vanadium pentoxide, g/L; Or sodium arsenate, g/L; Or potassium antimony tartrate, g/L		0.01 0.02 0.03						
Organic phosphate ester, mL/L		1						
Triethanolamine, mol/L						0.2		
Sodium citrate, mol/L			0.052					15.5 g/L
Sodium hypophosphite, mol/L			0.27					60 g/L
Formaldehyde (HCHO), mol/L				0.08				

Boric acid, mol/L			0.5					12.4 g/L
Nickel sulfate, mol/L			0.002					1 g/L
Polyethylene glycol, g/L								0.001
2,2′-Dipyridyl, mg/L				0–500				
EDTA, mol/L				0.036	0.11			
Formaldehyde; or glyoxylic acid, mol/L					0.05 0.15	0.06		
Stabilizer, g/L								0.001
DL-Malic acid, mol/L						0.02		
Sodium cyanide, mol/L							0.40	
Sodium carbonate, mol/L							0.43	
pH		12	9.2	12	12.5	12.5	11	9
Temp., °C	25	20–80	65	70	60	30	55–65	75
P, Ni, or Zn in deposit, w/o						Ni: 37.8 mol% when $CuSO_4$= 0.003 mol/L	Cu: 26.6–74.3 at%	1.55
References	[41]	[42]	[43]	[44]	[45]	[46]	[47]	[48]

TABLE 2.7 Representative Electroless Plating Bath Formulations for Au From Various Authors

Bath compositions and operating conditions	Au					
Gold potassium cyanide, g/L		5.8	1.45			
Gold cyanide, g/L	2.0					
Sodium gold sulfite, mL/L				25		0.079 mol/L
Gold sodium chloride, mol/L					0.01	
Sodium hypophosphite, g/L	10.0					
Potassium cyanide, g/L	0.2	13	11			
Sodium sulfite, g/L				35	0.32 mol/L	0.127 mol/L
Potassium hydroxide, g/L		11.2	11.2			
Potassium borohydride, g/L		21.6	10.8			
Sodium EDTA, g/L			5			
Ethanolamine, mL/L			50			
Sodiumtetraborate (borax), g/L				32		
Formaldehyde, mol//L						0.625
Thiourea (thiocarbamide), mol/L				0.33		
Sodium thiosulfate, mol/L					0.08	
Disodium phosphate, mol/L					0.32	
Sodium bicarbonate, mol/L						0.025
pH	13.5			9.0	9.0	8.0
Temp., °C	96	75	70–75	40	60	1–4
References	[49]	[50]	[51]	[52]	[53]	[54]

2.10 GROUP B METALS (Zn, Cd, In, Sn, Pb, As, Sb, AND Bi) AND A FEW BINARY ALLOYS OF THESE METALS

In Table 2.12, one bath formulation for electroless Zn metal; one for Cd; two for In; two for Sn; one for Pb; one for As; one for Sb; and one for Bi are listed, which are cited from different papers published in 1992, 1992, 1982 and 1992, 1980 and 1993, 1992, 1992, 1992, 1992, and 1993, respectively.

In Table 2.13, the bath formulations for seven binary alloys of group B metals are listed, which are cited from different papers published in 1992 (In–Fe, In–Ni, In–Sn, In–Pb, and In–As), and in 1994 (In–Sb and Sn–Pb), respectively. The bath for Bi_2Te_3 is cited from a paper published in 2005.

TABLE 2.8 Representative Electroless Plating Bath Formulations for Ag From Various Authors

Ag					
A, Bath compositions and operating conditions	**B, Bath compositions and operating conditions**				
Silver nitrate, g/L 20 Amonia solution proper quantity Water, L 1	Formaldehyde, mL/L 40 Water, mL/L 200				
A and B are mixed at the ratio of 5:1 [55]					
C, Bath compositions and operating conditions	D, Bath compositions and operating conditions				
Silver nitrate, g/L 454 Ammonia solution, mL/L 355 Water, L 5.45	Rochelle salt (potassium sodium tartrate), g/L 1590 Epsom salt (magnesium sulfates), g/L 114 Water, L 3.64				
C (256 mL) and D (256 mL) are mixed first, then distilled water of 3.61 L is added [56]					
Bath compositions and operating conditions	Ag				
Sodium silver cyanide, g/L	1.83				
Sodium cyanide, g/L	1.0				
Silver nitrate, mol/L		0.003		0.008–0.05	1
Sodium hydroxide, g/L	0.75				0.01 mol/L
Dimethylamine borane (DMAB), g/L	2.0				
Thiourea, g/L	0.25				
Rochelle salt (potassium sodium tartrate), mol/L		0.035			
Ethylenediamine, mol/L		0.018			
3,5-Diiodotyrosine, mol/L		0.00004			
Ammonia, mol/L				0.4–1.83	
Acetic acid, mol/L				0.16–0.73	
Hydrazine-hydrate, mol/L				0.1	
Additives (as a surfactant and to level and brighten the deposit), mol/L				<0.01	
Silver as silver (I) complexes, g/L			1.5–15		
Sodium thiosulfate, g/L			15–200		
Sodium sulfite, g/L			1–20		
Disodium EDTA, g/L			0.1–10		

(Continued)

TABLE 2.8 Representative Electroless Plating Bath Formulations for Ag From Various Authors *(cont.)*

Ammonium hydroxide, mol/L					2
Glucose, mol/L					0.1
Trisodium citrate, g/L					0.2
Ethanol					20% vol.
pH		10–10.5	7.5–9	10–10.6	>10
Temp., °C	55–65	35–40	55–75		65
References	[57]	[58]	[59]	[60]	[61]

From the above various electroless plating of metals and binary alloys, the following features can be seen.

Research into electroless nickel started interest into this topic, but use of most of the other electroless metals and binary alloys also started quite early. Today all of them are still studied, though the contents of the research are now broader and deeper.

Some of the electroless baths in the early stage of development included some toxic agents, for example, cyanide was used in the electroless plating baths for Au and Ag, and formaldehyde in the bath for Cu. Of course, those baths are not environmentally friendly. Today, people pay more attention to the so-called "green" electroless technology and are attempting to make the electroless plating industry more environmentally friendly. This will be discussed in more detail in Chapter 7.

Research into electroless plating was basically limited to the electroless plating process itself in the early stages, but today the situation is quite different. For example, in addition to the electroless plating process, extended range of associated processes are being investigated, such as the study of electroless Ni–P plating on stainless steel [24], the effects of deposition parameters on surface morphology and crystallographic orientation of electroless Ni–B coatings [30], electroless Ni–B supported on carbon for direct alcohol fuel cell applications [97], selective electroless nickel deposition on copper as a final barrier/bonding layer material for microelectronics applications [98], and electroless nickel deposition from methane sulfonate bath [99], to name just a few.

Similarly to the early investigations, most work still concentrates on the electroless plating of Ni, Co, Cu, Ag, Au, Pd, and their binary alloys. Pt, Rh, Ru metals and their binary alloys are still only studied a little, and Os has only one or two papers dedicated to it. In addition, only one or two papers for the electroless plating of group B metals (Zn, Cd, In, Sn, Pb, As, Sb, and Bi) and

TABLE 2.9 Representative Electroless Plating Bath Formulations for Pd and Palladium-Based Binary Alloys From Various Authors

Bath compositions and operating conditions	Pd				Pd–Cu*		Pd–Ru	Pd–P			Pd–B
					Pd bath	Cu bath					
Tetramminoplladous chloride, g/L	5.4										
Palladium chloride, g/L		2	5	3.2	5.45			3.6–4.4	0.01 mol/L	1	0.01 mol/L
Cooper sulfate, g/L						6.225					
Sodium hydroxide 38%, g/L		4				20		10–20			
Ethylenediamine, mol/L									0.08		0.08
Sodium EDTA, g/L	33.6		70		70	20.0988					
Ammonium hydroxide 28%, mL/L	350 g/L	160	500	320	389.6					120	
Hydrazine, mL/L	0.3 g/L		10	1%, 200	10 (1 mol)						
Formaldehyde(37), mL/L						14.039					
Triton X-100, mg/L						25					
2,2-Bipyridyl, mg/L						5					
Thiodiglycolic acid, mg/L									30		50
Ammonium chloride, g/L		27								20	
Sodium hypophosphite, g/L		10						10.6–21.2	0.06 mol/L	10	
Sodium thiosulfate, g/L								0.0372–0.0447			
Hydrochloric acid, mL/L				4.0	10.9					2	
Trimethylamine borane, mol/L											0.04
Activating: Palladium (II) Acetate, 99% pure Chloroform, HPLC grade							33 g/L 1 L/L				

(Continued)

TABLE 2.9 Representative Electroless Plating Bath Formulations for Pd and Palladium-Based Binary Alloys From Various Authors *(cont.)*

Bath compositions and operating conditions	Pd				Pd–Cu*		Pd–Ru	Pd–P			Pd–B
					Pd bath	Cu bath					
Reducing: Deionized H_2O 28–30 wt% NH_4OH Hydrazine (3 mol)							 650 mL/L 340 mL/L 10 mL/L				
Pd stock: Deionized H_2O 28–30 wt% NH_4OH 37 wt% HCl Palladium (II) Chloride, 99% pure							 602 mL/L 392 mL/L 6 mL/L 5.5 g/L				
Ru solution: Ruthenium (III) Chloride, 99.9% pure Deionized H_2O							 Up to 3.8 g/L 1 L/L				
Pd–Ru bath: Pd stock Ru solution							 800 mL/L 200 mL/L				
pH		9.8 ± 2	10					8–10	80	10	7.0
Temp., °C	80	25	60	20	65 ± 5		50	40–50	50	45	50
Cu, Ru, P, B in deposit, wt%					As thickness to 30 μm, Cu: 24		7.5 ± 2.1 for 10 wt% Ru bath		P:3–17 mol%	P:7.1–7.3 at%	B:18–22 mol%
References	[62]	[63]	[64]	[65]	[66]		[67]	[68]	[69]	[70]	[71]

**Such a process is called a two-step method. First, layers of Pd and Cu were sequentially deposited by electroless plating on activated substrates. Then, the composite membrane was annealed at high temperature (e.g., between 350°C and 700°C) in flowing H_2 for times ranging from 6 days to 25 days to induce intermetallic diffusion and obtain homogeneous Pd–Cu films. Other binary electroless Pd-based alloys such as Pd–Ag and Pd–Au were all prepared by this technology. Here, electroless plating Pd–Cu alloy is just acted as an example. From Fig. 2.1, we can see the two steps clearly.*

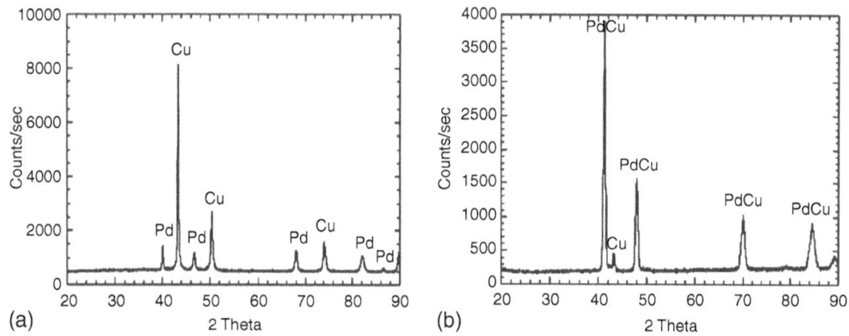

FIGURE 2.1 XRD results for the annealing of Golden Technologies Company (GTC) (now Coors Tech Oak Ridge, Oak Ridge, Tennessee) Pd–Cu membrane. Film thickness ≅ 22 μm. (a) As electrolessly deposited; (b) annealed at 600°C for 12 h under helium [66].

their binary alloys have been published to date. This situation reflects the level of application of electroless plating of the respective metal (Fig. 2.1).

2.11 ELECTROLESS PLATING OF TERNARY ALLOYS

As mentioned in Chapter 1, although initially there were few studies on the electroless plating of ternary alloys, gradually these alloys have been studied more and more. Currently, most electroless plating applications use ternary alloys today because they have superior properties to coatings of metals and binary alloys.

This book will divide the ternary alloys into Ni–Me–P, Ni–Me–B, Co–Me–P, Co–Me–B systems, and other metal-based ternary alloy systems. Of course, not all baths reported will be listed; three examples are chosen for each alloy system.

2.11.1 Ni–Me–P Alloy Plating Baths

Although few papers of Ni–Me–P ternary alloys (Me is a metal other than Ni) were published in the early stage, a few reviews of Ni–Me–P alloys did appear in that period of time [100–102]. After the 1990s, many more papers concerning Ni–Me–P alloys were published because of their superior properties and broad applications. However, it is difficult to find reviews of these alloys in the period of time. Here, some representative baths for producing Ni–Me–P alloys are listed. In general, only three baths of every alloy system are listed, from the early, rapid, and in-depth development stages. Among the ternary electroless Ni–Me–P alloys, Ni–Mo–P, and Ni–W–P alloys were paid more attention, so many papers about these systems have been published. However, only three examples will be listed here, for simplicity. In addition, though an electroless Ni–Mo–P coating was first proposed by Mallory in 1976 [103], no details of the bath were given in his paper, so it is not used as an example in Table 2.16.

TABLE 2.10 Representative Electroless Plating Bath Formulations for Pt and Platinum-Based Binary Alloys From Various Authors

Bath compositions and operating conditions	Pt				Pt–Pd	Pt–Rh	Pt–Ru
Sodium palatinate, g/L	10						
Platinum metal concentration, g/L		1.9					
Diamminebis(nitrito-N,N) platinum (II), g/L			0.1–5			0.1–5	
Chloroplatinic acid hexahydrate, mmol/L				4.0 g/L	0.15–1		
Palladium chloride, mmol/L					1–1.85		
Hexachloroplatinic acid and ruthenium (III) chloride were used as the source of Pt and Ru ions, respectively							C_{Pt}/C_{Ru}= 0.2–5.0 in bath compositions
Triamminetris(nitrito-N,N,N)rhodium(III), g/L						0.1–5	
5-Sulfosalycylic acid, g/L				2.0			2
Sodium hydroxide, g/L	5						
1,3,6-Sodium naphthalene trisulfonate tribasic hydrate, g/L				0.5			0.5
Sodium benzene 1,3 disulfonate, g/L				1.0			0.6
Ethylamine, g/L	10						
Ethylenediamine, mol/L		0.8					
Hydrazine dihydrochloride, g/L				2.0			
Hydrazine or hydrazine salts, g/L			0.1–10				
Formic acid, g/L							10

Ammonium hydroxide, mL/L			125			10–500	
Hydrazine, g/L	1	2–4 mol/L			0.035 mol/L	0.1–10	
Imidazol, mol/L		0.5					
Diarsenic pentoxide, mol/L		6.5×10^{-4}					
Hydrochloric acid, g/L				33	0.38 mol/L		30
pH		9	11			8–13	
Temp., °C	35	90	50	70 ± 0.2	55	60–90	70
Rh, Ru in deposit, at%					Pd: 52–91	Rh: 1–13 wt%	Ru: 75–6.8 as the C_{Pt}/C_{Ru}= 0.2–5.0 in bath compositions
References	[72]	[73]	[74]	[75]	[76]	[77]	[78]

TABLE 2.11 Representative Electroless Plating Bath Formulations for Ru, Rh, Os, and Cr–P Binary Alloy From Various Authors

Bath compositions and operating conditions	Ru				Rh	Os	Cr–P
Ruthenium chloride, g/L	1	0.014 mol/L	0.32	3.08 mmol/L			
Triamminetris(nitrito-N,N,N)rhodium(III), g/L					0.01–5		
Osmium tetraoxide, mol/L						0.01	
Chromic sulfate, mol/L							0.1
Sodium hypophosphite, mol/L		0.27				0.05	0.1
Di-ammoniumhydrogencitrate, mol/L		0.014					
Di-ammonium citrate, mol/L						0.003	
Ammonium oxalate, mol/L		0.07					
Sodium hydroxide, g/L	50	Small amount				Small amount	
Addition agent,		Small amount				Small amount	
Ammonium hydroxide (28%), mL/L	50		32		10–500		
Hydrazine, g/L	2		(5%)50 mL/L	30 mmol/L	0.1–10		
Hydrocloric acid			0.4 mL/L	0.12 mmol/L			
Glycine, mol/L						0.01	
Ammonia						Small amount	
Potassium thiocyanate, mol/L							0.1–0.2
Sodium chloride, g/L							10
Boric acid, g/L							20
Sodium nitrite, mmol/L							5
pH	12.6–13.6	9.5		11.5–14	8–13	10	3.0
Temp., °C	55–65	85	85	30–80	60–90	85	50
P in deposit, wt%							10
References	[79]	[80]	[65]	[81]	[82]	[83]	[84]

TABLE 2.12 Representative Electroless Plating Bath Formulations for Zn, Cd, and In Metals From Various Authors

Bath compositions and operating conditions	Zn	Cd	In		Sn		Pb	As	Sb	Bi
Zinc chloride, mol/L	0.08									
Cadmium chloride, mol/L		0.08								
Indium sulfate, mol/L			0.007							
Indium chloride, mol/L				0.08						
Stannous chloride, mol/L					1 mL/L	0.08				0.04
Trisodium citrate, mol/L		0.34			100 g/L					
Lead chloride, mol/L							0.04			
Sodium arsenate, mol/L								0.08		
Antimony trichloride, mol/L									0.08	
Bismuth trichloride, mol/L										0.08
Titanium chloride, mol/L	0.04	0.04		0.04		0.04	0.03	0.04	0.04	
EDTA, mol/L	0.08							0.08		
Na_2EDTA, mol/L		0.08	0.019		15 g/L	0.09	0.08		0.08	0.08
Nitrilo triacetic acid (NTA), mol/L	0.20	0.20		0.20		0.10	0.20	0.20	0.10	0.20
Triethanolamine (TEA), mol/L			0.02–0.03							
Sodium borohydride, mol/L			0.05							

(Continued)

TABLE 2.12 Representative Electroless Plating Bath Formulations for Zn, Cd, and In Metals From Various Authors *(cont.)*

Bath compositions and operating conditions	Zn	Cd	In		Sn		Pb	As	Sb	Bi
Sodium citrate, mol/L				0.32		0.24	0.35		0.32	0.34
Sodium acetate, g/L					10					
Citric acid, mol/L	0.34							0.34		
Benzenesulfonic acid (32% W/V), mL/L					1					
Titanium (III) chloride (30% W/V in 24% W/V HCl), mL/L					1.5					
pH	8–10	10.5	9.5–9.75	10	9	6.5–7.5	9.0	6–10	7.5	8.8
Temp., °C	70–90	70	80	80	90	50–70	60	70–90	20	60
References	[85]	[86]	[87]	[88]	[89]	[90]	[91]	[85]	[85, 92]	[93]

TABLE 2.13 Representative Electroless Plating Bath Formulations for a Few Binary Alloys of Group B Metals From Various Authors

Bath compositions and operating conditions	In–Fe	In–Ni	In–Sn	In–Pb	In–As	In–Sb	Sn–Pb	Bi_2Te_3
Indium chloride, mol/L	0.08	0.08	0.08	0.08	0.08	0.08		
Iron (II) sulfate, mol/L	0.04							
Nickel chloride, mol/L		0.04						
Stannous chloride, mol/L			0.08				0.079	
Lead chloride, mol/L				0.0005			0.001	
Sodium arsenate, mol/L					0.002			
Antimony trichloride, mol/L						0.0002		
Tellurium powder (99.99%), mol/L								5×10^{-3}
Bismuth nitrate (99.99%), mol/L								3.34×10^{-3}, 5×10^{-3}, 1×10^{-2}
Titanium trichloride, mol/L	0.04	0.04	0.04	0.04	0.04	0.04	0.04	
Na_2EDTA, mol/L							0.09	
Nitrilo triacetic acid (NTA), mol/L	0.20	0.20	0.20	0.20	0.20	0.14	0.10	
Sodium citrate, mol/L	0.32	0.32	0.32	0.32	0.32	0.34	0.24	
Nitric acid (11 mol), mL/L								87
pH	10.0	9.0	7.0	9.0	9.0	7.8	7.0	
Temp., °C	80	80	80	80	80	60	40–70	19–21
Composition in deposits, wt%	Fe: 0.16	Ni: 15.18	Sn: 99.50	Pb: 8.42	As: 2.81	Sb: 50 at%	Pb: 3.0–28 at% as lead chloride: 0.001–0.01 M	Bi_2Te_3*
References			[88]			[94]	[95]	[96]

*According to the authors [96], the details of the bath prepared should be as follows in order to obtain the electroless Bi_2Te_3 deposit. Three Te^{IV} solutions were prepared by dissolving 5×10^{-3} mol of Te in 87 mL of nitric acid 11 N, 3.34×10^{-3} mol, 5×10^{-3} mol, and 10^{-2} mol of bismuth nitrate pentahydrate were added to the different solutions. Finally, the volume was adjusted to 1 L in order to obtain respectively three solutions with $[Bi^{III}]/[Te^{IV}]$ ratio equal to 2/3, 1, and 2 and a $[Te^{IV}]$ concentration constant and equal to 5×10^{-3} mol/L.

In Table 2.14, three bath formulations for electroless Ni–Fe–P are listed, which are cited from different papers published in 1966, 2000, and 2011; three for Ni–Co–P from 1967, 1997, and 2011; and three for Ni–Re–P from 1968, 2000, and 2011, respectively.

In Table 2.15, only one bath formulation for each of electroless Ni–Ru–P, Ni–Rh–P, Ni–Pd–P, Ni–Pt–P, and Ni–Be–P are listed, cited from papers published in 1998, 1969, 1993, 1995, and 1969 and three for Ni–Cr–P from 1978, 2000, and 2010, respectively.

In Table 2.16, three baths for Ni–Mo–P are listed, which are cited from papers published in from 1988, 1996, and 2010; three for electroless Ni–W–P from 1983, 1996, and 2010; and three for Ni–Cu–P from 1974, 1992, and 2010, respectively.

In Table 2.17, three bath formulations for electroless Ni–Zn–P are listed, which are cited from papers published in 1990, 1999, and 2010; three for Ni–Sn–P from 1977, 2000, and 2010; only one for Ni–Tl–P and Ni–Sb–P from 2003 and 1975; and three for Ni–B–P from 1986, 1995, and 2009, respectively.

2.11.2 Co–Me–P Alloy Plating Baths

In Table 2.18, three bath formulations for electroless Co–Fe–P are listed, which are cited from papers published in 1976, 2003, and 2009; two for Co–Mo–P from 2002 and 2011; three for Co–W–P from 1967, 1997, and 2009; and only one for Co–Re–P from 1967, respectively.

In Table 2.19, one bath formulation for electroless Co–Ir–P alloy is listed, which is cited from a paper published in 2007; three for Co–Zn–P from 1966, 1991, and 2007; one for Co–Sn–P from 1998; and one for Co–Cu–P from 2003, respectively.

2.11.3 Ni–Me–B Alloy Plating Baths

In Table 2.20, one bath formulation for electroless Ni–Cr–B is listed, which is cited from a paper published in 1992; four for Ni–Mo–B from 1979, 1987, 2001, and 2009; and three for Ni–W–B from 1979, 2003, and 2009, respectively.

In Table 2.21, two bath formulations for electroless Ni–Fe–B are listed, which are cited from a paper published in 1994 and 2006; three for Ni–Co–B are from 1990, 1999, and 2004; and one for Ni–Cu–B from 1995, respectively.

In Table 2.22, two bath formulations for electroless Ni–Sn–B are listed, which are cited from a paper published in 1979 and 1993; three baths for Ni–Tl–B from 1983, 2005, and 2010; one for Ni–Pb–B from 2002, respectively.

2.11.4 Co–Me–B Alloy Plating Baths

In Table 2.23, one bath formulation for electroless Co–Mo–B is listed, which is cited from a paper published in 2010; two for Co–W–B from papers in 2002

TABLE 2.14 Representative Electroless Plating Bath Formulations for Ni–Me–P (Me = Fe, Co, and Re) Alloy Systems from Various Authors

Bath compositions and operating conditions	Ni–Fe–P			Ni–Co–P			Ni–Re–P		
Nickel chloride, mmol/L	56								
Nickel sulfate, g/L		1–20	0.01 mol/L	25	0.0167 mol/L	9.17–3.94	30	0.10 mol/L	38–3.8 mmol/L
Ferrous ammonium sulfate, mmol/L	20								
Cobalt sulfate, g/L				17.65	0.0083 mol/L	4.21–9.83			
Ferrous sulfate, g/L		2–22	0.07 mol/L						
Potassium rhenate, mol/L							1.5 g/L	0–0.0075	17–69 mmol/L
Sodium potassium tartrate, mmol/L	100–350	60 g/L							
Sodium hypophosphite, g/L	10	18–48	0.48 mol/L		0.200 mol/L	38.68	10	0.20 mol/L	190 mmol/L
Hypophosphorous acid, g/L				18.83					
Succinic acid, mol/L								0.20	
Ammonia, mol/L	3.6								
Sucrose, g/L		2							
Boric acid. g/L		5							
Sodium citrate, mol/L			0.20	80 g/L		44 g/L			
Ammonium sulfate, mol/L			0.10	40 g/L	0.250				

(Continued)

TABLE 2.14 Representative Electroless Plating Bath Formulations for Ni–Me–P (Me = Fe, Co, and Re) Alloy Systems from Various Authors *(cont.)*

Bath compositions and operating conditions	Ni–Fe–P			Ni–Co–P			Ni–Re–P		
Citric acid, g/L				51.4					
Ammonium chloride, g/L						26	50		
Glycine, mol/L							0.4		
Sodium succinate, mol/L								0.20	
pH		11	10	7.2–8.6	7.8–9.5	8.5	9	4.5	9.5–10
Temp., °C	75	78	65–70	50–95	25	75	98	90–95	90–95
[Re]/[Ni] in bath solution									0.45–18
Composition in deposits, wt%	Fe: 25 P: 0.5–1	Fe: 30–38 P: 1–3 at%	Fe: 3.85 P: 4.3 at%		Co: 42 P: 8 at%	Co: 11.01–45.06 P:9.346–6.49	Re: 45 P: 2.3	Re: 1–11 P: 8–7	Re: 30–73 P: 0–15 at%
References	[104]	[105]	[106]	[107]	[108]	[109]	[110]	[111]	[112]

TABLE 2.15 Representative Electroless Plating Bath Formulations for Ni–Me–P (Me = Ru, Rh, Pd, Pt, Be, and Cr) Alloy Systems From Various Authors

Bath compositions and operating conditions	Ni–Ru–P	Ni–Rh–P	Ni–Pd–P	Ni–Pt–P	Ni–Be–P	Ni–Cr–P		
Nickel chloride, g/L	37.5					30		
Nickel sulfate, g/L		7	0.1 mol/L	0.1 mol/L	7		13	15
Rhodium chloride, g/L		0.2						
Palladium chloride, mol/L			0.005					
Hydrogen hexachloroplatinate, g/L				0–1.25				
Beryllium fluoride, g/L					0.3			
Cromic chloride, g/L						40	10.7	18
Ruthenium chloride, g/L	0.31							
Sodium hypophosphite, g/L		10	0.1 mol/L	0.1 mol/L	10	20	42.4	15
Sodium citrate, g/L	80	20	0.3 mol/L	0.3 mol/L	20	10	20	
A solution of 10 g/L of mercapto-benzothiazole in 2N NaOH, mL/L		0.2			0.2			
Ammonium chloride, g/L	40							
Triethenolamine, mL/L	75							
Ammonium sulfate, mol/L			0.5	0.1				
Thiodiglycol Acid (TDG), mg/L			10					
Citric acid, g/L						7	34	
Lactic acid, g/L							18	
pH		8.2	10	10	8.2	3.83	4.6	13–13.5
Temp., °C		98	80 ± 2	90 ± 1	98	92	85	85–90
Composition in deposits, wt%	Ru: 0.2 P: 11.2 at%	Rh: 12	Pd: 32.8 P: 2	Pt: 0–9.8 P: 11–7.8	Be: <1	Cr: 4 at%	Cr: 9.52 P: 18.2 at%	
References	[113]	[114]	[115]	[116]	[114]	[117]	[118]	[119]

TABLE 2.16 Representative Electroless Plating Bath Formulations for Ni–Me–P (Me = Mo, W, and Cu) Alloy Systems From Various Authors

Bath compositions and operating conditions	Ni–Mo–P			Ni–W–P			Ni–Cu–P		
Nickel sulfate, mol/L	0.1	0.1	25 g/L	0.27	0.1	20 g/L	35 g/L	0.01–0.057	0.1
Sodium molybdate, mol/L	0–0.02	0–0.01	0.17–0.54 g/L						
Sodium tungstate, mol/L				0–0.12	0.05–0.4	15 g/L			
Cuprous chloride, g/L				.			0–200 ppm		
Copper sulfate, mol/L								0.0005–0.004	0.0008
Sodium hypophosphite, mol/L	0.2	0.2	20 g/L	0.066	0.4	40 g/L	15 g/L	0.11–0.68	0.28
Sodium citrate, mol/L	0.1	0.1	25 g/L	0.005–0.25	0.3	16 g/L		0.034–0.4	0.2
Sodium acetate, g/L							Do		
Ammonium sulfate, mol/L				0.227					
Thiodiethyleneglycol, ppm							25		
Acetate acid, g/L			2.1						
Citric acid, g/L						8			
Glycollic acid, mol/L	0.2	0.2							
Ammonium fluoride, g/L			0.004						
Sodium dodecyl sulfate, g/L			0.008						
Ammonium chloride, g/L						14		0–1.86 mol/L	
Ammonium Acetate, mol/L									0.5
pH	9	9	11.25–12.0	9.0	9	9.5	~5.0	5.8–9	5
Temp., °C	90	90	75–90	90	90 ± 1	92 ± 2	90.56–96.11	65–90	78
Composition in deposits, wt%	Mo: 0–14.9 P: 20–2	Mo: 0–14.37 P: 11.80–2.56	Mo: 6.3–5.5 P: 1.3–2.4	W: 0–5 P: 8.4–6	W: 32.5–4.0 P: 10.5–1.8 at%	W: 2.2 P: 8.3	Cu: 1–8 P: 5–15	Cu: 100–15 P: 0–15 As pH: 6.5–9.5	Cu: 11 P: 12
References	[120]	[121]	[122]	[123]	[5, 124]	[125]	[126]	[127]	[128]

TABLE 2.17 Representative Electroless Plating Bath Formulations for Ni–Me–P (Me = Zn, Sn, and Tl) Alloy Systems From Various Authors

Bath compositions and operating conditions	Ni–Zn–P			Ni–Sn–P			Ni–Tl–P	Ni–Sb–P	Ni–B–P		
Nickel sulfate, g/L			27	0.133 mol/L		26	0.1 mol/L	7.5			
Nickel chloride, g/L	7.5	47.5			55				0.10 mol/L	30	0.1 mol/L
Zinc chloride, g/L	Up to 1	8									
Zinc sulfate, g/L			4								
Sodium stannate, g/L				0–0.033 mol/L	5	0.875–3.5					
Thallous sulfate, mol/L							0.005–0.01				
Antimony gluconate, mg/L								100			
Sodium hypophosphite, g/L	3.5	45	32	0.095 mol/L	18	15	0.1–0.2 mol/L	20	0.25 mol/L	10	0.094 mol/L
DMAB, mol/L									0.01–0.025		
Sodium borohydride, g/L										2	
Potassium borohydride, mol/L											0.185
Sodium citrate, mol/L			59 g/L	0.289	80 g/L		0.02–0.1				
Sodium gluconate, g/L								60			
Ammonium sulfate, g/L			27								
Ammonium chloride, g/L	12.5	26		0.934 mol/L							

(Continued)

TABLE 2.17 Representative Electroless Plating Bath Formulations for Ni–Me–P (Me = Zn, Sn, and Tl) Alloy Systems From Various Authors *(cont.)*

Bath compositions and operating conditions	Ni–Zn–P			Ni–Sn–P			Ni–Tl–P	Ni–Sb–P	Ni–B–P		
Lactic acid, g/L						11					
Citric acid, g/L	19.8	4				6					
Lactic acid, g/L			10								
Glycine, g/L						7			0.40 mol/L		
Boric acid, g/L								15			
Ethylenediamine, mL/L										80	0.36 mol/L
Sodium hydrate, g/L										40	
Lead(II) nitrate, mg/L										10	
Cadmium sulfate, mol/L											5.68×10^{-4}
pH	8.2	7–10	9	9	9	7	5–6	5	5.5 ± 0.2		13
Temp., °C	60	85	80 ± 5	90 ± 1	90	85	80–90		70 ± 1	95	90
Composition in deposits, wt%	Zn: 10 P: 16 at%	Zn: 6–20 P: 6–10		Sn: 0–1.0 P: 3.4–5.4	Sn: 10.4 P: 4.5 at%	Sn: 0.3–1 P: 6.3–8.9		Sb: 2 P: 4.5	B: 0.28–0.70 P: 7.82–6.80	B: 15.2 P: 4.1	B: 1.43 P: 1.18
References	[129]	[130]	[131]	[132]	[21]	[133]	[134]	[135]	[136]	[137]	[138]

TABLE 2.18 Representative Electroless Plating Bath Formulations for Co–Me–P (Me = Fe, Mo, W, and Re) Alloy Systems From Various Authors

Bath compositions and operating conditions	Co–Fe–P			Co–Mo–P		Co–W–P			Co–Re–P
Cobalt sulfate, mol/L	0.089	0.09	0–0.2	0.087	0.05	30 g/L		0.1	30 g/L
Cobalt chloride, g/L							30		
Ferrous sulfate, mol/L	0–0.073	0.07	0–0.2						
Sodium tungstate, mol/L						0–30 g/L		0–0.2	
Sodium molybdate, mol/L				0.000413					
Ammonium tungstate, g/L							10		
Potassium perrhenate, g/L									0–0.8
Sodium hypophosphite, mol/L	0.377	0.3–0.6	0–0.5	0.239	0.1	20 g/L	20	0.5	20 g/L
Ammonium sulfate, mol/L	0.303	0.05–0.3							
Sodium citrate, mol/L	0.095	0.1–0.5	0.25	0.44	0.1	80 g/L	80 g/L		80 g/L
Maltose, mol/L			0.0058						
Boric acid, mol/L			0.5	0.5	0.2				
Polyoxyethylene nonylphenyl ether phosphate (RE-610), g/L				0.05			0.5		
Ammonium hydroxide, g/L						60			60
Glycine, mol/L								0.6	
Ammonium chloride, g/L						50			50

(Continued)

TABLE 2.18 Representative Electroless Plating Bath Formulations for Co–Me–P (Me = Fe, Mo, W, and Re) Alloy Systems From Various Authors *(cont.)*

Bath compositions and operating conditions	Co–Fe–P			Co–Mo–P		Co–W–P			Co–Re–P
Molybdic acid, mol/L					1×10^{-5}–1×10^{-3}				
Sodium molybdate, mol/L					1×10^{-5}–1×10^{-3}				
Ammonium molybdate, mol/L					1×10^{-5}–5×10^{-4}				
Maleic acid, mol/L					0.01				
Malonic acid, mol/L								0–0.5	
pH	8.1 ± 0.1	8–12	9.5–11.5	8.9–9	9.2	8.9	8.5–10.5	9–11	8.9
Temp., °C	80 ± 1	30–90	60–90	90–95	82	95 ± 1	55–90	80	95 ± 1
Composition in deposits, wt%	Fe: 0–25 P: 4.3–0.3	P: 1–2.5 at. %	Fe: 41.1–27.1 P: 1.3–6.5 As NaH_2PO_2 increase	Mo: 2.68 P: 3.17 O: 2.46 at%	Mo: 0–4.03 P: 8.84–0.53 Mo: 0–3.14 P: 9.60–0.91 Mo: 0–5.04 P: 7.93–2.53 at%*	W: 0–9 P: 6–3	W: 1.8–3.4 P: 8.9–10.3 at%	W: 4.2 P: 2.9 at%	Re: 0–30 P: 0–5
References	[139]	[140]	[141]	[142]	[143]	[144]	[145]	[146]	[144]

**When the concentrations of the molybdenum compounds used are rather larger, there are large amounts of oxygen in the composition of Co–Mo–P alloy deposits. O (at%): 28.96 and 62.84 when molybdic acid are 5×10^{-4} and 1×10^{-3} mol/L; 31.06 and 67.97 when sodium molybdate is 5×10^{-4} and 1×10^{-3} mol/L; and 43.90 when ammonium molybdate is 5×10^{-4} mol/L, respectively. According to the authors, it seems that rather adsorbed molybdate species or molybdenum oxide than elemental molybdenum was present in such deposits.*

TABLE 2.19 Representative Electroless Plating Bath Formulations of Co–Me–P (Me = Zn, Sn) Alloy Systems From Various Authors

Bath compositions and operating conditions	Co–Ir–P		Co–Zn–P		Co–Sn–P	Co–Cu–P
Iridium sulfate, g/L	0.551					
Cobalt chloride, g/L		7.5		15		20
Cobalt sulfate, mol/L	23 g/L		0.05		22 mmol/L	
Cuprous sulfate, g/L						1.5
Zinc chloride, g/L		0–1.0				
Zinc sulfate, g/L			0.02 mol/L	2		
Sodium stannate, mmol/L					4	
Sodium citrate, g/L	130					
Sodium hypophosphite, g/L	25	3.51	0.20 mol/L	20	95 mmol/L	15
Ammonium chloride, g/L		12.5				
Ammonium hydroxide, mol/L					0.1	
Sodium citric, mol/L			0.25		85 mmol/L	
Aminoacetic acid, mol/L			0.30			
Rochelle salt, g/L				50		
Citric acid, g/L		19.84				
Boric acid, g/L	31		0.25 mol/L			30
FC 98 surfactant, mmol/L					1	
EDTA, g/L						8.8
Hydroquinone, g/L						5
Nickel chloride, g/L						0.8
pH	8.5	8.2 ± 1	9.5	~9	8.6	4.5
Temp., °C	80	80 ± 0.5	80	80–85	75	70
Composition in deposits, wt%	Ir: 33.9 P: 43	Zn: 0–4 P: ~4		Zn: 1.6 P: 1.8	Sn: 1–3 P: few at%	
References	[147]	[148]	[149]	[150]	[151]	[152]

TABLE 2.20 Representative Electroless Plating Bath Formulations for Ni–Me–B (Me = Cr, Mo, W) Alloy Systems From Various Authors

Bath compositions and operating conditions	Ni–Cr–B	Ni–Mo–B				Ni–W–B		
Nickel sulfate, mol/L	0.04	0.1	0.05		0.1	0.1	0.10	0.11
Nickel chloride, g/L				30				
DMAB, mol/L	0.2	0.04	0.06		0.05	0.04	0.05	0.069
Cr-complex*, mol/L	0.3							
Mo-complex, mol/L		0.001–0.05	0–0.0125					
MoO_3, mol/L					0–0.01			
W-complex, mol/L						0.001–0.05		
Sodium molybdate, g/L				0.06–6.049				
Sodium tungstate, mol/L							0–0.10	
Ethylene diamine, mL/L				50–70				
Tungstic acid, mol/L								0.03
Sodium citrate, mol/L			0.10		0.2		0.2–0.4	
Hydroxyacetic acid, mol/L		0.3				0.3		
Ammonium acetate, mol/L								0.285
Potassium borohydride, g/L				0.6–1.4				
Sodium hydroxide, g/L				50				
Glycolic acid, mol/L			0.20					
pH	4–6	5, 7, 9	9	>13	9	5, 7, 9	9	8.5
Temp., °C	70	70 ± 2	90	86 ± 1	70	70 ± 2	70	70–75
Composition in deposits, wt%	Cr: 1.7–0.8 B: 22–24 at%	Mo: 2.9–31 B: 3.2–1.6	Mo: 0–21.1 B: 6.3–0.1	B: 4.26–0.41		W: 10–10.5 B: 5.1–0.8	W: 0–8.9 B: 7.4–2.5	W: 19–21 B: 8–9 at%
References	[153]	[100]	[154]	[155]	[156]	[152]	[157]	[158]

*Cr-complex contains equal molar quantities of Cr and citric acid.

TABLE 2.21 Representative Electroless Plating Bath Formulations for Ni–Me–B (Me = Fe, Co, Cu) Alloy Systems From Various Authors

Bath compositions and operating conditions	Ni–Fe–B		Ni–Co–B			Ni–Cu–B
Nickel sulfate, mol/L	0.05	0.02–0.06		0.1	0.07	
Nickel chloride, mol/L			Variable			0.1
Ferrous sulfate, mol/L	0.05	0.08–0.04				
Cobalt chloride, mol/L			Variable			
Cobalt sulfate, mol/L					0.03	
Ni + Co, mol/L			0.05			
$FeSO_4$ + $NiSO_4$ was kept as, mol/L		0.1				
DMAB, mol/L	0.03			0.03	0.024	
Cuprous chloride, mmol/L						0–4
Sodium tartrate, mol/L	0.15		0.4			
Potassium sodium tartarate, mol/L		0.2				0.06
Sodium citrate, mol/L	0.10	0.2			0.10	
Potassium borohydride, g/L		2.0–5.0				1.85
Ammonium sulfate, mol/L	0.50					
Sodium hydroxide, mol/L		0.125				1.5
Ethylenediamine, mL/L		5.0				1 mol/L
Hydrazine dihydrochloride, mol/L			0.8			
Thiourea, mol/L			1.3×10^{-5}			
Borax, mol/L			2×10^{-3}			
pH	7–11		12.5	6.5	7.5	
Temp., °C	90	40–55	80	70 ± 1	75 ± 1	90
Composition in deposits, wt%	Fe: 23.3–26.8 B: 22–24	Fe: 47–56 at%	Co: 55.8 B: 0.27	Co: 0–21.1 B: 6.3–0.1	Co: 73.2 B: 2.6 at%	Cu: 0–8.0 B: 22.8–21.5 at%
References	[159]	[160]	[161]	[162]	[163]	[164]

TABLE 2.22 Representative Electroless Plating Bath Formulations for Ni–Me–B (Me = Sn, Tl) Alloy Systems From Various Authors

Bath compositions and operating conditions	Ni–Sn–B		Ni–Tl–B			Ni–Pb–B
Nickel sulfate, mol/L	0.1					
Nickel chloride, mol/L		0.1	0.08	15–50 g/L	20 g/L	24 g/L
Dithallium sulfate, mg/L					1–60	
Lead tungstate, g/L						0.02
DMAB, mol/L	0.04	0.051	4.0 g/L			
Sn-complex, mol/L	0.001–0.05					
Tin tetrachloride, mol/L		0–0.12				
Thallic nitrate, g/L			0–0.5	0.012–0.042		
Sodium acetate, mol/L			0.4			
Sodium borohydride, g/L				0.5–1.8	1	0.48
Ethylene diamine, mL/L				32–72	100 g/L	59 g/L
Sodium citrate, mol/L		0.2				
Hydroxyacetic acid, mol/L	0.3					
Ammonium chloride, mol/L		0.5				
Sodium hydroxide, g/L				42	45	39
pH	5, 7, 9	7	6.5		14	
Temp., °C	70 ± 2	90 ± 1	60		85	95 ± 1
Composition in deposits, wt%	Sn: 8.9–3.5 B: 3.4–1.7	Sn: <18 B: 5.3–1.5	Tl: 0–7.8 B: 13.5–3.5 at%	Tl: ~7, B: ~3 as changing $NiCl_2$		Pb: 1.25 B: 5.75
References	[154]	[165]	[166]	[167]	[168]	[169]

TABLE 2.23 Representative Electroless Plating Bath Formulations for Co–Me–B (Me = Mo, W, Fe) and Co–P–B Alloy System From Various Authors

Bath compositions and operating conditions	Co–Mo–B*		Co–W–B		Co–Fe–B		Co–P–B
	A	B					
Cobalt chloride, g/L	50		0.1 mol/L				
Cobalt sulfate, mol/L				0.07	0.10–0.07	0.03–0.14	28.1 g/L
Sodium molybdate, g/L	12						
Sodium tungstate, mol/L			0.03	0.076			
Ferrous sulfate, mol/L					0.01	0.005–0.022	
Sodium borohydride, g/L		40					
DMAB, mol/L			0.06	0.12	0.025	0.085–0.68	4 g/L
Sodium citrate, mol/L			0.3	0.46	0.05	0.05	0.3 g/L
5-Aminotetrazole, mol/L				0.0004			
Sodium tartrate, mol/L					0.20	0.20	
Thallium acetate, mol/L						1.06×10^{-4}	
Thiourea, mol/L						1.76×10^{-6}–1.25×10^{-5}	
Sodium hypophosphite, g/L							0.0094/0.069
Hydrogen chloride, g/L							0.083
Ammonium chloride, g/L	50						

(Continued)

TABLE 2.23 Representative Electroless Plating Bath Formulations for Co–Me–B (Me = Mo, W, Fe) and Co–P–B Alloy System From Various Authors *(cont.)*

Bath compositions and operating conditions	Co–Mo–B*		Co–W–B		Co–Fe–B		Co–P–B
	A	B					
Sodium hydroxide, g/L		10		0.66			
Ammonium hydroxide, mL/L	45						
Ammonium sulfate, mol/L					0.20	0.20	
Phosphonic acid, mol/L					0.06	0.06	
Boric acid, mol/L				1.0			
Surfactant RE-610, cc/L							1.5×10^{-2}
pH			9.5		9	6.5–9.5	9.5–9.7
Temp., °C	25		75	70	70	78± 1	80–85
Composition in deposits, wt%			W: 20 B: ** at%	W: 15.5 B: **	Fe: 4 B: 2 at%	Fe:–10 B: 0.4–0.8 at%***	
References	[170]		[171]	[172]	[173]	[174]	[175]

Here the preparation of Co–Mo–B/Ni foam catalyst used the so-called modified electroless plating method [169], which is as follows. The pretreated Ni foam was immersed in solution A for 20 s at 25°C and then, an equal volume of solution B was poured into solution A. The mixture was kept undisturbed until bubble generation ceased (which generally took several minutes). This EP process was repeated twice to obtain the desired Co–Mo–B loading.

***The content is under the limit of detection.*

****Iron and boron atomic percentage of the Co–Fe–B alloy coating versus DMAB concentration. Plating bath.* parameters: sodium tartrate (0.2 M), sodium citrate (0.05 M), ammonium sulfate (0.2 M), phosphorous acid (0.06 M), cobalt sulfate (0.06 M), and iron sulfate (0.01 M). pH = 6.5 for $CoSO_4$ and $FeSO_4$ curves and pH = 5.5 for DMAB.

and 2009, two for Co–Fe–B from 2000 and 2008, and one for Co–P–B from 2007, respectively.

2.11.5 Other Ternary Alloy Plating Baths

In Table 2.24, one bath formulation for electroless Pd–Co–P is listed, which is cited from a paper published in 1973, one for Pd–Zn–P from 1969, and two for Cu–Sn–P from papers in 1995 and 2007, respectively.

TABLE 2.24 Representative Electroless Plating Bath Formulations for Other Ternary Alloys From Various Authors

Bath compositions and operating conditions	Pd–Co–P	Pd–Zn–P	Cu–Sn–P	
Palladium chloride, g/L	2	2		
Cobalt sulfate, g/L	29.6			
Zinc sulfate, g/L		36		
Cuprous sulfate, mol/L			1.5 g/L	0.020
Tin sulfate, g/L			2.0	
Tin chloride, mol/L				0.018
Nickel sulfate, mol/L				3.8×10^{-4}
Hydrogen chloride (38%), mL/L	4	4		
Sodium hypophosphite, g/L	10	10	14.4	0.10 mol/L
EDTA, g/L			8.8	0.070 mol/L
Hydroquinone, g/L			5.0	
Sodium acetate, g/L			6.0–10.0	
Ammonium chloride, g/L	27	27		
Paradioxybenzence, mol/L				2.7×10^{-5}
Ammonium hydroxide (28%), mL/L	160	160		
Tartaric acid, mol/L				0.40
Boric acid, g/L			30	
pH		9.8 ± 2	4.0–5.5	6–7
Temp., °C	60	50–60	50–70	48–52
Composition in deposits, wt%	Co: 10 P: 28	Zn: 35.99 P: 1.24	Sn: 12.7 P: 3.6	
References	[176]	[63]	[177]	[178]

2.12 ELECTROLESS PLATING OF QUATERNARY ALLOYS

2.12.1 Ni-Based Quaternary Alloy Plating Baths

In Table 2.25, one bath formulation for electroless Ni–Fe–B–P is listed, which is cited from two papers published in 2000 and 2010; two for Ni–Fe–Co–P from 1995 and 2008; one for Ni–Fe–Ru–P from 2010; one for Ni–Fe–Zn–P from 2009; one for Ni–Fe–La–B from 2006; one for Ni–Co–Tl–P from 1991; and one for Ni–Cr–Cu–P from 2005, respectively.

In Table 2.26, one bath formulation for electroless Ni–Cr–W–P is listed, which is cited from a paper published in 2010; one for Ni–Mo–W–P from 2008; one for Ni–W–Re–P from 2001; one for Ni–W–Cu–P from 2006; one for Ni–W–Sn–P from 2007; one for Ni–Re–B–C from 2000; and two for Ni–Sn–Cu–P from 1998 and 2000, respectively.

2.12.2 Co-Based Quaternary Alloy Plating Baths

In Table 2.27, one bath formulation for electroless Co–Fe–Mn–P is listed, which is cited from a paper published in 2007; two for Co–Ni–Mn–P from 1991 and 2009; one for Co–Ni–Re–P from 1985; two for Co–Ni–W–P from 2007 and 2009; one for Ni–W–Cu–P from 2002; and one for Co–Ni–Zn–P from 1993, respectively.

In Table 2.28, one bath formulation for electroless Co–Mo–W–P is listed, which is cited from a paper published in 2008; one for Co–W–Re–P from 2007; one for Co–W–Pt–P from 2010; one for Co–W–Zn–P from 2003; one for Co–Ni–Fe–B from 2000; one for Co–Ni–La–B from 2005; one for Co–W–Re–B from 2007; and one for Co–W–B–P from 2007, respectively.

2.13 ELECTROLESS PLATING QUINARY AND MULTIALLOYS

In Table 2.29, one bath formulation for electroless Ni–Fe–Zn–Cu–P is listed, which is cited from a paper published in 2009; one for Ni–Co–Fe–La–P from 2007 and a Chinese patent in 2008; one for Ni–Co–Re–Mn–P from 1985; and one for Co–W–Re–B–P from 2007, respectively.

2.14 SUMMARY

Because the plating bath is the basic technology of electroless plating this chapter describes and discusses the electroless plating baths for metals, binary alloys, and multicomponent alloys.

EP has been investigated for more than 70 years, and too many bath formulations of EP deposits have been published to be listed here. Therefore, only a few representative bath solutions for each metal or alloy are given.

The representative bath solutions for various systems include those of pure metals (Ni, Co, Cu, Ag, Au, Pd, Pt, Ru, Rh, Os, Zn, Cd, In, Sn, Pb, As, Sb, and

TABLE 2.25 Representative Electroless Plating Bath Formulations for Ni-Based Quaternary Alloy Systems

Bath compositions and operating conditions	Ni–Fe–B–P	Ni–Fe–Co–P		Ni–Fe–Ru–P	Ni–Fe–Zn–P	Ni–Fe–La–B	Ni–Co–Tl–B	Ni–Cr–Cu–P
Nickel sulfate, g/L	1–20	0.025 mol/L	0.04 mol/L	12				15
Nickel chloride, mol/L					35 g/L	0.1262	21.4 g/L	
Iron sulfate, mol/L	2–22 g/L	0–500 ppm	0.03	35 g/L	25 g/L	0.0252		
Zinc sulfate, g/L					20			
Cobalt sulfate, mol/L		0.025	0.03					
Cobalt chloride, g/L							9	
Ruthenium chloride, g/L				1.25				
Lanthanum(III) chloride, mol/L						1.0		
Thallium sulfate, g/L							0.01 + 0.1*	
Cooper sulfate, g/L								0.2
Chromium(III) chloride, g/L								10
DMAB, mol/L						0.035		
Sodium molybdate, g/L								0.5
Sodium hypophosphite, g/L	13–48	0.2 mol/L	0.20 mol/L	25	15			40
Potassium borohydride, g/L	0–1.4							
Sodium borohydride, g/L							0.528 + 99.5*	
Potassium sodium tartrate, g/L	60							
Sucrose, g/L	2							
Sodium citrate, g/L		0.1 mol/L	0.30 mol/L	60			36	20
Boric acid, g/L	5		0.50 mol/L	10		0.40		
Sodium tartrate, mol/L		0.15				0.08		

(Continued)

TABLE 2.25 Representative Electroless Plating Bath Formulations for Ni-Based Quaternary Alloy Systems *(cont.)*

Bath compositions and operating conditions	Ni–Fe–B–P	Ni–Fe–Co–P		Ni–Fe–Ru–P	Ni–Fe–Zn–P	Ni–Fe–La–B	Ni–Co–Tl–B	Ni–Cr–Cu–P
Ethylene diamine, g/L							60	
Sodium hydroxide, g/L							47.8 + 105.7*	
Multi-hydroxyl carboxylicacids (major) + polycarboxylic acid (minor) as complexant, g/L								40
Ammonium sulfate, mol/L		0.5						
Glycine, mol/L						0.07		
Stabilizer, mmol/L						0.1–0.2		
Ammonium hydroxide, g/L							18	
Fluorine-containing surfactant								appropriate amount
Nitro-organic compounds or iodide as stabilizer, mg/L								0.5
pH	11	9	7.0–8.5	10	8.8	7.8	13.5	4–5
Temp., °C	78	90	90 ± 0.5	90 ± 0.5	90	65 ± 1	87.8	80–90
Composition in deposits, wt%	Fe_{max}: 38 B_{max}: 1.4 P_{max}: 3 at%	Fe: 1.3 Co: 41.3 P: 9.4	Fe: 1.23–8.69 Co: 24.3–37 P: 26.41–17.3	Fe: 6.16 Ru: 2.34 P: 13.68 at%	Fe: 2.23 Zn: 6.99 P: 5.3		Co: 2 Tl: 5 B: 3	Cr: 8.2 Cu: 2.4 P: 8.2 Fe, Cd: minute amount
References	[105, 179]	[180]	[181]	[182]	[183]	[184]	[185]	[186]

**The second figure indicates the adding quantity after the substrates have immersed into the bath for some time, and the procedure is as follows. After immersing the substrates into the coating bath at 87.8°C for about 1 h, an additional 0.4 g of thallium sulfate and 10 mL of a sodium borohydride solution (0.83 pounds sodium borohydride in 1 gallon of water containing also about 400 g of sodium hydroxide). After 2 h the coated substrates were removed from the coating bath. The original units are gallon (USA) and pound, here all are converted into liter and gram.*

TABLE 2.26 Representative Electroless Plating Bath Formulations for Ni-Based Quaternary Alloy Systems

Bath compositions and operating conditions	Ni–Cr–W–P		Ni–Mo–W-P	Ni–W–Re–P*	Ni–W–Cu–P	Ni–W–Sn–P	Ni–Re–B–C	Ni–Sn–Cu–P
	A	B						
Nickel sulfate, g/L	30–40		26.2	0.05 mol/L	0.076 mol/L	20	0.1 mol/L	
Nickel chloride, g/L								55
Sodium molybdate, g/L			2.4					
Sodium tungstate, g/L	25–30		30–40	0.1 mol/L	0.09 mol/L	30		
Chromium chloride, g/L		10–15						
Potassium perrhenate, mol/L				1.047×10^{-4}				
Ammonium perrhenate, mol/L							0.0005–0.01	
Cooper sulfate, mol/L					0.003			
Cooper chloride, g/L								0.3
Sodium stannate, g/L						0.3		5
DMAB, mol/L							0.03	
Diethylenetriamine, mol/L							0.005	
Sodium hypophosphite, g/L	10–20	10–20	10.6	0.114 mol/L	0.14 mol/L	20		18
Sodium citrate, g/L	80–100		30	0.285 mol/L	0.119 mol/L	35	0.005	80
Lactic acid, g/L			9		0.05 mol/L	5		
Sodium acetate, g/L		10–20						
Ammonium chloride, g/L	40–60							

(Continued)

TABLE 2.26 Representative Electroless Plating Bath Formulations for Ni-Based Quaternary Alloy Systems *(cont.)*

Bath compositions and operating conditions	Ni–Cr–W–P							
	A	B	Ni–Mo–W-P	Ni–W–Re–P*	Ni–W–Cu–P	Ni–W–Sn–P	Ni–Re–B–C	Ni–Sn–Cu–P
Ammonium hydroxide, mL/L	10–20							
Ammonium sulfate, mol/L					0.20	30		
Stabilizer, mg/L	10–30	10–20						
Potassium oxalate, g/L		4–6						
Lead nitrate, ppm							2	
pH	8.8–9.2	4–6	9	8.5	8	9	6.5	9
Temp., °C	85–90		85–90	89 ± 1	88 ± 2	88 ± 2	70	90
Composition in deposits, wt%	A:B Cr W P 4:6 3.2 4.1 5.6 5:5 3.0 5.2 5.2 6:4 2.7 5.7 5.0 8:2 2.3 6.2 4.6		Mo: 12.4 W: 1.3 P: 4.5	W: 2.6 Re: 3.2 P: 13.3	W: 5.5 Cu: 4.08 P: 4.92	W: 6.9 Sn: traces P: 6.4	Re: 9–60 B: 1.6–1.2 C: 0.22–0.15	Sn: 1.22 Cu: 17.47 P: 15.26 at%
References	[187]		[188]	[189]	[190]	[191]	[192]	[193]

According to the authors [189], codeposition of refractory metals requires that the specific Me/Ni concentration ratio: W/Ni should be greater than unity, whereas Re/Ni should be very small, owing to the strongly preferred Re deposition. So, the following ratio of metal concentration (g/L) was used: W/Ni = 6.3; Re/Ni = 6:6 × 10^{-3}.

TABLE 2.27 Representative Electroless Plating Bath Formulations for Co-Based Quaternary Alloy Systems

Bath compositions and operating conditions	Co–Fe–Mn–P	Co–Ni–Mn–P		Co–Ni–Re–P	Co–Ni–W–P		Co–Ni–Cu–P	Co–Ni–Zn–P
Iron sulfate, mol/L	0.029							
Nickel sulfate, mol/L		0.01	10 g/L	0.08	0.03	5 g/L	0.04	0.02
Sodium tungstate, mol/L					0.0025	1.6–8.2 g/L		
Ammonium perrhenate, mol/L				0.003				
Cobalt sulfate, mol/L	0.078	0.035	10 g/L	0.06	0.1	5 g/L	0.06	0.08
Manganese sulfate, mol/L	0.04	0.05	6–12 g/L					
Cooper sulfate, mol/L							0.003	
Zinc sulfate, mol/L								0.02
Sodium malonate, mol/L		0.5		0.75				
Sodium hypophosphite, mol/L	0.3	0.2	40 g/L	0.20	0.2	20 g/L	0.5	0.3
Sodium malate, mol/L		0.05						
Sodium citrate, mol/L	0.3		60 g/L		0.1	60 g/L		
Boric acid, g/L			30			30		
Tartronic acid, mol/L				0–0.07				
Sodium tartrate, mol/L				0.20				
Sodium gluconate, mol/L				0.30				
Ammonium chloride, mol/L	0.3							
Ammonium sulfate, mol/L		0.5		0.5	0.5			0.5

(Continued)

TABLE 2.27 Representative Electroless Plating Bath Formulations for Co-Based Quaternary Alloy Systems *(cont.)*

Bath compositions and operating conditions	Co–Fe–Mn–P	Co–Ni–Mn–P		Co–Ni–Re–P	Co–Ni–W–P		Co–Ni–Cu–P	Co–Ni–Zn–P
Glycine, mol/L							0.6	
Sodium succinate, mol/L								0.4
Malic acid, mol/L								0.05
pH	8	9.6	8 ± 2	9.2	7–9.5	8 ± 2	10	9
Temp., °C	90 ± 5	80 ± 5	80 ± 2	80	50	80 ± 2	80	37
Composition in deposits, at%	Fe: 8.81 Mn: 0.71 P: 7.06	Ni: 12.7 Mn: 1 P: 6	Ni: 36.7–25 Mn: 0.3–2 P: 8–13	Ni: 42–45.2 Re: 5.2–5.9 P: 7.8–9.9	Ni: 47.99–6.36 W: 0.4 P: 5.5	Ni: 26.9–13 W: 0.1–2 P: 5–13	Ni: 4.0 Cu: 10 P: 5.5 wt%	Ni: 13 Zn: 2 P: 3
References	[194]	[195]	[196]	[197]	[198]	[196]	[199]	[200]

TABLE 2.28 Representative Electroless Plating Bath Formulations for Co-Based Quaternary Alloy Systems

Bath compositions and operating conditions	Co–Mo–W–P	Co–W–Re–P	Co–W–Pt–P	Co–W–Zn–P	Co–Ni–Fe–B	Co–Ni–La–B	Co–W–Re–B	Co–W–B–P
Iron sulfate, mol/L					0.01			
Nickel sulfate, mol/L					0–0.02*			
Nickel chloride, g/L						5		
Cobalt sulfate, mol/L			0.1	0.05	0.10–0.08*			28.1 g/L
Cobalt chloride, g/L	20–40	15–35				11.67	15–35	
Rhenium chloride, g/L		0.1–30					0.1–30	
Copper sulfate, as antibacterial additive		10 ppb–1000 ppm					10 ppb–1000 ppm	
Ammonium hexachloroplatinate, mol/L			0.004					
Zinc sulfate, mol/L				0–1.0				
Sodium stannate, mol/L				0–0.06				
Tungstic acid, g/L	2–6	5–15					5–15	
Molybdenum oxide, ppm	50–250							
La, g/L						0.8		
Sodium tungstate, mo/L			0.005					0–20 g/L
Potassium sodium tartrate, mol/L			0.3					
DMAB, mo/L					0.025		5–15 g/L	4 g/L
Sodium hypophosphite, mol/L			0.4	0.1–0.6				0–8.05 g/L
Hypophosphorous acid, g/L	5–15	10–50						
EDTA, g/L		10–20					10–20	

(Continued)

TABLE 2.28 Representative Electroless Plating Bath Formulations for Co-Based Quaternary Alloy Systems *(cont.)*

Bath compositions and operating conditions	Co–Mo–W–P	Co–W–Re–P	Co–W–Pt–P	Co–W–Zn–P	Co–Ni–Fe–B	Co–Ni–La–B	Co–W–Re–B	Co–W–B–P
Sodium borohydride, g/L						0.8		
Citrate acid, g/L	40–80				0.05 mol/L			
Sodium citrate, g/L								88.23
Boric acid, g/L	0–30	2–5	0.5 mol/L		0.06 mol/L		2–5	
Calfax 10LA-75, ppm	200–300							
Triton X-100, g/L		0.02–0.04					0.02–0.04	
Ammonium citrate, g/L		60–90					60–90	
Sodium tartrate, mol/L				0.5	0.20	65 g/L		
Ammonium chloride, g/L						12		
Ammonium sulfate, mol/L				0.5	0.20			
Hydrochloride, mL/L								3
Stabilizer, g/L						1		
Surfactant RE-610, cc/L								1.5×10^{-2}
pH	8.0–9.5	9	10–13.5	10.5	9	14	9	9.5–9.7
Temp., °C	65–80	50–95	65–90	75	70	70	50–95	80–85
Composition in deposits, wt%		W: 2–4 Re: 0.5–4 P: 8–15 at%	W: 3.4–4.6# Pt: 14.6–6.2# P: 4.4–5.5# at%,	W: 1.7 Zn: 0.2 P: 7 ± 1	Ni: 0–23 Fe: 8 B: 2		W: 2–4 Re: 0.5–4 B: 3–4 at%	
References	[201]	[202]	[203]	[204]	[173]	[205]	[202]	[206]

The total concentration of ($CoSO_4 \cdot 7H_2O$ + $NiSO_4 \cdot 6H_2O$) was kept constant at 0.10 M.
#*As the temperature changed.*

TABLE 2.29 Representative Electroless Plating Bath Formulation for Quinary and Multicomponent Alloys From Various Authors

Bath compositions and operating conditions	Ni–Fe–Zn–Cu–P	Ni–Co–Fe–La–P	Ni–Co–Re–Mn–P	Co–W–Re–B–P
Nickel sulfate, g/L		18–26	0.08 mol/L	
Nickel chloride, g/L	35			
Iron sulfate, g/L	25	5–17		
Ammonium perrhenate, mol/L			0.003	
Cobalt sulfate, mol/L		6–18	0.06	
Manganese sulfate, mol/L			0.03	
Lanthanum(III) oxide, g/L		0.2–1.8		
Copper sulfate, g/L	0.35			
Zinc sulfate, g/L	20			
Cobalt chloride, g/L				15–35
Tungstic acid, g/L				5–15
Rhenium chloride, g/L				0.1–30
Sodium hypophosphite, g/L	15	21–35	0.20 mol/L	
DMAB, g/L				5–15
Hypophosphorous acid, g/L				10–50
Sodium citrate, g/L		30–35		
Succinic acid, g/L		6–15		
Sodium malonate, mol/L			0.30	
Sodium tartrate, mol/L			0.20	
Ammonium sulfate, g/L		38–50	0.50 mol/L	
Malic acid, M		8–17		
Thiourea, g/L		0.08–0.5		
Urea, g/L		9–21		
Hydrochloride acid, g/L		25–55		
Tartronic acid, mol/L			0.05–0.10*	
EDTA, g/L				10–20
Boric acid, g/L				2–5
Triton X-100, g/L				0.02–0.04
Ammonium citrate, g/L				60–90
Copper sulfate, as antibacterial additive				10 ppb–1000 ppm
pH	8.8	8.0 ± 0.2	9.2	9

(Continued)

TABLE 2.29 Representative Electroless Plating Bath Formulation for Quinary and Multicomponent Alloys From Various Authors *(cont.)*

Bath compositions and operating conditions	Ni–Fe–Zn–Cu–P	Ni–Co–Fe–La–P	Ni–Co–Re–Mn–P	Co–W–Re–B–P
Temp., °C	90	70 ± 5	80	50–95
Composition in deposits, wt%	Fe: 1.55 Zn: 8.77 Cu: 10.86 P: 4.16	Co: 8–18 Fe: 4–15 La: 0.7–1.6 P: 3–17	Co: 24 Re: 5.9 Mn: 0.02 P: 10 at%	W: 2–4 Re: 0.5–4 B: 3–4 P: 8–15 at%
References	[183]	[207,208]	[209]	[202]

**The contents of Ni and P in the film composition increase from 40 at% to 60 at% and 7.4 at% to 10 at%, respectively, but Co decreases from 40 at% to 24 at% in the range of 0–0.05 M of tartronic acid. Then, they do not change till the concentration of tartronic acid reaches 0.1 M. However, the contents of Re and Mn do not change significantly in the whole range of 0–0.1 M of tartronic acid.*

Bi), binary alloys (Ni–P, Ni–B, Co–P, Co–B, Cr–P, and Cu-, Pd-, and Pt-based binary alloys), ternary alloys (Ni–Me–P, Co–Me–P, Ni–Me–B, Co–Me–B, and other ternary alloys), quaternary alloys (Ni- and Co-based quaternary alloys) and quinary and multialloys.

Each bath formulation not only lists its composition but also the operating conditions. Besides these, the composition of the deposits was listed as far as possible.

In most cases, several formulations for every bath are given, including examples published in the early, medium-term, and recent stages. This is to allow readers to understand the progress that has been made and to compare their similarities and differences.

Although research into EP has carried on for more than 70 years, the investigation of bath solutions still continues today. It can be expected that new bath formulations will be continue to appear. So far, a unified and recognized bath formula has not materialized.

REFERENCES

[1] Mallory O, Hajdu JB, editors. Electroless plating – fundamentals and applications. Orlando, FL: American Electroplaters and Surface Finishers Soc; 1990.

[2] Wolfgang R. Electroless nickel plating. Metals Park, OH: Finishing Publications Ltd & ASM International; March 1991.

[3] Jiang X, Shen W. The fundamentals and practice of electroless plating. Beijing: National Defence Industry Press; June 2000.

[4] Djokić SS, Cavallotti PL. In: Djokić SS, editor. Electroless deposition, theory and application. Modern aspects of electrochemistry, No. 48: electrodeposition theory, practice. New York: Springer; 2010 [chapter 6].

[5] Zhang B, Hu W, Qu X, Zhang Q, Zhang H, Tan Z. Trans Inst Metal Finish 1996;74(2):69.
[6] Takano O, Matsuda H. Metal Finish, 1986;84(60):60.
[7] Zhang H, Zhang B, Tan Z, Lei Q. Mater Mech Eng 1995;19(1):19.
[8] Dini JW, Coronado PR. Plating 1967;54(4):385.
[9] Zarnoch KP, Iacovangelo CD, US Patent 4,695,489 Electroless nickel plating composition and method, September 22, 1987.
[10] Tashiro K, Watanabe T, Inaba H, Honma H. Jpn J Surf Finish 2000;51(6):606.
[11] Nakao S, Kim DH, Obata K, Inazawa S, Majima M, Koyama K, Tani Y. Surf Coat Technol 2003;132:169.
[12] Haag S, Burgard M, Ernst B. Surf Coat Technol 2006;201:2166.
[13] Cheng SL, Peng HC. J Eectrochem Soc 2010;157(2):D81.
[14] Takano O, Shigeta T, Ishibash S. Jpn J Surf Finish 1967;18(8):299.
[15] Hong MZ, Bae SC, Lee HS, Lee HC, Kim Y-M, Kim K. Electrochim Acta 2003;48(28):4213.
[16] Ma L, Wang G, Liu L, Li B. J Alloys Comp 2010;505(1):374.
[17] Brenner A, Riddell GE. J Res Bur Stand 1946;37:31. Brenner A., Riddell G.E. P&SF 1998;85:54.
[18] Schwartz M. Proc AES 1960;47:176.
[19] Brenner A, Riddell GE. Surf Technol 1980;10:81.
[20] Lambert MR, Duquette DJ. Thin Solid Films 1989;177:207.
[21] Zhang B, Xie H. Mater Sci Eng 2000;A281:286.
[22] Ashassi-Sorkhabi H, Mirmohseni A, Harrafi H. Electrochim Acta 2005;50(28):5526.
[23] Balaraju JN, Rajam KS. Int J Electrochem Sci 2007;2:747.
[24] Ayoub IE, Online J. Math Stat (OJMS) 2010;1(1):13.
[25] Mallory GO. Plating 1971;58:319.
[26] Rajagopal I. Bull Mater Sci 1983;5(3–4):323.
[27] Hu W, Zhang B, Zhang H. Plat Finish 1988;10(6):10.
[28] Di Giampaolo AR, Ordofiez JG, Gugliemacci JM, Lira J. Surf Coat Technol 1997;89:127.
[29] Dervos CT, Vassiliou R, Novakovic J. Rev Metal Madrid Extr 2005;232.
[30] Bulbul F. Met Mater Int 2011;17(1):67.
[31] Ransom LD, Zentner V. J Electrochem Soc 1964;111:1423.
[32] Pearstein F, Weightman RF. J Electrochem Soc 1974;121(8):1023.
[33] Malik M, Joseph L, US Patent 4,659,605 Electroless deposition magnetic recording media process and products produced thereby, (April 21, 1987).
[34] Anjana L, Selvam M, Srinivasan KN, Vasudevan T. Metal Finish 1994;92(8):47.
[35] Zhang B, Chen B, Yi K, Wang L. Electroplat Pollu Control 1997;17(6):19.
[36] Rudnik E, Gorgosz J. Surf Coat Technol 2007;201:6953.
[37] Chang YH, Lin CC, Hung MP, Chin TS. J Electrochem Soc 1986;133(5):985.
[38] Saito T, Sato E, Matsuoka M, Iwakura C. J Appl Electrochem 1998;28:559.
[39] Xuan T, Zhang L, Huang Q. J Rare Earths 2002;20(5):512.
[40] Miyake H, Inoue F, Yokoyama T, Shimizu T, Tanaka1 S, Terui1 T, Shingubara S. Japan J Appl Phys 2011;50:05ED01.
[41] Saubestre EB. Proc AES 1959;46:264.
[42] Schneble FW, Jr, Bay O, McCormack JF, Heights R, Zeblisky RJ, Wlliamson JD, Place M, Polichette J, US Patent 3,259,559 Method for Electroless Copper Plating, July 5, 1966.
[43] Hung A. P&SF 1988;75(1):62.
[44] Oita M, Matsuoka M, Iwakurac C. Electrochem Acta 1997;42(9):1435.
[45] Wu X, Sha W. Appl Surf Sci 2009;255:4259.
[46] Nawafune H, Uegaki T, Mizumoto S, Ishikawa M, Nakamura T. Jpn J Surf Finish 1998; 49(7):796.

[47] Domínguez-Ríos C, Moreno MV, Torres-Sánchez R, Antúnez W, Aguilar-Elguézabal A, González-Hernández J. Surf Coat Technol 2008;202:4848.
[48] Faraji S, Rahima AA, Mohamed N, Sipaut CS, Raja B. Mater Chem Phys 2011;129:1063.
[49] Brookshire RR, US Patent 2,976,181 Method of gold plating by chemical reduction, March 21, 1961.
[50] Okinaka Y. Plating 1970;57(9):914.
[51] Ali HO, Christie IRA. Gold Bull 1984;17(4):118.
[52] Sato Y, Osawa T, Kaueda K, Kobayakawa K. P$SF 1994;81(9):74.
[53] Kato M, Sato J, Otani H, Homma T, Okinaka Y, Osaka T, Yoshioka O. J Electrochem Soc 2002;149:C164.
[54] Dumee L, Velleman L, Sears K, Hill M, Schutz J, Finn N, Duke M, Gray S. Membranes 2011;1:25. doi:10.3390/membranes1010025.
[55] Nakanishi K. Kagaku no Ryoiki 1950;4:604.
[56] Gupata SK. Cent Glass Ceram Res Int Bull 1967;14:18.
[57] Pearlstein F, Wightman RF. Plating 1971;58(10):1014.
[58] Kubota A, koura N. Jpn J Surf Finish 1986;37:131.
[59] Mandich NV, Krulik GA, Singh R, US Patent 5,322,553 Electroless silver plating composition, June 21, 1994.
[60] Shacham-Diamand Y, Inberg A, Sverdlov Y, Croitoru N. J Electrochem Soc 2000;147(9): 3345.
[61] Heuck FCA, Staufer U. J Microelectromech Syst 2011;20(2):451.
[62] Rhoda RN. Trans Inst Met Finish 1959;36:82.
[63] Pearlstein F, Weightman RF, Pearlstein F, Weightman RF. Plating 1969;56(10):1158.
[64] Dittmeyer R, Höllein V, Daub K. J Mol Catal A Chem 2001;173:135.
[65] Ryi S-K, Li A, Jim LC, Grace JR. Intern J Hydro Ener 2011;36:9335.
[66] Roa F, Douglas Way J, McCormick RL, Paglieri SN. Chem Eng J 2003;93:11.
[67] Sabina KG, Keeling MK, Davidson AP, Hatlevik O, Way JD. Inter J Hydrogen Ene 2009;34:6484.
[68] Zayats AI, Stepanowa IA, Gorodyskii AV. Zashch Metal 1973;9(1):116.
[69] Nawafune H, Mizumoto S, Uchida E, Haga M. Surf Technol 1996;7(12):1060.
[70] Byeon JH, Kim Y-W. ACS Appl Mater Interf 2011;3:2912.
[71] Nawafune H, Kamiura Y, Mizumoto S, Haga M, Uchida E. J Jpn Ins Interconn Pack Elect Circ 1996;11(4):262.
[72] Rhoda RN, Vines RF, U.S. Patent 3,486,928, Bath and process for platinum and platinum alloy, December 30, 1969.
[73] Steinmetz P, Alperine S, Friant-Costantini A, Josso P. Surf Coat Technol 1990;500:43–44.
[74] Koslov AS, Palanisamy T, Narasimhan D, US Patent 6,391,477 Electroless autocatalytic platinum plating, May 21, 2002.
[75] Mustain WE, Kim H, Prakash S, Stark J, Osborn T, Kohl PA. Electrochem Solid State Lett 2007;10(12):B210.
[76] Wang C, Peng B, Xie H-N, Zhang H-X, Shi F-F, Cai W-B. J Phys Chem C 2009;113:13841.
[77] Kozlov AS, Palanisamy T, Narasimhan D. US Patent 6,706,420 Electroless platinum-rhodium alloy plating, March 16, 2004.
[78] Mustain WE, Kim H, Narayanan V, Osborn T, Kohl PA. J Fuel Cell Sci Technol 2010;7: 041013.
[79] Okuno K. P&SF 1990;77(2):48.
[80] Chang YS, Chou ML. J Appl Phys 1991;69(11):7848.
[81] Braun F, Tarditi AM, Cornaglia LM. Memb J Sci 2011;382(1–2):252.

[82] Kozlov AS, Palanisamy T, Narasimhan D, US Patent 6,455,175 Electroless rhodium plating, September 24, 2002.
[83] Chang Y-S, Chou M-L. Mater Chem Phys 1989;24(1–2):131.
[84] Tharamani CN, Begum NS, Mayanna SM. Mater Chem Phys 2004;83:278.
[85] Senda A, Nakagawa T, Takano Y, US Patent 5,160,373 Electroless plating bath, November 3, 1992.
[86] Senda A, Takano Y, Nakagawa T. Jpn J Surf Finish 1992;43(10):967.
[87] Nishikawa H, Minami S. Jpn J Surf Finish 1982;33(3):114.
[88] Senda A, Takano Y, Nakagawa T. Jpn J Surf Finish 1992;43(7):694.
[89] Warwick MF, Shirley BJ. Trans Inst Met Finish 58 Part 1980;1:9.
[90] Senda A, Morita K, Takano Y. Jpn J Surf Finish 1993;44(11):966.
[91] Senda A, Nakagawa T, Takano Y, Kasanami T. Jpn J Surf Finish 1992;43(4):324.
[92] Senda A, Takano Y, Nakagawa T. Jpn J Surf Finish 1992;43(6):589.
[93] Senda A, Takano Y, Nakagawa T. Jpn J Surf Finish 1993;44(2):145.
[94] Senda A, Takano Y, Kanbayashi Y, Morita K. Jpn J Surf Finish 1994;45(9):923.
[95] Senda A, Takano Y, Morita K. Jpn J Surf Finish 1994;45(3):283.
[96] Scidone L, Diliberto S, Stein N, Boulanger C, Lecuire JM. Mater Lett 2005;59:746.
[97] Hassan HB, Hamid ZA. Intern J Hydr Ener 2011;36:849.
[98] Rohan JF, O'Riordan G, Boardman J. Appl Surf Sci 2002;185(3–4):289.
[99] Srinivasan KN, John S. J Alloys Comp 2009;486(1–2):447.
[100] Mallory GO, Horhn TR. P&SF 1979;66(4):40.
[101] Iwamatsu K. Metal Finish 1989;87(5):25.
[102] Gonzalez OM, White RE, Cocke DL. P&SF 1990;77(11):63.
[103] Mallory GO. P&SF 1976;63(6):34.
[104] Schmeckenbacher AP. J Electrochem Soc 1966;113(8):778.
[105] Wang L, Zhao L, Huang G-f, Yuan X, Zhang B, Zhang J. Surf Coat Technol 2000;126:272.
[106] Tierno P, Goedel WA. J Phys Chem B 2006;110:3043.
[107] Lawless GW, Fisher RD. P&SF 1967;54(6):709.
[108] Homma T, Sezai Y, Osaka T, Maeda Y, Donnet DM. J Magn Magn Mater 1997;173:314.
[109] Kumar A, Singh A, Kumar M, Kumar D, Barthwal S. J Mater Sci Mater Electron 2011;22:1495.
[110] Pearlstein F, Weightman RF. Electrochem Technol 1968;6:427.
[111] Mencer D. J Alloys Comp 2000;306:158.
[112] Duhin A, Inberg A, Eliaz N, Gileadi E. Electrochim Acta 2011;56:9637.
[113] Wu G, Liu B, Deng J-F. Chem Lett 1998;(9):889.
[114] Pearlstein F. US Patent 3,485,597 Electroless deposition of nickel-phosphrous based alloy, December 23, 1969.
[115] Aoki K, Takano O, Nakata T. Jpn J Surf Finish 1993;44(11):957.
[116] Aoki K, Takano O, Kim D-H. Jpn J Surf Finish 1995;46(1):79.
[117] Hiura T, Aoki H. Trans IEICE 1978;J61-C(8):517.
[118] Yang Y, Sun D, Yang D. Corr Sci Protec Technol 2000;12(3):138.
[119] Lou B-y, Yu K, Wu M-m. In: Proc. ICECE '10 of the 2010 international conference on electrical and control engineering. Washington, DC: IEEE Computer Society; 2010. p. 4834.
[120] Koiwa I, Usuda M, Yamada K, Osaka T. J Electrochem Soc Solid State Sci Technol 1988;135(3):718.
[121] Lo YL, Chou SC, Hwang BJ. J Appl Electrochem 1996;26:733.
[122] Liu D-l, Yang Z-g, Zhang C. Mater Sci Eng B 2010;166:67.
[123] Koiwa I, Osaka T, Sawai H. J Metal Finish Soc Jpn 1983;34(12):600.

[124] Zhang B, Hu W, Zhang Q, Qu X. Mater Charac 1996;37:119.
[125] Liu H, Guo R-x, Zong Y, He B-q, Liu Z. Trans Nonferrous Met Soc China 2010;20:1024.
[126] Gulla M. US Patent 3,832,168 Metal finishing alloy of nickel-copper-phosphrous, August 27,1974.
[127] Cherkaoui M, Srhiri A, Chassaing E. P&SF 1992;79(11):68.
[128] Cissé M, Abouchane M, Anik T, Himm K, Belakhmima RA, Touhami ME, Touir R, Amiar A. Intern J Corrosion Volume 2010; Article ID 246908, 9 pages.
[129] Schlesinger M, Meng X. J Electrochem Soc 1990;137(6):1858.
[130] Oulladj M, Saidi D, Chassaing E, Lebaili S. J Mater Sci 1999;34:2437.
[131] Ranganatha S, Venkatesha TV, Vathsala K. Appl Surf Sci 2010;256:7377.
[132] Aoki K, Takano O, Isnibashi S, Nayashi T. Jpn J Surf Finish 1977;28(12):629.
[133] Zou Y, Cheng YH, Cheng L, Liu W. Mater Trans 2010;51(2):277.
[134] Ryu I-K, Korean J. Sanitation 2003;18(3):35.
[135] Boose CA. Trans IMF 1975;53:49.
[136] Mitl CK, Shrivastava PB. Metal Finish 1986;84(10):67.
[137] Zhang H, Zhang B, Tan Z, Lei Q. Mater Mech Eng 1995;19(1):19.
[138] Jiang B, Xiao L, Hu S, Peng J, Zhang H, Wang M. Opt Mater 2009;31:1532.
[139] DePew JR, Speliotis DE. P&SF 1976;54(6):705.
[140] Zeng Q, Zhang Y, Bonder MJ, Hadjipanayis GC. J Appl Phys 2003;93(10):6498. Parts 2 and 3.
[141] Huang W-Q, Huang G-F, Liang B, Xie C-L. J MMM 2009;321(9):1177.
[142] Shacham-Diamand Y, Zylberman A, Petrov N, Sverdlov Y. In: Proceedings of 2001 6th international conference on solid-state and integrated circuit technology. Shanghai, China: IEEE; 2001. p. 410–5; Microelec. Eng 2002;64:315.
[143] Rudnik E, Mucha M. Surf Eng 2011;27(9):683.
[144] Pearstein F, Weightman RP. Plating 1967;54(6):714.
[145] Shacham-Diamand Y, Lopatin S. Microelectr Eng 1997;77. 37–38.
[146] Tarozaitė R, Sukackienė Z, Sudavičius A, Juškėnas R, Selskis A, Jagminienė A, Norkus E. Mater Chem Phys 2009;117:117.
[147] Liao W-K, Study on cobalt alloy thin films as Cu/Si diffusion barrier, Thesis for the Degree of Master, Dept. of Appl. Chem. Chaoyang University of Technology, Taiwan, China, June 2007.
[148] Fisher RD. IEEE Trans Magn, Mag-2 1966;(4):681.
[149] Goto F, Yamamoto T, Shiota N, Osaka T. Jpn J Surf Finish 1991;42(3):345.
[150] Yuan X, Du C, Sun G, Pan N. Appl Surf Sci 2007;253:4546.
[151] Brusic VA, Marino JR, Eugene J, Carlos J, Alejandro G, Cyprian E. US Patent 5,755,859 Cobalt-tin alloys and their applications for devices, chip interconnections and packaging, May 26, 1998.
[152] Zhang Y, Wang X, Li X. Mater Prot 2003;36(7):34.
[153] Nakayama I, Fukawa M, Homma T, Osaka T. Jpn J Surf Finish 1992;143(9):835.
[154] Osaka T, Koiwa I, Yamada K, Nishikawa M, Usuda M. Bull Chem Soc Jpn 1987;60:3117.
[155] Rao Q, Wang H, Zhou Y. Electroplat Pollu Cont 2001;21(2):22.
[156] Yoshino M, Masuda T, Wakatsuki S, Sasano J, Matsuda I, Shacham-Diamand1 Y, Osaka T. In: 208th ECS meeting. Abstract #587, MA2005-02(14), 2006. p. 587.
[157] Osaka T, Takano N, Kurokawa T, Kaneko T, Ueno K. Surf Coat Technol 2003;169–170:124.
[158] Duhin A, Sverdlov Y, feldman Y, Shacham-Diamand Y. Electrochim Acta 2009;54:6036.
[159] Aoki K, Kim D-H, Takano O. Jpn J Surf Finish 1994;45(12):1287.
[160] Wang S. Funct Mater 2006;7(37):1154.

[161] Saito T, Yoshimura S-i, Sato E. Jpn J Surf Finish 1990;41(1):53.
[162] Kim M, Iizuka A, Kubomiya T, Osaka T. Jpn J Surf Finish 1999;50(4):353.
[163] Sankara Narayanan TSN, Stephan A, Guruskanthan S. Surf Coat Technol 2004;179:56.
[164] Hu W, Wu L, Wang L, Zhang B, Guan H. Physica B 1995;212:195.
[165] Aoki K, Takano O, Sasaki Y. Jpn J Surf Finish 1993;44(10):821.
[166] Matsuoka M, Hayashi T. Bull Univ Osaka Prefecture Series A, Eng Na Sci 1983;32(1):19.
[167] Parvini Ahmadi N, Khosroshahi RA, Abdeli M. Inter J ISSI 2005;2(2):19.
[168] Yang L, Song D, Shi X. Heat Treat Metals 2010;35(4):38.
[169] Delaunois F, Lienard P. Surf Coat Technol 2002;160(2–3):239.
[170] Dai H, Gao L, Liang Y, Kang X-D, Wang P. J Power Sources 2010;195:307.
[171] Itabashi T, Nakano H, Akahoshi H. In: Interconnect technology conference, 2002. Proceedings of the IEEE 2002 international 2002. p. 285–7.
[172] Koo H-C, Cho SK, Kwon OJ, Suh M-W, Im Y, Kim JJ. J Electrochem Soc 2009;156(7):D236.
[173] Yokoshima T, Kaneko D, Akahori M, Nam H-S, Osaka T. J Electroanal Chem 2000;491:197.
[174] Dadvand N, Jarjoura G, Kipouros GJ. J Mater Sci Mater Electron 2008;19:51.
[175] Ofek-Almog R, Sverdlov Y, Goldfarb I, Shacham-Diamand Y. Microelectron Eng 2007;84(11):2450.
[176] Pearlstein F, Robert F. US Patent 3,754,939 Electroless deposition of palladium alloys, August 28, 1973.
[177] Zhang J, Li Y, Zhou D. Mater Protec 1995;28(1):11.
[178] Ma M-h, Li X-h. J Nanking Xiaozhuang Univ 2007;(6):39.
[179] Yu Z, Hao S, Rao Z. Chem Res Appl 2010;22(1):78.
[180] Kim D-H, Aoki K, Takano O. Jpn J Surf Finish 1995;46(1):81.
[181] Wang S, Zhan J. Chin J Nonferr Metals 2008;18(6):1105.
[182] Guo H, Liu H, Li W, Xie G. Electroplat Finish 2010;29(7):20.
[183] Ayoub IE, Ibrahim IS. Isotope Rad Res 2009;4(Suppl. 2):1561. 41.
[184] Yang L, Xuan T, Ju Z, Zhang L. Electroplat Polu Control 2006;26(4):20.
[185] Edward McComas C. US Patent 5,019,163 Corrosion/wear-resistant metal alloy coating compositions, May 28,1991.
[186] Xiao X, Zhong P, Xia H, Chen Y. Mater Protec 2005;38(4):28.
[187] Jin Y, Fan J, Mou K, Wang X, Ding Q, Li Y, Liu C. Inorg Mater 2010;46(6):631.
[188] Yu Z, Hao S, Sun Y, Huang X, Liu Q. Hot Working Technol 2008;37(22):52.
[189] Valova1 E, armyanov S, Franquet A, Hubin A, Steenhaut O, Delplancke J-L, Vereecken J. J Appl Electrochem 2001;31:1367.
[190] Balaraju JN, Ezhil Selvi V, William Grips VK, Rajam KS. Electrochim Acta 2006;52:1064.
[191] Balaraju JN, Millath Jahan S, Jain A, Rajam KS. J Alloys Comp 2007;436:319.
[192] Kim M, Yokoshima T, Kubomiya T, Osaka T. Jpn J Sur Finish 2000;51(6):617.
[193] Zhang B, Xie H. Rare Metal Mater Eng 1998;27(2):79. Mater Sci Eng A 2000;281:286.
[194] Wang S, Zhang Q. Rare Metal Mater Eng 2007;36(12):2224.
[195] Liu Z, Chen H, Yang C, He Y. J Phys Condens Matter 1991;3:9007.
[196] Uwazumi H, Tei Y, Kainuma K, Higuchi K, Iso A, Kurihara H. US Patent 7,622,205 Disk substrate for a perpendicular magnetic recording medium and a perpendicular magnetic recording medium using the substrate, November 23, 2009.
[197] Koiwa I, Okabe Y, Matsubara H, Osaka T, Goto F. IEEE Trans Mag Japan, TJMJ-1 1985;(4):443.
[198] Cheng H, Xu S, Zou S, Zhang Q, Zheng J, Zhao W, Guan J, Sun Z. IEEE Trans Magn 2007;43(6):2433.
[199] Tarozaitė R, Stalnionis G, Sudavičius A, Kurtinaitiene M. Surf Coat Technol 2002;161:150.

[200] Matsuda H, Takano O. J Appl Electrochem 1993;23:1083.
[201] Chen Q, Valverde C, Paneccasio V, Petrov N, Stritch D, Witt C, Hurtubise R. US Patent 7,410,899 Defectivity and process control of electroless deposition in microelectronics applications, August 12, 2008.
[202] Lopatin S, Shanmugasundram A, Lubomirsky D, Pancham IA. US Patent 7,205,233 Method for forming CoWRe alloys by electroless deposition, April 17, 2007.
[203] Wei G, Jin Y, Yu Y, Ge H, Wang X, Jiang L, Sun L. Thin Solid Films 2010;519:1259.
[204] Yae S, Kanki T, Fukumuro N, Yamada Y, Matsuda H. Trans Inst Met Finish 2003;81(2):45.
[205] Xuan T, Feng S, Zhu Y. Electroplat Finish 2005;24(9):11.
[206] Shacham-Diamand Y, Sverdlov Y, Bogush V, Ofek-Almog R. J Solid State Electrochem 2007;11:929.
[207] Qin X, Huang Y, Du Z, Yue Ying C. J Rare Met 2007;31(5):631.
[208] Huang Y, Qin X, Qi S. CN Patent 101311308A Bath and process of electroless plating of Ni-Co-Fe-La-P alloy on surface of glass fiber, November 26, 2008.
[209] Koiwa I, Okabe Y, Matsubara H, Osaka T. Jpn J Sur Finish 1985;36(5):204.

Chapter 3

Electroless Composite Plating

Chapter Outline

Electroless composite plating (ECP) is a method of intentionally introducing insoluble or sparingly soluble particulate matter at the micron level in size into conventional electroless plating (EP). Its development comes from a greater understanding of EP, and from demands to apply EP more broadly. Currently, ECP has become an important field of EP. The development of ECP not only enhances our understanding of the character of EP, but also extends its application.

The codeposition particle in ECP can be any of a wide range of materials, such as the oxide, carbide, nitride, or boride of Cr, Mo, W, V, Nb, Ta, Ca, Al, B and Si, or polytetraflouroeterylene (PTFE), diamond, MoS_2, CF_n, BN, B_4C, Si_3N_4, talc, and graphite.

Oderkerken was the first to study ECP. He was granted a GB Patent "improved method of depositing bright nickel/chromium" on September 7, 1966 [1] and an US Patent "method for covering objects with a decorative bright nickel/chromium grating, as well as objects covered by applying this method" on January 17, 1967 [2], in which he made a structure consisting of a composite electroless plating containing finely divided Al_2O_3 or silicate powders, set between electrodeposited layers to improve its corrosion resistance. Later, Metzger et al. extended the Oderkerken work in their two US Patents 3,617,363 [3] and 3,753,667 [4], issued on November 2, 1971 and August 21, 1973, respectively. They used several particles, including SiC, WC, SiO_2, MoS_2, and Al_2O_3, of dimensions in the range from about 0.01 μm to about 100 μm incorporated within an EN layer. However, the preferred particles are in the size range of 0.5–10 μm. The dates on these patents indicate that composite coatings were studied from around 20 years after the invention of EN. The main drive for

Amorphous and Nano Alloys Electroless Depositions. http://dx.doi.org/10.1016/B978-0-12-802685-4.00003-0

such a study is the complexity of electroless composite plating, which makes it more difficult to control the stability of the bath.

Some reports were published consecutively, such as Christini et al. and Lukschandel who published their work on diamond-containing electroless nickel coatings in 1976 and 1978 [5,6]. Then in 1981, Hussaina and Such [7] reported a process for the electroless deposition onto a metal substrate of a nickel coating containing dispersed particles. They investigated the effect of heat treatment on the hardness of the composite coating, and the surface morphologies of the deposits with a scanning electron microscope and a Quantimet 720 image analyzer. Tulsi studied electroless Ni–P–PTFE coatings in 1983 because he found that such composite coatings have a high dry lubricity, low friction, precise and uniform torque and tension, good wear, and corrosion resistance. These results encouraged studies in the field, including coating molds for rubber and plastic components, tools for pumps, valves, and butterfly valve for oil and gas industry, fasteners, precision instrument parts, aluminum air cylinders, carburetors, choke shafts, etc. [8]. In 1985 Spencer, Jr. [9] invented the processes for the preparation of composite electroless coatings which comprise a metal and/or metal alloy, plus particulate matter, the latter having at least two distinct nominal sizes ranging from 0.1 μm to 100 μm. In 1987, Kim reported that carbon fluoride and metal can be codeposited from an electroless plating bath onto a suitable solid. To achieve this, carbon fluoride particles having an average size from 0.2 μm to 8 μm are dispersed in the electroless bath solution [10]. Henry, and Ernest M. Summers in their US Patent 4,830,889, issued in 1989 [11], reported a process for codepositing fluorinated carbon with electroless nickel, which provides a uniformly dispersed codeposit.

It is interesting to note that even in the early study period of ECP, we can find few review papers on composite coatings. One example that was published was Hubbell in 1978, and his review paper "chemically deposited composites – a new generation of electroless coating" described the corrosion resistance of EN composite coatings [12]. In 1983, Feldstein et al. published a review "electroless composite plating," which reviewed the state of the art and its evolution, including results depicting the merits of these coatings in comparison to conventional ones [13]. In the famous book "Electroless Plating: Fundamentals and Applications" edited by Mallory and Hajdu [14], Feldstein wrote a special Chapter 11 on "composite electroless plating" which described studies on ECPs up to 1990. However, the development of ECPs before 1990 was very limited, and at that time the volume of composite plating applications was quite small and limited to niche applications. Feldstein described such a situation in his article "composite electroless plating is still in its infancy" [14]. This point was also made by Hajdu and Zabrocky in 2000 [15]. We pointed out in Chapter 1 that the rapid development and broad application of ECP, and a corresponding number of published papers, occurred in the last 20 years.

3.1 GENERAL CONSIDERATIONS ABOUT ECP

The fundamental difference between ECP and conventional EP lies in intentionally putting particulate matter into the bath solution. Therefore, there could be stability implications for the plating bath due to the insoluble particles. The particle size and shape, relative density of the particle, particle charge, inertness of the particle, the concentration of particles in the plating bath, the method and degree of agitation, the compatibility of the particle with the matrix, and the orientation of the part being plated, etc., all of these factors should be considered in the ECP in order to insure optimal particle codeposition, consistency, plating rate, stability, and bath life.

It is well known that the surface area generally increases with decreased particle size, so the surface area of particles with size in micrometers is very large. For instance, 1 g of 1.0 μm sized diamond particles contains 310,000,000,000 particles [16]. This creates a surface area loading of nearly 100,000 cm^2/L, approximately 800 times the preferred loading of a conventional EN bath [13]. The intentional introduction of insoluble particles into the bath solution therefore affects plating rate, stability, and bath life, and can even result in rapid decomposition of the bath if no any measures are taken. The key in ECP is to make the bath stable by adding stabilizers to the bath solution. These types of stabilizers are also known as "particulate matter stabilizer (PMS)" in much of the literature.

The functions of PMS are as follows: (1) PMSs can isolate the finely divided particulate matter, thereby maintaining and ensuring its "inertness" in participation in the actual conventional electroless plating mechanism; (2) PMSs tend to modify the charge on the particulate matter via some electrostatic interaction and alteration of the double layer; (3) PMS will cause a significant shift in the zeta potential of the particulate matter when dispersed in water.

A PMS is usually selected such that it does not affect the basic kinetics of the plating process. A surfactant such as anionic, cationic, nonionic and amphoteric types, dispersants of various charges, and emulsifying agents can be used as a PMS. According to Feldstein and Lindsay [17], the zeta potential shift is: anionic PMS at least 15 mV; cationic PMS at least 10 mV, although most are 70 mV and above, and nonionic PMS at least 5 mV.

Four features have been proved by experimental data. The first is that a stabilizer for a specific bath may, in fact, be of little effectiveness in another bath. In other words, a stabilizer, although it improves the plating in certain baths, does not provide the same level of effectiveness in all baths. People supposed that the competitive reactions of adsorption and/or absorption of the PMS onto the particulate matter may be opposite by the presence of certain complexing or chelating agents, which are part of conventional EP baths. The character of these agents may affect the degree of adsorption or absorption onto the particles and so that the degree of isolation of the particle from the active chemistry of the EP. Therefore, it is easy for people to anticipate that a particulate matter stabilizer for a specific bath may, in fact, be of little improvement in another bath.

The second is that, in general, the concentration for the PMS is from about 0.01 wt% to about 0.5 wt%. The amount of the PMS used in the ECP is much larger than that of stabilizers used in the conventional electroless plating, which is a few milligrams per liter or less.

The third is that the incorporation of the PMS not only improves the stability of the electroless plating stability but also produces superior deposits. For example, compared to deposits from conventional EP, the composite coatings are more homogeneous and smoother. In some instances, the composite coatings are glossier than coatings resulting from EP bath alone without the particulate matter stabilizers [17].

The last feature is that addition of any PMS, though it may lead to bath stabilization, will also reduce significantly the plating value(s) to lower and impractical values.

In order to make the ECP work properly, particular attention must be paid to the following points.

1. The article to be metalized would be generally pretreated, for example, by cleaning, strike, etc., prior to the actual deposition step.
2. Stirring is an obvious feature in ECP, usually with a magnetic stirrer.
3. In addition to topping up stabilizers during ECP, continuous or semicontinuous filtration is needed to remove the finely divided matter.

In order to make the matrix harder and/or to improve adhesion and cohesion of the coating, or to smooth the surface, further step(s) such as heat treatment may take place after the deposition of the ECP.

3.2 BATH SOLUTIONS OF ECP

We will describe the bath solution of ECP according to the various alloy systems. This should facilitate comparisons of the features between ECP and conventional electroless plating. To date, a lot of ECP systems have been studied. Not all of the bath solutions studied are listed here, just some representative cases for every kind of plating.

3.2.1 Bath for Binary Alloy-Based ECP

In Table 3.1, four bath formulations for electroless Ni–P + SiC deposits are listed, cited from papers published in 1971, 1991, 2000, and 2006, respectively. One bath is for Ni–P + WC from 2007; and the other two baths are for Ni–P + diamond from 2002 and 2007, respectively (Fig. 3.1).

In Table 3.2, one bath formulation for electroless Ni–P + TiC alloy is listed, which is cited from a paper published in 1990, and one is listed for Ni–P + Cr_3C_2 alloy from 1981. Three baths for Ni–P + Al_2O_3 alloys from 1971, 1991, and 2006, respectively. are given, along with two baths for Ni–P + SiO_2 alloys from 2000 and 2007. One bath for Ni–P + Cr_3O_2 alloy from 2007 is also cited (Fig. 3.2).

TABLE 3.1 Representative Bath Solutions for the Electroless Composite Plating for Ni–P Based Systems

Bath compositions and operating conditions		Ni–P + SiC				Ni–P + WC	Ni–P + diamond	
Nickel sulfate, g/L		0.08 mol/L	0.1 mol/L	20–30	15–20	35	26	30
Sodium hypophosphite, g/L		0.23 mol/L	0.1 mol/L	15–25	20–25	20	24	20
Lactic anion, mol/L		0.30					Lactic acid 27 g/L	
Propionic anion, mol/L		0.03					2.2 mL/L	
Lead ion, ppm		1					2	
Sodium acetate, mol/L			0.1			5	30 g/L	14 g/L
Sodium citrate, mol/L			0.1	25–30 g/L		15		
Thallium sulfate (additive), mol/L			0.1					
Lactic acid (88%), mL/L					30–33			10 g/L
Citrate acid, g/L								10
Sodium dodecyl sulfate (SDS), g/L								trace
SiC	Size, μm	0.5–25	1.0	3–5	1			
	Amount, g/L	15%*	1–10	0.5–10	5–15			
WC	Size, μm					1		
	Amount, g/L					0–40		
Diamond	Size, μm						8–12	10
	Amount, g/L						2–10	12

(Continued)

TABLE 3.1 Representative Bath Solutions for the Electroless Composite Plating for Ni–P Based Systems *(cont.)*

Bath compositions and operating conditions	Ni–P + SiC				Ni–P + WC	Ni–P + diamond	
Agitation	Mechanical	Air blow	Magnetic	Air	Magnetic	Not marked	Electric stirring
Stirring rate, rpm					0–200		225
pH	4.6	7	5–5.6	4.3–4.5	2–8	5	4.6
Temp., °C	93–98	80	80–90	90 ± 1	60–100	70–90	80
Concentration in deposit, wt%	P: 9 SiC: 5	SiC: 1–16 As air 15 mL/min	SiC: 8↗19, 19↘13 Vol% as pH = 5, see Fig. 3.1	SiC: 4.3 P: 10.5	WC_{max}: 53% at added WC = 20 g/L	Diamond: 10.9–18.4 vol%	Diamond: 14.12 As intermittent stirring method: stir 10 s stopped 10 s
References	[3]**	[18]	[19]	[20]	[21]	[22]	[23]

**Volume of the coating %.*

***1973 Metzger et al. pointed out evidently in their US patent that in order to obtain a deposit having the particles uniformly dispersed there through the intentional induced particles must be essentially insoluble in the plating bath, and must be noncatalytic, and must not interfere with or stop the plating reaction. The concentration of the particles in the plating bath is preferably in the range from about 0.1% by weight of the solids to about 2% by weight of the solids, but in any event not more than four times the weight of the metal in the bath expressed as free metal or reduced metal. In addition, the particles must be maintained in suspension throughout the bath during the plating process.*

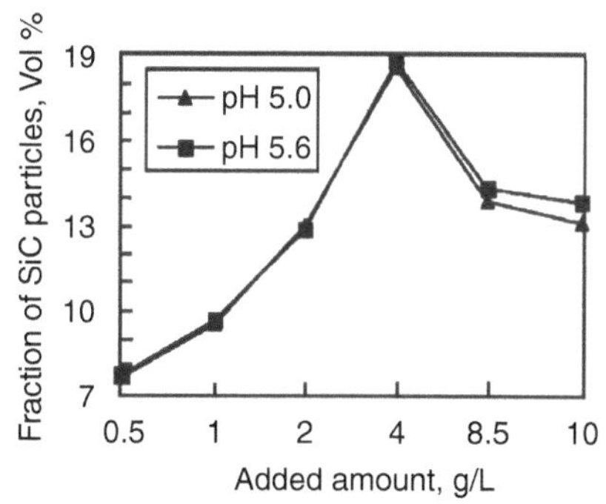

FIGURE 3.1 **Effect of SiC added amount on the fraction in the composite [19].**

In Table 3.3, one bath formulation for electroless Ni–P + TiO_2 alloy is listed, which is cited from a paper published in 2009. One bath for Ni–P + ZrO_2 alloy from 2008 is given, together with one bath for Ni–P + CeO_2 alloy from 2006, one bath for Ni–P + YSZ alloy from 2009, one bath for Ni–P + Fe_3O_4 alloy from 2009, one bath for Ni–P + $K_2Ti_6O_{13}$ alloy from 2007, and one bath for Ni–P + Si_3N_4 alloy from 2009.

In Table 3.4, two bath formulations for electroless Ni–P + MoS_2 alloys are listed, which are cited from papers published in 1999 and 2010. Two baths for Ni–P + PTFE alloy from 2002 and 2011 are also included, together with one bath for Ni–P + BN(h) alloy from 1998, and one bath for Ni–P + Talc alloy from 2006 (Fig. 3.2).

In Table 3.5, two bath formulations for electroless Ni–P + B_4C alloy are listed, which cited from a paper published in 2003 and 2011. Two baths for Ni–P + CaF_2 alloy from 1994 and 2009 are also given, along with two baths for Ni–P + polymer alloys from 2009 and 1998, and one bath for Ni–P + graphite alloy from 2006.

In Table 3.6, one bath formulation for electroless Ni–B + SiC alloy is listed, which is cited from a paper published in 1987. Alongside this, there is a Ni–B + diamond alloy from a US patent in 1976, one for Ni–B + graphite alloy from 1993, one for Ni–B + Al_2O_3 alloy from 1994, one for Ni–B + Si_sN_4 alloy from 2004, and one for Ni–B + BN alloy from 2006.

In Table 3.7, one bath formulation for electroless Co–P + SiC, Co–P + Al_2O_3, and Co–P + SiO_2 alloys, respectively, is listed, which is cited from a paper published in 2008. Further to this, two Co–B + PTFE alloys from 1994 and 2008 are included, respectively, along with a bath for Cu–P + SiC alloy from 2011, and one for Cu–P +graphite alloy from 2001.

3.2.2 Bath for Ternary Alloy-Based ECP

In Table 3.8, one bath formulation for electroless Ni–W–P + SiC, Ni–W–P + Al_2O_3 and Ni–W–P + ZrO_2 alloys, respectively, are listed. The formulations

TABLE 3.2 Representative Bath Solutions for the Electroless Composite Plating of Ni–P Based Systems

Bath compositions and operating conditions	Ni–P + TiC	Ni–P + Cr_3C_2*		Ni–P + Al_2O_3		Ni–P + SiO_2		Ni–P + Cr_2O_3
Nickel sulfate, g/L	Not indicated		0.08 mol/L	25	21	0.15 mol/L	25	25
Sodium hypophosphite, g/L	Not indicated		0.23 mol/L	18	24	0.24 mol/L	20	20
Sodium acetate, g/L				10				
HEDP (1-hydroxyethyliphosphonate), mol/L						0.08		
Lead acetate, mol/L						6×10^{-6}		
Lactic acid, mol/L			0.30		25 g/L		10	10
Propionic acid, mol/L			0.03		3 g/L			
Sodium gluconate, g/L							52	52
Sodium lauryl sulfate, g/L							0.05	0.05
OP-21 neutral surfactant, mol/L					30			
Lead acetate, ppm		1						
Lead(II) nitrate, mg/L				Small		0.5	0.5	
TiC — Size, μm	Not indicated							
TiC — Amount, g/L	10							
Cr_3C_2 — Size, μm		2–3						
Cr_3C_2 — Amount, g/L		10–200						

Al_2O_3	Size, μm			0.5	3.88	1			
	Amount, g/L			**	12	6			
SiO_2	Size, μm						7	Not indicated	
	Amount, g/L						0.25 mol/L	10–35	
Cr_2O_3	Size, μm								Not indicated
	Amount, g/L								15
Agitation		Not indicated	Air	Mechanical	Magnetic	Magnetic	Electrical	Mecahnical	Mecahnical
Stirring rate, rpm		Not indicated	Not indicated	Not indicated	Not indicated	2000	1 min at every 10 min interval	240	240
pH		4.6	4.5	4.6	5.6	4.7–4.9	4.5–5	5.2	5.2
Temp., °C		87	92	93–98	90	90 ± 2	88–95	82	82
Concentration in deposit, wt%			P: 7.2 Cr_3C_2: 27	P: 9	P: 13.82 (at%) Al_2O_3: 12	P: 7.8 Al_2O_3:8	P: 12, Si: 2.77, Fe: 2.32, O: 6.65% at%	P: 9.3$\searrow$6.18 SiO_2: 4.3$\nearrow$7.31	P: 9.374 SiO_2: 4.34
References		[24]	[25]	[4]	[26]	[27]	[28]	[29]	

*A commercia1 electroless nickel bath solution (Ni Foss 80) was used for the Ni–P bath.
**A concentration by weight of about 1% of the dry solids in the plating bath.

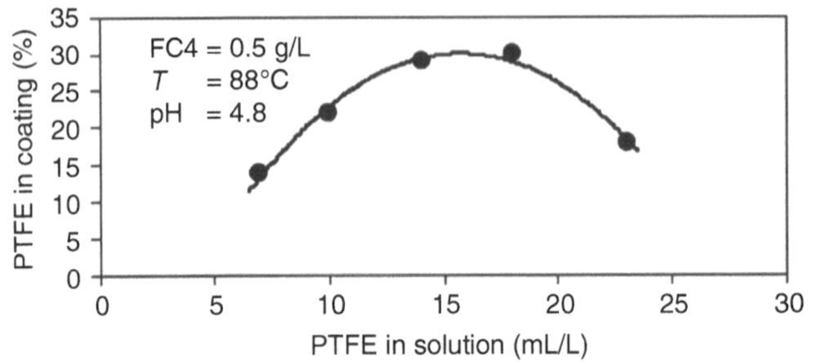

FIGURE 3.2 PTFE content in coating vs. PTFE concentration in solution [39].

are cited from papers published in 2004, 2010, and 2009. In addition, one formulation for Ni–WP + Si_3N_4 from 2004 and one for Ni–W–P + PrO_2, both from 2007, are also given.

In Table 3.9, one bath formulation for electroless Ni–Mo–P + Al_2O_3 and Ni–Mo–P + PPS alloys, respectively, is given, cited from a paper published in 2005. In addition, a formulation for Ni-Co–B + diamond from 1991 and one for Ni–Cu–P + PTFE from 2005 are included in the table.

3.2.3 Bath for ECP With Two Kinds of Particles

In Table 3.10, one bath formulation for electroless Ni–P + SiC + MoO_2 is given, cited from a paper published in 2008. Also one for Ni–P + SiC + graphite from 2006, one for Ni–P + SiC + PTFE from 1999, one for Ni–P + TiC/Al_2O_3 from 2007, one for Ni–P + Al_2O_3+ ZrO_2 from 2002, and one for Ni–P+ MoS_2 + PTFE from 2010 are included.

Table 3.11 includes one bath formulation for electroless Ni–P + diamond + CaF_2 alloy, which is cited from a paper published in 2007. Formulations for Ni–B + SiC + RE from 1995, Ni–W–P + diamond +(CF)n from 2008, Ni–Re–P+ Al_2O_3 + CaF_2 from 2008, Ni-W-B + SiC + CeO_2 from 1999, and Ni–P+ ZrO_2–Al_2O_3–Al_3Zr from 2005 are also given.

3.3 SUMMARY

The following conclusions can be drawn from the discussion in the chapter.

1. Although some reports about ECP were published before 1990, most were published after that date, especially after 2000.
2. To date, the published papers of ECP include binary-based EP + one kind of particle alloy systems, ternary-based EP + one kind of particle alloy systems, binary based EP + two kinds of particle alloy systems, and ternary based EP + two kinds of particle alloy systems. Most of the published data about binary based EP + one kind of particle alloy systems concentrate on

TABLE 3.3 Representative Bath Solutions for the Electroless Composite Plating of Ni–P Based Systems

Bath compositions and operating conditions		Ni–P + TiO_2	Ni–P + ZrO_2	Ni–P + CeO_2	Ni–P + YSZ *	Ni–P + Fe_3O_4	Ni–P + $K_2Ti_6O_{13}$	Ni–P + Si_3N_4
Nickel sulfate, g/L		21	0.10 mol/L	23		21.2	25	21
Sodium hypophosphite, g/L		24	0.20 mol/L	21		24	15	24
Sodium formate, mol/L			0.50					
Complexing agent				Proprietary				
Stabilizer				Proprietary				
Lactic acid, g/L		30				28 mL/L		25
Propionic acid, g/L		8				2.2 mL/L		3
Succinic acid, g/L						12		
Glycine, g/L			0.15 mol/L					
Sodium acetate, g/L							12	
Sodium citrate, g/L							10	
Surface agent (not indicated), mg/L							20–40	
Lead acetate, ppm		0.7–1.5						3
Lead(II) nitrate, mol/L			3.75×10^{-6}					
TiO_2	Size, μm	0.2–0.3						
	Amount, g/L	0.5–2						
ZrO_2	Size, μm		5					
	Amount, g/L		12					

(Continued)

TABLE 3.3 Representative Bath Solutions for the Electroless Composite Plating of Ni–P Based Systems *(cont.)*

Bath compositions and operating conditions		Ni–P + TiO_2	Ni–P + ZrO_2	Ni–P + CeO_2	Ni–P + YSZ *	Ni–P + Fe_3O_4	Ni–P + $K_2Ti_6O_{13}$	Ni–P + Si_3N_4
CeO_2	Size, μm			4–17				
	Amount, g/L			0–25				
YSZ	Size, μm				2			
	Amount, g/L				50			
Fe_3O_4	Size, μm					0.05–0.2		
	Amount, g/L					10		
$K_2Ti_6O_{13}$ Whiskers**	Size, μm						diameter: 1.5 length: 35	
	Amount, g/L						not indicated	
Si_3N_4	Size, μm							Sub micron
	Amount, g/L							1
Agitation		Electrical	Magnetic	Mechanical	Mechnical	Mechanical	Magnetic	Magnetic
Stirring rate, rpm		500	500	600	Not indicated	600	Not indicated	900 ± 25
pH		4.4–4.7	6	4.8–5.0	4.9–5.4	4.5	4.5–5	4.6 ± 0.2
Temp., °C		94 ± 1	75	88 ± 1	89	90 ± 1	80–90	85 ± 2
Concentration in deposit, wt%		P: 15.2↘13.2 TiO_2: 3.7↗8.8	P: 4.8 ± 0.3 ZrO_2: 14.4 ± 0.4	P: 10.18 CeO_2: 7.44	P: 6–9 YSZ: 30			P: 10.0 Si_3O_4: 40% by volume
References		[30]	[31]	[32]	[33]	[34]	[35]	[36]

The electroless Ni–P bath solution was manufactured by Schloetter Company Ltd under the Tradename of Slotonip 2010 and produced a bright mid-phosphorous (6–9%) nickel deposit.
***According to the authors [35], $K_2Ti_6O_{13}$ whiskers have specific tunneling structures and fine ceramic polycrystalline whiskers, therefore it is comparatively difficult to codeposit PTWs with Ni–P by electroless plating, and the pretreatment of $K_2Ti_6O_{13}$ whiskers is very important. As for how to do the pretreatment of $K_2Ti_6O_{13}$ whiskers, the interesting readers could read the original paper.*

TABLE 3.4 Representative Bath Solution for the Electroless Composite Plating of Ni–P Based Systems

Bath compositions and operating conditions	Ni–P + MoS_2		Ni–P + PTFE		Ni–P + BN(h)	Ni–P + Talc
Nickel sulfate, g/L	21	20	25	*		
Nickel chloride, g/L					21	
Nickel metal, g/L						6 ± 1**
Sodium hypophosphite, g/L	24	25	30		24	
Nickel hypophosphite, g/L						23 ± 3
Citrate acid, g/L			18			
Sodium citrate, g/L					45	
Sodium acetate, g/L			18			
Sodium orthophosphite, g/L						0–180
Lactic acid, g/L	30	30				
Propionic acid, g/L	3	2.2				
Ammonium chloride, g/L					30	
CTAB (Hexadecyl trimethyl ammonium Bromide), g/L		0.6		0.3		
PVP (Polyvinyl pyrrolidone), g/L		5		0.2		
Fluorine carbon surfactant, $C_{20}H_{20}F_{23}N_2O_4I$ (FC-4), g/L			0–1.0			
SDS (Sodium dodecyl sulfate), g/L				0.3		
Lead(II) nitrate, ppm	0.7–2					
Thiourea, ppm			1			

(Continued)

TABLE 3.4 Representative Bath Solution for the Electroless Composite Plating of Ni–P Based Systems *(cont.)*

Bath compositions and operating conditions		Ni–P + MoS_2		Ni–P + PTFE		Ni–P + BN(h)	Ni–P + Talc
MoS_2	Size, μm	1–5	1.5				
	Amount, g/L	10–25	6–8				
PTFE emulsion (60 wt%)	Size, μm			0.05–0.5	0.2		
	Amount, g/L			4–50 mL/L	7 mL/L		
BN (h)	Size, μm					7.39	
	Amount, g/L					1, 2.5	
Talc	Size, μm						not indicated
	Amount, g/L						40, 80
Agitation		Not indicated	Magnetic	Not indicated	Magnetic	Magnetic	Mechanical
Stirring rate, rpm		300	Not indicated	Not indicated	100	Not indicated	500
pH		4.4–4.7	4.5	4.9	4.5	8	5 (4.8–5.4)
Temp., °C		90	90 ± 1	88–90	88	60 ± 1	88 (86–92)
Concentration in deposit, wt%		MoS_2: 5		P: 8–10; PTFE see Fig. 3.2	P / PTFE 9.4 / 1.8 8.8 / 8.1 (CTAB) 9.4 / 1.6 (PVP) 10.1 / 0 (SDS)	P: 5.96,10.15 BN: 12.4, 45.2 vol%	P: 5.46, 5.65 O: 4.00, 6.17 Mg: 1.28, 1.51 Si: 1.17, 1.5
References		[37]	[38]	[39]	[40]	[41]	[42]

**A commercial electroless nickel bath (SLOTONIP 70A from Schlotter) was used to obtain the Ni–P coating, which provided Ni–P deposit with high phosphorous content, 9–11 wt%.*
***Europlate Ni216 commercial bath.*

TABLE 3.5 Representative Bath Solution for the Electroless Composite Plating of Ni–P Based Systems

Bath compositions and operating conditions	Ni–P + B_4C		Ni–P + CaF_2		Ni–P + polymer		Ni–P + graphite
Nickel carbonate, g/L	*	15					
Nickel sulfate, g/L			23–31	30	30	35	30
Sodium hypophosphite, g/L		15	32–40	30	25	15	22
Hypophosphorous acid, mL/L		32					
Sodium citrate, g/L				20		50	
Lactic acid, mL/L		32	10–20				
Acetic acid, g/L		18					
Propionic acid, mL/L		3					
Dimethylamine, g/L		1.7					
Propionic acid- lactic acid, g/L					10		
Citrate acid, g/L			8–12				
Succinic acid, g/L			2–8				28
D/L Malic acid, g/L							13
Stabilizer A (not indicated), g/L			15–25				
Sodium acetate, g/L				15	15		
Lead acetate, g/L							0.5
Anionic surfactant(not indicated)				Small			
Nonionic surfactant (not indicated)				Small			
Surfactant (not indicated), g/L					0.2–4		
Allyl alcohol ethoxylate, g/L						1–10	

(Continued)

TABLE 3.5 Representative Bath Solution for the Electroless Composite Plating of Ni–P Based Systems *(cont.)*

Bath compositions and operating conditions		Ni–P + B_4C		Ni–P + CaF_2		Ni–P + polymer		Ni–P + graphite
Perfluoro ethyl ammonium iodide, g/L								0.1
B_4C	Size, μm	5–11	2					
	Amount, g/L	1–8	5–15					
CaF_2	Size, μm			<2	1–5			
	Amount, g/L			2–5	15			
Polyvinyl pyrrolidone (PVP) – 10, 24.5, 40	Size, μm					Not indicated		
	Amount, g/L					0.2–4		
Polyvinyl chloride (PVC) Polyvinyl alcohol (PVA) Polyacrylamide (PAm)	Size, μm						Not indicated	
	Amount, g/L						4	
Graphite	Size, μm							Not indicated
	Amount, g/L							2–30
Agitation		Magnetic	Mechanical	Not indicated	Intermittent	Not indicated	Not indicated	Not indicated
Stirring rate, rpm		750	200	Not indicated	Not indicated	Not indicated	Not indicated	230±
pH		4.6	6.5	4.7–5.1	4–5	4.6	5–6	5.2
Temp., °C		90	83	87–92	85	80	80–85	89
Concentration in deposit, wt%		B_4C: 12–33 vol%	P: 9.96–8.84 B_4C: 2.32–6.18	P: 11–12.4 CaF_2: not indicated				P: 9.4↘6.83 Graphite: 2.4↗12.73
References		[43]	[44]	[45]	[46]	[47]	[48]	[49]

The commercial SH490L5 nickel–phosphorus electroless bath (Scholter, Germany) was used.

TABLE 3.6 Representative Bath Solution for the Electroless Composite Plating of Ni–B Based Systems

Bath compositions and operating conditions	Ni–B + SiC	Ni–B + diamond	Ni–B + graphitre	Ni–B + Al_2O_3	Ni–B + Si_3N_4	Ni–B + BN
Nickel chloride, g/L	30			35	30	35
Nickel acetate, g/L		50				
Nickel sulfate, mol/L			0.10			
Potassium borohydride, g/L	1.0			0.8		
Sodium borohydride, g/L					0.8	0.6
Sodium hydroxide, g/L	40					55
Ethylenediamine, g/L	36			58	90	62
DMAB, g/L		2.5	0.05 mol/L			
Sodium hydroxide, g/L				42	90	
Thallium acetate (stabilizer), mg/L					16	
Sodium lauryl sulfate, g/L					0.1	
Ammonium sulfate, mol/L			0.10			
Lactic acid, g/L		25				
Boric acid, mol/L			0.10			
Sodium citrate, g/L		25				
Thiodiglycolic acid (TDGA), g/L		0.1				

(Continued)

TABLE 3.6 Representative Bath Solution for the Electroless Composite Plating of Ni–B Based Systems *(cont.)*

Bath compositions and operating conditions		Ni–B + SiC	Ni–B + diamond	Ni–B + graphitre	Ni–B + Al_2O_3	Ni–B + Si_3N_4	Ni–B + BN
Santomerse S, g/L			0.1				
Potassium sodium tartrate, mol/L				0.10			
Ammonium hydroxide			Appropriate				
Polyoxy ethylene nonyl phenylether, mg/L				500			
Perfluore alkyl ammonium iodide, mg/L				500			
Lead chloride, ppm		20					
Lead nitride, g/L							0.06
Stabilizer (not indicated)					Appropriate		
Silicon nitride, g/L						0–25	
SiC	Size, μm	1.5–2.5					
	Amount, g/L	≈40					
Diamond	Size, μm		2–8				
	Amount, g/L		2				
Graphite	Size, μm			2			
	Amount, g/L			0.1–1			

Al_2O_3*	Size, μm				1		
	Amount, g/L				10		
Si_3N_4	Size, μm					Not indicated	
	Amount, g/L					0–25	
BN	Size, μm						≤5
	Amount, g/L						3
Agitation		Magnetic	Electric	Not indicated	Electric	Mechanical	Not indicated
Stirring rate, rpm		600	350	Not indicated	30	600	20–40
pH		13.5	6.5	6–10	13	14	13–14
Temp., °C		75 ± 1	50–55	50–80	80	95 ± 1	85 ± 7
Concentration in deposit, wt%		B: 4.1 SiC: 4.8	diamond: 20 vol%	graphite: 2–5.5	Al_2O_3: 4	B: 6.5 Si_3N_4: 2 Tl: 0.3**	
References		[50]	[51]	[52]	[53]	[54]	[55]

In order to prevent the Al_2O_3 particles agglomerate and precipitation or floating in the bath, the Al_2O_3 particles must be carried out with hydrophilic treatment before adding them to the bath solution. Anionic surfactant (not indicated): 1–1.5 g/L; nonionic surfactant (not indicated): 0.5–1.0 g/L.

***The incorporation of Tl along with Ni and B in the composite deposit due to the use of thallium acetate as the stabilizer in the bath solution.*

TABLE 3.7 Representative Bath Solution for the Electroless Composite Plating of Co–P Based Systems

Bath compositions and operating conditions		Co–P + SiC	Co–P + Al_2O_3	Co–P + SiO_2	Co–P + PTFE		Cu-P + SiC	Cu-P + graphite
Cobalt sulfate, mol/L		0.05	0.05	0.05				
Cobalt chloride, g/L					27	27		
Copper sulfate, g/L							25	25
Sodium hypophosphite, mol/L		0.1	0.1	0.1	30 g/L	20–35 g/L	125 g/L	125 g/L
Nickel sulfate, g/L							3.125	3.125
Sodium citrate, mol/L		0.1	0.1	0.1		40 g/L	50 g/L	50 g/L
Ammonium citrate, g/L					40			
Ammonium chloride, g/L						22.5		
Boric acid, mol/L		0.3	0.3	0.3			25 g/L	
Fumaric acid, mol/L		0.05	0.05	0.05				
Glycine, g/L					10	10		
Surfactant FC-400, g/L					0.04	0.02–0.05		
Polyglycol, g/L							0.001	
Stabilizer (not indicated), g/L							0.001	
SiC	Size, μm	4					4–10	
	Amount, g/L	1–10					5	
Al_2O_3	Size, μm		4					
	Amount, g/L		1–10					

SiO_2	Size, μm			4				
	Amount, g/L			1–10				
PTFE	Size, μm				Not indicated	Not indicated		
	Amount, g/L				4	4*		
Graphite	Size, μm							4–10
	Amount, g/L							5***
Agitation		Magnetic	Magnetic	Magnetic	Not indicated	Not indicated	Magnetic	Mechanical
Stirring rate, rpm		150	150	150	Not indicated	Not indicated	Not indicated	Not indicated
pH		9.2	9.2	9.2	7–8	9	9	9
Temp., °C		82	82	82	80–90	90	90	90
Concentration in deposit, wt%		P: 6 ± 1 SiC: 45 ± 2	P: 6 ± 1 Al_2O_3: 5 ± 1	P: 6 ± 1 SiO_2: 2.5 ± 2	P: 3 PTFE: 10	P: 4.34 C: 7.69 Fe: 1.96**		P: 0.23 C_g: 11.65 O: 11.03 Fe: 11.97
References			[56]		[57]	[58]	[59]	[60]

*PTFE emulsion was milled for 30 min in 200 r/min before adding them into the bath solution.
**From analyzing the composition of the composite deposit there is no F element except P, C, Fe, and Co elements. The reason may cause from that SEM cannot measure the F element because of its amount rather few. As for Fe element appeared in the deposit may result from that the subtract of A3 steel was exposure or very small amount of Fe rust left behind during processing of the substrate and plating. C element in the coating indicated obviously that PTFE particles have embedded in the composite deposit. In addition, microstructure of the composite coatings shows evidently that PTFE particles uniformly embedded in the cobalt, phosphorus coating.
***The SEM micrographs of the electroless Cu–P composite coatings show that superfine C_g particles were successfully codeposited in the Cu–P alloy matrix.

TABLE 3.8 Representative Bath Solutions for Ternary Alloy-Based ECP Systems

Bath compositions and operating conditions		Ni–W–P + SiC	Ni–W–P + Al_2O_3	Ni–W–P + ZrO_2	Ni–W–P + Si_3N_4	Ni–W–P + PrO_2
Nickel sulfate, g/L		120	35	0.10 mol/L	26	20
Sodium hypophosphite, g/L		25	15	0.20 mol/L	20	30
Sodium tungstate, g/L		90	28	0.005–0.0125 mol/L	40	20
Complexing agent (not indicated), g/L		180				40
Additive (not indicated), ml/L		30				
Sodium citrate, g/L			90		100	
Ammonium chloride, g/L			50			
Ammonium acetate, g/L					13	
Buffering agent (not indicated), g/L						25
Glycine, mol/L				0.15		
Sodium formate, mol/L				0.50		
Lead acetate, ppm			Small amounts			
Lead(II) nitrate, mol/L				3.75×10^{-6}		
Thiourea, mg/L					4	
Stabilizer, g/L						0.08
SiC	Size, μm	2–3				
	Amount, g/L	50				

Al_2O_3	Size, μm		1			
	Amount, g/L		1–6			
ZrO_2	Size, μm			5		
	Amount, g/L			6		
Si_3N_4	Size, μm				0.1	
	Amount, g/L				10	
PrO_2	Size, μm					Not indicated
	Amount, g/L					0.04
Agitation		Not indicated	Magnetic	Magnetic	Mechanical	Air
Stirring rate, rpm		300–500	800 ± 50	500	Not indicated	Not indicated
pH		8–8.5	9	6	9	5.5
Temp., °C		65–90	88 ± 2	75	85–90	90 ± 2
Concentration in deposit, wt%			P: 10.3 W: 1.01 Al_2O_3: 6.3	P: 4.34↗4.83 W: 0.69↗1.15 ZrO_2: 13.45↘12.2		
References		[61]	[62]	[63]	[64]	[65]

TABLE 3.9 Representative Bath Solutions for Ternary Alloy-Based ECP Systems

Bath compositions and operating conditions		Ni–Mo–P+ Al_2O_3	Ni–Mo–P + PPS	Ni–Co–B + Diamond	Ni–Cu–P + PTFE
Nickel acetate, g/L		25	25	44	
Nickel sulfate, g/L					50
Copper sulfate, g/L					1
Cobaltous acetate, g/L				6	
DMAB, g/L				2.5	
Sodium hypophosphite, g/L		20	20		25
Sodium tungstate, g/L		0.6	0.6		
Sodium citrate, g/L		32	32	25	60
Lactic acid, g/L				25	
Thiodiglycolic acid, g/L				0.1	
Santomerse S, g/L				0.1	
Ammonium acetate, g/L					40
Dodecane, mg/L		5	5		
Thiourea, mg/L		4	4		
Fluorine carbon surfactant, (FC-4), g/L					0–0.6
Al_2O_3	Size, μm	Not indicated			
	Amount, g/L	6			
Polyphenylene sulfide (PPS)	Size, μm		Not indicated		
	Amount, g/L		1		

Diamond	Size, μm			9	
	Amount, g/L			2	
PTFE	Size, μm				0.05–0.5
	Amount, mL/L				4–18
Agitation		Not indicated	Not indicated	Mechanical	Without using
Stirring rate, rpm		Not indicated	Not indicated	350	
pH		8.0–8.5	8.0–8.5	6.4	7–8.5
Temp., °C		85 ± 3	85 ± 3	60	77~85
Concentration in deposit, wt%				Diamond: 11 vol/%	P: 10.5↘4.3 Cu: 6.3↗11.5 C: 3.5↘2.7 F: 7.7↘5.4
References		[66]		[67]	[68]

TABLE 3.10 Representative Bath Solution for ECP with Two Kinds of Particles

Bath compositions and operating conditions		Ni–P + SiC +MoS_2	Ni–P + SiC +graphite*	Ni–P + SiC +PTFE	Ni–P +TiC/Al_2O_3****	Ni–P + Al_2O_3 + ZrO_2	Ni–P+ MoS_2 +PTFE
Nickel sulfate, g/L		25	20–25	21	35	Not indicated	20
Sodium hypophosphite, g/L		30	25–30	24	15	Not indicated	25
Sodium citrate, g/L		15			15	Not indicated	
Glycine, g/L		25		25			
Succinic acid, g/L		15					
Sodium acetate, g/L		15	30–40	0.1	5		
Lactic acid, g/L			20–30	30			30
Propionic acid, g/L				3			2.2
Surfactant (not indicated), mg/L		40–90	100–400				
Stabilizer (not indicated), mg/L			1–3				
Ammonium sulfate and chliride						Not indicated	
Lead nitrate, ppm				0.7–1.5			
CTAB, g/L							1.5
PVP, g/L							5
SiC	Size, μm	not indicated	4	1–5			
	Amount, g/L	2–8	6–10	15–20			
MoS_2	Size, μm	not indicated					
	Amount, g/L	2–8					

Graphite	Size, μm		4				
	Amount, g/L		5–8				
PTFE	Size, μm			≅0.5			
	Amount, mL/L			10–30			
TiC/Al_2O_3	Size, μm				90		
	Amount, g/L				10		
Aluminum chloride	Size, μm					5–10	
	Amount, g/L					Not indicated	
Zirconium oxychloride	Size, μm					5–10	
	Amount, g/L					Not indicated	
MoS_2	Size, μm						1.5
	Amount, g/L						6–8
PTFE	Size, μm						1
	Amount, g/L						6–8
Agitation		Not indicated	Mechanical	Mechanical	Magnetic	Not indicated	Magnetic
Stirring rate, rpm		240	Not indicated	Not indicated	150	Not indicated	No indicated
pH		3.8–5.4	4.6–4.9	3.8–4.2** 4.4–4.7***	5.5	9	4.5
Temp., °C		86–90	89–91	90** 80***	85	90	90 ± 1

(Continued)

TABLE 3.10 Representative Bath Solution for ECP with Two Kinds of Particles *(cont.)*

Bath compositions and operating conditions	Ni–P + SiC +MoS_2	Ni–P + SiC +graphite*	Ni–P + SiC +PTFE	Ni–P +TiC/Al_2O_3****	Ni–P + Al_2O_3 + ZrO_2	Ni–P+ MoS_2 +PTFE
Concentration in deposit, wt%			SiC: 25 vol% PTFE: 20vol%	TiC/Al_2O_3: 61 vol%	P: 9.5 Al_2O_3: 0.025 ZrO_2: 0.287	P: 8.64 Mo: 3.16 F: 1.86
References	[69]	[70]	[37]	[71]	[72]	[38]

The innovative Ni–P + SiC + PTFE composite deposit was obtained by first immersing the specimen in the bath for the codeposition of SiC and, subsequently, in the one for the codeposition of PTFE.
***SiC bath.*
****PTFE bath.*
*****The powders TiO_2, Al, and C were weighed in the stoichiometric ratio of 3.0:4.0:2.7, respectively, to form TiC/Al_2O_3 composites (1:1.14 wt%) and then blended in a slow rotating mill with alumina ball for 5 h. The powders were then pressed into a cylindrical shape of 1.3 cm diameter and 3 cm long at 100 MPa in a metal mould to form a green body with −60% relative density using. The reaction between these powders was carried out in SHS reactor where the upper surface of the specimen was ignited in argon atmosphere using thin graphite electrode. The combustion product was then crushed and pulverized to pass ~90 μm.*

TABLE 3.11 Representative Bath Solutions for ECP with Two and Three Kinds of Particles

Bath compositions and operating conditions	Ni–P + CaF_2+ diamond	Ni–B+ SiC + RE	Ni–W–P+ diamond + (CF)n	Ni–Re–P+ Al_2O_3 + CaF_2	Ni–W–B+ SiC + CeO_2	Ni–P + ZrO_2 +Al_2O_3 + Al_3Zr *
Nickel chloride, g/L		30			30	Not indicated
Nickel sulfate, g/L	25		25	20		
Sodium borohydride, g/L		0.8				
Potassium borohydride, g/L					2–3	
Ethylene diamine, g/L		60				
Sodium hydroxide, g/L		40				
Sodium hypophosphite, g/L	30		20	20		Not indicated
Sodium tungstate, g/L			50		50–70	
Potassium perrhenate, g/L				1.5		
Sodium citrate, g/L	100		90	80		Not indicated
Complexing agent 1 (not indicated), g/L					40–80	
Complexing agent 2 (not indicated), g/L					0–15	
Lactic acid, mL/L	5			5		
Additive (not indicated), mg/L	10			10		
Accelerate impregnant (not indicated), mg/L	8			8		

(Continued)

TABLE 3.11 Representative Bath Solutions for ECP with Two and Three Kinds of Particles *(cont.)*

Bath compositions and operating conditions		Ni–P + CaF_2+ diamond	Ni–B+ SiC + RE	Ni–W–P+ diamond + (CF)n	Ni–Re–P+ Al_2O_3 + CaF_2	Ni–W–B+ SiC + CeO_2	Ni–P + ZrO_2 +Al_2O_3 + Al_3Zr *
Stabilizer (not indicated), mL/L			9–11	Appropriate			
Ammonium sulfate, g/L				30			Not indicated
Ammonium chloride, g/L *							Not indicated
Zirconium oxychloride, g/L *							Not indicated
Diaomond	Size, μm	3–5		Not indicated			
	Amount, g/L	6		Not indicated			
CaF_2	Size, μm	≤2			2–3		
	Amount, g/L	8			18		
SiC	Size, μm		Not indicated			3–3.5	
	Amount, g/L		5–20			60	
RE (oxide)	Size, μm		Not indicated				
	Amount, g/L		0–8				
(CF)n	Size, μm			Not indicated			
	Amount, mL/L			not indicated			
Al_2O_3	Size, μm				Not indicated		
	Amount, g/L				8		
CeO_2	Size, μm					0.5–1	
	Amount, g/L					not indicated	

Agitation	Not indicated	Mechanical	Magnetic	Not indicated	Not indicated	Not indicated
Stirring rate, rpm	Not indicated	Not indicated	Intermitted	Not indicated	Not indicated	Not indicated
pH	7	13–13.5	8.5	4.5	13.5	9 ± 0.25
Temp., °C	80	82 ± 1	88 ± 2	80	40	90 ± 2
Concentration in deposit, wt%		B: 4.2 SiC: 6.5 RE (oxide): 2	P: 12.41 W: 5.21 diamond + (CF) n: 8–9		W: 0.66 SiC: 10.74 CeO_2: 1.12	P: 8.2 Al: 0.50 Zr: 4.44 Ni: 86.9 **
References	[73]	[74]	[75]	[76]	[77]	[78]

All of the three kinds of insoluble particles are not introduced intentionally in the bath, but the authors believe that they all are produced during the electroless plating by adding chemicals aluminum chloride and zirconium oxychloride through a coprecipitation reaction as follows: $8AlCl_3 + ZrOCl_2 \cdot 8H_2O + 26NH_4OH \rightarrow 4Al_2O_3 + ZrO_2 + 26\ NH_4Cl + 4H_2O$. And they also sure that when three atoms of Al and one of Zr come in the vicinity of each other, it is expected to form an intermetallic compound Al_3Zr. Their experimental observations from SEM and selected area diffraction (SAD) show that Al_3Zr, Al_2O_3, and ZrO_2 phases have been confirmed in the composite coating.

***The results are from quantitative analysis of the $Ni-P-X$ ($X = ZrO_2-Al_2O_3-Al_3Zr$) coating by point scan using EDXA. It is very strange that there is no O element in their composite coating.*

Ni–P, second is Ni–B and Co–P, whereas very few papers have been published about Cu-P. Many reports of binary based EP + two kinds of particle alloy systems, and ternary based EP + two kinds of particle alloy systems have been published. No papers can be found that report on quaternary based EP + one or two kinds of particles alloy systems. However there is a report for Ni–P + ZrO_2 + Al_2O_3 + Al_3Zr ECP though the three kinds of particles are not intentionally introduced to the bath, but they all are produced during the electroless plating by adding chemicals aluminum chloride and zirconium oxychloride through a coprecipitation reaction.

3. Twenty-three kinds of composite particles have been studied, including SiC, WC, diamond, TiC, Cr_3C_2, Al_2O_3, SiO_2, Cr_2O_3, TiO_2, ZrO_2, CeO_2, YSZ, Fe_3O_4, $K_2Ti_6O_{13}$, Si_3N_4, MoS_2, PTFE, BN, Talc, B_4C, CaF_2, polymer, and graphite.
4. As discussed in Section 3.1, the PMS should be added into the bath solution of ECP. Most of the bath solutions do this, but some of them do not as the reasons are described in Section 3.1.
5. All of the representative bath solutions need to be stirred with a kind of stirrer (mechanical, electrical, magnetic, or air flow). The stirring rate usually is several hundred rpm.

REFERENCES

[1] Odekerken JM, GB Patent 1,041,753 Improved method of depositing bright-nickel/chromium, September 7, 1966.

[2] Odekerken JM, US Patent 3,298,802 Method for covering objects with a decorative bright nickel/chromium grating, as well as objects covered by applying this method, January 17, 1967.

[3] Metzger W, Ott R, Pappe G, Schmidt H, US Patent 3,617,363 Process for electroless metallizing incorporating wear-resisting particles, November 2, 1971.

[4] Metzger W, Ott R, Pappe G, Schmidt H, US Patent 3,753,667 Articles having electroless metal coatings incorporating wear-resisting particles therein, August 21, 1973.

[5] Christini TP, Eustice AL, Graham AH, US Patent 3,936,577 Method for concomitant particulate diamond deposition in electroless plating, and the product thereof, February 3, 1976.

[6] Lukschandel J, Trans Inst Metal Finish, 1978;56(3):118.

[7] Hussaina MS, Such TE. Surf Technol 1981;13(2):119.

[8] Tulsi SS. Finishing 1983;7(11):14.

[9] Spencer RA Jr, US Patent 4,547,407 Electroless metal coatings incorporating particulate matter of varied nominal sizes, October 15, 1985.

[10] Kim JT, US Patent 4,716,05 Composites of metal with carbon fluoride and method of preparation, December 29, 1987.

[11] Henry JR, Summers EM, US Patent 4,830,889 Co-deposition of fluorinated carbon with electroless nickel, May 16, 1989.

[12] Hubbell FN. P&SF 1978;65(12):58.

[13] Feldstein N, Lancsek T, Lindsay D, Salerno L. Metal Finish 1983;81(8):35.

[14] Mallory GO, Hajdu JB. Electroless plating: fundamentals and applications. Orlando, FL: AESF; 1991.

[15] Hajdu JB, Zabrocky S. Metal Finish 2000;98(5):42.
[16] Mypolex®, Micropolycrystalline diamond powder, E. I. du Pont de Nemours and Company, Wilmington, DE, USA p. 17.
[17] Feldstein N, Lindsay DJ, US Patent 6,306,466 Stabilizers for composite electroless plating, October 23, 2001.
[18] Onda T, Okayasu N, Yoshuhara S, Sato E. J Surf Finish Soc Jpn 1991;42(1):79.
[19] Wu YC, Li GH, Zhang LD, Song GM. J Mater Eng Perfor 2000;9(4):379.
[20] Gao J, Liu L, Wu Y, Shen B, Hu W. Surf Coat Technol 2006;200:5836.
[21] Abdel Hamid Z, El Badry SA, Abdel Aal A. Surf Coat Technol 2007;201:5948.
[22] Sheela G, Pushpavanam M. Metal Finish 2002;100(1):45.
[23] Zhou H, Wei H, Wang J. Surf Technol 2007;36(5):50.
[24] Hansen PL, Mqller P. J. Mater Sci Lett 1990;9:152.
[25] Dennis JK, Sheikh ST, Silverstone EC. Trans IMF 1981;59:118.
[26] Zhang H, Zhang B, Tan Z, Tang H. Mater Mech Eng 1990;14(4):36.
[27] Balaraju JN, Kalavati, Rajam KS. Surf Coat Technol 2006;200:3933.
[28] Yu X, Wang H, Yang Z, Yi P, Xin X. Appl Surf Sci 2000;158:335.
[29] Karthikeyan S, Srinivasan KN, Vasudevan T, John S. Electroplat Finish 2007;28(1):1.
[30] Novakovic J, Vassiliou P. Electrochim Acta 2009;54:2499.
[31] Szczygiel B, Turkiewicz A, Serafińczuk J. Surf Coat Technol 2008;202:1904.
[32] Balaraju JN, Sankara Narayanan TSN, Seshadri SK. Mater Res Bull 2006;41:847.
[33] Bahiyah Baba N, Waugh W, Davidson AM. World Acad Sci Eng Technol 2009;49:715.
[34] Zuleta AA, Galvis OA, Castaño JG, Echeverría F, Bolivar FJ, Hierro MP, Pérez-Trujillo FJ. Surf Coat Technol 2009;203:3569.
[35] Jin Y, Hun L. J Mater Sci Technol 2007;23(3):387.
[36] Balaraju JN, Rajam KS. Int J Electrochem Sci 2007;2:747.
[37] Straffelini G, Colombo D, Molinari A. Wear 1999;236:179.
[38] Mohammadi M, Ghorbani M, Azizi A. J Coat Technol Res 2010;7(6):697.
[39] Zhaoa Q, Liua Y, Müller-Steinhagenb H, Liu G. Surf Coat Technol 2002;155:279.
[40] Mafi IR, Dehghanian C. Appl Surf Sci 2011;257:8653.
[41] Leóna OA, Staiaa MH, Hintermann HE. Surf Coat Technol 1998;108–109:461.
[42] Etcheverry B, Le Coz F, Alexis J, Paris JY, Petit JA, Bonino JP, Mechanical properties of electroless codeposited Ni–P + Talc composite coatings. In: Sudarshan TS, Jeandin M, Stigich JJ, editors. Surface modification technologies XVIII. Materials Park, OH: ASM International, and UK: IOM Communications Ltd.; 2006. p. 355–63; Etcheverry B, Adhesion, tribology and mechanical composite coatings NiP-Talc multifunctional reduced ecological footprint, PhD Thesis, The National Institute Polytechnic of Toulouse, France, 2006.
[43] Monir Vaghefi SM, Saatchi A, Ebrahimian-Hoseinabadi M. Surf Coat Technol 2003;168(2–3):259.
[44] Karthikeyan S, Venkatachalam G, Srinivasan KN, Vasudevan T, Narayanan S. Electroplat Finish 2011;30(6):18.
[45] Li N, Huang Y, Wang C, Li M, Furukawa N. Mater Protec 1994;27(5):1.
[46] Song D, Yang L. Lubric Eng 2009;34(11):20.
[47] Khalifa ORM, Sakr E. Open Corrosion J 2009;2:211.
[48] Shoeib MA, Mokhtar SM, Abd El-Ghaffar MA. Metal Finish 1998;96(11):58.
[49] Karthikeyan S, Neelakandan MA. Electroplat Finish 2006;25(4):1.
[50] Zhao G, Lu W, Wang F. Electroplat Pollu Cont 1987;7(3):1.
[51] Christini TP, Graham AH, US Patent 3,936,577 Method for concomitant particulate diamond deposition in electroless plating, and the product thereof, February 3, 1976.

[52] Nishira M, Yamagishi K, Takano O. Jpn J Surf Finish 1993;44(11):961.
[53] Wu F, Chu S. Surf Technol 1994;23(4):154.
[54] Krishnaveni K, Sankara Narayanan TSN, Seshadri SK. In: International symposium of research students on materials science and engineering. December 20–22, 2004, Chennai, India.
[55] Shen R, Chen K, Liu J. J Southwest Uni Sci Technol 2006;21(4):33.
[56] Rudnik E. Appl Surf Sci 2008;255:2613.
[57] Tang J, Xie Y, Tong G. Lubri Eng 1994;19(6):27.
[58] Yang Q, Tang J. Lubri Eng 2008;33(1):85.
[59] Faraji S, Rahim AA, Mohamed N, Sipaut CS, Raja B. Mater Chem Phys 2011;129:1063.
[60] Faraji S, Rahim AA, Mohamed N, Sipaut CS. Surf Coat Technol 2011;206:1259.
[61] Zhang H, Wang Q, Dai Y. Surf Technol 2004;33(4):35.
[62] Balaraju JN, Kalavati, Rajam KS. Surf Coat Technol 2010;205:575.
[63] Szczygiel B, Turkiewicz A. Appl Surf Sci 2009;255:8414.
[64] Du K, Ge R, Gao Y. Plat Finish 2004;26(2):8.
[65] Huang Y, Liu F, Zhao Y. Electroplat Finish 2007;26(2):1.
[66] Li J, Jiang B. Corr Protec 2005;26(8):326.
[67] Christini TP, Eustice AL, Graham AH, US Patent RE33,767 Method for concomitant particulate diamond deposition in electroless plating, and the product thereof, December 19, 1991.
[68] Zhao Q, Liu Y. Surf Coat Technol 2005;200:2510.
[69] Du CP, Liu JY, Wei Z, Zhang H. Electroplat Finish 2008;27(1):17.
[70] Wu YT, Lei L, Shen B, Hu WB. Surf Coat Technol 2006;201:441.
[71] Abdel Aal A, Zaki ZI, Abdel Hamid Z. Mater Sci Eng A 2007;447:87.
[72] Sharma SB, Agarwala RC, Agarwala V, Satyanarayana KG. J. Mater Sci 2002;37:5247.
[73] Liu C, Zhong L, Lei Y. J Synth Cryst 2007;36(1):210.
[74] Kuo Z, Liu H, Wang Z, Wang M. Acta Metall Sinica (English Lett) 1995;8(2):118.
[75] Guo L, Zhang Z, Yin X. China Mater Technol Equip 2008;(No. 2):52.
[76] Liu C, Zhong L, Lei Y. China Surf Eng 2008;21(5):46.
[77] Zhu C, Guo Z. Eng Chem Metall 1999;20(3):225.
[78] Sharma SB, Agarwala RC, Agarwala V, Satyanarayana KG. Metall Mater Trans B 2005;36 B:23.

Chapter 4

Nano Electroless Plating

Chapter Outline

Amorphous and Nano Alloys Electroless Depositions. http://dx.doi.org/10.1016/B978-0-12-802685-4.00004-2

Nano science and technology has developed very recently, and spread of the use of this technology indicates that human society entered the nanomaterials age at the very beginning of the new century [1,2]. Today, not only has nanoscience penetrated and extended into nearly all other macro disciplines [3], but nanoscience and technology have started to affect the world's economies, science and technology, the military, and human life. Numerous industries have turned to nano science and technology to solve their problems and move forward. Naturally, EP has also followed this trend. It is usually agreed that the field of nanotechnology was applied to EP at the end of the last century. In fact, the so-called nano EP has been studied since the beginning of 2000, just as we pointed out in Chapter 1 (see Fig. 1.1b and explanation). Today, nano EP has become the most active field in EP.

Why has nano EP appeared and developed so rapidly? The main reason is that the convenient EP coatings could not fully meet the needs of critical applications in rapidly advancing science and technology. Nano EP deposits perform superlatively, and therefore can be used in such applications. In this section, the metallization process of integrated circuits (IC) production will be outlined briefly. The technology of nano EP metallization has developed to the point at which the metallization process is much simpler. EP nanotechnologies are now better and simpler than other expensive and more complicated methods such as e-beam, X-ray lithography, or the production of devices using light phase shift photomasks.

There is a strict definition for nanomaterials (NMs), which is that the length in one direction must be in the range of 1–100 nm regardless of whether the NM is zero-dimensional (0D), one-dimensional (1D), two-dimensional (2D), or three-dimensional (3D). Three-dimensional NMs are bulk NMs, that is, they are a bulk material containing NMs, so an EP deposit that does not contain any NMs cannot be called a nano EP material. If the EP material itself or inclusions in the EP material exceeds the size range of NMs, the material cannot be classified as a nano EP material. Only EP nano materials are discussed in this chapter. If the size of the EP material itself or inclusions in it exceed the size range of NMs in some cases; this will be pointed out clearly.

Nano EPs can be divided into four types. The first is zero-dimensional nano EP materials, which includes nanonanoparticles, nanocubes, and core-shell nanonanoparticles. The second is one-dimensional nano EP materials, which includes nanotubes, nanowires, and nanofibers, and nano EP coatings or particles deposited onto nanotubes, nanowires, and nanorods. The third is 2D nano EP materials, which includes nanofilms deposited onto the surface or interface of substrates, as well as nanofilms, nanoplates, naodisks, and nanowalls. The final category is the nano EP bulk materials, which includes the electroless nanocomposite and EP nanoarchitecture materials.

EP deposits usually are amorphous materials, and these will be discussed more in detail in Chapter 8 (composition and microstructure). It is known that amorphous alloy materials are metastable states. They in fact are a supercooled liquid, and will change into a more stable or equilibrium state under certain conditions. The technical parameters in such a process of transformation can be chosen to promote the rapid nucleation of a very large number of nuclei, with very slow growth of the new phase, in which case the size of the new phase grain must be very small, even as small as nano size. This is the so-called amorphous nanocrystallization method (ANM). There are three kinds of ANM. The most popular method is heat treatment, the second is laser annealing, and the third is the amorphous nanocrystallization produced by deformation. We will introduce the first one here [4]. In 1988, Yoshizawa et al. prepared amorphous magnetic $Fe_{74.5-x}Cu_xNb_3Si_{13.5}B_9$ ($x = 0$, 0.5, 1.0, and 1.5) alloy ribbons by the single roller method. The width and thickness of these ribbons were 5 μm and from 15 μm to 20 μm, respectively. Then, the ribbons were wound into 19 mm outer diameter and 15 mm inner diameter specimens. These were then thermally annealed at 673 K to 923 K for 1 h under a nitrogen gas atmosphere. The authors found that the annealed alloy is composed of bcc Fe solid solution with a grain diameter of about 10 nm [5]. Then in 1991, Wang Jintang's group prepared amorphous $Ni_{80}P_{20}$ alloy ribbons by using the single roller method. The amorphous sample was heated rapidly at 300 K/min to 325°C for 10 min and then rapidly quenched to room temperature. Two crystalline phases, one of the Ni_3P compound and the other of fcc Ni(P) austenite solid solution, were observed. The average grain sizes were 13 and 9 nm for Ni_3P and Ni(P) solid solution, respectively, as estimated by X-ray diffraction and electron microscopy [6]. From these two examples, it is evident that the amorphous ribbons had really transformed to nanocrystalline phases after appropriate heat treatment. Now ANM has been studied thoroughly, and a large number of papers about it have been published. In addition, some reviews and book chapters on this subject have also been prepared [4, 7–10]. We will not discuss it more in detail here.

In this chapter, the technology for the preparation of nano EP materials, including the bath solution composition and operational conditions, will be described and discussed in detail for every EP nanomaterial. Unlike the arrangement in the second and third chapters, we will not list a whole table for all EP nano materials giving their bath solution composition and operational conditions. In contrast, each respective EP nano system will be described separately. The manufacturing principle, chemical composition and structure, and nano features of the EP nano system to be mentioned are also important pieces of information given in the chapter. Moreover, some properties will be referred to for some nano EP materials. However, the details of the properties of the nano EP materials will not be described here as pointed out at the beginning of the book.

Four kinds of nano EP will be discussed in the following sections.

4.1 BULK NANO EP MATERIALS

As mentioned above, bulk nano EP materials refer to the bulk materials produced by nano electroless composite plating (nano ECP) method and EP nanoarchitecture bulk materials.

Similarly to conventional ECP, discussed in Chapter 3, a nano ECP can be defined as a method of introducing intentionally insoluble or sparingly soluble particulate matter in the nano size range into conventional electroless plating. The main difference between nano and conventional ECP is the different size of the intentionally insoluble particles, where the former is nanonanoparticles but the latter is micron particles. Of course, other parameters such as the additive surfactants, and/or stirring method and intensity could differ from the counterpart conventional ECP simply because nanonanoparticles had been added. As for the bath composition, this is the same as that used for conventional metals, hence we just use simple words to explain it but not list the name and amount of chemicals in detail.

EP nanoarchitectural materials are just the 3D NMs prepared by EP.

4.1.1 Nano ECP

As of now, many nano ECP materials have been prepared, including metal based, binary alloy based, and ternary alloy based nano ECP materials.

4.1.1.1 Metal-Based Nano ECP Materials

Zhang et al. studied electroless composite coatings of Ag doped with nanonanoparticles of $A1_2O_3$, Si_3N_4, and ZrC [11]. They prepared these Ag nano ECP films on the surface of polyamide fabrics by introducing corresponding nanonanoparticles into a conventional electroless Ag plating bath with the assistance of ultrasound. The three kinds of nanonanoparticles are of purity greater than

99.0% and an average diameter of less than 30 nm. The bath solution composition and operational conditions were: silver nitrate 5.0 g/L, ammonium hydroxide 2.5 g/L, glucose 4.5 g/L, tartaric acid 0.4 g/L, potassium hydroxide 3.5 g/L, pH = 13, temperature = 30°C, and an ultrasonic oscillator operating at 50 kHz, 100 W. It was found that the composite coatings were smoother and denser, and had smaller grain sizes than a conventional electroless Ag coating. All the coatings had similar crystallinity, but their grain sizes were different, being 25.1 nm [Ag (111) crystal surface] for the conventional EP Ag film, but 22.1, 19.4, or 17.2 nm for the nano ECP Ag films codeposited with Al_2O_3, Si_3N_4, or ZrC nanoparticles, respectively. The composite coatings possessed better wear resistance, adhesion strength, and oxidation resistance than the Ag coating. In addition, the composite coatings had similar resistance to salt solution to the Ag coating, but their electromagnetic shielding effectiveness was slightly lower.

Ling et al. have studied electroless plating of a Ag + TiO_2 nanoparticle composite on PET fabric by using different dispersants [12]. The dispersants used were sodium alginate (SA), sodium tripolyphosphate (STPP), polyvinylpyrrolidone (PVP), and tween 80 (T80). The average diameter of the TiO_2 nanonanoparticles was less than 20 nm. The experimental data indicated that the introduction of intentionally insoluble TiO_2 nanonanoparticles had no effect on the structure of the Ag film when PVP or T80 was used as a dispersant. However, the crystal size increased. The abrasion resistance of the nano Ag + TiO_2 plated fabric was greater and its air permeability was lower. In addition, it has been found that the electromagnetic shielding effectiveness was improved when the mass concentration of TiO_2 nanonanoparticles was below 0.2 g/L.

In 2011, Zhang and Feng compared electroless copper composite plating on polyester (PET) fabric by using different nanonanoparticles (ZnO, Si_3N_4, Fe_3O_4, ZrC, and SiO_2) [13]. The EP bath for Cu includes copper sulfate (20 g/L), sodium hypophosphite (75 g/L), sodium citrate (30 g/L), boric acid (35 g/L), and nickel sulfate (7.9 g/L). The operation temperature is 80°C, and the pH is 9.8. The average size of the nanonanoparticles is less than 25 nm (ZnO, Si_3N_4, Fe_3O_4, and SiO_2) and 50 nm (ZrC). The bath solution was stirred by ultrasonic vibration. They found that compared with the Cu-plated fabric, the surface roughness of Cu composite plating increases but the holes decrease distinctly. The grain sizes of Cu for the Cu+ ZnO, Cu + Si_3N_4, Cu + Fe_3O_4, Cu + ZrC, and Cu + SiO_2 composite coatings are 19.5, 24.0, 21.9, 26.7, and 28.0 nm, respectively. The shielding effectiveness changes slightly but the abrasion resistance decreases. In addition, the shielding effectiveness increases with an increase in weight, but the surface resistance decreases first, and then reaches a certain level. Figure 4.1 compares the SEM photographs of conventional and five kinds of ECP Cu + nanoparticle films. It can be seen that the conventional electroless copper plated fabric coating surface is relatively smooth, but has more stomata; after adding nanonanoparticles the surfaces of ECP coatings become relatively rough, but their pores are significantly reduced or disappear, and they become denser.

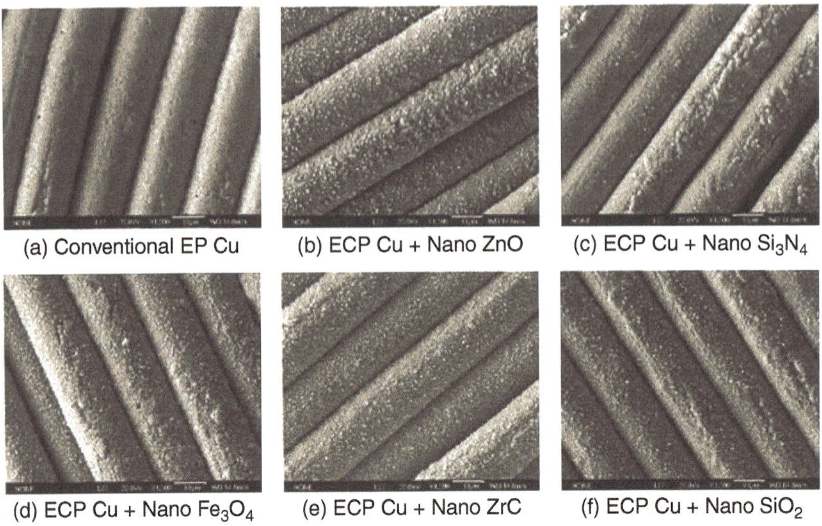

FIGURE 4.1 SEM photographs of the conventional EP Cu and ECP Cu +nano particles films on polyester fabric [13].

It should be noted that only a few papers on metal-based nano ECP materials have been published to date.

4.1.1.2 Binary Alloy-Based Nano ECP Materials

A number of papers have been published which discuss binary alloy-based nano ECP materials, especially Ni–P + nanonanoparticles ECP materials. We will not describe all of the results, but will just give examples of nano ECP materials for every kind of nanoparticle.

The category nano ECP Ni–P, codeposited with nanonanoparticles, includes Ni–P + nano metal, Ni–P + various kinds of nanonanoparticles such as carbide (TiC, Ti(C,N), WC, Cr_3C_2, SiC), oxide (TiO_2, ZrO_2, CeO_2, Cr_2O_3, Al_2O_3, ZnO, SiO_2), and others (TiN, Si_3N_4, MoS_2, PTFE, C, Diamond, CNT), and Ni–P + two kinds of nanonanoparticles (Ni–P + TiO_2 + PTFE, Ni–P + Al_2O_3 + MoS_2, Ni–P + Al_2O_3 + PTFE, Ni–P + C + CaF_2).

Other binary-based nano ECP materials include Ni–B + nanoparticle deposits (Ni–B + ZrO_2, Ni–B + Al_2O_3, Ni–B + Diamond, Ni–B + Diamond like carbon, Ni–B + C_{60}), and Co–P + Al_2O_3 nano ECP deposits.

All of these nano ECP materials will be discussed in the following sections.

4.1.1.2.1 Ni–P + Nano Metal Particle Deposits

In 2008, Ma et al. studied the nanocomposite electroless coating Ni–P + Au [14]. The composition of the electroless plating bath that they used is nickel sulfate (35 g/L), sodium hypophosphite (10 g/L), sodium citrate (10 g/L), and sodium acetate (7 g/L). The pH is 4.0–5.0, and temperature is 85°C. 20# steel was used as the substrate. 2.0×10^{-5} mol/L gold nanonanoparticles were added to the basic electroless solution; however, the authors did not indicate the size of the Au nanonanoparticles. They found that the Ni–P + Au composite coating had 9.85 wt% P and 2.38 wt% Au. The composite coating with gold nanonanoparticles had a larger hardness value than the Ni–P coating. In addition, the composite coatings exhibited better corrosion resistance and uniformity than the particle-free coatings. In 2009, these authors published results of a study on the nanocomposite electroless coating Ni–P + Ag [15]. The basic composition and the parameters of their electroless plating bath are: nickel sulfate (25 g/L), sodium hypophosphite (30 g/L), sodium citrate (30 g/L), sodium acetate (20 g/L), lead (II) acetate (0.002 g/L), thiourea (0.003 g/L), pH 4.0–5.0, and bath temperature 85°C. The composite plating solution is prepared by adding silver nanonanoparticles (1×10^{-7} mol/L) to the basic solution. Figure 4.2 shows that the Ag particles really are nanonanoparticles. The composite coating of Ni–P + Ag contains 11.17 wt% P and 0.24 wt% Ag. The hardness of the composite coating is greater than that of a Ni–P alloy coating.

In 2009, Haijima Yuki et al. reported electroless nickel composite plating film codeposited niobium nanopowder [16]. The electroless Ni–P bath composition and operating conditions are nickel sulfate (0.075 mol/L), sodium hypophosphite (0.10 mol/L), sodium citrate (0.20 mol/L), ammonium sulfate (0.50 mol/L), bath temperature 60°C, pH 8.50 ± 0.05, agitation180 rpm (magnetic stirrer). The amount of Nb powder added is 20 g/L. The size of nanonanoparticles is on average several hundred nm, small ones are the most spherical particles having a particle diameter of tens of nanometers. X-ray photoelectron spectroscopy (XPS) quantitative analysis of Nb 3d on Ni–P composite film shows that Ni, P, O, and Nb are 69.7, 10.2, 8.9, and 11.2 wt%, respectively. Composite plating improves functionalities

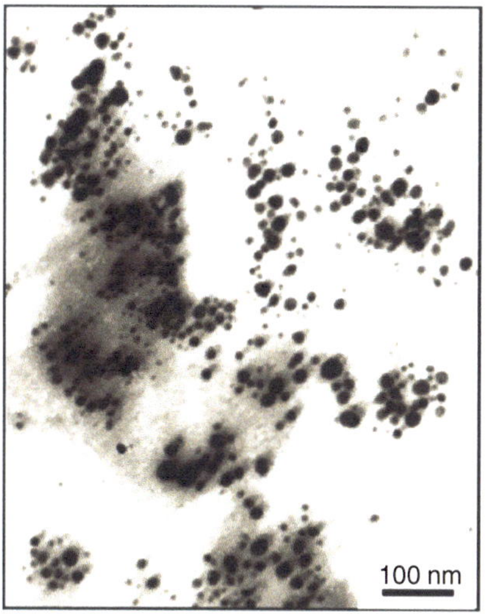

FIGURE 4.2 **Transmission Electron Microscopy (TEM) image for the silver nanoparticles [15].**

of wear resistance, corrosion resistance, lubricity, etc. through codeposition with suitable particles. Using the nanopowder, much of the composite plating film formed a reactive composite plating film alloy during heat treatment.

4.1.1.2.2 Ni–P + Various Kinds of Carbide Nanoparticle Deposits

Several kinds of carbide nanonanoparticles added to Ni–P coatings have been reported, namely TiC, WC, ZrC, Ti(C,N), and SiC.

In 2012, Afroukhteh et al. prepared a Ni–P composite coating codeposited with nanoTiC particles and evaluated its anticorrosion properties [17]. The substrate they used was low carbon steel plates with surface area of 30 mm × 30 mm. A commercial Ni-P electroless plating bath with a high phosphorous content (9–11 wt%) was used. TiC nanonanoparticles with an average particle size of 40 nm were used for composite coating, from 0.3 g/L to 0.1 g/L. The operating parameters were a temperature of 90°C, a stirring rate of 100 rpm and a pH of 4.5. The experimental results show that adding TiC nanonanoparticles to a Ni–P electroless bath not only changes the surface morphology of the Ni–P coating, but also improves the corrosion resistance of the steel substrate in comparison with Ti-free Ni–P electroless coating. Figure 4.3 shows the surface morphology of Ni–P+ nanoTiC coatings particles with (a) 0 g/L, and (b) 0.3 g/L, demonstrating the very different structure of the two coatings. The composition of the nano ECP coating of Ni–P + 0.3 g/L TiC is Ni 84.7 ± 0.2 wt%, P 8.1 ± 0.2 wt%, and TiC 1.7 ± 0.2 wt% respectively. It was also found that the improvement in corrosion resistance largely depends on the phosphorous and TiC concentrations in the coatings.

In 2005, Zhao Fangxia et al. reported a Ni–P + nano Ti(C,N) electroless composite depositing on T10A steel [18]. The optimal ECP process is: nickel sulfate 25 g/L, sodium hypophosphite 29.9 g/L, sodium citrate 50 g/L, lactic

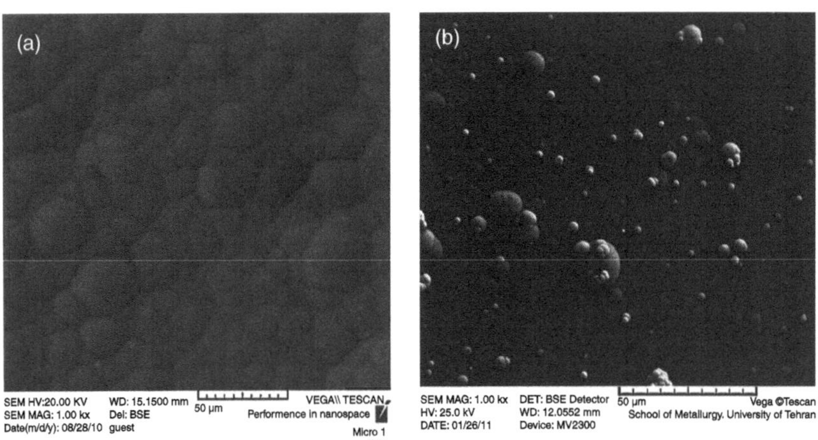

FIGURE 4.3 Surface morphology of Ni–P coatings plated from baths with different concentrations of nanosized TiC particles, (a) 0 g/L, and (b) 0.3 g/L [17].

acid 34 g/L, lead (II) chloride 0.001 g/L, nanoTiC(C,N) 6 g/L, pH 4.0 and a bath temperature of 80°C. The experimental data show that the significance of the factors influencing the nano ECP processing decreased in the following order: pH> nickel-phosphorus ratio > temperature > concentration of nano Ti(C,N). They found that the highest hardness of nano Ti(C,N) ECP coating was 975.2 HV, which is 3.8 times as hard as that of T10A.

In 2007, Xu et al. [19] synthesized a Ni–P + WC nanometer diamond composite coating. The main composition of the solution and some experimental conditions for the electroless plating bath are as follows: nickel sulfate 28 g/L, sodium hypophosphite 25 g/L, lactic acid 10 mL/L, pH 4.5, temperature 90 ± 1°C, and WC nanonanoparticles (almost spherical in shape, with a diameter in the range of 30–80 nm) 10 g/L. A mechanical stirrer (200 rpm) was used during the plating. It has been found that Ni–P + WC nanocoatings, when compared to the conventional Ni–P coating, exhibited a low friction coefficient, high hardness, and heat treatment of these coatings at 400°C for 1 h could further increase the microhardness and wear resistance. This improvement was attributed to the existence of WC nanophase grains and the crystallization of a hard Ni_3P phase. The abrasive wear was superior for the Ni–P + WC nanocoatings, while the combined adhesive and fatigue mechanism was better for the conventional Ni–P coating. Figure 4.4 shows the surface morphology of Ni–P + nano WC [19].

In 2011 Zhang Hui, and Feng Yiwei reported electroless Ni–P + nano ZrC composite plating on PET fabrics [20]. The bath solution they used was nickel sulfate 30 g/L, sodium hypophosphite 30 g/L, sodium citrate 20 g/L, and ammonium chloride 25 g/L. The pH of the bath is 9, and the temperature was 60°C. The average size of ZrC nanonanoparticles is less than 50 nm with purity of

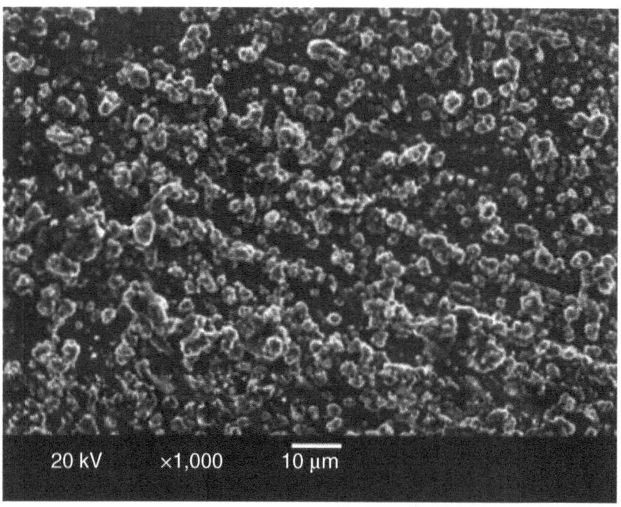

FIGURE 4.4 **Surface morphology of Ni–P+nano WC [19].**

over 99.0%. A supersonic stirrer was used during the plating. It was found that nanometer-sized crystals of Ni–P were formed in the coating. The decomposition temperatures of electroless Ni–P and Ni–P + ZrC composite coating on PET fabrics were lower than that of PET fabric. The shielding effectiveness gradually decreased with increasing levels of nano ZrC particles. Compared with the Ni–P plated fabric, the abrasion resistance of Ni–P + ZrC coated fabric was increased.

Several papers have been published on ECP Ni–P + SiC coatings in recent years [21–26], although this text will mention only one. In 2011, Allahkaram et al. [24] reported the characterization and corrosion behavior of electroless Ni–P + nanoSiC coating on the X70 steel substrate without any surfactant. A commercial electroless nickel bath (SLOTONIP 70A from Schlotter) with sodium hypophosphite as reducing agent was used to obtain the coatings. This bath provided Ni–P deposits with a high phosphorous content, 9–10% P. Temperature was set at 90 ± 2°C and pH was fixed at 4.6 ± 0.1. 4 g/L of nanosized SiC particles (average size 40 nm) were added to the plating bath by ultrasonic dispersion for 15 min, followed by deposition process for 2 h under magnetic agitation condition at a speed of 300 rpm. The XRD pattern of Ni–P + nano-SiC coating was very similar to that of a Ni–P coating. Energy dispersive spectroscopy (EDS) results demonstrated the presence of SiC particles in the coating, and SEM images confirmed the presence of SiC nanonanoparticles with almost uniform distribution in the coating, as seen in Fig. 4.5. The corrosion current density was less in the Ni–P and Ni–P + nano-SiC coated samples than uncoated X70 steel. The Ni–P + nano-SiC coated sample had the most corrosion resistance because there was less effective metallic area available for corrosive media to react with. The overall protection mechanism of Ni–P and Ni–P + nano-SiC coatings was achieved by the formation of a layer of adsorbed hypophosphite anions ($H_2PO_2^-$).

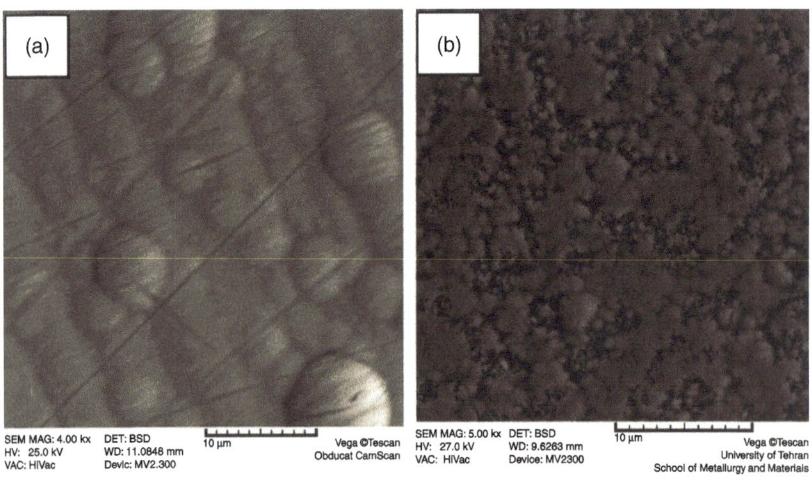

FIGURE 4.5 **Surface morphology of (a) Ni–P and (b) Ni–P+nano SiC coatings [24].**

4.1.1.2.3 Ni–P + Various Kinds of Oxide Nanoparticle Deposits

Several kinds of oxide nanonanoparticles introduced in Ni–P coating have been reported up today, namely TiO_2, ZrO_2, Cr_2O_3, CeO_2, Al_2O_3, ZnO, and SiO_2.

Several reports of Ni–P + nano TiO_2 particles coatings published in recent years [27–30]. This chapter takes only one paper as an example. In 2008, Abdel Aal et al. prepared Ni–P + TiO_2 nanocomposite coatings with various contents of TiO_2 nanoparticulates [27]. The plating bath consists of nickel sulfate 35 g/L as a source of nickel, sodium hypophosphite 15 g/L as reducing agent, and sodium citrate 15 g/L, sodium acetate 5 g/L as a complexing agent. The bath operated at pH 4.7–4.9 and a temperature of 90 ± 2°C. The TiO_2 particles used were 25–40 nm in size, and were obtained from Degussa Company. These nanonanoparticles were added to the solution at concentrations of 1 to 5 g/L. Magnetic stirring was applied to mix the TiO_2 particles and the electrolyte for 8 h for complete dispersion of powder into the plating solution. Subsequently, the suspension was stirred by ultrasonic agitation for 30 min just before plating. It has been indicated that the incorporation of TiO_2 nanoparticulates was found to be improved by the addition of TiO_2 powder to the plating bath. However, it has no significant effect on the P content in the deposit. SEM images showed that the morphology of Ni–P + TiO_2 nanocomposite coating is finer and smoother than that of Ni–P coating. The catalytic activity of the prepared electrodes toward electro-oxidation of small organic molecules was found to vary with the amount of TiO_2 embedded into the Ni deposit.

In 2011 Yang et al. studied fabrication and characterization of electroless Ni–P + ZrO_2 nanocomposite coatings [31]. In order to synthesize EP Ni–P + ZrO_2 nanocomposite coatings with highly dispersed TiO_2 particles, these author used a novel technique to produce nanoparticle oxide. First, 11.3 mL zirconium (IV) n-propoxide (70 wt% solution in 1-propanol) was dissolved in a mixture containing 30.9 mL anhydrous ethanol and 2.8 mL diethanolamine (DEA). This solution was then stirred continuously to achieve a complete chelation between $Zr(OPr)_4$ and DEA. Second, a mixture of 0.46 mL deionized water and 4.5 mL ethanol solution was mixed with the solution under vigorous stirring to form a fresh sol. The bath composition and plating parameters are: nickel sulfate 15 g/L, sodium hypophosphite 14 g/L, sodium acetate 13 g/L, hydrogen fluoride (40%), ammonium bifluoride 8 g/L, thiourea 1 mg/L, ZrO_2 sol 12 mL/L, temperature 80 ± 2°C, and pH value 6.4 ± 0.2. The solution was kept homogeneous by magnetic stirring at a speed of 200 rpm. The data showed that an as-plated nanocomposite coating exhibits much increased microhardness up to 1045 HV200 and remarkably improved wear resistance. X-ray and electron diffraction patterns show a phase transformation in the Ni matrix of the coating from amorphous to nanocrystalline when ZrO_2 sol is introduced.

Eslami Rad and Dehghanian in 2010 reported EP Ni–P composite coatings with embedded Cr_2O_3 nanonanoparticles with average size 100 nm [32]. They used a commercial electroless nickel bath (SLOTONIP 70A from Schlotter) to obtain the coating, which provides Ni–P deposit with high phosphorous content,

9–10 wt%. The amount of Cr_2O_3 added was 5 g/L. In addition, a nonionic polymeric surfactant of 15 g/L was added to the bath. Plating conditions were pH 4–5, and temperature 88 ± 1°C. The nano powders were added to the solution, stirred by magnetic stirrer or ultrasound for 20 and 10 min, respectively. Then this solution was added to the main bath and the substrate was placed in the bath when the solution temperature reached 88°C. The results showed that both Ni–P and Ni–P nanocomposite coatings had significant improvement on corrosion resistance in comparison to the substrate. The codeposition of Cr_2O_3 in the coating improved its corrosion characteristics.

In 2008, Jin Huiming et al. studied the microstructure and corrosion behavior of electroless deposited Ni–P + nano CeO_2 coating [33]. The electroless bath Ni–P composition is 32 g/L nickel sulfate, 25 g/L sodium hypophosphate, 12 g/L sodium citrate, 5 mL/L phosphoric acid, and 0.1 g/L sodium laurel sulfate. The temperature is set at 85°C and the pH is kept at 4.5 using ammonia solution. Nanometric CeO_2 powders of 20 nm diameter are added to the bath at a concentration of 20 g/L and ultrasonically suspended for 20 min. The results show that the Ni–P coating has a partially amorphous structure mixed with nanocrystals, while the Ni–P + CeO_2 coating has a perfectly amorphous structure. Figure 4.6 shows TEM bright field (TEM BF) images of (a) Ni–P and (b) Ni–P + CeO_2 coatings.

Ni–P + nano CeO_2 coating's pure amorphous structure was the result of Ni's hindered crystal-typed deposition and P's promoted deposition. The anticorrosion properties were better in the nano CeO_2-containing coating, due to its less liability to undergo local-cell corrosion than its CeO_2-free counterpart.

Several groups have reported their results on EP Ni–P + nano Al_2O_3 coatings in recent years [34–39]. In 2006, Balaraju et al. studied the microstructure, hardness, and corrosion resistance of electroless Ni–P + nano Al_2O_3 composite coatings [35]. The size of the alumina powders was 50 nm. The electroless bath contains nickel sulfate 21 g/L, sodium hypophosphite 24 g/L, lactic acid

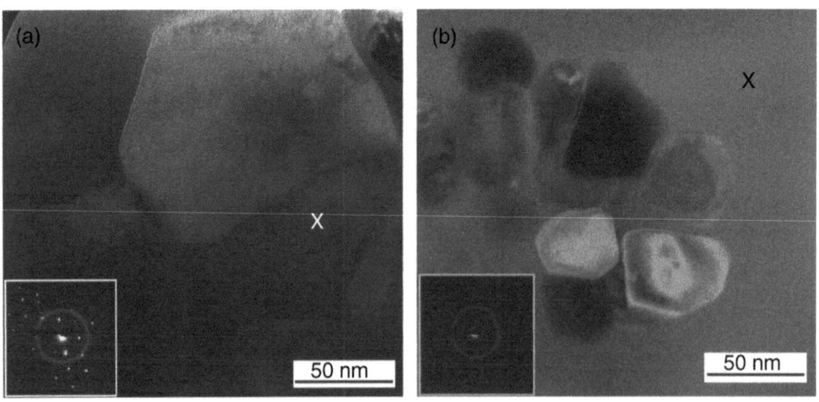

FIGURE 4.6 TEM BF images of Ni–P (a) and Ni–P/CeO_2 (b) coatings, the selected area diffraction is tested at point "X" in the image [33].

25 g/L, propionic acid 3 g/L, and small amounts of lead nitrate and operated at pH 4.7–4.9 and a temperature of 90 ± 2°C. About 50 mL of electroless nickel solution containing 6 g/L of alumina powder was thoroughly mixed using mortar and pestle. A magnetic stirrer was used to get uniform suspension of particles in the solution. About 400 mL of the solution containing alumina particles was taken in a 500 mL glass beaker and stirred for more than 12 h on a magnetic stirrer using a PTFE coated magnetic paddle to get better wettability of the particles. The data showed that a marginal improvement in hardness was noticed in as-plated composite coatings. A 15% increase in micro hardness was observed in the heat treated (400°C for 1 h) composite coatings. Potentiodynamic polarization measurements made on the deposits in 3.5% sodium chloride solution showed that uniform corrosion occurred in the composite coatings.

In 2006, Shibli et al. reported incorporation of nano zinc oxide for improvement of electroless nickel plating [40]. The substrate used was mild steel. The average size of the ZnO particles was less than 25 nm. The alkaline bath included the composition of nickel sulfate 30 g/L, sodium hypophosphite 10 g/L, sodium citrate 84 g/L, and ammonium chloride 50 g/L. The plating was operated at pH 8–10 and temperature 90 ± 8°C. The electrochemical and metallurgical properties of the Ni–P + nano ZnO plates were evaluated and compared with that of the unreinforced Ni–P plates. It was found that the incorporation of the nano ZnO particles resulted in improvement of metallurgical and corrosion resistance characteristics of the plates.

Some papers on ECP Ni–P + nano SiO_2 particles films have published in very recent years [41–43]. In 2010, de Hazan et al. reported homogeneous electroless Ni–P + SiO_2 nanocomposite coatings with improved wear resistance and modified wear behavior [42]. Disks with a diameter of 40 mm made of 100Cr6 steel were used as substrates. The SiO_2 nanoparticle was used with a mean particle size of 40–50 nm. The initial bath contained 6 g/L nickel and 35 g/L sodium hypophosphite. The pH of the mixed bath was 4.7 ± 0.1. The mixed composite bath was heated to plating temperatures of 86–87°C. The bath was mixed by a magnetic stirrer (100–200 rpm). Tribological measurements of the heat-treated coatings indicated that while the hardness of the coating was decreased compared to that of the Ni–P coating without particles, an improvement in Pin-on-Disk wear resistance up to a factor of 2 could be achieved. The aspect ratio of the wear track of samples heat treated above 270°C was found to increase with the concentration of nanonanoparticles in the coatings. This observation indicates that the nanonanoparticles also modify the wear behavior of the coatings.

In 2005, Agarwala [44] reported EP nanocomposite coatings Ni–P + $BaZn_{2-y}Co_yFe_{16}O_{27}$ (y = 0.0, 0.4, 0.8, 1.2, 1.6, and 2.0). The EP composite coatings were obtained using a bath having the metal salt (nickel sulfate 30 g/L), the reducing agent (sodium hypophosphite 20 g/L), the complexing agent (sodium acetate 10 g/L), the stabilizer (thiourea few particles (if needed)), pH controlling reagent (ammonia solution (a few drops)) for maintaining the pH at 9.0 ± 0.25 at temperature 90 ± 2°C. The EP bath was kept at a constant temperature and periodically stirred so as to keep the EP bath composition and temperature uniform. SEM micrographs show the presence of agglomerated particles prior to microwave processing, and particles in the range of 20–60 nm could be seen after microwave treatment. It is evident from the SEM micrograph of Ni–P + ferrite coating that globule size of Ni–P is reduced when ferrite particles are trapped into Ni–P matrix. The present Ni–P composite coating is found to have the composition of Ni 95.165 wt% with P 3.335 wt% composition.

4.1.1.2.4 Ni–P + Various Kinds of Other Nanoparticle Deposits

Besides the carbide and oxide nanonanoparticles mentioned already, other kinds of nanonanoparticles are also incorporated in the binary alloy-based nano ECP coatings. They are TiN, Si_3N_4, MoS_2, PTFE, C, diamond, and CNT.

A few reports have been published for ECP Ni–P + nano TiN particles coatings [45–47]. Mafi and Dehghanian in 2011 reported studying the effects of the addition of TiN nanonanoparticles to Ni–P electroless coatings [46]. A low carbon steel disk with a diameter of 20 mm and thickness of 5 mm was used as a substrate. The average particle size of TiN nanopowder is approximately 20 nm. A nickel bath (SLOTONIP 70A from Schlotter) was used to obtain the coating which provided Ni–P deposit with high phosphorous content, 9–11 wt%. The TiN nanonanoparticles and the surfactant (polyethylene glycol) were added to a separate portion of the bath and stirred by magnetic stirrer and ultrasonic mixer for 20 and 10 min, respectively. Then, the mixture was added to the main bath and the specimen was placed in it when the solution temperature reached 88°C. 1 g/L TiN was added to the bath. The ECP was operated under conditions with a pH of 4.5 and a temperature of 88°C and a stirring rate of 100 rpm. The results show that the incorporation of TiN particles into the coating has an adverse effect on the corrosion properties of the specimens. The hardness of the specimens increases dramatically by adding TiN. Furthermore, the hardness of the specimens increases after conducting the heat treatment. The corrosion and hardness behavior of the Ni–P system after heat treatment largely depended on the temperature of heat treatment.

In 2011, two Chinese scholars studied ECP Ni–P + nano Si_3N_4 coating on polyamide fabric [48]. The average size of Si_3N_4 particles is less than 20 nm. The composition of the acidic bath and operated condition are nickel sulfate 30 g/L, sodium hypophosphite 25 g/L, sodium citrate 5.5 g/L, and sodium acetate 50 g/L, pH 8–10 and temperature 80°C. The experimental results showed that with the incorporation of Si_3N_4 nanonanoparticles the composite coating prepared in acidic bath was in an amorphous state. After deposition by acidic Ni–P + nano Si_3N_4 composite coating, the decomposition temperature of polyamide fabric was lower. With the increase of Si_3N_4 nanonanoparticles content in the coating, the shielding effect of fabrics plated in acidic bath was almost unchanged. Compared with traditional Ni–P electroless coating, the wear resistance

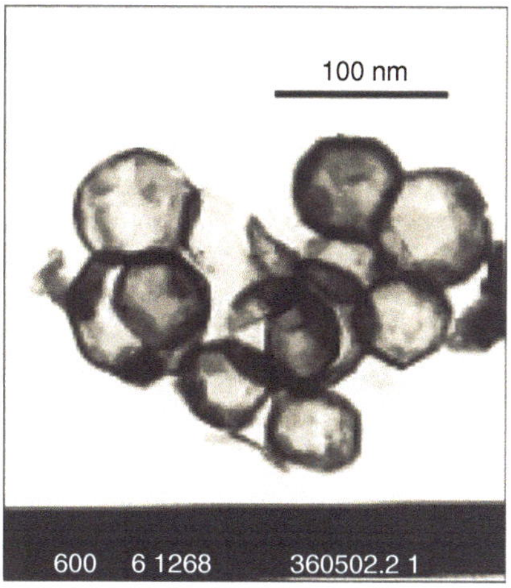

FIGURE 4.7 TEM image of nano-MoS_2 [51].

of fabrics having Ni–P + nano Si_3N_4 electroless composite coatings prepared in acidic bath was reduced. In 2006, Sarret et al. reported ECP Ni–P + nanoSi_3N_4 coating as well [49].

There have been two reports on ECP Ni–P + nano Mo_2S coating [50, 51]. In 2009, Hu et al. studied the corrosion and friction reduction of electroless Ni–P coating with nano molybdenum disulfide nanonanoparticles [51]. MoS_2 nanonanoparticles with an average diameter of 80–100 nm were used in the plating. Figure 4.7 shows the TEM image of nanoMoS_2. A medium carbon steel plate (C: 0.45 wt%, 10 mm diameter, 20 mm thickness) and a cylinder (C: 0.45 wt%, 40 mm diameter, 30 mm thickness) were used as substrates. The basal EP Ni–P bath had the following compositions: 20 g/L nickel sulfate, 24 g/L sodium hypophosphate, 33 mL/L lactic acid and 2.2 mL/L propyl acid, NaOH 12 g/L, proper dispersant and additive, and nano MoS_2 nanonanoparticles 1–4 g/L. The bath temperature was maintained at 90 ± 2°C and the pH of the bath was controlled at 4.3–4.6 by NaOH. MoS_2 nanonanoparticles were well dispersed in the bath solution by ultrasound for 1 h. During the plating, the bath solution was agitated using a magnetic stirrer. The results indicated that the anticorrosion ability of the composite coating was decreased because of the addition of nano-MoS_2 particles. In addition, the friction coefficient of the Ni–P + nano MoS_2 composite coating decreased greatly compared with those of Ni–P electroless coatings.

One paper reported an ECP Ni–P + nano PTFE coating in 2008 [52]. PTFE particles were in the range of 50–200 nm in size. The constituents of the electroless plating bath and plating conditions were 28 g/L nickel sulfate, 30 g/L sodium hypophosphate, 20 g/L sodium citrate, 20 g/L sodium acetate, 10 g/L succinic acid, 5 g/L lactic acid, 20 g/L potassium iodate, 6 g/L PTFE, pH 5, and a temperature of 85°C. The results showed that lubricating PTFE particles have been uniformly incorporated into the Ni–P + PTFE coating. Measuring the friction behavior and wear resistance of the surface modification shafts in rolling contact with bushing showed that the Ni–P + nano PTFE coating deposited on the shaft plays an excellent role in reducing the wear of the bushing.

A synthesis, characterization, and infrared-emissivity study of Ni–P + CB (carbon black) nanocomposite coatings by electroless process was published in 2010 [53]. Composite electroless Ni–P + CB coatings were deposited on an ABS (22% acrylonitrile, 15% butadiene, 63% styrene) substrate. The composition and deposition parameters for the Ni–P + CB plating solutions were 30 g/L nickel sulfate, 26 g/L sodium hypophosphate, 25 g/L sodium citrate, 25 g/L ammonium chloride, 10 g/L triethanolamine, 120 mg/L surfactant, 1–8 g/L CB (18 nm), temperature at 50°C, and pH 8.5–9.5. The results show that the nano CB particles are distributed uniformly in the Ni–P coatings. When the CB concentration is lower than 4 g/L, there is a sharp increase in the CB content of the coating with increasing CB concentration in the solution, whereas above 4 g/L the CB content levels off. The composite possesses lower infrared-emissivity value than those of pure ABS plastic, CB, and a Ni–P coating.

Many papers on ECP Ni–P + nano diamond (ND) coatings have been published [54–61]. In 2003, Matsubara et al. studied the codeposition of cluster diamond into electrolessly plated nickel films with 5 nm in size of ND [54]. They found that the ND content in the deposit increased with an increase of the ND concentration in plating bath, but decreased as the plating bath temperature increased. In 2004, Matsubara et al. reported the codeposition characteristics of nano diamond particles in electrolessly plated nickel films [55]. The size of the ND is also 5 nm. The authors found that the content of ND particles in the film increased with decreasing deposition rate and decreasing hydrogen evolution rate accompanied by plating reaction. The content did not exceed 2% when using an ordinary plating solution that was as used for the ammoniacal-citrate nickel plating bath. In 2005, Xu Hui et al reported the tribological properties, hardness, and corrosion behaviors of the ECP Ni–P + ND coatings on medium carbon steels [56]. The diamond powder is close to spherical with a diameter in the range of 3–8 nm. Figure 4.8 shows the Atomic Force Microscope (AFM) images of Ni –P (a) and Ni–P + ND (b) electroless composite coating. In 2006 Matsubara et al. reported the codeposition behavior of ND with electrolessly plated nickel films [57]. The size of the ND particles is 5 nm. These authors found that the ND content in deposit increases or decrease with increase of the ND concentration in plating bath with or without ammonium of sulfate, respectively. In 2007 Matsubara et al. studied the codeposition mechanism of ND with electrolessly plated nickel films [58]. They successfully fabricated composite films with 0–14% ND, semibrilliant from the bath which contained only citrate as a complexing agent. They believe that the adsorbed Ni-complex precipitates the incorporation of nanodiamond particles into the nickel matrix. In addition, incorporation becomes more probable when the adsorbed amount of Ni-complex on the ND particles was large. In 2010 another Chinese group also reported Ni–P-nanodiamond composite electroless

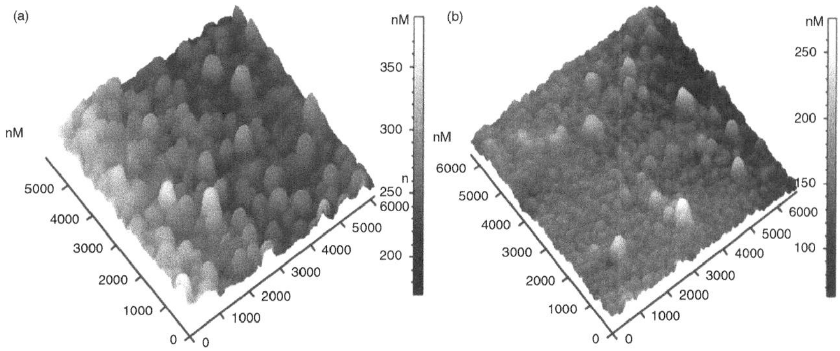

FIGURE 4.8 AFM images of Ni –P (a) and Ni–P+ND (b) electroless composite coating [56].

plating [60]. Their results show that the thickness of the composite coating decreases with the rotational speed of the stirrer, while the micro hardness of the coating and the ND content of the coating increase with increasing ND levels in the bath solution. The wear resistance of the composite coating deposited in a bath with 6 g/L ND increases by 50% in contrast with the pure Ni–P coating. In 2012 a group of Iranian scholars reported deposition, characterization, and electrochemical evaluation of Ni–P + nano diamond composite coatings [61]. The Ni–P + ND composite depositions were coated on API-5L X65 steel substrates (30 mm × 25 mm × 15 mm). Ni–P + ND was deposited by Ni–P electroless bath containing various concentrations of 20 nm average size ND particles as reinforcement. A commercial electroless nickel bath (SLOTONIP 70 A from Schlotter) with sodium hypophosphite as reducing agent was used to produce the coatings. The temperature of electroless bath varied between 88 and 93°C and the pH was kept constant at 4.6 during the process. The incorporation of diamond nanometer particles significantly increased the hardness of depositions as compared to the Ni–P as-deposited matrix. Corrosion behavior, consisting of corrosion tendency and resistance of Ni–P + ND depositions, improves by embedding nanometer particles within the Ni–P matrix due to decreasing of porosity density by accommodation of nanonanoparticles to them.

In 2000, a Chinese Group published a paper on Ni–P + carbon nanotube (CNT) nanocomposite depositions [62]. CNTs used there were synthesized by the catalytic decomposition of acetylene over mesoporous silica containing cobalt nanonanoparticles, which were prepared by a sol–gel process. In order to improve the dispersion of CNTs in the electroless bath, the purified CNTs were milled for 10 h with a planetary ball mill machine in an ether liquid at a rotating speed of 900 rpm. The CNTs have central hollow tubes with diameters of 20–40 nm, which is a typical multiwalled nanotube structure. The diameters of CNT after ball milling are still less than 100 nm. Bath compositions and conditions for electroless plating are nickel sulfate 20–25 g/L, sodium hypophosphite 20–25 g/L, sodium acetate 10–15 g/L, acetate acid 5–10 mL/L, surfactant (cetyl-trimethylamine bromide) 200–400 mg/L, ball-milled CNTs 2.5 g/L, pH 4.5–5.1, and temperature

80–85°C. The resultant Ni–P + CNT composite coatings exhibited not only high wear resistance but also low friction coefficients compared with the Ni–P + SiC and Ni–P + graphite composite coatings. The favorable effects of CNTs on the tribological properties of the electroless coatings are attributed to improved mechanical properties and unique topological structure of the hollow nanotubes. In addition to preventing the rough contact between the two mating metal surfaces, the shorter CNTs can easily slide or roll within the surfaces of friction pairs. Therefore, the wear rate and friction coefficient of Ni–P + CNT composite coatings can be substantially reduced. A paper in 2005 [63] addressed the synthesis and characterization of Ni–P + CNTs nanocomposite film for microelectromechanical systems (MEMS) applications. The incorporation of multiwalled carbon nanotubes (average diameter: 10–20 nm) into the electroless Ni matrix can greatly enhance its mechanical properties. When the concentration of CNTs embedded in the nickel matrix is over a critical point, the Young's modulus and hardness of the composite film can be increased up to 665.9 and 28.9 GPa, respectively, which are four times larger than that of the pure Ni–P film. Furthermore, the E/ρ ratio of the cantilever beam with 28.2 vol% CNTs concentration is approximately 3.9 times that of the pure Ni one. Two groups from Iran published two papers in 2012 [64, 65]. One paper reported the effect of CNTs on the corrosion and tribological behavior of electroless Ni–P + CNT composite coating [64]. Pristine chemical vapor deposition-grown multiwalled carbon nanotubes (MWCNT), 10–30 μm in length and 10–20 nm in diameter, were used in the plating. The results indicated that the incorporation of CNTs in the coating improved both tribological behavior and corrosion resistance. These improvements have been attributed to the superior mechanical properties, unique topological structure, and high chemical stability of nanotubes. Another paper [65] studied the effect of heat treatment on the properties of electroless Ni–P + CNTs composite coatings. The results show that as-plated Ni–P + CNT composite coating has a nanocrystalline structure. Proper heat treatment (400°C for 1 h) significantly improves the coating density and its structure, giving rise to an enhanced corrosion resistance.

4.1.1.2.5 Ni–P + Two Kinds of Nanoparticle Deposits

There are only a few reports published on ECP Ni–P + two kinds of nanoparticle deposits, and these are Ni–P + SiO + CeO composite coatings [66], Ni–P + Al_2O_3 + PTFE nanocomposite films [67], Ni–P + TiO_2 + PTFE nanocomposite coatings [68], and Ni–P + PTFE + biocide polymer nanocomposite coatings [69]. Here only the Ni–P + TiO_2 + PTFE nanocomposite coatings [68] will be mentioned. In 2011, Liu and Zhao studied the influence of surface-energy components of Ni–P + TiO_2 + PTFE nanocomposite coatings on bacterial adhesion. A stainless-steel 316 L plate (25 mm × 26 mm × 1 mm) was used as the substrate. Bath composition and operating conditions were nickel sulfate 15–35 g/L, sodium hypophosphate 20–40 g/L, sodium citrate 10–30 g/L, sodium acetate 15 g/L, lactic acid 0–25 mL/L, succinic acid tittle, thiourea 0–3 ppm, surfactant 0.2–0.4 g/L, PTFE (60 wt%) 3–9 mL/L, TiO_2 0.1–1.5 g/L, pH 4.5–6.0, temperature 80–95°C, and stirring 60 rpm. The TiO_2 powder was obtained from Aldrich, with a particle size of 25 nm, and the 60 wt% PTFE emulsion was obtained from Aldrich, with a particle size in the range of 50–500 nm. The experimental results demonstrated that there was a strong correlation between bacterial attachment (or removal), and the CQ ratio, which is defined as the ratio of Lifshitz van der Waals (LW) apolar to electron donor surface-energy components of substrates, was strongly correlated with bacterial adhesion. The coatings with the lowest CQ ratio had the lowest bacterial adhesion or the highest bacterial removal.

4.1.1.2.6 Ni–P + Three Kinds of Nanoparticle Deposits

Only one paper to date has reported a Ni–P + three kinds of nanoparticle deposit, this being by Sharma et al. in 2005 [70]. The main constituents of the bath used for the synthesis of electroless composite coating, Ni–P + ZrO_2 + Al_2O_3 + Al_3Zr include nickel sulfate, sodium hypophosphite, trisodium citrate, ammonium chloride, ammonium sulfate, aluminum chloride, and zirconium oxychloride. Samples were coated at 90°C ± 2°C at a pH of 9.0 ± 0.25 to obtain constant phosphorus content in the coatings. The pH of the bath solution was adjusted by adding ammonia solution periodically as and when required. The composition of electroless Ni–P–X coated sample obtained by area scan under scanning electron microscopy (SEM)–energy-dispersive X-ray analysis (EDXA) is (wt%) Ni 90.7, P 8.9, Zr 0.21, and Al 0.16. The bright field TEM micrograph reveals the distribution of very fine particles in the matrix, which have nano-sized islands or particles of the order 5–8 nm.

4.1.1.3 Other Binary Alloy-Based Nano ECP Materials

Several papers on other than Ni–P binary alloy based nano ECP materials have been published, including Ni–B + ND particles [71, 72], Ni–B + nano diamond-like (NDL) particles [73], and Ni–B + C_{60} nanonanoparticles [74].

In the paper published in 2008 concerning Ni–B + ND particles [71], the composition of the bath and the experimental parameters used for electroless Ni–B coatings are main solution (nickel chloride 24 g/L, ethlenediamine 60 mL/L), reducing solution (potassium hydroxide 26.5 g/L, sodium borohydride 120 g/L, sodium hydroxide 250 g/L), stabilizing solution (sodium hydroxide 13 g/L, lead tungstate 2.6 g/L, ethlenediamine 40 mL/L, EDTA 13 g/L), temperature 88 ± 1°C and pH 13.5. The main solution was heated to about 88–90°C with agitation using magnetic stirrer. 2.6 mL reducing solution and 2.6 mL stabilizing solution were added into the heated bath every 30 min during the coating procedure. Then cleaned substrates (10 × 10 × 60 mm plain carbon steel) were placed in the bath. The average diameter of the diamond particles was 5 nm. Two different surfactants and different compositions were used to suspend these particles into the Ni–B coating baths. One is (Rhodafac Re-610 nonyl phenoxy (polyethyleneoxy) ethanol, Rhone-Poulenc), another is nonionic Surfactant-2, which is not indicated in the paper. The results show that 5–6 wt% B of the ECP nanocoatings was observed. Heat treatment (1.5 h at 385°C) increased the hardness of the Ni–B coatings at 1175 HV levels. The best resistance to wear was obtained by using nano diamond particle suspension with surfactant 2. Reinforcing Ni–B coatings with nano size diamond using electroless method increased the microhardness and wear resistance properties of coated steel substrates.

In a paper studying alleviating fretting damage properties of Ni–B + C_{60} as a nanocomposite coating on titanium alloy [74], the composition and deposition parameters for the bath were nickel chloride 20 g/L, ethylenediamine 100 g/L, sodium borohydride 1 g/L, sodium hydroxide, 40 g/L, nano C_{60} particles 8 g/L, stabilizer and surfactant small amount, temperature 85°C, pH 14, and a magnetic force agitator stirs the plating solution. From the experimental data, it can be seen that the hardness of the Ni–B + C_{60} nanocomposite coating is lower than that of Ni–B coating, but the roughness value of the Ni–B + C_{60} nanocomposite coating surface is higher than that of Ni–B coating. However, the self-lubricating performance and the fretting wear performance of the Ni–B + C_{60} nanocomposite coating are superior to that of the Ni–B coating. In addition, the Ni–B + C_{60} nanocomposite coating can bear more cycle period and has lower coefficient of friction in its running-in stage compared with Ni–B coating.

4.1.1.4 Ternary Alloy-Based Nano ECP Materials

A number of papers on ternary alloy-based nano ECP materials have been published [75–81, 29], and most of these from 2008 to 2012 [75–81]. The coatings in question are: Ni–W–P + MoS_2 [75], Ni–Ce–P + TiO_2, [76], Ni–Cu–P + ZrO_2 [77], Ni–Cu–P + CNT [78], Ni–Zn–P + TiO_2 [79], Ni–Sn–P + $ZnCr_2S_4$ [80], Ni–Sn–P + SiC [81], and Ni–Sn–P + TiO_2 [29]. In three papers [76, 80, 29], the size of nanonanoparticles was not given by the original authors, so these results will not be described further.

In a paper published in 2012 and reporting studies on the preparation and properties of electroless Ni–W–P alloy coatings and its nano MoS_2 composite, Ranganatha and Venkatesha used the following bath composition: nickel sulfate 20.05 g/L, sodium hypophosphate 14.84 g/L, sodium citrate 29.41 g/L, ammonium sulfate 26.42 g/L, and lactic acid 4.23 g/L [75]. In order to minimize the agglomeration of 40 nm-sized MoS_2, the bath solution was subjected to magnetic stirring for a period of 12 h prior to the plating experiments. Electroless plating was performed under constant stirring at 500 rpm for 1 h at a temperature of 80°C while keeping the bath pH at 8. The results show that an increase in codeposition of the alloying metal tungsten changed the structure and morphology of the deposits and enhanced the corrosion resistance. Ni–W–P deposits with increasing amounts of tungsten metal were found to possess higher friction coefficients and greatly improved microhardness. The incorporation of MoS_2 nanonanoparticles into Ni–W–P alloy coatings strongly influenced the reduction in the corrosion and friction.

In the paper "Process and properties of electroless Ni–Cu–P + ZrO_2 nanocomposite coatings" published in 2012, Ranganatha et al. used the following

FIGURE 4.9 SEM image of Ni–P–Cu+ZrO_2 nano composite coating from the bath: BB+ copper sulfate 1 g/L and zirconium sulfate 1 g/L [77].

basic bath solution for Ni–P coating (BB): nickel sulfate 50 g/L, sodium hypophosphate 25 g/L, sodium citrate 60 g/L, and ammonium acetate 40 g/L [77]. The bath compositions used to generate quaternary Ni–Cu–P + ZrO_2 composites were BB with concentrations of copper sulfate 1 g/L and zirconium sulfate 1 g/L, and other concentrations. Electroless plating experiments were performed under constant magnetic stirring at 500 rpm for 1 h at a temperature of 80 ± 5°C and a pH of 7. In order to minimize the agglomeration of 30–35 nm-sized ZrO_2 the bath solution was subjected magnetic stirring for a period of 12 h prior to plating experiments. It has been found that copper changes the surface morphology of the deposits significantly.

Figure 4.9 is the SEM image of Ni–P–Cu+ZrO_2 nanocomposite coating from the bath: BB+ copper sulfate 1 g/L and zirconium sulfate 1 g/L. The Ni–Cu–P + ZrO_2 coatings were more resistant toward corrosion than a plain Ni–P deposit. In addition, the Ni–Cu–P + ZrO_2 specimen was the most noble of all with a seven-fold decrease in the corrosion rate and twice the Ni–P–ZrO_2 in 3.5 wt.% NaCl solution and contrasted significantly with the Ni–P coating.

Wu et al. studied the preparation and characterization of a Ni–Cu–P + CNTs quaternary electroless composite coating in 2008 [78]. The substrate is low carbon steel (15 mm × 40 mm × 2 mm). The bath components and operating conditions for Ni–Cu–P + CNTs electroless coating are as follows: nickel sulfate 32 g/L, copper sulfate 2.0 g/L, sodium hypophosphite 16 g/L, potassium sodium tartrate 4 g/L, sodium citrate 16 g/L, potassium iodide 1 mg/L, surfactants 1 mg/L, CNTs 0.8 g/L, 1.6 g/L, 2.4 g/L, 3.2 g/L, pH value 5–9, and temperature 75°C. The CNTs used in this work were multiwalled carbon nanotubes, 10–30 nm, synthesized by CVD technology. Ball milling of the purified CNTs was carried out using a GN-2 type ball mill operating at 200 rpm for 15 h for better dispersion characteristics and faster processing. Air agitation was employed during the deposition process. The experimental data show that the mass percent content of the CNTs in the composite coating reaches the highest value of 3.68 wt% when the CNTs' concentration is 1.6 g/L. Moreover, the CNT content in the composite coatings increases first and then decreases with increasing concentration of CNTs in the bath. The hardness of as-deposited composite coating increases first, and then decreases with increasing CNTs concentration in the bath and peaks at 1200 HV0.1 when the CNTs concentration was 1.6 g/L. In contrast, after heat treatment at 400°C in a hydrogen protective atmosphere for 1 h, the hardness value increases to about 1500 HV0.1. The corrosion resistance of Ni–Cu–P + CNT composite coating is excellent compared with that of Ni–Cu–P coating.

Ranganatha et al. reported the development of an electroless Ni–Zn–P + nano TiO_2 composite coatings and their properties in 2010 [79]. The bath of Ni–Zn–P ternary alloy deposits is nickel sulfate 27 g/L, zinc sulfate 4 g/L, sodium hypophosphite 32 g/L, sodium citrate 59 g/L, ammonium sulfate 27 g/L, lactic acid 10 g/L. The operating conditions for Ni–Zn–P + TiO_2 electroless coating are pH value 9, temperature 80 ± 5°C, and magnetic stirring at 500 rpm. TiO_2 nanonanoparticles (100–200 nm) of 1.5 g/L were added to the bath. The agglomeration of TiO_2 was minimized by subjecting the electrolyte to magnetic stirring for a period of 12 h prior to plating experiments. The authors found that codeposited particles did not influence the structure or the phase transformation behavior. The presence of metastable and stable phases was identified in the deposits heated to 400°C temperature. The composite coatings exhibited higher microhardness than pure alloy. The annealed Ni–Zn–P + TiO_2 coatings have good corrosion resistance in 3.5 wt% NaCl solution compared to Ni–Zn–P coatings.

In a paper published in 2010 [81], Khalifa et al. studied the corrosion behavior of an electroless Ni–Sn–P–SiC nanocomposite coating. The bath compositions and operational conditions were: nickel sulfate 30 g/L, sodium hypophosphite 125 g/L, sodium acetate 120 g/L, lactic acid 10 mL/L, propionic acid 110 mL/L, surfactant 0.2 g /L, tin (II) chloride 0.05 g/L, SiC nano powders (a mean size of 129.4 nm) 4–10 g/L, pH 5, plating temperature 80–83°C, and magnetic stirring 3 rpm. The data show that the structure of the composite coatings as deposited was amorphous. After heating to 400°C, the final products of the crystallization and reaction of composite coatings were Ni, Ni_3P, Ni_3Si, $Ni_{31}Si_{12}$ and $Ni_{2.97}Sn_2$. The corrosion protection follows the sequence Ni–Sn–P + SiC> Ni–Sn–P > Ni–P > Ni–P + SiC in 3.5% NaCl solution.

4.1.1.5 Summary of Nano ECP Materials

From the above description and discussion, the following conclusions can be arrived at.

Most of the papers on nano ECP materials were published after 2000, with only a very few being reported in the 1990s. That means the field of nano ECP is new to EP.

Nearly all types of conventional ECP alloy systems have been studied as a nano ECP coating even though they have been known for less than 10 years. This indicates that the field is generating much research interest. The reasons for this interest are as discussed at the beginning of this chapter.

Most of the nano ECP bath solutions contain some surfactants because of the presence of intentionally insoluble nanonanoparticles in the bath. The large specific surface area of nanonanoparticles causes an 800 times larger loading of conventional electroless bath [82]. Incorporation of surfactants can overcome this extraordinary loading and wetting problem.

All of the nano ECP bath solutions used a kind of agitation method (air, electric, or magnetic stirrer) to stir the solution during the plating process in order to keep the nanonanoparticles in suspension, and stop them agglomerating.

The various properties of nano ECP coatings are briefly introduced for nearly every example system. The results show that they are indeed superior to those of the conventional counterpart ECP coatings. Since they are similar to the properties of the plain EP alloys, those of the nano ECP alloy coatings will not be described more in detail in this book, for the reasons we presented in the preface.

4.1.2 EP Three-Dimensional Nanostructured Materials (3D NSMs)

EP nanoarchitecture bulk materials also are called EP 3D NSMs. Can EP really prepare 3D MSMs as well as nano ECP materials? Yes, using EP 3D ordered NSMs can be easily fabricated at low temperatures. Why have 3D NSMs recently attracted intensive research interest? The first reason is the aim of synthesizing 3D NSMs with a controlled structure and morphology. The second is that the 3D NSMs have higher surface area and supply enough absorption sites for all involved molecules in a small space [83]. The third is that these materials with porosity in three dimensions could lead to a better transport of molecules [84]. The fourth reason is that EP is a simple and inexpensive fabrication technique, and its use allows 3D NSMs to be fabricated relatively quickly and effectively [85].

4.1.2.1 Nanoflowers

In recent years, some work has been carried out on fabricating metal nanoflowers using electroless deposition. In 2010, Bechelany et al. [86] prepared Au nanoflowers and nanocrowns with single crystalline tips based on electroless deposition and natural lithography. The fabrication process of gold nanoflowers is as follows. A monodisperse suspension of polystyrene (PS) microspheres was applied on the surface of the substrate, and spheres were self-assembled into a close-packed, three-dimensional ordered lattice via attractive capillary forces. Specimens were then placed in hexamethyldisilazane (HMDS) vapor. After removal of the PS (used as a mask), the Si substrate was immersed in an electroless Au plating solution containing 1 mmol/L $NaAuCl_4$, 0.15 mol/L HF, and 12% saccharin at room temperature. The obtained Au flowers, with long-range hexagonal ordering, have a diameter of 150 nm and are composed of nanoneedles of Au with a diameter of less than 10 nm and a length of about 50 nm (Fig. 4.10). The authors found that surface enhancement Raman spectroscopy (SERS) study of gold nanoflowers showed massive and reproducible enhancements of the Raman signal.

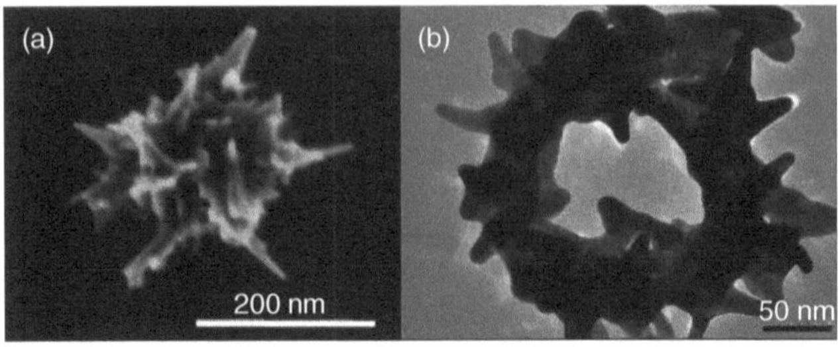

FIGURE 4.10 **SEM image of an individual nanoflower structure composed of nanoneedles of Au with diameters less than 10 nm and a length of 50 nm (a), TEM BF image of an Au nanocrown (b) [86].**

In 2011, Ding et al. [87] also prepared flowerlike and snowflake-like gold hierarchical nanostructures at carbon coated copper grid/Langmuir monolayer/$HAuCl_4$ aqueous solution interfaces under different Langmuir monolayers through an electroless deposition process. Figure 4.11 shows the TEM micrographs of gold nanoflowers formed under THTPyPzCo monolayers under different conditions.

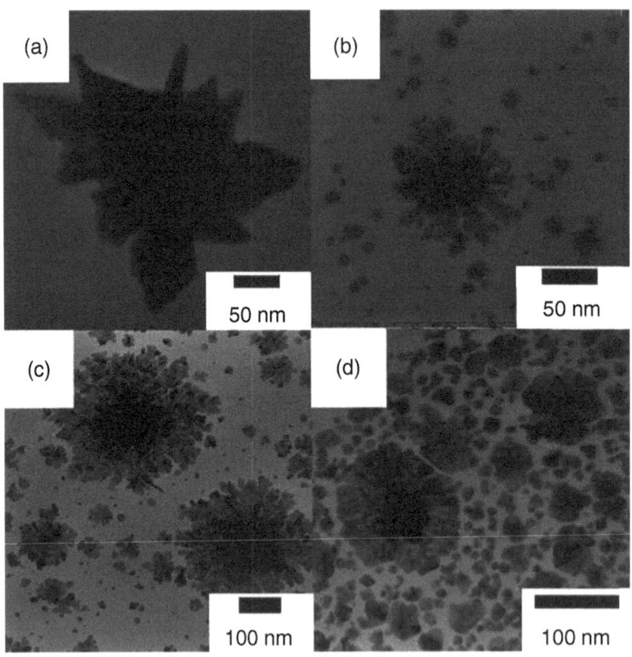

FIGURE 4.11 **TEM micrographs of gold nanoflowers formed under THTPyPzCo monolayers under different conditions. (1) The subphase concentration is 1 × 10^{-4} mol/L, the reaction time is 10 min, and the temperature is 45°C (a). (2) The subphase concentration is 1 × 10^{-4} mol/L, the temperature is 35°C, and the reaction times are 2 (b), 10 (c), and 30 min (d) [87].**

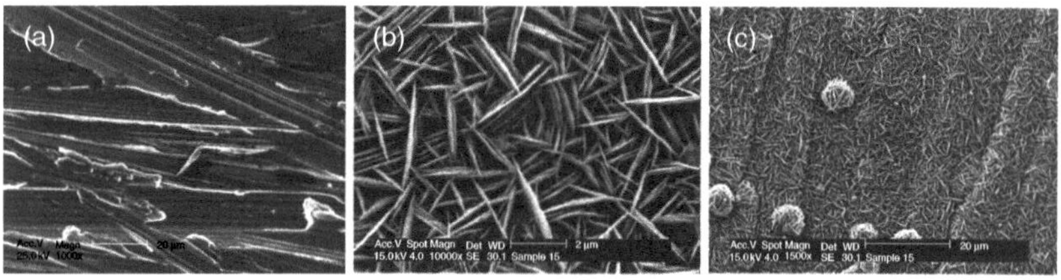

FIGURE 4.12 **SEM images of Cu foils rubbed with the 180-C sandpaper, and silver nanostructures from the silver mirror reaction on the corresponding Cu foils, see the text for more details [88].**

In 2005, Qu and Dai fabricated [88] silver nanostructures through a silver mirror reaction on reactive substrates. Silver nanoclusters with various morphologies ranging from the leaflike to flowerlike hierarchical structures can be produced from the silver mirror reaction on commercially available copper foils after being treated with a dilute aqueous HCl solution under different conditions. The silver mirror solution the authors used contains a freshly prepared silver complex of $Ag(NH_3)_2^+$ (6.375×10^{-3} mol/L), glucose (3.405×10^{-2} mol/L), and tartaric acid (4.005×10^{-3} mol/L). In a typical experiment, silver microstructures were produced by immersing a Cu foil into the silver mirror solution for a certain time at room temperature. They found that on the surface of copper foil rubbed by a rough sandpaper (GatorGrit 180-C, Fig. 4.12a), densely packed silver nanostructures in somewhat scattered spherical microclusters formed. Close inspections of the surface and spherical microcluster show similar "building blocks" of perpendicularly aligned silver nanosheets in both areas (Fig. 4.12b). Then, the randomly interconnected silver nanosheets within the spherical microcluster stand out from the surface like a silver "flower" (Fig. 4.12c). These nanosheets are very thin with a thickness less than 100 nm.

4.1.2.2 Dendritic Nanostructures

Dendritic nanostructures prepared via the EP method have attracted much research interest recently. In 2005, Qiu et al. [89] reported the synthesis of unique Ag dendritic nanostructures, with stems, branches, and leaves, via a simple electroless metal deposition method. A p-type, B-doped silicon (100) (1–5 Ωcm) wafer was used as a substrate. After cleaning by acetone to degrease the Si surface, followed by etching in diluted aqueous HF solution for 10 min, the cleaned silicon wafer then was etched in a 5.0 mol/L HF solution containing 0.02 mol/L silver nitrate at 50°C for 30 min in a conventional teflon-lined stainless steel vessel. Finally, the wafer was rinsed with deionized water and blown dry in air. The thick silver film covering the silicon wafer was detached for microstructural observation. SEM showed images of tree-like silver dendrites on the thick silver film. The diameters of stems for silver dendritic nanostructures are ~100 nm, the same as the sizes of the silver nanoclusters formed at the initial stage. Shapely stems, symmetrical branches, and leaves can be clearly observed (Fig. 4.13).

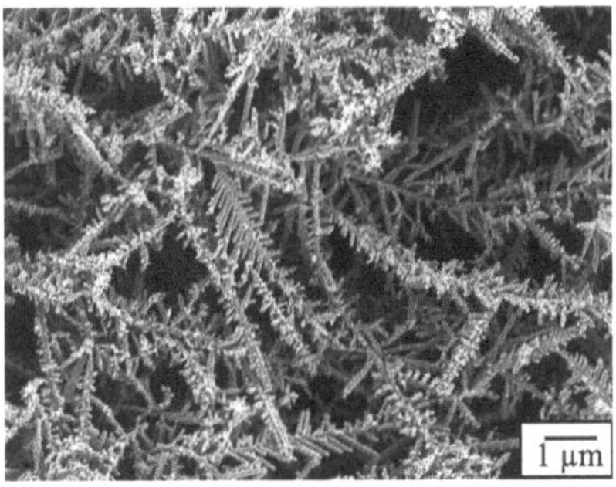

FIGURE 4.13 SEM image of the tree-like Ag dendrite grown at 50°C for 30 min, the diameters of stems for silver dendritic nanostructures are ~ 100 nm [89].

The formation process and mechanism of these Ag dendritic nanostructures obtained from EP can be schematically described as

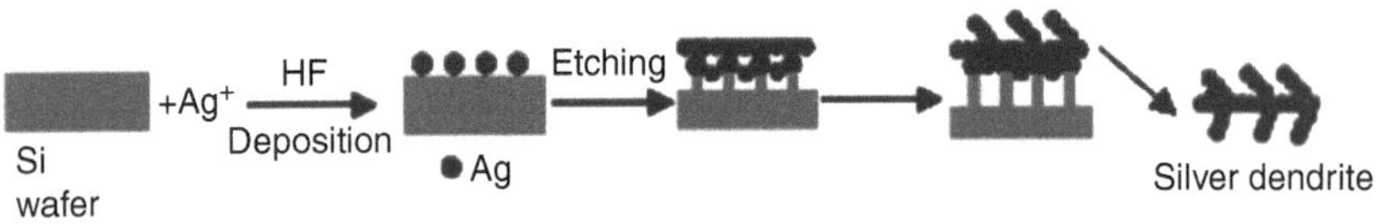

In 2010, Drozdowicz-Tomsia et al. [90] also studied the Ag dendritic nanostructures from EP. They pointed out that silver nuclei are formed initially, with relatively uniform size in the range 50–100 nm. When the tree-like three-dimensional (3D) silver structures are formed with clearly defined branches, the thickness of these branches is almost constant over the entire aggregate and each section is composed from agglomerated small particles of similar size to the initial nuclei.

Some researchers [89–93] like to call the aforementioned EP a simple electroless metal deposition (SEMD) method because the deposition is only carried out in an HF solution containing a metallic salt, for example, $AgNO_3$ which is much simpler than a conventional EP bath. In addition, for the formation of the silver dendritic nanostructures, Ye et al. [93] calculated the anodic potential difference, ΔE, between the NH_4F–$AgNO_3$ solution and HF–$AgNO_3$ solution under the identical conditions is 0.177 V. Therefore, electroless silver deposition in NH_4F solution is thermodynamically favored.

In 2007, Fang et al. [94] also reported the synthesis of dendritic silver nanostructures using a similar simple electroless deposition method. From the SEM image (Fig. 4.14a) and TEM images (Fig. 4.14c and d) of silver dendrites, and an optical micrograph of silver trees (Fig. 4.14b), it can be seen that the Ag dendrites are really in the range of nano size. The authors found that above an $AgNO_3$ concentration of 100 mmol/L, the silver dendrites show a prefect single crystalline structure. So a silver dendrite with well-defined morphology and

FIGURE 4.14 **SEM image (a) and TEM images (c and d) of the Ag dendrites, and optical micrograph of Ag trees (b). They all are prepared at RT in different silver nitrate concentrations for different reaction times: (a and b) 200 mM for 10 s, (c) 60 mM for 30 s, and (d) 100 mM for 10 s [94].**

crystalline structure could be obtained by controlling the concentration of silver ions in the solution.

In 2009, Lv et al. [95] also reported the synthesis of Ag nanodendrities by EP.

In 2008 Yan et al. [96] fabricated Cu dendritic nanostructures via electroless deposition. The bath solution these authors used is: 0.12 g of copper chloride was dissolved in a deionized water/hydroperoxide solution system. The pH value of the mixture solution was adjusted to around 2.6 using acetic acid solution. Then, the cleaned zinc foil was immersed in the mixed solution and put into an autoclave. After that, the autoclave was sealed to an electric oven and the temperature was maintained at 120°C for 4–18 h. The autoclave was cooled to room temperature. Finally, the resulting products were taken out, and the precipitate was collected, filtered off, and washed with water, then with absolute ethanol several times and dried in air at 80°C for several hours. Figure 4.15 shows a perfect Cu magnification SEM image (B) of the individual Cu dendrite, which clearly displays their detailed structure characteristics, composed of a long central trunk with the secondary branches lying parallel to each other. The authors pointed out that the secondary branches are composed of many small nanonanoparticles showing a rough surface structure, many branches are produced through attachment of small nanonanoparticles to the branches. More interestingly, some tertiary branches develop from the secondary branches in more complex dendritic structures. However, the authors did not state the size of the nanonanoparticles.

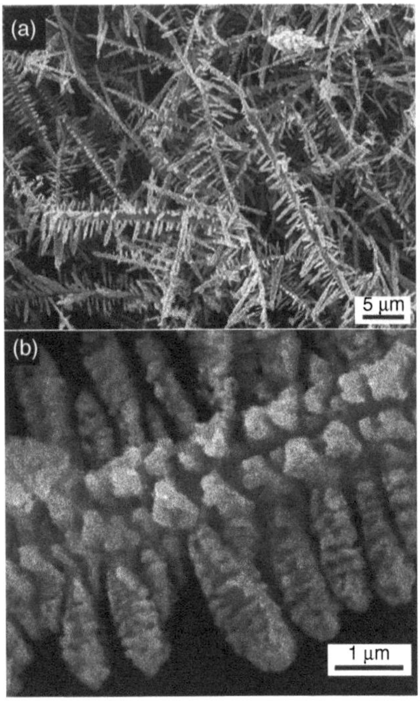

FIGURE 4.15 The perfect Cu dendritic structures at 18 h growth (a), and a higher magnification SEM image (b) of the individual Cu dendrite [96].

In 2006, Fang et al. [97] synthesized Au dendritic nanostructures via a simple and fast electroless metal deposition route at room temperature and ambient pressure for different reaction times. A zinc plate (99.9%) was used as a substrate. Only three ingredients of $HAuCl_4$, water and zinc are needed during the electroless metal deposition. The zinc plate is placed in the container of $HAuCl_4$ solution, and the dendritic gold product can be obtained on the zinc plate within several seconds. From the SEM images of gold dendrites synthesized with 1 mmol/L $HAuCl_4$ (Fig. 4.16), the features of the dendrites can be seen clearly, having threefold symmetric characteristics composed of a trunk (black line), branches (dashed line), and leaves (white line). The authors found through detailed TEM investigations that the Au dendrites with a threefold symmetric characteristic were built up of numerous nanonanoparticles roughly 5–10 nm in size.

In 2011, Zhang Yanling et al. [98] reported Pt nanonanoparticles on three-dimensional dendrite-like Au supported on Si (100) by electroless displacement deposition. They dipped etched silicon wafer into a solution containing $HAuCl_4$/HF for some time to obtain the dendritic Au on the silicon surface. The prepared Au/Si wafer then was immersed into the solution containing H_2PtCl_6/HF to gain the Pt/Au/Si. In such a way, Pt particles can be distributed on a dendritic Au surface in the H_2PtCl_6/HF solution with higher electro-catalytic activity and

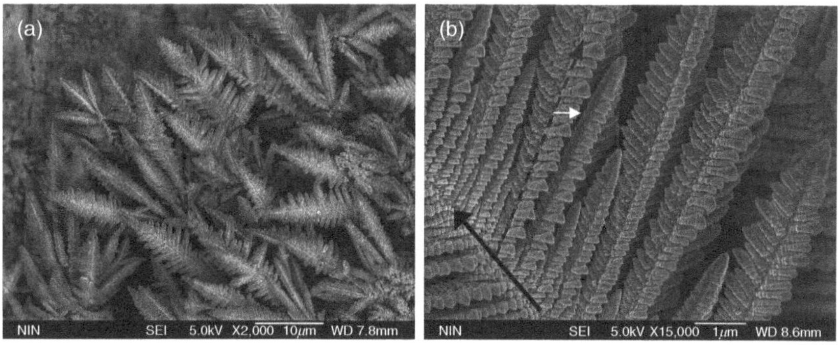

FIGURE 4.16 SEM images of gold dendrites. (a) Gold dendrites synthesized with 1 mM $HAuCl_4$. (b) A gold dendrite composed of a trunk (black line), branches (dashed line) and leaves (white line) [97].

excellent stability in acid solution. The formation process of Pt/Au/Si catalyst can be represented as

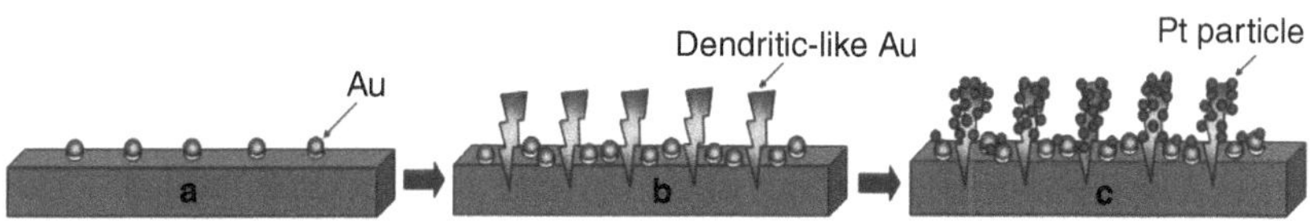

The HR-TEM images of Pt/Au catalysts scratched from the Pt/Au/Si wafer show that all the Pt particles appear to be well dispersed on the surface of branches of Au and have a relatively narrow size of 3–5 nm. In addition, they found that the Pt nanonanoparticles are highly crystallized on the Au support, and the d spacing in the Pt(111) is 0.226 nm

4.1.2.3 Other Shapes of 3D NSMs

Besides the nanoflowers and dendritic nanostructures other shapes of 3D NSMs can be synthesized by EP as well. Various kinds of shapes of 3D NSMs are including Ag nanoinukshuks [99], Au nano mushrooms [100], Ag and Au 3D spindle-shaped nanostructures [101], Au–Ag alloy 3D open-ended, cylindrical nanostructures [102], CuO 3D nano dandelions [103], highly ordered 3D polyaniline (PANI) nanostructures [104], Ni–NiO 3D core-shell nanoarrays [105], and 3D tin-coated nanoporous copper [106]. Out of these, this chapter will discuss three studies in more detail.

Aizawa et al. in 2005 [99] fabricated Ag nanoinukshuks on germanium. They used a dip-and-rinse galvanic displacement approach to synthesizing complex, highly regular crystalline Ag inukshuk architectures directly on Ge surfaces. Besides the substrate Ge only two ingredients, $AgNO_3$ and water, are required. The temperature of reaction was maintained at RT. The results show that the structures, termed nanoinukshuks, are based on stacks of hexagonal

FIGURE 4.17 (a) A stone inukshuk in Northern Canada. (b) SEM image of Ag nano-inukshuks formed by immersion of Ge (111) p-type in a 1.0 × 10^{-3}–M $AgNO_3$ (aq) for 4 h. (c) SEM image of close-up of facets on the tips of Ag metallic nano-inukshuks formed on p-type Ge(111) with 1.0 × 10^{-3} M $AgNO_3$ for 24 h [99].

metallic structures that grow, initially, parallel to the surface normal of the germanium. The nanoinukshuks appear as 300 nm diameter stacked hexagons with facets, and can grow up to 10 or more micronmeters in length, perpendicular to the (111) planes of the silver hexagons. The hexagonal plates themselves are 50–75 nm thick, and in the first 4–6 h of deposition, lie approximately parallel to the surface normal of the germanium, leading to growth of nanoinukshuks out from the surface (Fig. 4.17).

In 2009, Wang et al. [100] used a combination of block copolymer (BCP) lithography and a simple galvanic displacement reaction to prepare nanostructured gold films for surface-enhanced Raman scattering. The plating process was arranged such that a monolayer of inverse PS-b-P2VP micelles is deposited on a Si wafer by spin coating. Morphology reconstruction by treatment using acetic acid leads to the formation of a nanoporous BCP template. Immerging the Si wafer into a $NaAuCl_4/HF$ plating solution yields first an array of Au nanopillars connected to a continuous Au film on the Si wafer. Then caps grow on top of the Au nanopillars as soon as they completely fill the nanopores in the BCP template. Finally, a thick gold film with surface features characterized by length scales that are similar to those accessible by top-down lithography and self-assembly of colloids results. The formation of arrays of Au nanostructures for SERS by bcp-templated galvanic displacement can be schematically summarized (PS, blue; P2VP, orange; Si, black; Au, yellow) as follows:

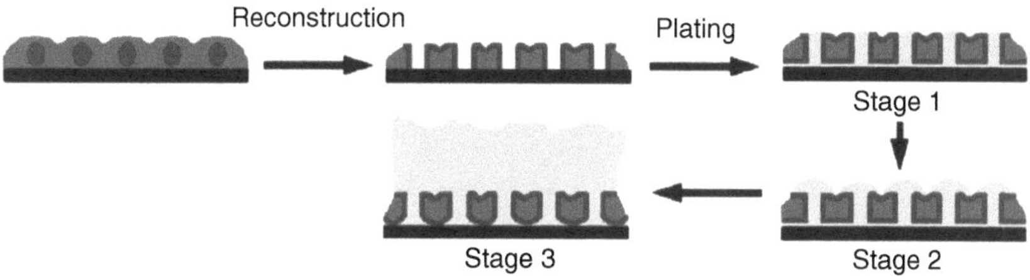

The authors found that after plating for 15 s (stage 1), Au nanopillars with diameters of about 30 nm arranged in an array with a period of ~50 nm, corresponding to that of the BCP template. After plating for 30 s (stage 2), mushroom-shaped entities form by the growth of caps on top of the nanopillars, which impinge on each other when their diameter approaches the period of the BCP template. A regular array of stems connects the caps with the Si wafer. The thickness of mushroom layer is ~50 nm. Finally, after the plating for 90 s, the gold layer is 700–800 nm thick, whereas the peak-to-valley distances of the surface features are of the order of 100 nm and above. In a word, high throughput and large-area patterning by BCP lithography can be exploited to generate high-performance SERS substrates with periods of a few tens of nanometers consisting of dense arrays of metal nanostructures with shapes and spacings optimized for high SERS performance. Figure 4.18 illustrates the TEM image of mushroom layer in stage 2 sample obtained by plating for 30 s.

Liu and Zeng in 2004 [103] prepared "dandelions" composed of CuO nanoribbons. The plating solution that they used is of copper (1.00 mol/L) by mixing $Cu(NO_3)_2$ in pure ethanol solvent. Typically, 15–25 mL/L of the above solution was added with 15–25 mL/L of ammonia solution (32%), followed by an addition of 4–5 mL/L of aqueous NaOH (1.00 mol/L) and 0–5 g of solid $NaNO_3$. Putting the solution mixture into a Teflon-lined stainless steel

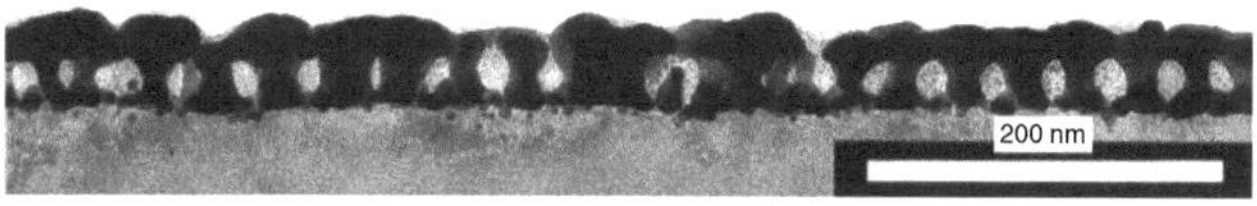

FIGURE 4.18 TEM image of cross-sectional mushroom layer cut normal to the surface of the Si wafer in stage 2 [100].

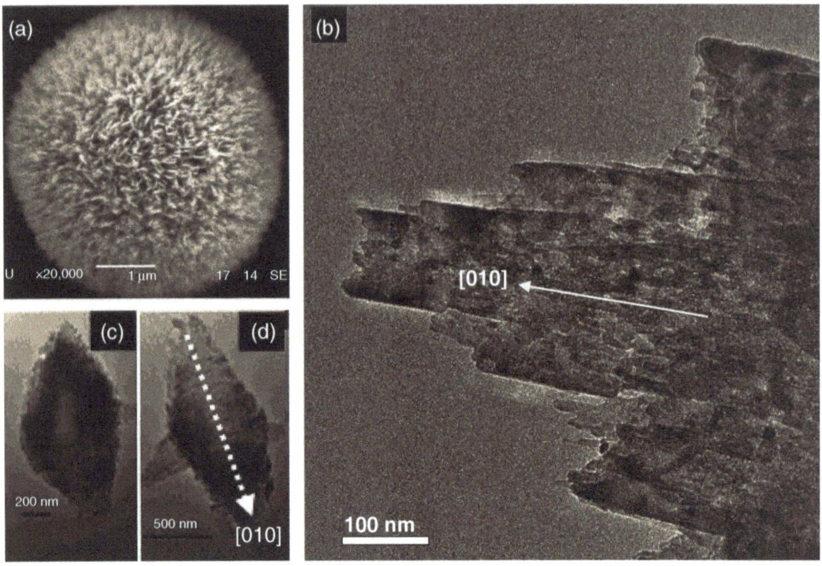

FIGURE 4.19 **(a) SEM image of a detailed view on an individual CuO microsphere (prepared at 100°C, 24 h); (b) HRTEM image of oriented aggregation of tiny CuO nanoribbons (in the tip area of a rhombic CuO crystal strip); (c and d) TEM images of two rhombic CuO crystal strips formed from smaller one-dimensional nanoribbons [103].**

autoclave, then the autoclave was heated at 100–180°C for 2–24 h in an electric oven. After reaction, the autoclave was cooled in tap water, and CuO products were washed via centrifugation–redispersion cycles, with each successive supernatant being decanted and replaced with deionized water and ethanol. The experimental results show that the SEM images of as-prepared CuO samples are just like "dandelions," which are the CuO crystallites self-organized into spherical assemblies. From the SEM/TEM/SAED investigation it has been found that the nanoribbons that construct the rhombic CuO crystal strips (Fig. 4.19c and d) have breadths of about 10–20 nm and lengths not exceeding the strip dimension in [010] (HRTEM/TEM, Fig. 4.19b). In addition, the nanoribbons are aligned with one another along some main crystallographic axes of CuO via an "oriented attachment" process, which shows that the overall assembly is essentially single crystalline. The dimension of CuO crystal strips has the following hierarchical order [010] > [100] ≫ [001], while the thickness of the crystal strips in [001] is 20–30 nm. Figure 4.19a shows the SEM image of an individual CuO dandelions microsphere.

4.2 2D NANO EP MATERIALS

It should be noted that EP is one of the most convenient methods for fabricating 2D nano materials (2D NMs). This is because the EP method has been used to prepare 2D bulk thin films since the very beginning of its discovery. In principle, 2D nanofilm materials of all the metals, alloys, and composite alloys discussed in Chapters 2 and 3 could be synthesized through EP while controlling the thickness of plating films in the range of nano size. Of course, not all of such 2D nanofilm materials have been fabricated as yet, though many 2D nano films of metals and alloys prepared by EP have been reported in the literature [107–145]. Besides the 2D nanofilm materials, other shapes of 2D NMs such as nanoplates, nanodisks, nanoshells and nanosheets, nanowalls, nano circles and rings, nano honeycomb, nanoline and nanofin pattern, nanorod arrays, a nano grating, and nano irregular pattern have been reported. Of course, we cannot describe and discuss all of these 2D NMs in detail, but instead this section will give details about a selected few.

4.2.1 EP 2D Nano Films

4.2.1.1 EP 2D Au Nano Films

EP 2D Au nano films are the most studied of the EP 2D nano films [107–116] some detail will be given about them here. In 2002, Porter et al. [107] investigated electroless nanoparticle film deposition compatible with photolithography, microcontact printing, and dip-pen nanolithography (DPN) patterning technologies. They found that nanonanoparticles of Au formed spontaneously as thin, morphologically complex metallic films upon various semiconducting or metal substrates such as Ge(100), Cu, Zn, and Sn. For example, DPN was demonstrated through the writing of a 0.55 μm long gold line with a width of 30 nm and height of 10 nm through the spontaneous electroless deposition of $AuCl4^{-}$, delivered via the AFM tip, upon an untreated Ge(100) wafer. In a typical experiment, a silicon tip was dipped in an aqueous 20 mmol/L $AuCl4^{-}$ solution, diluted with acetonitrile (1:10 v/v) to reduce surface tension, and dried for 5 min. Writing was accomplished at a rate of 0.2 μm/s in a constant humidity environment of 50%. The DPN can be outlined as

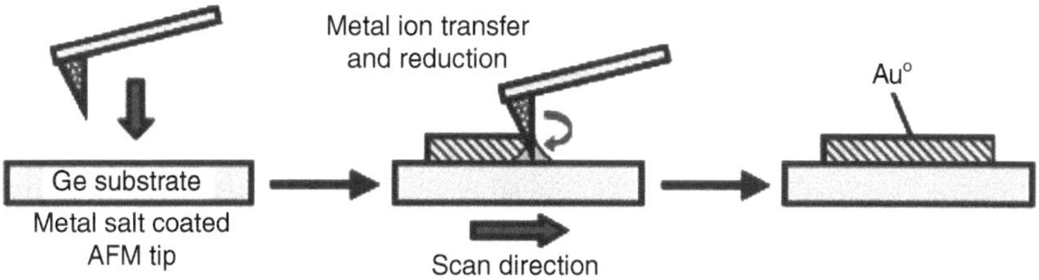

Horiuchi et al. in 2010 [113] reported Pt colloid catalyzed etchingless Au electroless plating with strong adhesion to polymers. Because Pt colloidal nanoparticles have excellent catalytic activity for echingless electroless plating (ELP), the Pt colloidal nanoparticles can be immobilized via electrostatic interactions on a substrate simply by dipping it into a Pt colloid. Pt colloid was synthesized by the reduction of metal salt H_2PtCl_6 by $NaBH_4$ in the presence

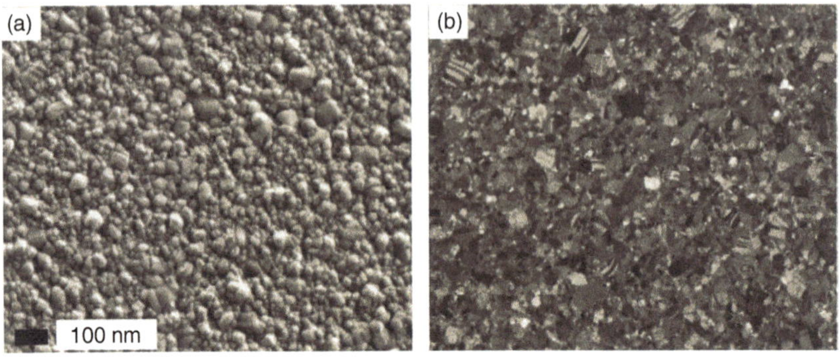

FIGURE 4.20 SEM micrographs showing the surfaces of gold films plated for 1 min (a) and 5 min (b). (a) is secondary electron image and (b) is back-scattered electron images [113].

of polyvynlypyrrodidone (PVP) at room temperature for one day, which gave a stable aqueous suspension containing 100 ppm metal nanoparticles. Owing to the excellent catalytic property of the Pt nanoparticles, continuous gold thin films can be produced at room temperature using a simple cyanide-free gold electroless plating solution composed of chloroauric acid and hydrogen peroxide. Experiments indicate that after 10 s of the immersion of the Pt-catalyzed film into the plating bath, Au particles with the diameter of 10–30 nm are deposited on the Pt nanoparticles. The number of the Au particles increases with the plating time, and finally the entire surface is covered with the gold deposits forming a continuous film. After plating for 5 min, the secondary electron (SE) image presents the topographic features of the surface fully covered with the gold grains with the size of 50–100 nm, while the back-scattered electro (BS) image presents the variation of the crystal orientations of the grains densely packed with no voids at the boundaries (Fig. 4.20). Therefore, it is indicated that the autocatalytic reaction not only contributes to the thickness increment but it may also facilitate the interconnection between the neighboring grains. The authors believe that this process is an environment-friendly thin metal film deposition process that avoids the use of toxic and hazardous substances.

Ultrasmooth Au thin films on Si were investigated by Gutés et al. in 2011 [114]. They used galvanic displacement (GD), a type of electroless deposition, to obtain ultrasmooth gold thin films on silicon <111 >. These authors said that the novel aspect of the method presented herein is the absence of fluoride ions in the liquid phase, and its principal advantage when compared to previous efforts is that the process is inherently self-limiting. Therefore, no silicon oxide is removed during the process. And the maximum gold film thickness is achieved when elemental silicon is no longer available once the surface is oxidized completely during the GD displacement process. They used three gold plating solutions with different concentrations of $KAuCl_4$ (2, 0.2, and 0.02 mmol/L) The results show that the Au film is to confirm surface smoothness, suggesting that the monolayer growth does not follow the Volmer–Weber growth mode, in contrast

FIGURE 4.21 **AFM image of the sample from 0.2 mmol/L solution and after 60 s immersion time. Scanned area of 2 μm >× 2 μm, z scale of 10 nm, and rms of ~0.72 nm [114].**

to the GD process from aqueous conditions with fluorinated species. The maximum thickness of Au films is ~35 nm. Figure 4.21 shows AFM image of the sample from 0.2 mmol/L solution and after 60 s immersion time.

4.2.1.2 EP 2D Ag Nano Films

Ag nanofilm prepared by EP has also been the focus of recent research interest [117–125]. In 2006, Park et al. [118] used a very simple electroless plating method to prepare optically tunable nanostructured Ag films. Cleaned glass was dipped in the reaction mixture and incubated for 30 min at 45 ± 1°C with vigorous stirring. The concentration of $AgNO_3$ in absolute ethanol was kept at 0.5 mM for all experiments, while that of butylamine was subjected to change. AFM images of the silver films prepared at various silver ion-to-butylamine molar ratios are shown in Fig. 4.22. The mean grain sizes are determined to be 25 ± 4, 46 ± 6, 71 ± 19, and 110 ± 37 nm for the surfaces a–d, respectively. That is to say that the grain size of silver can be readily controlled to range from 20 nm to 150 nm. Such nanostructural features correlated well with their UV/vis absorption characteristics, as well as with their surface enhanced Raman scattering (SERS) activities.

In 2009, Gutés et al. [121] studied Ag nanostructures on silicon based on a galvanic displacement process. The Ag plating solution composed of dissolving the appropriate amount of AgF and KF in deionized (DI) water to the desired final Ag^+ and F^- concentrations. Silicon chips were immersed and left in solution for the desired time. Average thicknesses and root mean square (RMS) roughness values for the Ag films are from 18.8 nm to 81.4 nm and 5.0 nm to 45.4 nm corresponding to the immersion time from 5 s to 60 s when the Ag^+ and F^- concentrations are 1 and 20 mmol/L, respectively. It is obvious from the AFM results that both Ag film thickness and RMS roughness increase with deposition time for samples prepared from 1:20 Ag:F ratio.

In 2011, Sabayev et al. [124] studied electrical properties of nanofilm Ag prepared by EP. Ammonia acetate solution containing [mol/L]: $AgNO_3$, 0.05, NH_4OH (25%), 1.83, CH_3COOH, 0.73 and hydrazine-hydrate as reducing agent, 0.1, at pH 10.5 was used for Ag EP. The authors measured the influence

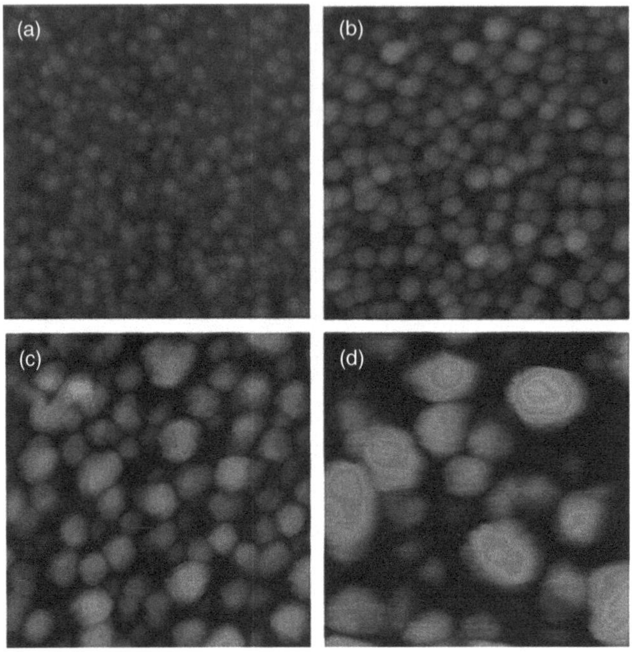

FIGURE 4.22 **AFM images (500 nm × 500 nm) of silver films prepared by electroless deposition at different molar ratios of $AgNO_3$ and butylamine: (a) 1:0.1, (b) 1:0.2, (c) 1:0.5, and (d) 1:1. The mean particle sizes are 25, 46, 71, and 110 nm, respectively; the Z scale is 100 nm [118].**

of the metal concentration in solution on the film thickness and surface roughness of Ag nanofilm. The content of metal ions was varied by dilution of the base solutions by deionized water to the ratios of 1:10, 1:6, and 1:3 named as low, medium, and high concentration, respectively. When at medium concentration the thickness of Ag nanofilm increases from 25 nm to 110 nm while the deposition time enhances from 25 s to 115 s. The surface RMS of nano Ag films varies from 4.3 nm at low concentration to 13.7 nm at high concentration. The resistivity measurements of Ag nano films indicated that films with a thickness lower than 20 nm demonstrate high initial resistivity, near up to 1000 μΩ-cm for the film thickness of about 15 nm. However, in the vicinity of electrical percolation threshold the resistivity starts to drop at the Ag film thickness of 30 nm, then reduces to 2.5 μΩ-cm at thickness of 60 nm, which is near to the pure Ag bulk resistivity value.

It is interesting to note that Hao et al. in 2011 prepared EP multilayer Ag nanofilms for surface-enhanced Raman scattering sensing of arsenate [125]. The bath for depositing the Ag nanofilms on glass slides is a solution of $AgNO_3$ with $BuNH_2$ in anhydrous ethanol. The operation was without stirring at room temperature (approximately 22°C). Their experiments show that newly reduced Ag atoms would grow on the top of the well-defined first-layer Ag nanonanoparticles (AgNPs) to form second layer AgNPs (second-layer) just by continuously raising the reaction time to 5–40 h. The second layer AgNPs were much larger than those in the first layer and kept growing in particle size and coverage degree as the reaction time was prolonged. When the reaction time was >9 h, some dendrites formed on the top of the double layer with a statistically uniform distribution, which could be regarded as a third layer though its coverage degree was not high. As for the formation mechanisms, the authors belived that a AgNPs double layer was formed by directly reducing Ag+ ions on the glass surfaces, while a top layer (third-layer) of Ag dendrites was deposited on the double layer by self-assembling AgNPs or AgNPs aggregates which had already formed in the suspension. The mean sizes of the AgNPs in the first-layer and second-layer Ag films prepared in 5/2.5 mM $AgNO_3/BuNH_2$ solution increased from 20 ± 5 nm to 50 ± 10 nm and from 60 ± 5 nm to 180 ± 30 nm as the reaction time enhanced from 1.5 h to 40 h and from 5 h to 40 h, respectively. The Ag dendrites consist of small AgNPs of 20–50 nm.

4.2.1.3 EP 2D Cu Nano Films

ED Cu nano films [127–132] have also attracted research interest. Electroless plating nanocrystalline copper films on glass was reported in 2007 [129]. The bath used to obtain the EP copper films consisted of 10 g/L $CuSO_4$, 40 g/L $NaKC_4H_4O_6$, 8 g/L NaOH, 2 g/L Na_2CO_3, 1 g/L NiC_{12}, 20 mL/L HCHO, and 0.2–0.5 g/L sodium dodecyl benzene sulfonate (SDBS). The pH value of the bath was 12.0–12.5. The temperature of the bath was maintained at 40°C and agitated by magnetic stirring. The mean grain size increases from 15.5 to 27.2 nm when the deposition time increases. The Cu film thickness is from 52.2 nm to 107 nm when the deposition time is from 1 min to 5 min. Then the thickness is beyond the nanometer range as the deposition time is larger than 5 min. The film surface roughness is 11.656–16.684 RMS as the deposition time is from 1 min to 5 min.

Scudiero et al. in 2011 [131] studied electroless deposition of copper thin film on Ge and Si surfaces. Cu nanofilm was deposited on Ge and Si substrates by galvanic displacement at room temperature. The film thickness was estimated from AFM imaging to be 95 nm for films deposited on Si and about 600 nm for films grown on Ge. Even for relatively brief immersion times, coverage is fairly uniform even though the average particle size is very small. As the deposition time increases, the particle diameter rapidly grows and large clusters appear. After deposition for 30 min, the average roughness RMS was estimated to be about 15 nm and the mean size of the Cu particle to be about 100 nm. At this point, the film appears granular and relatively dense.

4.2.1.4 EP 2D Pd and Pt Nano Films

Compared with the previously mentioned Ag and Au ED nano films, relatively few studies have been carried out on EP 2D Pd and Pt nano films [133–135]. In 2006, Shi et al. [134] reported the synthesis of Pd nanostructures by spontaneous electroless deposition. The composition of the electroless plating bath

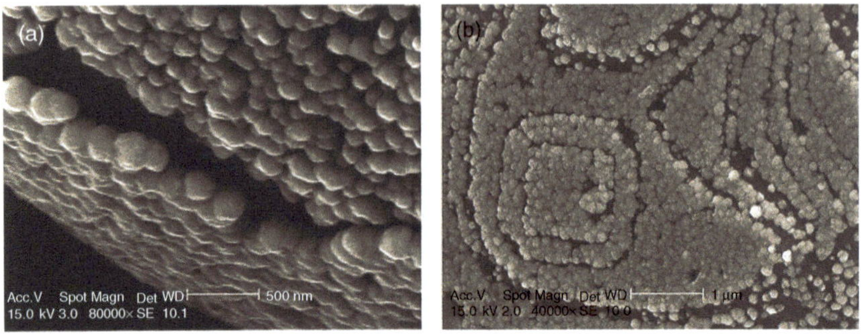

FIGURE 4.23 **SEM micrographs of the nanoparticle-monolayer Pd membrane packed by Pd nanoparticles (a), and the Pd nanostructural arrays formed on the irregular steps at the surface of the porous stainless steel substrate (b) [134].**

and other experimental parameters there used were: $PdCl_2$ 2 g/L, Na_2EDTA 40.1 g/L, NH_3 (28%) 198 mL/L, N_2H_4 (1 M) 5.6 mL/L, pH 10–10.4, and temperature 60°C. Figure 4.23 shows the nanoparticle monolayer palladium membrane packed by palladium nanonanoparticles (a), and the palladium nanostructural arrays formed on the irregular steps at the surface of the porous stainless steel substrate (b). SEM micrograph in Fig. 4.23a shows the nanoparticle monolayer Pd membrane is deposited on the wall of a pore at the stainless steel substrate. The thickness of the Pd layer is around 80–130 nm, which is the same as the diameters of the Pd nanonanoparticles. The palladium nanowires arrays in Fig. 4.23b is formed due to the aggregation of Pd nanonanoparticles and the Pd nanostructure arrays are wholly followed the irregular steps at the surface of the substrate and their diameters are inhomogeneous.

In 2010, Pernstich et al. [135] put electroless Pt nanofilms onto polymers, metals, and ceramics. In their preparation, dissolved *cis*-$[PtCl_2(sty)_2]$ is in equilibrium with *trans*-$[PtCl_2(sty)_2]$ and $[Pt_2Cl_4(sty)_2]$. The formation of the trans isomer appears important for further reaction: experiments with $[Pt_2Cl_4(sty)_2]$ dissolved in toluene did not yield a platinum film, while the addition of styrene, which formed *trans*-$[PtCl_2(sty)_2]$ *in situ*, led to film formation after a variable induction period. The proposed chemical reaction, leading to a reduction of the precursor to metallic platinum and finally to a film, is given in the following equation:

$$2cis-[PtCl_2(sty)_2] \rightleftarrows \left\{ \begin{array}{l} trans-[PtCl_2(sty)_2] \\ [Pt_2Cl_4(sty)_2]+2(sty) \end{array} \right. \xrightarrow[\text{of toluene}]{\text{methyl group}} 2Pt+4HCl\uparrow+4(sty)$$

TEM images of a cross section of a Pt film deposited onto Kapton reveal the presence of two distinct layers. A first layer, referred to as the primary film, has a thickness of 10–15 nm and is comprised of particles with ≈1 nm diameter. The secondary film has a thickness of 60–65 nm with particle size increased

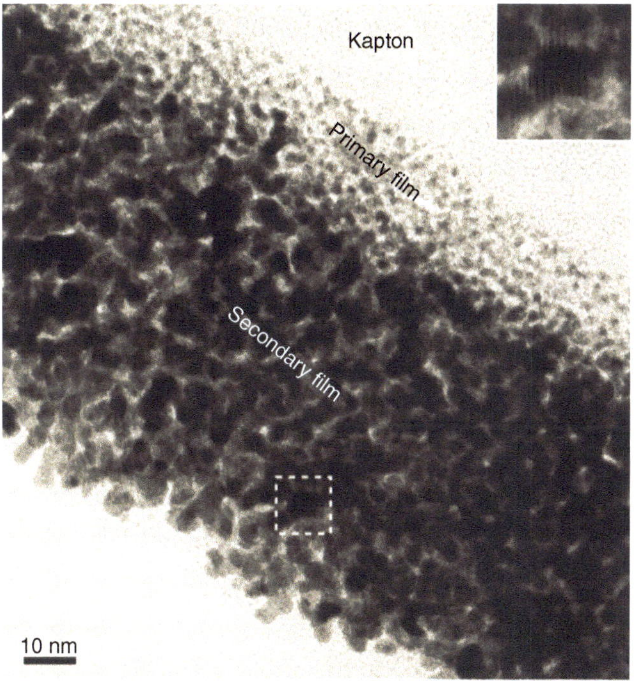

FIGURE 4.24 **TEM images of a cross section of a platinum film deposited onto Kapton [135].**

to 3–5 nm. The porosity of the film can be seen in Fig. 4.24. Also, crystalline grains (dashed square) can be identified as shown in the inset of the figure.

4.2.1.5 EP 2D Nano Alloy Films

Some EP 2D nano alloy films have been reported [136–145], including Ni–P [136, 140, 141], Co–W–P [137, 138], Ag–W [139], Pd–Au, Pd–Ag, Pd–Ni, Pd–Pb, and ternary system Pd–Au–Ni [142], Pd-Ni–P [143], Ni–P/Cu–Ni [144], and Ni-Co–P [145].

In 2008, Crobu et al. prepared electroless nanocrystalline Ni–P alloy coatings and studied their corrosion resistance [140]. The chemical composition of the electroless bath and its operational conditions are: nickel sulfate 5.5–6.5 g/L, sodium hypophosphite 18–22 g/L, complexing agents and stabilizers proprietary, traces of PbC_{12}, pH 4.6–5, temperature 85–90°C, and plating duration 1–2 h. The deposition film was rather thick, but it was composite of a face centered phase of nickel with a crystalline size of 1.2 nm. This nanocrystalline phase is a consequence of the difference in atomic radius between nickel and phosphorus leading to a chemically ordered structure. In addition, the roughness in the flat areas was found to be RMS ± 1.5 nm from AFM measurements. Chemical composition of the coatings showed a phosphorus content of 18.4%–19 wt%. The formation of a phosphorus enriched surface layer is responsible for

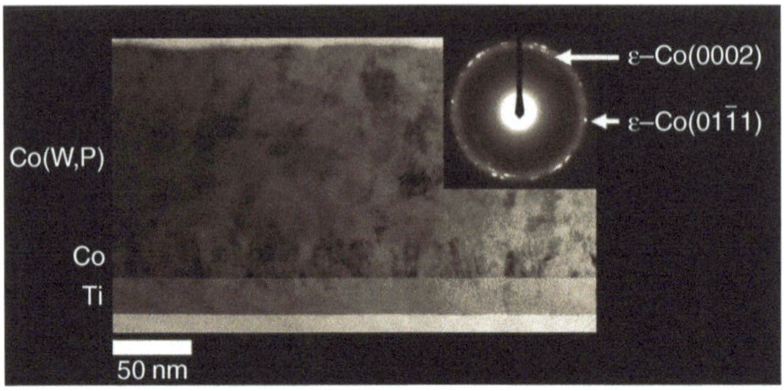

FIGURE 4.25 Cross section bright field TEM micrograph of electroless deposited Co(W,P) on a cobalt catalytic seed layer. Inset, selected area electron diffraction (SAD) of the layer [138].

the high stability and corrosion resistance of the nano Ni–P alloys. And surface analytical techniques provide direct evidence of this P-enriched layer.

In 2001, Kohn et al. [138] reported electroless deposited Co(W,P) nano films as diffusion barriers for copper metallization. The solution composition, pH, and temperature for electroless deposition of Co(W,P) are: cobalt sulfate 0.082 mol/L, sodium hypophosphite 0.169 mol/L, sodium citrate 0.492 mol/L, boric acid 0.502 mol/L, sodium tungstate 0.03 mol/L, surfactant RE-610 0.05 g/L, pH 9, and T 85–95°C. The composition of the deposited Co(W,P) films has been determined to be Co: 88–90 at%, P: 8–10 at%, W: 1.5–2.5 at%. The typical microstructure of the as-deposited films from bright field (BF) TEM micrograph is relatively homogeneous with grains sized 10–20 nm, approximately. Figure 4.25 shows the cross section bright field TEM micrograph of electroless deposited Co(W,P) on a cobalt catalytic seed layer. The Co–W–P films have been found to be effective diffusion barriers for copper after thermal anneal at 400–500°C according to C–t measurements of metal oxide semiconductor (MOS) capacitors and secondary ion mass spectrometry (SIMS) depth profiles of stack samples. These barrier properties may be explained by the microstructure of the nano films. The phosphorus and tungsten impurities are most probably enriching the grain boundaries, forming a "stuffed" barrier. These impurities have also been found to stabilize the hcp structure, delaying the phase transition to fcc by more than 80°C compared with bulk pure cobalt.

Strukova et al. in 2010 [142] reported nanocrystalline Pd alloy films coated by electroless deposition, which included Pd–Au, Pd–Ag, Pd–Ni, Pd–Pb, and the ternary system Pd–Au–Ni. These Pd alloy films were produced by immersion plating through heating a metallic substrate with a solution of metals halogenides and ammonium chloride in an aprotonic-dipolar organic solvent. The aprotonic-dipolar solvent (ADS) possesses the required combination of donor and ionizing capabilities and forms coordination compounds with many transition metals. The experiments show that the coatings are dense, pore-free 0.005–1 μm thick films with high adhesive strength to the substrate surface. The deposited films consist of nanocrystalline grains with sizes in the range of 11–35 nm. The coatings of Pd–Au, Pd–Ag, and Pd–Ni have a solid solution structure, whereas Pd–Pb is an intermetallic compound. SEM and TEM investigations reveal the existence of clusters formed by nanocrystalline grains.

4.2.2 EP 2D Nanoplates

Some reports of EP 2D nanoplates have been published recently [146–150]. In 2005, Ah et al. [146] synthesized Au nanoplates in an aqueous solution of 250 mL/L of 1 mM $HAuC_{14}$, which was heated for 30 min with stirring. Then various amounts of 250 mmol/L poly (vinyl pyrrolidone) (PVP) (molar ratio between the repeating unit of PVP and $HAuC_{14}$ was from 0 to 8) were added, and then 3.75 mL/L of 34 mmol/L sodium citrate (the molar ratio of sodium citrate to $HAuC_{14}$ was 0.5) was quickly added. The size of the Au nanoplates was controlled by changing the concentration of PVP, while the molar ratio of the reducing agent to $HAuC_{14}$ was kept at 0.5 (Fig. 4.26). Field emission-scanning electron microscopy (FE-SEM) images of the Au nanoplates show that the average width and thickness are of 450 ± 29 and 39 ± 4 nm (a), 310 ± 25 and 28 ± 3 nm (b), 158 ± 13 and 18 ± 3 nm (c), and 96 ± 12 and 14 ± 2 nm (d).

There is significant interest in the direct growth of metal nanoplates with well-controlled dimensions on semiconductor substrates because of their broad potential use. Sun did just such work in 2007 [148]. He found that placing a drop (~6 μL) of aqueous solution of $AgNO_3$ (concentration of 1 M) on the surface of the cleaned n-AsGa wafer initiates the growth of Ag nanoplates at room temperature (~22°C) via a galvanic redox reaction with different reaction times. The SEM images indicated that the Ag nanoplates have smooth surfaces and a thickness of 28 ± 5 nm and edge lengths of 400–450 nm (Fig. 4.27), respectively. Figure 4.27b presents a TEM image of an individual Ag plate sitting

FIGURE 4.26 (a–d) FE-SEM images of Au nanoplates. Molar ratio of the repeating unit of PVP (MW ≈ 55,000, n ≈ 500) to $HAuC_{14}$ was 8:1 (a), 5:1 (b), 2:1 (c), and 0.1:1 (d) [146].

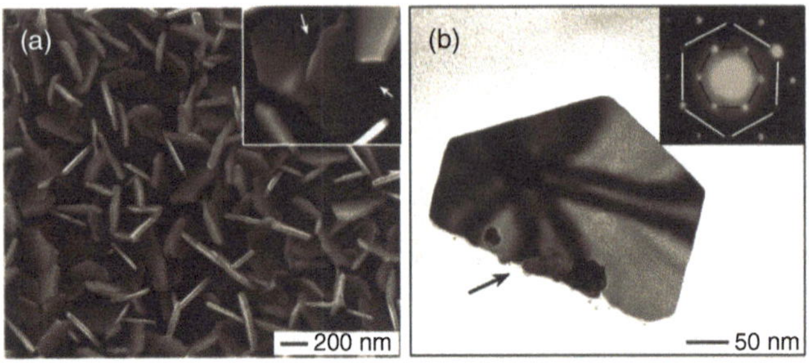

FIGURE 4.27 Characterization of Ag nanoplates formed at a reaction time of 2 min. (a) High-magnification SEM image; and (b) low HRTEM image. The arrows in (b) and inset of (a) highlight the roughness of the edges of the Ag nanoplates, which stick to the GaAs substrate [148].

on a TEM grid against one of its flat surfaces. The edge (indicated by the arrow) that sticks to the GaAs wafer is rough, whereas the edges out of the wafer surface are straight and sharp. This profile of the edges of the Ag nanoplates is also confirmed by the SEM observations (inset of Fig. 4.27a). The inset of Fig. 4.27b is the selected area electron diffraction pattern obtained by focusing the electron beam perpendicular to the flat surfaces of the plate. The formation of a hexagonal pattern indicates that the Ag nanoplates are crystallized in a face-centered cubic (fcc) lattice with the (111) flat surfaces. The author also used a (100) n-type Si wafer as substrate to grow electroless Ag nanoplates.

4.2.3 EP 2D Nanodisks

Very few studies have been reported for the EP 2D nanodisks [151,152]. Wolcott et al. in 2006 [151] fabricated the ultrathin WO_3 nanodisks using a wet chemical route with poly(ethylene glycol) (PEG) as a surface modulator. The WO_3 nanostructures formed are dominated by very thin disks with dimensions on the nanometer to micrometer scale. SEM and TEM images reveal the structures to have dimensions on the order of 350–1000 nm in length, 200–750 nm in width, and 7–18 nm in thickness and possessing textured single crystalline features.

Aherne et al. in 2009 [152] synthesized Ag nanoprisms using electroless deposition. Typically, nanoprisms are produced by combining 35 mL/L of distilled water, ascorbic acid (0.525 mL/L, 10 mmol/L), and 3.5 mL/L of Ag nanoparticle seed solution, followed by the addition of $AgNO_3$ solution (21 mL, 0.5 mmol/L) at a rate of 4 mL/min while stirring vigorously. After the reaction was completed, tri-sodium citrate solution (17.5 mL/L, 25 mmol/L) was added. The size of the formed Ag nanoprisms is 21 nm. The Ag nanoprisms are single crystals with their basal plane as the (111) lattice plane. By reducing the volume of seed solution, it is possible to manipulate the size of Ag nanoprisms. Figure 4.28 shows TEM images of edge-on-oriented and flat-lying Ag nanoprisms.

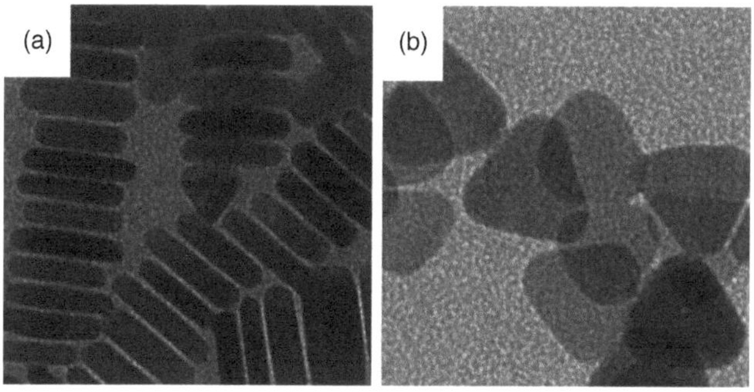

FIGURE 4.28 **TEM images of edge-on-oriented and flat-lying Ag nanoprisms [152].**

4.2.4 EP 2D Nanoshells and Nanosheets

Three groups from China studied EP 2D nanoshells and nanosheets from 2006 to 2012 [153–155]. Here just the ordered array of gold nanoshells interconnected with gold nanotubes prepared by one of the groups is briefly described. Dong et al. in 2006 [153] dipped the macroporous polystyrene (PS) nanomold into a plating solution that was prepared by mixing solutions A and B in a volume ratio of 10:3. Solution A consisted of hydrogen tetrachloroaurate (1.0 g), sodium chloride (0.5 g), and deionized water (80 mL); solution B consisted of tartaric acid (1.125 g), sodium hydroxide (15.0 g), ethanol (19 mL/L), and deionized water (50 mL/L). Then, the PS nanomold was rinsed with deionized water. After the sample had dried, the PS matrix was removed by dissolving in a solution of chloroform for 12 h. During this procedure, gold is electroless deposited on the void surface of the PS template. Continuous gold shells can be formed within the pores of the polymer membrane. Finally, the whole structure is immersed in a solution of chloroform, wherein the PS membrane is dissolved, and a gold nanoshell/ nanotube network is left. Figure 4.29 is an SEM image of a multilayered array of non-close-packed (NCP) gold nanoshells interconnected with gold nanotubes. The wall thickness of the Au structure is estimated

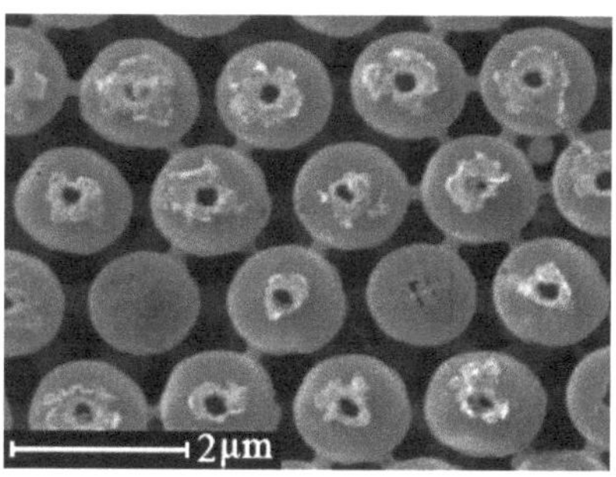

FIGURE 4.29 **SEM image of a multilayered array of non-close-packed (NCP) gold nanoshells interconnected with gold nanotubes [153].**

to be about 40 nm by TEM analysis. Thus, the hollow Au nanospheres in the 2D metal networks are not complete shells but look like nanocups that are interconnected with nanotubes at their equator, as is illustrated in Fig. 4.29.

4.2.5 EP 2D Nanowalls

Three Chinese groups investigated the EP 2D nanowalls from 2006 to 2011 [156–158]. Jiang et al. in 2006 [156] used the electroless process to fabricate Cu microstructures assembled with nanowalls. In their experimental procedure, 0.01 mol of CuCl and 0.01 mol of NaOH were dissolved in 20 mL of distilled water; then 0.03 mol of NaH_2PO_2 was added to the above solution under vigorous stirring for 5 min. After that, the solution was transferred into a 50 mL Teflon-lined stainless-steel autoclave, which was then filled with distilled water up to 90% of the total volume. The autoclave was sealed and maintained at 200°C for 24 h. After completion of the reaction, the resulting solid products were filtered off, washed with absolute ethanol, and distilled water several times, and then dried in a vacuum at 60°C for 6 h. Figure 4.30 shows the field-emission scanning electron microscopy (FE-SEM) patterns of copper microstructures with net shape consisting of nanowalls (a) and SEM images of copper microstructures with radial shape (b) prepared at 200°C for 24 h. In Fig. 4.30a, the high-resolution FE-SEM image indicates that the thickness of the nanowalls is about 50 nm with a length of about 1–3 μm, and the nanowalls construct a network on the surface of the microstructures. Figure 4.30b proves that the microstructure mainly consists of ordered nanowalls with a thickness of about 50 nm and a length of 15 μm radiating from the center of the microstructure.

In 2011, Nien et al. [158] prepared Cu_2S nanowalls by immersing cleaned and polished Cu substrates in an aqueous solution containing Na_2S (1 mol/L) and HCl (1 mol/L) for 5 and 40 min at around 4°C. After the above immersion process, the samples were dried in air for various characterizations. XRD analysis revealed the formation of the monoclinic phase of Cu_2S nanostructures on

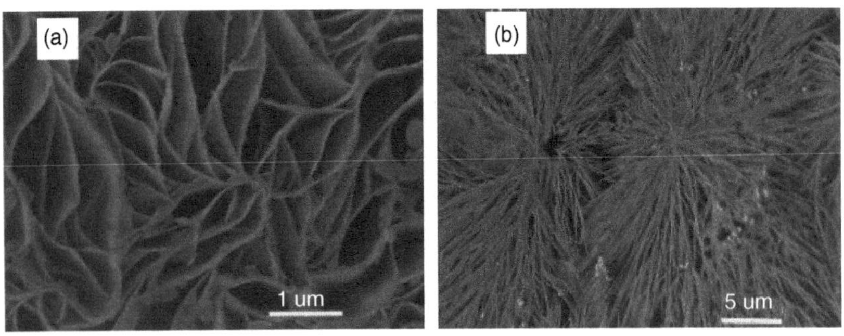

FIGURE 4.30 **FE-SEM patterns of copper microstructures with net shape consisting of nanowalls (a) and SEM images of copper microstructures with radial shape (b) prepared at 200°C for 24 h [156].**

Cu substrates, which exhibited a thickness of less than 100 nm and a height of several μm of sheets (nanowalls). The measurements from TEM and EDS indicated that the Cu_2S nanowalls were thinner near the sheet edge and had voids inside the sheets, supposedly due to Na_2S/HCl etching.

4.2.6 EP 2D Nano Circles and Rings

Two groups from USA and Japan reported EP 2D nano circles and rings [159,160]. In 1999, Zehner and Sita [159] prepared nanoscale copper circles via electroless deposition. They used an ultrathin film of microphase-separated poly(styrene-block-methyl methacrylate) (PS-*b*-PMMA) as a template, to generate nanometer-scale patterns of tetraalkylammonium passivated palladium colloids. This pattern serves, in turn, as the template for the electroless deposition of semicontinuous 20 nm wide copper nanostructures. Figure 4.31 indicates the TEM image of microphase-separated PS-b-PMMA. In addition, the two-dimensional Fourier transform (2D FT) of the image in the figure provides a measure of the repeat distance of the phase-separated pattern, which in this case is 41 ± 1 nm, as determined from the one ring that is visible in the 2D FT spectrum (see the inset of the figure).

In 2011, Sugimura et al. [160] fabricated concentric Au circles with a nm-scale line thickness on Si substrates by electroless plating and by using a scanning probe microscope as a lithographic tool. First, a microtemplate was fabricated on a hexadecyl self-assembled monolayer (HD-SAM) covalently bonded to a Si(111) substrate by scanning probe anodization lithography in which the

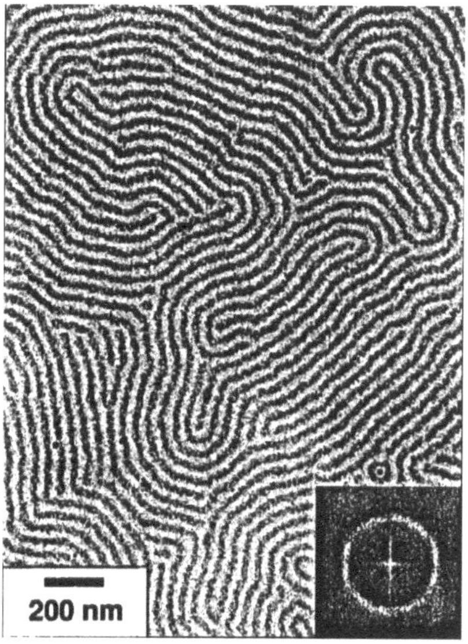

FIGURE 4.31 TEM of microphase-separated PS-b-PMMA. The inset shows a 2D FT spectrum of the image where the bright ring corresponds to a spacing of 41 ± 1 nm [159].

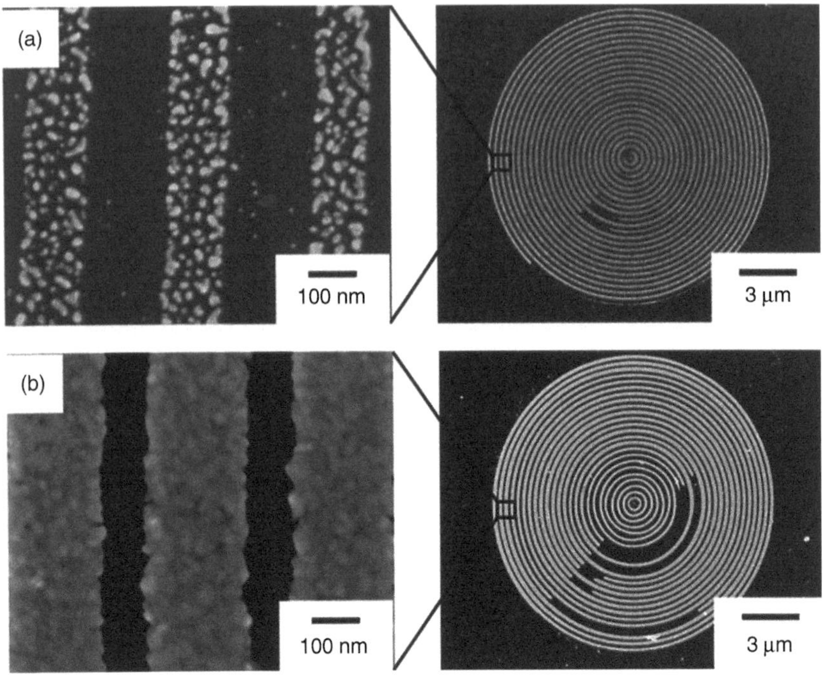

FIGURE 4.32 FE-SEM images of Au nano circles plated for 1 and 10 min on the template fabricated from the sample anodized circles served for 1-min plating (a), and for 10-min plating (b) respectively [160].

HD-SAM was locally degraded and the substrate Si was anodized along a trace of a conductive AFM tip biased negatively to the grounded Si substrate. Anodic Si oxide formed by this local anodization process was then etched with HF in order to expose the Si surface under the anodic oxide layer. Consequently, a Si nanopattern surrounded with the HD-SAM identical to the anodic oxide pattern was formed. This Si nanopattern served as a template for fabricating an Au nanostructure by means of electroless plating. As the ED reaction progressed, Au started to nucleate on the exposed Si surface, while no nuclei were formed on the surrounding HD-SAM. Figure 4.32 illustrates the FE-SEM images of Au nano circles. At a plating time of 1 min, the deposited Au does not form a continuous film as can be seen in a magnified FE-SEM image shown in Fig. 4.32a. At this plating time, the nucleation area is identical to the area where anodized by AFM, that is, lines of 120 nm wide. The height of the Au deposits is about 12 nm as confirmed by AFM. After plating for 5 min, Au nuclei have further grown so as to merge each other although gaps remain between the merged Au islands. The deposit height increased up to about 16 nm. In contrast to this slight growth of the Au deposits in the vertical direction, the Au islands have grown more markedly horizontally. The width of the Au deposition lines becomes to be 190 nm. At a plating time of 10 min, continuous Au lines could be fabricated, as demonstrated in Fig. 4.32b. Their height and width are 25 and 220 nm, respectively. In such a way, a nano Au multiring has been fabricated successfully.

4.2.7 EP 2D Nanohoneycomb

Between 2007 and 2011 four groups from Japan and Italy synthesized EP 2D nanohoneycomb patterns [161–164]. In 2007, Ono et al. [161, 162] prepared the patterning of an ordered honeycomb structure and a hexagonally arranged convex array of Cu and Ag on Si. They first fabricated the hole arrays on Si using self-organized anodic alumina. Then, they performed Cu or Ag dot arrays on Si by electroless deposition using anodic porous alumina as mask. Third, they prepared the metal honeycomb structure on Si by EP using colloidal spheres (polystyrene beads, PS) as mask. The EP Cu or Ag was laid down by dipping in 0.02 wt% $CuSO_4{\cdot}5H_2O$/47 wt% HF solution for 2 min or 1×10^{-4} mol /L $AgClO_4$/1×10^{-3} mol/L NaOH solution for 20 min. The authors evaluated the size of the deposited dots on the Si surface using AFM. The average height of the dots was in around 25 nm and the width was 80–90 nm. Each dot appears to consist of two or three particles. STEM images of Cu deposits shown in Fig. 4.33a indicate that the copper particles, with an average size of 20–40 nm, were coagulated. Figure 4.33b shows SEM images of Cu particles deposited on Si substrate composing a honeycomb structure obtained using 3 μm sized PS beads as masks for electroless plating. The framework of the Cu honeycomb pattern was composed of coagulated Cu particles with a size of approximately 50 nm. As for the Ag honeycomb structure, they found that the average size of the Ag particles is approximately 100 nm but it ranges from 10 nm to 250 nm, as shown in Fig. 4.33c.

In 2011, Ouchi et al. [164] reported a Co-Ni–P alloy electroless deposition process for fabricating ferromagnetic nanodot arrays. The bath solution and condition for electroless deposition are $NaH_2PO_2{\cdot}H_2O$ 100 mmol/L, N_2H_4 50 mmol/L, $(NH_4)_2SO_4$ 250 mmol/L, $NiSO_4{\cdot}6H_2O$ 16.3 mmol/L, $CoSO_4{\cdot}7H_2O$ 8.7 mmol/L, pH (adjusted with NH_4OH) 8.6, and temperature 50°C. In order to improve the crystallinity of the Co–Ni–P deposits in the initial deposition stage, the authors used a fcc-Cu (111) underlayer with low lattice mismatch to hcp-Co (0002) as substrate. Then nanopatterned substrates were formed by UV-nanoimprint lithography. Co–Ni–P films were electroless deposited on the Cu surface of the nanopatterned substrates by using dual reducing agents, $H_2PO_2^-$ and N_2H_4. Co–Ni–P layers demonstrated high perpendicular magnetic coercivities in the initial deposition stage since the highly crystalline hcp (0 0 0 2) Co–Ni–P layers were grown parallel to the Cu underlayers. Figure 4.34 shows representative SEM images (a–c) of Co–Ni–P nanodot arrays with diameters of 150 nm, pitches between dots of 300 nm, and heights of 50 nm fabricated by the autocatalytic electroless deposition process. In Fig. 4.34a, the Co–Ni–P nanodot arrays were clearly observed with no defect. Furthermore, it was confirmed that Co–Ni–P was deposited only on the Cu surface as shown in the cross-sectional SEM image of the Co–Ni–P nanodots (Fig. 4.34b).

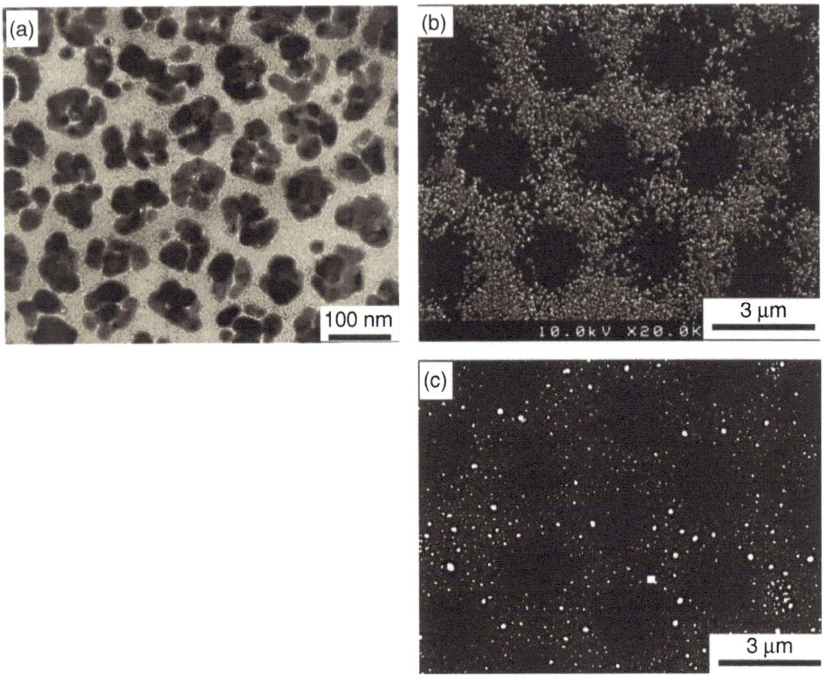

FIGURE 4.33 (a) STEM images of Cu dot arrays deposited on Si surface. (b) SEM images of Cu honeycomb structure on Si substrate using 3 μm sized PS beads as masks for electroless plating. (c) SEM images of Ag honeycomb structure on Si substrate using 3 μm sized PS beads as mask for electroless plating [161,162].

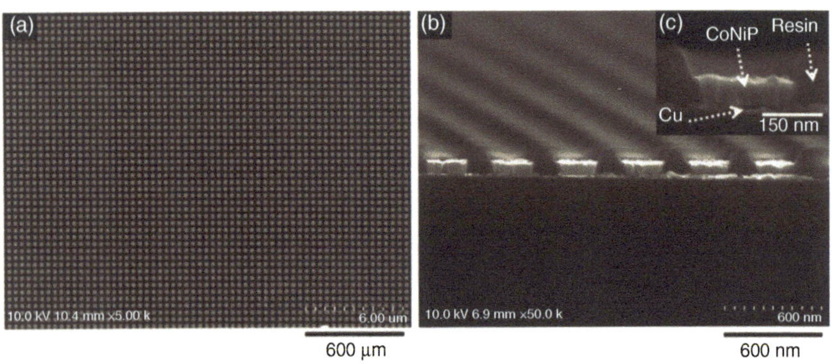

FIGURE 4.34 SEM images of CoNiP nanodot arrays with diameters of 150 nm, pitches between dots of 300 nm, and heights of 50 nm: (a) top view of SEM image, (b) cross-sectional view of SEM image, (c) highly magnified cross-sectional view of SEM image [164].

4.2.8 EP 2D Nanoline, Nanofin Pattern, and 2D Nano Grating

In 2008, Miyoshi et al. reported the fabrication of nanoline arrays of noble metals [165] and Ag [166] by an electroless plating and selective etching process. The fabrication procedure of nanoline arrays and metal nanofin arrays is resist line-pattern template→surface catalysis→metal EP→removal of top coat→removal of template. The bath solution and conditions for EP Cu are 0.05 M $CuCl_2$, 0.1 M H_3BO_3, 0.05 M ethylenediaminetetraacetic acid disodium salt dehydrate, and 0.1 M DMAB as a reductant, and kept at 50°C for 2 min. The bath solution and condition for EP Au are 3.5 mmol/L potassium dicyanoaurate dissolved in HGS-500 at 70°C for 10 min. The steps for fabricating the layer of Cu on Ag are as follows. First, EP the Ag layer was deposited on the template. Aqueous ammonia (0.2 mol/L) was added dropwise to an aqueous silver nitrate (0.15 mol/L) solution until a brown precipitate was formed and then dissolved. 1.9 mol/L glucose in a mixture of methanol/water (3:7, v/v) was used as reductant. The Ag film is deposited by adding the glucose solution to the silver nitrate solution, and then putting the template substrate into this mixture for 5 min. Second, this substrate was immersed in the Cu plating bath to form a Cu layer on the Ag layer. If just the step mentioned first is carried out, electroless Ag deposition can occur. Figure 4.35 shows cross-sectional SEM images of the metal nanolines and nanofins. The Cu nanoline is ca. 288 nm in height and ca. 67 nm in width. The finger-like structures between Cu nanolines were observed. The fabricated Au nanoline pattern was 364 nm in height and 31 nm in width. Experimental data show that the Ag and the Cu layers in the Ag/Cu bimetallic nanolines were not overlapped with each other, but have a clear-cut bilayer structure consisting of Ag and Cu metals. Measurement from the SEM image (Fig. 4.35) indicates the height and width of the Ag/Cu bimetallic nanolines are 337 and 118 nm,

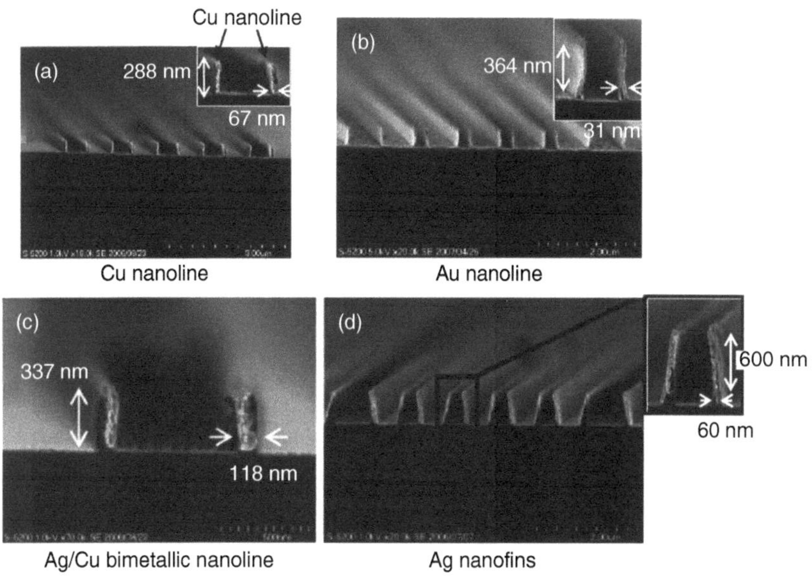

FIGURE 4.35 Cross-sectional SEM images of metal nanolines: (a) Cu nanolines, (b) Au nanoline, and (c) Ag/Cu bimetallic nanolines. (d) Cross-sectional SEM images of the Ag nanofins. The inset images in (a), (b), and (d) are magnified images [165,166].

respectively. As for the Ag nanofins, the height and width are ca. 600 and 60 nm, respectively.

In 2011, Han et al. investigated how to completely fill a 41 nm trench pattern with EP Cu [167]. First, the authors formed an amine-self-assembled monolayer (SAM) on the 41 nm trench pattern and subsequently attached the 4 nm AuNPs catalyst by dipping the substrate in a AuNPs suspension maintained at pH 6. Then, the trench pattern was filled with Cu by the EP method using a polyethylene glycol (PEG) bath additive. The bath composition and condition are $CuSO_4 \cdot 5H_i$ O 0.04 mol/L, EDTA 0.08 mol/L, Glyoxilic acid 0.08 mol/L, PEG 4000 1ppm, and conditions adjust pH to 12.5, at 70°C. Figure 4.36a illustrates the TEM image of Ta (barrier) coated 41 nm trench pattern activated by ca. 3 nm AuNPs. After 40 s of deposition (Fig. 4.36b), Cu filling was completed without defects or seams.

In 2007, Cui et al. [168] developed an approach to fabricating large area 50 nm period gratings having a 22 nm linewidth. They used a fabrication cycle twice, each combining nanoimprint lithography with a spatial frequency doubling based on electroless plating, lift-off, and reactive ion etching. The electroless Ni plating solution formulation is $NiSO_2 \cdot 6H_2O$ 16 g/L, Na citrate 10 g/ L, DMAB 0.2 g/ L, thiourea 0.8 mg/L, pH(NH_4OH) 8.5, and temperature 65°C. The plating rate is ~30 nm/min. Boron in deposit is <0.7% (unit is not indicated in the paper). After having the 100 nm period grating, they used it to carry out another cycle of nanoimprint and frequency doubling process to achieve a 50 nm period grating. The completed 50 nm period grating with about 22 nm

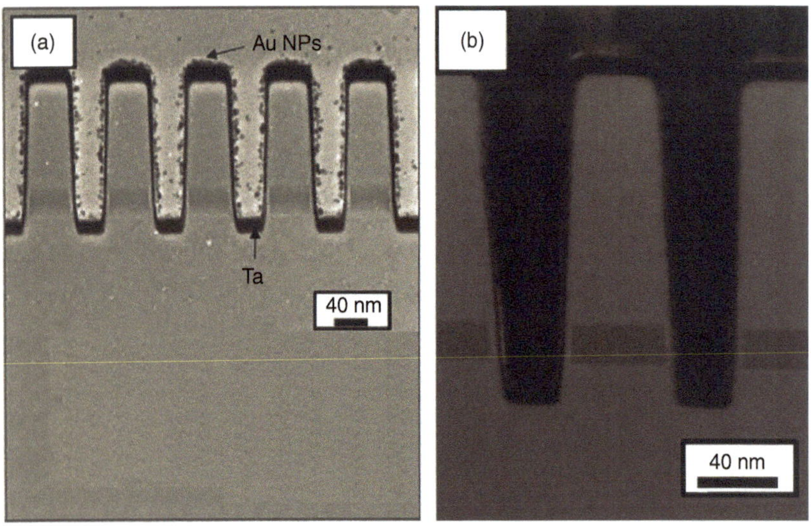

FIGURE 4.36 **(a) TEM image of Ta (barrier) coated 41 nm trench pattern activated by ca. 3 nm AuNPs. (b) TEM image of cross-sectional features of Cu deposited in 41 nm trenches from PEG 4000 1ppm bath at deposition times of 40 s [167].**

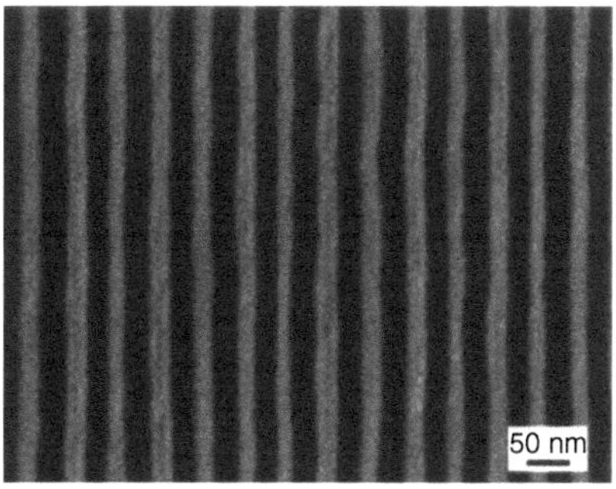

FIGURE 4.37 **50 nm period grating with 22 nm linewidth over a large area fabricated by doubling twice from a 200 nm period grating [168].**

linewidth is shown in Fig. 4.37. The line roughness and irregularity of the master 200 nm period grating are further amplified in the 50 nm period case.

4.3 LINEAR (1D) NANO EP MATERIALS

Because of their importance in research and developments the wide range of potential applications that they have, 1D NMs have attracted a lot of interest during the last decade. 1D NMs are ideal systems for exploring a large number of novel phenomena at the nanoscale and for investigating the size and dimensionality dependence of functional properties. In addition, it is just as pointed out by Kuchibhatla et al. in their review paper of 1D NMs [169] that "1D NMs has a profound impact in nanoelectronics, nanodevices, and systems, nanocomposite materials, alternative energy resources, and national security."

EP is the most effective method for manufacturing 1D NMs because of its convenience and low cost. To date, several kinds of 1D NMs have been reported, including nanotubes, nanowires, nanorods, nanobelts, nanoribbons, and hierarchical nanostructures. As pointed out in the beginning of this chapter, nano EP coatings or particles deposited onto 1D NMs such as the nanotubes, nanowires, nanorods, etc., also belong to the 1D NMs.

4.3.1 EP Nanotubes

CNTs were discovered early in 1991 [170], but it was not fabricated by an EP method. To date, there have been no reports of fabricating CNT by EP; however, due to their excellent properties, CNTs are very useful for supporting metal nanoparticles in many potential applications, ranging from advanced catalytic systems through very sensitive electrochemical sensors to highly efficient fuel cells. Therefore, functionalization of CNT with metal/alloy nanoparticles has received ever-increasing interest in recent years. This chapter will just describe briefly the materials such as nano metals and alloys + CNT composites by EP. In addition, metal, alloy, or compound nanotube and nanotube arrays are also discussed. Of course, only the 1 D NMs of such materials will be discussed in this section.

Use EP nano metals and alloys to reinforce CNTs can be traced back to 1994. Tsang et al. [171] used a simple chemical method to fill carbon nanotubes. Satishkumar et al. in 1996 [172] studied acid-treated CNT decorated with nanoscale Au, Pt, and Ag clusters. In 1997, Li et al. reported the coating of CNT with nickel by electroless plating method [173]. Li et al. in 1998 [174] studied nanoscale electroless Ni and Co deposition in aligned carbon nanotubes. Ang et al. in 1999 [175] investigated the electroless plating of metals onto carbon nanotubes activated by a single-step activation method. They believed that "No previous research has been published on carbon nanotubes activated by the single-step procedure." In 2000, Chen et al. [176] prepared the EP CNT + Co metal-matrix composite*. Since then, many papers looking at CNT + metals, alloys, and other materials, such as CNT + Cu [177–182], CNT + Ag [183–188], CNT + Au [189–193], CNT + Ni [194–196], CNT + Pd [175, 179, 197], CNT + Pt [198–201], CNT + Sn [202], CNT + Ni–P [175, 203–214], CNT + Ni–B [215], CNT + Co–P [216–218], CNT + Co-B [219], CNT + binary alloys [220–222], CNT + ternary alloys [223–225], and CNT + oxide [226] have been published.

Other than on reinforced CNT, few papers on EP nano metal/alloy/compound nanotubes and nanotubes arrays have been published.

Only one example of every kind will be discussed below.

4.3.1.1 EP Nano Metals Coated on CNT

As mentioned in the last paragraph, many EP nano metals such as Cu, Ag, Au, Ni, Pd, Pt, and Sn coated on the CNT have been published. Of these metals, more papers for Cu, Ag, Au, and Pt have been published, but for Pd and Sn only a few papers have been published.

4.3.1.1.1 EP Nano Cu Coated on CNT

Ang et al. in 2000 [177] reported the decoration of activated carbon nanotubes with copper. In order to obtain copper-decorated nanotubes, they immersed the activated and sensitized CNTs into 100 mL of Cu bath ($CuSO_4$ 3.1 g/L,

---

*Eder in his review paper of Carbon Nanotube-Inorganic Hybrids [Eder D, Chem Rev, 110 (2010) 1348–1385] said: "Electroless metal deposition was first mentioned by Qu et al., who observed that Au and Pt nanoparticles formed spontaneously on SWCNTs immediately after immersing the CNTs in a solution of $HAuC_{14}$ or Na_2PtC_{12}." It is obvious that this is not the case because the paper of Qu et al. published in 2005 (Qu L, Dai L, J Am Chem Soc, 127 (2005) 10806), which was later than the mentioned papers for many years. In another review paper by S R Bakshi, D Lahiri and A Agarwal [Intern Mater Rev, 55 (1) (2010) 41–64], the authors said, "The very first report of electroless deposition technique was by Chen and coworkers on a Co-CNT system to prepare 1D nanostructures for magnetic recording." The paper of Chen et al. was published in 2000 (Chen X, Xia J, Peng J, Li W, Xie S: Compos Sci Technol, 2000, 60, (2), 301–306). It is obvious as well that, this is also not the case.

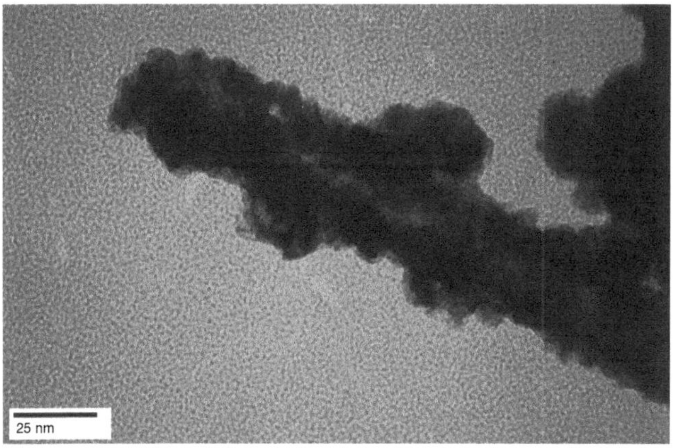

FIGURE 4.38 **A TEM micrograph of a copper-coated nanotube at higher magnification [176].**

disodium salt of EDTA 20.0 g/L, HCHO (added later) 20.0 g/L, Na_2SO_4 40.0 g/L, HCOONa 20.0 g/L, polyethyleneglycol 6.6 g/L, pH =7; temperature = 25°C) with stirred for 0.5 h, followed by addition of 2 g of HCHO with vigorous stirring. The CNT was well coated with a layer of copper electroless deposits. The coating appears to consist of Cu particles less than 10 nm in diameter. Figure 4.38 is a TEM micrograph of a copper-coated nanotube at higher magnification.

In 2011, Li et al. [181] electrolessly deposited a uniform and complete Cu coating onto MWCNTs. Three categories of nonmetallic materials, ceramics, plastics, and carbon were used as substrates. The substrate was dipped into a 60°C Cu electroless plating bath, which contained 6–8 g/L $CuSO_4$, 35–45 g/L ethylenediamineteraacetic acid (EDTA), 75 ppm 2′2′-dipyridyl, and 10–20 g/L glyoxylic acid. The experimental data show that there are numerous 10–20 nm Cu particles deposited on the CNT surface. The Cu film had an average thickness of 125 nm. Figure 4.39 shows the SEM of Cu electroless coated on organic Pd–Cu activated MWCNTs.

4.3.1.1.2 EP Nano Ag Coated onto CNT

In 2008, Byeon and Hwang [184] carried the electroless deposition of Ag on MWCNTs. After carrying out the pretreatment processes, the MWCNTs were immersed into a solution for the EP of silver. The silver EP bath was a mixture of solutions A and B at a 1:1 (v/v) ratio. Solution A contained 2 g of $AgNO_3$, 60 g of disodium ethylenediamine-tetraacetate (Na2-EDTA), 88 mL of isopropyl alcohol, 12 mL of acetic acid, and 400 mL of $NH_3 \cdot H_2O$ in 1 L of deionized (DI) water. Solution B contained 3 mL of hydrazine, 2 mL of mercerine, and 400 mL of ethyl alcohol in 1 L of DI water. They found that for EP of 5 and 10 min, the mean outer diameters of the coated MWCNTs were ~85 and ~105 nm, respectively. The experimental measurements show that the average diameters of the

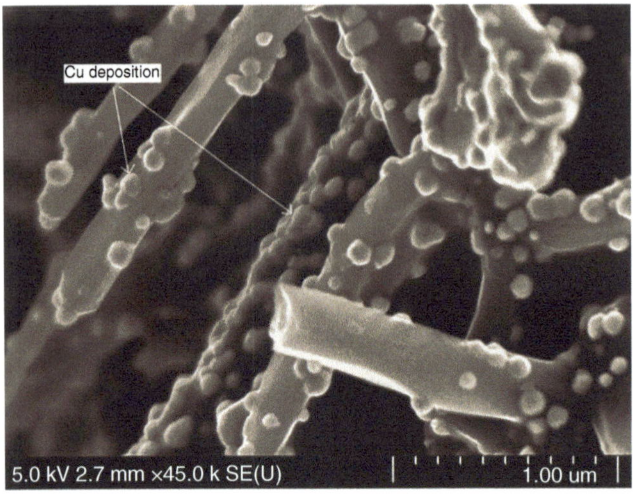

FIGURE 4.39 SEM of Cu electroless coated on organic Pd–Cu activated MWCNTs [180].

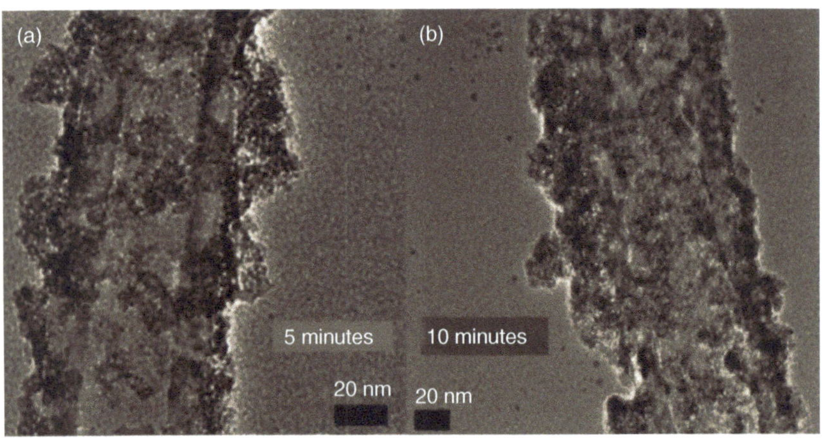

FIGURE 4.40 HRTEM images of Ag-coated MWCNTs, (a) for 5 min, (b) for 10 min [183].

metallic silver after EP for 5 and 10 min were 1.9 ± 0.33 and 2.6 ± 0.34 nm, respectively. Figure 4.40 is the high resolution transmission electron microscope (HRTEM) images of silver coated MWCNTs, (a) for 5 min, (b) for 10 min.

4.3.1.1.3 EP Nano Au Coated on CNT

Ma et al. in 2006 [191] studied the synthesis of Au nanoparticles supported on CNTs by using EP. The bath used for EP of CNTs in whiden gold consists of $KAu(CN)_2$ 0.7 g/L, KCN 5.5 g/L, KOH 5.6 g/L, KBH_4 5.0 g/L, Na_2EDTA 2.5 g/L, and glycolic amide 25 mL/L. The reaction was carried under 25°C for 10–20 min with a proper agitation. Figure 4.41 illustrates the HRTEM micrographs of the

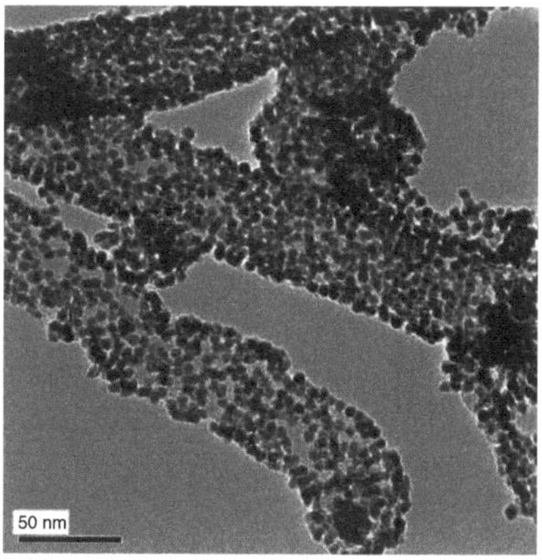

FIGURE 4.41 HRTEM micrograph of Au nanoparticle supported on a carbon nanotube [190].

Au nanoparticles supported on CNTs, which reveals the vary dense, and homogeneous dispersion of spherical gold nanoparticles on the surface of the CNTs, with a sharp particle size distribution centered at around 3–4 nm of diameter.

4.3.1.1.4 EP Nano Ni Coated on CNT

Lin in 2010 [196] studied decorating carbon nanotubes with Ni particles using an electroless deposition technique for hydrogen storage applications. The Ni electroless deposition bath that they used was composed of nickel chloride (11.9–23.7 g/L), DMAB (7 g/L), sodium acetate (13.2 g/L), and linear alkylbenzene sulfonate (0.1 g/L). The pH was adjusted to 5.5 using dilute HCl. The solution volume (mL) to CNT mass (mg) ratio was set at two. The plating temperatures were at 5–55°C for different periods of time (5 and 10 min). In addition, various Ni^{2+} concentrations in the plating bath (0.05–0.10 mol/L) were used. The Ni, C content, and particle size in the CNT/Ni nanocomposites prepared with various electroless deposition conditions (Ni^{2+} concentrations in the plating bath 0.05–0.10 M, temperature 5–25, and plating time 5–10 min) are from 1.9 wt% to 51.5 wt%, 98.1 wt% to 48.5 wt%, and 2 nm to 30 nm. Their energy dispersive spectrometer (EDS) analytical data show evidently that the particle mainly consists of Ni (B is too light to be detected with EDS) even though reducing agent of DMAB was added in the plating bath solution. However, from the insert in Fig. 4.42a, it can be seen that the nanoparticle is a single crystal because its high resolution lattice image represents the clear lattice fringes, and the distance between the lattice planes is 2.07 Å, which is slightly larger than that of pure Ni (2.034 Å). The lattice expansion can be attributed to a small amount of the B incorporation into the Ni crystal. Compared with Fig. 4.42a and b, it can be seen that the diameter of tiny nanoparticles decorated on CNT sidewall increases from approximately 5 nm to about 10 nm when the Ni^{2+} in the plating solution enhances from 0.07 mol/L (a) to 0.1 mol/L (b).

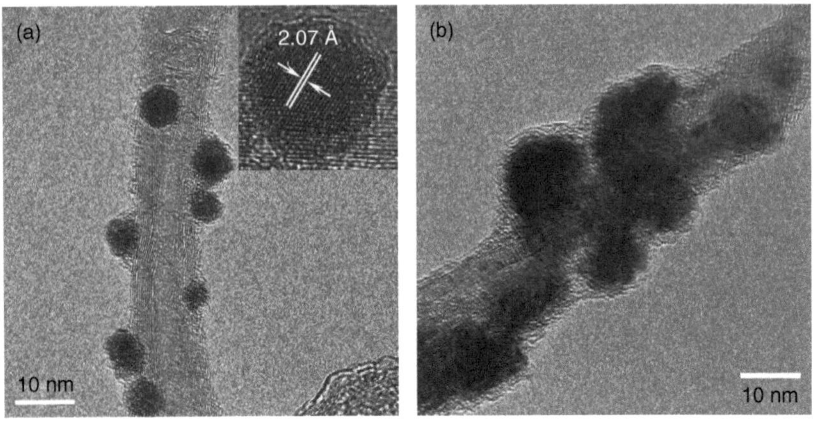

FIGURE 4.42 TEM micrographs of the CNT/Ni nanocomposites prepared from the plating solution with [Ni^{2+}] of 0.07 mol/L (a) and 0.1 mol/L (b), respectively, when the deposition time was 5 min and the reaction temperature was 5°C [195].

4.3.1.1.5 EP Nano Pd Coated on CNT

Ang et al. in 1999 [175] reported the electroless plating of Pd onto CNT activated by a single-step activation method. These authors believed that no previous research has been published on carbon nanotubes activated by the single-step procedure. The palladium electroless plating bath there used was prepared by dissolving 0.05 g of PdC_{12} in 0.1 mL of HCl and 10 mL of Millipore water. Sixteen milliliters of NH_4OH (28%) and 2.7 g of NH_4Cl were added, and the solution was left to stand for 4 h. One gram of hypophosphite NaH_2PO_2•$2H_2O$ was stirred into the solution, and the volume was made up to 100 mL with Millipore water. The pH of the bath was adjusted to 8.25 with ammonium hydroxide, and it had a bath temperature of 25°C. Figure 4.43 shows that Pd nanoparticles with ~5 nm in diameter coated the CNT densely.

4.3.1.1.6 EP Nano Pt Coated on CNT

In 2008, a group from China and the USA [201] prepared crystalline Pt nanoparticles electrolessly deposited on poly(acrylic acid) (PAA) grafted MWCNTs (PAA-g-MWCNTs). They deposited Pt nanoparticles on the PAA-g-MWNTs from the platinum precursor, chloroplatinic acid H_2PtC_{16}•$6H_2O$, using ethylene glycol (EG) as solvent and reducing agent. A typical experiment is as follows: 0.05 g PAA-g-MWNTs (PAA grafting amount, 15 wt%) was charged in a 25 mL flask containing 10.0 mL EG and 0.025 g H_2PtC_{16}•$6H_2O$. The reaction temperatures were different, depending on the addition of ions. The HRTEM images of Pt/MWCNTs nanocomposites (Fig. 4.44) show that the average size of Pt particles increased with the concentration of Pt precursors from ~1.3 nm (H_2PtC_{16}•$6H_2O$/EG) 0.1 mg/mL), 3.2 nm (0.2 mg/mL) to 8.7 nm (1 mg/mL), as determined by Sherrer's formula through line broadening of the Pt (220) peak.

FIGURE 4.43 TEM micrograph of activated and accelerated nanotubes coated with palladium nanoparticles ~5 nm in diameter. Copper holey carbon support films were used [175].

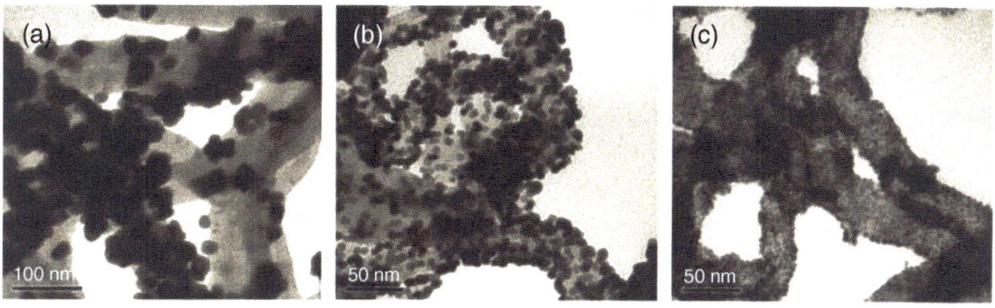

FIGURE 4.44 HRTEM images of Pt/MWCNTs nanocomposites prepared from different $H_2PtCl_6 \cdot 6H_2O$/EG ratios: (a) 1 mg/mL, (b) 0.2 mg/mL, and (c) 0.1 mg/mL at 160°C for 30 min [201].

4.3.1.1.7 EP Nano Sn Coated on CNT

In 2007, Liu et al. [202] deposited Sn nanoparticles on CNT by electroless plating. Cu plates were used as substrates for the growth of carbon nanotubes (CNTs) and deposition of Sn. The sample needed to be sensitized and activated before electroless Sn deposition. The plating bath composition and operating parameters for electroless Sn deposition were $SnCl_2 \cdot 2H_2O$ (0.20 mol/L), HCl (0.6 mol/L), $NaH_2PO_2 \cdot H_2O$ (0.8 mol/L), $Na_3C_6H_5O_7 \cdot 2H_2O$ (0.1 mol/L), $CO(NH_2)_2$ (1.4 mol/L), temperature (30°C), and time (10 min). Figure 4.45 shows a FE-SEM image of CNTs deposited on Sn by a displacement method. The diameters of CNTs displacing Sn are about 100 nm, not larger than that before Sn displacement. The EDS spectrum of this sample shows that the sample contains elements including C, Sn, Cu, O, and Pd. The electroless Cu on the surface of CNTs cannot be completely displaced by Sn. The Sn in the sample also is partly oxidized by oxygen in air.

4.3.1.2 EP Nano Ni–P, Ni–B, Co–P, and Co–B Coated on CNT

Many reports about EP of nano Ni–P coated on CNT have been published [175, 203–214], but only a few about CNT + Ni–B [215], CNT + Co–P [176, 216–218], CNT + Co-B [219].

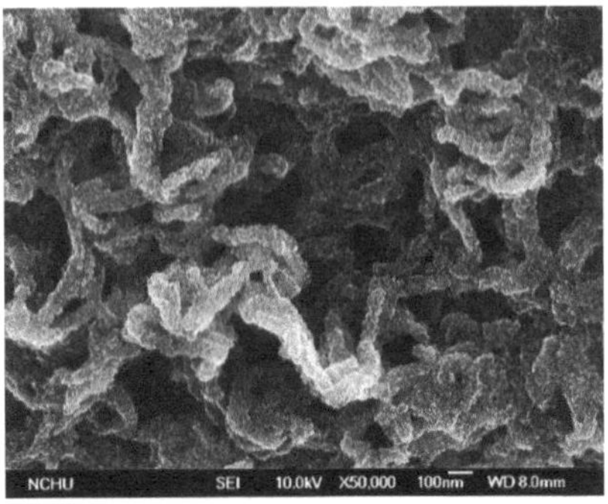

FIGURE 4.45 **FESEM image of CNTs with displacing Sn [202].**

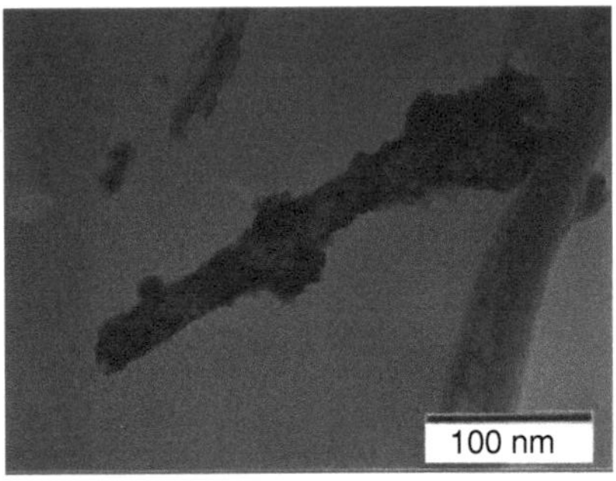

FIGURE 4.46 **TEM micrograph of an activated and accelerated tube coated with Ni–P electroless deposit [175].**

We discuss EP CNT + Ni–P first. In 1999, Ang et al. [175] reported the electroless plating of Ni–P onto CNT activated by a single-step activation method. The Ni–P electroless plating bath they used was composed of 30 g/L $NiCl_2$•6H2O, 10 g/L $NiSO_4$•$6H_2O$, 15.5 g/L $Na_3C_6H_5O_7$•$2H_2O$†, 100 g/L NaH_2PO_2•H_2O†, 100 g/L NH_4Cl, and 2.5 g/L $Pb(NO)_3$, the pH was adjusted to 8.25 with NH_4OH, and the bath temperature was 25°C. The activated and accelerated CNTs were stirred in 50 mL of the bath solution for 30 s in order to obtain the decorated Ni–P on them. Under the conditions employed, electroless plating was rapid, and resulted in tubes which were almost fully covered with a layer of the deposit. The thickness and density of the coating can be varied by adjusting the concentration of $H_2PO_2^-$ reductant. The experimental data revealed the presence of Ni and codeposited P in the coating on CNT, and the structure was essentially amorphous. Figure 4.46 shows a carbon nanotube densely decorated with a Ni–P electroless deposit of particles of ~2 nm in diameter.

In 2006, Wang et al. [207] reported the synthesis of CNT-supported Ni–P nanoparticles by electroless plating. The CNTs with diameter of 100–200 nm

†The original paper stated $Na_3C_6H_5O_7$•$1.5H_2O$ and NaH_2PO_2•$2H_2O$, it should be $Na_3C_6H_5O_7$•$2H_2O$ and NaH_2PO_2•H_2O, respectively.—The author.

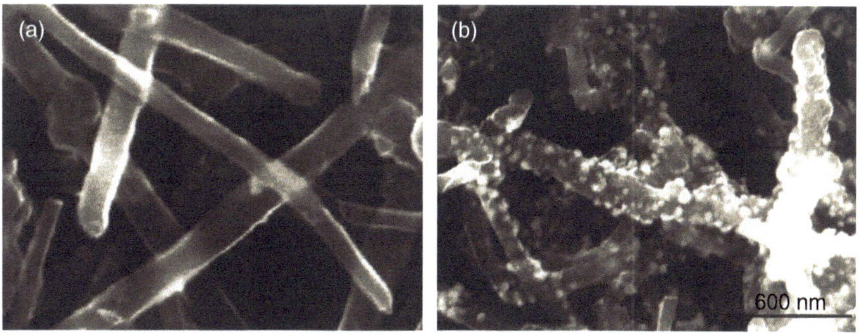

FIGURE 4.47 FE-SEM images of CNTs before (a) and after (b) the electroless deposition for Ni–P nanoparticles [207].

were used in the study. After a pretreatment procedure was carried out on the surface of CNTs, the electroless Ni–P deposition was conducted into Ni–P plating bath at the temperature of 60°C for 1 min, followed by a rapid cooling into iced water. As seen in Fig. 4.47b, the uniformly dispersed Ni–P nanoparticles are firmly attached on CNT surfaces. The size of the well-dispersed nanoparticles is estimated to be around 40 nm in diameter. Electron probe X-ray microanalyzer (EPMA) quantitative analysis shows that the P content in the CNT-supported Ni–P nanoparticles is about 20–21 wt%, and the loading of Ni–P nanoparticles on the CNTs is estimated at about 14.0 wt%.

In 2012, Liu et al. [215] studied the electroless preparation and characterization of Ni–B alloy nanoparticles supported on multiwalled carbon nanotubes. The MWCNTs used in the study were typically 50–80 nm in diameter and 10–20 mm in length. The acid treatment of the high crystalline MWCNTs was conducted in order to introduce functional groups onto the inert nanotube surface prior to the electroless deposition. The activated MWCNTs were placed into 10 mL of deionized water containing polydimethylsiloxane (PDMA, 0.01 g) at 50°C. Then, the mixture was placed into a 20 mL-plating bath comprising $NiC_{12}•6H_2O$ (6.86×10^{-2} mol/L), ethylenediamine ($4.14 - 10^{-1}$ mol/L), $NaBH_4$ (2.18×10^{-1} mol/L), and NaOH (1.5 mol/L) with the temperature of the plating bath kept at 50°C. Electroless plating was conducted with the temperature held steady for 10 min with magnetic stirring. Figure 4.48 shows the SEM image of some N–B +MWCNT samples. It can be seen that the deposited Ni–B NPs are homogenously distributed on the surfaces of the MWCNTs with the size distribution centered at around 60 nm. The inset in Fig. 4.48a shows the high-magnification SEM image of a single Ni–B particle connected to the MWCNT, in which many subunits with the size of ~5 nm can be observed at the surface of the single spherical particle. The authors believe that the subunit feature is quite special and has never been reported as far as they know. They argued that it might be contributing to the larger surface area of Ni–B NPs.

Considering EP CNT + Co–P deposition, Chen Xiaohua et al. reported its preparation rather early in 2000 [176]. The composition of the plating solution

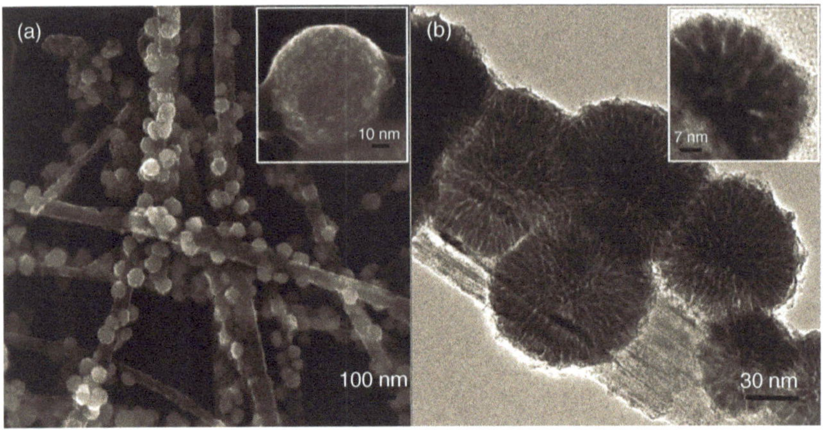

FIGURE 4.48 SEM image of the Ni–B+ MWCNT samples, inset is high magnification image (a) and (b) HR-TEM images of Ni–B+ MWCNT samples, inset is high magnification image [215].

and the reaction conditions are: 20 g/L $CoSO_4{\bullet}7H_2O$, 25 g/L $NaH_2PO_2{\bullet}H_2O$**, 24 g/L $(NH4)_3C_6H_5O_7$, 4.5 × 10^{-2} $Pb(NO_3)_2$, temperature 35°C and pH adjusted using NH_4OH. From the SEM/EDX analyses, it can be seen that in Fig. 4.49a the Co–P deposit is formed as spherical grains, which appear as very closely packed crystallites; in addition, the Co–P tended to aggregate as nanoparticles on the outer surfaces of the nanotubes, but the nanotubes are not completely or uniformly coated. It must be pointed out that the authors did not measure either the content of P or the size of the deposited Co–P nanoparticles. Lu et al. in 2008 [218] also reported the preparation of CNT + Co–P nanoparticles. The plating solution and the reaction conditions they used are a little different: 0.036 mol/L $CoSO_4{\bullet}7H_2O$, 0.236 mol/L $NaH_2PO_2{\bullet}H_2O$, 0.485 mol/L H_3BO_3, 0.496 mol/L $C_4H_4KNaO_6{\bullet}4H_2O$, temperature 70°C, pH 9 adjusted using NaOH, and plating time 30 min with stirring. The results show that the structure of particles was a solid solution of phosphorus dissolved in a microcrystalline cobalt matrix with the size of 15–40 nm as seen in Fig. 4.49b. Comparing the results from Fig. 4.49a [176], the cover of Co–P nanoparticles on CNTs was rather densely and uniformly. However, they still did not measure the content of P in the nanoparticles.

In 2003, Liu et al. [219] studied a simple method for coating CNTs with a layer of Co–B amorphous alloy. In order to obtain the CNT + Co–B composite, the authors prepared a saturated $Na_2[Co(OH)_2]$ solution by dissolving $Co(OH)_2$ to a 6 mol/L NaOH aqueous solution. The preoxidized CNTs were sonicated in the saturated $Co(OH)_2$ solution. Then, a 1.0 mol/L KBH_4 solution was added slowly into the above solution. The plating reaction lasted 10–30 min (typically 25 min) with continued vigorous stirring. Figure 4.50 shows the HRTEM image of a coated Co–B/CNT having an inner diameter of 25 nm. There is a Co–B coating layer apparently covering the surface of the coated nanotubes. The coating

**The original paper wrote as $NaH_2PO_2{\bullet}2H_2O$, it should be $NaH_2PO_2{\bullet}H_2O$.—The author.

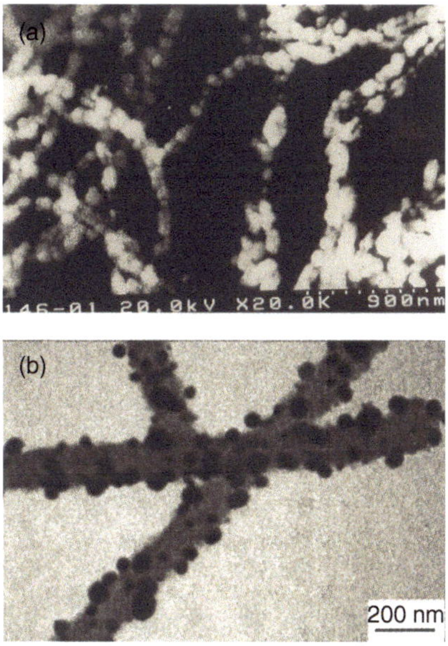

FIGURE 4.49 (a) SEM image of the cobalt-coated carbon nanotubes by electroless plating: pH = 10.0, t = 18 min, cobalt deposits are formed by grains on the surface of carbon nanotubes [176]; (b) TEM image of carbon nanotubes plated with Co–P nanoparticles [218].

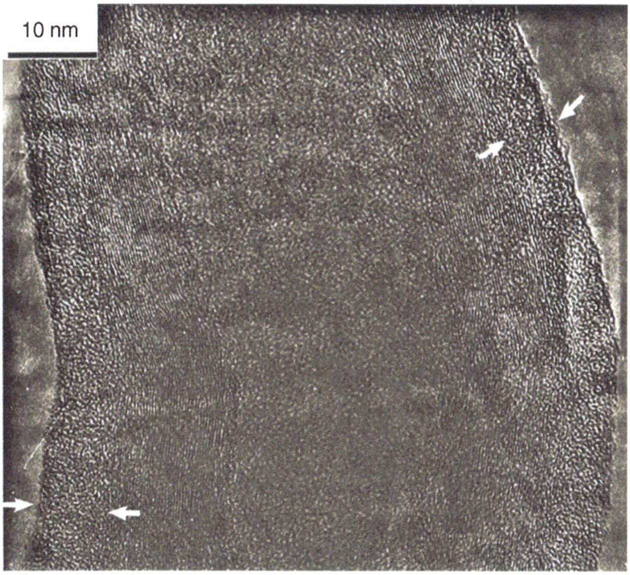

FIGURE 4.50 A typical HRTEM image of a Co–B-coated nanotube. Arrows indicate a coating layer with an image contrast pattern typical of an amorphous structure [219].

layer is around 7 nm in thickness in this case. The measured results from X-ray diffraction (XRD) and induced coupled plasma (ICP) spectroscopy indicate that the coated Co–B layer was amorphous character, and that the coating layer contained both cobalt and boron with the atomic composition of $Co_{73.9}B_{26.1}$.

4.3.1.3 EP Nano Binary Metallic Alloys Coated on CNT

Only a few papers reported EP nano binary metallic alloys coated on CNT. In 2007, Cai et al. [220] prepared one-dimensional nanoseale composite based on CNTs with magnetic Ni–Co alloy coating by EP. The EP bath solution is composed of $CoSO_4$•$7H_2O$, $Na_3C_6H_5O_7$•$2H_2O$, NH_4Cl, NaH_2PO_2•H_2O. The pH of the bath is ~8, and the plating temperature is 75°C. The plating process is mechanically stirred thoroughly. The compositions of the Ni–Co alloy coating are from $Ni_{80}Co_{20}$ to $Ni_{80}Co_{20}$ as the ratio of Ni:Co (molar ratio) in the bath depletes from 8:2 to 2:8. Figure 4.51 shows the SEM (a) and TEM (b) images of $Ni_{80}Co_{20}$. The CNT is covered near completely with Ni–Co nanoparticles. The thickness of the Ni–Co coating is around 60 nm.

In 2010, Zhao et al. [221] prepared a deposit Co–Fe alloy on the surface of MWCNTs by electroless plating. The EP Co–Fe alloy was produced in an alkaline solution, containing 0.099 mol/L cobaltous sulfate, 0.011 mol/L ferrous sulfate, 0.31 mol/L sodium hypophosphite NaH_2PO_2 as reducing agent and 0.32 mol/L sodium citrate $Na_3C_6H_5O_7$ as complexing agent. The pH of the bath was 9 adjusted by ammonia solution, and the process was optimized at 60°C. The experimental data show that a layer of Co–Fe alloy deposits on the surface of MWCNTs and the contents of Co and Fe are fairly high. But the authors did not show the contents of Co and Fe, in addition their EDS results (original Fig. 1c) show some P content exists in the Co–Fe alloy deposit. An field-emission transmission electron microscope (FETEM) image of MWCNTs after electroless plating is shown in Fig. 4.52, which indicates the size of the alloy particles must be less than 100 nm, but the authors still did not illustrate how large are they.

In 2009, Huang et al. [222] studied the CNTs + Co–Sn nanocomposites produced by EP. The electroless bath consists of 75 g/L Na_2SnO_3•$3H_2O$, 3.5 g/L CoC_{12}•$6H_2O$, 150 g/L $C_4H_4O_6KNa$ •$4H_2O$, and 20 g/L $K_3C_6H_5O_7$•$4H_2O$. In order to enhance the dispersion of the CNT powder, the suspension was sonicated by

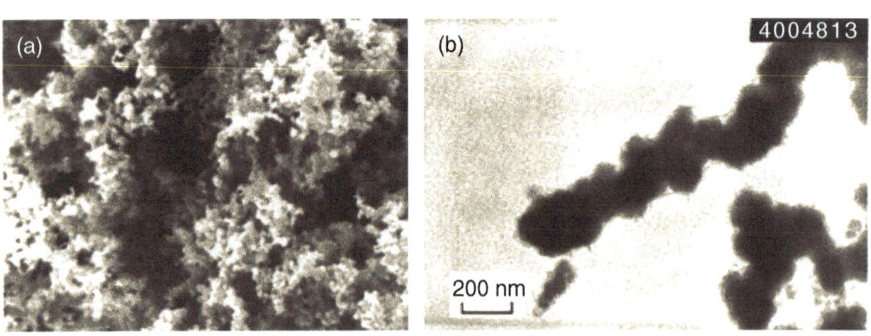

FIGURE 4.51 **SEM (a) and TEM (b) images of composite $Ni_{80}Co_{20}$ [220].**

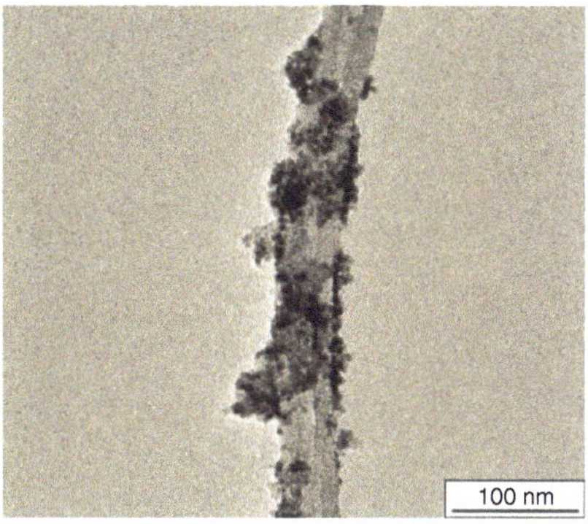

FIGURE 4.52 **FETEM images of MWCNTs after electroless plating Co–Fe alloy [221].**

ultrasonic vibration for 28 h, followed by gently adding $NaBH_4$ solution under continuous and vigorous stirring until the air bubbles no longer emerged. The CNTs + Co–Sn composites were then filtrated and subsequently washed with large amounts of distilled water till the pH of filtrate near 7. Figure 4.53a presents typical TEM image of the CNTs + Co–Sn nanocomposite morphology, which clearly exhibit the distribution of Co–Sn alloy nanoparticles with a size of about 15 nm on average on the surface of the CNTs. Figure 4.53b shows EDS spectrum and the contents of C, Co, Sn, and O of CNTs + Co–Sn nanocomposites. It can be seen that the as-coated material on CNTs is an alloy of Sn and Co rather than simple mixture.

4.3.1.4 EP Nano Ternary Alloys Coated on CNT

Some papers of EP nano ternary alloys coated on CNT have been published in recent years mainly from few groups in China [223–225]. He et al. in 2008 [223] prepared an electroless Ni–Co–P alloy plating on CNT. The bath solution and operation condition are 12–15 g/L $NiSO_4$•$7H_2O$, 13–16 g/L $CoSO_4$•$7H_2O$,

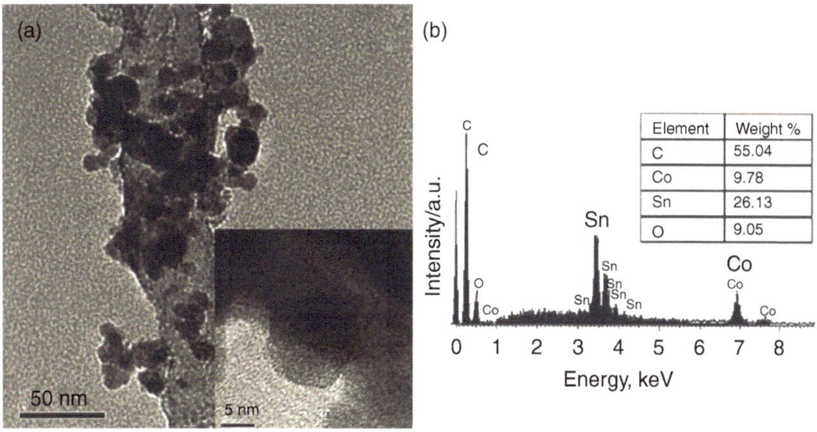

FIGURE 4.53 **(a) TEM image spectrum of CNT +Co–Sn nanocomposite at a high magnification level, and (b) EDS of Co–Sn/CNT nanocomposite [222].**

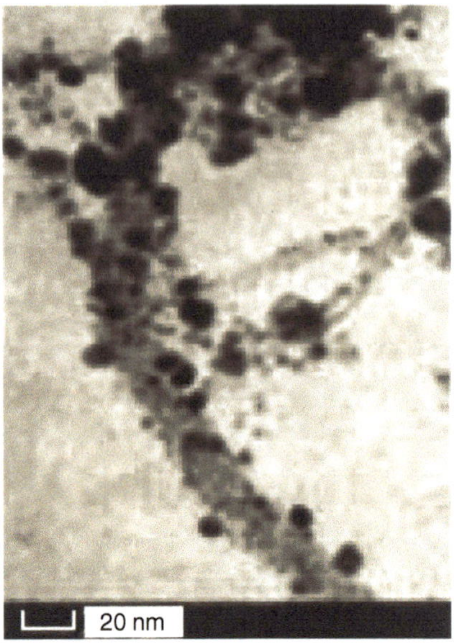

FIGURE 4.54 TEM image of Ni-Co–P alloy coated CNTs [223].

34–37 g/L $NaH_2PO_2 \cdot H_2O$, 80–82 g/L $Na_3C_6H_5O_7 \cdot 2H_2O$, 48–51 g/L $(NH_4)_2SO_4$, 1 mg/L surfactant, the pH value 8.5–8.8, the plating temperature 20°C, and the plating time 2 h. Before plating, the CNTs in the plating solution were subjected to ultrasonic dispersion for 5 min. The whole plating process has been mechanical stirred. Figure 4.54 shows a TEM image of Ni–Co–P alloy coated CNTs. It can be seen that a thickness of 8–15 nm of Ni–Co–P alloy layer evenly covered the surface of the CNTs. The measured results show that the structure of the as-plated Ni–Co–P alloy coating is amorphous, and the contents (mass fraction) are Ni 19.98 wt%, Co 21.62 wt%, and P 3.28 wt%. In addition, there is some O in the coating. Wang et al. in 2011 [224] and Ye et al. in 2009 [225] prepared Ni–Co–P alloys coated carbon nanotubes as well.

4.3.1.5 EP Nano Oxide on CNT

Just one report on EP CNT + nano oxide has been published. In 2008, Reddy et al. [226] used a new one-step synthesis method for coating multiwalled carbon nanotubes with cuprous oxide nanoparticles. In a typical procedure, 150 mg of purified MWCNTs were first suspended in 30 mL of 1 wt% copper sulfate solution containing a mixture of tartaric acid and sodium hydroxide, and then stirred with ultrasonication for 30 min. Next, 0.5 mL formaldehyde was added dropwise to the suspension, and the reaction mixture was stirred at 60°C for 1.5 h. It can be seen from the TEM images of Cu_2O nanoparticles that the particles were well dispersed onto the CNTs (Fig. 4.55a). And the size of the Cu_2O nanoparticles increased from 21 nm to 32 nm as the concentration of copper sulfate was increased from 1 wt% to 3 wt%. A high-magnification high-voltage

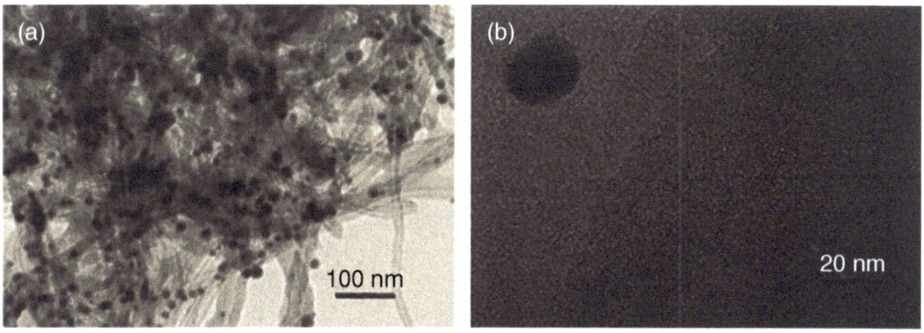

FIGURE 4.55 **TEM images of Cu_2O-coated MWCNTs prepared using (a) 1% solutions of copper sulfate and (b) high-magnification highvoltage TEM image of the composites in (a) [226].**

TEM image (Fig. 4.55b) also clearly shows the walls of the CNTs and the Cu_2O nanoparticles coated on the surface of the nanotubes in the composite. The authors found that the interaction between the Cu_2O nanoparticles and the nanotubes was quite strong because thorough washing did not.

4.3.1.6 EP Nano Metal/Alloy/Compound Nanotubes and Nanotubes Arrays

4.3.1.6.1 EP Nano Metal Nanotubes and Nanotubes Arrays

Electroless deposition of open-end Cu nanotube arrays was undertaken by Li et al. in 2004 [227]. First, the sensitized and activated anodic Al oxide (AAO) template was stirred in 50 mL of Cu bath at room temperature for various deposition times. The composition of the plating bath they used was 7 g/L of $CuSO_4$•$5H_2O$, 25 g/L of potassium sodium tartrate, and 9.3 g/L of formaldehyde. The pH value was adjusted to 7–8 by sodium hydroxide. From the SEM image, the outer diameter of the Cu nanotubes is about 300 nm, corresponding well to the pore size of the AAO template. The inner diameter of the hollow tubes is related to the electroless deposition conditions, which can be simply controlled by the deposition time. The wall thickness of the tubes increased with deposition time, and solid nanowires were obtained eventually. However, the authors did not indicate the thickness of the tubes. EDX analysis of the Cu nanotubes containing an AAO membrane confirms the presence of copper element; in addition, a multiple Pd peak is observed due to the activity procedure, which confirms the deposition of Pd nanoparticles on the membrane surface.

Early in 1994, Brumiik et al. [228] prepared EP Ag nanotubes. Electroless silver was deposited using a two-part (A and B) bath. Part A consisted of 454 g silver nitrate, 355 mL ammonium hydroxide, and 4.54 L water. Part B consisted of 1,590 g Rochelle salt (potassium sodium tartrate), 114 g Epsom salt (magnesium sulfate), and 3.64 L water. Equal volumes of A and B were mixed with 14.1 parts water. The temperature of the electroless bath was 35°C. They have prepared Ag nanotubules with diameters as small as 30 nm. They have proved that tubules of this extremely small diameter are hollow because liquid flows

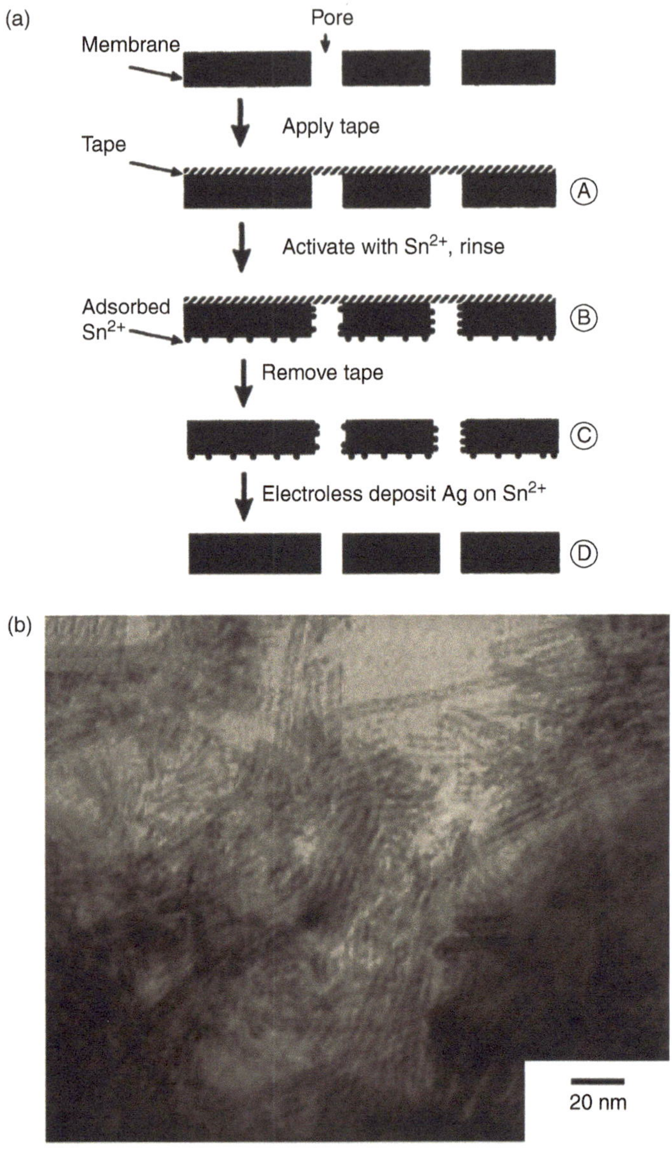

FIGURE 4.56 **(a) Schematic of the electroless deposition for forming Ag tubules [228], and (b) TEM image of an aggregate of Ag nanotubes with inner and outer diameters 4 and 7 nm, respectively obtained by reducing their precursory mixed surfactant LC phases with hydrazine [230].**

freely through the tubule impregnated polycarbonate membrane. Membranes of this type are commercially available with pores as small as 15 nm.

Zhang et al. and Kijima et al. also prepared the Ag nanotubes [229, 230]. Figure 4.56 shows (a) the schematic of the electroless deposition for forming Ag tubules [228], and (b) TEM image of Ag nanotubes obtained by reducing their precursory mixed surfactant LC phases with hydrazine.

Just a few groups reported the fabrication of EP Au nanotube or array of nanotubes [231–235]. Wen and Yang in 2002 [231] used a simple redox reaction to deposit continuous Au films on the Cu_2S nanowire surfaces, forming a novel type of Cu_2S/Au. The Cu_2S core was typically ~60 nm thick, and the Au coating was from a few nm up to 100 nm thick depending on the coating time period and the concentration of Au^{3+} in the coating solution. The Au coating layer was found to be polycrystalline. For relatively thick Au coatings, pure gold nanotubes can be successfully fabricated by etching away the Cu_2S materials core. Figure 4.57 shows the TEM image of a typical gold nanotube. It has diameter of ~300 nm and a wall thickness of 50 nm. Clearly, one end of the gold tube is closed and the other end is open. The ED pattern showed the polycrystalline fcc structure of the gold tube and the absence of the Cu_2S crystals originally in the core.

In 2012, Yu et al. [235] prepared gold nanotube (GNT) membranes using gold deposition on a porous AAO template. Electroless gold deposition on AAO was performed via several steps. The first step is called sensitization. Then, the sensitized AAO was immersed into 100 cm^3 ammoniacal solution of 0.029 M $AgNO_3$ for 30 min, in which a redox reaction leads the formation of a thin layer of silver nanoparticles on the membrane surface to form catalytic sites for the oxidation of formaldehyde and the concurrent reduction of Au(I) to Au(0). The

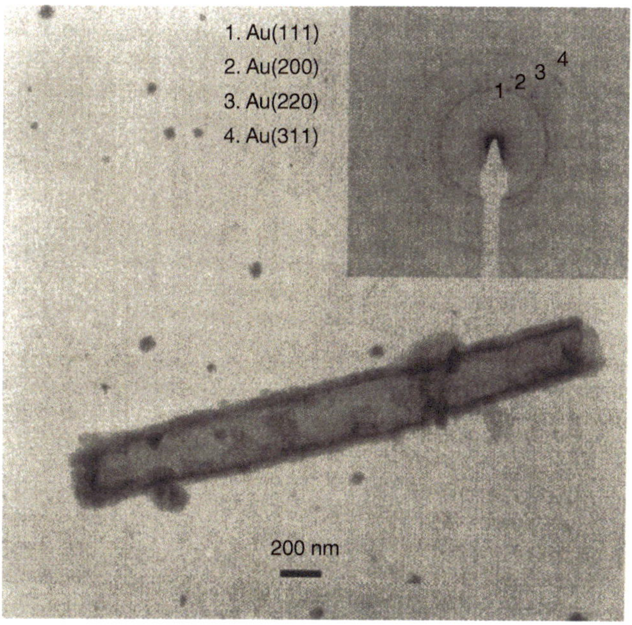

FIGURE 4.57 TEM image of a Au nanotube after the removal of the Cu_2S core. Inset: ED pattern of the Au nanotube [231].

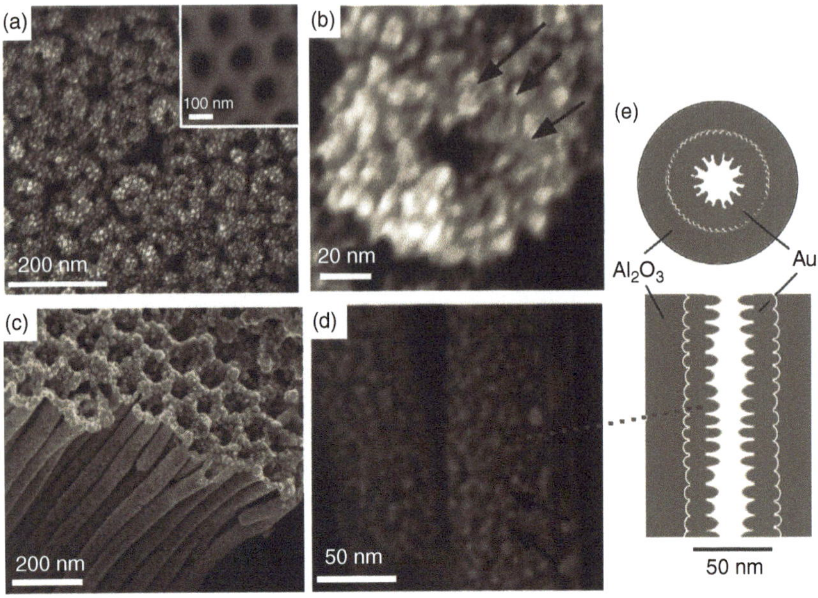

FIGURE 4.58 **HR SEM images of the GNT structures; (a) the top view of GNT–AAO membranes showing clustered top surface. The surface of AAO pores before deposition is shown in the inset; (b) the image of single pore with gold clusters (arrows) about 5 nm in diameter; (c) GNT structure after removal of AAO showing their intake structure across whole membrane; (d) details of fractured Au nanotube structure with clustered Au surface (arrows) inside of nanotubes; (e) schematic diagram of single GNT–AAO structure (top and profile) which shows Au clusters on inner and outer nanotube surface (the AAO layer is only partially presented) [235].**

third step is called displacement deposition (electroless deposition step), the silver coated membrane was washed with Milli-Q water, and then was immersed into a standard gold plating solution containing 0.079 mol/L $Na_3Au(SO_3)_2$, 0.127 mol/L Na_2SO_3, 0.625 mol/L formaldehyde, and 0.025 mol/L $NaHCO_3$. The pH was adjusted to pH = 8 and pH = 9 in two sets of preparation experiments and the temperature was kept at ~1°C for the required deposition time (24 h). Figure 4.58 shows the details of the plating GNT structures and their nano nature.

EP Ni nanotubes were reported in 2006 [236]. The AAO membranes used for fabricating the Ni nanotubes were obtained from Whatman Inc. with the diameter of 250–300 nm and thickness of 60 mm. The AAO membranes were pretreated and sensitized before carrying out the electroless deposition. The nickel-plating bath is composed of nickel sulfate, nickel chloride, lead nitrate, sodium hypophosphite, sodium citrate, and ammonium chloride. Figure 4.59 indicates the top-view SEM image of EP Ni nanotubes. However, the authors did not report the size of the Ni nanotubes. The EDX spectrum of the Ni nanotubes containing the AAO membrane shows the presence of nickel element.

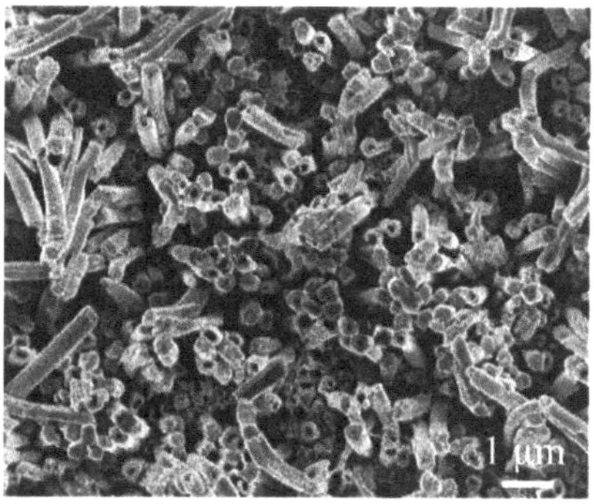

FIGURE 4.59 **Top-view SEM images of EP Ni nanotube arrays [236].**

Integrated arrays of Pd and Pt nanotubules with high aspect ratios on Al sheets were synthesized by Chu et al. in 2008, using a critical-potential anodization and an electroless deposition [237]. The critical-potential anodization provided ordered porous anodic alumina (PAA) films as supporting materials, with controllable pore intervals of 50–980 nm by anodizing potentials and adjustable pore sizes in 20–900 nm by chemical dissolution in acidic or alkali solutions. The Pd electroless deposition they used was carried out in a neutral bath of pH 6–7 using sodium hypophosphite NaH_2PO_2 as a reducing reagent. The bath was composed of 0.02 M PdC_{12}, 0.05 M Na_2HPO_2, 0.08 M $NH_2CH_2CH_2NH_2$, and 30 ppm 2,2-S($CH_2COOH)_2$, operated at 40–60°C for 30 s to 2 min with magnetic stirring. The Pt electroless solution the authors used was an alkali bath pH 10–12 using hydrazine as a reducing reagent, operated at 60–80°C for 3–7 min with magnetic stirring, which mainly contained 0.02 M H_2PtC_{16}, 0.3 M NH_2NH_2, and 0.8 M $HONH_3Cl$, with an appropriate amount of ammonia solution 25–28% for pH adjustment. After EP, Pd was uniformly deposited on the surface and pore walls of PAA films as a thin film, giving a nanoporous Pd film (pore diameter 80–200 nm (Fig. 4.60a) or Pd nanotubule arrays (tube diameter 320 nm, tube wall thickness 60–120 nm, tube length 4–5 μm (Fig. 4.60b)). In the case of short-time deposition, integrated arrays of Pd nanotubules with open ends were formed because deposition near the pore entrance was faster than the inside and the base of pores. With prolonged deposition time and proper activating treatment, uniform Pd coating down to the pore bases of the PAA films (Fig. 4.60b) can be achieved, thus leading to an integrated array of Pd nanotubules with one end closed, which completely shuts in the anodic alumina films from the solutions. Similarly to Pd nanotubules, Pt was also coated uniformly on and inside the PAA films in a hollow type with open entrances without blocking inside, thus producing ordered nanoporous Pt films (pore sizes 200–400 nm (Fig. 4.60c)) or united Pt nanotubule arrays embedded in the porous alumina matrix (tube diameter 600 nm, tube wall thickness 100–200 nm, tube length 10–12 μm (Fig. 4.60d)). In particular, the walls of Pt nanotubules were not as dense as those the Pd nanotubules. The Pt tube

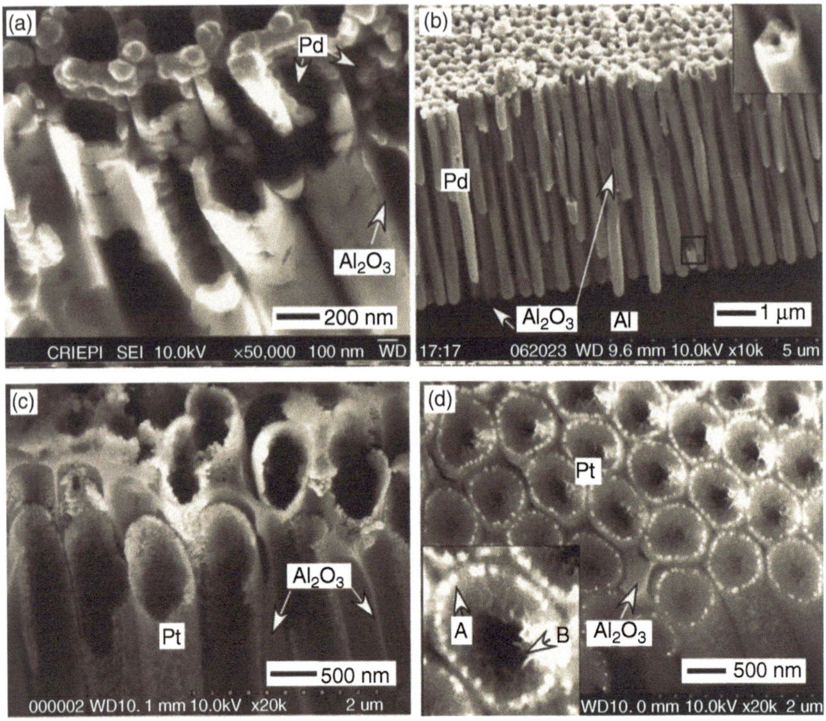

FIGURE 4.60 FESEM images of Pd nanotubules embedded in porous alumina films and Pd electroless deposition for 40 s. (a) Top view and (b) vertical fracture sectional images. FESEM images of Pt nanotubules embedded in porous alumina films for 3 min in an alkali bath with hydrazine as a reducing reagent: (c) top view, and (d) a transverse fracture sectional image. The Pt tube walls, presented a two-layer structure, that is, a chain ring of nanoparticles (~20 nm) as seen the inset in Fig. 4.60 d, in which arrow A pointed the formed egglike composite nanoparticles, and the arrow B pointed Pt grew on the Pt/Pd nanoparticles by a homonuclear catalysis in dendrite crystals [237].

walls, interestingly, possessed a two-layer structure, see the inset in Fig. 4.60d, that is, a chain ring of nanoparticles (~20 nm) and an assembly of inward-pointed nanoflakes or nanoneedles (50–100 nm long). The unique structure of the Pt nanotubules embedded in the PAA films could be ascribed to the reaction mechanism of Pt electroless deposition. Pt may have first deposited on the Pd colloids by heteronuclear catalysis in a dense layer and formed egglike composite nanoparticles (arrow A in the inset of Fig. 4.60d), and then grew on the Pt/Pd nanoparticles by a homonuclear catalysis in dendrite crystals (arrow B in the inset of Fig. 4.60d).

As a result, the Pt nanotubule arrays may possess larger surface areas than the Pd nanotubule arrays with the same tube dimensions. The nanotubule arrays gave a good mechanical strength and electrical conductivity; in addition, the Pd and Pt nanotubule arrays as electrodes were stable and durable even in a sulfuric acid solution with a high concentration of 50 wt%. From the EDX spectra of the Pd and Pt nanotubules embedded in PAA films on Al sheets, a small amount of phosphor (5–8 at% P) was detected out from the Pd nanotubule arrays, indicating the formation of Pd–P alloys instead of the pure Pd. However, pure Pt was formed for the Pt nanotubules with hydrazine as the reducing reagent. Moreover, the inclusion of a small amount of Pd (3–8 at%) was detected out in the Pt nanotubule arrays, which could be ascribed to the Pd catalytic colloids formed in the activating step. The XRD patterns of as-deposited Pd and Pt nanotubule arrays embedded in alumina films on Al indicated both Pd and Pt nanotubules exhibited the polycrystalline structures. The calculated crystal sizes of the Pd and Pt nanotubules from Scherrer's formula were evaluated to be ~5 and ~10 nm, respectively. This indicates that the Pd nanoparticles and Pt nanoflakes observed in the FE-SEM images were secondary particles that were formed by coagulation of numerous nanocrystals.

4.3.1.6.2 EP Nano Alloy Nanotubes and Nanotubes Arrays

Only a few groups have published studies of Ni–P nanotubes arrays recently [238–240]. In 2005, Lin et al. [238] synthesized highly arrayed Ni–P nanotubes. The original AAO templates with a pore diameter of about 50–200 nm were used. One side of the AAO template was sputtered with 200 nm Au film to act as a catalytic layer. The bath solution was prepared from $NiSO_4$ (0.05 mol/L), NaH_2PO_2 (0.04 mol/L), $NaC_2H_3O_2$ (0.01 mol/L), and $Na_3C_6H_5O_7$ (0.005 mol/L). The AAO template was immersed into the bath solution at 90°C and pH of 4.5, and Ni–P nanotube arrays were formed. After removing the AAO in aqueous NaOH, the Ni–P nanotube arrays were obtained. Judging by the investigations in the literature, it is widely known that the solid nanowires structure can be obtained via an AAO template. However, a Ni–P tube structure was formed in this study. The reasons may be that as the Au was sputtered into the pores, hollow structure could be formed as shown in Fig. 4.61d. Then, the deposited Ni–P could cause preferential growth on the pore wall to form nanotubes. From the SEM image in Fig. 4.61 the nanotubes were estimated to have average dimensions of 200–300 nm and length of about 2 μm, and with a wall thickness of about 40 nm. The corresponding electron diffraction pattern (c) indicates it is an amorphous structure. EDS analysis indicates that the nanotubes are composed of Ni and P. The XRD patterns show that the Ni–P nanotubes exhibit amorphous structure with ~5 nm ultrafine grains. The grain size in the Ni–P nanotubes increases with increasing annealing temperature. At 700°C, the average grain size was estimated to be 35 nm.

In 2011, Chen et al. [241] prepared Sn and Sb separately deposited into TiO_2 nanotubes (TNTs) by electroless plating. The pretreated TNTs were immersed in an aqueous solution containing 0.3 M SnC_{12}, 3.0 mol/L NaOH, 0.9 mol/L NaH_2PO_2, and 0.6 mol/L sodium citrate for 20 min at 65°C for electroless deposition of Sn. The electroless deposition of Sb was carried out in a solution containing 0.1 mol/L SbC_{13}, 0.3 mol/L NaH_2PO_2, and 0.3 mol/L sodium citrate

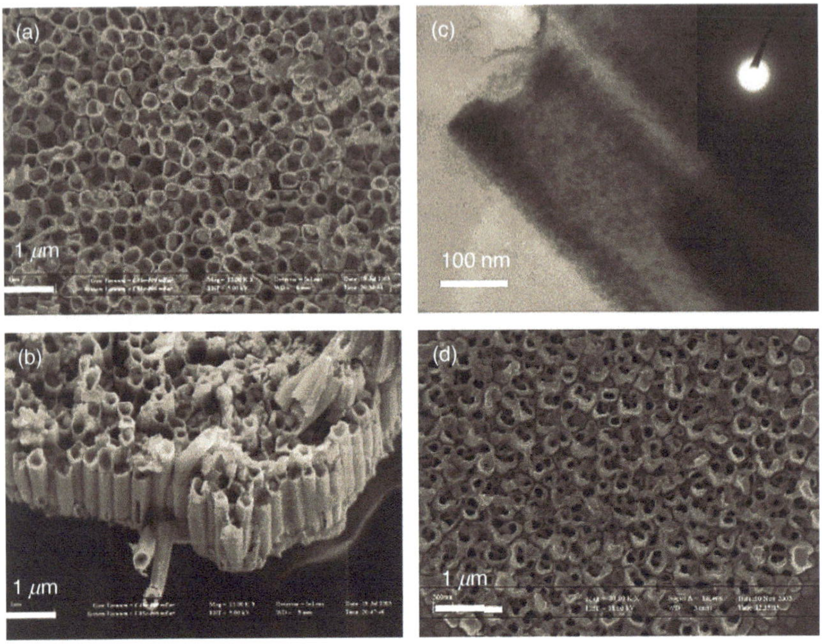

FIGURE 4.61 **SEM and TEM images of Ni–P alloy nanotube arrays formed by EP (a, b) Up view and side view of the SEM image of Ni–P nanotube arrays controlling bath solution at 90°C and pH = 4.5. (c) TEM image and diffraction pattern of Ni–P nanotubes. (d) The hollow structure of catalyst Au film after removing the AAO in NaOH (aq) [238].**

for 20 min at 65°C. Then, electroless deposition of Sn was repeated for 30 min at 65°C to form the TNT/Sn–Sb nanotubes. The average diameter of these nanotubes was found to be ~120 nm, and the tube length is about 420 nm (Fig. 4.62). The surface morphology is dense and rough. The EDX spectrum confirms the presence of Ti, O, Sn, and Sb in the TNT/SnOx–Sb nanotubes, with a molar ratio of Ti:Sn:Sb:O of 11.8:29.2:1.8:57.2.

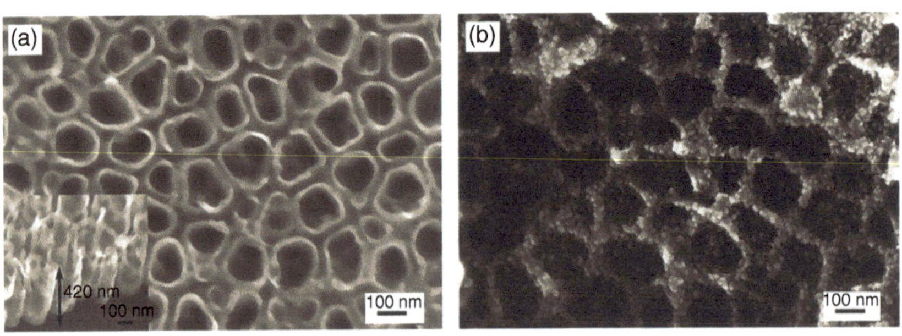

FIGURE 4.62 **SEM image of (a) the TNTs (inset shows a cross-sectional view of TNTs); and the structure of SnOx–Sb in the TNTs; (b) the TNTs after Sn was deposited for 20 min and annealed for 3 h [241].**

FIGURE 4.63 (a) Top view FE-SEM and (b) TEM image of the end-closed NiCoFe-B nanotube arrays [242].

In 2011, Azizi et al. [242] fabricated end-closed NiCoFe–B alloy nanotube arrays by an electroless method. The template AOA was pretreated before the fabrication of Ni–CoFe nanotube arrays. The bath solution for the formation of the arrays contained $8.5 \times 10 - 3M$ $NiSO_4{\bullet}6H_2O$, 2.5×10^{-2} mol/L $CoSO_4{\bullet}7H_2O$, 1.3×10^{-2} mol/L $FeSO_4{\bullet}7H_2O$, 0.143 mol/L lactic acid and 7×10^{-2} mol/L DMAB and its pH was fixed at ~7–8. The TEM image (Fig. 4.63b) of the NiCoFe–B nanotubes indicated about 110 nm for their outer diameter, corresponding well to the average internal diameter of the pores on the AAO. According to the TEM image, the average wall thickness of the nanotubes is about 15 nm.

Besides the alloy nanotubes, in 2011, Neetzel et al. [243] prepared lepidocrocite (c-FeOOH) nanotubes in ion track etched polycarbonate templates using an electroless method. The authors used a two-step method to synthesize the lepidocrocite nanotubes. First, electroless precipitation of lepidocrocite particles was performed by using 0.4 mol/L of $FeC_{12}{\bullet}4H_2O$ in 100 mL of deionized water. The pH of the solution (pH = 4) was adjusted with HCl. While the solution was stirred rapidly, a 0.1 mol/L NaOH solution was added dropwise until a pH value of 6.5 was reached. The oxidation of the initial rust was performed in an open vessel by bubbling air at a flow rate of 50 mL/min. Finally, the solid product was washed and filtered 10 times and then dried at 70°C under atmospheric pressure for several days. The second step was to produce the lepidocrocite nanotubes. The previously prepared and dried iron oxide particles were transferred into 10 mL of hexane and then put in an ultrasonic bath for 5 min. Meanwhile, the template was put inside an open vessel and cooled down to 0°C. Then, hexane containing iron oxide particles was added to the template. The reaction takes about 12 h, and the solution has to be put in an ultrasonic bath every hour. Afterward, the template was removed from the solution and dried at room temperature and atmospheric pressure. Figure 4.64a–b shows an SEM of produced nanotubes unhinged from the polycarbonate template. Their lengths and diameters are from about 2 to 4 µm and 100 nm, respectively. Their wall thickness is about 20 nm. Energy-dispersive X-ray spectroscopy affirms that the

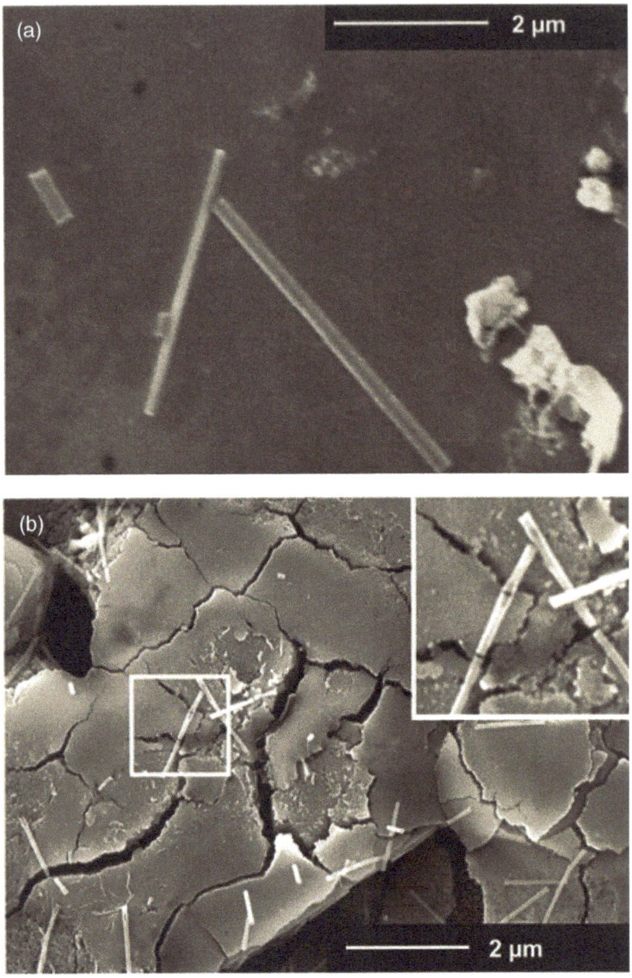

FIGURE 4.64 **(a, b) SEM pictures of produced lepidocrocite nanotubes unhinged from polycarbonate template (length: about 2–4 μm, diameter: 100 nm, wall thickness: 20 nm) [243].**

material consists of iron and oxide. The detection of carbon can be traced back to residues of polycarbonate.

4.3.2 EP Nanowires

Nanowires and nanofibers can be considered to be synonymous. The term refers to a solid nanostructure with a diameter in the range of nanometers (1–100 nm). Nanowires are used in this book although nanofiber is still used in some places. They can be catalogued as several groups, namely metal nanowires, alloy nanowires, semiconductor nanowires, oxide nanowires, and other nanowires.

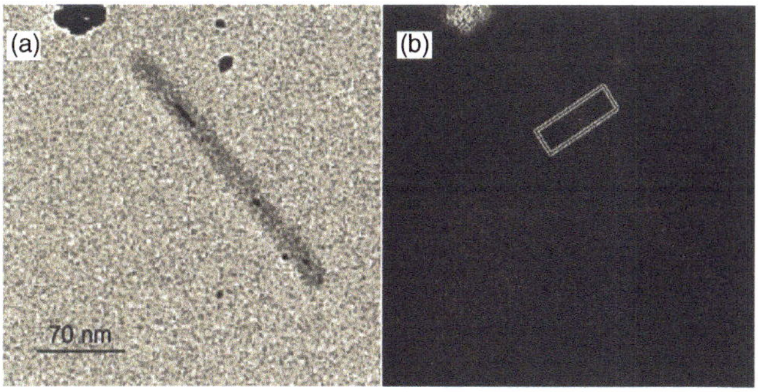

FIGURE 4.65 **A TEM BF image of (a) Cu nanowires within the central channel of TMV particles and the corresponding elemental distribution images (b) recorded with the three-window technique (Cu $L_{2,3}$ edge at 931 eV, Ni $L_{2,3}$ edge at 854 eV, and Co $L_{2,3}$ edge at 779 eV) [244].**

4.3.2.1 EP Nano Metal Nanowires

4.3.2.1.1 EP Cu Nanowires

Several papers have reported EP Cu nanowires [244–246]. In 2006, Balci et al. [244] described the synthesis of copper nanowires of 3 nm in diameter with electroless deposition within the 4 nm wide channel of tobacco mosaic virus (TMV) particles. An EP copper solution was used to metalize the TMV with copper at lower pH level (pH ~7.5). A Pd (II)-incubated TMV suspension was mixed with an equal volume of a solution containing 0.032 mol/L $CuSO_4{\bullet}5H_2O$, 0.04 mol/L Na_2H_2EDTA (ethylenediaminetetraacetic acid disodium salt), 0.05 mol/L TEA (triethanolamine), and 0.067 mol/L DMAB in water. The metallization step was carried out at room temperature. Figure 4.65a and b shows bright-field TEM images of copper nanowires in the virions, ~3 nm in diameter. In four separate experiments, the authors found that between 40 and 50% of the TMV particles were metallized with Cu, and that the length of the wires varied from 5 nm to 150 nm. They never found a 300 nm long copper wire within the virus central channel.

4.3.2.1.2 EP Ag Nanowires

Some groups studied EP Ag nanowires [247–251]. In 2004, Barbic et al. [247] presented a method of fabricating single-crystal Ag nanowires based on the electroless deposition of silver into the pores of polycarbonate membranes by the metal amplification process. The fabrication procedure for preparing single-crystal Ag nanowires is as follows. A polycarbonate membrane with an Au layer on one side is placed in a holder so that the Ag enhancement solution can only contact the Au layer through the pores of the membrane. The reduction of the free Ag ions is initiated at the Au interface spontaneously. This initial Ag layer on the Au film acts further as the catalyst for the continual growth of Ag nanowires inside the pores. Free Ag nanowires can be obtained by removing the Au film and dissolving the membrane. Figure 4.66 shows a typical bright-field TEM

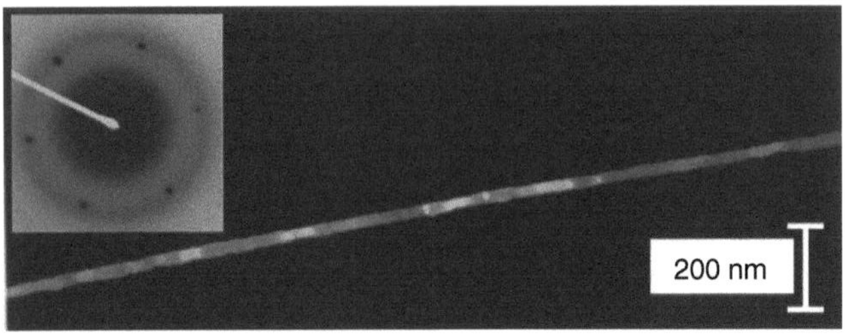

FIGURE 4.66 TEM image of a 30 nm diameter Ag nanowire. The inset shows a selected area diffraction pattern from the nanowire. The ring features in the diffraction pattern are from the silicon nitride TEM membrane substrate [247].

micrograph of a silver nanowire with 30 nm diameter and –5 nm deviation in the diameter size due to the nonuniformity of the membrane pores used to guide the nanowires' growth. From a selected area diffraction pattern of the nanowire, as shown in the inset of the figure, one can see the single-crystal diffraction of the Ag nanowires. In 2004, Sun and Xia [250] prepared Ag nanowires also by EP. The authors synthesized Ag nanowires using a polyol process. A typical procedure is to simultaneously inject 3 mL of an ethylene glycol solution of $AgNO_3$ (0.085 mol/L) and 3 mL of an ethylene glycol solution of PVP (0.13 mol/L) into ethylene glycol (5 mL) that had been preheated to 160°C. This mixture was then heated at 160°C for 1.5 h under vigorous magnetic stirring. Figure 4.67a shows that the Ag nanowires obtained had a pentagonal cross-section, with five straight side edges parallel to its long axis, five flat side surfaces bounded by {100} facets, 10 {111} end facets, and a twinned crystalline structure of five-fold symmetry. The authors have also investigated the reaction between these silver nanowires and various volumes of the 1 m mol/L $HAuCl_4$ solution. When 0.3 mL of $HAuCl_4$ was used (Fig. 4.67b), each Ag nanowire developed interior cavities (i.e., tubular strips). All the Ag nanowires were transformed into nanotubes with smooth and uniform sheaths when the volume of the $HAuCl_4$ solution was increased to 0.6 mL (Fig. 4.67c). The nanotubes essentially inherited all the morphological features associated with silver nanowires, each tube also had a pentagonal cross-section, five straight side edges, and five flat side surfaces (see the inset picture). Observations confirmed that the elemental gold generated from the replacement reaction was epitaxially deposited on the surfaces of Ag templates. Dealloying of these Au–Ag alloyed nanotubes also led to the formation of pinholes in their surfaces (Fig. 4.67d).

4.3.2.1.3 EP Au Nanowires

Only a few groups have published their work on EP Au nanowires [252–254]. Earlier in 1994, Menon and Martin [252] described an EP Au electroless deposition procedure for filling the pores in nanoporous filtration membranes with Au nanowires. They immersed an Ag-coated membrane in an Au plating bath that contained 7.9×10^{-3} mol/L of $Na_3Au(SO_3)_2$, 0.127 mol/L of Na_2SO_3, and

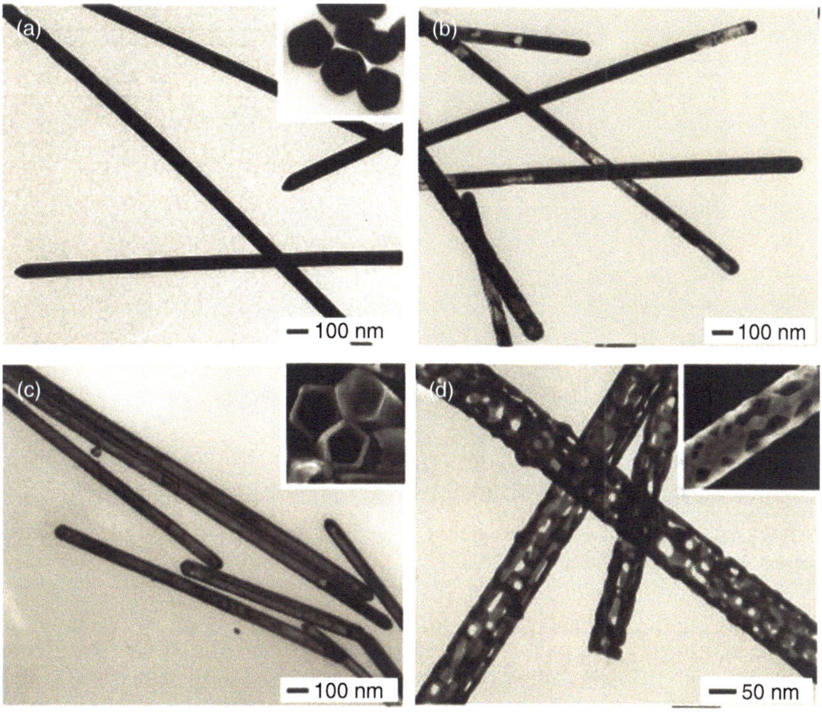

FIGURE 4.67 **TEM images taken from Ag nanowires (a) before and (b) after they had reacted with volume of 1 mM $HAuCl_4$ solution: 0.3 mL. The inset of (a) shows TEM image taken from the microtomed sample of silver nanowires. The inset of (b) is SEM images of the Ag nanowires B [250].**

0.625 mol/L in formaldehyde; the temperature of this bath was ~2°C. The Ag particles are galvanically displaced by Au since gold is a more noble metal. As a result, the pore walls and faces become coated with Au particles. Au plating was continued for a period of 24 h, resulting in the deposition of Au nanowires into the pores. In addition, both faces of the membrane become coated with thin (~500 nm thick) Au films. This membrane is referred to as Au films on both faces and Au nanowires running through the pores of the Au/Au-PC/Au membrane, where Au-PC stands for the Au-nanowire-containing polycarbonate. After plating, the Au/Au-PC/Au membrane was thoroughly rinsed with water and then immersed in 25% nitric acid for 12 h to dissolve any residual Sn or Ag that might be left in the membrane. Figure 4.68 indicates the SEM of Au nanowires (50 nm diameter) obtained after dissolving away the polycarbonate template membrane.

4.3.2.1.4 EP Ni Nanowires

Several papers on EP Ni nanowires have been published [255–261]. In 2007, Srivastava et al. [258] prepared Ni nanowires using an alumina membrane as a template. First, the membrane was treated with fluorosilanes to make the

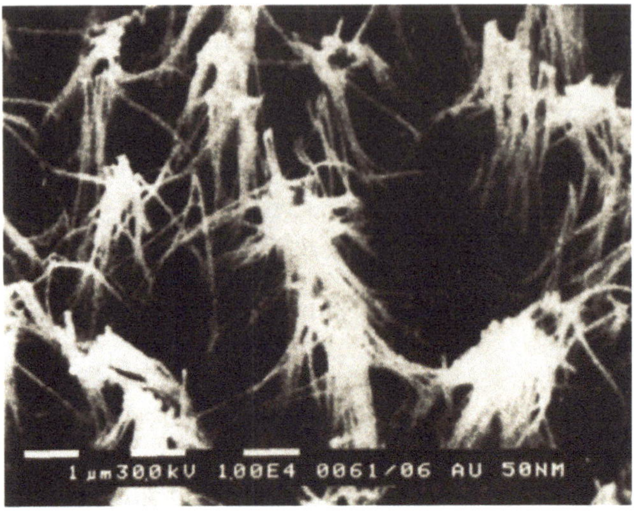

FIGURE 4.68 SEM of Au nanowires (50 nm diameter) obtained after dissolving away the polycarbonate template membrane [252].

surface completely hydrophobic. These freshly cleaved hydrophilic edges were used as templates for the deposition of polyelectrolyte multilayers (PEMs) and nickel. They selectively deposited polyelectrolyte multilayers on those hydrophilic edges. Then, PEMs were built on the hydrophilic edge of the pore membrane and an electroless nickel bath was used for the deposition of nickel onto the multilayers to form nickel nanowires. After dissolving the membrane, free standing nickel nanowires were obtained. The EP bath was prepared using nickel sulfate 40.0 g/L, sodium hydroxide, sodium citrate 20.0 g/L, lactic acid 10.0 g/L, and DMAB 2.0 g/L. Before plating, the pH of the nickel bath was adjusted to 6.5 ± 0.1 by adding small amounts of 1.0 M NaOH. The SEM images of Ni nanowires are shown in Fig. 4.69. The length of the resulting nanowires

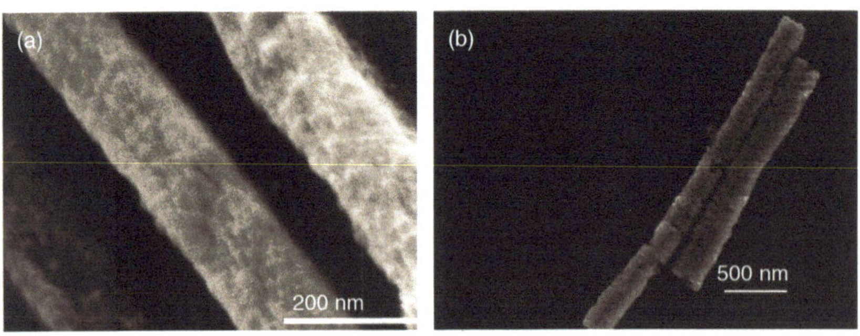

FIGURE 4.69 Images of nickel nanowires (a) Nickel nanowires growing on the edge of the alumina membrane pore wall. (b) A group of nickel nanowires after dissolution of the alumina membrane [258].

is 2–4 μm although the membranes are 60 μm thick. From the quartz crystal microbalance (QCM) data, the PEMs-Ni nanowire has a thickness of around 47 nm; 7 nm from the PEMs and 40 nm from the nickel.

4.3.2.1.5 EP Pd Nanowires

Some reports on EP Pd nanowires have been published [262–265]. In 2008, Shi et al. [264] believed that they could control Pd nanowire size through an electroless deposition process. The composition of the EP bath and experimental parameters used are $PdCl_2$ (g/L) 4.2, 3.0 and 1.8–2.0, $Na_2EDTA{\bullet}2H_2O$ (g/L) 40.1 $NH_3{\bullet}H_2O$ (28%) (mL/L) 198, N_2H_4 (1 M) (mL/L) 5.6, pH 10–10.4, and temperature 60°C. Figure 4.70 illustrates the deposition of Pd on porous stainless-steel substrate with a deposition time from 90 s to 150 s when the concentration of $PdCl_2$ in the plating solution was 3.0 g/L. A continuous Pd nanowire was built, as shown in Fig. 4.70a. The nanowire made of nanoparticles was easily recognized at a deposition time of 90 s and the diameter of these nanowires was measured at ~125 nm, but the size of most nanoparticles distributed on the substrate surface ranged from 50 nm to 70 nm. The diameter of the Pd nanowire increased up to

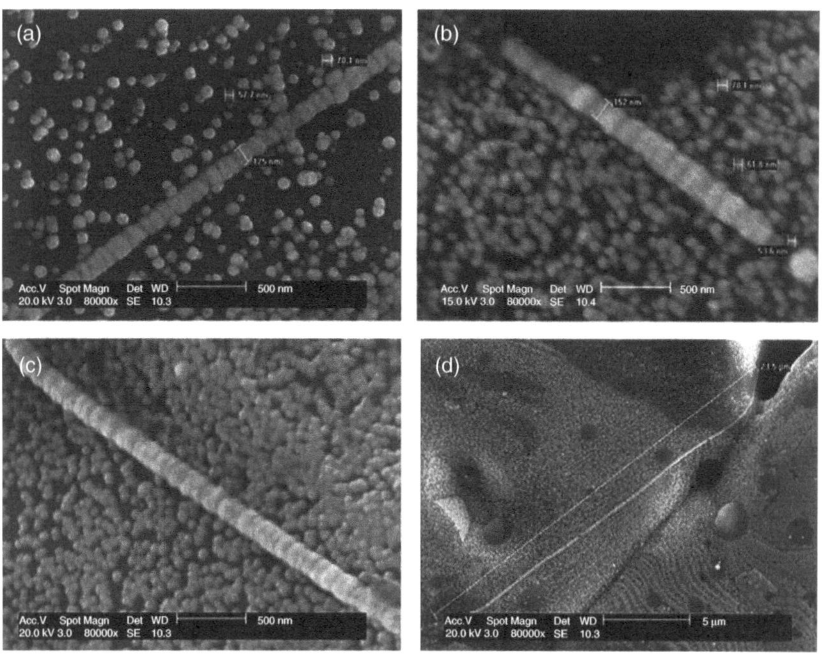

FIGURE 4.70 SEM images of Pd nanowires deposited on porous stainless steel substrate from 90 s to 150 s, the concentration of $PdCl_2$ in the bath solution was 3.0 g/l. (a) 90 s, the length of Pd nanowire over 23.5 μm in d, (b) 120 s, the length of Pd nanowire over 23.5 μm in d, (c) 150 s, the length of Pd nanowire over 23.5 μm in d, (d) the length of a nanowire built on the substrate surface was over 23.5 mm and was not located along a grain-boundary of the porous stainless steel substrate [264].

152 nm after the deposition time was extended to 120 s, as shown in Fig. 4.70b. By increasing the deposition time to 150 s, the nanowire was built homogeneously and its diameter was about 200 nm, as illustrated in Fig. 4.70c. The data indicated that the size of most nanoparticles distributed on the surface did not change and still remained in the same range of about 50–70 nm with the extension of the deposition time from 90 s to 150 s in Figs. 4.70a–c. Figure 4.70d shows that the length of a nanowire built on the substrate surface was over 23.5 mm and was not located along a grain boundary of the porous stainless-steel substrate.

4.3.2.1.6 EP Co Nanowires

Balela et al. [266] reported EP nanowires in 2011. The authors prepared the metal ion solution and reducing agent solution separately. First, 10 mmol Co(II) acetate and 0.125 mmol poly(ethylene glycol) [PEG, $H(OCH_2CH_2)_nOH$, $M_w = 20{,}000$] were dissolved in 50 cm^3 propylene glycol (PG, $C_3H_8O_2$) while nitrogen gas (N_2) was bubbled through it at 50 cm^3/min. Then, 25 cm^3 of 1.0 mol/L NaOH-PG solution was added to hydrolyze the Co(II) ions and adjust the pH to about 12. Next, 10 cm^3 of 2.5 mmol/L H_2PtCl_6-PG solution was added to the metal ion solution as a nucleating agent to provide heterogeneous nucleation sites for Co. As the reducing agent solution, 25 cm^3 deaerated PG solution containing 4.0 M N_2H_4 was prepared. The reducing agent solution was then mixed thoroughly with the metal ion solution in a 200 cm^3 Pyrex beaker by mechanical stirring. The total solution had a volume of 115 cm^3, with final concentrations of 0.087 M Co(II) acetate, 1.1 mM PEG, 0.22 M NaOH, 0.22 mM H_2PtCl_6, and 0.87 M N_2H_4. The final pH remained about 12. The resulting solution was placed in a superconducting magnet for 1 h. The synthesized nanowires were magnetically separated from the solution and ultrasonicated for 10 min in ethanol. This washing process was repeated 10 times to remove organic contaminants on the surface of the nanowires. Under a field of 1 T, Co nanowires with an average diameter of about 240 nm and lengths up to 30 mm were obtained (Fig. 4.71b). When the strength of magnetic field is increased, the mean diameter of Co nanowires is decreased. Co nanowires about 190 nm in mean diameter and 160 mm long are prepared under 10 T (Fig. 4.71c). The authors also found that crystallite sizes of the Co nanoparticles (about 15 nm) and nanowires (10–12 nm) determined from the broadening of the 111 fcc Co peaks by Scherrer's equation are much smaller than the apparent diameters of the samples. This indicates that the products are polycrystals, which corresponds well to the morphology observed by electron microscopy in Fig. 4.71.

4.3.2.1.7 EP Pb Nanowires

One paper has been reported for fabricating EP Pb nanowires [267]. Pb nanowires were prepared on zinc substrate by using 10 mL of $Pb(OAc)_2$ solution in a 20 mL vial. After the growth, the zinc foil was cleaned with deionized water and was dried with N_2 gas. Figure 4.72a shows the bright-field TEM images of Pb nanowires. From an analysis of the electron diffraction pattern and EDS, it was confirmed

FIGURE 4.71 SEM and TEM images of products fabricated by electroless deposition of Co in a PG solution containing 0.087 M Co(II) acetate, 0.22 M NaOH, 0.22 mM H_2PtCl_6, and 0.87 M N_2H_4 at room temperature under a magnetic field of (a) 0, (b) 1, (c) 10 T [266].

that the 1D nanostructure consists of a single-crystalline Pb nanowire core and a thin amorphous ZnOx shell. The thicknesses of the shell of Pb/ZnOx nanocables were as little as 10–20 nm. From the inset electron diffraction pattern in Fig. 4.72b, one can see that the growth of the Pb nanowire is in the [100] direction.

4.3.2.2 EP Alloy Nanowires

Besides pure metal EP nanowires, the binary and ternary alloy nanowires have been studied. Ni–P [268], Co–P [269], Co–Fe [270], Co–Ni [271], and Pb–Se

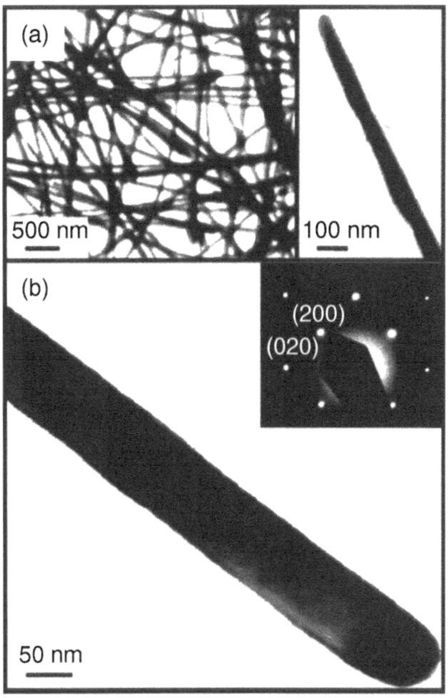

FIGURE 4.72 (a) TEM BF image of Pb/ZnOx nanocables with the high-magnification image of a nanocable as the inset. (b) TEM image showing the Pb NWs after the ZnOx shells were stripped with a dilute HCl dip. The inset electron diffraction pattern in (b) reveals that the growth direction of Pb nanowire is in [100] direction [267].

[272] binary alloy nanowires have been published. As for EP ternary alloy nanowires, Ni–W–P [273], Ni–Co–Fe [274], Co–Fe–P [275], Co–Fe–B [276], Co–Ni–P [277], Co–Zn–P [278] have been reported. Only few samples will be discussed in the following sections.

4.3.2.2.1 EP Nano Binary Alloy Nanowires

EP Ni–P nanowires were reported in 2009 [268]. The bath solution used by Ren et al. was 25 g/L $NiSO_4 \cdot 6H_2O$, 30 g/L $NaH_2PO_2 \cdot H_2O$, 15 g/L $CH_3COONa \cdot 3H_2O$, 15 g/L $Na_3C_6H_5O_7 \cdot 2H_2O$, and 1 mg/L $(NH_2)_2CS$. Electroless deposition was carried out at room temperature for 1 h. These experiments show that the nanowires are about 60 mm in length and 70 nm in diameter, that is, the aspect ratio is beyond 800. The small-angle electron diffraction (SAED) pattern shown in the inset of Fig. 4.73b indicates that the deposited nanowires are amorphous. EDS reveals that the nanowires are composed of Ni and P, the atomic ratios of the nanowires are 88.89:11.1, and the mass ratios are 93.81:6.19.

In 2012, Balci et al. [270] synthesized 3 nm wide Co–Fe alloy nanowires by simple wet chemical electroless deposition inside tubular tobacco mosaic virus particles. The EP bath is based on DMAB as reductant, and succinate as

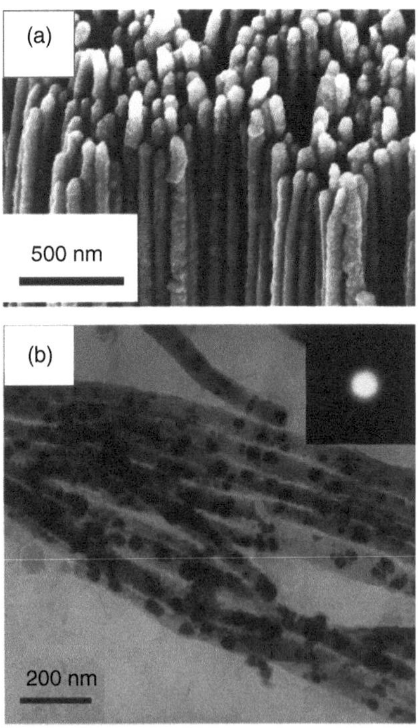

FIGURE 4.73 **(a) FE-SEM images of Ni–P nanowires deposited for 30 min in AAO templates, (b) TEM images of Ni–P nanowires deposited for 10 min in AAO templates. The inset is SAED pattern [268].**

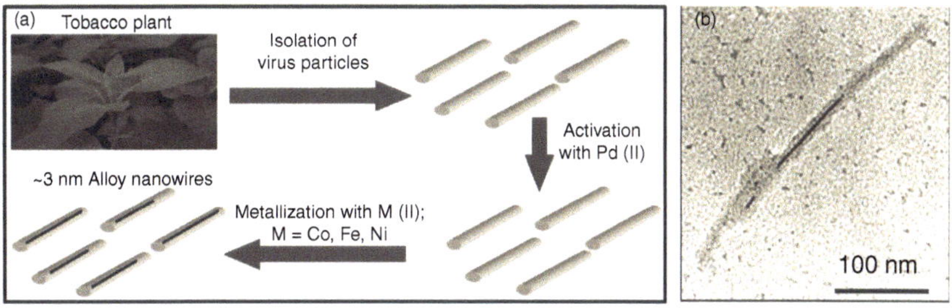

FIGURE 4.74 (a) Scheme of synthesis of Co–Fe alloy nanowires in the tubular channel of TMV particles. (b) TEM micrograph of Co–Fe alloy nanowires within the channel of a TMV particle (dark gray) [270].

complexant. Figure 4.74a indicates the scheme of synthesis of extremely small diameter (~3 nm) Co–Fe alloy nanowires in the tubular channel of tobacco mosaic virus (TMV) particles. The particles are isolated from plants, wet chemically sensitized/activated with Pd(II) catalyst, and metallized with Co(II) and Fe(II) in suspension. Finally, ~3 nm Co–Fe alloy nanowires can be obtained in the TMV particles (virions). Figure 4.74b just shows the TEM micrograph of Co–Fe alloy nanowires within the channel of a TMV particle (dark gray). The black line in the center corresponds to an alloy wire of ~3 nm in diameter and approximately 100 nm in length. In the lower part of the same virion a second small nanowire (~10 nm long) is located. Another short virion (~100 nm long) is attached to the lower portion of the metallized virion.

Cho et al. [272] in 2005 synthesized single-crystal Pb–Se nanowires in solution through oriented attachment of nanocrystal building blocks. They dissolved 0.76 g of lead acetate trihydrate and 2 mL of oleic acid in 10 mL of diphenyl ether. Heating to 150°C for 30 min under a nitrogen flow forms lead oleate *in situ* and dries the solution. After cooling to 60°C, the lead oleate solution was mixed with 4 mL of 0.167 M trioctylphosphine selenide (TOPSe) solution in trioctylphosphine (TOP) and injected under vigorous stirring into a hot (250°C) solution containing 0.2 g of *n*-tetradecylphosphonic acid (TDPA) dissolved in 15 mL of diphenyl ether. The injection of stock solution initiated the temperature drop as shown in the reaction temperature profile (Fig. 4.75). After 50 s of heating, the reaction mixture was cooled to room temperature using a water bath. The reaction conditions for Pb–Se straight nanowires are summarized as stabilizing agents, solvent OA (oleic acid)-TDPA-TOP, Ph_2O (phenyl ether); OA:Pb ratio 3.15:1; Pb:Se ratio 3:1; injection/growth temperature (°C) 250/190; and growth time 30 s-1 min. Figure 4.76 shows HR (a) SEM and (b) TEM images of Pb–Se nanowires grown in solution in the presence of oleic acid. The shape and morphology of the nanowire can be controlled by introducing cosurfactants into the reaction mixture. Figure 4.76c–f indicates that long, straight wires with a narrow diameter distribution form in the presence of TDPA. The average length of the nanowires is typically 10–30 μm, while the average diameter can be tuned from ~3.5 to ~18 nm with a 10% standard

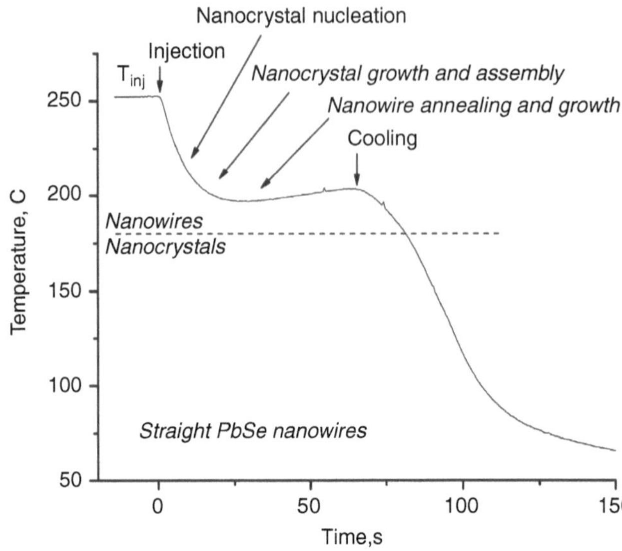

FIGURE 4.75 Temperature profile of the reaction mixture during synthesis of straight 6 nm diameter PbSe nanowires [272].

deviation by adjusting the temperature profile during growth and controlling the reaction time. From Fig. 4.76g and h, one can see the evolution of a single batch of Pb–Se nanowires grown at 170°C in the presence of oleic acid and an excess of selenium precursor (Pb:Se = 1:3). The wires evolve from chainlike assemblies, 8 nm in diameter, to straight thicker wires, 12 nm in diameter. The steady growth of the (111) planes results in elimination of the {111} facets, producing straight smooth wires.

4.3.2.2.2 EP Ternary Alloy Nanowires

Ni–W–P nanowire arrays fabricated by EP were reported in 2004 by Yuan et al. [273]. Ni–W–P nanowire arrays were deposited in the pores of the anodic alumina membrane (AAM) from a solution of 15 g/L $NiSO_4$•$6H_2O$, 10 g/L Na_2WO_4•$2H_2O$, 22 g/L NaH_2PO_2•H_2O, and 40 g/L sodium citrate at 80–85°C. The AAM with Ni–W–P nanowire arrays was dissolved in 5 wt% NaOH solution at 25°C for 5 min and lightly washed several times with distilled water to remove the dissolved AAM and the remaining NaOH solution, and then the isolated Ni–W–P nanowire arrays can be obtained. An SEM image of EP Ni–W–P alloy nanowire arrays (Fig. 4.77a) reveals that they are uniform in size, all pores of the AAM are filled up, the nanowires are parallel to each other, and they are about 3 μm in length. Figure 4.77b shows that the Ni–W–P nanowires are about 60 nm in diameter and 3 μm in length. The SAED (inset to (b)) and the XRD spectrum (not shown here) of the Ni–W–P alloy nanowire illustrate that the structure of the nanowire is amorphous. The EDS reveals that the nanowires

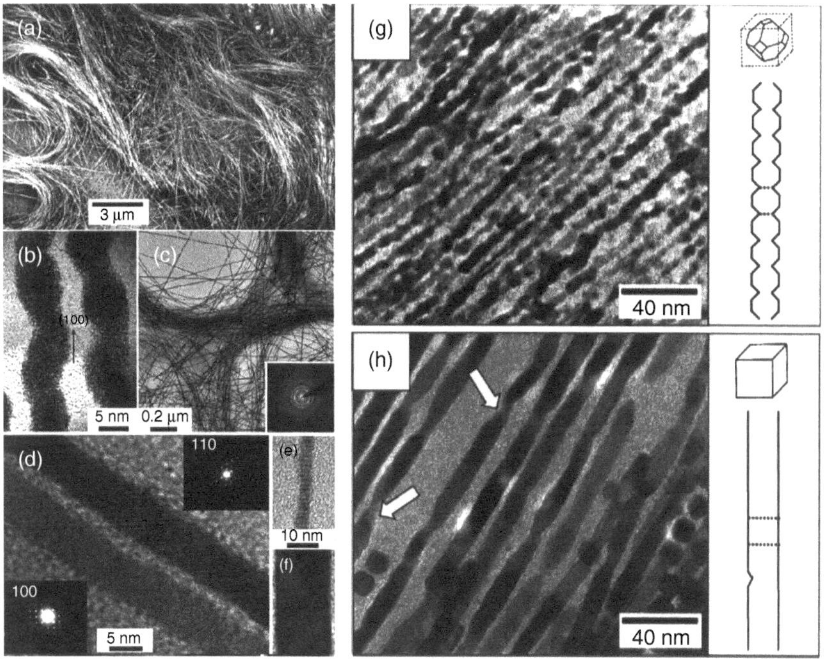

FIGURE 4.76 HR (a) SEM and (b) TEM images of PbSe nanowires grown in solution in the presence of oleic acid. (c) Overview and (d–f) HR TEM images of PbSe nanowires formed in the presence of oleic acid and *n*-tetradecylphosphonic acid. SAED from a film of PbSe nanowires (inset to c) and single nanowires imaged along the (100) and (110) zone axes (insets to d). The diameter of PbSe nanowires can be tuned from (e) ~4 nm to (f) ~18 nm. Shape evolution of PbSe nanowires grown for (g) 30 s and (h) 10 min at 170°C using a molar ratio of Pb:Se precursors of 1:3 and oleic acid as the stabilizing agent. Arrows in (h) point to the (111) face defects. The average wire morphology and the structure of the hypothetical nanocrystal repeat unit are depicted to the right of the TEM images [272].

are composed of Ni, W, and P, the atomic ratio is 65.79:6.51:27.70, and the mass ratio is 65.28:20.22:14.50.

In 2007, Dadvand and Kipouros [276] fabricated electroless of Co–Fe–B alloy nanowires within an alumina template. The electroless plating solution contained cobalt (II) sulfate heptahydrate (0.06 mol/L) as a source of cobalt ions, iron (II) sulfate heptahydrate (0.01 mol/L) as a source of iron ions, dimethylaminoborane (0.26 mol/L) as a reducing agent and boron source, sodium tartrate (0.2 mol/L) and sodium citrate (0.05 mol/L) as complexing agents, ammonium sulfate (0.2 mol/L) as a buffering agent, phosphorous acid (0.06 mol/L) as a slow-rate reducing agent, and thallium acetate (1.06×10^{-4} mol/L) as a stabilizer. The pH of solution was pH 6.5 adjusted by a concentrated solution of sodium hydroxide and the operating temperature was maintained at 75 ± 1°C with a thermostatically controlled water bath. Figure 4.78 shows that the length of $Co_{90}Fe_3B_7$ nanowires increase from 10 μm to 25 μm as the

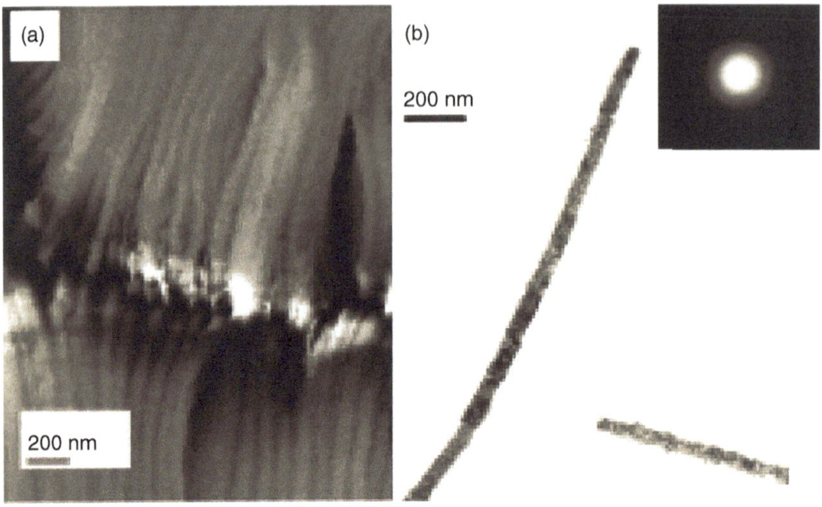

FIGURE 4.77 **(a) SEM images of Ni–W–P nanowire arrays. View from the top. (b) TEM image of Ni–W–P alloy nanowires with diameter about 60 nm. Inset to (b) is the SAED of Ni–W–P alloy nanowires [273].**

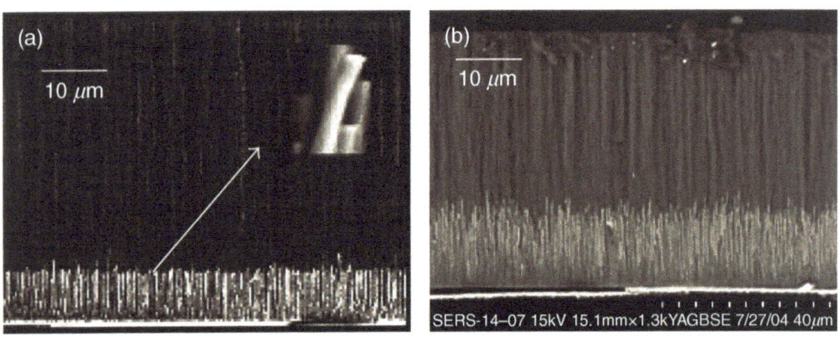

FIGURE 4.78 **SEM micrograph of EP $Co_{90}Fe_3B_7$ nanowires within the nanopores of alumina template at various deposition times: (a) 45 min. The nanowires are 10 μm in length and 200 nm in diameter; (b) 1-hour and 33-min. Nanowires are 25 μm in length and 200 nm in diameter [276].**

deposition time increased from 45 min to 1 h and 33 min, but their diameters are maintained at 200 nm.

In 2010, Azizi et al. [274] synthesized Ni–Co–Fe nanowires by electroless deposition. The chemical composition of the electroless bath is 0.017 mol/L $NiSO_4{\bullet}6H_2O$, 0.017 mol/L $CoSO_4{\bullet}7H_2O$, 0.013 mol/L $FeSO_4{\bullet}7H_2O$, 0.143 mol/L lactic acid, and 0.07 mol/L DMAB. Figure 4.79 shows Ni–Co–Fe nanowires fabricated in one stage on activated AAO under vacuum. The diameter of the nanowires corresponds well to the pore size of the AAO template. Thus, the diameter of the nanowires can be adjusted by choosing an AAO

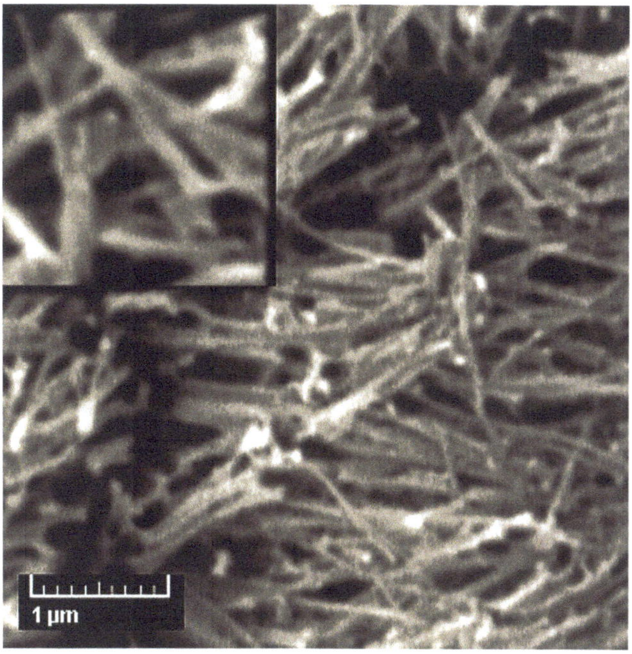

FIGURE 4.79 SEM image of the Ni-Co–Fe nanowires fabricated in one stage activated AAO under vacuum [274].

template with different pore size. EDS analysis indicates that these nanowires are an alloy with composition of $Ni_{84.5}Co_{9.25}Fe_{6.25}$.

4.3.2.3 EP C Nanowires

Carbon nanowires (CNWs) similar to CNTs have not as yet been fabricated by an EP process. Composite CNWs coated with metals or alloys have, however, attracted some research interest. For example, CNWs + Cu [279–282], CNWs + Ag [283, 284], CNWs + Au [285, 286], CNWs + Ni–P [287–290], CNWs + Ni–B [291], CNWs + Ni–Fe–P [292,293], CNWs + Ni–Co–B [294], CNWs + Ni–Fe–Co–P [295], CNWs + Ni–P–ZrO_2–Al_2O_3, CNWs + Ni–Fe [296], glass fibers + Ag [297].

4.3.2.3.1 CNWs + Nanometals Composite

In 2008, Gómez de Salazar et al. [279] studied the preparation of a CNW surface to coat copper by electroless process. Electroless plating solution used was prepared with $CuSO_4•5H_2O$ 5 g/L, ($HO(C_2H_4O)_nH$) 7.5 mL/L, HCOONa 10 g/L, (HCHO) 9 mL/L, $Na_2H_2EDTA•2H_2O$ 22 g/L. NaOH was added to the reaction solution until pH = 12.5 and maintained there for 30 min because the pH value of the solution decreased gradually during the reaction progress. The electroless process was carried out at room temperature and under magnetic stirring. Experiments show

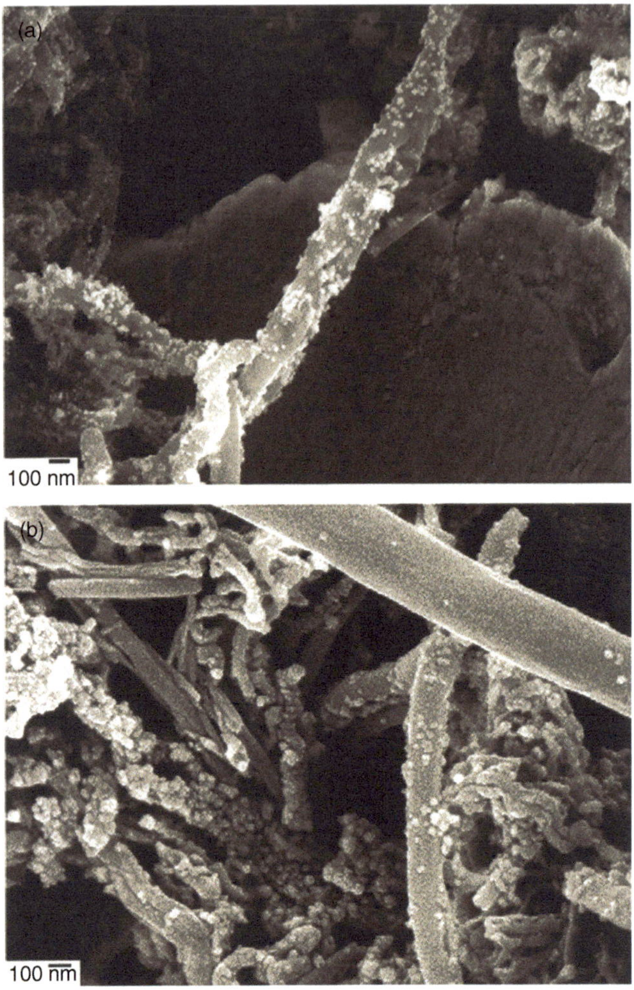

FIGURE 4.80 SEM images of Cu coating over CNWs obtained by coarsening treatments, (a) $K_2Cr_2O_7/H_2SO_4$ and (b) HNO_3 [279].

that when CNWs were pretreated with $K_2Cr_2O_7$ solution, a nucleation mechanism was favored, the Cu coating obtained had a particle size between 15 and 25 nm (Fig. 4.80a) and it was heterogeneous all over the CNWs surface. However, when the HNO_3 solution was used in pretreatment step, the particle size Cu coating was between 40 and 60 nm, because this mineral acid solution favored the growth mechanism (Fig. 4.80b). In this case, in addition to copper coating homogeneous over CNMs, nanoparticles copper aggregates were obtained between CNWs.

In 2011, Lei et al. [284] prepared Ag-coated polyacrylonitrile nanofiber membranes. First, they used electrospinning to fabricate polyacrylonitrile (PAN)/ $AgNO_3$ composite nanofibers. After reduction, Ag nanoparticles dispersed along

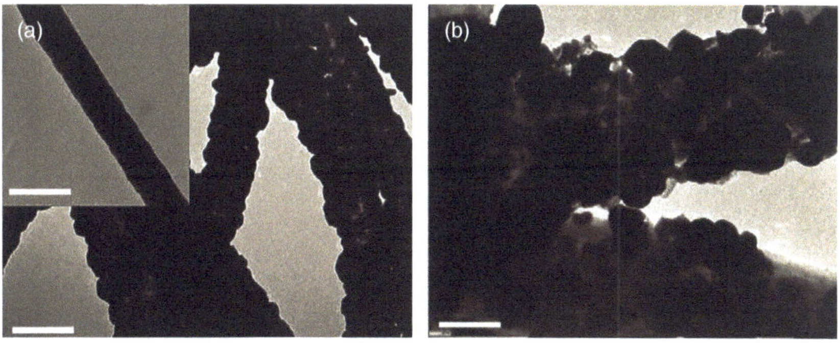

FIGURE 4.81 **TEM images of (a) Ag-coated PAN nanofibers, (b) the Ag-coated PAN nanofibers are mixed with $FeCl_3$ solution for 1 h. The inset in (a) represents the TEM images of PAN nanofibers containing Ag nanoparticles. Scale bar: 200 nm [284].**

the nanofibers act as seeds in the following metal electroless plating step for the growth of a continuous Ag shell. Then an *in situ* oxidation reaction between Ag shells and $FeCl_3$ solution was carried out to prepare Ag/AgCl coated PAN nanofiber membranes. The as-prepared materials exhibited excellent photocatalytic activity under visible-light, long-term stability, flexibility, as well as easy separation from the liquid. From the TEM images in the inset of Fig. 4.81a, the uniformity of the Ag nanoparticles can be seen. They have a diameter less than 5 nm and are densely distributed along the PAN nanofibers (a.v. 120 nm). When the Ag-coated PAN nanofibers are mixed with $FeCl_3$ aqueous solution, the *in situ* oxidation reaction occurs immediately. When the reaction time increases to 1 h (Fig. 4.81b) the shells of the composite structures become much coarser consisting of larger and nonuniform particles. This is due to the tendency for agglomeration into larger particles when the amount of AgCl increases.

In 2006, Metz et al. [285] fabricated a continuous film of highly nanotextured gold, completely coating the CNWs by EP. EP was achieved by immersing the carboxylic acid-modified nanofibers in a tin-sensitizing solution and then activating it with silver. This leaves neutral atomic silver (AgO) bound to the surface. Finally, the samples were placed in individual electroless deposition baths at 6°C for durations of 1–22 h. Figure 4.82 shows SEM mages of the nanofibers after various immersion times in the gold bath. After 1 h (Fig. 4.82a), small gold clusters are visible. After 5 h (Fig. 4.82b), they appear to be fully coalesced, forming a gold sheath that encapsulates each nanofiber, with protruding spikes up to 50 nm long. Upon continued immersion, the surface develops an even finer nanoscale texture. Images (c) and (d) of Fig. 4.82 show a sample after 7.5 h, exhibiting spikes up to 100 nm long.

4.3.2.3.2 CNWs + Nanoalloy Composite

In 2004, Arai et al. [287] prepared Ni–P-coated carbon nanofibers by electroless deposition. The electroless Ni–P plating bath is $NiSO_4 \cdot 6H_2O$ 0.08 mol/L,

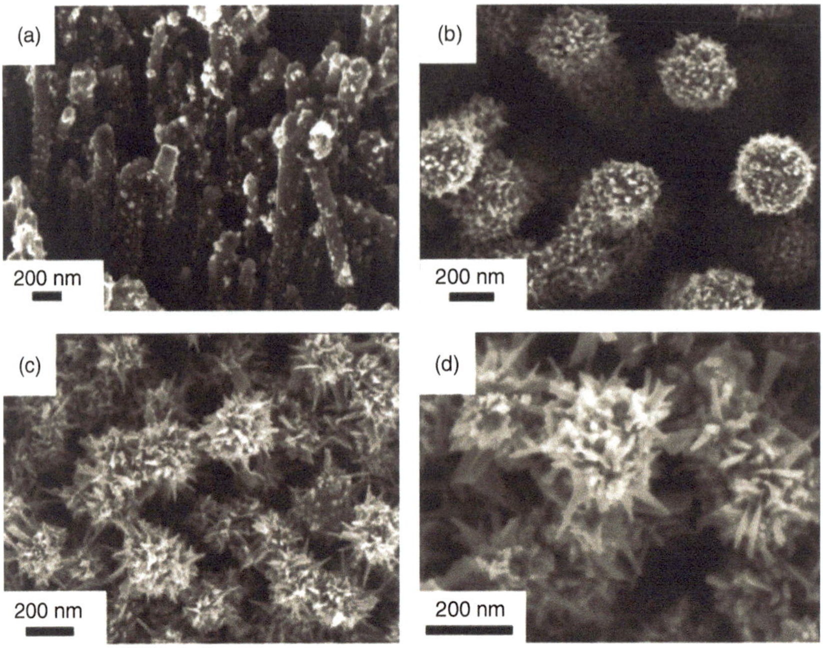

FIGURE 4.82 SEM images of gold-coated VACNWs after various times in the gold electroless deposition bath: (a) 1, (b) 5, (c), 7.5 (d), 7.5 h [285].

$NaPH_2O_2{\bullet}H_2O$ 0.2 mol/L, and $C_6H_5Na_3O_7{\bullet}2H_2O$ 0.08 mol/L. The pH of the bath was adjusted to 9 by the addition of NH_3 solution. Electroless plating was performed for 15 min at 35°C under ultrasonic treatment. Figure 4.83 shows SEM micrographs of the raw vapor-grown carbon nanofibers (VGCFs) (a, c) and the electroless Ni–P-plated VGCFs (b, d). Nickel was deposited on all the VGCFs homogeneously, resulting in Ni–P-coated VGCF powder, as shown in Fig. 4.83b. The thickness of the Ni–P coating was about 25–50 nm and the morphology of the Ni–P coating was not smooth, as seen in Fig. 83d. EPMA qualitative analysis showed that phosphorus, about 2–3 wt%, was present in the Ni–P coatings on the VGCFs.

EP Ni–B-coated CNWs were prepared by Arai et al. in 2011 [291]. Commercially available VGCNFs were used in this study. The VGCFs had a diameter of 150 nm and a length of 10 μm. The electroless plating bath composition used was 0.1 mol/L $NiSO_4{\bullet}6H_2O$ + 0.1 mol/L $Na_3C_6H_5O_7$ + x mol/L DMAB. The bath pH was eight. Electroless plating was performed with stirrer agitation at 30°C. Experimental data show that the boron content in the alloy deposits increased with increasing boron concentration in the plating bath, ranging from 14 to 24 at% B (2.9–5.5 mass%). Figure 4.84 shows the growth of the Ni–B alloy using the plating bath with 0.01 M DMAB. The Ni–B alloy (Ni–14 at% B) thickness on the VGCNFs increased with increasing reaction time. Ni–B alloy particles are observed on the VGCNFs at 30 s (Fig. 4.84 b). These particles grew on the VGCNFs (Fig. 4.84c) with increasing time and finally

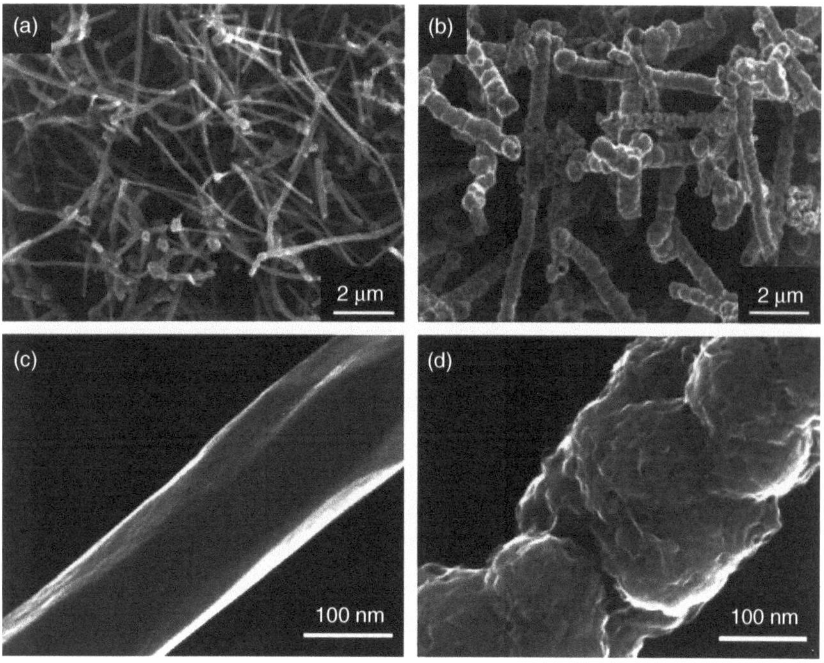

FIGURE 4.83 **SEM images of (a) raw VGCFs and (b) nickel-coated VGCFs obtained by electroless deposition. Enlarged images corresponding to (a) and (b) are shown in (c) and (d), respectively [287].**

perfectly coated the VGCNFs at a reaction time of 300 s (Fig. 4.84d). This reaction time corresponds to a mean film thickness of about 10 nm (see the inset in Fig. 4.84b). After a reaction time of 300 s, the alloy films grew autocatalytically on the deposited Ni–B alloy on the VGCNFs. Thus, perfect coating of each VGCNF with semiamorphous Ni–B alloy (Ni–14 at% B) was achieved at film thicknesses above 10 nm using a plating bath with 0.01 M DMAB.

In 2007 and 2011, Park et al. [292, 293] studied the hybrid composites containing Ni–Fe coated CNWs prepared by electroless plating. The typical diameter and length of CNWs are 100–200 nm and 10–20 μm. Ni–Fe–P plating solution is composed of $FeSO_4•7H_2O$, $NiSO_4•6H_2O$, $KNaC_4H_4O_6•4H_2O$, $NaPH_2O_2•H_2O$‡, NaOH, and deionized water. Electroless Ni–Fe–P plating is progressed at 75°C for 8 min. Figure 4.85a shows the SEM image of Ni–Fe–P coated CNWs. On the whole, the Ni–Fe–P deposited layers are homogenously distributed on the each surface of CNFs. Figure 4.85b shows the TEM BF images of Ni–Fe–P coated CNWs. Light rods are CNWs and dark points of the edge are Ni–Fe–P coated layers. The thickness of typical Ni–Fe–P deposition is about 50–100 nm. From the EDS the composition of the Ni–Fe–P coating is 26.20 wt% Ni, 70.05 wt% Fe, and 3.75 wt% P.

‡The origin is $NaPH_2O_2•6H_2O$, it should be $NaPH_2O_2•H_2O$ —The author.

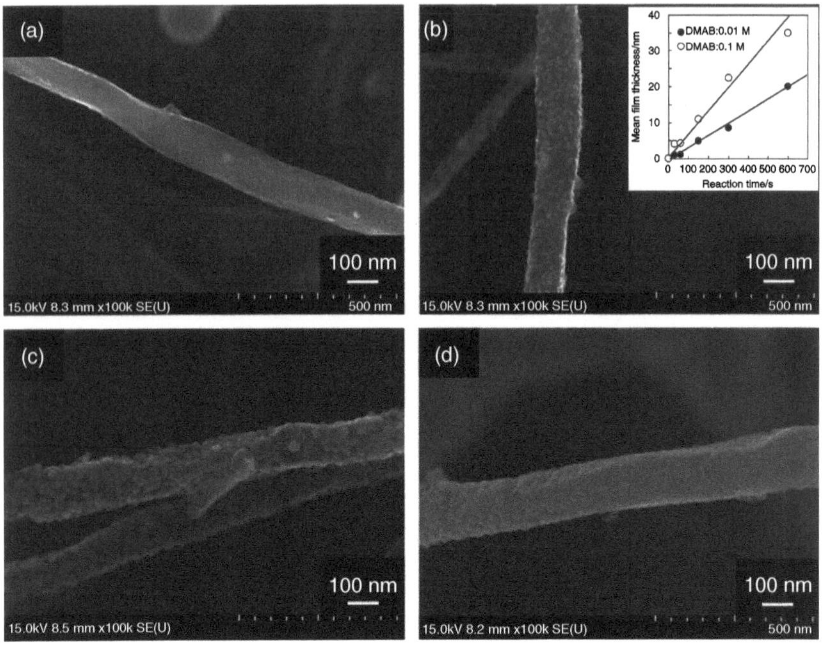

FIGURE 4.84 SEM images showing the growth of Ni–B alloy deposits on VGCNFs using an electroless plating bath with 0.01 M DMAB (a) before electroless deposition and (b) after 30, (c) 150, and (d) 300 s. The inset in (b) shows relationship between reaction time and mean film thickness of the Ni–B alloy on the VGCNFs [291].

Using EP, Xie et al. [294] prepared amorphous Ni–Co–B alloy supported on CNWs in 2008. They also prepared EP Ni–Fe–Co–P coatings on coiled CNWs in 2005 [295]. We just discuss the later more in detail. The coiled CNWs used in the study were synthesized by a low temperature (300°C) nano-copper catalytic CVD method. The electroless deposition of Ni–Co–Fe–P alloy was performed in an alkaline solution, containing 10 g/L nickel chloride, 12 g/L cobalt sulfate heptahydrate, 10 g/L ammonium ferrous sulfate, 20 g/L sodium hypophosphite monohydrate as reducing agent, and 80 g/L potassium sodium tartrate tetrahydrate as stabilizer. The pH was controlled by the addition of ammonia solution. The process was optimized at 80°C. Figure 4.86 shows the SEM images of the uncoated and coated CNWs. It can be seen from the SEM images that the electroless coatings are continuous and uniform. The thickness of the coatings is about 100 nm. The chemical composition of the coatings determined by EDX was Ni 52%–70%, Fe 5%–13%, Co 10%–32%, and P 3%–8% (wt%).

A group from India fabricated Ni–P–ZrO_2–Al_2O_3 composite coatings on carbon fiber woven fabric by the EP technique in 2002 [296]. The bath components and operating conditions for EP coating are nickel sulfate, sodium hypophosphite, sodium citrate, ammonium sulfate, and chloride, aluminum

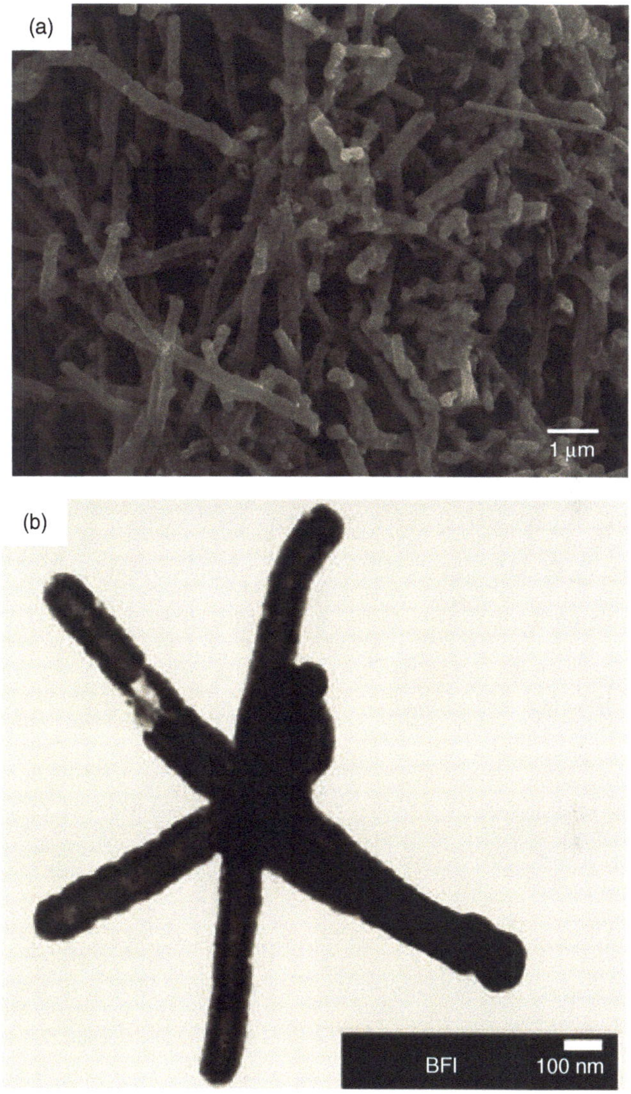

FIGURE 4.85 **SEM image (a) and TEM BF image (b) of Ni–Fe–P/CNWs [292].**

chloride, zirconium oxychloride, pH 9.0, temperature 90°C, and time of coating 5 min. The experimental data indicate that the coating on carbon fabric is amorphous in nature; the amorphous nature of the EP coating is due to the presence of phosphorus atoms distorting the crystal lattice of nickel. Quantitative analysis of EP Ni–P–ZrO_2–Al_2O_3 coating determined by EPMA shows that Ni 90.188, P 9.5, Al_2O_3 0.025, and ZrO_2 0.287 (wt%). From Fig. 4.87d fine grains of 5–15 nm size uniformly distributed in the matrix are identified as Ni, Al_2O_3, ZrO_2.

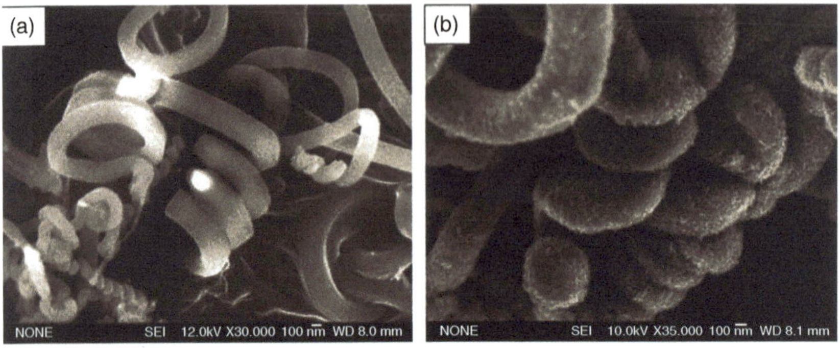

FIGURE 4.86 **SEM images of CNWs: (a) uncoated wires and (b) coated wires [295].**

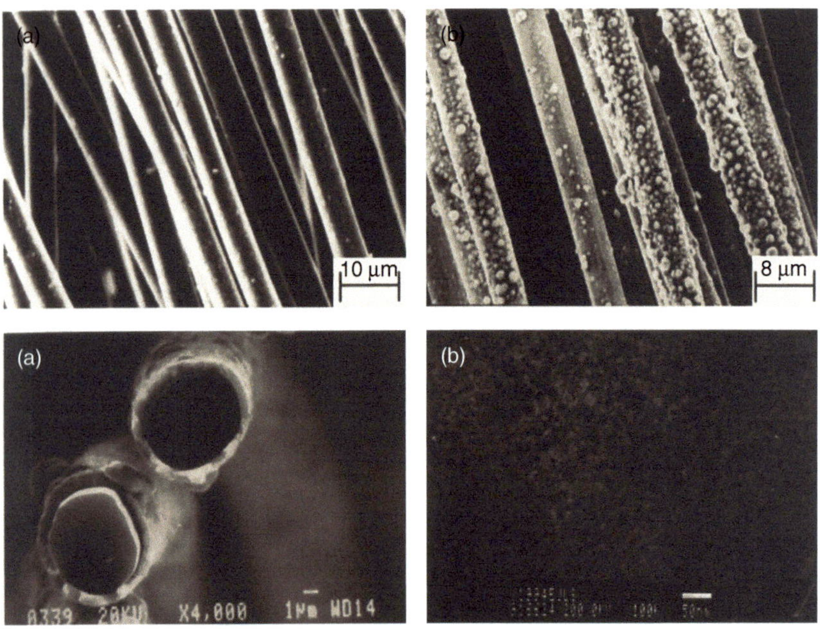

FIGURE 4.87 **SEM micrographs of uncoated carbon fabric in: (a) longitudinal position. SEM micrograph of Ni–P-ZrO_2–Al_2O_3 coated carbon fabric in: (b) longitudinal position and (c) cross-sectional position. TEM micrographs of Ni–P--ZrO_2–Al_2O_3 coating on carbon fabric: (d) bright-field image [296].**

In addition, fabrication and properties of silvered glass wire (GW) by dopamine functionalization and electroless plating were studied in 2012 [297]. Figure 4.88 indicates the experimental results. Figure 4.88a is the SEM images of GF-polymerized poly(dopamine) (PDA)/Ag without extra reducing agent at magnifications 5000×. It can be seen that uniformly dispersed discrete Ag clusters are deposited on the GF surface. However, the Ag particles are dispersive

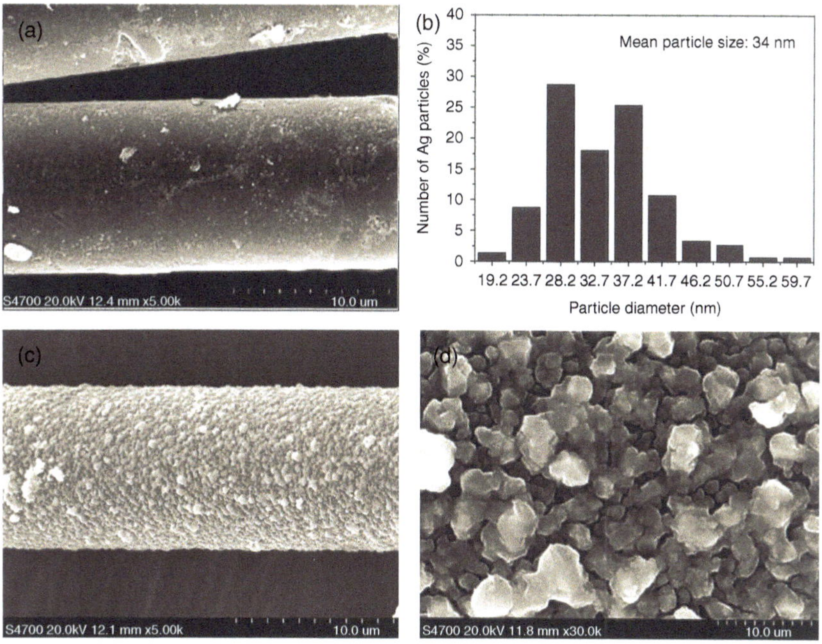

FIGURE 4.88 **(a) SEM images of GF-PDA/Ag without extra reducing agent at magnifications ×5000, (b) particle size distribution of Ag nanoparticles reduced by PDA only, (c) and (d) SEM photographs of GF-PDA/Ag at silver nitrate concentrations: 10 g/L (dopamine concentration = 2 g/L) [297].**

and do not form a continuous silver layer in this case. Part (b) is the particle size distribution of Ag nanoparticles reduced by PDA only, the discrete Ag clusters have a narrow size distribution with a mean diameter of 34 nm. Parts (c) and (d) are the SEM photographs of GF-PDA/Ag at silver nitrate concentrations: 10 g/L (dopamine concentration = 2 g/L). Right now, the Ag particles form a continuous silver layer, covering the whole GF. The experimental data indicate the presence of Ag nanoparticles in a crystalline metal form.

4.3.2.4 EP Si Nanowires

Silicon nanowires (SiNWs) are not similar to the CNT; they can be fabricated using Ag-assisted electroless etching processes [298–307]. In addition, people studied the composite SiNWs coated with metals, for example, SiNWs + Au [308,309], SiNWs + Pt [310], SiNWs + Ni [311–313], SiNWs + Ni–Pd [314].

EP SiNWs: Fabricated SiNWs using Ag-assisted electroless etching processes could be dated back to 2002 [298]. Since then, many reports have been published on this subject [299–307]. The technique is based on the galvanic displacement of Si by $Ag^{+} \rightarrow Ag^{0}$ reduction on the wafer surface. Usually p-type (100) and (111) wafers were used. Silicon substrates were cut into $1.5 \times 1.5\ cm^2$ squares and washed with DI water. Then, SiNWs were synthesized at the

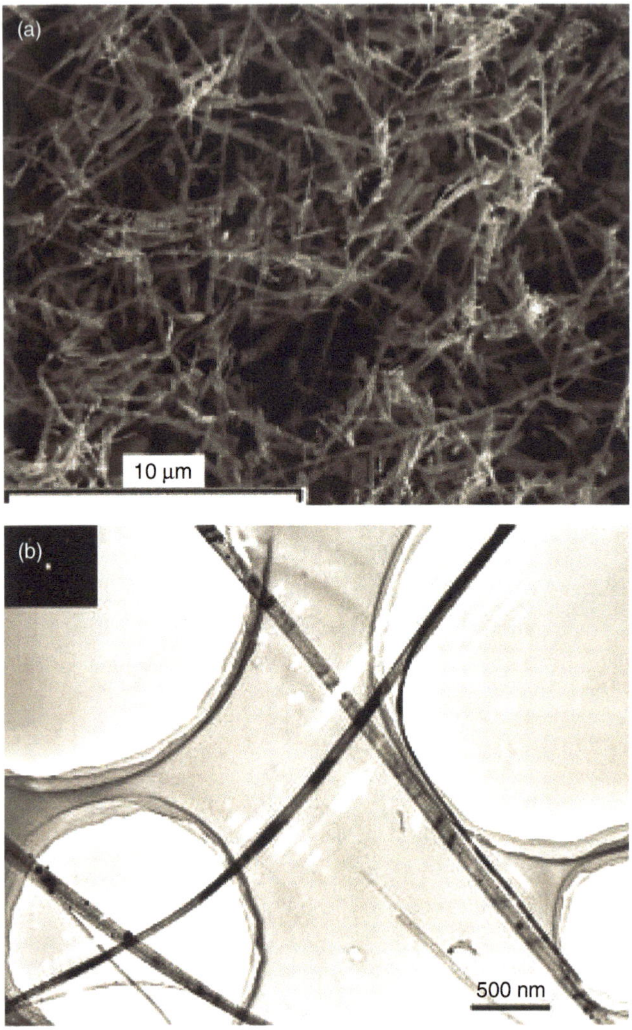

FIGURE 4.89 **(a) TEM image of Si nanowires. (b) TEM image of a single Si nanobelt. The upper left inset in (b) shows the electron diffraction pattern of one nanowire [298].**

concentration ratio, for example, 0.02M: 4.6M for $AgNO_3$/HF at 50°C with different chemical etching times from 15 min to 60 min. After the chemical etching process, the SiNWs arrays were covered with Ag dendrites on the top of the prepared SiNWs array. Therefore, the as-prepared samples were dipped in 30% HNO_3 solution for some times to remove all residual Ag from the nanowire surfaces. Peng et al. [298] found that Ag dendrite film was wrapped the Si wafer (Fig. 4.89a). Fig. 4.89b indicates the TEM image of the SiNWs. The diameters of the SiNWs normally range from 30 nm to 50 nm and their lengths are 20–50 µm. The upper left inset in Fig. 4.89b shows that a typical small-angle electron diffraction (SAED) pattern of one nanowires could be indexed for [110] zone axis of single-crystal silicon. Interestingly, the authors used other oxidizing agents such as K_2PtCl_6, $KAuCl_4$, $Cu(NO_3)_2$, $Fe(NO_3)_3$, $Mn(NO_3)_3$, and $Co(NO_3)_3$ instead of $AgNO_3$ in HF solution to see if metal Pt, Au, Cu, etc., also like Ag a dendrite film can be formed, wrapping the Si wafer. The results confirmed that continuous metal grain films were usually formed in the EP processes of Pt, Au, Cu in ionic metal solution. These continuous metal grain films will lead to uniform silicon etching; therefore, no nanowires could be formed in these cases. As for $Fe(NO_3)_3$, $Mn(NO_3)_3$, and $Co(NO_3)_3$ the experiments show that strong luminescent porous Si layers could also be formed, but no corresponding metal deposition occurred. In these instances, numerous micrometer-sized pillar-like, cone-like, or crater-pit-like microstructures could be observed. That is to say that only Ag dendrite film can be formed with $AgNO_3$ in HF solution. The growth of Ag dendrites plays an important role in the formation of SiNWs. However, Huang et al. [302] in 2010 proved from both theory and experiment that the similar as Ag, Cu can be utilized to catalyze the etching of Si into nanowires as well. Comparing the metal Ag, Cu will greatly lower the cost of Si nanowires from the metal-assisted etching method. Figure 4.90a and its inset show typical plan view SEM images of a Cu-particle-loaded Si substrate. Cu-assisted electrochemical etching of Si results in nanowires or nanopores on the Si substrate (Fig. 4.90b and c), which look similar to the structures resulting from Ag-assisted electrochemical etching (see the original figure 2(b) and (c) in [302]). The Cu particles also locate at the etching front, and nanowires with a diameter as small as 30 nm are obtained. Usually, Cu particles can be oxidized by oxidants (e.g., H_2O_2, Fe^{3+}) and vanish immediately after a short-time incubation in the etchant. Consequently, Cu-assisted chemical etching of Si usually results in shallow pits on the surface of the Si (Fig. 4.90d) due to the loss of Cu, instead of nanowires or nanopores.

4.3.2.4.1 EP SiNWs + Nano Metal and Alloy Composite

Sayed et al. studied galvanic gold deposition on SiNWs [308,309]. First, they prepared SiNWs. After cleaning, Si wafers were first immersed into a hot solution of $H_2O:NH_4OH:H_2O_2$ (5:1:1) for 15 min, and after rinsing with excess water, they were immersed into a hot solution of $H_2O:HCl:H_2O_2$ (6:1:1) for 15 min. Following another cleaning procedure, the wafers were immersed into 1% HF (aq) for 7 min to remove the native surface oxide film. The wafer was divided into smaller pieces. To hydrogen passivate the SiNWs, the substrates were immersed in 1% HF (aq) for 3 min, and then rinsed with water and dried under nitrogen. After that, the silicon wafer shard was immersed in the desired aqueous gold salt and hydrofluoric acid solution in a Teflon beaker. The gold salt/acid solutions were prepared by mixing 0.1 mL of 0.01 M $KAuCl_4{\bullet}xH_2O$ and 9.9 mL of 1% HF (aq) for a given time. After metal deposition, the sample was thoroughly rinsed with water and dried under a nitrogen stream. Figure 4.91 shows the formation of nanostructured Au films on single-crystal shards of

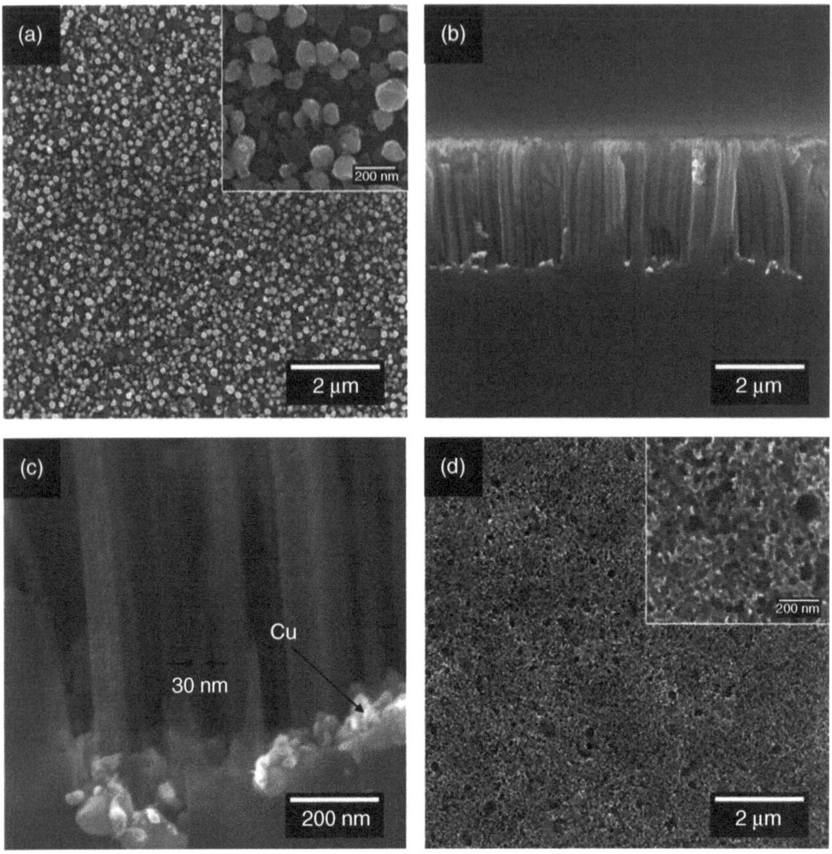

FIGURE 4.90 **(a) Plan view SEM images of Cu particles loading on a Si substrate. (b, c) Cross-sectional view SEM images of a Cu-loaded Si substrate subjected to electrochemical etching in a HF solution. (d) A plan view SEM image of a Cu-loaded Si substrate subjected to chemical etching in a HF/H_2O_2 solution [302].**

Si(111) and SiNWs, by immersion in dilute $KAuCl_4$ (aq) and HF (aq) for short periods of time (seconds to minutes). The plan view (top view) SEM images reveal more continuous growth of gold on Si(111), and cross-section SEM image shows that the thickness of the gold layer is ~10 nm. On silicon nanowires, Au nanoparticles of different sizes (20–30 nm) are observed.

In 2009, Peng et al. [310] studied aligned n-type silicon nanowire (SiNW) arrays decorated with discrete 5–10 nm Pt nanoparticles (PtNPs) fabricated by aqueous electroless Si etching followed by an electroless platinum deposition process. The diameter distribution of the SiNWs is in the range of 20–300 nm, and the axial wire length is about 17 μm. After the SiNWs arrays were chemically etched and cleaned, they were immediately immersed into a solution containing 50:1 HF solution and H_2PtCl_6 (0.001 M) to decorate SiNW surface with PtNPs via electroless plating. Figure 4.92a and c shows a schematic of

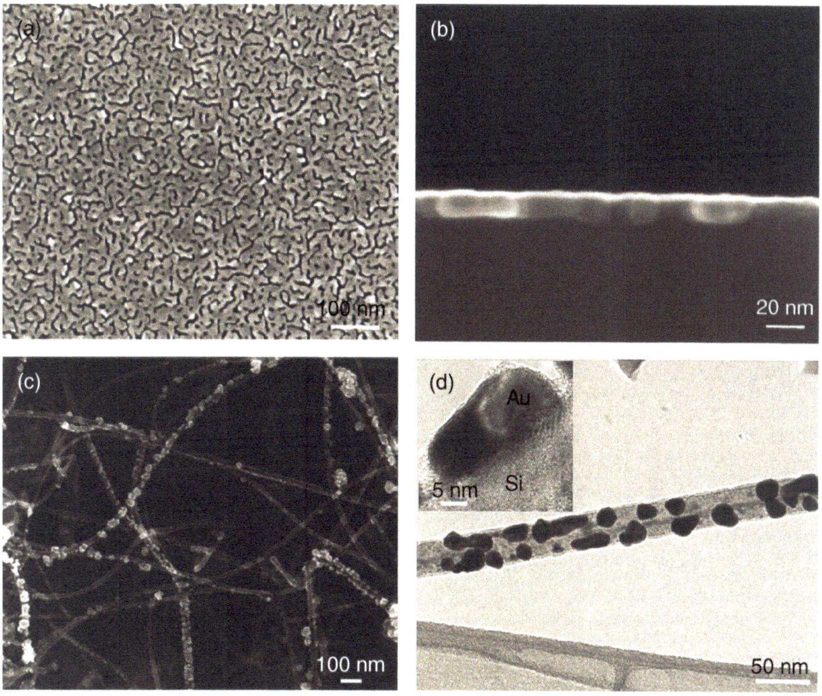

FIGURE 4.91 **(a, b) Plan view and cross-section SEM images of Au films on Si(111), formed from 0.1 mM $KAuCl_4$ (aq) and 1% HF (aq) for 7.5 min. (c) SEM image of SiNWs functionalized with Au nanoparticles. Conditions: 1 mM $KAuCl_4$ (aq) and 1% HF (aq) for 30 s. (d) TEM image for Au nanoparticles on a SiNW following removal of the SiNWS in panel c from their substrate via sonication in 100% ethanol, followed by spotting the supernatant layer on a lacy-carbon grid. The inset in (d) shows HRTEM image of a SiNW with a single Au nanoparticle [308].**

PtNP-decorated n-SiNW arrays being prepared by aqueous electroless etching of silicon and electroless plating. Figure 4.92b shows the cross-sectional SEM of the heat-prepared Si nanowire array; dendritic Ag can be observed inside the SiNWs array. Figure 4.92d shows the low-magnification TEM image of NP-decorated SiNWs, indicating that most of the PtNPs are primarily located at the wire tips. The mean diameter of PtNPs is typically 5–10 nm. Both electron diffraction (ED) patterns and HRTEM images confirmed the high-quality single-crystal characteristics of the nanowires.

Few reports refer to SiNWs decorated with Ni nanoparticles by EP [311–313]. In 2012, Wang et al. [313] prepared Ni nanoparticles by electroless deposition on silicon nanowires. The Ni electroless plating bath used was $NiSO_4{\bullet}6H_2O$ 1 mol/L, $(NH_4)_2SO_4$ 0.5 mol/L, sodium dodecyl sulfate 0.001 mol/L, NH_4F 2.5 mol/L, sodium succinate 0.2 mol/L, sodium citrate 0.2 mol/L, and pH 8.5–9.5 adjusted by ammonia. Figure 4.93 shows SEM cross-sectional images of silicon nanowires at different magnifications. Figure 4.93a shows that the EP

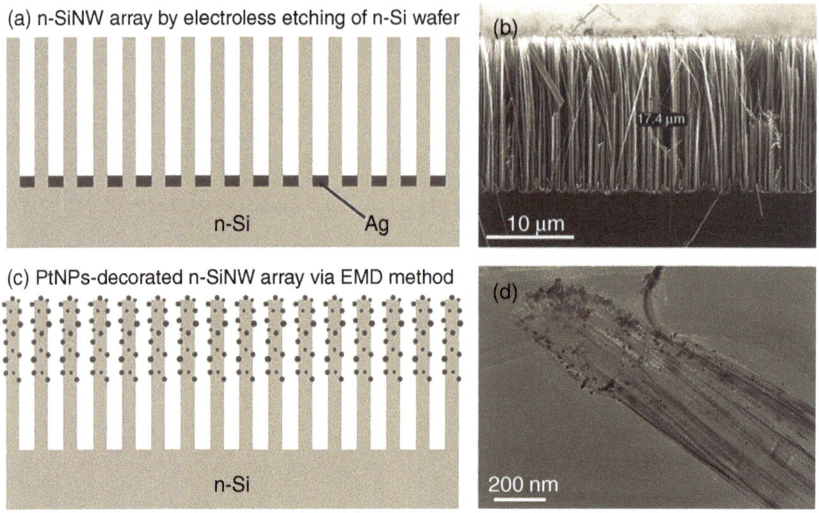

FIGURE 4.92 (a, c) Schematic of PtNP-decorated n-SiNW arrays prepared by aqueous electroless etching of silicon and electroless metal deposition. (b) Cross-sectional SEM of as-prepared Si nanowire array; dendritic Ag can be observed inside the SiNWs array. (d) Low-magnification TEM image of PtNP-decorated n-SiNWs showing PtNPs on the wire surface [310].

Ni particles were distributed on the surface of SiNWs. Figure 4.93b is a partial enlargement of Fig. 4.93a, in the region of the red frame. It was shown that Ni particles were noninteracting and monodisperse in Fig. 4.93b. Figure 4.93c is the enlargement of the red frame in Fig. 4.93b. It can be observed that the Ni particles exhibited homogeneous spherical structure and the diameter of Ni particles was ~35 nm. The XRD patterns of Ni/SiNWs indicate that the as-deposited Ni had a face-centered-cubic (fcc) structure. No other peaks that could be attributed to impurities, for example, P were detected, which indicated that a phase-pure nickel film had been obtained. However, several very weak peaks of nickel oxide were also detected by XRD, indicating nickel oxide formed on the surface.

The Ni–Pd/SiNWs nanocomposites were prepared by coating Ni and trace Pd simultaneously onto the SiNWs by electroless coplating [314]. The recipe for the Ni electroless plating bath is the same as that mentioned for plating Ni/SiNWs. Electroless coplating of Pd–Ni was conducted by adding $PdCl_2$ 100 mg/L into the Ni electroless plating bath. The experiment has shown that the diameter of nanowires ranges from 60 nm to 300 nm, and the thickness of the plated film onto the SiNWs' surface is about 10 nm, covering the silicon nanowires uniformly.

4.3.3 EP Nanorods

What is the difference between nanorods and nanowires? There is no serious definition. Usually, people use aspect ratios (length-to-width ratio) to separate them. Nanorods (NRs) are the materials that have a width of 1–100 nm and

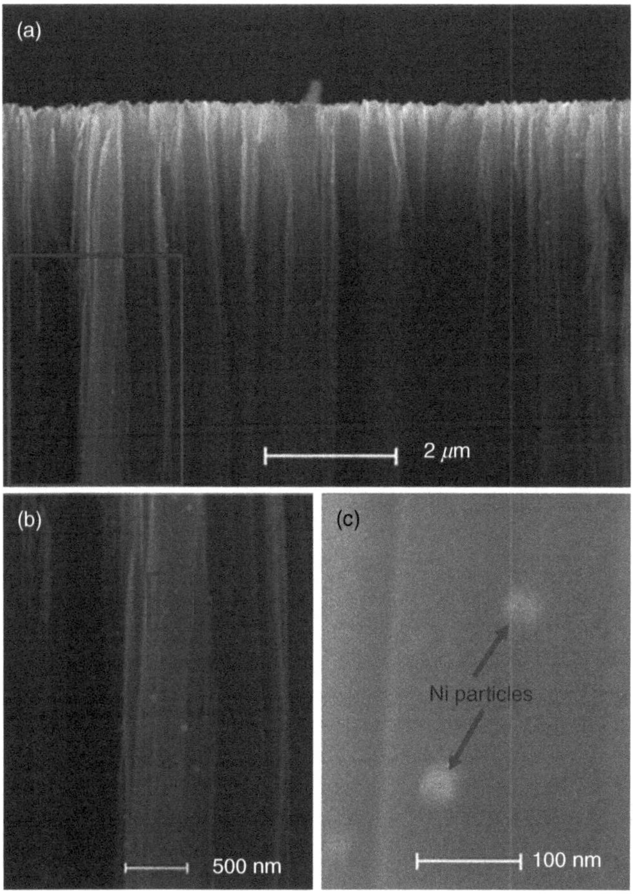

FIGURE 4.93 SEM cross sectional images of silicon nanowires with different magnifications. (a) By the electroless plating, Ni particles were distributed on the surface of SiNWs. (b) Partial enlargement of the red frame in (a). (c) Partial enlargement of the red frame in (b) and the diameter range of Ni particles was ~35 nm [313].

aspect ratios greater than 1 but less than 20; and NWs are the materials that have aspect ratios greater than 20 [315]. But, this distinction is not absolute, researchers may use other methods than the standard of aspect ratios to separate them.

Compared to nanotubes and nanowires, rather fewer reports of EP nanorods have been published as yet. The following papers reported the preparations and features of nanorods: AgNRs [316, 317], AuNRs [318, 319], NiNRs [320], ZnONRs [321–323], ZnO NRs + Ag [324], C_{60}NRs + Ni [325], Rod-like viruses for the synthesis of CuNRs, and PdNRs [326].

AgNRs were first fabricated in 2004 [316]. The authors used 6 nm silver particles to prepare AgNRs with high aspect ratios. The silver seeds were prepared by reduction of silver nitrate with sodium borohydride in the presence of citrate

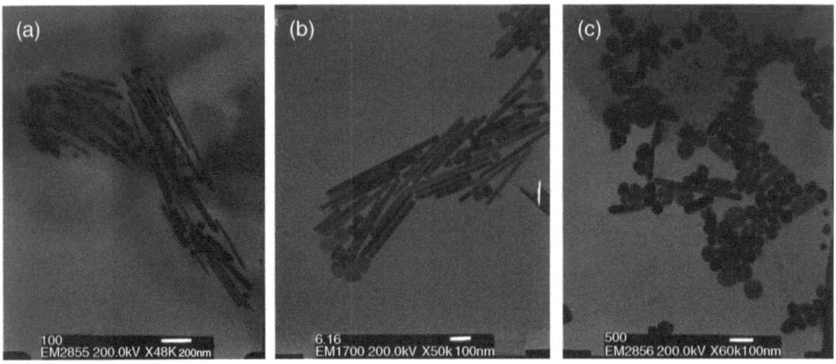

FIGURE 4.94 TEM images of AgNRs that were obtained at different pHs (a) 6.95 (b) 10.65, and (c) 12.24 [316].

as stabilizer. Secondary nucleation during the growth stage was inhibited by carefully controlling the growth conditions using a weak reducing agent (ascorbic acid) and CTAB in aqueous solution. It can be seen from the TEM images of AgNRs (Fig. 4.94) that the lengths of AgNRs are 600, 423, 138 nm, the widths are 17.6, 27, and 38 nm, respectively. As a result, the aspect ratios of the AgNRs are 34 (6.95), 15.6 (10.56), and 3.6 (12.24). The figures in brackets are the pH values. That is to say that aspect ratio of AgNRs decreased with an increase of pH. The control of pH could increase and then decrease the length of a nanorod. This effect was observed for temperature as well.

In 2011, Krishnamoorthy et al. [319] prepared AuNRs using a route as follows. First, the authors used self-assembly amphiphilic copolymers to create high-resolution molds for nanoimprint lithography (NIL) spanning entire 100 mm Si wafers. Then, the nanoporous poly(methyl methacrylate) (PMMA) templates obtained through NIL were subjected to galvanic displacement reactions to create gold nanorod arrays. The electroless growth of gold was carried out in a solution mixture containing 2.3 mmol/L $HAuCl_4$ and 0.9% HF. The electroless deposition occurs through the *in situ* reduction of Au(III) ions in solution to Au(0) driven by the oxidation of silicon in the presence of HF. The immersion of templates into the plating bath for duration of 1 min was sufficient to create discrete and well-defined gold nanorods. Figure 4.95 shows the scheme

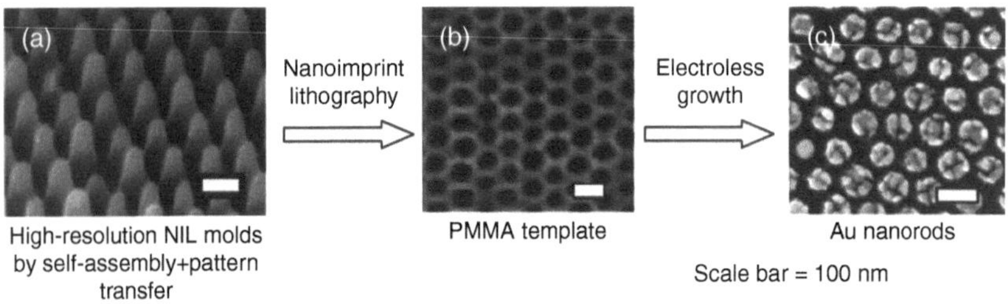

FIGURE 4.95 Scheme of fabricated steps [319].

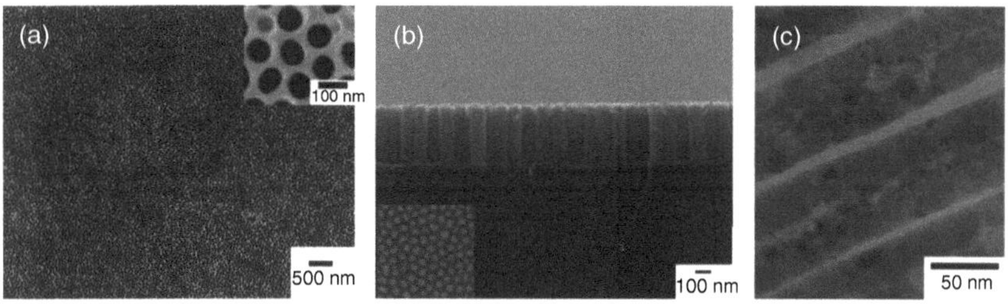

FIGURE 4.96 (a) SEM top-view image of AAO template. (b) SEM cross-sectional image of NiNRs on a Si substrate. The inset picture in (b) is SEM top-view image of NiNRs. (c) Cross-sectional TEM image of the as-deposited NiNRs [320].

of fabricated steps. AuNR arrays with a diameter of 40 nm and an aspect ratio of ~6 were obtained through this approach.

Ni nanorod arrays were fabricated on Si substrates using electroless deposition in 2009 [320]. The bath composition and operation parameters for the EP Ni used by Liu et al. are as follows: $NiSO_4{\bullet}6H_2O$ 20 g/L, $NaH_2PO_2{\bullet}H_2O$ 27 g/L, $Na_2C_4H_4O_4{\bullet}6H_2O$ 16 g/L, $Pb(NO_3)_2$ 1 ppm, pH 5.0, and temperature 65°C. Figure 4.96 shows the SEM and TEM images of AAO template and NiNRs.

It can be seen from the figure that as shaped by the AAO, NiNRs were constructed with a dimension of ~330 nm in height and ~70 nm in diameter (aspect ratio = ~4.7). The cross-sectional TEM image of the as-deposited NiNRs indicates that the rods are composed of a columnar structure that commences growth along the AAO walls and merges at the center of the pores, with an average diameter of ~25 nm. The HRTEM image shows that the columnar structure contains several nanocrystalline grains with an average size of about 2–3 nm, as highlighted, with the grains having random orientation and embedded in an amorphous matrix.

In 2003, Harnack et al. [322] reported the preparation of ZnONRs. The solution they used was 29.5 g (0.13 mol) of zinc acetate dihydrate dissolved in 125 mL of methanol at 60°C. Then, a solution of 14.8 g (0.23 mol/L) of potassium hydroxide in 65 mL of methanol was added. The reaction mixture was stirred for several days at 60°C. During this time, the nanoparticles precipitate. The length and the width of the resulting nanorods depend on the reaction time. Figure 4.97 shows a TEM image of ZnONRs after a reaction time of 3 days. The nanorods have a diameter of 15–30 nm and a length of 200–300 nm. Ye et al. also studied the coating of ZnO nanorods with nanosized Ag particles by electroless plating process [324]. The solution they used to fabricate ZnONRs is 40 mL of an equimolar aqueous solution of zinc nitrate and methenamine (10 mM), added to a flask and heated at a constant temperature of 95°C for 1.5 h with vigorous stirring. The diameter of ZnO nanorods obtained is about several hundred nanometers. The manufacturing procedures for EP Ag nanoparticles on the surfaces of ZnO nanorods are as follows. The ZnO nanorods (40 mg) were added to 20 mL of a mixture of methanol and water

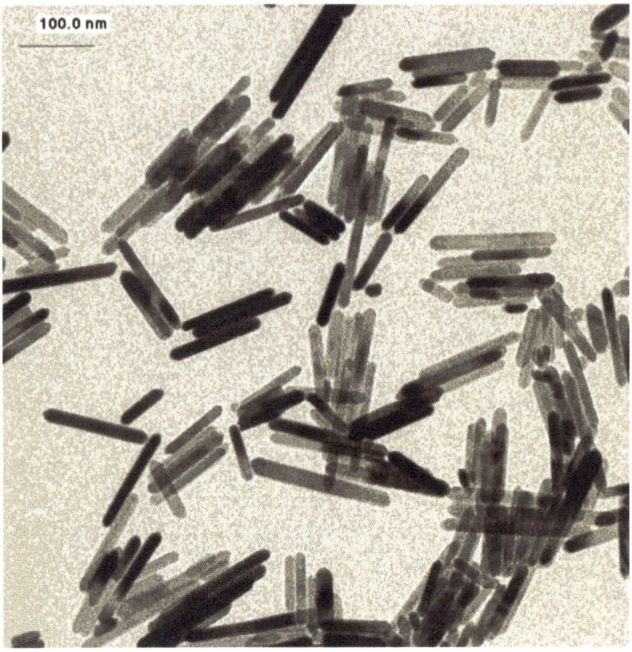

FIGURE 4.97 **TEM image of ZnO nanorods after a reaction time of 3 days [322].**

(1:2, w/w) containing $SnCl_2$ (0.029 M) and CF_3COOH (0.072 M). After agitation for 30 min, the suspension was centrifuged, the supernatant removed, and 10 mL of water added followed by 10 mL of ammonical $AgNO_3$ solution. When the reaction was finished, the resulting Ag-coated ZnO nanorod composites, which were present as a dark brown precipitate, were centrifugally separated from the suspension and ultrasonically washed with water. Figure 4.98 shows TEM and SEM micrographs of the ZnONRs and Ag-coated ZnO nanorod composites. It can be seen that the sharp edges of NRs are hexagonal (Fig. 4.98b) and their diameter ranges from 300 nm to 600 nm. Figure 4.98c and d represents TEM images of the ZnO nanorod after the first and second silver depositions. The second deposition repeats the coating step. It is obvious that the surface coverage of Ag nanoparticles is increased (Fig. 4.98d), resulting from not only the fresh formation of Ag nanoparticles but also the growth of previously existing Ag nanoparticles. After the first deposition, a number of remarkably monodisperse Ag NPs with a diameter of about 10 nm are uniformly distributed on the surface of the ZnO nanorod. However, after the second deposition, the size of Ag NPs reaches around 60 nm.

In 2009, Chen et al. [325] investigated the coating of C_{60}NRs with Ni–P by an electroless plating method. C_{60}NRs were synthesized by a simple solution method [326]. The treated C_{60} rods were then dipped in an alkaline plating bath solution for nickel deposition. The bath solution and operational conditions are

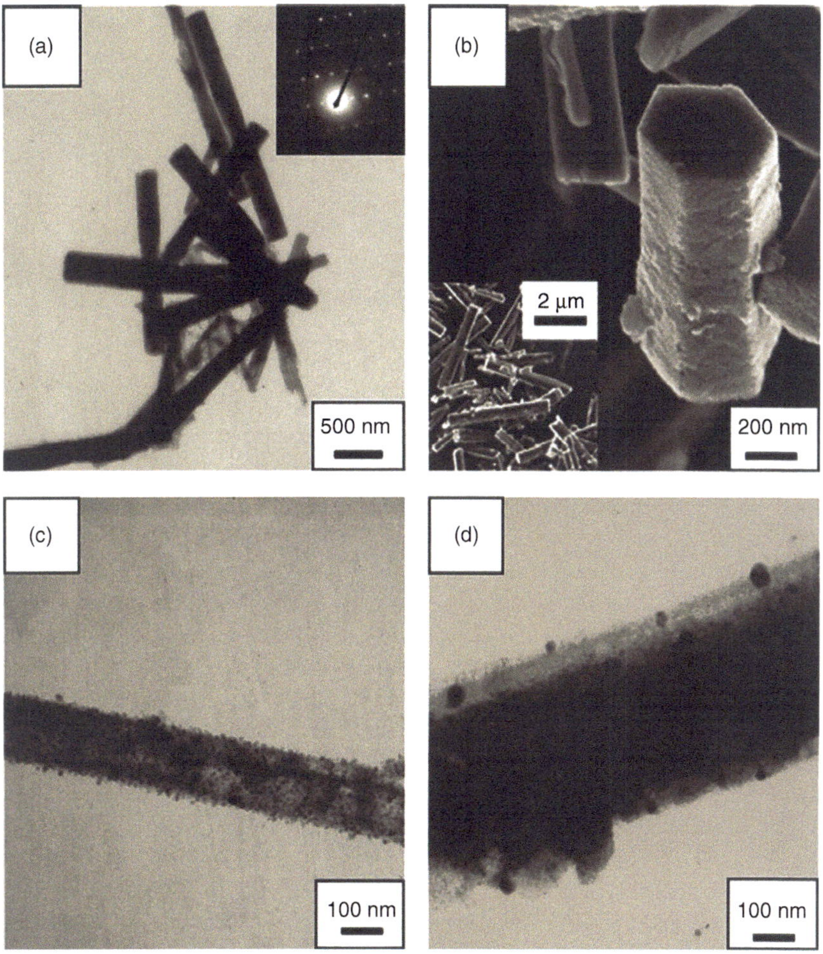

FIGURE 4.98 **SEM and TEM images: (a, b) TEM and SEM images of the ZnO nanorods (insets: corresponding SAED pattern in a; the low-resolution image in b). (c, d) TEM images of ZnO nanorod after the first and the second deposition of Ag nanoparticles, respectively [324].**

$NiSO_4{\bullet}7H_2O$ 30 g/L, $NaH_2PO_2{\bullet}H_2O$ 30 g/L, $Na_3C_6H_5O_7{\bullet}2H_2O$ 10 g/L, NH_4Cl 30 g/L, pH adjusted by NH_4OH to 8.0, and temperature maintained at 45°C. Figure 4.99 shows TEM images of a typical nickel-coated C_{60}RD with different plating times, 2 and 4 h, respectively. It can be seen that the C_{60}RDs are coated with a lot of particles. From the EDX analysis and TEM images, the authors estimated the average size of Ni particles on the coated C_{60}NRs to be 5.3 nm. The chemical composition and structure of the sample can be determined by EDX and XRD. It was found that the structural phase is amorphous, and the coating layer mainly consists of Ni and P. The weight ratio of Ni/P is estimated to be 1:0.09, corresponding to a composition as $Ni_{0.85}P_{0.15}$.

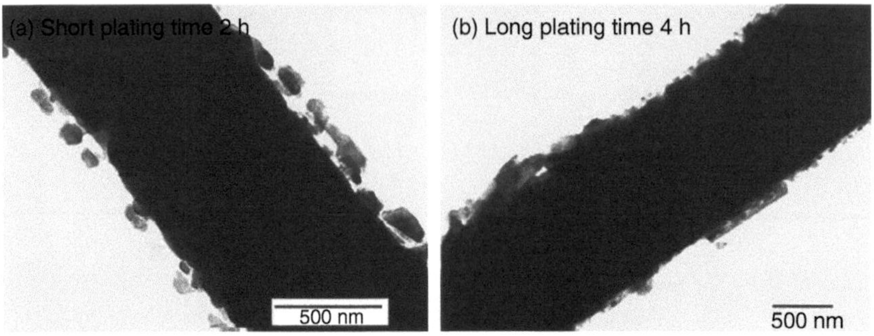

FIGURE 4.99 **TEM images of the nickel-coated C_{60} nanorods [325].**

It is interesting to note that in 2012 a group from USA used biotemplating rod-like viruses to synthesis CuNRs by EP [327]. Three rod-like plant virus tobacco mosaic virus (TMV) and bacteriophages such as fd and M13 are used by the authors because they are ideal templates for producing high aspect ratio nanorods. There are no reports to date on the metallization of the fd bacteriophage. First, Pd-activated virus outer surfaces were prepared, and then straight, continuous, and dispersed CuNRs by electroless deposition of Cu on such surfaces were synthesized. Electroless Cu plating was performed by mixing 1 μg/mL Pd-TMV into a Cu plating bath containing 1 mmol/L copper (II) sulfate ($CuSO_4$), 0.03 mmol/L EDTA, and 1 mmol/L DMAB. From the TEM images, it can be seen that the resulting Pd-TMV nanorods were straight and uniform in thickness with an average diameter of 35 ± 4 nm. Most of the Pd-TMV nanorods were structurally intact with length of 305 ± 119 nm. SEM image (inset in Fig. 4.100a) shows that Pd-TMV nanorods on a Si wafer are well dispersed. As for the Cu-TMV, during the time taken by the experiment, after 6 min reaction time, the change in diameter is not significant. Therefore, reaction time was set to 6 min, and the Cu-TMV was an average diameter of 46 ± 5 nm. Similar to Pd-TMV, Pd-fd Y21M and Pd-M13 were densely and continuously coated with 2–3 nm Pd nanoparticles. The average diameters of Pd-M13 and Pd-fd Y21M were 23 ± 4 and 20 ± 3 nm, respectively. TEM analysis indicates that Pd-M13 nanorods were slightly more entangled than the Pd-fd Y21M, reflecting the more flexible nature of the M13 template prior to metallization. After Cu plating, the average diameter of the Cu coated nanorods was 24 ± 4 nm for Cu-M13 and 35 ± 8 nm for Cu-fd Y21M. The Cu-M13 nanorods had aggregates larger than 10 μm and the Cu-fd Y21M nanorods formed 2–3 μm clusters. Overall, based on TEM images the coating on Cu-M13 (Fig. 4.100e) was less uniform in comparison to Cu-fd nanorods (Fig. 4.100f).

4.3.4 EP Nanobelts

Only few papers on EP nanobelts (NBs) were published resently.

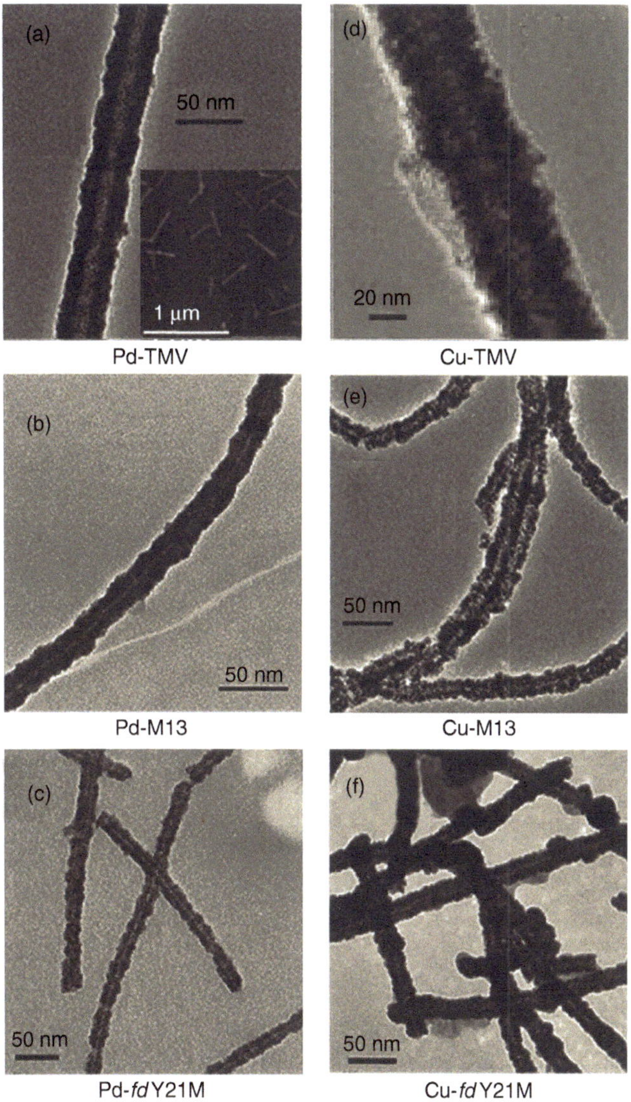

FIGURE 4.100 TEM images of metalized TMV, M13 and fd Y21M. (a) a Pd-TMV produced with 0.4 mM $Na_2Pd_2Cl_4$, inset in (a) is the SEM image of Pd-TMV dispersed on Si wafer. (b) Pd-M13. (c) Pd-fd Y21M. (d) Cu-TMV was taken from the plating bath for 9 min. (e) Cu-M13. And (f) Cu-fdY21M [327].

Sun et al. prepared Ag NBs in 2003 [328]. First, they synthesized Ag NPs in an aqueous solution by reducing $AgNO_3$ with $NaBH_4$ in the presence of poly(vinyl pyrrolidone) (PVP) and sodium citrate. TEM image of these Ag NPs indicated that they were spherical in shape with a diameter of

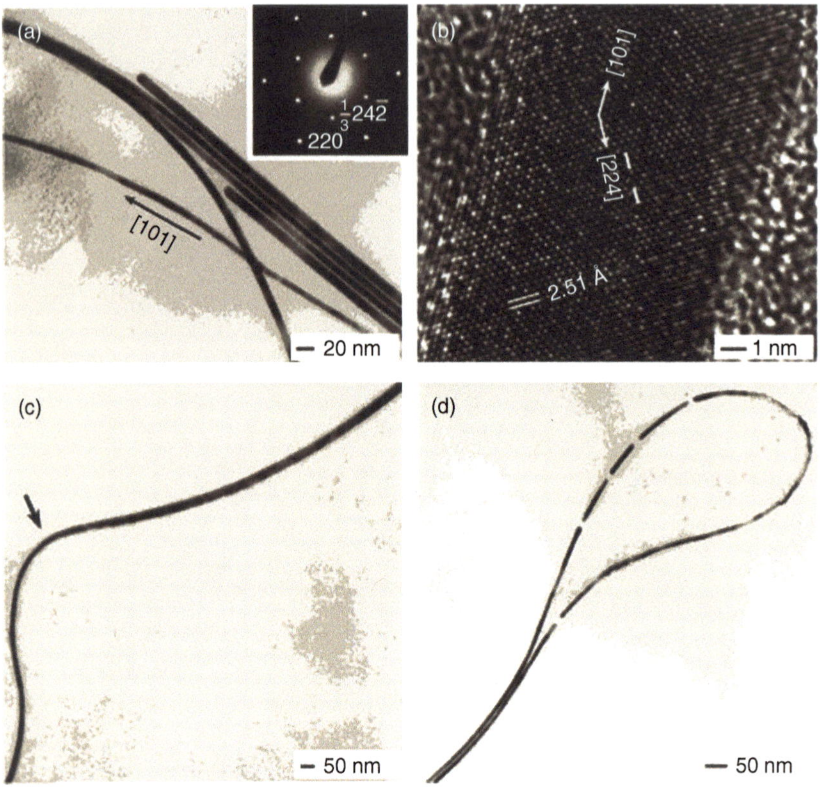

FIGURE 4.101 (a) TEM image of Ag NBs, with the inset showing a typical EDP taken from an individual nanobelt. (b) HRTEM image of a nanobelt indicating that the nanobelt grew along the [101], together with a highly crystalline structure. (c) TEM image of an individual nanobelt showing a twisting in the middle portion, whereby its width can be measured. (d) TEM image of a Ag nanobelt –10 nm in diameter that broke into short segments during sample preparation [328].

3.5 ± 0.8 nm. Electron diffraction patterns (EDP) confirmed that these Ag NPs were crystallized in the fcc structure. After refluxing an aqueous dispersion of spherical colloids of silver under ambient conditions of a chemical laboratory for 10 h, electron microscopic studies revealed that the Ag NPs had been transformed from spheres into a mixture consisting of triangular nanoplates (~95%) and wire-like nanostructures (~5%). The wire-like nanostructures could be separated from the nanoplates using centrifugation. Figure 4.101a shows the TEM image of some such nanostructures, indicating the uniformity in both lateral dimension and surface smoothness. These NBs had a mean diameter of ~9 nm and lengths up to 25 μm. The belt-like morphology could be easily identified when the lateral dimensions of these nanostructures were wider than 20 nm. Figure 4.101c shows the TEM image of one such silver

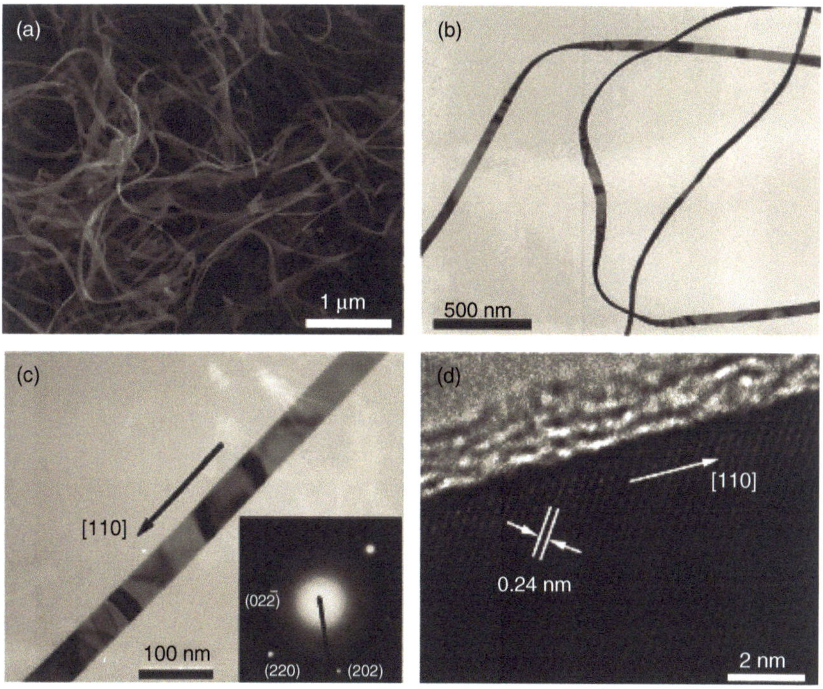

FIGURE 4.102 **(a) SEM, (b, c) TEM, and (d) HRTEM images of AuNBs obtained at 4°C. (Inset) ED pattern corresponding to the whole region of (c) [329].**

NBs whose lateral dimension was ~25 nm. Figure 4.101d shows the TEM image of a 10 nm Ag NB that broke into a number of fragments along its longitudinal axis.

EP Au NBs were reported in 2008 [329]. Zhao et al. first mixed at room temperature 3.45 mL of water, 0.65 mL of 0.05 M cetyltrimethylammonium bromide (CTAB), 0.50 mL of 0.01 M sodium dodecylsulfonate (SDSn), and 0.10 mL of 0.01 mol/L hydrochloroauric acid trihydrate ($HAuCl_4$•$3H_2O$) to give a yellowish solution. Then, 0.30 mL of 0.1 mol/L ascorbic acid was mixed with the solution at either 4 or 27°C, and kept at this temperature under static conditions for ~12 h, leading to formation of Au nanobelts. Figure 4.102a and b shows representative SEM images of the Au product obtained at 4°C, which suggest that the product exclusively consists of very flexible, 1D belt-like nanostructures typically several tens of micrometers in length, ranging from 40 nm to 150 nm in width, and about 15–20 nm in thickness. The XRD and EDS patterns indicate that the nanobelts are composed of pure gold with characteristic of fcc Au.

In 2008, Sun and Hagner [330] studied the rapid formation of ultralong nanobelts of water-soluble inorganic salt of $Na_3Au(SO_3)_2$ on a large scale via organic solvent-induced crystallization, carried out by adding a diluted solution

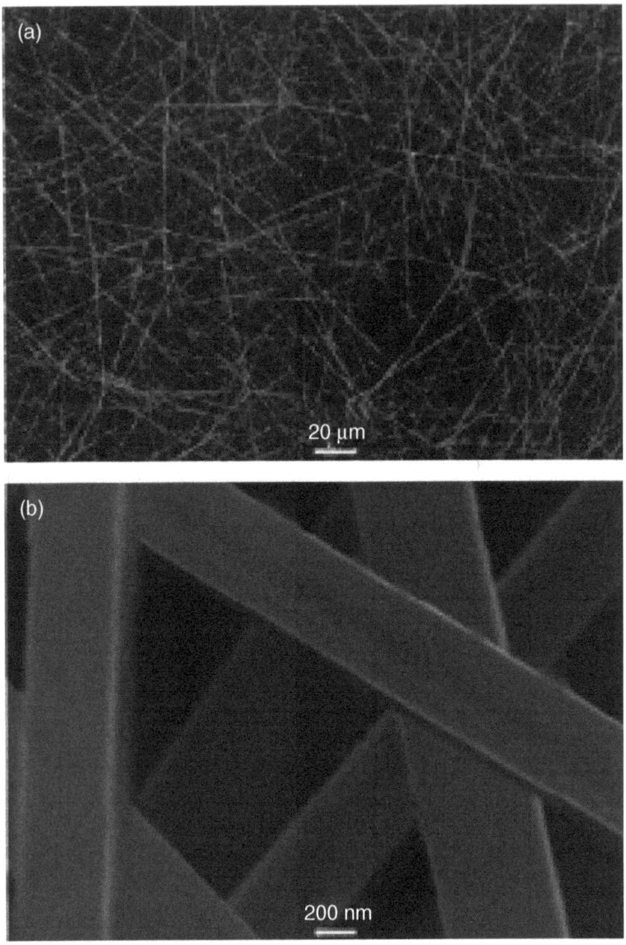

FIGURE 4.103 (b) Low-magnification SEM, and (a) high-magnification SEM image of the $Na_3Au(SO_3)_2$ nanobelts obtained by adding a diluted solution of 5 μL of gold bath in 295 μL of H_2O into 700 μL of DMF at room temperature [330].

of commercial gold bath, an aqueous solution of $Na_3Au(SO_3)_2$. It can be clearly seen from Fig. 4.103 that these 1D structures are uniform and belt-shaped about 500 nm in width and 200 nm in height. Strictly speaking, this is beyond the nanometer range (1–100 nm).

4.4 ZERO-DIMENSIONAL NANO EP MATERIALS

As mentioned in the very beginning of this chapter, 0D NMs include nanoparticles, nanocubes, nanoparticles array, and core-shell nanoparticles. We will discuss their synthesis, composition and structure, and features as follows.

4.4.1 EP Nanoparticles

Usually, NPs are nanospheres, and the sphericity of the NPs is prefect in some cases. However, the shapes of NPs other than spheres are also included in NPs. EP NPs have been reported as metal NPs, including Cu NPs [331, 332], Ag NPs [333–341], Au NPs [342–347], Ni NPs [348–349], Pd NPs [350–352], Pt NPs [353–355], Co NPs [356–358], and alloy NPs, such as Ni–P NPs [359], Ag–Au NPs [360], Ag–Pd NPs [361], Fe–Ni NPs [362], Fe–Pt NPs [363], Ni–Co NPs [364–367], Co–Pt NPs [368], Fe-Co–Ni NPs [364,365,369], Sm–Co NPs [370].

4.4.1.1 Nano Metal PNs

4.4.1.1.1 Cu NPs

Only a few reports on Cu NPs have been published [331,332]. In 2008,Yagi et al. [332] reported the formation of Cu NPs by electroless deposition using aqueous CuO suspension. The first step was to prepare a CuO colloidal aqueous suspension by dispersing CuO powder of 0.060 mols in 42.0 cm^3 of distilled water. The second step was to add 18.0 cm^3 of 10 wt% gelatin aqueous solution into the solution as a dispersing agent. The pH of the mixed solution was adjusted to 12.0 at 298 K by adding 1.0 mol/L sodium hydrate aqueous solution. The solution temperature was kept at 323–353 K and the solution was magnetically stirred at a rate of 500 rpm. 42.0 cm^3 of 1.43 mol/L hydrazine aqueous solution was added into 18.0 cm^3 of 10 wt% gelatin aqueous solution as a reducing solution. Then, the hydrazine solution was added to the CuO aqueous suspension as a reducing agent to deposit Cu particles. The total amount of reaction suspension was 120.0 cm^3, and thus the reaction suspension was 0.50 mol/L hydrazine aqueous solution with 0.50 mol/L dispersed CuO. Figure 4.104 indicates SEM images of precipitates after liquid-phase reduction for 6 h at 323–353 K. It has been found that the fraction of smaller particles tended to increase with increasing reaction temperature. When the reaction temperature at 323 K (Fig. 4.104a), only particles around 300 nm in diameter are observed. However, when the temperatures are up to 333 and 353 K, particles of two different sizes, less than 50 and 300 nm in diameter, are observed (Fig. 4.104b and c). Figure 4.104d shows the SEM image of precipitates obtained by reduction for 2 h at 353 K after centrifugal classification. The smaller particles around 30 nm were readily separated from the larger ones by centrifugal classification without aggregation.

4.4.1.1.2 Ag NPs

A lot of papers about EP Ag NPs have been published [333–341]. We will go into depth about just of one, published in 2012 [341]. Bastide et al. [341] studied electroless deposition of Ag nanoparticles on PECVD SiNx:H dielectric layers. They used P-type (100) oriented c-Si wafers as substrates for the growth of 300 nm thick SiNx:H and SiOx layers. Electroless deposition was performed by putting the dielectric layers in contact with an aqueous solution of HF containing $AgNO_3$ at 0.14 mol/L or 0.5 mmol/L, respectively. The Si substrate was isolated

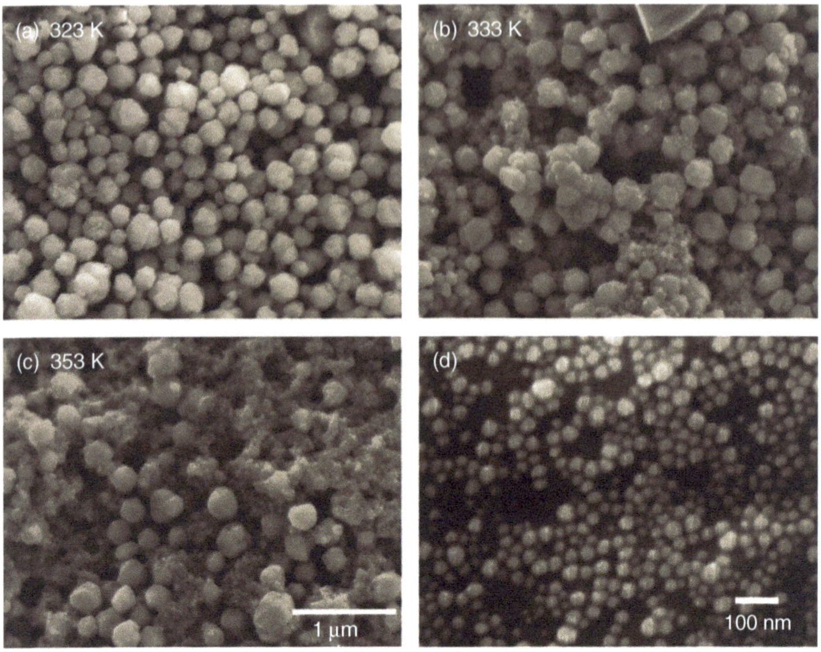

FIGURE 4.104 **SEM images of precipitates obtained at (a) 323, (b) 333, and (c) 353 K by liquid-phase reduction for 6 h. SEM image of precipitates (d) after centrifugal classification obtained at 353 K by liquid phase reduction for 2 h [332].**

from the solution by an O ring seal. After metallization, the samples were rinsed thoroughly in deionized water and dried with nitrogen. Figure 4.105 illustrates AFM images of SiNx:H surfaces after Ag deposition in HF-$AgNO_3$ for 30 s to 60 min. The inset in 30 s picture shows size and density of AuNPs as a function of deposition time. It can be seen that the NP size increases almost linearly with time (right axis in inset) to reach a mean value of ~100 nm after 60 min. That is to say, the NP size is less than 100 nm when the reaction time is less than 1 h. In addition, during the first 15 s the number of AgNPs rapidly increases, reaches a maximum value at 475 μm^{-2} after 2 min, and then decay exponentially to ~70 μm^{-2} after 60 min.

4.4.1.1.3 Au NPs

A small number of papers on EP Au NPs have been published [342–347]. In 2012, Yagi et al. investigated electroless growth of size-controlled gold nanoparticles using hydroquinone [346]. The first step was to prepare Au seed. An aqueous $HAuCl_4{\bullet}4H_2O$ stock solution (29.5 mmol/L) was prepared. Ultrapure water (30 cm^3) and $HAuCl_4$ stock solution (29.5 mmol/L, 0.300 cm^3) were mixed. This solution was immersed in an oil bath at 403 K and stirred vigorously. An aqueous solution of trisodium citrate (Na_3Cit, 38.7 mmol/L, 0.900 cm^3) was

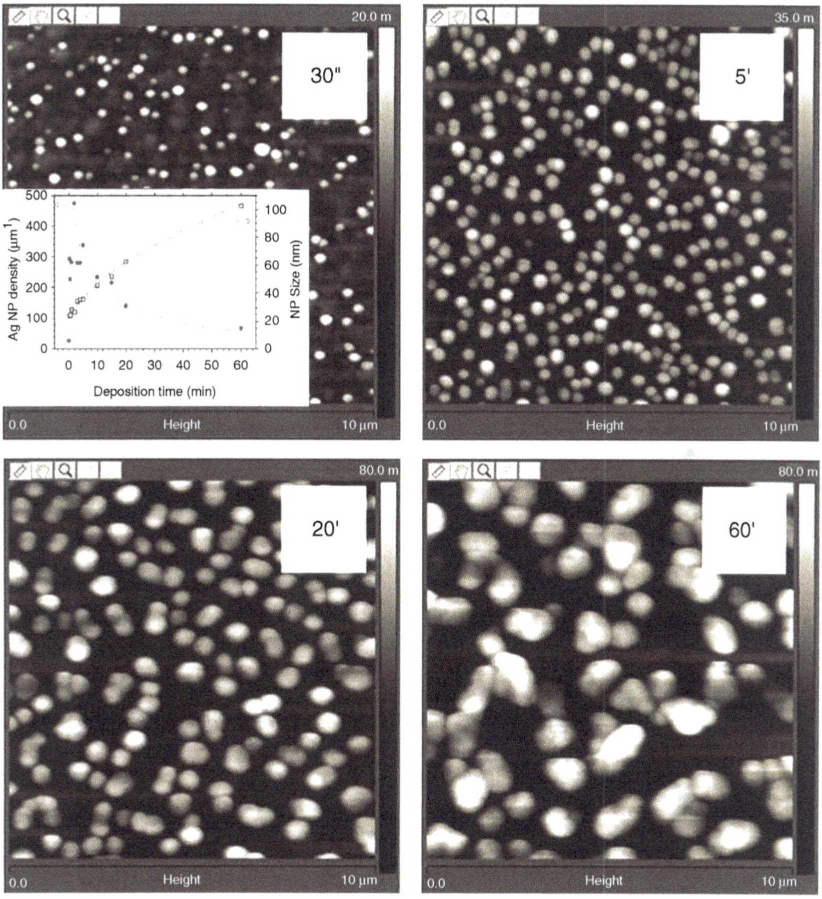

FIGURE 4.105 AFM images (tapping mode) of SiNx:H surfaces after Ag deposition in HF-$AgNO_3$ for 30 s to 60 min. The inset in 30 s picture shows size and density of AgNPs as a function of deposition time [341].

added to the solution just after it started boiling. After 10 min, the solution was cooled in air. The final seed solution contained Au atoms with a concentration of 0.284 mmol/L. The second step was to prepare Au growth solution. The concentration of the growth solution was 0.295 mmol/L $HAuCl_4$, 0.0851 mmol/L Na_3Cit, and 0.300 mmol/L hydroquinone. The initial pH of the growth solution was to be about 3.0 at 298 K. Three Au growth solutions (named growth solutions 1–3) containing different concentrations of seeds were prepared by mixing the $HAuCl_4$ stock solution (29.5 mmol/L, 0.100 cm^3) with 0.406, 0.200, 0.100 cm^3 of the seed solution. Therefore, the ratio of the number of Au seeds was about 4:2:1 in growth solutions 1, 2, and 3, respectively. Figure 4.106 shows TEM images and particle distributions of Au seeds and Au nanoparticles grown in the growth solutions 1–3 with an initial pH of 3.0. Spherical, monodisperse

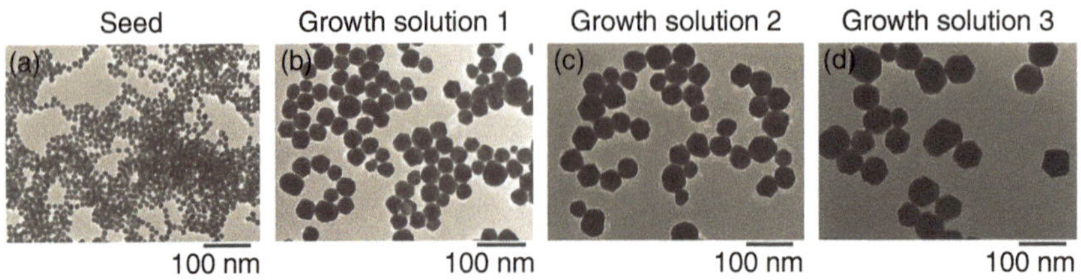

FIGURE 4.106 TEM images and size distribution of (a) Au seeds, and Au nanoparticles grown in (b) growth solution 1, (c) growth solution 2, and (d) growth solution 3 for 168 h. The concentration of growth solutions 1–3 was the same and only differed in their seed concentration [346].

Au seeds with a mean diameter of about 12.5 nm were successfully formed. The size of Au NPs grown from the seeds increased from 38.7 nm to 63.5 nm with decreasing seed concentration.

4.4.1.1.4 Ni NPs

Only a few papers on EP Ni NPs have been published [348,349]. Wu et al. in 2008 [349] synthesized Ni nanoparticles supported on metal oxides using electroless plating. The Ag/support was synthesized first. The electroless nickel plating was performed by adding the Ag/support to the plating solution with the following composition: Nickel(II) sulfate hexahydrate $NiSO_4{\bullet}6H_2O$, 12.0 g/L; KBH_4, 5.5 g/L; $H_2NCH_2CH_2NH_2$, 10 g/L; and NaOH, the amount needed to reach pH 13.5. The temperature of the bath solution was 323 K. The reaction lasted until no significant bubbles were observed (about 20–40 min). Figure 4.107 shows the TEM images of Ag/TiO_2 samples and their resulting Ni/TiO_2 samples. The particle size of the Ni NPs supported on the Ag/TiO_2 with 0.02 wt% Ag loading is ~100 nm (Fig. 4.106a and c). With the Ag loading increased from 0.02 wt% to 0.05 wt%, the Ni particle size decreases from ~100 nm to ~40 nm (Fig. 4.106b and d) and the Ni loading increases from 4.5 wt% to 8.1 wt%. It indicates that more new nuclei for the deposition of Ni are formed by promoting the Ag loading.

4.4.1.1.5 Pd NPs

Few papers on EP Pd NPs have been published [350–352]. In 2006, Lee et al. [351] prepared mesoporous Pd nanospheres using EP. The procedures for preparing Pd NPs are as follows. First, ~0.5 mL pure HNO_3 was added into 0.1 M SDS aqueous solution of 50 mL, keeping the concentration ratio of H^+ to SDS (C_{H^+}/C_{SDS}) at ~1.5. Now, the pH value of the mixed solution was ~1.14. Then, ~0.24 mmol palladium acetate ($Pd(OAc)_2$) was gradually dissolved in SDS/HNO_3 aqueous solution. After stirring slowly for 5 min the solution became orange, 10% hydrazine aqueous solution of 0.5 mL was slowly dropped in the $Pd(OAc)_2$/SDS/HNO_3 aqueous solution that was still stirred. The solution turned from orange to dark brown and then Pd nanospheres were obtained. Figure 4.108 indicates the HRTEM image of self-assembled, mesoporous Pd nanospheres.

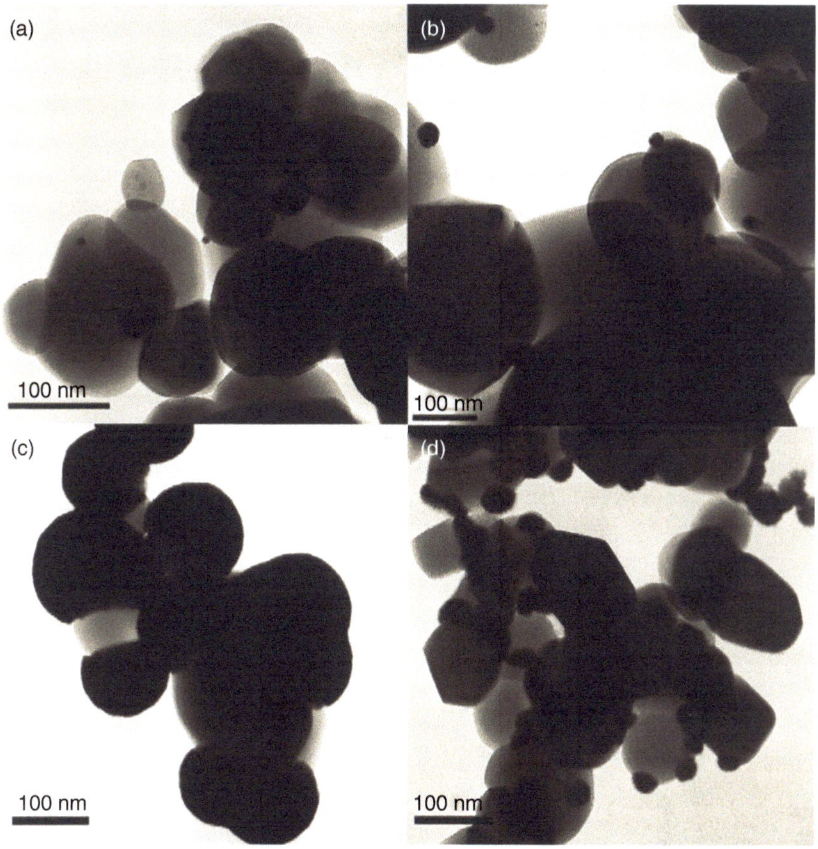

FIGURE 4.107 TEM images of samples with different Ag weight loadings. Ag/TiO_2: (a) 0.02%, and (b) 0.05%. Ni/TiO_2: (c) 0.02%, and (d) 0.05% [349].

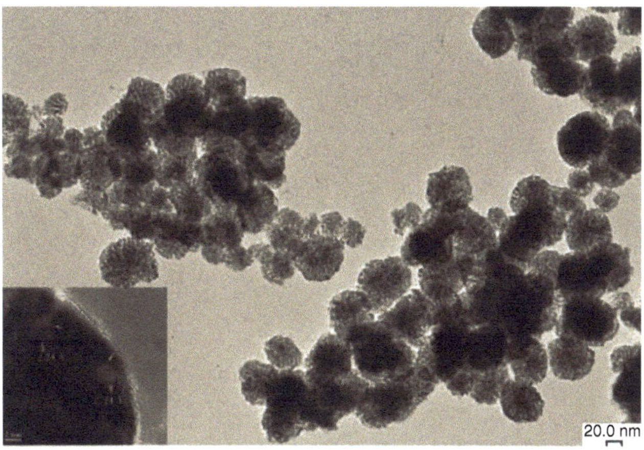

FIGURE 4.108 The HRTEM image of self-assembled, mesoporous Pd nanospheres [351].

The mean size of Pd is 41.5 nm. Larger Pd nanospheres can be obtained if $Pd(OAc)_2$ concentration increases, for example, the mean size of Pd will increase to ~55.8, and ~56.2 when the $Pd(OAc)_2$ concentration increases to 0.48 and ~0.72 mmol, respectively.

4.4.1.1.6 Pt NPs

There are very few published papers on EP Pt NPs [353–355]. Kokkinidis et al. in 2001 [354] reported electroless deposition of Pt on Ti. The substrate (titanium discs) they used was cut from a titanium wire and fixed into a teflon cylinder. Electroless Pt was deposited by immersing the substrate (a freshly polished titanium plate) in an aqueous solution of 0.1 M $HClO_4 + 2 \times 10^{-4}$M K_2PtCl_6 for deposition times from 5 s to 60 s and the solution was kept at constant temperature $T = 300$ K and was stirred using a magnetic stirrer. From the TEM image (Fig. 4.109), one can see that the deposited Pt forms islands on the titanium substrate. The islands were most likely built up from Pt nanoparticles, which were homogeneously dispersed on the titanium surface. The amount of platinum deposited increases with increasing immersion time. Due to coalescence and agglomeration processes, the Pt particles form aggregates with linear sizes 30–40 nm and a polycrystalline structure (see the inset in Fig. 4.109). In fact, such structures are known as "cryptocrystalline" or "pseudocrystalline." It has been found that the blocks, which start forming aggregates, have sizes 4–12 nm.

4.4.1.1.7 Co NPs

A few papers on EP Co NPs have been published [356–358]. In 2010, Balela et al. [358] synthesized metallic Co NPs by electroless deposition in an aqueous

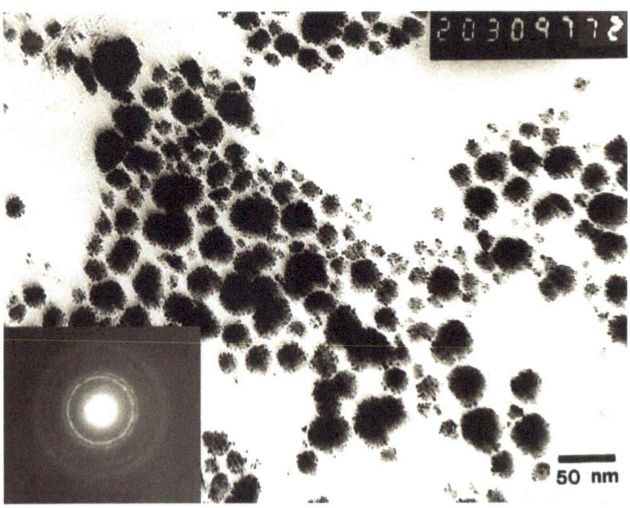

FIGURE 4.109 TEM image and the corresponding diffraction pattern of electroless deposited Pt crystals on freshly polished Ti after immersing it in 0.1 M $HClO_4 + 2\times10^{-4}$ M K_2PtCl_6 solution for 60 s [354].

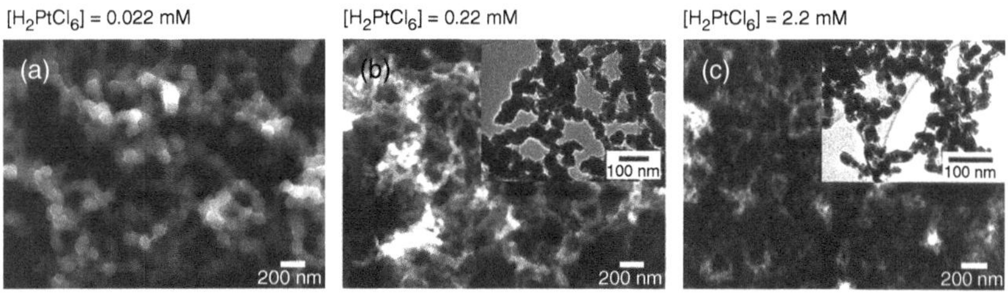

FIGURE 4.110 **SEM images of Co NPs prepared using (a) 0.02, (b) 0.22, and (c) 2.2 mmol/L H_2PtCl_6 nucleating agent. The mean diameters of the Co NPs are (a) 112, (b) 54, and (c) 24 nm. Insets are the TEM images of the Co NPs [358].**

solution at room temperature. The bath solution consisted of 0.087 M Co (II) acetate tetrahydrate, 1.1 mmol/L poly(ethylene glycol), 0.22 M NaOH, 0.022–2.2 mmol/L $H_2PtCl_6{\bullet}6H_2O$, and 0.87 mol/L $N_2H_4{\bullet}H_2O$. The solution was continuously stirred and nitrogen gas (N_2) was bubbled at 50 mL/min. The pH value of the total solution was about 12. After the solution was kept for 1 h, Co NPs were magnetically separated and washed several times with ethanol. Figure 4.110 shows SEM images of Co NPs prepared using (a) 0.02, (b) 0.22, and (c) 2.2 mmol/L H_2PtCl_6 nucleating agent.

The mean diameters of the Co NPs are (a) 112, (b) 54, and (c) 24 nm. Insets are the TEM images of the Co NPs [358].

4.4.1.2 Alloy PNs

4.4.1.2.1 Ni–P Alloy NPs

Liu et al. [359] in 2008 reported electroless deposition of Ni–P nanodots. The substrates used are a-Si coated glass plates. The thickness of a-Si is 200 nm. The plating solution of Ni–P composed of $NiSO_4{\bullet}6H_2O$ 30 g/L, $NaH_2PO_2{\bullet}H_2O$ 30 g/L, $C_3H_5O_3Na$ 40 g/L, $C_2H_5O_2N$ 10 g/L, and surfactant 20,000 ppm. The pH value of the bath was controlled at the range of 9.0 ± 0.2. The bath temperature was set at 20°C, and various values of temperatures between 100 and 200°C were used. The experimental results show that the Ni–P NPs density increases monotonically as the deposition temperatures increase. At temperature = 100°C, the Ni–P dots were quite sparsely distributed, but the dot densities for 160 and 200°C were up to 1.0×10^{10} and $5.1 \times 10^{10}/cm^2$, respectively. However, the dot size remains in the same range of 15–30 nm. Figure 4.111 shows the SEM pictures of the Ni–P NPs uniformly distributed on a-Si film. The measured composition of deposited Ni–P nanodots and films indicates that the content of phosphorus in the deposited Ni–P film is around 8–10 at%.

4.4.1.2.2 Ag–Au alloy NPs

EP hollow Au–Ag nanospheres (NSs) were first prepared in 2012 [360]. Choi et al. first prepared Ag nano seeds by reducing $AgNO_3$ using $NaBH_4$. Then,

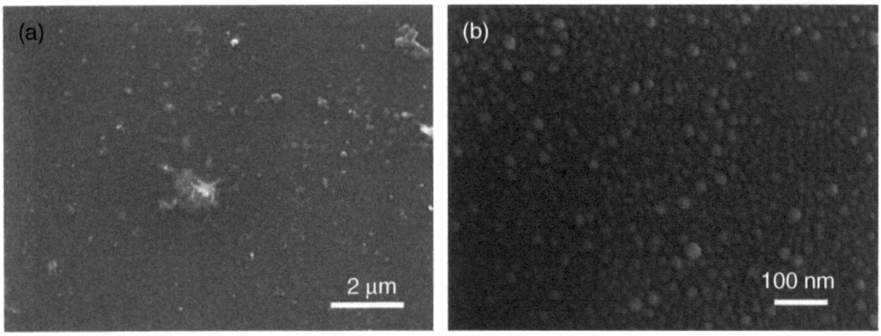

FIGURE 4.111 SEM morphologies of Ni–P NPs deposited on a-Si coated glass by EP at Ts = (a) 100, and (b) 160°C. The deposition time is 30 s [359].

a $HAuCl_4$ solution was prepared by adding 100 μL of a 20 mmol/L aqueous $HAuCl_4$ solution to 10 mL of a 0.05 mol/L CTAB solution. Using the Ag nano seeds and $HAuCl_4$ solutions, hollow nanostructures can be synthesized using the galvanic replacement reaction for about 12 h at room temperature. The SEM images (Fig. 4.112a and b) show the morphology and size of Ag and Au−Ag nanospheres. They are really sphere shape. The sizes of Ag and Au–Ag nanospheres are 40 (± 3) and 45 (± 3) nm. The hollow interiors of Au–Ag NSs were visible in the HRTEM images of Fig. 4.112c.

4.4.1.2.3 Ag–Pd Alloy NPs

Silver et al. in 1996 [361] prepared Ag–Pd NPs from ethylene glycol solutions of polyvinyl pyrrolidone (PVP) and silver and palladium nitrates. The authors first prepared a 14 nm Ag–Pd alloy colloid by dissolving 7.0 g PVP in 75 cm^3 ethylene glycol at room temperature. Afterward, 0.43 g $AgNO_3$ and 0.25 g $Pd(NO_3)_2$ were added to this solution, and the formed suspension was mechanically stirred for 20 min to allow dissolution of the precursors. The solution was then heated up to 120°C at a heating rate of 1°C/min. The optimum reaction time was found to be 4 h at 120°C. Ethylene glycol acts both as a solvent and as a reducing agent, while the role of PVP is to prevent interparticle sintering and to stabilize the colloidal dispersion. The TEM in Fig. 4.113 shows the shapes and sizes of Ag–Pd alloy particles. These particles exhibit a quasispherical shape, mean size around 14 nm, and do not show significant interparticle sintering. Chemical analysis of the Ag–Pd alloy colloids indicates the following composition (wt%): 5.3% C, 0.9% H, 1.0% N, 4.9% O, 64% Ag, and 24% Pd. It is noticeable that the initial Ag/Pd ratio (70/30) of the starting solution has been conserved in the final powder.

4.4.1.2.4 Fe–Ni Alloy NPs

In 1996, Viau et al. [362] prepared Fe–Ni monodisperse NPs in polyols. First, iron and nickel hydroxide are precipitated *in situ* by adding NaOH to a solution in a polyol of highly soluble precursors. Thus, iron (II) chloride tetrahydrate

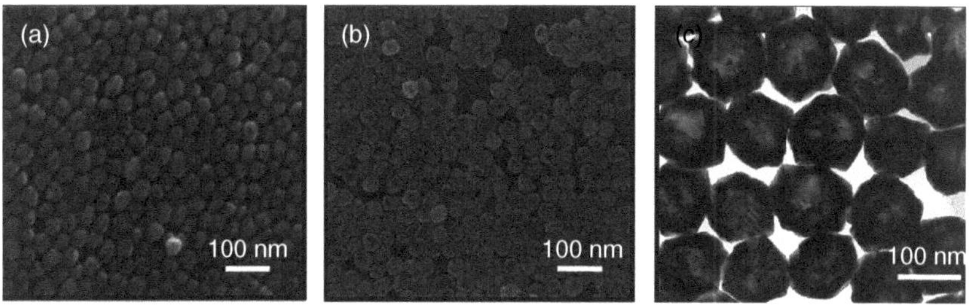

FIGURE 4.112 SEM images of Ag (a) and Au–Ag (b) nanospheres. (c) HRTEM image of hollow Au–Ag [360].

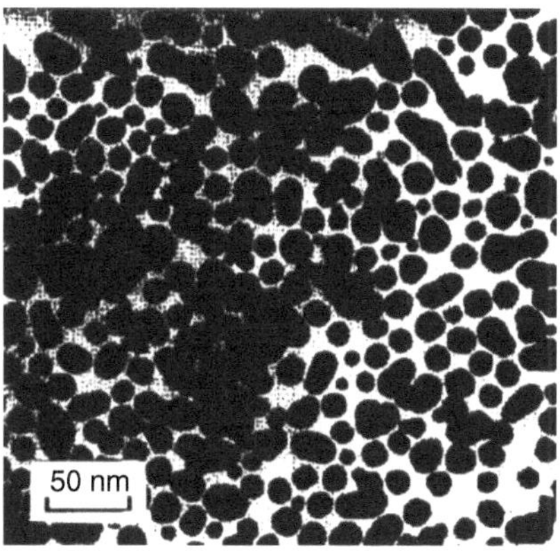

FIGURE 4.113 TEM of Ag–Pd (70/30 at wt%) alloy NPs [361].

and nickel chloride hexahydrate (Prolabo) are dissolved in 200 cm^3 of ethylene glycol or propylene glycol and sodium hydroxide is dissolved in a same volume of polyol. The concentrations of the reactants are adjusted to get after mixing the two solutions a metallic precursor concentration typically 0.2 mol/L and a NaOH concentration in the range 1–2 mol/L. Then, the Fe–Ni NPs can be obtained from the reaction. In the whole composition range available Fe–Ni particles prepared in polyols present a spherical shape, a very narrow size distribution and a mean diameter in the range 0.1–0.25 μm. TEM micrographs evidenced secondary particles built-up by aggregation of small primary particles. The SAED pattern of such an isolated secondary particle ascertains the polycrystalline character of these particles. The average crystallite size from X-ray diffraction and TEM analysis is ~5 nm.

4.4.1.2.5 Fe–Pt NPs

In 2004, Hinotsu et al. [363] synthesized Fe–Pt monodisperse by using a modified polyol process. First, they dissolved specified amounts of metal precursors in 100 mL of polyol like ethylene glycol or trimethylene glycol or tetraethylene glycol as the case may be. Then, the solution was transferred to a vessel with reflux attachment, placed in an oil bath/mantle heater, and heated at a constant rate under gentle mechanical stirring. During this stage, the metal precursor undergoes dissolution and is reduced to metal ions, and then the formation of nucleus and particle growth takes place. Fe–Pt nanoparticles with particle diameter between 5 and 10 nm and with fcc and a mixture of fcc and fct phases were obtained by decelerating the reaction.

4.4.1.2.6 Ni–Co Alloy NPs

Few papers reported the synthesis of EP Ni–Co NPs [364–367]. Toneguzzo et al. in 2000 [365] parepared Ni–Co NPs by polyol process. Co_xNi_{100-x} NPs were prepared from cobalt(II) and nickel(II) acetate tetrahydrates and dissolved with sodium hydroxide in 3.5 mol/L of 1,2-propanediol. The sodium hydroxide concentration is 0.25 mol/L. The solution containing all the reactants and the nucleating agent was slowly heated up to the boiling point under mechanical stirring at 250 rpm and then maintained at this temperature during the whole reaction process. Chemical analysis of the as-made powders indicates that some O (0.08–1.2 wt%) and C (0.27–0.48 wt%) contents are involved in the NPs. Dark-field TEM images show that Co_xNi_{100-x} particles are made up of a few crystallites with a more irregular shape. Mean crystallite sizes are 40 nm ($Ni_{80}Co_{20}$ and $Ni_{20}Co_{80}$ alloy NPs) and 65 nm ($Ni_{50}Co_{50}$ NPs) respectively. SAED patterns confirm that these NPs are fcc phase.

4.4.1.2.7 Co–Pt Alloy NPs

In 2003, Chinnasamy et al. [368] reported the synthesis of Co–Pt NPs by using the polyol process. Cobalt acetylacetonate 0.01 mol/L/platinum acetylacetonate 0.01 mol/L were dissolved either separately or together in trimethylene glycol (TMEG) 200 mL in a three-neck flask. The appropriate amounts of NaOH were then dissolved in this solution. The solution was placed in an oil bath, heated to 195°C with constant mechanical stirring, and allowed to reflux for a maximum period of 3.5 h at this temperature. Then the suspension was allowed to cool to RT. The black precipitated particles were isolated by centrifuging, and were then washed three times with ethanol to remove the byproducts. The TEM image of C–Pt NPs in Fig. 4.114 shows an average particle size of 4 nm. The XRD pattern and electron diffraction pattern reveal that the as-synthesized particles have chemically disordered fcc structure.

4.4.1.2.8 Fe–Co–Ni Alloy NPs

In 2003, Vemagiri et al. [369] studied the synthesis of Fe–Co–Ni NPs deposited by a polyol electroless process. The substrate they used is a (100)-oriented Si wafer sputter coated with 5 nm Ti and 25 nm Cu layers. A partially soluble 0.05 mol/L nickel acetate, 0.05 mol/L cobalt acetate, and 0.05 mol/L iron acetate in ethylene glycol are prepared. 0.2 mol/L NaOH dissolved completely in ethylene glycol is separately prepared. 50 mL of the nominal plating solution

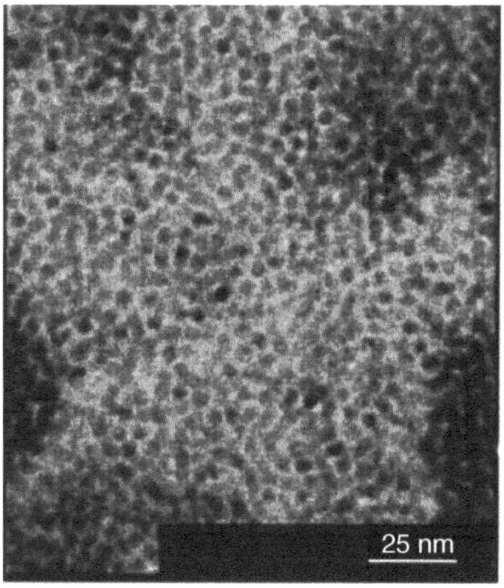

FIGURE 4.114 TEM micrograph of the as synthesized Co–Pt NPs [368].

was prepared by mixing 25 mL of the NaOH–ethylene glycol solution and proportional volumes of nickel, cobalt, and iron solutions corresponding to the film composition to be prepared. Nitrogen is bubbled through the plating solution to prevent oxidation of the deposited film. After reaching the refluxing temperature of 194°C, the silicon substrate was introduced into the reaction solution. The deposition was carried out for 15 min. The measured results show that the atomic compositions of the NPs films deposited from different nominal compositions are Fe 1.56, Co 71.95, Ni 26.5 at% for the nominal composition $Fe_{0.1}Co_{0.4}Ni_{0.5}$ and Fe 4.57, Co 35.93, Ni 59.51 for $Fe_{0.3}Co_{0.2}Ni_{0.5}$, respectively. The average grain size is 13.0 nm for $Fe_{0.1}Co_{0.4}Ni_{0.5}$, and 12.1 nm for $Fe_{0.3}Co_{0.2}Ni_{0.5}$.

4.4.1.2.9 Sm–Co Alloy NPs

Sm–Co magnetic NPs were prepared by polyol synthesis in 2012 [370]. Tian et al. added the mixture of $Sm(NO_3)_3$•$6H_2O$ (0.004 mol/L) and $Co(NO_3)_2$•$6H_2O$ (0.02 mol/L) to 50 mL/L of tetraethylene glycol in a flask under stirring, and then heated the mixture to 300°C gradually. During this process, acetic acid CH_3COOH at different levels (0.03–0.07 mol/L) was injected into the solution at 100°C and then PVP (10 mL of 0.1 or 0.2 mol/L) was injected into the solution to protect the particles from oxidation at 200°C. The mixture solution was refluxed at 300°C for 1–2 h and became black. Then, the resulting products can be obtained. It can be seen from Fig. 4.115a that the synthesized Sm–Co alloy NPs have a narrow size distribution (5–8 nm) and disperse well. The particles' morphology is similar to equiaxed grain, and their surface contours are irregular. HRTEM and EDP images (Fig. 4.115b and c) show that the product consists of two different nanoscale domains, one with lattice fringe distance at 0.21 nm and the other at 0.246 nm, corresponding to the (111) planes and (110) planes in the structured $SmCo_5$, respectively. The interplane distances of

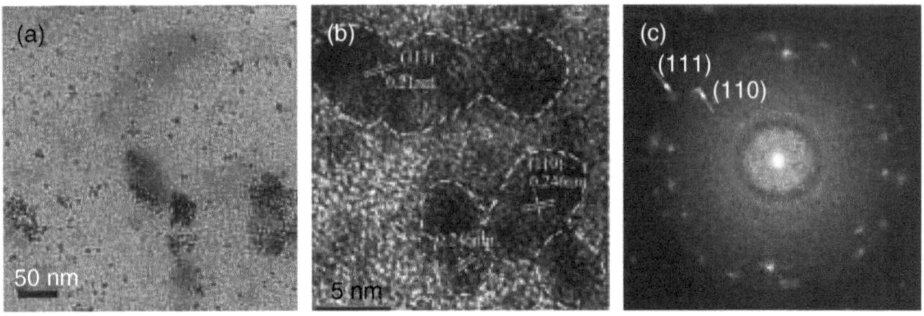

FIGURE 4.115 TEM micrographs of Sm–Co alloy NPs (a) TEM, (b) HRTEM images, and (c) diffraction pattern of the Sm–Co alloy NPs prepared by polyol [370].

the nanoparticles are less than those of typical $SmCo_5$ bulks (0.2115 nm and 0.2498 nm). In addition, one can see that each particle is a single-crystal structure, and its growth direction is along the <111> or <110>. The XRD pattern of the nanoparticles prepared by polyol synthesis with different CH_3COOH contents shows that the diffraction peaks of Sm–Co alloy phases increase with the increasing of CH_3COOH contents. The reason is that the additive of CH_3COOH will increase the concentration of H^+ and decrease the reaction velocity, which shortens the gap of the reduction time of Co^{2+} and Sm^{3+} and realizes the coreduction. Besides, the ionized CH_3COO^- can combine with Sm^{3+} and decrease the stability of Sm^{3+}, which promotes the reduction of Sm^{3+}. So the additive of CH_3COOH can promote the formation of Sm–Co phases during synthesis. That is to say, Sm–Co NPs cannot be directly synthesized by polyol from nitrates precursors.

4.4.2 EP Nanoparticle Arrays

Very few reports on EP NP arrays have been published [371, 372] to date. In 2003, Homma et al. [371] studied the maskless and electroless fabrication of Cu nanodot arrays. The substrate they used is P-type Si(1 0 0) wafers (9–18 Ω-cm). Nanoscopic patterns were formed on the surface by a nanoindentation technique using a scanning probe microscope (SPM) with a diamond probe. By controlling the parameters such as the indentation force (within the order of 1/10 mN), nanoscopic patterns with different degrees of defects were formed. By immersing the patterned wafer into 2.0% dilute fluoride solution (DHF) containing 40 mmol/L of Cu (a) for 1 min, formation of nanostructures of Cu such as arrays of nanodots has been obtained. Figure 4.116a shows representative tapping mode atomic force microscopy (TMAFM) images of a 5 × 5 array of Cu nanodots fabricated on the Si(1 0 0)wafer surface formed by the nanoindentation process. Indentation forces for the rows A (left) to E (right) are 15.8, 22.2, 28.6, 34.7, and 41.2 mN, respectively. Figure 4.116a indicates the diameter and pitch of Cu nanodots are approximately 50 and 80 nm, respectively.

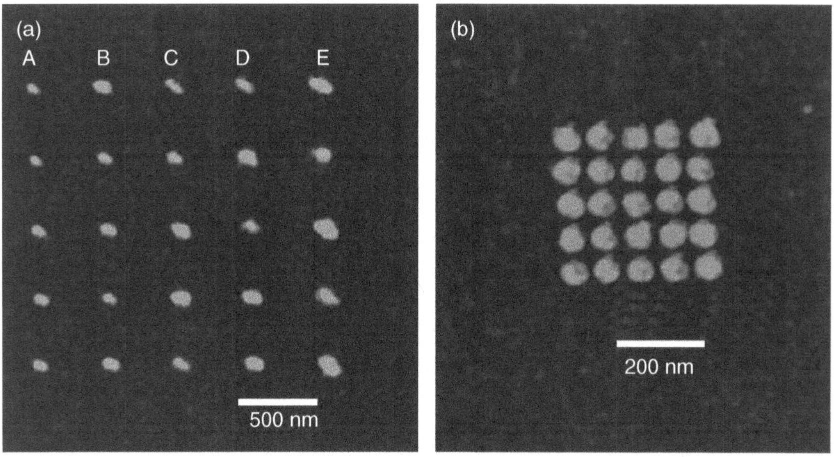

FIGURE 4.116 **(a) TMAFM image of 5 × 5 Cu nanodot array fabricated on the Si(1 0 0) wafer surface formed by the nanoindentation process. Indentation force for the rows A (left) to E (right) are 15.8, 22.2, 28.6, 34.7 and 41.2 mN, respectively. (b) TMAFM image of 5 X 5 Cu nanodot array. Indentation force: 23.8 mN [371].**

Arrays of EP spherical Au NPs were synthesized by Blake et al. in 2010 [372]. The substrate is a conductive indium-tin-oxide (ITO)-coated polished glass slide. EP Au plating was performed in three steps: (1) immersion in a solution of tin (Sn^{2+}) for 3 min, (2) ammoniacal $AgNO_3$ for 2 min, and (3) sodium gold sulfite [$Na_3(Au(SO3)_2)$] for 10 min. After Au plating, the sample was lifted off by immersion in 50°C acetone for 5 min followed by 50 min of gentle sonic agitation in 50°C acetone, resulting in regular arrays of Au nanostructures. Figure 4.117 shows SEM images of regular arrays of Au NPs created on a Sn presensitized ITO substrate. Original image magnification was changed from (a) 20,000× to (b) 80,000× to show top-down images of regular NP patterns. AuNPs with a radius (*r*) of 47.1 ± 7.4 nm were patterned into regular square arrays with an interparticle distance (d) of 646.0 ±12.4 nm.

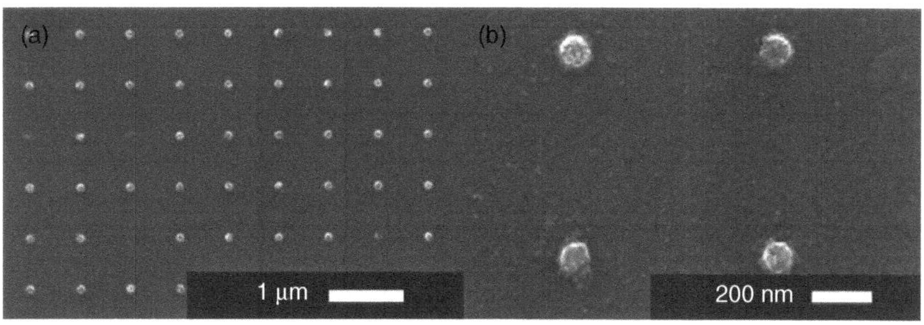

FIGURE 4.117 **SEM images of regular Au NP arrays on a Sn presensitized substrate in order of increasing magnification, (a) 20,000 ×, (b) 80,000 × [372].**

4.4.3 EP Nanoparticles Other Than Spherical Shape

Besides the mentioned EP NPs with basically spherical shape in Sections 4.4.1 and 4.4.2, NPs with other shapes such as triangular plates [328, 373], octapods and tetrapods [355], cubes and cages [255, 374–376] have also been reported.

4.4.3.1 Nano Metal Triangular Plates

Triangular nanoplates of silver with sharp corners have been synthesized by refluxing aqueous dispersions of spherical silver colloids with an average diameter of 3.5 nm [328]. According to the authors' typical operation procedure, Ag NPs were first synthesized in an aqueous solution by reducing $AgNO_3$ with $NaBH_4$ in the presence of PVP and sodium citrate. Figure 4.118a shows a TEM image of these Ag NPs, indicating that they were spherical in shape with a diameter of

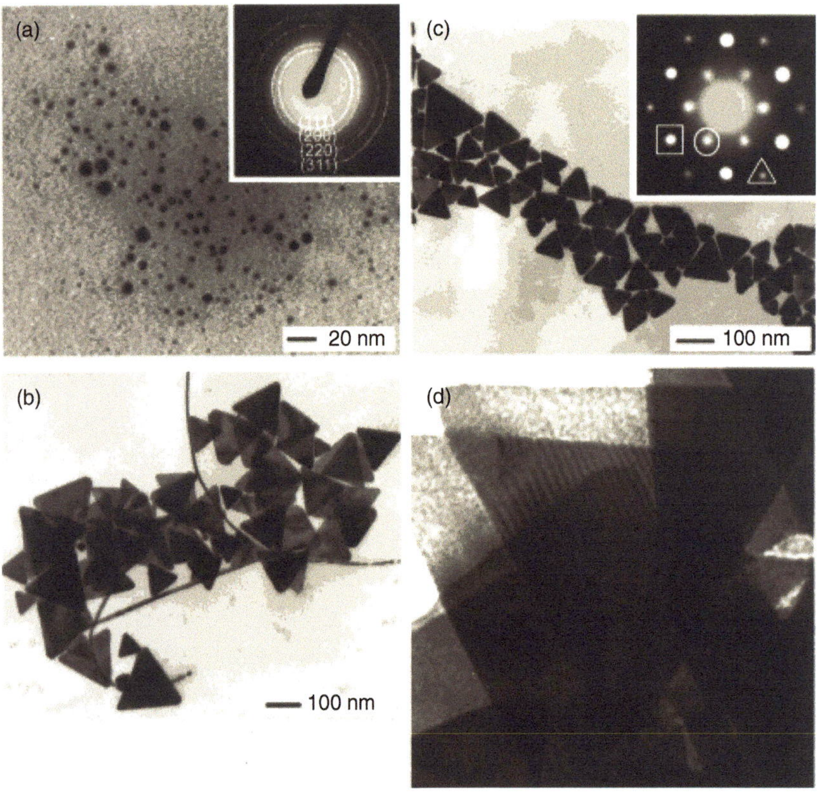

FIGURE 4.118 (a) TEM image of the Ag NPs before refluxing. The inset is their SAED pattern. (b) TEM image of this sample after refluxing, indicating the formation of a mixture of triangular nanoplates (with sharp corners) and nanobelts of silver. (c) TEM image of triangular nanoplates self-assembled into a stripe of monolayer. The inset gives a typical ED pattern taken from an individual nanoplate supported on the TEM grid against one of its triangular faces. (d) A typical Moiré fringe pattern formed by stacking two or more nanoplates against their triangular faces [328].

3.5 ± 0.8 nm. Electron diffraction patterns (the inset of Fig. 4.118a) indicate that they were the fcc structure. After the dispersion had been refluxed under the ambient condition for 10 h, TEM study (b) revealed that the Ag NPs had been transformed from spheres into a mixture consisting of triangular nanoplates (~95%) and wire-like nanostructures (~5%). Because of their surface flatness, corner sharpness, edge smoothness, and shape regularity, the Ag nanoplates tended to self-assemble into monolayered structures (Fig. 4.118c) on the surfaces of carbon-coated TEM grids. The inset in Fig. 4.118c shows a typical ED pattern taken from an individual nanoplate supported on the TEM grid against one of its triangular faces. The hexagonal symmetry of these patterned spots implied that the two triangular faces of each nanoplate were bounded by the {111} planes. The strongest spots (squared) could be indexed to the allowed {220} reflection, the outer spots (triangled) with the weakest intensity could be assigned to the allowed {422} reflection, and the inner spots (circled) with a weaker intensity corresponded to the formally forbidden (1/3){422} reflection. Figure 4.118d shows a typical example of Moiré patterns, which was an array of parallel lines. The formation of the pattern can be attributed to the difference in orientation between the two silver nanoplates.

In 2009, Lee et al. [373] fabricated porous Ag–Pd triangle nanoplates by electroless deposition. Porous triangular Ag–Pd nanoplates of various bimetallic ratios were prepared by a galvanic displacement reaction in which $Pd(OAc)_2$ reacted with Ag nanoplates. In order to reduce the interaction of free hexadecyltrimethylammonium bromide (CTAB) with the synthesized Ag–Pd nanoplates during this reaction, Ag nanoplate/CTAB solution was precipitated by high-speed centrifugation before synthesizing the porous bimetallic nanotemplates. It can be seen from the TEM images in Fig. 4.119 that the size of the triangular plates is really in the nanometer range.

4.4.3.2 Nano Metal Octapods and Tetrapods

In 2004, Herricks et al. [355] employed a polyol process with sodium nitrate to control the morphology of the synthesized Pt nanoparticles. Ethylene glycol served

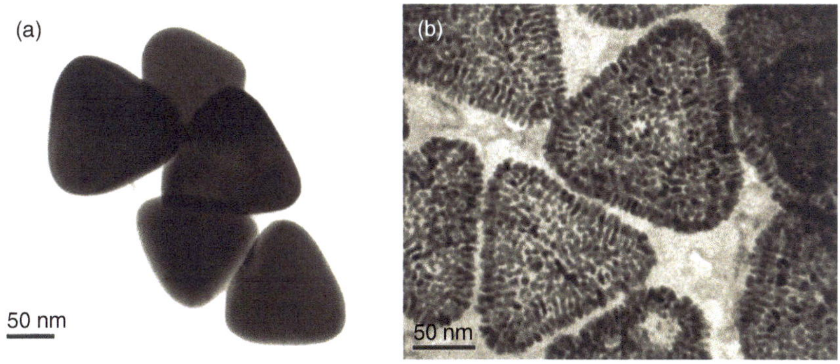

FIGURE 4.119 TEM images of triangular Ag nanotemplate (a) and porous triangular $Ag_{18}Pd_1$ nanoplate (b) [373].

as both the reducing agent and solvent. A solution containing 1 mL of 80 mmol/L H_2PtCl_6 was rapidly added to 7 mL ethylene glycol (held at 160°C) that contained both $NaNO_3$ and poly(vinyl pyrrolidone) (PVP, $M_w = 55{,}000$). The ratio of $NaNO_3$/H_2PtCl_6 was varied from 0 to 11. Figure 4.120a–d shows TEM images of Pt NPs synthesized with different $NaNO_3$/H_2PtCl_6: (a) 0.3, (b) 3.3, (c) 5.5, (d) 11.0. The morphology of the synthesized Pt nanomaterials varies greatly with dimensions but lies between 3 and 30 nm. It is clear that the exact ratio ($NaNO_3$/H_2PtCl_6) had a profound influence on the morphology of resultant nanoparticles. When the concentration of $NaNO_3$ increases from 0.3 to 3.3, the NPs exhibited a transition from irregular particles (Fig. 4.120a) to uniform spherical colloids (Fig. 4.120b). TEM images from octahedral and tetrahedral Pt particles (Fig. 4.120e–h) illustrate their subtle change in morphology. In Fig. 4.120e, an octahedral particle is displayed in the [110] orientation with the corners pointing toward the <100> directions, so this octahedral crystal is bounded by {111} facets. As increasing the ratio of $NaNO_3$/H_2PtCl_6 from 5.5 to 11.0, Fig. 4.120f shows that the particle is also oriented along the [110] axis with the corners pointing toward the {100} directions. The same trend was also observed for the tetrahedral nanoparticles. Figure 4.120g shows that the corners of the tetrahedron point toward the [111] directions. Figure 4.120(2h) illustrates a tetrahedral particle synthesized at a much higher concentration of $NaNO_3$, clearly showing enhanced growth in the <111> directions as the arms of this particle become more prominent to appear like a tetrapod. The variation in the structure of the Pt nanomaterials was due to the concentration of the nitrate ion, which was responsible for slowing the reduction of the Pt(II) and Pt(IV) ions in solution by ethylene glycol. This effect may be attributed to the formation of stable complexes between the nitrate and Pt(II) and Pt(IV) ions in solution, thus altering the reaction kinetics. The experimental results indicate that at the highest concentration (Fig. 4.120h), both octapod and tetrapod particles with dimensions >30 nm were found with relatively high yields in the final product.

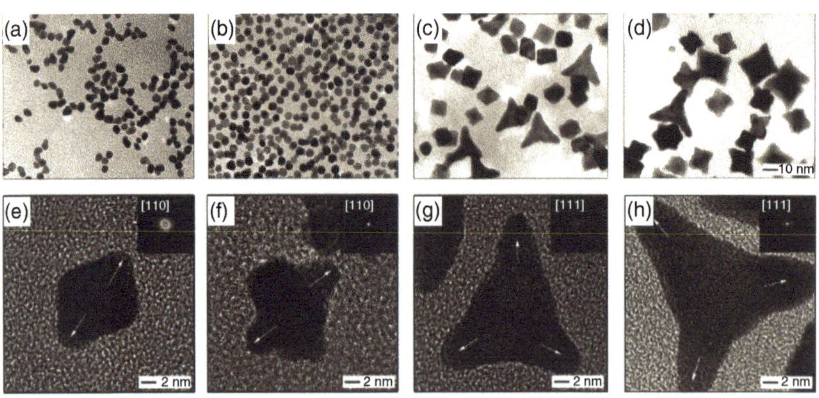

FIGURE 4.120 **TEM images of Pt NPs synthesized with different $NaNO_3$/H_2PtCl_6: (a) 0.3, (b) 3.3, (c) 5.5, (d) 11.0. The 10-nm scale bar is applicable to (a) to (d). HRTEM images of Pt NPs obtained with the ratio $NaNO_3$/H_2PtCl_6 at (e, f) 5.5 and (g, h) 11.0 respectively [355].**

4.4.3.3 Nano Metal Cubes and Cages

Several papers on nano metal cubes [250, 374] and cages [375–382] have been published to date.

In 2004, Sun and Xia [250, 374] investigated the morphological, structural, and compositional changes involved in the galvanic replacement reaction between silver nanocubes and $HAuCl_4$ in aqueous medium. The typical synthesis process is that 5 mL of anhydrous ethylene glycol was heated at 160°C for 1 h. A 3-mL aliquot of an ethylene glycol solution of $AgNO_3$ (0.25 M, 99 + %) and poly-(vinyl pyrrolidone) (PVP, 0.19 M, M_w = 55,000) and 3 mL of an ethylene glycol solution of PVP (0.19 M) were simultaneously added to the hot ethylene glycol at a rate of 0.375 mL/min. The reaction mixture was then heated at 160°C for 40 min. Magnetic stirring (at a rate of ~400 rpm) was maintained throughout the synthesis. The product was dominated by cubic NPs, with a small fraction (<5%) of nanostructures with other morphologies that included, for example, rods, cubooctahedrons, tetrahedrons, and spheres. Figure 4.121 is a typical SEM image of the silver nanocubes, indicating that monodispersity could be achieved using the modified polyol process. These Ag nanocubes were characterized by smooth surfaces and a mean edge length of around 111 ± 13 nm. The electron diffraction pattern in the inset displays the square symmetry, indicating that each Ag nanocube was a single crystal bounded by {100} facets.

Xia Younan's group have published many papers on Au and other metals nanocages in recent years [375–382]. They not only studied the synthesis of Au nanocages, but also their properties and theranostic applications. Gold nanocages are hollow nanostructures with porous walls. These authors fabricated Au nanocages by simply using the galvanic replacement reaction between Ag

FIGURE 4.121 The SEM image is taken from silver nanocubes [250].

nanocubes and chloroauric acid. The reaction process is as follows. When the Ag nanocubes were mixed with aqueous $HAuCl_4$, the galvanic replacement reaction occurred immediately. The reaction was initiated randomly from any one of the six {100} faces of an Ag nanocube through a pitting process. The Ag will give up its electrons and dissolve into the solution as ions while a thin layer of gold is being deposited on the outer surface of the cube. The driving force comes from the different electrochemical potentials between these two metals. As the reaction continues, more and more Ag from the interior of the cube will be removed through small pits in the wall, forming a hollow Au–Ag alloyed nanobox. When all the pure Ag in the interior of the Ag cubes has been removed, Ag from the Au–Ag alloy walls will begin to be removed by the $HAuCl_4$. Finally, a highly porous structure known as a nanocage has been formed. The dimensions, wall thickness, pore sizes, and pore densities of the resultant Au nanocages can all be controlled by varying the ratio Ag nanocubes/$HAuCl_4$. Figure 4.122 shows the SEM and TEM images of the Ag nanocubes, Au–Ag nanoboxes, and Au nanocages. From the figure the conversion of an Ag nanocube into an Au/Ag nanobox, then a predominately Au nanocage, can be seen evidently.

4.4.4 EP Core-Shell Nanoparticles

Metal core-shells NPs are a new class of NPs and have received significant attention in recent years because of their highly tunable optical properties [383]. In 1994, Zhou et al. [384] studied the first metal core-shell NPs, consisting of an Au_2S dielectric core surrounded by a gold shell. They used a two-step process to prepare these NPs. First, they prepared a $HAuCl_4$ solution and a Na_2S solution by dissolving chloroauric acid ($HAuCl_4{\cdot}4H_2O$) and sodium sulfide ($Na_2S{\cdot}9H_2O$) into super pure water at room temperature. Then, they mixed controlled amounts of the two solutions together to get the unstable gold sulfide, Au_2S, particles. Second, they inject a little Na_2S solution into the Au_2S solution. During the reaction, the gold atoms in the Au–S bonds on the surface of Au_2S were reduced by the S ion and the surface became gold coated. The gold-surface layer grows thicker and the Au_2S core becomes smaller with time, until all the particles become Au particles. Before the addition of more Na_2S solution in the second step, they measured the average particle diameter by TEM. They found that the average diameters of the Au_2S core are about 4 or 5 nm, and the diameters of whole core (Au_2S)-shell (Au) large particles are about 25 nm.

Since then, more work on EP core-shell NPs has been published. We just give a brief discussion in the following sections.

4.4.4.1 EP Core-Metal Shell PNs

If element A is coated with B to form A–B core-shell NP, then we use symbol A@B to denote it. There are two types of core-shell NPs. One is EP core-metal shell NPs, which means that the shell is a metal. Another is EP core-alloy shell NPs,

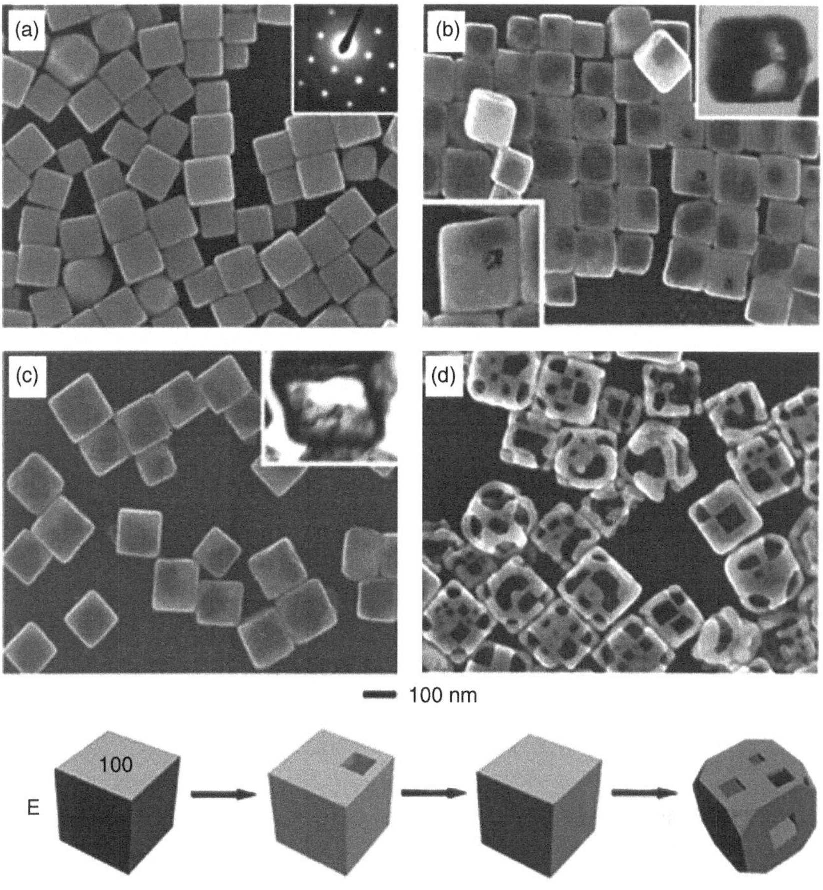

FIGURE 4.122 **(a) SEM of Ag nanocubes; inset indicates that they are single crystals. (b) SEM of product with pinhole (lower inset) on the exposed face of ~1 in 6 nanocubes. TEM (upper inset) shows a microtomed sample reveals early hollowing out. (c) SEM of Au–Ag nanobox; TEM (inset) of a microtomed sample reveals the hollow interior of the nanobox. (d) SEM of porous Au nanocages. (e) Illustration indicates the conversion of an Ag nanocube into a Au/Ag nanobox then a predominately Au nanocage [379].**

which means that the shell is an alloy. Several kinds of EP core-metal shell NPs have been reported, including Co@Cu [385], Pt@Cu [386], βSiC@Cu [132], SiO_2@Ag [387,388], Cu@Au [389], SiO_2@Au [390], SiC@Ni [391], $BaTiO_3$@Ni [392], Co@Pd [393], Co@Pt [394–396], Al_2O_3@Pt [397], and Fe_3O_4@Co [398]. Also, several kinds of EP core-alloy shell NPs have been published, including Cu@FeNi [399], Al_2O_3@Ni–P [400], $SrFe_{12}O_{19}$@Ni–P [401], $BaNi_{0.4}Ti_{0.4}Fe_{11.2}O_{19}$@Ni–P [402], B_4C@Ni_3P [403], Fe_3O_4@Ni–B [404], B_4C@Ni–B [405], and Al_2O_3@Co–P [406]. Not all kinds of EP core-shell (metal) NPs and EP core-shell (alloy) NPs will be discussed, but some of them will be mentioned in more detail.

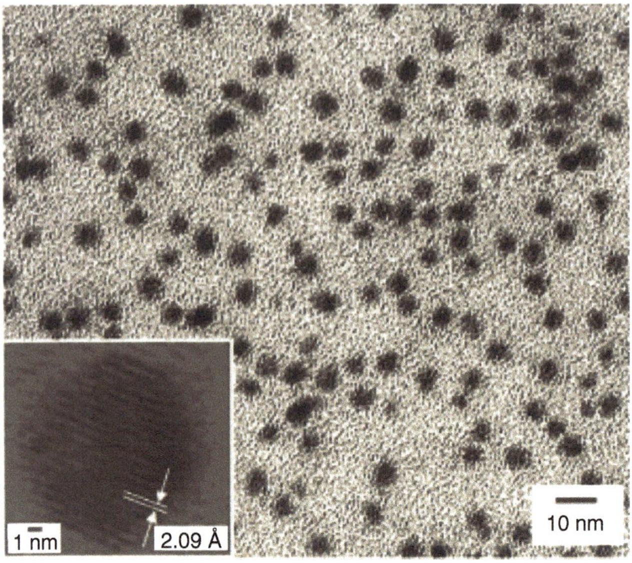

FIGURE 4.123 TEM and HRTEM images of 6.5 nm Co@Cu core-shell NPs. Measured lattice spacing (inset figure) of 2.09 Å by HRTEM is clearly matched to the shell lattice parameter known for the (111) plane of Cu [385].

In 2005, Lee et al. [385] fabricated Co@Cu and other various types of core-shell nanomaterials such as Co@Au, Co@Pd and Co@Pt through redox-transmetalation reactions. Co@Cu core-shell NPs were obtained from $Cu(hfac)_2$ (48 mg, 0.1 mmol/L) and Co nanoparticles (14 mg, 0.2 mmol/L). The reaction proceeded at 140°C for 2 h in *o*-dichlorobenzene (ODCB) solution containing 0.1 mmol/L (44.5 mg) of NaAOT (AOT= bis(2-ethylhexyl)sulfosuccinate) as a capping molecule. Figure 4.123 shows a TEM image of Co@Cu NPs with a particle size of 6.5 nm. From EDS the composition ratio of Co/Cu is 54:46. HRTEM analysis shows a lattice distance of 2.09 Å due to the lattice parameter of a fcc copper (111) plane, and these core-shell NPs are roughly estimated as 5.2 nm Co core and 0.7 nm Cu shell.

SiO_2@Ag NPs was studied by Kobayashi et al. in 2001 [387]. They deposit Ag NPs on SiO_2 spheres by pretreatment steps in electroless plating. First, monodispersed colloidal SiO_2 particles were prepared by the Stöber method. Second, to 1 mL of the silica colloid in water/methanol, with a silica concentration of 1.1 wt%, was added 9 mL of a solution of $SnCl_2$ (0.029 mol/L) and CF_3COOH (0.072 mol/L) in a 50% (v/v) methanol/water mixture. This resulted in deposition of the "sensitizer" (Sn^{2+}) onto colloid surfaces (SiO_2@Sn^{2+}). Finally, 9 mL of an aqueous solution of ammonical $AgNO_3$ (0.039 mol/L) was added to 1 mL of the colloid. The deposition of Ag nanoparticles on the colloid surfaces (SiO_2@Ag-1) can be obtained. The morphology of these nanoparticles comprises a myriad of individual, remarkably monodisperse Ag NPs with

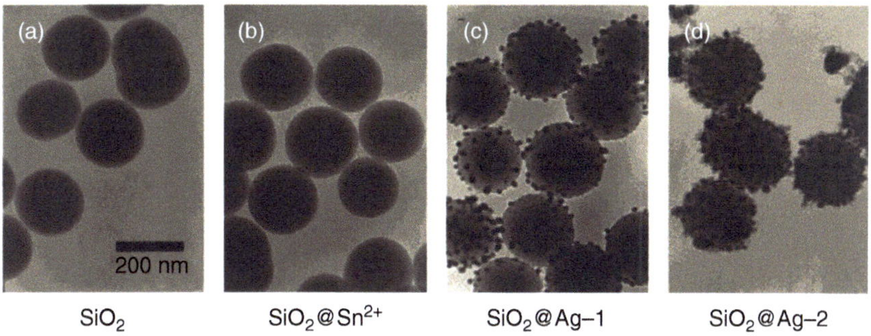

FIGURE 4.124 TEM micrographs of SiO_2 spheres during the successive deposition steps. Scale bars correspond to all four photos [387].

a diameter of 21.9 ± 2.4 nm uniformly distributed on the surface of the silica spheres (Fig. 4.124c). Repeating the last step, a larger number of Ag nanoparticles can be deposited (Fig. 4.124d). In the end, the surface coverage by Ag NPs was increased, either through the formation of new, small particles or through the growth of previously existing Ag NPs. The Ag particle size ranges in this sample from 13.3 nm to 60.0 nm.

Cu@Au NPs was prepared by Nadagouda et al. in 2007 [389] using aqueous ascorbic acid (vitamin C). They also synthesized core (Cu)-shell (Pt, Pd, and Ag) and core (Fe)-shell (Au, Pt, Pd, and Ag) by the same method, but here we will just mention Cu@Au NPs. Synthesis of core-shell Cu@Au NPs is as follows: 10 mL of 0.1 N ascorbic acid (vitamin C) was reacted with core $CuCl_2$ (2 mL, 0.1 N) at room temperature, and then shell $HAuCl_4•3H_2O$ (2 mL, 0.01N) was added and allowed to react at room temperature for 1 h without using any capping or dispersing agents. The schematic diagram of formation of the Cu@ Au core-shell NPs is as follows: Fig. 4.125 is the TEM image of Cu@Au NPs. The size of nanoparticles was in the range of 5~50 nm.

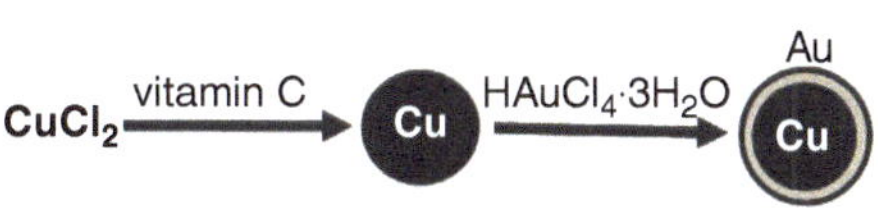

In 2012, Wang et al. [392] fabricated $BaTiO_3$@Ni NPs by electroless plating, and investigated the microstructure and electromagnetic characteristics of the NPs. $BaTiO_3$ powders were prepared by the sol–gel method. After pretreatment, the $BaTiO_3$ powders were introduced into an electroless plating solution to deposit nickel. The plating solution (0.1 mol/L $NiSO_4•6H_2O$, 60 mL/L $N_2H_4•H_2O$, 50 g/L $C_4H_4KNaO_6•4H_2O$, 0.5 mg/L CH_4N_2S) with pH adjusted by NaOH to 11.0 was maintained at a constant temperature of 80°C. The plating time was 60 min. The TEM image of $BaTiO_3$@Ni in Fig. 4.126 clearly reveals that the $BaTiO_3$ particles are fully enwrapped with a Ni coating layer of approximately 30 nm in thickness. The deposit's growth was initiated on the absorbed Pd catalytic

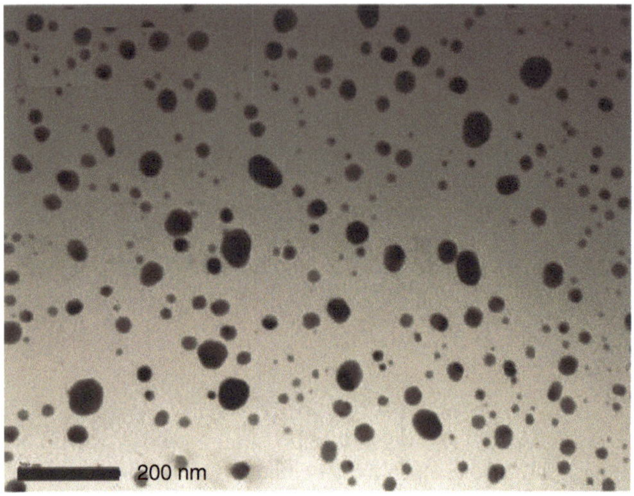

FIGURE 4.125 TEM image of core (Cu)-shell (Au) NPs [389].

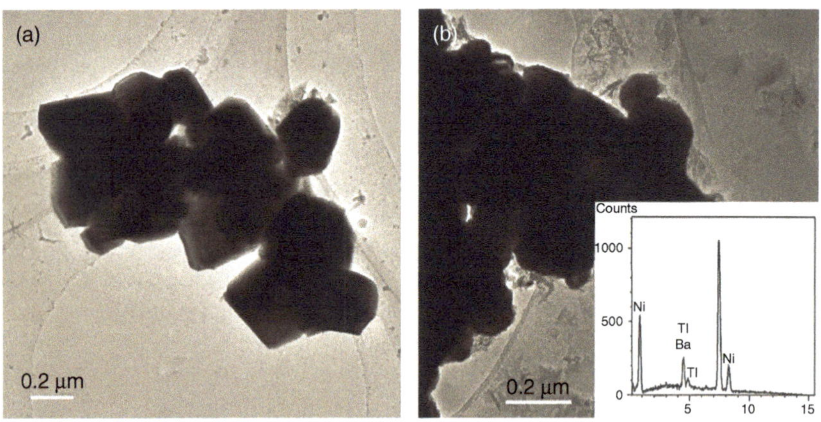

FIGURE 4.126 TEM images of (a) $BaTiO_3$ particles (b) $BaTiO_3$@Ni particles, inset is the corresponding EDX spectrum, confirming the existence of Ni on the $BaTiO_3$ particles [392].

centers and further grew vertically and laterally to form a continuous coating layer due to the autocatalytic activity of Ni. The EDS spectrum of the $BaTiO_3$@Ni particles in the inset of Fig. 4.126b shows the expected presence of Ni, Ba, Ti, and O elements, which further proves the occurrence of electroless Ni plating.

In 2007, Shao et al. [393] prepared carbon-supported Co–Pd core-shell NPs using a redox-transmetalation (electroless deposition) method involving the oxidation of Co by Pd cations, yielding a Pd shell around a Co core. The procedure for synthesizing core-shell Co@Pd NPs is based on the tetraoctylammonium bromide (TOAB) surfactant. 0.15 mmol/L $CoCl_2$ and 0.1 mmol/L TOAB were dissolved in 30 mL tetrahydrofuran (THF). Forty milligrams of carbon particles was added to the mixture under sonication, and 10 mL of superhydride (1.0 mol/L $LiB(C_2H_5)_3H$ in THF) saturated with nitrogen was added drop-wise into the mixture to reduce Co^{2+} to Co^0. The mixture was sonicated for 30 min. The product of Co/C particles was dispersed in 30 mL of THF via sonication in a nitrogen atmosphere for 10 min. A solution of 20 mL THF containing 0.05 mmol/L $Pd(NO_3)_2$ was added drop-wise to the Co/C suspension; the replacement reaction was completed after another 30 min sonication. The Co@Pd/C particles then were obtained. During the redox-transmetalation synthesis of the Co–Pd core-shell NPs, the Co nanoparticle, acting as the seed, is oxidized while the Pd^{2+} is reduced simultaneously, so forming the metal Pd due to the higher redox potential of Pd than of Co. Figure 4.127 is a TEM image of Co@Pd/C NPs, which reveals that the average particle size of the Co@Pd NPs after heat treatment at 180°C was 3.5 nm. In addition, experimental results found that the particle's size could be controlled by changing the concentration of $CoCl_2$ and the superhydride. The inset of Fig. 4.127 is an HRTEM image of a single Co@Pd nanoparticle. The measured distance between the two nearest atom rows is 2.22 Å, that is, close to the (111) interplanar distance of Pd (2.25 Å). Rows of Pd atoms can be observed on the surfaces of most Co@Pd particles, indicating the uniformity of the Pd shell. The analysis from ICP-AES measurements shows Co@Pd NPs consist of 34 mol% Pd and 66 mol% Co, which is consistent with the expected Pd/Co molar ratio (33/67). The authors believed that, on the basis of their experimental results, the Co@Pd NPs are spherical, and the estimated thickness of the Pd shell was 0.54 nm, that is, about 2.4 atomic layers.

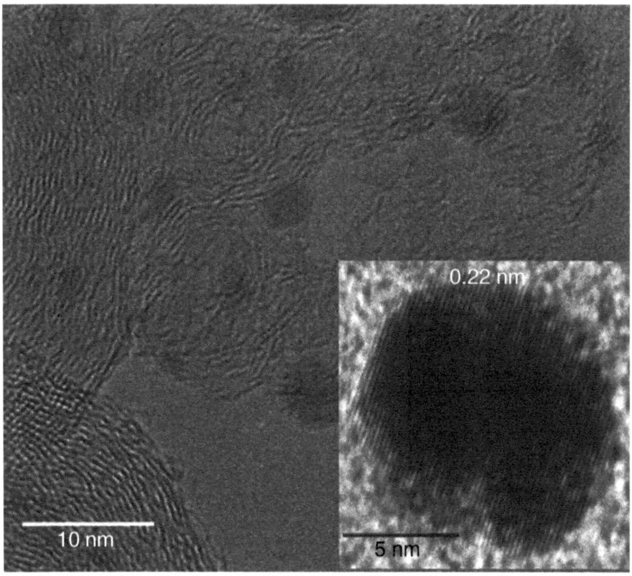

FIGURE 4.127 **TEM image of Co@Pd/C NPs. Inset is an HRTEM image of a single core-shell nanoparticle. The rows of Pd atoms can be seen on the surfaces of most particles [393].**

Carbon-supported Co@Pt NPs were prepared by Beard et al. in 2009 [396] using electroless deposition methods. Cobalt was impregnated on the support by dissolving an appropriate quantity of $CoCl_2$ in approximately 20 mL of ethanol and then slurrying with the pretreated carbon support. The Co was then reduced by either room temperature exposure to an aqueous sodium borohydride bath (molar ratio of $BH_4^-/Co^{2+} > 10$) or in flowing H_2 at 400°C in a tubular reactor. These authors used a bath solution containing H_2PtCl_6, DMAB, and sodium citrate to electroless deposit Pt on the cobalt. All electroless deposition baths were composed of a 1:10:5 molar ratio of Pt/DMAB/citrate at an initial pH of 11; half of the DMAB was added initially and the remaining DMAB added 20 min later. The deposition temperature was maintained at 80°C, 60°C, or room temperature, and time was kept at 40 min. HRTEM images of Co/C@Pt NPs are shown in Fig. 4.128. A darker center obviously

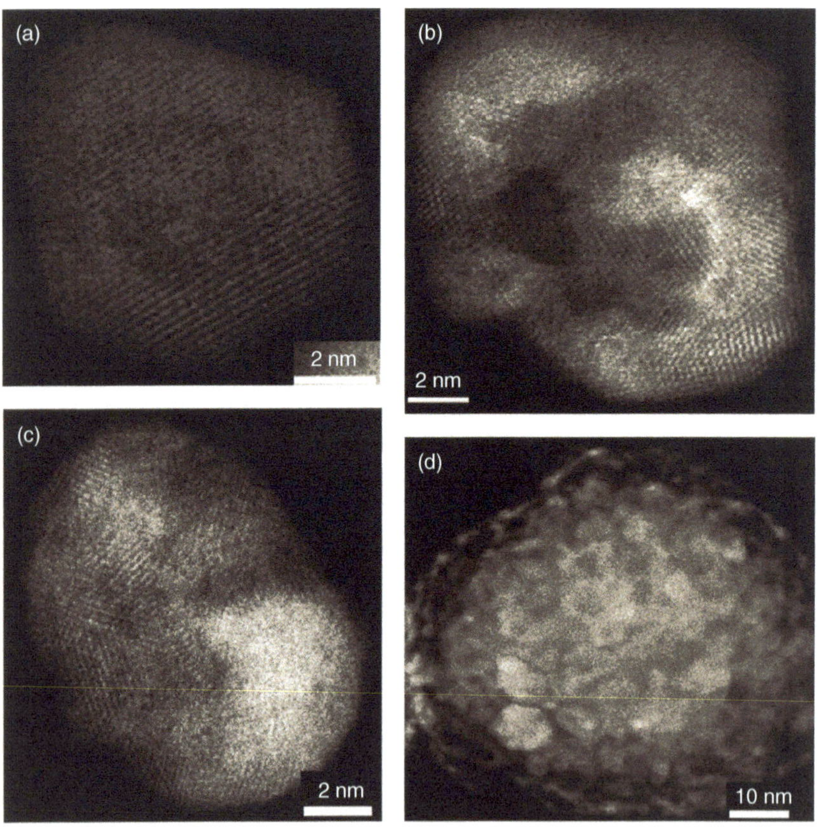

FIGURE 4.128 HRTEM images of Pt–Co/C samples prepared by the electroless plating of Pt at 80°C: (a) 8.0% Pt on 1.0% Co/C; and (b) 4.5% Pt on 1.0% Co/C. HRTEM images of 4.8% Pt5.6% Co/C (c, d) prepared by EP of Pt at room temperature: (c) complete encapsulation of a smaller Co particle compared to extensive Pt aggregation on a larger Co particle in (d) [396].

exists in the middle of each particle, indicating the presence of a lower atomic weight species, Co in this case. Surrounding the Co core is a brighter shell of Pt (higher atomic weight).

4.4.4.2 EP Core-Alloy Shell PNs

Cu@FeNi core-shell NPs were first prepared by Wang et al. in 2012 [399]. The preparation for Cu0.4@Fe0.1Ni0.5 NPs is as follows: dissolving 3.2 mg of $CuCl_2$, 1.3 mg of $FeSO_4$, 5.7 mg of $NiCl_2$, and 80.0 mg of PVP into 8 mL of distilled water, and then introducing the solution into a flask that contains 40.0 mg of ammonia borane (NH_3BH_3) with shaking for several minutes. During the process of fabrication, magnetic stirring is not needed because of the presence of aggregation of the prepared magnetic particles. Figure 4.129a indicates clearly the obvious contrast of the inner black core and the outer gray shell of the core-shell NPs. The HRTEM image of one particle (Fig. 4.129b) indicates that the core is crystallized while the shell is amorphous. The clear lattice fringes confirm that the core part is cubic Cu, which is confirmed by its Fourier

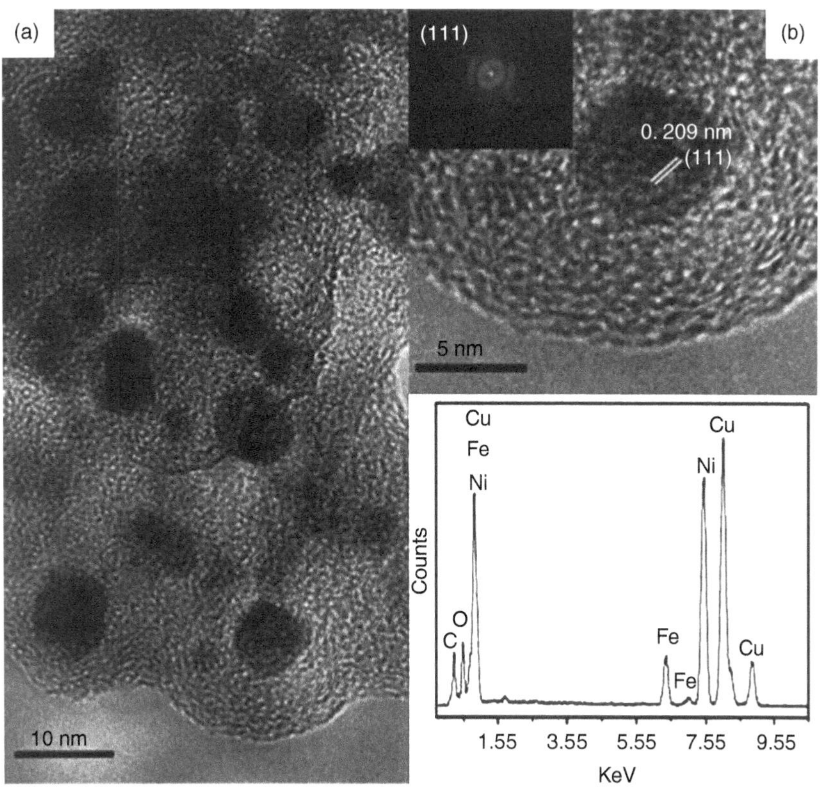

FIGURE 4.129 **TEM (a), HRTEM (b) images, FT pattern (b inset) and EDS spectrum (c) for Cu0.4@Fe0.1Ni0.5 NPs [399].**

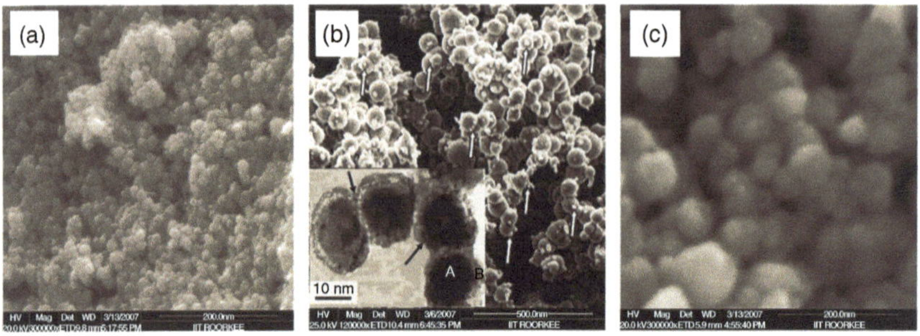

FIGURE 4.130 **FESEM micrographs of (a) nanocrystalline RAM; (b) fully or partially grown (marked with arrows) EP Ni–P nanoglobules, and (c) annealed EP (Ni–P)/RAM nanocomposite at 400°C for 4 h. Inset in (b) is HRTEM image of Ni–P nanoglobules [402].**

Transform (FT) pattern (inset in Fig. 4.129b). The EDS analysis (Fig. 4.129c) confirms that the outer amorphous shell is composed by bimetallic Fe–Ni. And the atomic ratio for Fe:Ni is 0.16:0.84, agreeing with the appointed atomic ratio (0.1:0.5). Figure 4.129 shows clearly that the size of Cu@FeNi particle is less than 10 nm.

In 2009, Sharma et al. [402] developed an electroless $BaNi_{0.4}Ti_{0.4}Fe_{11.2}O_{19}$/(Ni–P) nanocomposite powder for enhanced microwave absorption. The core is Ni-, Ti-substituted nanocrystalline M-type barium hexaferrite powder ($BaNi_xTi_xFe_{12-2x}O_{19}$ ($x = 0.4$), a radar absorbing material (RAM)). Activated RAM powder was coated uniformly with Ni–P alloy layer by the EP technique using alkaline bath (pH 9) at 90°C for a deposition time of 60 s. The field emission scanning electron microscope (FESEM) micrograph of as-synthesized RAM powder (Fig. 4.130a) shows the presence of agglomeration of spherical nanosized particles (10–15 nm). Figure 4.130b indicates fully or partially grown (marked with arrows) EP Ni–P nanoglobules. The inset in Fig. 4.130 b is HRTEM image of Ni–P coating, in which the coating can be easily seen (marked with arrow). The grain size within the particles of annealed EP RAM@Ni–P is increased to 25–50 nm range due to simultaneous growth of EP Ni–P and RAM powders (Fig. 4.130c). Therefore, the deposition of EP Ni–P on RAM powder has increased the size range to 15–25 nm, thus indicating that the deposition thickness of Ni–P coatings is in the range of 5–10 nm. The FESEM-EDAX analysis shows that the composition of the as-synthesized EP RAM@Ni–P core-shell NPs (% wt) is O 26.20, Ti 6.98, Ba 3.95, Ni 30.32, Fe 27.90, and P 4.65, respectively.

In 2008, Zhu et al. [405] reported electroless coating of Ni–B onto B_4C particles. The B_4C particles used have average particle size about 2.27 μm. The EP bath solution used $NiSO_4 \cdot 6H_2O$ (99.0%) as the Ni^{2+} source, ethylenediamine ($C_2H_8N_2$) as complexing agent. The compositions of EP bath are $Ni:C_2H_8N_2 = 1:9$, $Ni:NaBH_4 = 1:1$, and $Ni:B_4C = 0.0671–3.3536$. The plating temperature was kept at 85 ± 2°C. The pH value was 12–14 adjusted with

FIGURE 4.131 SEM images of B_4C particles coated with varying Ni–B thickness: (a) 5 nm thickness, (b) 50 nm thickness. SEM cross-section images of B_4C particles coated with varying Ni–B thickness: (c) 5 nm thickness, and (d) 50 nm thickness [405].

sodium hydroxide. The plating time was 30 min under mechanical stirring. Figure 4.131 shows the SEM images of the B_4C@Ni–B NPs with two targeted Ni–B nanolayer thickness. Figure 4.131a indicates that when the targeted Ni–B nanolayer thickness is 5 nm, the sharp edge of the B_4C particles can still be observed, but the coating layer almost fully covers the B_4C particle surfaces. When the targeted Ni–B nanolayer thickness increase to 50 nm, Fig. 4.131b shows the original B_4C morphology completely disappears and small nodules appear on the B_4C particles. Figure 4.131c and d shows the SEM cross-section images of the B_4C particles coated with varying Ni–B thickness: 5 nm (c) and 50 nm (d). It can be seen that the bright Ni–B nanolayers distribute all around the B_4C particle surfaces, which means Ni–B coverage on the B_4C particles is complete. As the targeted Ni–B nanolayer thickness increases, the actual Ni–B nanolayer thickness increases accordingly. When the targeted Ni–B thicknesses are 5 and 50 nm, the cross-section thicknesses are 43 and 90 nm respectively (Fig. 4.131c and d). Generally, the Ni–B layer thickness is higher than the targeted thickness.

Co–P coated Al_2O_3 powders NPs were prepared by electroless plating in 2003. Zhang et al. [406] used the following bath solution: $CoSO_4{\cdot}7H_2O$ 14.3–33.3 g/L, $Na_3C_6H_5O_7{\cdot}2H_2O$ 39.3– 91.6 g/L, $NaH_2PO_2{\cdot}H_2O$ 5–45 g/L, H_3BO_3 5–45 g/L. The original Al_2O_3 powder was produced by a sol–gel process and

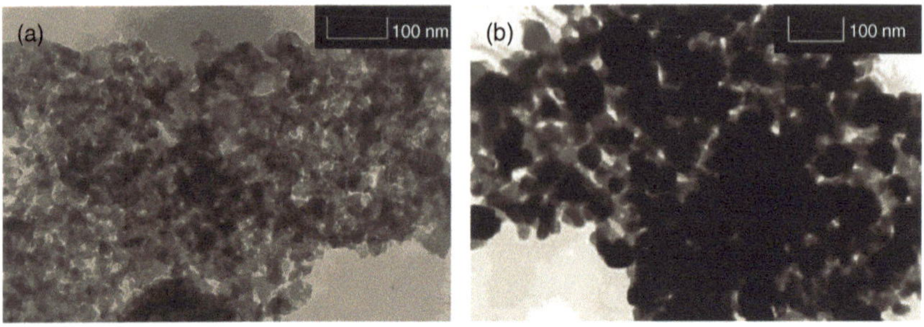

FIGURE 4.132 TEM images of (a) initial alumina powder, and (b) of Al_2O_3@Co–P core-shell NPs with NaH_2PO_2: 35 g/L [406].

is of 10–20 nm in diameter. The pH value of the solution was adjusted to 8.0–12.0 by the addition of 5 mol/L NaOH. The plating temperature was at 45–50°C. Figure 4.132 shows TEM images of (a) initial alumina powder, and (b) of Al_2O_3@Co–P core-shell NPs with NaH_2PO_2: 35 g/L. The average size of the Al_2O_3@Co–P NPs is 50–60 nm. The experiments found that the average particle size becomes larger when prepared in higher NaH_2PO_2 content; meanwhile the uniformity becomes better. As the NaH_2PO_2 content increases, both the weight gain and the particle size of Al_2O_3@Co–P NPs increase. In addition, the reaction time increases with NaH_2PO_2 content as well, which makes the powder of better uniformity. Analysis from energy spectrum of Al_2O_3@Co–P NPs indicates that Co, P, and Al are all found at different positions. Moreover, XRD pattern reveals the coexistence of peaks of Al_2O_3 and Co–P phases. So it is believed that the Al_2O_3@Co–P NP is the composite of Al_2O_3 and Co–P.

4.5 SUMMARY

After discussing EP nanomaterials in detail, we can make some conclusive remarks as follows.

Recently, and particularly during the last 10 years, nano electroless plating has been developing rapidly. The speed of development was considerably more rapid than the development of ordinary EP which took over 70 years. Nanomaterials have superior properties over ordinary EP materials, which accounts for their rapid growth. This is also why scientists pay a lot of attention to nano EP and attempt to fabricate more and more EP nanomaterials.

Not only have nano EPs developed very rapidly, but also the research field that is involved in is very broad. Nearly all the aspects of ordinary electroless plating have been studied from a nano electroless point of view. From bulk nano EP materials to 2D, 1D, and 0D EP nanomaterials, all of these areas have been studied in recent years. Although in some subjects only very few or just one to two papers have been published to date, for example, EP nanobelts, this seems to be a good start.

In this chapter, the preparation technology, including the bath solution composition and operational conditions, has been discussed in detail for every EP nanomaterial mentioned. Although only one or two typical samples for every kind of nanomaterials have been mentioned, rather than all the nanomaterial systems published, still a large number of them reported in nearly 160 papers have been described. The manufacturing principles, chemical composition and structure, and nano features of the EP nanosystem are very important, and are introduced in this chapter. Of course, detailed properties of nano EP materials will not be described, for the reasons pointed out at the beginning of the book.

It seems that the formation mechanism of nano EPs is basically not referred to in papers published to date. Is the problem too complicated or is the study time not long for nano EP? It needs scientists to answer this question.

Nano EP is the newest field in electroless plating, and it is in its ascendant stage. People expect more and better papers and reports to be published, and they should cover areas such as better and more convenient methods of preparation, a deep understanding of the nature of nano EP, an exploration of the superior properties of nano EP, and especially development of more environmentally-friendly green nano chemical plating methods.

It is fully expected that nano EP will have a bright future.

REFERENCES

[1] Zhang B, Yang Y. Arts Soc Sci J 2011;1. ASSJ-28.
[2] Yang Y, Zhang B. J South China Uni Technol (Soc Sci Edi) 2011;(13):101. (6).
[3] Porter AL, Jan Y. Nature Nanotechnol 2009;4:534.
[4] Zhang B. Physical fundamentals of nanomaterials. Beijing, China: Chemical Industry Press; 2009.
[5] Yoshizawa Y, Oguma S, Yamauchi K. J Appl Phys 1988;64:6044.
[6] Lu K, Wang JT, Wei WD. J Appl Phys 1991;69(1):522.
[7] Lu K. Mater Sci Eng 1996;R16:161.
[8] Hono K, Ping DH. Mater Charact 2000;44:203.
[9] Murty BS, Hono K. Mater Sci Eng 2001;A312:253.
[10] Clavaguera-Mora MT, Clavaguera N, Crespo D, Pradell T. Prog Mater Sci 2002;47:559.
[11] Zhang H, Shen L, Chang J. J Mater Protec 2011;44(3):1.
[12] Ling M, Zhang H. J Textile Res 2011;32(12):103.
[13] Zhang H, Feng Y. J Qindao Uni (Eng Technol Edi) 2011;26(1):42.
[14] Ma H, Liu Z, Tian F, Sun H. J Alloys Comp 2008;450:348.
[15] Ma H, Tian F, Li D, Guo Q. J Alloys Comp 2009;474:264.
[16] Haijima Y, Matsumura A, Sugiyama T, band KI. Jpn J Surf Finish 2009;60(2):123.
[17] Afroukhteh S, Dehghanian C, Emamy M. Appl Surf Sci 2012;258:2597.
[18] Zhao F, Zhang Z, Qiu T, Shi B. Mater Mech Eng 2005;29(3):59.
[19] Liu YY, Yu J, Huang H, Xu BH, Liu XL, Gao Y, Dong XL. Surf Coat Technol 2007;201:7246.
[20] Zhang H, Feng Y. Electropla Poll Cont 2011;31(2):22.
[21] Gao J, Liu L, Wu Y, Shen B, Hu W. Surf Coat Technol 2006;200:5836.
[22] Bigdeli F, Allahkaram SR. Mater Desi 2009;30:4450.
[23] Xu H, Li C, Wang W. J Heilongjiang Inst Technol 2010;24(1):43.

[24] Allahkaram SR, Honarvar Nazari M, Mamaghani S, Zarebidaki A. Mater Desi 2011;32:750.
[25] Zarebidaki A, Allahkaram SR. Micro Nano Lett 2011;6(11):937.
[26] Zarebidaki A, Allahkaram SR. Key Eng Mater 2011;203:471.
[27] Abdel Aal A, Hassan HB, Abdel Rahim MA. J Electroanaly Chem 2008;17:619.
[28] Chen W, Gao W, He Y. Surf Coat Technol 2010;204:2493.
[29] Khalifa ORM, Abd El-Wahab E, Tilp AH. Austra J Basic Appl Sci 2011;5(6):136.
[30] Zeng H, Wei Z, Hu G, Tang F, Pan Z, Luo G, Xu B. Plat Finish 2011;33(6):16.
[31] Yang Y, Chen W, Zhou C, Xu H, Gao W. Appl Nanosci 3 2011;1:19.
[32] Eslami Rad N, Dehghanian Ch. Iranian J Mater Sci Eng 2010;7(4):1.
[33] Jin H, Jiang S, Zhang L. Chin Chem Lett 2008;19:1367.
[34] Gao J, Liu L, Shen B, Hu W, Ding W. P&SF 2004;91(5):74.
[35] Balarajua JN, Kalavatib, Rajam KS. Surf Coat Technol 2006;200:3933.
[36] Fang NG, Yu H, Sun D, Wang X, Meng H, Yang D. Electroplt Finish 2007;26(3):1.
[37] Hamdy AS, Shoeib MA, Hady H, Abdel Salam OF. J Appl Electrochem 2008;38:385.
[38] de Hazan Y, Werner D, Z'graggenc M, Groteklaes M, Graule T. J Colloid Interf Sci 2008;328:103.
[39] Allahkaram SR, Zoughi M, Rabizadeh T. Micro Nano Lett 2010;5(5):262.
[40] Shibli SMA, Jabeera B, Anupama RI. Appl Surf Sci 2006;253:1644.
[41] Dong D, Chen XH, Xiao WT, Yang GB, Zhang PY. Appl Surf Sci 2009;255:7051.
[42] de Hazan Y, Zimmermann D, Z'graggenc M, Roos S, Aneziris C, Bollier H, Fehr P, Graule T. Surf Coat Technol 2010;204:3464.
[43] Rabizadeh T, Allahkaram SR. Mater Des 2011;32:133.
[44] Agarwala RC. Pramana 2005;65(5):959.
[45] Yu L, Huang W, Zhao X. Surf Technol 2009;28(5):17.
[46] Mafi IR, Dehghanian C. Appl Surf Sci 2011;258:1876.
[47] Yu L, Huang W, Zhao X. J Alloys Comp 2011;509:4154.
[48] Zhang H, Feng Y. Plat Finish 2011;33(1):1.
[49] Sarret M, Müller C, Amell A. Surf Coat Technol 2006;201:389.
[50] Zou TZ, Tu JP, Zhang SC, Chen LM, Wang Q, Zhang LL, He DN. Mater Sci Eng A 2006;426:162.
[51] Hu X, Jiang P, Wan J, Xu Y, Sun X. J Coat Technol Res 2009;6(2):275.
[52] Liu D, Yan Y, Lee K, Yu J. In: Proc. CIST2008 & ITS-IFToMM2008 Beijing, China, p. 482.
[53] Liu X, Wu C, Wang X. J Coat Technol Res 2010;7(5):659.
[54] Matsubara H, Tazawa H, Saito N, Nishiyama H, Inoue Y, Mayuzumi M. Jpn J Surf Finish 2003;54(5):374.
[55] Matsubara H, Kobayashi M, Nishiyama H, Saito N, Inoue Y, Mayuzumi M. Electrochemistry 2004;72(6):446.
[56] Xu H, Yang Z, Li M, Shi Y, Huang Y, Li H. Surf Coat Technol 2005;191:161.
[57] Matsubara H. Jpn J Surf Finish 2006;57(7):484.
[58] Matsubara H, Abe Y, Chiba Y, Nishiyama H, Saito N, Hodouchi K, Inoue Y. Electrochim Acta 2007;52:3047.
[59] Matsubara H. J Surf Sci Jpn 2009;30(5):279.
[60] Zhu Y, Chen Y, Zhu C, Shen X. Acta Metall Sin (Engl Lett) 2010;23(6):409.
[61] Mazaheri H, Allahkaram SR. Appl Surf Sci 2012;258:4574.
[62] Chen WX, Tu JP, Gan HY, Xu ZD, Wang QG, Lee JY, Liu ZL, Zhang XB. Surf Coat Technol 2002;160:68.
[63] Shen G, Cheng Y, Tsai L. IEEE Trans Nanotechnol 2005;4(5):539.
[64] Alishahi M, Monirvaghefi SM, Saatchi A, Hosseini SM. Appl Surf Sci 2012;258:2439.
[65] Zarebidaki A, Allahkaram SR. Micro Nano Lett 2012;7(1):90.

[66] Yu X, Wang H, Yang Z, Yin P, Xin X. 158. Appl Surf Sci 2000;335.
[67] Xu X, Zhong L, Liu J. Electroplat Poll Cont 2005;25(3):18.
[68] Liu C, Zhao Q. Langmuir 2011;27:9512.
[69] Su XJ, Zhao Q, Wang S, Zhang S, Noormofidi N, Kreutzwiesner E, Slugovc C. In: Malayeri MR, Watkinson AP, Müller-Steinhagen H, editors. Proceedings of the international conference on heat exchanger fouling and cleaning, June 5–10, 2011, Crete Island, Greece.
[70] Sharma SB, Agarwala RC, Agarwala V, Satyanarayana KG. Metall Mater Trans B 36b 2005;23.
[71] Kaya B, Gulmez T, Demirkol M. In: Proceeedings of the world congress on engineering and computer science 2008, WCECS 2008, October 22–24, 2008, San Francisco, CA.
[72] Gurga A, Mochalin V, Pepe D, Picardi C, Gogotsi Y. Adv Technol Mater Mater Proc 2009;10(1):47.
[73] McComas E, US Patent Application Publication US 2006/0024514 Al Electroless plating with nanometer particles, February 2, 2006.
[74] Xiao S, Song D, 978-1-4244-7161-4/10 (2010 IEEE, 2010 International Conference on E-Product E-Service and E-Entertainment (ICEEE), Nov. 7–9, 2010.
[75] Ranganatha S, Venkatesha TV. Phys Scr 2012;85:035601. (10pp).
[76] Zhao B, Wang R, Li Z, Li H. Mater Heat Treat 2010;39(24):173.
[77] Ranganatha S, Venkatesha TV, Vathsala K. Mater Res Bull 2012;47:635.
[78] Wu Y, Ren R, Wang F, Yuan Z, Wang T, Hu X. Mater Res Bull 2008;43:3425.
[79] Ranganatha S, Venkatesha TV, Vathsala K. Appl Surf Sci 2010;256:7377.
[80] Li X, Ma M, Zhang C. Surf Technol 2002;31(6):24.
[81] Khalifa ORM, Abd El-Wahab E, Tilp AH. J Appl Sci Res 2010;6(12):2280.
[82] Mallory GO, Hajdu JB. Electroless plating: fundamantals and applications. Florida: American Electroplaters and Surface Finishers Society; 1990.
[83] Shen Q, Jiang L, Zhang H, Min Q, Hou W, Zhu JJ. J Phys Chem C 2008;112:16385.
[84] Teng XW, Liang XY, Maksimuk S, Yang H. Small 2006;2:249.
[85] Qiu T, Wu XL, Siu GG, Chu PK. J Electr Mater 2006;35(10):1879.
[86] Bechelany M, Brodard P, Elias J, Brioude A, Michler J, Philippe L. Langmuir 2010;26(17):14364.
[87] Ding H, Wang M, Chen L, Fan W, Lee Y, Qian D, Hao J, Liu H. Colloids Surf A 2011;387:1.
[88] Qu L, Dai L. J Phys Chem B 2005;109:13985.
[89] Qiu T, Wu XL, Mei YF, Chu PK, Siu GG. Appl Phys A 2005;81:669.
[90] Drozdowicz-Tomsia K, Xie F, Goldys EM. J Phys Chem C 2010;114:1562.
[91] Gorostiza P, Kulandainathan MA, Diaz R, Sanz F, Allongue P, Morante JR. J Electrochem Soc 2000;147(3):1026.
[92] Peng K, Yan Y, Gao S, Zhu J. Adv Funct Mater 2003;13:127.
[93] Ye W, Shen C, Tian J, Wang C, Bao L, Gao H. Electrochemi Commun 2008;10:625.
[94] Fang J, You H, Kong P, Yi Y, Song X, Ding B. Cryst Growth Des 2007;7(5):864.
[95] Lv S, Suo H, Zhou T, Wang C, Jing S, Fu Q, Xu Y, Zhao C. Solid State Commun 2009;149:227.
[96] Yan C, Xue D. Cryst Growth Des 2008;8(6):1850.
[97] Fang J, Ma X, Cai H, Song X, Ding B. Nanotechnology 2006;17(23):5841.
[98] Zhang Y, Tong H, Sun Z, Yang S, Wen Z, Zhu J, Hao L, Han Y, Zhang X. J Solid State Electrochem 2011;15:2231.
[99] Aizawa M, Cooper AM, Malac M, Buriak JM. Nano Lett 2005;5(5):815.
[100] Wang Y, Becker M, Wang L, Liu J, Scholz R, Peng J, Gösele U, Christiansen S, Kim DH, Steinhart M. Nano Lett 2009;9(6):2384.
[101] Xu P, Mack NH, Jeon S, Doorn SK, Han X, Wang H. Langmuir 2010;26(11):8882.
[102] Zhang H, Jin R, Mirkin CA. Nano Lett 2004;4(8):1493.

[103] Liu B, Zeng H. J Am Chem Soc 2004;126:8124.
[104] Vlad A, Dutu CA, Guillet P, Jedrasik P, Fustin C-A, Södervall U, Gohy J-F, Melinte S. Nano Lett 2009;9(8):2838.
[105] Lo C, Huang C, Liu C, Chen C, Kuo C, Lin H, Tseng Y. J Magn Magn Mater 2011;323:1950.
[106] Zhang S, Xing Y, Jiang T, Du Z, Li F, He L, Liu W. J Power Sources 2011;196:6915.
[107] Porter LA Jr, Choi HC, Schmeltzer JM, Ribbe AE, Elliott LCC, Buriak JM. Nnno Lett 2002;2(12):1369.
[108] Miyake H, Ye S, Osawa M. Electrochem Commun 2002;4:973.
[109] Hrapovic S, Liu Y, Enright G, Bensebaa F, Luong JHT. Langmuir 2003;19:3958.
[110] Hu J, Li W, Chen J, Zhang X, Zhao X. Surf Coat Technol 2008;202:2922.
[111] Ahn W, Taylor B, Dall'Asén AG, Roper DK. Langmuir 2008;24:4174.
[112] Flavel BS, Yu J, Ellis AV, Shapter JG. Electrochim Acta 2009;54:3191.
[113] Horiuchi S, Nakao Y. Surf Coat Technol 2010;204:3811.
[114] Gutés A, Carraro C, Maboudian R. ACS Appl Mater Interf 2011;3:1581.
[115] Thangavel S, Ramaraj R. J Coll Interface Sci 2011;355:293.
[116] Chen D, Xu B, Sun S, Tong Y. J Catal Today 2012;182:46.
[117] Shacham-Diamand Y, Inberg A, Sverdlov Y, Bogush V, Croitoru N, Moscovich H, Freeman A. Electrochim Acta 2003;48:2987.
[118] Park HK, Yoon JK, Kim K. Langmuir 2006;22:1626.
[119] Schaefers S, Rast L, Stanishevsky A. Mater Lett 2006;60:706.
[120] Ron I, Friedman N, Cahen D, Sheves M. Small 2008;4(12):2271.
[121] Gutes A, Laboriante I, Carraro C, Maboudian R. J Phys Chem C 2009;113:16939.
[122] Hao J, Xu Z, Han M, Xu S, Meng X. Collo Surf A: Physicochem Eng Aspe 2010; 366:163.
[123] Campione M, Parravicini M, Moret M, Papagni A, Schr€oter B, Fritz T. Langmuir 2011;27:12008.
[124] Sabayev V, Croitoru N, Inberg A, Shacham-Diamand Y. Mater Chem Phys 2011;127:214.
[125] Hao J, Han M, Xu Z, Li J, Meng X. Nanoscale Res Lett 2011;6:263.
[126] Hao J, Han M-J, Li J, Meng X. J Coll Interface Sci 2012;377:51.
[127] Boontongkong Y, Cohen RE, Rubner MF. Chem Mater 2000;12:1628.
[128] Azzaroni O, Zheng Z, Yang Z, Huck WTS. Langmuir 2006;22:6730.
[129] Zhang H, Jiang Z, Lian J, Hou X. Trans Nonferrous Met Soc China 2007;17:s736.
[130] Ye J, Chen Q, Xiong Y, Chai Y, Chen P. Phys Stat Sol (a) 2008;1–5. DOI 10.1002/pssa.200723305.
[131] Scudieroa L, Fasasib A, Griffiths PR. Appl Surf Sci 2011;257:4422.
[132] Zhu SL, Tang L, Cui ZD, Wei Q, Yang XJ. Surf Coat Technol 2011;205:2985.
[133] Porter LA Jr, Choi HC, Ribbe AE, Buriak JM. Nano Lett 2002;2(10):1067.
[134] Shi Z, Wu S, Szpunar JA. Chem Phys Lett 2006;422:147.
[135] Pernstich KP, Schenker M, Weibel F, Rossi A, Caseri WR. Appl Mater Interf 2010;2(3):639.
[136] Martyak NM. Chem Mater 1994;6:1667.
[137] Shacham-Diamanda Y, Lopatin S. Electrochim Acta 1999;44:3639.
[138] Kohn A, Eizenberg M, Shacham-Diamand Y, Israel B, Sverdlov Y. Microelectro Eng 2001;55:297.
[139] Bogusha V, Inberg A, Croitoru N, Dubin V, Shacham-Diamand Y. Thin Solid Films 2003;426:288.
[140] Crobu M, Scorciapino A, Elsener B, Rossi A. Electrochim Acta 2008;53:3364.
[141] Bi H, Kou KC, Rider AE, Ostrikov K, Wu HW, Wang ZC. Appl Surf Sci 2009;255:6888.

[142] Strukova GK, Strukov GV, Batov IE, Sakharov MK, Kudrenko EA, Mazilkin AA. Mater Chem Phys 2010;119:377.
[143] Yoshida N, Tasaki T, Chang T-FM, Shibata A, Ishiyama C, Sone M. Microelectro Eng 2011;88:2401.
[144] Jiang SX, Guo RH. Surf Coat Technol 2011;205:4274.
[145] Balaraju JN, Parashar C, Rajam KS. In: 2010 12th Electronics packaging technology conference, p. 418–22.
[146] Ah CS, Yun YJ, Park HJ, Kim W-J, Ha DH, Yun WS. Chem Mater 2005;17:5558.
[147] Kan C, Zhu X, Wang G. J Phys Chem B 2006;110:4651.
[148] Sun Y. Chem Mater 2007;19:5845.
[149] Sun Y, Qiao R. Nano Res 2008;1:292.
[150] Sun Y, Pelton M. J Phys Chem C 2009;113:6061.
[151] Wolcott A, Kuykendall TR, Chen W, Chen S, Zhang JZ. J Phys Chem B 2006;110:25288.
[152] Aherne D, Charles DE, Brennan-Fournet ME, Kelly JM, Gun'ko YK. Langmuir 2009;25(17):10165.
[153] Dong W, Dong H, Wang Z, Zhan P, Yu Z, Zhao X, Zhu Y, Ming N. Adv Mater 2006;18:755.
[154] Zhuo Y, Sun W, Dong L, Chu Y. Appl Surf Sci 2011;257:10395.
[155] Feng Q, Liu S, Wang X, Jin G. Appl Surf Sci 2012;258:5005.
[156] Jiang CL, Zhang WQ, Liu YK, Qian YT. Cryst Growth Des 2006;6:2603.
[157] Hu Y, Liu J, Ding R, Wang K, Jiang J, Ji X, Li Y, Huang X. Thin Solid Films 2010;518:6876.
[158] Nien Y-T, Chang Y-H, Chen I-G. Mater Lett 2011;65:1201.
[159] Zehner RW, Sita LR. Langmuir 1999;15:6139.
[160] Sugimura H, Kanda M, Ichii T, Murase K. J Photochem Photobio A: Chem 2011;221:209.
[161] Ono S, Oide A, Asoh H. Electrochim Acta 2007;52:2898.
[162] Asoh H, Sakamoto S, Ono S. J Collo Interf Sci 2007;316:547.
[163] Coluccio ML, Das G, Mecarini F, Gentile F, Pujia A, Bava L, Tallerico R, Candeloro P, Liberale C, De Angelis F, Fabrizio ED. Microelectr Eng 2009;86:1085.
[164] Ouchi T, Shimano N, Homma T. Electrochim Acta 2011;56:9575.
[165] Miyoshi K, Fujikawa S, Kunitake T. Collo Surf A: Physicochem Eng Aspects 2008;321:238.
[166] Miyoshi K, Aoki Y, Kunitake T, Fujikawa S. Langmuir 2008;24:4205.
[167] Han WK, Hwang GH, Hong SJ, An HH, Yoon CS, Kim JH, Lee MJ, Hong G, Park KS, Kang SG. Mater Chem Phys 2010;123:401.
[168] Cui B, Yu Z, Ge H, Chou SY. Appl Phys Lett 2007;90:043118.
[169] Kuchibhatla SVNT, Karakoti AS, Bera D, Seal S. Prog Mater Sci 2007;52:699.
[170] Iijima S. Nature 1991;354:56.
[171] Tsang SC, Chen YK, Harris PJF, Green MLH. Nature 1994;372(10):159.
[172] Satishkumary BC, Voglz EM, Govindarajx A, Rao CNR. J Phys D: Appl Phys 1996;29:317.
[173] Li Q, Fan S, Han W, Sun C, Liang W. Jpn J Appl Phys 1997;36:L501.
[174] Li J, Moskovits M, Haslett TL. Chem Mater 1998;10:1963.
[175] Ang LM, Hor TSA, Xu Guoqin, Tung CH, Zhao SP, Wang JLS. Chem Mater 1999;11:2115.
[176] Chen X, Xia J, Peng J, Li W, Xie S. Compos Sci Technol 2000;60:301.
[177] Ang LM, Hor TSA, Xu GQ, Tung CH, Zhao SP, Wang JLS. Carbon 2000;38:363.
[178] Xu C, Wu G, Liu Z, Wu D, Meek TT, Han Q. Mater Res Bull 2004;39:1499.
[179] Qu L, Dai L. J Am Chem Soc 2005;127:10806.
[180] Peng Y, Chen Q. Collo Surf A: Physicochem Eng Aspe 2009;342:132.
[181] Li J, O'Keefe MJ, O'Keefe TJ. Surf Coat Technol 2011;205:3134.
[182] Peng Y, Chen Q. J Electrochem Soc 2012;159(2):D72.
[183] Feng Y, Yuan H. J Mater Sci 2004;39:3241.

[184] Byeon JH, Hwang J. Surf Coat Technol 2008;203:357.
[185] Li C, Wan B, Qia Y, Lu W, Lian J. Inter J Min Metall Mater 2009;16(51):598.
[186] Arai S, Fujii J. J Electrochem Soc 2011;158(8):D506.
[187] Wu C-S, Lee C-Y, Chen J-K, Kuo S-W, Fan S-K, Cheng C-C, Chang F-C, Ko F-H. Inter J Electrochem Sci 2012;7:4133.
[188] Peng Y, Chen Q. Nanosca Res Lett 2012;7:195.
[189] Choi HC, Shim M, Bangsaruntip S, Dai H. J Am Chem Soc 2002;124:9058.
[190] Li X, Ma X, Li S, Wen S, Lun N. J Mater Sci Eng 2004;22(1):48.
[191] Ma X, Li X, Lun N, Wen S. Mater Chem Phys 2006;97:351.
[192] Shi Y, Yang R, Yuet PK. Carbon 2009;47:1146.
[193] Dobrzanski LA, Pawlyta M, Krzton A, Liszka B, Tai CW, Kwasny W. Acta Phys Poloni A 2010;118(3):483.
[194] Liu H, Cheng G, Zheng R, Zhao Y, Liang C. J Mol Catal A: Chem 2005;230:17.
[195] Liu H, Cheng G, Zheng R, Zhao Y, Liang C. Diam Rela Mater 2006;15:15.
[196] Lin K-Y, Tsai W-T, Chang J-K. Inter J Hydro Ener 2010;35:7555.
[197] Karousis N, Tsotsou GE, Evangelista F, Rudolf P, Ragoussis N, Tagmatarchis N. J Phys Chem C 2008;112:13463.
[198] Liu Z, Lin X, Lee JY, Zhang W, Han M, Gan LM. Langmuir 2002;18:4054.
[199] Mu Y, Liang H, Hu J, Jiang L, Wan L. J Phys Chem B 2005;109:22212.
[200] Qu L, Dai L, Osawa E. J Am Chem Soc 2006;128:5523.
[201] Guo G, Qin F, Yang D, Wang C, Xu H, Yang S. Chem Mater 2008;20:2291.
[202] Liu WL, Hsieh SH, Chen WJ. Appl Surf Sci 2007;253:8356.
[203] Chen X-h, Wang J-x, Deng F-m, Li H-j, Wang M, Lu X-n, Peng J-c, Li W-z. New Carbon Mater 2000;15(4):39.
[204] Chen WX, Tu JP, Wang LY, Gan HY, Xu ZD, Zhang XB. Carbon 2003;41:215.
[205] Chen C-S, Chen X-H, Li X-Q, Zhang G, Yi G-J, Zhang H, Hu J. Acta Phys Sinica 2004;53(2):531.
[206] Wang F, Arai S, Endo M. Carbon 2005;43:1716.
[207] Wang F, Arai S, Park KC, Takeuchi K, Kim YJ, Endo M. Carbon 2006;44:1298.
[208] Hu G-h, Wu H-h, Yang F-z. Electrochemistry 2006;12(1):25.
[209] Zhang H, Zhao G, Zhou Y. Surf Technol 2009;38(6):13.
[210] Arai S, Sato T, Endo M. J Electrochem Soc 2010;157(11):D570.
[211] Lin K-Y, Tsai W-T, Chang J-K. Inter J Hydro Ene 2010;35:7555.
[212] Li W, Jin H, Hao Y, Chen T, Dai J, Wang Q. J Nanomater 2011;5. Article ID 348958.
[213] Zarebidaki A, Allahkaram S-R. J Alloys Comp 2011;509:1836.
[214] Wang Y, Liu Y. J Mater Metall 2012;11(2):116.
[215] Liu Z, Li Z, Wang F, Liu J, Ji J, Park KC, Endo M. Mater Res Bull 2012;47:338.
[216] Zhen Z, Li S, Chen H, Zhang Y, Yu T. New Technol New Proc No 2005;8:63.
[217] Chen Y, Qiu C, Su H, Cao M, Zhu J. Chi Surf Eng 2003;16(2):29.
[218] Lu Z, Zhao D, Shen Z. Mater Mech Eng 2008;32(4):62.
[219] Liu Z-J, Xu Z, Yuan Z-Y, Chen W, Zhou W, Peng L-M. Mater Lett 2003;57:1339.
[220] Cai D, Gu J, Huang Z, Kang F. J Mater Sci Eng 2007;25(5):712.
[221] Zhao Y, Hao Z, Yajuan X, Yuexin D. Chi J Aerona 2010;23:377.
[222] Huang L, Cai J-S, He Y, Ke F-S, Sun S-G. Electrochem Commun 2009;11:950.
[223] He B, Wu Y, Ye M, Xie T. Eectropla Fini 2008;27(9):13.
[224] Wang G, Huang Y, Sun X. Devel Appl Mater 2011;26(3):34.
[225] Ye M, Xie T, He B-l, Wu Y-c, Meng G-w, Zhang L-d. Chin J Chem Phy 2009;22(4):411.
[226] Reddy KR, Sin BC, Yoo CH, Park W, Ryu KS, Lee J-S, Sohn D, Lee Y. Scr Mater 2008;58:1010.

[227] Li N, Li X, Yin X, Wang W, Qiu S. Solid State Commun 2004;132:841.
[228] Brumiik CJ, Menon VP, Martin CR. J Mater Res 1994;9:1174.
[229] Zhang SH, Xie ZX, Jiang ZY, Xu X, Xiang J, Huang RB, Zheng LS. Chem Commun (Camb) 2004;7(9):1106.
[230] Kijima T, Yoshimura T, Uota M, Ikeda T, Fujikawa D, Mouri S, Uoyama S. Angew Chem Int Ed 2004;43:228.
[231] Wen X, Yang S. Nano Lett 2002;2(5):451.
[232] Harrell CC, Lee SB, Martin CR. Anal Chem 2003;75:6861.
[233] Velleman L, Guillaume F, Brunee JL, Losic D, Shapter JG. In: Nanosci. nanotechnol., (ICONN), 2010 international conference on date of conference, Feb. 22–26, 2010. p. 98–101.
[234] Ma C, Han R, Qi S, Yeung Edward S. J Chromatogr A 2012;1238:11.
[235] Yu Y, Kant K, Shapter Joe G, Jonas A-M, Losic D. Micropo Mesopo Mater 2012;153:131.
[236] Wang W, Li N, Li X, Geng W, Qiu S. Mater Res Bull 2006;41:1417.
[237] Chu SZ, Kawamura H, Mori M. J Electrochem Soc 2008;155(5):D414.
[238] Lin S-C, Chen S-Y, Cheng S-Y, Lin J-C. Solid State Sci 2005;7:896.
[239] Ren X, Huang X, Zhang H. Acta Phys Chim Sin 2006;22(1):102.
[240] Xie G, Wang Z, Li G, Shi Y, Cui Z, Zhang Z. Mater Lett 2007;61:2641.
[241] Chen Y, Ni Q, Han W, Wang L, Kirk DW, Thorpe SJ. Scr Mater 2011;65:986.
[242] Azizi A, Mohammadi M, Sadrnezhaad SK. Mater Lett 2011;65:289.
[243] Neetzel C, Gasi T, Ksenofontov V, Felser C, Ionescu E, Ensinger W. Nucl Instrum Methods Phys Res B 2011;282:96.
[244] Balci S, Bittner M, Hahn K, Scheu C, Kneza M, Kadri A, Wege C, Jeske H, Kern K. Electrochim Acta 2006;51:6251.
[245] Kudo H, Fujihira M. IEEE Trans Nanotechnol 2006;5(2):90.
[246] Valenzuela K, Raghavan S, Deymier PA, Hoying J. J Nanosci Nanotechnol 2008;8(7):1.
[247] Barbic M, Mock JJ, Smith DR, Schultz S. J Appl Phys 2002;91(11):9341.
[248] Sun Y, Mayers BT, Xia Y. Nano Lett 2002;2(5):481.
[249] Sun Y, Mayers B, Herricks T, Xia Y. Nano Lett 2003;3(7):955.
[250] Sun Y, Xia Y. J Am Chem Soc 2004;126:3892.
[251] Yan G, Wang L, Zhang L. Rev Adv Mater Sci 2010;24:10.
[252] Menon VP, Martin CR. Anal Chem 1995;67:1920.
[253] Song JH, Wu Y, Messer B, Kind H, Yang P. J Am Chem Soc 2001;123:10397.
[254] Gu J, Shi J, Xiong L, Chen H, Li L, Ruan M. Solid State Sci 2004;6:747.
[255] Knez M, Bittner AM, Boes F, Wege C, Jeske H, Maiâ E, Kern K. Nano Lett 2003;3(8):1079.
[256] Hui L, Hong L, Songhai X, Weilin D, Minghua Q, Yufeng L, Hexing L. Chem Mater 2008;20(12):3936.
[257] Mani S, Saif T, Han Jong H. IEEE Trans Nanotechnol 2006;5(2):138.
[258] Srivastava D, Hendricks Troy R, Lee I. Nanotechnology 2007;18:245305. (6pp).
[259] Kawamori M, Yagi S, Matsubara E. J Electrochem Soc 2011;158(8):E79.
[260] Shi Z, Walker AV. Langmuir 2011;27:11292.
[261] Kawamoria M, Yagib S, Matsubara E. ECS Trans 2012;41(30):1.
[262] Gu J-L, Shi J-L, Xiong L-M. Chem J Chin Uni 2004;25(11):1986.
[263] Shi Z, Wu S, Szpunar Jerzy A. Chem Phys Lett 2006;422:147.
[264] Shi Z, Wu S, Fairbridge C, Szpunar JA. Micro Nano Lett 2008;3(2):50.
[265] Lim J-S, Kim S-M, Lee S-Y, Stach EA, Culver JN, Harris MT. Nano Lett 2010;10:3863.
[266] Balela Mary Donnabelle L, Yagi S, Matsubara E. J Electrochem Soc 2011;158(4):D210.
[267] Wang C-Y, Lu M-Y, Chen H-C, Chen L-J. J Phys Chem C 2007;111:6215.
[268] Ren X, Jiang CH, Huang XM, Li DD. Physica E 2009;41:349.
[269] Yuan XY, Wu GS, Xie T, Lin Y, Zhang LD. Nanotechnology 2004;15:59.

[270] Balci S, Hahn K, Kopold P, Kadri A, Wege C, Kern K, Bittner AM. Nanotechnology 2012;23:045603. (7pp).
[271] Kawamori M, Yagi S, Matsubara E. J Electrochem Soc 2012;159(2):E37.
[272] Cho K-S, Talapin Dmitri V, Gaschler W, Murray CB. J Am Chem Soc 2005;127:7140.
[273] Yuan XY, Xie T, Wu GS, Lin Y, Meng GW, Zhang LD. Physica E 2004;23:75.
[274] Azizi A, Mohammadi M, Sadrnezhaad SK. In: Proc. 3rd conference on nanostructures (NS2010) March 10–12, 2010, Kish Island, I.R. Iran.
[275] Yuan XY, Wu GS, Xie T, Geng BY, Lin Y, Meng GW, Zhang LD. Solid State Sci 2004; 6:735.
[276] Dadvand N, Kipouros Georges J. J Nanomater Vol 2007;6. Article ID 46919.
[277] Yuan XY, Wu GS, Xie T, Lin Y, Meng GW, Zhang LD. Solid State Commun 2004; 130:429.
[278] Yuan X, Du C, Sun G, Pan N. Appl Surf Sci 2007;253:4546.
[279] Gómez de Salazar JM, Barrena MI, Merino C, Merino N. Mater Lett 2008;62:494.
[280] Córdoba JM, Odén M. Surf Coat Technol 2009;203:3459.
[281] Silvain JF, Vincent C, Heintz JM, Chandra N. Comp Sci Technol 2009;69:2474.
[282] Tamayo-Ariztondo J, Córdoba JM, Odén M, Molina-Aldareguia JM, Elizalde MR. Comp Sci Technol 2010;70:2269.
[283] Jang J, Ryu SK. J Mater Proc Technol 2006;180:66.
[284] Lei J, Wang W, Song M, Dong B, Li Z, Wang C, Li L. Reac Func Polym 2011;71:1071.
[285] Metz KM, Tse KY, Baker SE, Landis EC, Hamers RJ. Chem Mater 2006;18:5398.
[286] Metz KM, Colavita PE, Tse K-Y, Hamers RJ. J Power Sour 2012;198:393.
[287] Arai S, Endo M, Hashizume S, Shimojima Y. Electrochem Commun 2004;6:1029.
[288] Tzeng S-S. Carbon 2006;44:1986.
[289] Xie G, Lü Z, Wang G. Mater Chem Phys 2009;118:281.
[290] Heo G-y, Hong Y-t, Park S-j. Macromole Res 2012;20(5):503.
[291] Arai S, Imoto Y, Suzuki Y, Endo M. Carbon 2011;49:1484.
[292] Park K-Y, Lee S-B, Kim J-B, Yi J-W, Lee S-K, Han J-H. In: 16th International conference on composite materials, Kyoto, Japan, 2007.
[293] Park K-Y, Han J-H, Lee S-B, Yi J-W. Composites: Part A 2011;42:573.
[294] Xie G, Sun W, Li W. Catal Commun 2008;10:333.
[295] Xie G, Wang Z, Cui Z, Shi Y. Carbon 2005;43:3181.
[296] Sharma SB, Agarwala RC, Agarwala V. J Mater Sci 2002;37:5247.
[297] Xu C, Tian M, Liu L, Zou H, Zhang L, Wang W. J Electrochem Soc 2012;159(4):D217.
[298] Peng K, Yan Y, Gao S, Zhu J. Adv Mater 2002;14:1164.
[299] Hochbaum AI, Chen R, Delgado RD, Liang W, Garnett EC, Najarian M, Majumdar A, Yang P. Nature 2008;451:163.
[300] Dalchiele EA, Martín F, Leinen D, Marotti RE, José Ramón R-B. J Electrochem Soc 2009;156(5):K77.
[301] Xu W, Palshin V, Flake JC. J Electrochem Soc 2009;156(7):H544.
[302] Huang ZP, Geyer N, Liu LF, Li MY, Zhong P. Nanotechnology 2010;21:465301. (6pp).
[303] Sivakov VA, Brönstrup G, Pecz B, Berger A, Radnoczi GZ, Krause M, Christiansen SH. J Phys Chem C 2010;114:3798.
[304] Kato Y, Adachi S. J Electrochem Soc 2011;158(7):K157.
[305] Chen C-Y, Phan DH, Wong C-C, Yen T-J. J Electrochem Soc 2011;158(5):D302.
[306] Chiou AH, Lin IF, Su CK, Hsu CY, Wu WF, Chou CP. In: 2011 11th IEEE International Conference on Nanotechnology, Portland Marriott, August 15–18, 2011, Portland, Oregon.
[307] Sun X, Tao R, Lin L, Li Z, Zhang Z, Feng J. Appl Surf Sci 2011;257:3861.

[308] Sayed Sayed Y, Wang F, Malac M, Meldrum Al, Egerton Ray F, Buriak JM. ACS Nano 2009;3(9):2809.
[309] Sayed Sayed Y, Wang F, Malac M, Li P, Wang D, Buriak JM. Cryst Eng Commun 2012;14:5230.
[310] Peng K-Q, Wang X, Wu X-L, Lee S-T. Nano Lett 2009;9(11):3704.
[311] Tao B, Zhang J, Miao F, Li H, Wan L, Wang Y. Sens Actuators B 2009;136:144.
[312] Yan Q, Wang Z, Zhang J, Peng H, Chen X, Hou H, Liu C. Electrochim Acta 2012;61:148.
[313] Wang Z, Zhu J, Chen X, Yan Q, Zhang J, Chen Y. J Alloys Comp 2012;511:257.
[314] Tao B, Zhang J, Hui S, Wan L. Sens Actuators B 2009;142:298.
[315] Murphy CJ, Jana NR. Adv Mater 2002;14(1):80.
[316] Lee G-J, Shin S-I, Kim Y-C, Oh S-G. Mater Chem Phys 2004;84:197.
[317] Sławiński GW, Zamborini FP. Langmuir 2007;23:10357.
[318] Mohl M, Kumar A, Reddy ALM, Kukovecz A, Konya Z, Kiricsi I, Vajtai R, Ajayan Pulickel M. J Phys Chem C 2010;114:389.
[319] Krishnamoorthy S, Krishnan S, Thoniyot P, Low HY. ACS Appl Mater Interfaces 2011;3:1033.
[320] Liu C-M, Tseng Y-C, Chen C, Hsu M-C, Chao T-Y, Cheng Y-T. Nanotechnology 2009;20:415703. (7pp).
[321] Pacholski C, Kornowski A, Weller H. Angew Chem Int Ed 2002;41(7):1188.
[322] Harnack O, Pacholski C, Weller H, Yasuda A, Wessels JM. Nano Lett 2003;3(8):1097.
[323] Pacholski C, Kornowski A, Weller H. ZnO nanorods: growth mechanism and anisotropic functionalization. In: Hartland GV, Zhu X-Y, editors. Proceedings of SPIE 5513, physical chemistry of interfaces and nanomaterials III, 232 (October 14, 2004); doi:10.1117/12.559645.
[324] Ye X, Zhou Y, Chen J, Sun Y, Wang Z. Mater Lett 2008;62:666.
[325] Chen A, Liu B, Hou Y, Wen G, Liu D, Lu S, Li Z, Li X, Zou Y, Zou B, Li D, Cui T, Zou G. Chin Sci Bull 2009;54:2539.
[326] Wang L, Liu B, Liu D, Yao M, Hou Y, Yu S, Cui T, Li D, Zou G, Iwasiewicz A, Sundqvist B. Adv Mater 2006;18:1883.
[327] Zhou Jing C, Soto Carissa M, Chen M-S, Bruckman MA, Moore Martin H, Barry E, Ratna Banahalli R, Pehrsson Pehr E, Spies Bradley R, Confer Tammie S. J Nanobiotechnol 2012;10(18):12.
[328] Sun Y, Mayers B, Xia Y. Nano Lett 2003;3(5):675.
[329] Zhao N, Wei Y, Sun N, Chen Q, Bai J, Zhou L, Qin Y, Li M, Qi L. Langmuir 2008;24:991.
[330] Sun X, Hagner M. Chem Mater 2008;20:2869.
[331] Choi J, Chen Z, Singh RK. J Electrochem Soc 2003;150(8):C563.
[332] Yagi S, Nakanishi H, Matsubara E. J Electrochem Soc 2008;155(6):D474.
[333] Jin R, Cao Y, Mirkin CA, Kelly KL, Schatz GC, Zheng JG. Science 2001;294:1901.
[334] Li D, Chen S, Zhao S, Hou X, Ma H, Yang X. Appl Surf Sci 2002;200:62.
[335] Brevnov DA, Olson TS, López GP, Atanassov P. J Phys Chem B 2004;108:17531.
[336] Wiley B, Herricks T, Sun Y, Xia Y. Nano Lett 2004;4(9):1733.
[337] Kim D, Jeong S, Moon J. Nanotechnology 2006;17:4019.
[338] Lu X, Tuan H-Y, Chen J, Li Z-Y, Korgel BA, Xia Y. J Am Chem Soc 2007;129:1733.
[339] Blackwood DJ, Khoo SM. Solar Ener Mater Solar Cells 2010;94:1201.
[340] Chang S, Combs Zachary A, Gupta MK, Davis R, Tsukruk VV. Appli Mater Interf 2010;2(11):3333.
[341] Bastide S, Nychyporuk T, Zhou Z, Fave A, Lemiti M. Solar Ener Mater Solar Cells 2012;102:26.
[342] Swami A, Kumar A, D'Costa M, Pasricha R, Sastry M. J Mater Chem 2004;14:2696.

[343] Asefa T, Lennox RB. Chem Mater 2005;17(10):2481.
[344] Dinga J, Wang H, Lin T, Lee B. Synth Metals 2008;158:585.
[345] Huang C, Jiang J, Lu M, Sun L, Meletis Efstathios I, Hao Y. Nano Lett 2009;9(12):4297.
[346] Yagi S, Oeda N, Kojima C. J Electrochem Soc 2012;159(7):H668.
[347] Liu S, Liu J, Wang L, Zhao F. Bioelectrochemistry 2010;79:37. Au NPs.
[348] Hegde MS, Larcher D, DuPont L, Beaudoin B, Tekaia-Elhsissen K, Tarascon J-M. Solid State Ionics 1997;93:33.
[349] Wu Z, Ge S, Zhang M, Li W, Tao K. J Coll Interf Sci 2009;330:359.
[350] Johansson A, Lu J, Carlsson J-O, Boman M. J Appl Phys 2004;96(9):5189.
[351] Lee C-L, Kuo L-C, Huang Y-C, Yen Y-W. Electrochem Commun 2006;8:697.
[352] Chen L-J, Wan C-C, Wang Y-Y. J Coll Interface Sci 2006;297:143.
[353] Kokkinidis G, Papoutsis A, Stoychev D, Milchev A. J Electroanaly Chem 2000;486:48.
[354] Kokkinidis G, Stoychev D, Lazarov V, Papoutsis A, Milchev A. J Electroanaly Chem 2001;511:20.
[355] Herricks T, Chen J, Xia Y. Nnao Lett 2004;4(12):2367.
[356] Zhao Y-W, Zheng RK, Zhang XX, Xiao JQ. IEEE Trans Magn 2003;39(5):2764.
[357] Balela Mary Donnabelle L, Yagi S, Lockman Z, Aziz A, Amorsolo AV Jr, Matsubara E. J Electrochem Soc 2009;156(9):E139.
[358] Balela Mary Donnabelle L, Yagi S, Matsubara E. Electrochem Solid-State Lett 2010;13(2):D4.
[359] Liu Y-M, Sung Y, Pu N-W, Chou Y-H, Yeh K-C, Ger M-D. Thin Solid Films 2008;517:727.
[360] Choi Y, Hong S, Liu L, Kim SK, Park S. Langmuir 2012;28:6670.
[361] Silver P-Y, Vijayakrishnan V, Vibert P, Herrera-Urbina R, Elhsissen KT. Nanostruc Mater 1996;7(6):611.
[362] Viau G, Fiévet-Vincent F, Fihet F. Solid State Ionics 1996;84:259.
[363] Hinotsu T, Jeyadevan B, Chinnasamy CN, Shinoda K, Tohji K. J Appl Phys 2004;95(11):7477.
[364] Toneguzzo Ph, Acher O. IEEE Trans Mag 1999;35(5):3469.
[365] Toneguzzo Ph, Viau GV, Acher O, Guillet F, Bruneton E, Fievet-Vincent F, Fievet F. J Mater Sci 2000;35:3767.
[366] Chow GM, Zhang J, Li YY, Ding J, Goh WC. Mater Sci Eng A 2001;304–306:194.
[367] Yagi S, Kawamori M, Matsubara E. J Electrochem Soc 2010;157(5):E92.
[368] Chinnasamy CN, Jeyadevan B, Shinoda K, Tohji K. J Appl Phys 2003;93(10):7583.
[369] Vemagiri JK, Murthy SK, Gunasekaran RA, Coane P, Varahramyan K. Mater Lett 2003;57:4098.
[370] Tian J, Zhang S, Qu X. SciencePaper Online, September 25, 2012, http://www.paper.edu.cn/index.php/default/releasepaper/content/201209-279.
[371] Homma T, Kubo N, Osaka T. Electrochim Acta 2003;48:3115.
[372] Blake P, Ahn W, Roper DK. Langmuir 2010;26(3):1533.
[373] Lee C-L, Tseng C-M, Wu R-B, Wu C-C, Syu C-M. Collo Surf A: Physicochem Eng Aspects 2009;352:84.
[374] Sun Y, Xia Y. Science 2002;298:2176.
[375] Chen J, Wiley B, Li Z-Y, Cambell D, Saeki F, Cang H, Au L, Lee J, Li X, Xia Y. Adv Mater 2005;17:2255.
[376] Chen J, McLellan JM, Siekkinen A, Xiong Y, Li Z-Y, Xia Y. J Am Chem Soc 2006;128(46):14776.
[377] Chen J, Wang D, Xi J, Au L, Siekkinen A, Warsen A, Li Z-Y, Zhang H, Xia Y, Li XD. Nano Lett 2007;7(5):1318.
[378] Au L, Zheng D, Zhou F, Li Z-Y, Li X, Xia Y. ACS Nano 2008;2(8):1645.
[379] Skrabalak SE, Chen J, Sun Y, Lu X, Au L, Cobley LM, Xia Y. Acc Chem Res 2008;41(12):1587.

[380] Yavuz MS, Cheng Y, Chen J, Cobley CM, Zhang Q, Rycenga M, Xie J, Kim C, Song KH, Schwartz AG, Wang LV, Xia Y. Nat Mater 2009;8:935.
[381] Cobley CM, Au L, Chen J, Xia Y. Expert Opin. Drug Deliv 2010;7(5):577.
[382] Xia Y, Li W, Cobley LM, Chen J, Xia X, Zhang Q, Yang M, Cho EC, Brown P. Acc Chem Res 2011;44(10):914.
[383] Hirsch LR, Gobin AM, Lowery AR, Tam F, Drezek RA, Halas NJ, West JL. Ann Biomed Eng 2006;34(1):15.
[384] Zhou HS, Honma I, Komiyama H. Phys Rev B 1994;50(16):12052.
[385] Lee W-r, Kim MG, Choi J-r, Park J-Il, Ko SJ, Oh SJ, Cheon J. J Am Chem Soc 2005;127:16090.
[386] Sarkar A, Manthiram A. J Phys Chem C 2010;114:4725.
[387] Kobayashi Y, Verónica S-M, Liz-Marzán Luis M. Chem Mater 2001;13:1630.
[388] Kobayashi Y, Tadaki Y, Nagao D, Konno M. J Coll Interf Sci 2005;283:601.
[389] Nadagouda MN, Varma RS. Cryst Growth Des 2007;7(12):2583.
[390] Zhang L, Feng Y-G, Wang L-Y, Zhang J-Y, Chen M, Qian D-J. Mater Res Bull 2007;42:1457.
[391] Uysal M, Karslioğlu R, Alp A, Akbulut H. Appl Surf Sci 2011;257:10601.
[392] Wang G, Ma L, Chang Y, Liu C. Appl Surf Sci 2012;258:3962.
[393] Shao M, Sasaki K, Marinkovic NS, Zhang L, Adzic Radoslav R. Electrochem Commun 2007;9:2848.
[394] Park J-I, Cheon J. J Am Chem Soc 2001;123:5743.
[395] Park J-I, Kim MG, Jun Y-w, Lee JS, Lee W-r, Cheon J. J Am Chem Soc 2004;126:9072.
[396] Beard KD, Borrelli D, Cramer AM, Blom D, Van Zee JW, Monnier JR. ACS Nano 2009;3(9):2841.
[397] Shi G, Shan Y, Huang Y. J Shenyang Jianzhu Uni (Natural Science) 2007;23(6):965.
[398] Wang G, Chang Y, Wang L, Liu C. Adv Powder Technol 2012;23:861.
[399] Wang H-L, Yan J-M, Wang Z-L, Jiang Q. Intern J Hydro Ener 2012;37:10229.
[400] Ling G-p, Li J, Wang S-j. J Zhejiang Uni (Science) 2001;2(4):376.
[401] Pan X, Shen H, Qiu J, Gu M. Mater Chem Phys 2007;101:505.
[402] Sharma R, Agarwala RC, Agarwala V. J Alloys Comp 2009;467:357.
[403] Hajizamani M, Alizadeh A, Ehsani N. Appl Nanosci 2012;. DOI 10.1007/s13204-011-0055-7.
[404] Li X, Han X, Tan Y, Xu P. J Alloys Comp 2008;464:352.
[405] Zhu X, Dong H, Lu K. Surf Coat Technol 2008;202:2927.
[406] Zhang C, Ling GP, He JH. Mater Lett 2003;58:200.

Chapter 5

Electroless Plating Fe-Based Alloys

Chapter outline

As the name suggests, Fe-based alloys must have more than 50 at% Fe, and the total amount of other constituents in it must be less than 50 at%. All of the EP Fe-based alloys are such alloys.

5.1 WHY ELECTROLESS PLATING Fe–B ALLOYS?

China began her EN research and application at the end of 1970s and at the beginning of 1980s because of the cultural revolution. This meant that EN in China started about 20 years later than that in the rest of the world. The author's group was one of the units that started research early in China. After studying

Amorphous and Nano Alloys Electroless Depositions. http://dx.doi.org/10.1016/B978-0-12-802685-4.00005-4

EN for several years, we wondered that why there were no reports of Fe-based electroless-plated coatings. Fe, Co, and Ni all belong to the iron group of elements in the periodic table. Electroless-plated Ni–P (B) and Co–P (B) alloys had been developed and used in many industries for nearly 40 years when we raised this question. Can Fe-based alloys not be used for EP? And if not, why not? If they can be electroless-plated, how can we prepare them? If this is possible, then what are the differences between the technology of EP Ni-based alloys and that of Fe-based alloys? We thought these problems over and over. We wanted to get answers to these questions for ourselves, and the only way of doing this is to carry out detailed experimental research. We therefore considered that EP Fe–B alloys would be a good subject for investigation.

We chose EP of Fe–B as a research subject not only because it was a new area that no one had investigated before, but also because of its importance. At that time, my group was researching amorphous alloys prepared by various methods, for example, splat quenching and mechanical alloying. We understood that Fe–B alloys would have extreme magnetic properties, such as large magnetostriction and high permeability, and even Invar-type behavior over a wide temperature range [1–5]. From the application point of view, Fe–B alloys were important. For example, Fe–B films are very useful for various electroacoustic applications [6].

5.2 DISCOVERY OF EP Fe–B ALLOYS

After selecting this subject, the group started the journey of exploring the possibility of electroless-plated Fe-based alloys through a project. "The electroless plating Fe–B amorphous alloys" was chosen for a graduate student research thesis on this topic.

A comprehensive literature search found no papers on EP Fe–B alloys up to the middle of the 1980s, and just one paper on electroless Fe–P deposit, "Chemical iron–phosphorus films" written by Ruscior and Croialã in 1971 [7]. It should be pointed out that this paper was published more than 20 years after the discovery of EP with Ni-based alloys by Brenner and Riddel [8]. It has always been a surprise to us that no subsequent papers by these authors have been published, and no similar papers refer to EP Fe–P alloys until the middle of the 1980s.

Considering all of these factors including the issues mentioned in Section 5.1, we felt that we have chosen a very difficult subject for our research.

In the 1971 paper, the authors did not analyze the plating mechanism in detail, nor did they indicate the electroless deposition rate. In addition, the preparation method for Fe–P is rather simple, and no structure and properties were measured except the very simple magnetic property. However, the authors pointed out an important point in the paper, namely that the electroless Fe–P films can only be obtained by using a combination of an electrochemical (galvanic) process caused by Cu–Al coupling and chemical deposition resulting from reduction. From the very beginning, we considered this point in our experiments. That is to say, a Cu–Al couple (a Cu plate is the substrate, and an Al plate touching the substrate is suspended in the bath solution with a wire) similar to Ruscior and Croială's was used in our experiments.

Because Ni–P (B) and Co–P (B) deposit as EPs very easily, early in the 1960s scientists began to add ferrous salts into these bath solutions to see if Fe–Ni–P (B) and Fe–Co–P (B) ternary Fe-based EP deposits could be obtained. In 1966, Schmeckenbecher [9] used the following bath solution to electroless plate Ni–Fe–P films. The bath he used is nickelous chloride 56 mmol/L, ferrous ammonium sulfate 20 mmol/L, 100–350 mmol/L sodium potassium tartrate, 10 g/L sodium hypophosphite, and 3.6 mol/L ammonia. The temperature of the bath solution was 75°C. His experimental results indicated that the maximum content of Fe in the EP Ni–Fe–P deposits is 30%. In 1971, Schmeckenbecher [10] also studied EP Ni–Fe–B. Also in 1971 DePew and Speliotis [11] studied EP Co–Fe–P, and in 1997 Yazima et al. [12] studied EP Co–Fe–B. All three groups have found that the rates of all three alloy deposits decrease rapidly as the content of Fe^{2+} in the bath solution increases, as shown in Fig. 5.1. The bath solution and operating conditions for each ternary alloy are indicated in the figure that relates to them. In addition, these authors have found that all of the three EP Fe-based ternary alloy deposits cannot be obtained. The Fe content of the deposits are all less than 50%. For example, the Fe content of the EP Ni–Fe–B deposit reaches a maximum value of about 30% [10]. The maximum Fe content (47%) only appears in the EP Co–Fe–B film [12]. This is because of the low catalytic activity of iron. This phenomenon can also be seen in recent similar research reports, for example, see EP Ni–Fe–P in 2000 ([13], original Fig. 2. Effect of metallic salt ratio on Ni–Fe–P film deposition rate and composition) and EP Co–Fe–B in 2008 ([14], original Fig. 1 Co–Fe–B deposition rate versus $FeSO_4$ concentrations). All of these experimental results clearly indicate that EP ternary Fe-based alloys cannot be obtained by adding ferrous salts into the baths of Ni–P (B) and Co–P (B). Or does this require an alternative innovative solution?

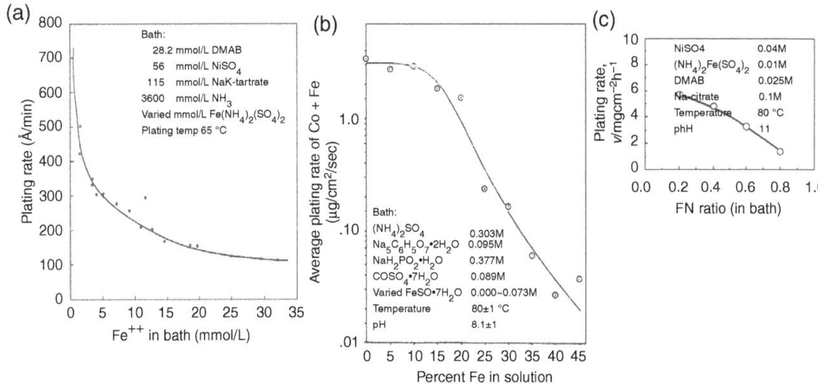

FIGURE 5.1 EP deposition rates versus ferrous salts in bath solutions. (a) Ni–Fe–B bath [10], (b) Co–Fe–P bath [11], and (c) Co–Fe–B bath [12]. The FN ratio = $Fe^{2+}/(Fe^{2+} + Ni^{2+})$ in (c).

As mentioned above, the Cu–Al couple (a Cu plate is the substrate, and an Al plate touching the substrate is suspended in the bath solution with a wire) similar to that used by Ruscior and Croialã was used from the very beginning of our experiments. However, no Fe–B deposits were obtained, not even after doing experiments for several months. The research team discussed and made many attempts. By changing each parameter of the bath solution one by one, and paying attention to every detail in the experiment, Fe–B deposits were finally obtained after more than a year. Since then, the deposition parameters, the structure and morphology, the crystallization process, and the mechanical and magnetic properties of amorphous Fe–B alloy deposits have been obtained and investigated systematically. An excellent master thesis was produced from this project, and several papers were published in national and international journals [15–19].

5.2.1 The Plating Bath and Affective Parameters

The plating bath and operating conditions of the EP Fe–B alloys that were finally obtained are as follows: $FeSO_4{\bullet}7H_2O$ 20 g/L, $KNaC_4H_4O_6{\bullet}4H_2O$ 90 g/L, KBH_4 3–20 g/L; NaOH 12–20 g/L, plating at 40°C for 20 min [18].

To find out the above plating bath and operating conditions, a number of affective parameters have been investigated repeatedly, including ligand, reduction agent KBH_4, pH, temperature, and plating time. The experimental results are as follows.

Although the strong reducing agent KBH_4 is used, its concentration is not the main rate-determining factor. Figure 5.2a shows that the deposition rate increases slightly with increasing KBH_4 concentration until saturation. It is well known that potassium borohydride is stable only at room temperature and in aqueous alkaline solution; otherwise, it will hydrolyze [20,21]. Therefore, the bath solution of EP Fe–B alloys should be alkaline.

Since the bath must be alkaline, a complexing agent for Fe^{2+} must be added to prevent precipitation of hydroxides or basic salts. To find the most effective

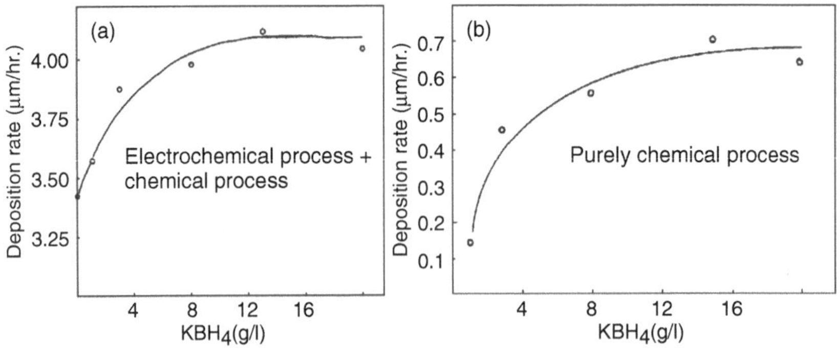

FIGURE 5.2 (a) Relationship between deposition rate and concentration of KBH_4. (b) Relation between deposition rate and concentration of KBH_4 for purely chemical deposition [18].

ligands the following was investigated: sodium citrate, ethylenediamine, sodium potassium tartrate, sodium pyrophosphate, and sodium acetate. The results indicated that sodium potassium tartrate was the most appropriate. The concentration of the ligand strongly affects the deposition rate. If the concentration is too low, the solution is unstable and the plating process proceeds with difficulty. On the other hand, if the concentration is too high, the deposition rate will also decrease due to decreasing activity of Fe. Therefore, to get the maximum rate, the best complexing conditions must be found. Our experimental results show that when the ligand concentration is at 90 g/L, the highest deposition rate is attained.

The pH of the plating solution has a considerable effect on the reduction process. As mentioned above, since the deposition takes place in alkaline solution only, sodium hydroxide was used for controlling the pH value. The measurements indicate that the deposition rate increases with increasing bath alkalinity. A plating rate of 1.8–5.0 μm/h is obtained as the concentration of NaOH varies from 12 g/L to 40 g/L.

As for the effect of temperature, the experimental results show that the maximum deposition rate occurs at the middle temperature range (40–50°C). Studies reveal that the rate of reduction of the iron cation is related to the bath temperature.

Effect of plating time: as the concentrations of reactants decrease during plating and some of the products also strongly affect the chemical and electrochemical reactions, the deposition rate changes. As described previously, the iron film formed would dissolve in the alkaline bath, so that although the total weight gain increases, the weight gain per unit time decreases with increase of plating time, as shown in Fig. 5.3.

5.2.2 Analysis of the Difficulty in Obtaining EP Fe–B Alloys

As mentioned above, Fe–B deposits result from the superimposition of an electrochemical process due to the metallic couple on a chemical process, and which amounts to the reduction of the Fe^{2+} ions by BH_4^- anions, accompanied by boron codeposition. The plating process of the bath without KHB_4 is purely

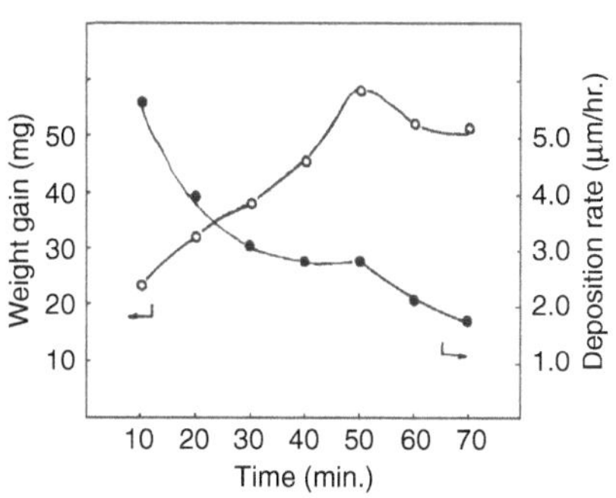

FIGURE 5.3 **The dependence of weight gain and deposition rate on deposition time [18].**

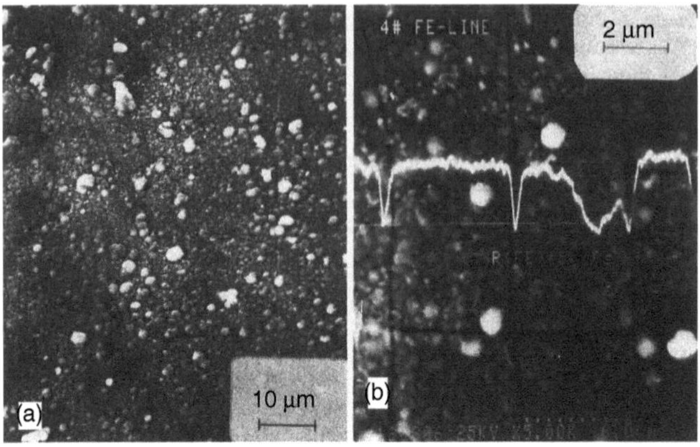

FIGURE 5.4 (a) SEM image for $Fe_{76}B_{24}$ alloy (1000×). (b) The graph from line scanning of the electron probe for the Fe of etched $Fe_{76}B_{24}$ alloy (5000×). [15]

electrochemical and there is a constant contribution to the plating process in baths containing KBH_4. Then by subtracting the deposition rate for the purely electrochemical process from the value of each point in Fig. 5.2a, the deposition rate of the purely chemical process can be obtained. The first point in Fig. 5.2a is obtained from the bath without KBH_4, so the deposition rate at this point is a purely electrochemical process. Using this approach the deposition rate of the purely chemical process is given by Fig. 5.2b. It can be seen from the figure that the rate is very small, the maximum rate being less than 0.7 μm/h, so the metal couple is necessary to amplify the chemical process.

When the concentration of KBH_4 is at 1 g/L, the purely chemical deposition rate of Fe–B is 0.14 μm/h from Fig. 5.2b. From the literature [22,23], the EP deposition rates of Co–B and Ni–B are 3.82 and 25 μm/h, respectively. Considering the redox potentials are 0.409 V for Fe^{2+}, 0.28 V for Co^{2+}, and 0.23 V for Ni^{2+} respectively [24], we can see that the EP rate increases sharply with decrease of the redox potential of the plated element.

From the measured relationship between the boron content and the concentration of KBH_4 and NaOH (see Table 1 and Fig. 6 in [18]), the reducing efficiency of KBH_4 can be calculated (see the detail given in [18]), and is found to be 1% for the purely chemical process. However, the reducing efficiency for the corresponding electroless-plated Ni–P using hypophosphite is 39%, EP Ni–B using borohydride is 6–45% [25], which is much larger than the reducing efficiency of KBH_4 in the EP Fe–B alloys. That is why, EP Fe–B alloys are so difficult to prepare.

5.2.3 Composition, Structure, and Properties of EP Fe–B Alloys

5.2.3.1 Morphology

The resultant surface appearance of EP Fe–B films was uniform and they adhered to the substrate. They were also smooth and bright. However, some rusty spots appeared on the films after several weeks in a humid atmosphere. The SEM micrographs of these alloys revealed that the deposit is composed of islands that are circular in shape and of different sizes. Figure 5.4a shows an SEM image of a $Fe_{76}B_{24}$ alloy [15]. After etching by 5% HNO_3 + C_2H_5OH, many etch pitches appeared which were low in iron – as demonstrated by line scanning of an electron probe for elemental Fe (see Fig. 5.4b). These pitches may be caused by internal stress or the evolution of hydrogen.

5.2.3.2 Composition and Structure

Figure 5.5 shows the XRD patterns of as-deposited specimens ($x = 2.2, 10.6, 18$) [15]. The substrates were nickel strips. It can be seen that only the α-Fe phase

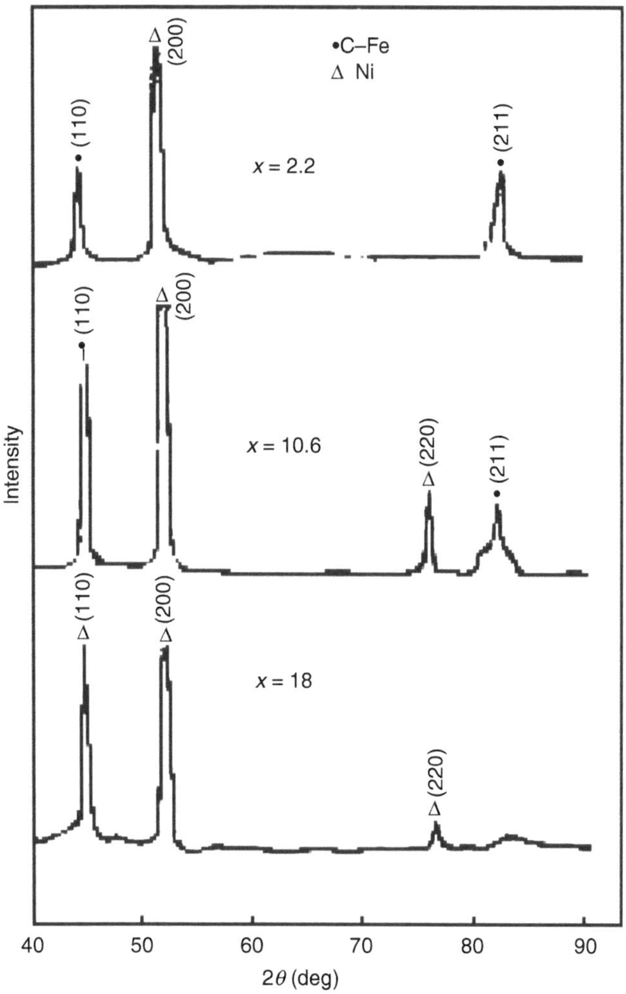

FIGURE 5.5 The XRD patterns of as-deposited $Fe_{100-x}B_x$ alloys ($x = 2.2, 10.6, 18$) [15].

exists for $x = 2.2$. For $x = 10.6$, the main phase is α-Fe, and some amorphous phases may exist as the XRD peak of α-Fe widens. When the boron content is not less than 18 at%, the diffraction patterns of the samples have no peak except the peaks of the substrate, which implies that their structures are amorphous.

The XRD patterns of the three heat-treated amorphous specimens ($x = 10.6$, 18, 27.4) have been obtained. The treatment temperature was 400°C. After heat treatment, step scanning measurements show that the diffraction peaks of α-Fe shift to the high angle side in these samples. The lattice parameter of the BCC lattice was determined by an extrapolation method. The result is 2.8494 Å for $Fe_{89.4}B_{10.6}$, which is much smaller than that of pure iron (2.8663 Å and 298 K) [26]. This value is also much smaller than that of the samples prepared by splat quenching (2.8610 Å for $x = 10$) [26] and there is no α-Fe, B phase. After heat treatment at 400°C for 1 h, the lattice parameter of $Fe_{89.4}B_{10.6}$ increases to 2.8630 Å as a result of precipitation of boron from the solid solution. This reveals that the boron atoms in the present case occupy substitutional sites in the BCC lattice. We note that for the results of the liquid-quenched Fe–B alloys by other authors, the boron atoms may occupy both substitutional and interstitial sites [2] or only substitutional sites in the BCC lattice [26].

5.2.3.3 Crystallization and Activation Energy

To research the crystallization processes of three Fe–B amorphous alloys, they were annealed at 400, 500, 550, and 600 °C for 1 h, respectively. The results show that the crystallization processes of Fe–B amorphous alloys are as follows:

$$
\begin{aligned}
x = 18 \quad & \text{amorphous} \\
& \rightarrow \text{amorphous} + \alpha\text{-Fe}(\text{B}), \\
& \rightarrow \alpha\text{-Fe} + \text{t-Fe}_3\text{B}, \\
x = 24 \text{ or } 27.4 \quad & \text{amorphous} \\
& \rightarrow \text{amorphous} + \text{t-Fe}_3\text{B}, \\
& \rightarrow \alpha\text{-Fe} + \text{t-Fe}_3\text{B}.
\end{aligned}
$$

The differential thermal analysis (DTA) measurements of the as-deposited Fe–B amorphous samples show that the crystallization temperature ranges from 730 K to 735 K for the composition range of $18 \leq x \leq 27.4$ at a heating rate of 20 K/min, which is nearly the same range for the splat quenching samples reported by Ray et al. [27].

The measured activation energy of crystallization E_a of the EP Fe–B amorphous alloys decreases sharply from 184.5 kJ/mol to 129.1 kJ/mol as the boron content increases from 18 at% to 27.4 at%, which is caused by differences in their internal structure and the degree of difficulty or ease of crystallization of the alloys. The activation energies of crystallization for $Fe_{80}B_{20}$ amorphous alloy prepared by rapid quenching from different authors range from 194 kJ/mol to 240 kJ/mol [28–30]; these are much higher than those of the EP Fe–B amorphous samples. Thus, the thermal stability of the present samples is lower than that of the rapidly quenched samples, which may be due to internal stress and homogeneities of composition and structure of the samples.

5.2.3.4 Mechanical and Magnetic Properties

The hardness of the Fe–B deposits plated on a mild steel substrate was determined by a Vickers indenter with diamond pyramid and a load of 5 g. The thickness of the deposits was in the range of 8 μm. Figure 5.6a shows that the hardness of as-plated deposits decreases with increasing boron content. This phenomenon arises from the internal structure of the Fe–B deposits. As mentioned above, for $x = 2.2$, only the α-Fe phase (BCC lattice) exists. It is a supersaturated solid solution, in which the boron atoms occupy the interstitial sites of the BCC lattice. It is a hard alloy. For $x = 10.6$, the main phase is α-Fe, and some amorphous phases may exist. Some B atoms do not occupy the interstitial sites, they move out, occupying the substitutional sites, so the hardness is lower. When the boron content rises to 18 at%, only amorphous phase exists. All of the B atoms have moved out from the interstitial sites, occupying the substitutional sites. Such phases are softer; therefore, the hardness reduces further.

After isothermal annealing for 1 h, the hardness of samples of $Fe_{76}B_{24}$ varies with treatment temperature as shown in Fig. 5.6b. The increase of hardness below 250°C is due to structural relaxation of short range order at the lower temperature. At higher temperatures, (250–350°C), the relaxation of the internal stresses and coagulation of boride particles results in a decrease in hardness. A very high hardness (1030 kg/mm^2) appears after treatment at 400°C as a result of the dispersion hardening due to Fe_3B phase formation. When the temperature is higher than 400°C, the sample crystallizes perfectly in its equilibrium structure that removes the hardening effect resulting from the structural changes or precipitation during crystallization. The hardness then decreases rapidly. But the hardness increases again due to the formation of a second precipitant α-Fe phase during annealing at 550°C for 1 h.

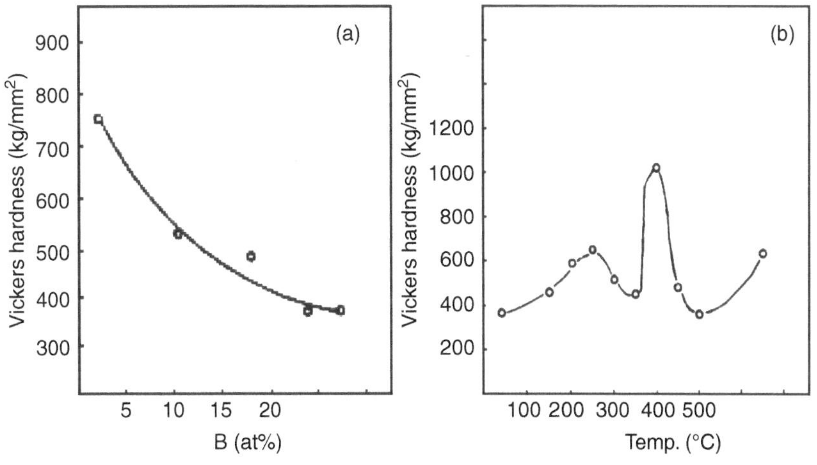

FIGURE 5.6 (a) Relationship between microhardness and boron content. (b) Relation between microhardness and heat-treatment temperature for amorphous $Fe_{76}B_{24}$ alloy [15].

The magnetic properties of EP amorphous samples of $Fe_{100-x}B_x$ ($x = 18$, 24, 27.4) and crystalline samples of $Fe_{100-x}B_x$ ($x = 2.2$, 10.6) have been studied. This chapter will concentrate on the magnetic moment μ and the Curie temperature T_c, for a discussion of other properties, please see [17].

From the measured saturation magnetization σ_s at room temperature the magnetic moment per Fe atom at temperature 0 $\mu(0)$ was derived. The results are shown in Fig. 5.7. The magnetic moment $\mu(0)$ decreases with increasing *B* concentration. For comparison, two sets of experimental data are also shown in Fig. 5.7, obtained on liquid-quenched ribbons [1,3] and on sputtered thin-film samples [31–33]. Although the magnetic moments for the three sets of samples all decrease with increasing *B* concentration, the decreases show considerable individual differences. In addition, Fig. 5.7 shows that in some cases the moment variation is linear and in other cases is nonlinear.

The experimental results can be understood from the simple rigid-band model [1,34] and the calculations of the magnetic moments of amorphous Fe–B alloys by first principles [35] (for details please see the original paper [17]). The rigid-band model supposes that the majority of the spin subband is fully occupied and that the sp electrons transfer from the B atoms to a minority of the spin subband that leads to a reduction of the magnetic moment. Therefore, $\mu(0)$ reduces with increasing B content in all three sets of samples. However, from the slopes of the curves in Fig. 5.7 the electrons donated by boron should be very different, which is difficult to explain using the model. From the calculated results of first principles, one can speculate that the electron transfer from B to Fe atoms would be different because of the different structure and different B coordination of Fe atoms for the electroless-plated samples compared with the other two sets of samples. Thus, the electron transfer upon increasing

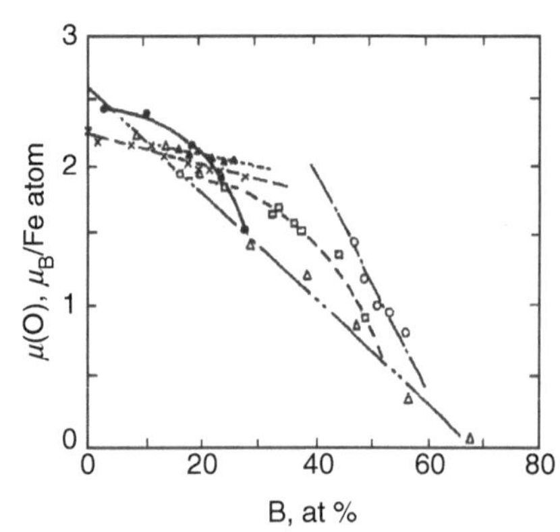

FIGURE 5.7 Saturation magnetic moments as a function of B content for Fe–B crystalline and amorphous alloys. The results for the samples from different preparing techniques are included: ● for the electroless-plated samples in the present work; × and ▲ for the liquid-quenched samples from Hasagawa et al. [3] and Luborsky et al. [1]; △, □, and ○ for the sputtered samples from Buschow et al. [32], Bayreuther et al. [32], and Webb et al. [34], respectively [17].

the B concentration would be different for the different samples. So, the different reduction of the magnetic moments for the three different sets of samples can be understood. It is obvious that further research is still needed for a quantitative explanation of the experimental data.

The measured values of T_c of the crystalline and amorphous alloys $Fe_{100-x}B_x$ are shown in Fig. 5.8 as a function of boron concentration. The value of T_c decreases from 903 K to 872 K upon increasing the B concentration from 2.2 at% to 10.6 at% for the crystalline alloys, but increases from 613 K to 776 K when the B concentration is 18–27.4 at% for the amorphous alloys. For comparison the data of T_c by other authors with two kinds of samples obtained by liquid quenching and sputtering are also given in the figure. It is obvious from Fig. 5.8 that the agreement of the measurements for the three different amorphous samples prepared by liquid quenching, sputtering, and EP is rather good, and the scatter of the data is within ±20°C from 10 at% to 30 at% B. It is interesting to note that from the curve of the experimental data of T_c for the three different types of samples in Fig. 5.8, a T_c value of about 230 K can be extrapolated for pure amorphous Fe. This value is very close to that obtained by extrapolation of T_c for the liquid-quenched samples [5].

How do we explain the T_c dependence of the B concentration for the amorphous alloys? From the molecular-field theory, we have

$$T_c = \frac{2n_{FeFe}J_{FeFe}S(S+1)}{3k_B} \quad (5.1)$$

where S is the spin, k_B is the Boltzmann constant, n_{FeFe} is the average coordination number of the FeFe nearest neighbors, and J_{FeFe} is the average exchange interaction. The experimental results by neutron scattering [38] indicate that

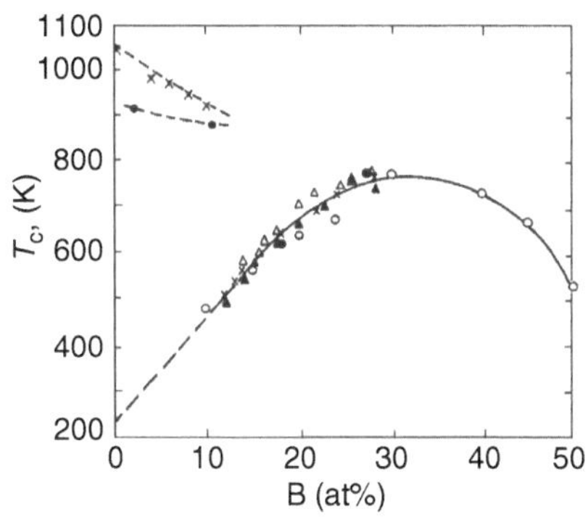

FIGURE 5.8 Curie temperatures as a function of B content for $Fe_{100-x}B_x$ crystalline and amorphous alloys. The results for the samples from different preparation techniques are included: ● for the electroless-plated samples in the present work; ×, ▲, and △ for the liquid-quenched samples from Hasagawa et al. [3], Luborsky et al. [1], and Chien et al. [36], respectively; ○ for the sputtered samples from Chien and Unruh [17, 37].

the coordination number for nFeFe is approximately 12 (11.96–12.33), independent of concentration, in $Fe_{100-x}B_x$ amorphous alloys with $14 \leq x \leq 25$. The same results are also obtained by the isotope-substitution method in amorphous alloys with $12 \leq x \leq 22$ [39]. So, we just need to consider the parameter J_{FeFe}. It is well known that J_{FeFe} depends sensitively on the Fe–Fe distance, which increases from 2.6 Å for $Fe_{86}B_{14}$ to 2.65 Å for $Fe_{75}B_{25}$, as determined by neutron scattering [38]. The conclusion therefore is that the increase of T_c with increasing *B* concentration is primarily caused by the increase of the average exchange interaction with increasing Fe–Fe distance.

As for the crystalline alloys of $x = 2.2–10.6$, all are stable in the α-phase, which indicates that n_{FeFe} remains constant, that is, 8 because α-Fe has a body-centered cubic (BCC) lattice structure. *S* decreases with increasing *B* concentration because of the reduction of the magnetic moment (Fig. 5.7). In addition, the lattice constant decreases from 2.8540 Å for $Fe_{97.8}B_{2.2}$ to 2.8494 Å for $Fe_{89.4}B_{10.6}$ [15], which means that the Fe–Fe distance and also J_{FeFe} decrease with increasing *B* concentration for the crystalline alloys. It is obvious from Eq. (5.1) that T_c should decrease with increasing *B* concentration for the crystalline alloys. Please see the original paper for a more detailed explanation [17].

The composition dependence of the coercive force H_c for the as-cast films of $Fe_{100-x}B_x$ alloys is shown in Fig. 5.9. It can be seen that the coercive force H_c first decreases, and then increases with increasing *B* concentration. However, the variation in H_c is not very large. The comparison of this functional relation to the results for the sputtered [40] and liquid-quenched [1] samples is also shown in Fig. 5.9. Also for these two types of samples the variation of H_c as a function of *B* concentration is not very large. However, the values of H_c are

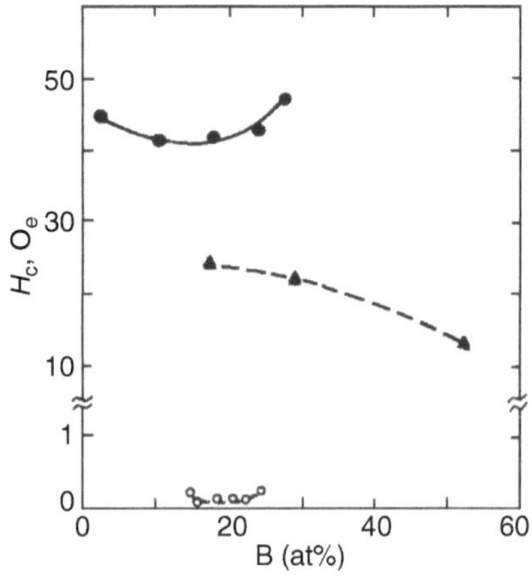

FIGURE 5.9 Coercive force as a function of B content for the crystalline and amorphous Fe–B alloys prepared by electroless plating (●). ○ represents the results of Luborsky et al. [1] for the liquid-quenched samples. ▲ represents the results of Stobiecki et al. [40] for the sputtered samples [17].

very different for the different types of samples. H_c amounts to 41–46 Oe for the electroless-plated amorphous samples, to 13–24 Oe for the sputtered samples [41], and only 0.08–0.21 Oe for the liquid-quenched samples [1]. Why is there such large difference in H_c for the three sets of samples? It is well known that the coercive force depends on structural inhomogeneities and the magnetic anisotropy of the sample. Structural inhomogeneities can be used to explain the difference of H_c for the three set samples in Fig. 5.9. In fact, even though the other properties were the same, Luborsky et al. [1] found that there is a large difference in the coercive force H_c between samples cast in air and in vacuum. The only difference for the two samples is that the samples cast in air have serrated sg edges and generally rougher upper surfaces, but those cast in vacuum have no such defects. The liquid-quenched samples have better structural homogeneity, which results in a very low value of H_c. However, the structural inhomogeneities for the samples prepared by the EP technique are the worst [15] of the three sets of samples so that their H_c values are very large.

5.2.4 Formation Mechanism of EP Fe–B Alloys

Figure 5.10 shows a schematic representation of EP Fe–B in the bath with a Cu–Al couple. First, just consider the plating process from the Cu–Al couple and ignore the electroless process due to the reduction of the Fe cations by BH_4^- anions. Because of the standard electrode (redox) potentials in aqueous solution at 25°C of metals Al^{3+}, Fe^{2+}, and Cu^{2+} are −1.66, −0.409, and 0.34 V, respectively [24], the Cu–Al couple must form a galvanic cell with Al as anode and Cu as cathode. Thus, the Al cations (Al^{3+}) and an equal number of electrons ($3e^-$) left the aluminum anode pass into solution. Because of the loss of Al during the process, Al usually is the so-called sacrificial anode. This is an oxidation process, which can be expressed as follows:

$$Al \xrightarrow{\text{oxidation}} Al^{3+} + 3e^- \text{ in the sacrificial anode} \tag{5.2}$$

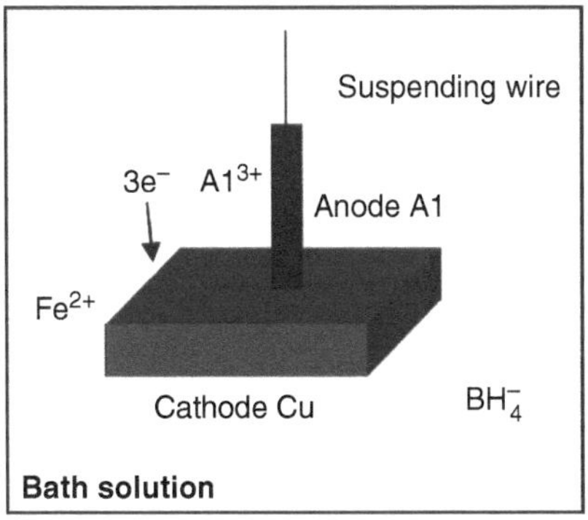

FIGURE 5.10 Schematic representation of EP Fe–B in the bath with a Cu–Al couple.

Meanwhile, the Fe cations are reduced at the copper cathode forming the Fe film because Al metal is more electronegative than Fe. This is a reduction process, which can be expressed as

$$Fe^{2+} + 2e^{-} \xrightarrow{\text{reduction}} \text{Fe in the cathode} \tag{5.3}$$

Now consider the formation of Fe–B films. The analysis and discussion of this in Section 5.2.2 that Fe–B deposits result from the superimposition of an electrochemical process which is due to the metallic couple on a chemical process which amounts to reduction of the Fe^{2+} ions by BH_4^- anions, accompanied by boron codeposition. Just because the metal couple amplifies the chemical process, the very small rate from pure chemical process with KBH_4 can be improved greatly, becoming practice and meaningful. The chemical reaction equation for the Fe–B films can be represented as follows:

$$BH_4^- + 4OH^- \rightarrow B(OH)_4^- + 2H_2 + 4e^- \tag{5.4}$$

$$BH_4^- + 8OH^- \rightarrow B(OH)_4^- + 2H_2O + 8e^- \tag{5.5}$$

These reactions provide electrons for the reduction of iron and boron atoms, and lead to the Fe–B films on the Cu substrate:

$$Fe^{2+} + 2e^{-} \xrightarrow{\text{reduction}} \text{Fe on the substrate} \tag{5.6}$$

$$B(OH)_4^- + 3e^{-} \xrightarrow{\text{reduction}} B + 4OH \text{ on the substrate} \tag{5.7}$$

As for the formation mechanism of Fe–B amorphous phase, it can be understood as follows. At the very beginning of plating, the number of the B atoms is rather few and that of the Fe atoms is large in the deposits, the reaction between B and Fe atoms is rather strong, the B atoms are easy to become an interstitial atom in the Fe lattice, thereby reducing the lowest total energy promoting the films stability. As the number of B atoms is very large, some reaction between the B atoms produced. Due to two reasons that one is Fe atoms difficult to perform long-distance running caused by the impeding and containing between the atoms, and another is the little driving force for atomic long-distance diffusion resulted from the temperature of the bath quite low, atomic groups are difficult to rearrange collectively in order to produce a regular atomic configuration suitable for crystallization. This situation means that crystallization is hindered and thereby it leads to easy formation of amorphous structure in the deposits. In such a way, the films are ensured in the lowest energy and become stable.

5.2.5 Problems and Worthwhile Improvements for EP Fe–B Alloys

First, even the EP amorphous Fe–B deposits got rusty after a period of time, which is not seen with electroless Ni or Co-based alloy deposits. This is an obvious problem that needs future work to progress.

Next, comparing the EP Ni, and Co-based alloys, little research work has been done with the EP Fe–B and Fe–P alloys. Further investigations of EP Fe-based alloys are needed to understand their processes and nature more deeply.

Third, even though EP Fe–B alloys were discovered a long time ago, their applications are rather few. In particular, their application in industry is rarely seen. This indicates that further deep research into EP Fe–B alloys and their application in industrial processes is needed. In addition, their application in industry needs to be promoted.

5.3 EP BINARY Fe–B ALLOYS

Some years after the discovery of EP Fe–B alloys, scholars from other countries began to investigate electroless Fe–B alloy deposits. In 1997, Fujita et al. [41] fabricated amorphous iron–boron films by EP. The authors used two types of baths and substrates to prepare the EP Fe–B depositions. The first kind of film they fabricated is called a "contact-plated film." The bath (A) contains $FeSO_4 \cdot 7H_2O$ (72 mmol/L), $C_4H_4KNaO_6 \cdot 4H_2O$* (320 mmol/L), NaOH (400 mmol/L**), and KBH_4 (20–370 mmol/L), and the substrate (1) is Cu foil in contact with an Al wire. It is obvious that this is nearly the same as reported in [15,18], except the amounts of the chemicals are slightly different. The second kind of film they fabricated is called a "chemically-plated film." Bath (B) is $FeSO_4 \cdot 7H_2O$ (200 mmol/L), $C_4H_4KNaO_6 \cdot 4H_2O$ (50 mmol/L), NH_3 (1.58 mol/L), and $NaBH_4$ (140–300 mmol/L), and the substrate (2) is Cu foil with sensitization and activation pretreatments. Their experimental results show that only the contact-plated films are effective, and amorphous $Fe_{100-x}B_x$ films with $x \sim 22$–28 at% are actually obtained. It is also confirmed that the key to obtaining such films with high thicknesses is the use of the substrate, with the copper–aluminum configuration forming the local battery. However, the chemically deposited films were found to be less attractive owing to their small thicknesses. But, these films had high concentrations (from 30 at% to 48 at%) of boron, the reason for which remained unclear.

In 1998, these authors studied [42] amorphous Fe–B films with soft magnetic properties. To reduce the coercivity of EP amorphous Fe–B films, they investigated the effects of three kinds of agents [sodium L-ascorbate ($C_6H_7NaO_6$), thiourea (NH_2CSNH_2), and ammonium sulfate ($(NH_4)_2SO_4$)] on magnetic and magnetoelastic properties of films. The first two agents, $C_6H_7NaO_6$ and NH_2CSNH_2, were found to be ineffective for improving the soft magnetic properties of films. On the contrary, however, $(NH_4)_2SO_4$ is very effective for magnetic softening: amorphous Fe–B plated films with a coercivity of less than 4 Oe were obtained by employing the bath composition of $FeSO_4 \cdot 7H_2O$ (71.9 mmol/L), KBH_4 (296.6 mmol/L), NaOH (400 mmol/L), $KNaC_4H_4O_6 \cdot 4H_2O$ (600 mmol/L), and $(NH_4)_2SO_4$ (80 mmol/L). These soft films exhibited a favorable large magnetostriction of 26×10^{-6}, being adequate for micromagnetoelastic devices.

---

* Original error in the paper [41] has been corrected.

** It is 400 M in original paper. It should be 400 mM, i.e. 400 mmol/L.

In 2007, Schoen et al. [43] reported a new electroless iron bath capable of depositing a ferromagnetic Fe–B coating onto Pd/Sn-catalyzed substrates at room temperature without the need for an accompanying galvanic couple. They used the bath to fabricate magnetic cellulose microfibers. The electroless Fe–B bath utilizes Fe(II) (0.025–0.050 mol/L) complexed by citrate ligand (0.125–0.250 mol/L) in alkaline solution as a metal source, with borate buffer (0.250 mol/L total borate, pH 9.7–10.2) as a pH controller and borohydride (0.025–0.100 mol/L) as the reductant. The fibrous cellulose they used was medium length, ~20 μm diameter × ~200 μm length, $\rho_c = 0.6\ \mathrm{g\ cm^{-3}}$. They used two-level factorial design statistical methods, which are rather complex, to characterize the effects of plating variables (i.e., bath pH and concentrations of each bath component) on bath behavior. The results show that only bath pH affects the quantity of metal deposited, with somewhat larger amounts of metal plating at lower pH values. Plating of Pd/Sn-catalyzed cellulose microfibers is essentially complete (>98%) and is not influenced by any of the variables under these conditions. In addition, surface analysis confirms the deposition of an adherent amorphous ferromagnetic Fe–B coating of estimated thickness ~140–346 nm and composition ~$Fe_{10}B$ onto the microfiber surface. Figure 5.11 is an optical micrograph of the samples. It can be seen that the fibers were nearly completely plated (> ~98%), becoming gray-black fibers. The samples were relatively free of debris (e.g., plated, nonfiber material). Although surface features consistent with limited nonhomogeneous plating (e.g., nodules, pits) were occasionally observed, the level of such defective fibers never exceeded 5% of the fibers examined for a given experiment.

In 2007, these authors reported further structural and magnetic properties of EP Fe–B microfibers [44] on the basis of their work in 2006 [43]. It is interesting to note that last time these authors wanted to use SEM or TEM to directly assess plating thickness and homogeneity, but their attempts were confounded by image distortions due to the fibers' magnetic fields [43]. However, this time, they succeeded in obtaining SEM images of the samples, as shown in Fig. 5.12a. We can see from the picture that the Fe–B-deposited films are made of nanoparticles that form around Pd/Sn electroless catalyst particles on

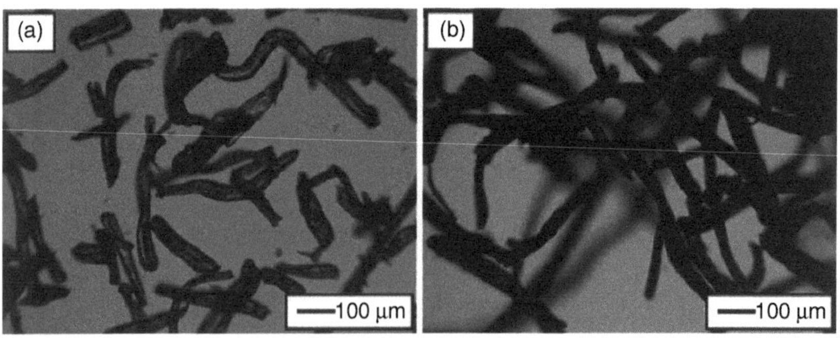

FIGURE 5.11 Optical micrographs (20×) of cellulose fibers. (a) Catalyzed, unplated fibers, (b) dried, plated fibers. The variable levels are $[Fe^{II}] = 0.05$ mol/L, $[BH4^-] = 0.10$ mol/L, [citrate] = 0.25 mol/L, and pH 10.2 [43].

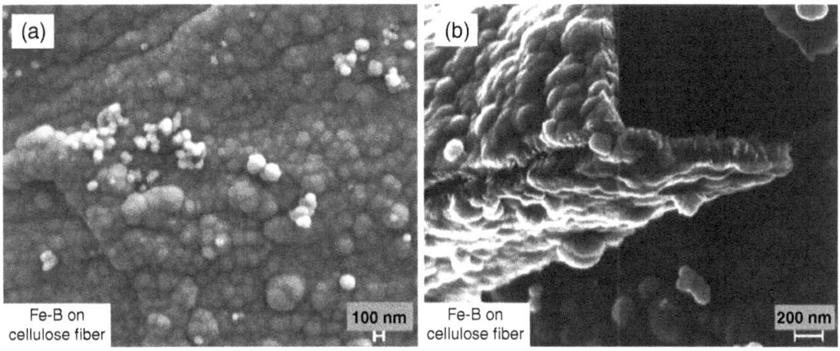

FIGURE 5.12 SEM micrographs of plated Fe–B films on cellulose fibers. (a) Left micrograph shows details of the surface. (b) Right micrograph shows the thickness of the coating [44].

the surface of the cellulose and grow together to cover the fibers uniformly. The nanoparticles have a diameter of 100–200 nm. The picture in Fig. 5.12b shows Fe–B that has broken free from the fiber surface, revealing the thickness of the coating, here about 150 nm.

XRD showed the presence of an amorphous phase. Extended X-ray absorption fine structure spectroscopy (EXAFS) studies confirmed an amorphous-like structure with the nearest coordination numbers around Fe atom to be 8.7 for Fe and 3.5 for B, giving the concentration of boron around 29 at%. The measured magnetic moment of 2.12 Bohr magnetons/Fe atom is consistent with composition obtained from EXAFS measurements. In addition, a coercive field of about 35 Oe was measured, which was a little less than ours measured results of 41–46 Oe for the EP amorphous Fe–B samples [17].

It is interesting to note that these authors applied for a US patent in May 2009 using the title of "Room temperature electroless iron bath operating without a galvanic couple for deposition of ferromagnetic amorphous Fe–B films" (2009/0117,285).

In 2006, Huang et al. [45] studied the effect of coupled aluminum on EP Fe–B deposition. The results show that coupled aluminum induces the Fe–B electroless deposition by shifting the potential negatively and decreasing the polarization resistance of anodic and/or cathodic reaction. The variations of the bath composition and temperature shift the mixed potential and change deposition rate of Fe–B electroless deposition. Due to the interactions between the two half reactions on the catalytic surface, the plating rate determined by electrochemical measurement in the complete bath is higher than that determined from the simulation half-reaction cells.

To date, not many papers on EP Fe–B have been published, and we look forward to this number increasing in the future.

5.4 EP FE–B-BASED MULTICOMPONENT ALLOYS

Because even the EP amorphous Fe–B deposits rusted easily after a period of time, overcoming this shortcoming therefore has become our most urgent task after finishing the EP Fe–B alloy depositions. In our thinking about ways to overcome this difficulty, it seems necessary to add extra elements into the Fe–B deposits. After adding W or Mo into Fe–B alloys, the EP Fe–W–B and Fe–Mo–B deposits do not rust so easily [46–48]. Following this routine, a graduate student in author's group chose "Electroless Fe-based multi-component alloy deposits" as her MS and PhD project. Multiple alloy systems such as Fe–Mo–B, Fe–W–B, Fe–Sn–B, and Fe–Mo–W–B were successfully deposited, and various properties of the deposited coatings were systemically studied. Thus, more papers of electroless Fe–B-based alloy deposits from the author's group were published [49–56].

5.4.1 EP Fe–W–B Alloy Deposits

In 1997, Yi et al. [46] prepared Fe–W–B alloy deposits by EP. The bath and the operating conditions are ferrous sulfate 20 g/L, sodium potassium tartrate 90 g/L (50–130 g/L), sodium hydroxide 20 g/L (10–40 g/L), potassium borohydride 8.5 g/L (4–20 g/L), sodium tungstate 33 g/L (20–60 g/L), temperature 40°C (32–60°C), and plating time 30 min (5–160 min). To obtain the deposits, an aluminum sheet must be in contact with the substrates, which leads to an electrochemical process. The effects of various parameters such as ligand, potassium borohydride, sodium hydroxide, sodium tungstate, plating temperature, and plating time on the deposition rate were investigated in detail. The composition, structure, surface appearance, and micro hardness were all measured. Table 5.1 indicates the relationship between the composition of deposits and the concentration of potassium borohydride. It was seen that the boron content in the deposits is <12 at%. The deposits are a type of α-Fe base solid solution. Only the (111) diffraction peak of α-Fe can be seen in the patterns. All the boron atoms and tungsten atoms exist in the alloys as interstitial atoms and substitution atoms, respectively. The Fe–W–B deposits do not have an amorphous phase. The reason may be that the addition of sodium tungstate in the bath will consume too many electrons to be reduced when WO_4^{2-} is reduced to metal W atom according to the following reaction formula:

$$WO_4^{2-} + 4H_20 + 6e^- \rightarrow W + 80H^- \quad (5.8)$$

Therefore, the addition of sodium tungstate in the bath decreases the boron content in the alloy films, which results in deposits of α-Fe solid solution and causes the Fe–W–B deposits to crystallize.

The surface morphology of the deposits listed in Table 5.1 was studied. The surface appearance of the deposits changed from bright to matte when the concentration of potassium borohydride went from 4 g/L to 12 g/L, and then from matte to bright. The morphology of the deposits is shown in Fig. 5.13. There are some particles on the smooth deposit. It can be seen from Fig. 5.13 (W3-1,000X), which shows the appearance of W3 under 1,000X, that the particles have nearly a spherical shape. The spot scanning electron probe microanalyzes the smooth area and the particles show that the particles have more tungsten

TABLE 5.1 The Relation Between the Composition and the Concentration of Potassium Borohydride of EP Fe–W–B Alloy Deposits

No.	Concentration (g/L)	Fe	W	B
		at%		
W1	20	90.16	3.51	6.33
W2	30	93.16	4.32	2.01
W3	40	85.59	2.50	11.91
W4	50	92.11	3.61	4.26
W5	60	66.20	2.51	9.29

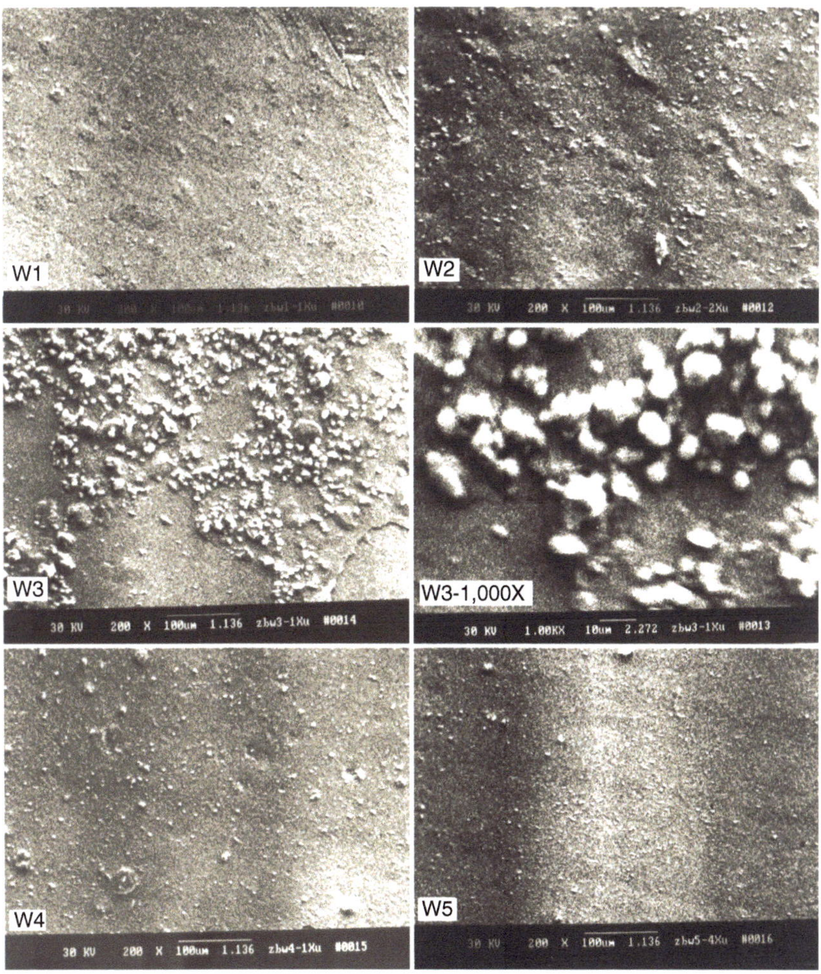

FIGURE 5.13 The surface appearance of specimens listed in Table 5.1 by standard electron microscopy (200×) [46].

content (particle: W 3.56 at%, smooth area: 2.77 at%). The quantity and size of the particles changed with the concentration of potassium borohydride.

The dependence of the microhardness for as-prepared deposits upon the boron content is given in Fig. 5.14. The microhardness increases as the boron content increases. Naka et al. [57] used the average outer-electron concentration (e/a) of the alloys perfectly to explain the microhardness of iron-based alloys prepared by rapid quenching. The average outer-electron concentration for the Fe–W–B deposits, by weighted means according to the atomic percentage of the elements, is also given in Fig. 5.14. It is clear that the microhardness increases and the e/a decreases as the boron content increases. This observation results from the strengthening of partial bonds by s–p hybrid orbits of boron atoms with the s–p–d orbits of the metal atoms. The higher the e/a, the lower the strength; hence, the microhardness decreases.

5.4.2 EP Fe–Mo–B Alloy Deposits

We have published two papers on EP Fe–Mo–B alloy films [47,48]. The influence of process parameters such as ligand, Na_2MoO_4, KBH_4, NaOH, plating temperature, and plating time, composition, and structure, glass-forming region (GFR), surface appearance, microhardness, corrosion, and crystal lization processes, and products of EP Fe–Mo–B alloys are all investigated. The bath solution and operating conditions are as follows, $FeSO_4$•$7H_2O$ 20 g/L, $KNaC_4H_4O_6$•$4H_2O$ 90 g/L, $NaMoO_4$•$2H_2O$ 48.4 g/L, KBH_4 10 g/L, NaOH 20 g/L, plating time 30 min, and plating temperature 40°C. To obtain the deposits, an aluminum sheet, which introduces an electrochemical process, must be in contact with the substrate. The results show that as the $NaMoO_4$ changed from 24.2 g/L to 84.7 g/L, the contents of Mo and B in the Fe–Mo–B deposits range from 9.9 at% to 16.9 at% and from 6.6 at% to 33.8 at%, respectively. To compare the Fe–W–B deposits, the contents of W and B range only from 0.8 at%

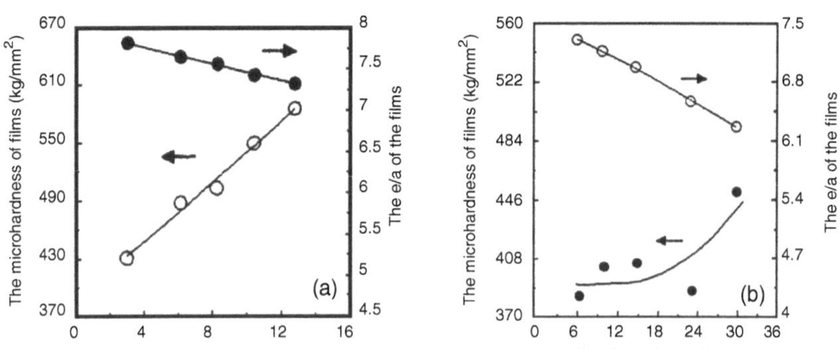

FIGURE 5.14 The dependence of microhardness and the average outer-electron concentration (e/a) on the boron content. (a) EP Fe–W–B deposits, (b) EP Fe–Mo–B deposits [46].

to 2.21 at% and from 2.95 at% to 12.71 at%, respectively, when the $NaWO_4$ is changed from 20 g/L to 60 g/L, which are much less than those in the Fe–Mo–B. Fukamich et al. [58] and Ray et al. [27] studied the Fe–B alloys obtained by rapid quenching, and drew the conclusion that Fe–B can form amorphization only when the boron content ranged from 9 at% to 26 at%. In our percent Fe–Mo–B work, with alloying the element of Mo as well, the boron content region for amorphization has been extended (from 6.63 at% to 30.1 at%). This may result from two kinds of effects of alloying Mo in the deposits. First, the addition of a suitable amount of molybdenum would improve the glass-forming ability (GFA) of the alloy. Second, the addition of the Mo atom makes the atoms' arrangement tend to become more chaotic during the deposition process, which improve the possibility of forming amorphization. (Please see the original paper for a more detailed explanation [48]). Figure 5.15a, b shows the surface appearance for $Fe_{71.1}Mo_{14.0}B_{14.9}$ and $Fe_{63.4}Mo_{13.3}B_{23.3}$. The surface of the as-deposited $Fe_{71.1}Mo_{14.0}B_{14.9}$ specimens is smooth and brightness. From Fig. 5.15b, on a lot of uniform homogeneous smaller particles, which consist of deposits, there are some bigger particles. The scanning electron probe microanalysis of the bigger particles and other areas shows that the bigger particles have somewhat higher molybdenum contents, which indicates that there is segregation of Mo in the particles. The measured results of microhardness of EP Fe–Mo–B films indicate similar behavior to Fe–W–B, that is the microhardness increases as the boron content increases. And it also can be understood with the average outer-electron concentration e/a, as seen in Fig. 5.14b. The XRD patterns and corresponding compositions for the 10 EP Fe–Mo–B samples are shown in Fig. 5.16. It can be seen that the samples FeMo12 and FeMo25 are crystal; others all have perfectly amorphous structures. As far as why the $Fe_{55.1}Mo_{11.1}B_{33.8}$ (sample FeMo25) forms a crystalline structure is concerned, the reason is that it has too higher boron atom content and lower Mo atom content, which results in a lot of number for Fe–B and Mo–B bonds

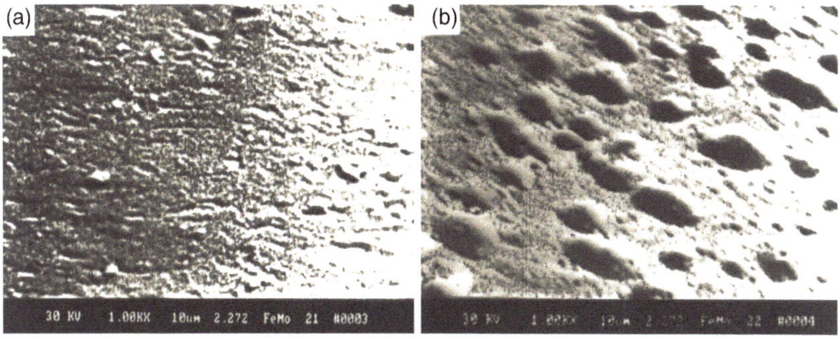

FIGURE 5.15 SEM images of EP Fe–Mo–B deposits. (a) $Fe_{71.1}Mo_{14.0}B_{14.9}$ and (b) $Fe_{63.4}Mo_{13.4}B_{23.2}$ [48].

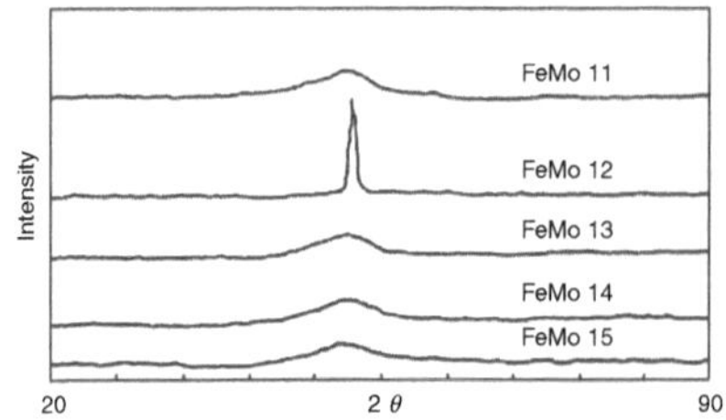

No.	Concentration of $NaMoO_4 \cdot 2H_2O$	Fe at%	Mo at%	B at%	Structure*
FeMo11	24.2 g/L	83.5	9.9	6.6	A
FeMo12	36.3 g/L	55.1	11.1	33.8	C
FeMo13	48.4 g/L	57.4	15.3	27.3	A
FeMo14	60.5 g/L	64.4	16.5	19.1	A
FeMo15	84.7 g/L	67.5	16.9	15.6	A

*A–amorphous, C–Crystal

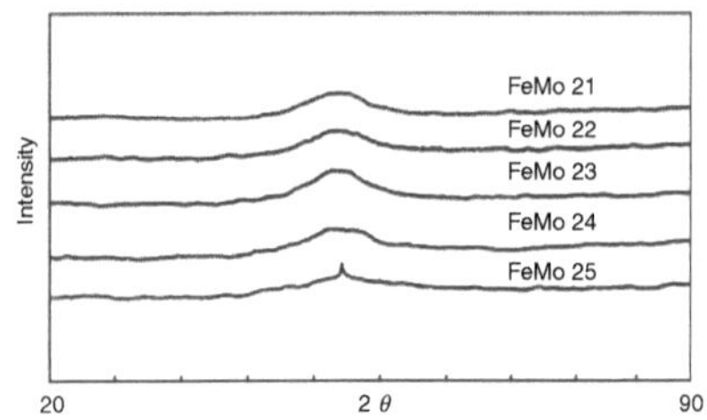

No.	Concentration of KBH_4	Fe at%	Mo at%	B at%	Structure*
FeMo21	2 g/L	71.1	14.0	14.9	A
FeMo22	6 g/L	63.4	13.3	23.3	A
FeMo23	10 g/L	58.7	11.3	30.0	A
FeMo24	15 g/L	73.6	16.3	10.1	A
FeMo25	20 g/L	74.7	18.9	6.4	C

*A–amorphous, C–Crystal

FIGURE 5.16 The XRD patterns of as-prepared deposits, the right tables are the corresponding composition and structures [48].

and a very few number for Fe–Mo bond, and impedes amorphization by very strong interaction of these bonds. The crystallization processes of the amorphous Fe–Mo–B alloys are as follows:

$$Fe_{71.1}Mo_{14.0}B_{14.9} \text{ amorphous} \rightarrow \text{amorphous} + \alpha\text{-Fe} \rightarrow \alpha\text{-Fe} + Mo_2B,$$

$$Fe_{63.4}Mo_{13.3}B_{23.3}(Fe_{58.7}Mo_{11.3}B_{30.0}, Fe_{73.6}Mo_{16.3}B_{10.1})\text{amorphous} \rightarrow \alpha\text{-Fe} + Mo_2B \rightarrow \alpha\text{-Fe} + Mo_2B + Fe_3B$$

The crystallization temperatures of the amorphous samples range from 748 K to 771 K for the B content from 10.1 at% to 30.0 at%, which is higher than 730–735 K in amorphous Fe–B films with the B content from 18 at% to 27.4 at%. The higher crystallization temperatures illustrate better thermal stability, which of course comes from the addition of metal Mo in the deposits.

5.4.3 EP Fe–Sn–B Alloy Deposits

Three papers of EP Fe–Sn–B alloy films have been published by our group [49–51]. The effects of various parameters, compositions, surface appearances, structures, formation range of amorphous phases, crystallization processes, microhardness, and corrosions are studied fully. The bath composition and operating condition for electroless Fe–Sn–B deposits are ferrous sulfate 20 g/L, sodium potassium tartrate 90 g/L, potassium borohydride 1 g/L, sodium, hydroxide, 20 g/L, sodium stannate, 7 g/L, temperature 40°C, and time 60 min. To obtain a deposit it was necessary for the copper sheet to be in contact with an aluminum sheet. Figure 5.17 shows the SEM images of EP Fe–Sn–B deposits fabricated by different Na_2SnO_3 concentrations: (a) 5 g/L; (b) 11 g/L; (c) 13 g/L; (d) 16 g/L;

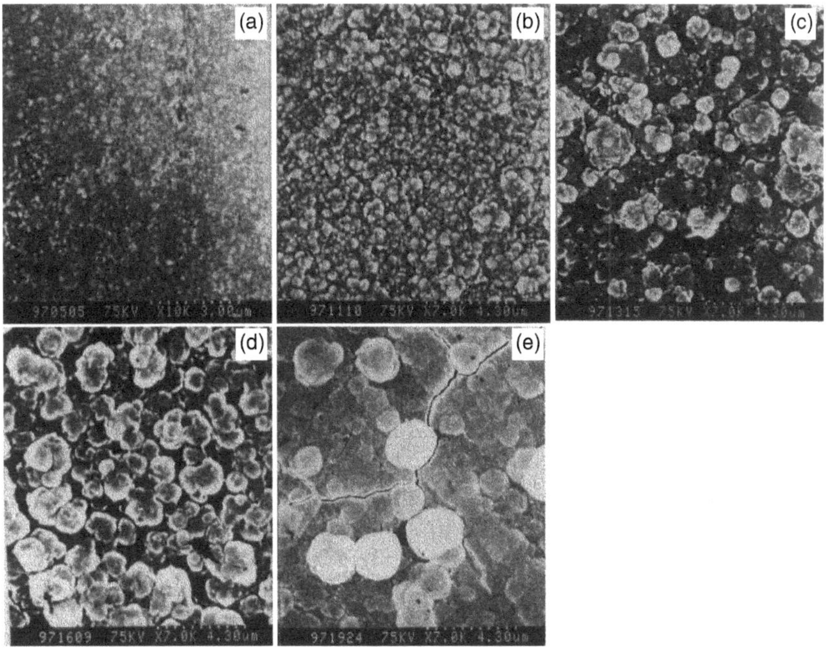

FIGURE 5.17 SEM images of Fe–Sn–B deposits fabricated by different Na_2SnO_3 concentrations. (a) 5 g/L; (b) 11 g/L; (c) 13 g/L; (d) 16 g/L; and (e) 20 g/L [49].

and (e) 20 g/L, respectively. It can be seen that there are some particles on the smooth Fe–Sn–B deposits. The particles have a nearly spherical shape. The size of the particles increases with the concentration of Na_2SnO_3 from 5 g/L to 20 g/L. The spot scanning electron probe microanalyzes of the smooth area and the particles show that the particles have higher tin content. The Sn–Fe atomic percentage ratio increases with increasing the concentration of Na_2SnO_3, from 3 g/L to 20 g/L. It has been found that the deposits, which have a high tin content, tend to crack at high Sn–Fe atom percentage ratios. This implies that the concentration of Na_2SnO_3 of the bath cannot be higher than 16 g/L. The formation range of amorphous phases in the Fe–Sn–B alloy deposits ranges from 12.1 at% to 27.0 at% of the boron content. The crystallization processes of different alloys vary with the composition. Initially, the crystallization phase is α-Fe, Fe_3Sn_2, and Fe_3B, and then it is $FeSn_2$ and Fe_3Sn for high temperatures. The crystallization temperatures of the amorphous samples range from 719 K to 729 K for a B content range of 12.1–27.0 at%, which is lower than that for amorphous Fe–Mo–B films but is the same as that for amorphous Fe–B films.

5.4.4 EP Fe–W–Mo–B Alloy Deposits

Two papers about EP Fe–W–Mo–B deposits have been published [52,53]. The standard parameters and their variation ranges in bath of Fe–W–Mo–B are as follows: $FeSO_4•7H_2O$ 20 g/L (3–35 g/L), $KNaC_4H_4O_6•4H_2O$ 90 g/L (60–120 g/L), $NaMoO_4•2H_2O$ 40 g/L (20–60 g/L), $NaWO_4•2H_2O$ 40 g/L (27–60 g/L), KBH_4 10 g/L (1–20 g/L), NaOH 20 g/L (0–24), plating time 30 min (10–60 min), and plating temperature 40°C (27–90°C). Effects of various process parameters such as ligand, metal salt, reductant, pH value, and temperature on the plating rate were studied. Composition, structure, and crystallization process of the deposits were all measured in detail. Table 5.2 shows the composition and structure of the seven Fe–W–Mo–B samples. From the table, it can be seen that the B content increases with increase of the concentration of KBH_4. The only exception is sample No. 4 because its composition is little higher than that of sample 5. The formation range of the amorphous structure in the EP Fe–Mo–W–B deposits is from 9.8 at% to 27.3 at% of boron content. As mentioned above, adding Mo into Fe–B deposits extends the boron content region for amorphization significantly. When incorporating W into Fe–Mo–B ternary films, the formation range of the amorphous structure in the EP Fe–Mo–W–B deposits (9.8 to 27.3 at% B) is not larger but is slightly less than that for the Fe–Mo–B deposits (6.6 to 30.0 at% B). This is because Mo facilitates the formation of amorphous structures, whereas W makes the formation of amorphous structures for Fe-based deposits difficult. The reason why the amorphous structures did not appear in the Fe–W–B alloy films was confirmed from another angle. The crystallization processes of all the samples were measured, but that for only one sample (sample 5 $Fe_{58.9}Mo_{18.2}W_{1.3}B_{21.6}$) is illustrated below.

$$\text{Amorphous deposit} \xrightarrow{300^\circ C} \alpha\text{-Fe} + Fe_3B + (Mo,W)_2B$$

$$\xrightarrow{400^\circ C} \alpha\text{-Fe} + Fe_3B + (Mo,W)_2B + Fe_2B$$

$$\xrightarrow{500,600^\circ C} \alpha\text{-Fe} + Fe_3B + (Mo,W)_2B + Fe_2B + (Mo,W)B + Fe_2(Mo,W)$$

TABLE 5.2 The Composition and Structure of the Seven EP Fe–W–Mo–B Samples

No.	KBH_4 (g/L)	Fe (at%)	Mo (at%)	W (at%)	B (at%)	Structure
1	2	73.9	16.7	0.6	8.8	C
2	6	71.9	17.7	0.6	9.8	A
3	8	73.3	13.5	0.7	12.5	A
4	10	63.7	12.1	1.3	22.9	A
5	12	58.9	18.2	1.3	21.6	A
6	15	62.7	10.0	Very low	27.3	A
7	20	58.2	13.5	Very low	28.3	C

A, amorphous; C, crystal.

It should be pointed out that, just as was mentioned in Section 5.2 the Fe–B films easily become rusty because some rusty spots appeared on the films after several weeks in a humid atmosphere. The way to overcome such a drawback in our thinking is to add one or two metallic elements to it. The experimental results from the above ternary and quaternary Fe–B-based deposits indicate that this is true, and the rusty phenomenon has been improved.

5.4.5 EP Fe–Ni–B Alloy Deposits

Very few reports about EP Fe-based Fe–Ni–B and/or Fe–Ni–P–B deposits have been published. In 2006, Wang [59] reported the EP and magnetic properties of an Fe–Ni–B alloy. The bath composition is: $FeSO_4{\bullet}7H_2O$ 0.08–0.04 mol/L, $NiSO_4{\bullet}6H_2O$ 0.02–0.06 mol/L. (The total concentration of $FeSO_4$ and $NiSO_4$ was kept as 0.1 mol/L, so the mole ratio of $FeSO_4/NiSO_4$ is the ratio of the concentration of $FeSO_4$ vs. $NiSO_4$ in the plating bath), $Na_3C_6H_5O_7$ 0.2 mol/L, $KNaC_4H_4O_6{\bullet}4H_2O$ 0.2 mol/L, KBH_4 2.0~5.0 g/L, NaOH 0.125 mol/L, $C_2H_8N_2$ 5.0 ml/L. The temperature was kept at 40–55°C. The plating time was 2.0 h. The load capacity of the EP was 40 cm^2/L. The influence of mole ratio of $FeSO_4/NiSO_4$ on the deposit composition is illustrated in Fig. 5.18. Besides the boron content in the deposit, the content of iron decreases linearly from 60.51 to 40.39 and that of nickel increases linearly from 39.49 to 59.61 as the mole ratio of $FeSO_4/NiSO_4$ decreases from 0.08/0.02 to 0.041/0.06. It is notable that high-iron deposits with iron content of 60.51 atom percentage can be prepared in this solution. When the mole ratio of $FeSO_4/NiSO_4$ around ≥ 2, the content of iron is larger than 50 at%, that is the deposits are Fe-based alloys. The paper found that with the decrease of the concentration of $FeSO_4$, the saturation magnetization (M_s) decreases from 1166.6 kA/m to 384.3 kA/m. These facts can be easily understood, because the atomic magnetic moment of iron is more than that of nickel, and the saturation magnetization of iron and nickel is 1707 and 485 kA/m, respectively; therefore, as the decrease of the iron content and increase of the nickel content in the deposit, the less the content of Fe, the less the saturation magnetization of the deposit is.

5.5 EP Fe–P ALLOYS

Early in 1971, Ruscior and Croialã [7] reported chemically made iron–phosphorus films. In the electroless bath, they used an Al strip to contact with the substrate Cu, forming a Cu–A1 galvanic couple for the first time. The bath composition is: ferrous ammonium sulfate 30 g/L (10–40 g/L), sodium citrate 45 g/L (25–45 g/L), sodium hypophosphite 30 g/L (25–40 g/L). The bath pH is 9.5–11.4, plating time 10–60 min, and bath temperature is 25–95°C. In the paper, the authors only studied the effects of bath temperature, plating time, and bath pH on the film thickness, composition, and coercivity, and magnetization of the deposits. It has been found since then that the film thickness increases linearly as increasing pH value and plating time. The deposition takes place in alkaline solutions only, and when pH = 9.5, the plating rate is small but it increases with increasing bath alkalinity. However, there are two distinct temperature intervals with respect to plating thickness change, as seen in Fig. 5.18a. The first one, including temperatures from 25°C to ~60°C, is marked by a relatively small change of the deposition rate (0.02 μm/h/deg). In this interval, Fe films free of phosphorus are deposited. The second interval, comprising temperatures from ~60°C to 95°C, is characterized by a greater plating rate change (0.07 μm/h/deg). In this temperature range, the Fe–P films are obtained. Figure 5.19b indicates further that at temperatures less than 60°C, there is only the electrochemical process and no chemical process. However at temperatures in the range of ~60–95°C the chemical process comes into play, and the superimposition of the electrochemical process with the purely chemical process results in the reduction of the Fe cations by $PO_2H_2^-$ anions, accompanied by phosphorus codeposition. So, the Fe–P films are obtained. From these facts the authors believed that the continuous electrochemical reduction of iron is necessary only to provide catalytic iron. That is why removing the aluminum foil results in immediate dissolution and/or oxidation of the iron which eliminates the necessary catalyst for the autocatalytic reduction of iron ions by hypophosphite.

The films obtained in the thickness range of 0.03–3 μm are uniform and adherent to the copper substrate. They are gray, but their aspect changes with increasing film thickness. The films below ~0.4 μm are smooth, bright, and do not oxidize in open air (over a period of many months, they did not lose their metallic luster); the films in the thickness range of ~0.4 to ~1 μm are less bright, whereas those over ~1 μm are dull and become oxidized in a few weeks. This situation is similar to that of the EP Fe–B deposits above. The films deposited at 80°C and pH = 10.5 contain 3% P, whereas those obtained at the same temperature but at a pH = 11.4 contain only 1% P. The Fe–P films deposited are ferromagnetic.

In 2006, Huang et al. [60] investigated the inducing effect of coupled aluminum on Fe–P deposition. The bath compositions and plating conditions are $FeSO_4{\bullet}7H_2O$ 9 g/L, $NaH_2PO_2{\bullet}H_2O$ 15 g/L, $KNaC_4H_4O_6{\bullet}4H_2O$ 36 g/L, H_3BO_3

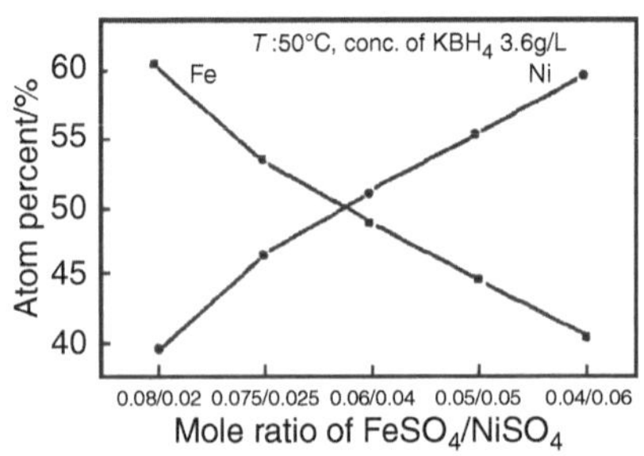

FIGURE 5.18 **Effect of mole ratio of $FeSO_4/NiSO_4$ on the deposit composition [59].**

3 g/L $C_{12}H_{12}O_{11}$ 1.2 g/L, pH 8–11.5, and T 50–85°C. During deposition, the substrate was connected with aluminum foil through a copper wire. The authors measured *in situ* mixed potentials of the process for electrolytes with different pH. The plating rate of iron alloys was measured gravimetrically and calculated from the deposition current density. The results show that coupled aluminum induces the Fe–P deposition by shifting the potential negatively and decreasing the polarization resistance of anodic and cathodic reaction. Therefore, the authors confirmed that coupled aluminum is necessary for Fe–P electroless deposition just like that Ruscior and Croialã did for EP Fe–P films [7] and just like Hu and Zhang did for EP Fe–B deposits [15]. In 2008, Huang et al. [61] investigated and discussed the deposition mechanism of electroless Fe–P films. They measured and analyzed the deposition rates in a bath containing hypophosphite and in the bath without hypophosphite, obtaining the same results (Fig. 1 in the paper [61] is similar to Fig. 5.19(b)) and conclusions. During the process the connected Al keeps the substrate surface active, and provides the necessary catalyst for the chemical reduction of iron ions by hypophosphite. The corrected aluminum cannot be removed during the whole plating process. In addition, they also investigated the effects of bath component and process parameters on the electroless Fe–P deposition. The results show that the bath component and process parameters obviously effect the formation and the composition of Fe–P film. The effect of hypophosphite on the deposition rate is more pronounced than that of $FeSO_4$ due to the higher polarization resistance of anodic reaction, indicating that the rate-determining step for the chemical iron reduction is the oxidation of hypophosphite.

5.6 EP Fe–P-BASED TERNARY-COMPONENT ALLOYS

Only very few reports have been published on EP Fe–P-based ternary alloy deposits, and none on EP Fe–P-based quaternary alloy films. In 2005, Chen and Zhong [62] reported the fabrication of Fe–W–P by EP. The plating bath was

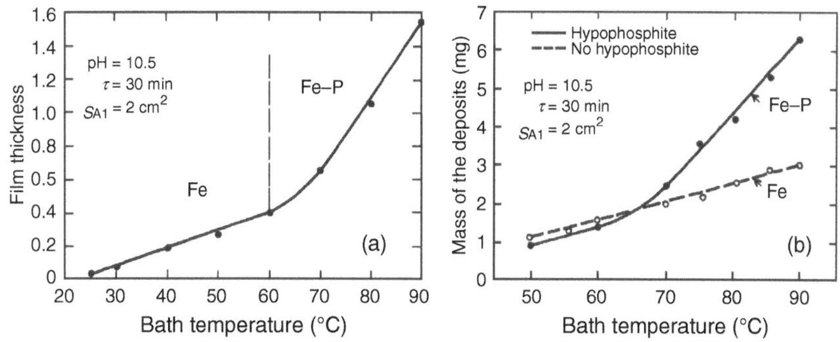

FIGURE 5.19 (a) Fe–P film thickness versus bath temperature. (b) Moss of Fe–P end Fe deposits vs. bath temperature [7].

composed of $Fe(NH_4)_2(SO_4)_2{\bullet}6H_2O$ 10–100 g/L, $Na_2WO_4{\bullet}2H_2O$ 10–100 g/L, $Na_3C_6H_5O_7{\bullet}3H_2O$ 38 g/L, $NaH_2PO_2{\bullet}H_2O$ 10–100 g/L, and a small amount of additive. The pH = 10.5, $T = 80°C$, and plating time = 0.5 h. During deposition, the copper substrate remained in contact with an aluminum strip. In the paper, the effects of concentration of $Fe(NH_4)_2(SO_4)_2$, Na_2WO_4, and NaH_2PO_2 on the depositing rate have been measured. In addition, the effects of concentration of $Fe(NH_4)_2(SO_4)_2$ and Na_2WO_4 on the morphology were also investigated, respectively. Figure 5.20 shows SEM images versus $Fe(NH_4)_2(SO_4)_2{\bullet}6H_2O$ of the EP Fe–W–P deposits. It can be seen that when the concentration of ferrous ammonium sulfate is low, the deposits are rather uniform, and the size of microparticles is about micrometers. As the concentration increases to 20 g/L, the size of the particles increases, and surface uniformity deteriorates. When the concentration is 30 g/L, the surface of deposits is composed of submicrometer particles. That is to say, the suitable concentration can produce surface grain refinement, obtaining more uniform and dense coating.

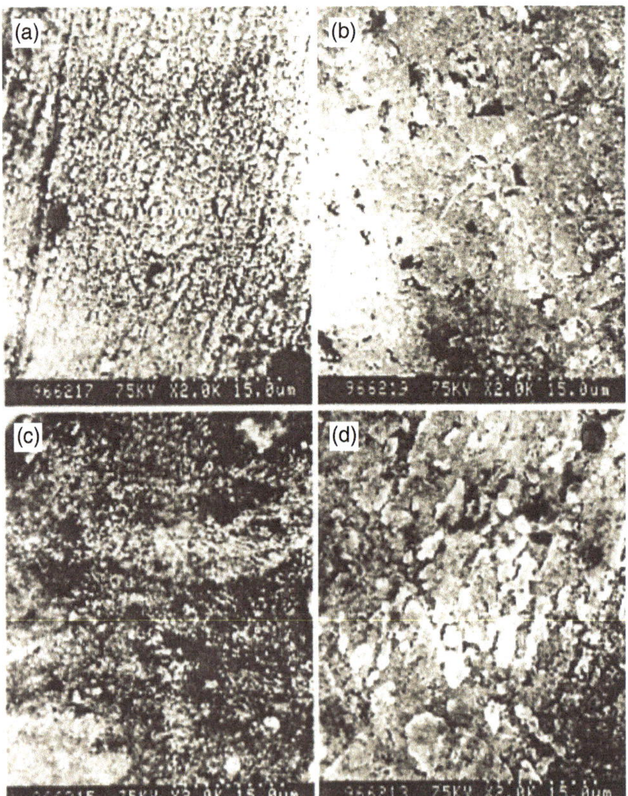

FIGURE 5.20 SEM morphology versus $Fe(NH_4)_2(SO_4)_2{\bullet}6H_2O$ ((a) 10 g/L, (b) 20 g/L, (c) 30 g/L, (d) 40 g/L) for EP Fe–W–P films [62].

5.7 SUMMARY

After describing, analyzing, and discussing the EP Fe–B-based and Fe–P-based deposits more in detail, the following remarks can be made.

Nearly 40 years after the discovery of chemical nickel plating by Brenner and Riddel, the EP Fe–B-based alloy deposits appeared. The reasons and the process for studying the EP Fe–B-based alloys have been described in this chapter.

Collectively, many papers on EP Fe–B-based alloys have been published, including several ternary and quaternary Fe–B-based alloys. Not only the manufacturing technique (bath composition and operating conditions), but also the fabricating principle, structure, composition, formation mechanism, and important properties of the Fe–B-based alloy films have all been studied systematically. However, papers investigating EP Fe–P deposits are rather fewer to date, for example, only Fe–W–P in the ternary but no quaternary Fe–P-based alloys have been reported. However, it can be said that currently we already have a considerable understanding of EP Fe-based alloy films.

Knowledge of the nature and characteristics of both EP Fe–B and Fe–P-based alloys are needed to carry out in-depth and extensive research. This is especially so for research on EP nano Fe-based alloy films, which we have not yet begun to study. Such a field will undoubtedly yield relevant ideas and will be well worth studying.

Unfortunately, there are few applications of EP Fe-based alloy films, particularly in industry. Of course, looking for such applications is not only necessary but also very meaningful. Studies in such field are definitely welcomed.

REFERENCES

[1] Luborsky FE, Liebermann HH, Becker JJ, Walter JL. In: Cantor B, editor. Proc ser int conf on rapidly quenched metals 3, vol. 2. London: Metals Society; 1978. p. 188.

[2] Hasegawa R, Ray R. J Appl Phys 1978;49:4174.

[3] Hasagawa R, Ray R. Phys Rev B 1979;20:211.

[4] Fukunaga T, Misawa M, Fukamichi K, Masumoto T, Suzuki K. In: Cantor B, editor. Proc. ser. int. conf. on rapidly quenched metals 3, vol. 2. London: Metals Society; 1978. p. 325.

[5] Chien CL, Musser D, Gyorgy EM, Sherwood RC, Chen HS, Luborsky FE, Walter JL. Phys Rev B 1979;20(1):283.

[6] Fujita N, Tanaka A, Makino E, Squire PT, Lim PB, Inoue M, Fujii T. Appl Surf Sci, 1997;14(13/l):61.

[7] Ruscior C, Croialã E. J Electrochem Soc 1971;18(5):696.

[8] Brenner A, Riddell GE. J Res Nat Bur Stand 1946;37:31–4. Proc Amer Electropla Soc 1946;33:23–9.

[9] Schmeckenbecher AF. J Electrochem Soc 1966;113(8):778.

[10] Schmeckenbecher AF. Plating 1971;58(12):905.

[11] DePew JR, Speliotis DE. Plating 1967;54(6):705.

[12] Yazima H, Numata H, Ohno I. Jpn J Surf Finish 1997;48(1):62.

[13] Wang L, Zhao L, Huang G, Yuan X, Zhang B, Zhang J. Surf Coat Technol 2000;126:272.

[14] Dadvand N, Jarjoura G, Kipouros GJ. J Mater Sci Mater Electron 2008;19:51.

[15] Hu W, Zhang B. Physica B 1991;175:396.
[16] Zhang B, Hu W. Rare Met 1991;10(4):301.
[17] Zhang B, Hu W. Physica B 1993;183:396.
[18] Hu W, Zhang B. Trans IMF 1993;71(1):30.
[19] Hu W, Zhang B. Electropla Protec 1993;13(1):13.
[20] Şahin Ö, Dolaş H, Özdemir M. Intern J Hydro Ener 2007;32:2330.
[21] Minkina VG, Shabunya SI, Kalinin VI, Martynenko VV, Smirnova AL. Intern J Hydro Ener 2008;33:5629.
[22] Chang YH, Lin CC, Hung MP, Chin TS. J Electrochem Soc 1986;133(5):985.
[23] Hu W, Zhang B, Zhang H. Plat Finish 1988;10:10.
[24] Weast RC, editor. Handbook of chemistry and physics. Cleveland, OH: CRC Press; 1977.
[25] Van den Mearakker JEAM. J Appl Electrochem 1981;11:395.
[26] Nakajima T, Kita E, Ino H. J Mater Sci 1988;23:1279.
[27] Ray R, Hasegawa R, Chou CP. Scr Metall 1977;11:973.
[28] Koster U, Herold U. Glassy metals 1. In: Guntherodt HJ, Beck H, editors. Berlin, Germany: Springer; 1981. p. 245.
[29] Luborsky FE. Mater Sci Eng 1977;28:139.
[30] Luborsky FE, Liebermann HH. Appl Phys Lett 1978;33:1717.
[31] Buschow KHJ, van Engen PG. J Appl Phys 1981;52:3557.
[32] Bayreuther G, Enders G, Hoffman H, Korndorfer U, Oestreicher W, Roll K, Takahashi M. J Magn Magn Mater 1983;31:1535.
[33] Webb DJ, Bhagat SM, Moorjani K, Poehler TO, Satkiewick FG. IEEE Trans Magn 1983;19:1892.
[34] Maksymowicz AZ, Stobiecki T, Jacocki E, Karas W. Phys Stat Sol B 1984;126:191.
[35] Cowlam N, Carr GE. J Phys F 1985;15:1109.
[36] Chien CL, Musser D, Gyorgy EM, Sherwood RC, Chen HS. Phys Rev B 1979;20:283.
[37] Chien CL, Unruh KM. Phys Rev B 1981;24. 1556;25(1982):5790.
[38] Ze X, Ishikawa Y, Fukunaga T, Watanabe N. J Phys F 1985;15:1799.
[39] Nold E, Lamparter P, Olbrich H, Rainer-Harbach G, Steeb S. Z Naturforsch A 1981;36:1032.
[40] Stobiecki F, Dubowik J, Baszynski J. In: Steeb S et al., editors. Proc ser int conf on rapidly quenched metals 5. Amsterdam: Elsevier; 1985. p. 1211.
[41] Fujita N, Tanaka A, Squire PT, Lim PB, Inoue M, Fujii T. Appl Surf Sci 1997;113/114:61.
[42] Fujita N, Inoue M, Arai KI, Lim PB, Fujii T. J Appl Phys 1998;83:7294.
[43] Dinderman MA, Dressick WJ, Kostelansky CN, Price RR, Qadri SB, Schoen PE. Chem Mater 2006;18:4361.
[44] Qadri SB, Dinderman MA, Dressick WJ, Schoen PE, Lubitz P, He JH, Tonucci RJ, Cross J. Appl Phys A 2007;89:493.
[45] Huang G-F, Huang W-Q, Wang L-L, Zhu Y-C, Zhang J-H, Wang Q-L. Int J Electrochem Sci 2006;1:354.
[46] Yi G, Wang L, Ouyang Y, Mao L, Guo Y, Zhang B, Xie H. Met Finish 1996;94(11):31.
[47] Wang L, Zhang B, Yi K, Ouyang Y, Hu Wangyu J. J Alloys Comp 1997;255(12):231.
[48] Yi G, Zhang B, Wang L, Ouyang Y, Xie H. J Electrochem Soc 2013;160(9):D403.
[49] Wang L, Zhao L, Hu W, Zhang B, Shu X, Sun X. Met Finish 2001;99(6):92.
[50] Wang L, Zhao L, Zhang B, Hu W, Shu X, Sheng X, Fang Z. J Alloys Comp 1999;287:234.
[51] Wang L, Zhao L, Hu W, Shu X, Zhang B. Adv Mat Eng Tech 2 (SPM 1) 2001;635.
[52] Wang L, Zhao L, Ouyang Y, Liao S, Zhang B. Rare Metals 2000;19(4):261.
[53] Wang L, Zhao L, Zhang B, Liao S, Ouyang Y, Hu W. Z Metallkd 1997;88(12):945.
[54] Wang L, Zhao L, Zhang B, Liao S, Hu W, Ouyang Y. P&SF 1998;85(12):96.

[55] Wang L, Zhao L, Zhang B, Hu W, Shu X, Sheng X. Z Metallkd 1999;90(5):338.
[56] Wang L, Zhao L, Zhang B, Liao S, Ouyang Y, Hu W, Shu X, Yuan X. P&SF 1999;86(12):84.
[57] Naka M, Tomizawa S, Watanabe T, Masumoto T. In: Grant NJ, Giessen BC, editors. RQM II, section 1. Boston, MA: Massachusetts Institute of Technology; 1976. p. 273.
[58] Fukamich K, Kikuchi M, Arakawa S, Masumoto T. Solid State Commun 1977;23:955.
[59] Wang S. Funct Metal Mater 2006;37(7):1153.
[60] Huang G-F, Huang W-Q, Wang L-L, Meng Y, Xie Z, Zou B-S. Electrochim Acta 2006;51:4471.
[61] Huang G-F, Huang W-Q, Wang L-L, Zou B-S, Wang Q-L, Zhang J-H. Int J Electrochem Sci 2008;3:145.
[62] Chen H, Zhong X. J Guangxi University (Nat Sci Ed) 2005;30(4):306.

Chapter 6

Impact Parameters and Deposition Rate

Chapter Outline

The electroless plating (EP) process is very complex. Each parameter of the bath composition, and all the operating conditions such as pH value, plating temperature, plating time, and bath loading all affect the deposition rate, and hence the composition and structure of the deposits. Studying the impact of parameters can give a better understanding of the plating technological

Amorphous and Nano Alloys Electroless Depositions. http://dx.doi.org/10.1016/B978-0-12-802685-4.00006-6

process and so give better control of the EP and in turn produce good-quality deposits. In other words, the influence of several parameters may be investigated, so that not only the nature and characteristics of EP can be comprehended correctly and thoroughly, but also good-quality EP products can be fabricated. In fact, the influence factors on EP have been studied continuously from the initial discovery of electroless Ni–P alloy films to today. Early in 1957, de Minjer and Brenner measured the deposition rate of EN plating [1], and examined the effect of the addition of thiourea on this rate. Today, studies of the impact of different parameters in EP still appear in many papers. Of course, such research is not comprehensive, and usually just presenting factors relevant to their own research work.

It should be pointed out that experimental measurements for a given parameter in EP are not always consistent between studies, but differences, even conflict in some cases, appear because of the complexity of the EP process. Therefore, there is a problem of selecting and screening data, that is, we do not use all the data, just the most suitable and typical. In addition, we use the most recently published data as much as possible, even when the data are nearly the same as older studies. Of course, the reader will correctly identify this selection and screening.

In Chapters 1 and 2 of the book "*Electroless Plating: Fundamentals and Applications*" by Mallory and Hajdu [2] and in Chapter 4 of the book "*Electroless Nickel Plating*" by Riedel [3], the impact of a range of parameters on EP is discussed in considerable detail. However, the material in the books is all taken from references published before 1990. Reviews devoted to discussing the influence of parameters in EP are seldom found. The author's group published such a review in 1999 [4] based on our research into the EP process, combined with the work of other authors. Of course, the material published more recently cannot be included in the paper.

We will describe and discuss the impact of a range of parameters in EP as detailed and comprehensive a way as possible in this chapter. We will pay particular attention to the explanation of experimental data.

6.1 EFFECTS OF PLATING BATH COMPONENTS ON DEPOSITION RATE

6.1.1 Effect of Metal Salts

Metal salts contribute the metal element to the bath. Without a metal salt there will be no electroless metal or alloys deposition. Hence, the metal salts are the most important constituent in the bath solution. Naturally, the amount of metal salts added must affect the plating rate and the characteristics of the deposit formed. Usually, only one kind of metal salt is added for the EP of pure metal and binary alloy systems. Ternary EP alloy systems need two kinds of metal salts. That is why researchers pay special attention to the study of the effects of metal salts in electroless plating – even after over 70 years since the discovery of EN; papers discussing the effect of metal salts can still be found in the literature today.

In order to illustrate and discuss this effect more closely, we will describe the effects of metal salts on deposition rate for pure metals, binary alloys, ternary alloys, and quaternary alloys separately.

6.1.1.1 Effect of Metal Salts on the Deposition Rate in the EP of Pure Metals

Figures 6.1–6.3 show the effects of metal salts for the EP of the pure metals Pb [5], Sn [6], Sb [7], Cu [8], Ag [9], Au [10], Ni [11], Co [12], and Pd [13] on deposition rates, respectively. It can be seen that the shapes of the influence

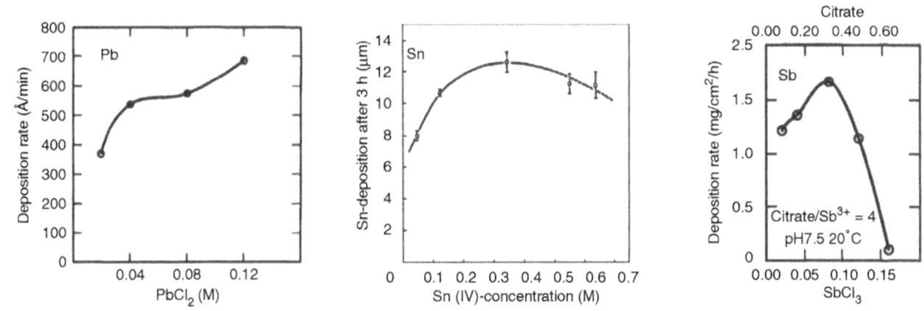

FIGURE 6.1 Deposition rates of EP pure metals Pb [5], Sn [6], and Sb [7] versus metal salts.

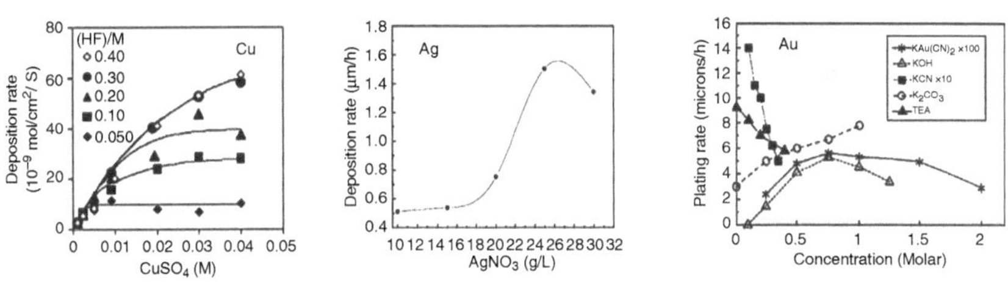

FIGURE 6.2 Deposition rates of EP pure metals Cu [8], Ag [9], and Au [10] versus metal salts.

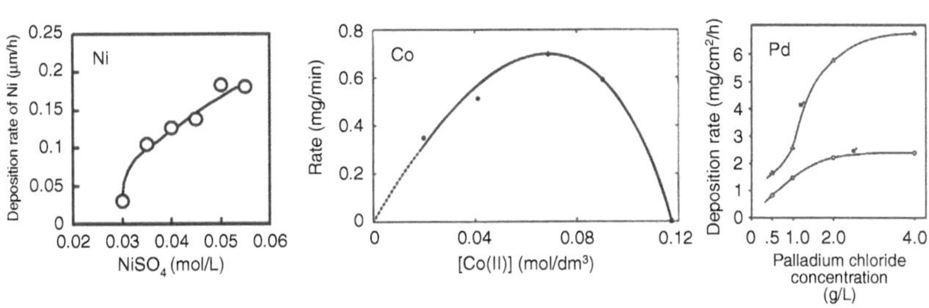

FIGURE 6.3 Deposition rates of EP pure metals Ni [11], Co [12], and Pd [13] versus metal salts.

curves are not the same. In order to see more clearly, we will first definite the shapes. The curve with maximum value is termed a mountain shape. A mountain shape has two sides, the side before maximum is called the rising side, and the side after maximum is called the falling side. Although the curves in Figs 6.1–6.3 are different, they fall into just two classes, one is the mountain shape, and the other is the shape of the rising side. The curve shapes of Pb, Cu, Ni, and Pd belong to the rising side; others are the mountain shape. In the following discussion, we will see that the mountain shape and the shape of the rising side can be classified as the same type. The mountain shape and the shape of rising side are, therefore, known as class I shapes, and the shape of the falling side will be termed a class II shape for simplicity hereafter.

The deposition rates of EP of Pb, Cu, Ni, and Pd all increase with increasing metal salt concentrations. But above a given concentration of metal salts, no deposition rate data can be obtained, that is, the plating bath cannot work normally because the bath solution is not stable or other reason. Of course, some details may not be the same but different metals may have different characteristics, for example, in the EP plating of Cu, when [HF] ≥ 0.30 M, Cu deposition rates are independent of [HF]. In the cases of other five EP metal platings (Sn, Sb, Ag, Au, and Co), their deposition curves all increase to a maximum value, and then decrease with increasing metal salt concentration.

6.1.1.2 *Effect of Metal Salt on Deposition Rate in EP Binary Alloys*

When considering EP binary alloy deposits, six alloy systems Ni–P [14–16], Co–P [17,18], Fe–P [19,20], Ni–B [21,22], Co–B [23,24], and Fe–B [25] will be discussed. The effects of metal salts on the deposition rate in these alloy films are shown in Figs 6.4 and 6.5. It can be seen that the curve of the plating rate versus metal salt follows a rising side shape only in the Fe–B film. The curves of the plating rate versus metal salt of the other five alloy deposits are all mountain shaped with a definite maximum value. It should be pointed out that Han et al. [15] in 1996 and Małecki and Micek-Ilnicka [16] in 2000 obtained the mountain shape with a definite maximum value for the plating rate versus metal salt in their studies of EP Ni–P. Shi et al. [18] in 2005, Akimoto et al. [20] in 1974, Shrivastava et al. [22] in 1985, and our group in 2007 [24] also obtained

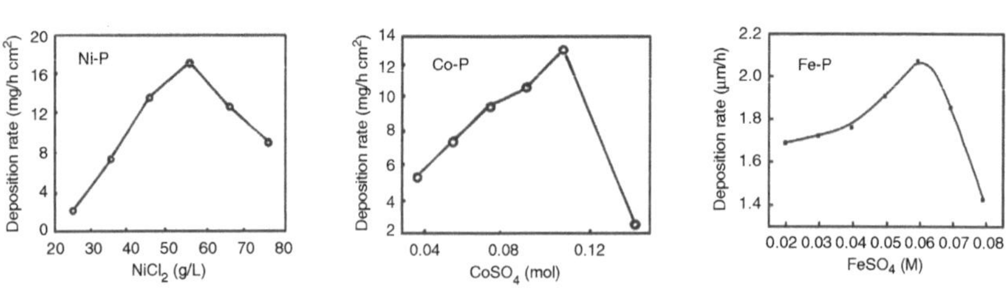

FIGURE 6.4 Deposition rates of EP binary Ni–P [14], Co–P [17], and Fe–P [19] alloys versus metal salts.

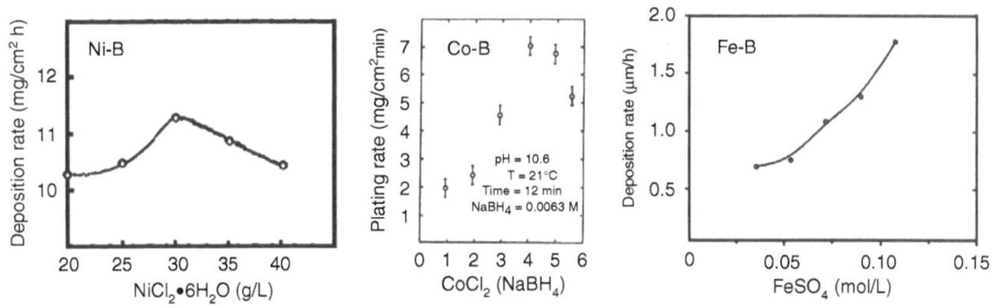

FIGURE 6.5 Deposition rates of EP binary Ni–B [21], Co–B [23], and Fe–B [25] alloys versus metal salts.

the mountain shape with a maximum value for the plating rate versus metal salt concentration in the studies of EP Co–P, Fe–P, Ni–B, and Co–B, respectively.

6.1.1.3 Effect of Metal Salt on Deposition Rate in the EP Ternary Alloys

There are a lot of EP ternary alloy deposits. This chapter only covers some typical EP ternary alloy deposits, namely Ni–W–P [26–30], Co–W–P [31–34], Fe–W–P [35], Ni–Mo–P [36,37], Ni–Sn–P [38], Co–Mo–P [39], Co–Fe–P [40,41], Ni–Fe–B [42,43], Co–Fe–B [44,45], Ni–Fe–P [46], Fe–W–B [47], Fe–Mo–B [48], Fe–Sn–B [49], Ni–Mo–B [50], Ni–Cu–B [51], and Ni–Sn–B [52]. Figures 6.6–6.11 show the effects of metal salts on plating

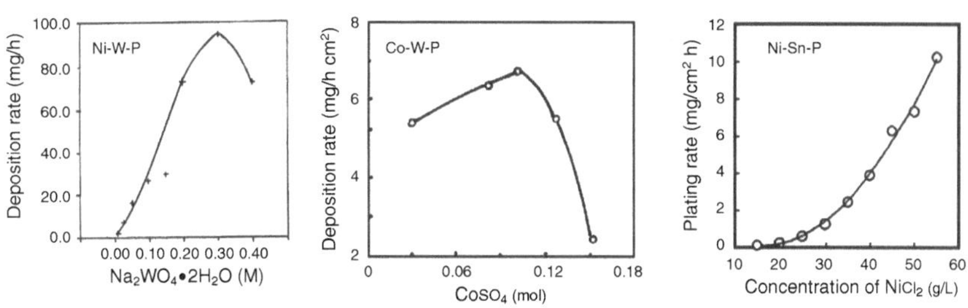

FIGURE 6.6 Deposition rates of EP ternary Ni–W–P [26], Co–W–P [31], and Ni–Sn–P [38] alloys versus metal salts.

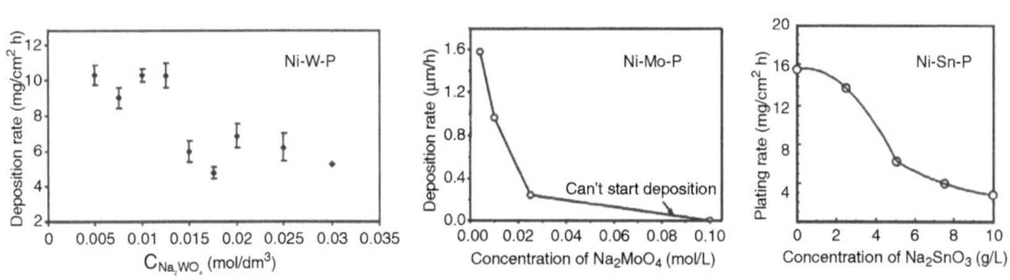

FIGURE 6.7 Deposition rates of EP ternary Ni–W–P [29], Ni–Mo–P [28], and Ni–Sn–P [38] alloys versus metal salts.

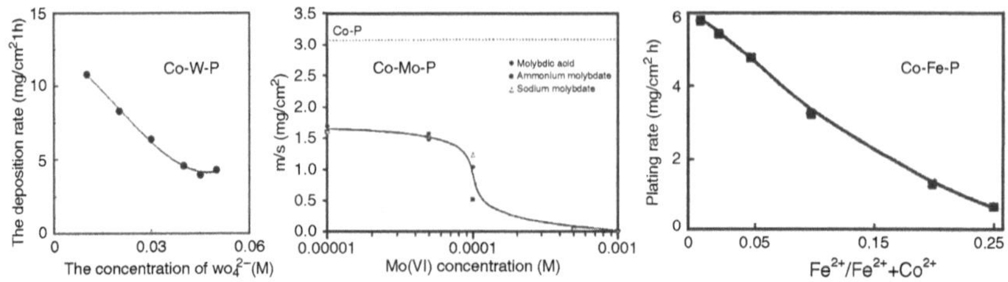

FIGURE 6.8 **Deposition rates of EP ternary Co–W–P [31], Co–Mo–P [39], and Co–Fe–P [41] alloys versus metal salts.**

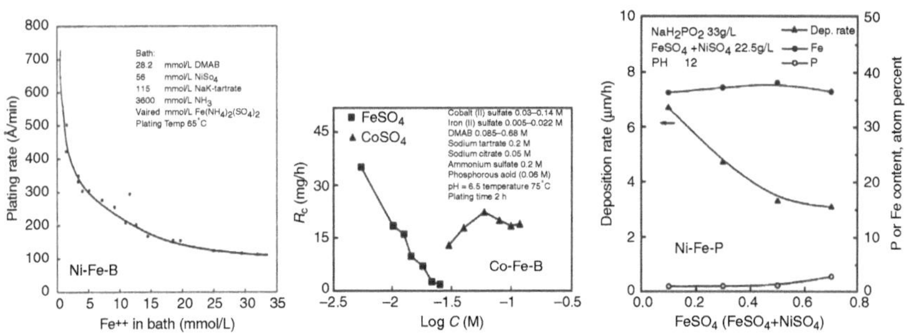

FIGURE 6.9 **Deposition rates of EP ternary Ni–Fe–B [42], Co–Fe–B [45], and Ni–Fe–P [46] alloys versus metal salts. R_c is the deposition rate and C is the concentration in Co–Fe–B.**

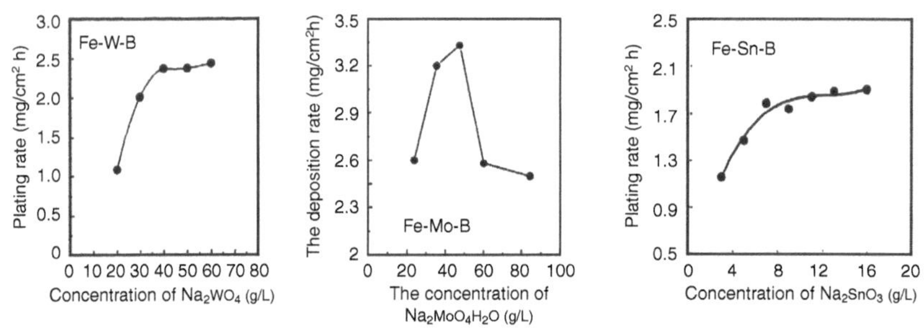

FIGURE 6.10 **Deposition rates of EP ternary Fe-W-B [47], Fe-Mo-B [48] and Fe-Sn-B [49] alloys versus metal salts.**

rates in these 18 ternary alloy systems. It can be seen from these figures that nine alloy systems (Ni–W–P, Co–W–P, Ni–Sn–P, Fe–W–B, Fe–Mo–B, Fe–Sn–B, Ni–Mo–B, Ni–Cu–B, and Ni–Sn–B) have the class I shape for the plating rates versus metal salts, and the other nine alloy systems (Ni–W–P, Ni–Mo–P, Ni–Sn–P, Co–W–P, Co–Mo–P, Co–Fe–P, Ni–Fe–B, Co–Fe–B, and Ni–Fe–P) have the class II shape. It also can be seen that the shapes of the curve of plating rate versus metal salt can be either class I or class II shapes when using

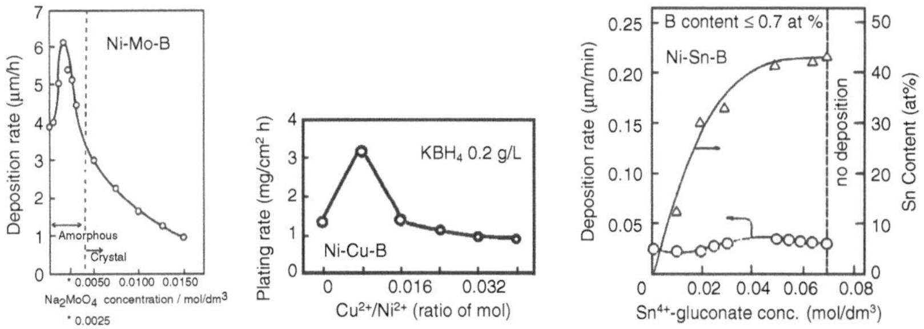

FIGURE 6.11 Deposition rates of EP ternary Ni–Mo–B [50], Ni–Cu–B [51], and Ni–Sn–B [52] alloys versus metal salts.

different metal salts in the same alloy system, for example, this is the case in the Co–W–P system. The class I and class II shapes are obtained when the metal salt is $CoSO_4$ [31] and Na_2WO_4 [31], respectively. Even more interesting is that different authors obtain conflicting results for the same system using the same metal salt. For example, this is evident in the EP Ni–W–P deposits. The author's group [26] and two other groups [27, 28] obtain the class I shape, but Szczygieł and Turkiewicz [29], and Younan and Shoeib [30] obtain the class II shape even though they all use the same metal salt (Na_2WO_4) in the EP Ni–W–P deposition (the bath compositions are all nearly the same).

6.1.1.4 Effect of Metal Salts on Deposition Rate in the EP Multicomponent Alloys

Many EP quaternary alloy depositions have been reported, as is indicated in Chapters 2 and 3. However, there are not that many reports of EP quaternary alloy depositions that cover systematic research into the plating process or that give effective parameters about the plating rate. The typical effects of metal salts on plating rate in EP quaternary alloy deposits such as Ni–Cu–Sn–P [53, 54], Fe–W–Mo–B [55, 56], Co–Fe–Mn–P [57], and Ni–Fe–Ru–P [58] will be described here. Until now, few quinary EP alloy deposits have been reported (see Table 2.29 in Chapter 2), but the study of effect of metal salts on plating rate can only be found in the EP quinary alloy Co–Ni–Mn–Re–P [59].

It can be seen from that the curves in Fig. 6.12 are the class I shape, and those in Fig. 6.13 are the class II shape.

Besides these six EP alloy systems, the effect of metal salts on plating rates has been also reported for other EP quaternary alloy systems. The effect of Na_2WO_4 concentration on plating rate in Co–Ni–W–P alloy films follows a class II shape (Fig. 1 in the original paper) [60]. The influence of Ni ion content in the bath on the deposition rate for the Co–Ni–Fe–B deposited alloys belongs to the class I shape (Figs 2 and 3 in the original paper) [61, 62]. The effect of ferrous sulfate concentration on the plating rate of electrolessly plated Ni–Co–Fe–P alloy films is the class II shape (Fig. 1 in the original paper) [63].

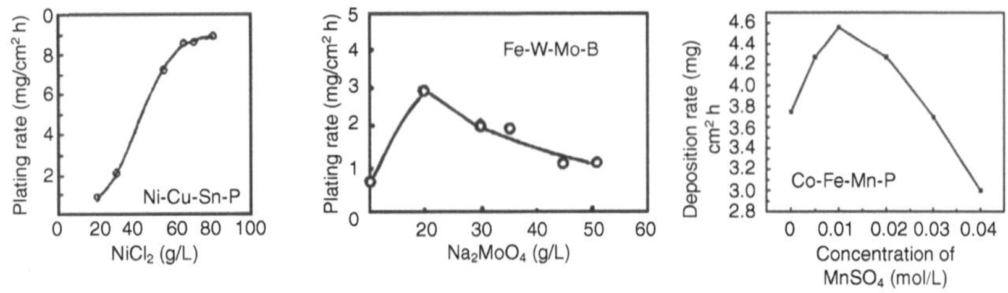

FIGURE 6.12 Deposition rates of EP ternary Ni–Cu–Sn–P [54], Fe–W–Mo–B [56], and Co–Fe–Mn–P [57] alloys versus metal salts.

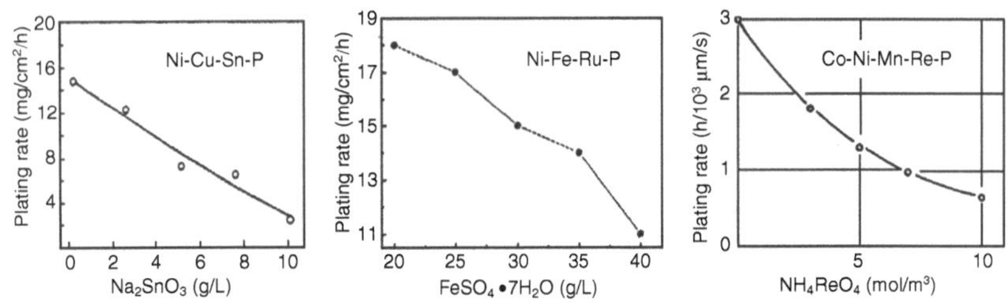

FIGURE 6.13 Deposition rates of EP ternary Ni–Cu–Sn–P [54], Ni–Fe–Ru–P [58], and Co–Ni–Mn–Re–P [59] alloys versus metal salts.

The considerable volume of experimental data on the effect of metal salts on deposition rates from the EP pure metals to quinary alloy systems has been summarized in the preceding sections. The key problem is how to understand and explain these facts including the complex phenomenon mentioned above. It must be pointed out that most authors gave no explanation of their experimental results but just set out the experimental facts; only a small number of authors have offered any explanation for their experimental results, among which is the work by Lu and Zangari [28] in 2003. They used cyclic voltammograms to measure the oxidation and reduction processes involved in the same electroless deposition solution systems, while varying the concentrations of solution constituents. Because they found that the plating rate versus Na_2WO_4 was class I shape, but the plating rate versus Na_2MoO_4 was class II shape, they wanted to use their measurements to explain the very different behaviors for Ni–Mo–P and Ni–W–P deposition. They found that introduction of Na_2MoO_4 into the Ni–P plating solution suppresses both the anodic and cathodic reactions. Therefore, the addition of Na_2MoO_4 inhibits the oxidation of hypophosphite and, through the coupling between anodic and cathodic processes, decreases the deposition rate. However, the introduction of Na_2WO_4 into the Ni–P electroless solution can accelerate both the anodic and cathodic reactions. So, the deposition rate of Ni–W–P increases with increasing concentration of Na_2WO_4. Further, these authors found that Mo(VI) species exist mainly as simple MoO_4^{2-} in the ternary electroless deposition system, whereas W(VI) species mostly appear as large cluster complex ions $[(WO_4)(Cit)(H)]^{4-}$ in the solution. The simple MoO_4^{2-} ion is easy to adsorb on the substrate, to interact with Ni, and to be discharged. Entirely because of its easy adsorption on the substrate, the preferentially adsorbed MoO_4^{2-} blocks the adsorption/reaction sites for Ni^{2+} gradually (shielding effect) and hinders Ni adsorption and reduction, so lowering the deposition rate or even completely stopping the reaction. Contrarily, although the large W(VI) complex ion is difficult to adsorb and discharge on the substrate, there is no such shielding effect as there is for Mo(VI) species. In addition, the formation of the W(VI) complex consumes some citrate ions. As a result, both the anodic and cathodic reactions are promoted, and the overall deposition rate increases. This is a rare explanation for the difference of the plating rates between the EP Ni–W–P and Ni–Mo–P alloy systems, through experimental measurement (CV measurement). This rarity is the reason for citing these authors' results in such detail. However, it is obvious that these authors did not use this explanation for all of the mentioned experimental facts. In fact, this explanation cannot explain all of the experimental data, especially the complex phenomenon mentioned above. Therefore, more in-depth research needs to be done. Also, in 2011 Rudnik and Mucha [39] commented: "In the electroless plating systems the adsorption hypothesis seems to be the most reasonable; however, more detailed studies are necessary."

In order to better understand and explain the experimental facts, we need comprehensive experimental data. That is why we described and classified the experimental data in detail in the above contents and figures.

First, we will discuss the class I shape of plating rates versus metal salts. Metal salts are the source of the metal ions; therefore, their availability must greatly affect the plating rate. All of the above experimental data, from EP pure metal to quinary alloy systems, are obtained under the condition that the other components of the bath solution and the plating conditions are not changed but fixed when considering the influence of metal salts on the plating rate. Of course, the reactants in the bath are not replenished during the whole experimental process, which differs from the large-scale production process in which the reactants would be replenished.

An increase in metal salt concentration means an increase in metal ions and the ratio of metal ions to reductant increases. As the metal ions are released from the ionic complexes, they are reduced by the electrons donated by the reductant, and thereby become neutral atoms. These neutral metal atoms gradually increase in number and then deposit onto the autocatalyzed or catalyzed substrate, increasing the plating rate. However, the amount of reductant is limited. This means that as the amount of reductant decreases there will be fewer electrons to release; thereby, the deposition rate rises to a maximum value and then decreases. That is why the mountain shape of plating rate versus metal salt can appear. Chang et al. [23] use their measured facts to explain the effect of $CoCl_2$ on plating rate in EP Co–B alloy films (the middle plot in Fig. 6.5).

These authors found that when the rate of release of Co^{2+} ions from complexed Co ions $[Co(NH_3)_6{}^{2+}]$ is equal to the deposition rate, the concentration of free Co^{2+} ions can remain constant. In such a case, the plating rate increases with the Co^{2+}/BH_4^- ratio. Increasing the $CoCl_2$ concentration increases the amount of free Co^{2+}. Once the Co^{2+}/BH_4^- ratio reaches 4 the plating rate reaches a maximum value. This situation just corresponds to the main reaction equation for the electroless deposition of metallic cobalt with a molar ratio of Co^{2+} and BH_4^- of 4:1.

$$4Co^{2+} + BH_4^- + 8OH^- \rightarrow 5H_2O + H_2BO_3^- + 4Co\downarrow \tag{6.1}$$

However, at large Co^{2+}/BH_4^- ratios, the concentration of Co^{2+} at the interface is too high to obstruct the arrival of the reducing agent, and thus, the deposition rate is retarded. At a Co^{2+}/BH_4^- ratio of about 5.5, the precipitated products of the reaction result in the bath solution failing, and thereby no deposits can be obtained. Shimachi et al. [52] illustrated this situation clearly, that is, no deposition in their measured plot of the plating rate versus Sn^{4+}-gluconate concentration of Ni–Sn–B depositions (the right plot in Fig. 6.11) because their measured results show that at the concentration above 0.07 mol/L, there was no deposition.

The situation of having only an existing rising edge but no falling side, which appears in some EP systems, is due to the bath solution becoming ineffective and even decomposing. This resulted from the precipitated product of the reaction when the plating rate reaches up to some maximum value; thereby, the normal reaction will be stopped after that point. Xie et al. [38] used their data to explain the plating rate versus $NiCl_2$ concentration in Ni–Sn–P deposits. They found that when the nickel chloride concentration is small, the nickel reduced by reductant is insufficient to form an autocatalyzed film on the substrate, so that no deposit is formed. As the nickel chloride concentration increases, so does the $Ni^{2+}/H_2PO_2^-$ ratio. The Ni^{2+} ions are reduced by electrons released from the reductant and become neutral atoms deposited on the autocatalyzed or catalyzed substrate. Therefore, the plating rate increases with the $Ni^{2+}/H_2PO_2^-$ ratio. However, if the nickel chloride concentration is too high (over 55 g/L), the reaction products are precipitated at the high operating temperature (90°C). The precipitation coats the bath, resulting in its failure.

From these discussions, it is obvious that the mountain shape and the shape of just the rising side are not substantially different; hence, they can be classified as one class.

Now, we discuss the class II shape of plating rates versus metal salts, that is, those that exhibited only a falling side. As mentioned previously, the class II shape is not like the class I shape, but illustrates a complex situation. It is easy to understand why the plating rates versus metal salts have the class II shape, such as Co–Fe–P (in Fig. 6.8), Ni–Fe–P, Ni–Fe–B, and Co–Fe–B (all three in Fig. 6.9). From the kinetics study earlier, it can be seen that adding ions of a catalytically less active metal (i.e., its standard potential is more negative than the nickel plating bath) will be more difficult to be reduced and to be converted to metals. The standard potentials of Fe and Ni are −0.44 V and −0.25 V, respectively, so the deposition rate of Fe, therefore, must be slower than that of Ni. Once the Fe^{2+} concentration in the bath increases, the reduction reaction of Fe controls the whole deposition process, because the Ni^{2+} concentration is less than the Fe^{2+} concentration in the bath. Therefore, the plating rate must decrease gradually and never increase until the bath solution becomes ineffective. Of course, the maximum in plating rate cannot be appeared in this case. This is just the case in the Ni–Fe–P and Ni–Fe–B systems (in Fig. 6.9).

Similarly, because the standard potential of Co is −0.28 V which is also larger than that of Fe, the same situation also appears in the Co–Fe–P and Co–Fe–B systems (Figs 6.8 and 6.9).

As for the value of the starting point of the plating rate versus metal salt in these ternary systems, it should be the plating rate value of the binary alloy system when the Fe salt $FeSO_4$ (Fe^{2+}) concentration is zero and the same concentration of Ni or Co salt is used in the bath. For example, the plating rate at the starting point on the curve of the plating rate versus metal salt (Na_2SnO_3) in Ni–Sn–P system (in Fig. 6.7) is about 16 mg/cm^2•h. The concentration of $NiCl_2$ is 55 g/L in the bath solution of Ni–Sn–P [38]. From the plating rate versus metal salt of the binary Ni–P system (in Fig. 6.4), we can find the plating rate is about 16.5 mg/cm^2•h when the $NiCl_2$ is 55 g/L. This is just the plating rate value of the starting point of the plating rate versus metal salt in Ni–Sn–P system (in Fig. 6.7) though there is a slight deviation of the numerical value.

Though Lu and Zangari [28] explained the different behaviors for Ni–Mo–P and Ni–W–P deposition used their CV measurement results, but they cannot explain the behaviors of the class II shape of either Ni–W–P and Ni–Mo–P systems (in Fig. 6.7), or Co–W–P and Co–Mo–P systems (in Fig. 6.8). Of course, using the difference of standard potentials as mentioned above the behaviors of these four systems cannot be explained as well. In addition, the class II shape in Ni–Sn–P system (in Fig. 6.7) cannot be understood with the difference of standards of Ni and Sn because the standard potential of Sn is −0.13 V, which is more positive than that of Ni. Comparing the behaviors of the class I (in Fig. 6.6) and class II (in Fig. 6.7), it is easy to understand that Na_2SnO_3 should suppress deposition, decreasing the plating rate. But up today no one has proposed a detailed mechanism to explain it and it is not known why the Na_2SnO_3 causes the plating rate decrease in the Ni–Sn–P deposition.

Moreover, the class II shapes of Ni–Cu–Sn–P, Ni–Fe–Ru–P, and Co–Ni–Mn–Re–P have not been explained at all. The complex phenomenon and conflicting evidence mentioned in Section 6.1.1.3 become more difficult for people to understand and interpret. Therefore people need to work more hard, proposing more explanations; the above effects of metal salts on plating rates from the pure metals to quinary alloy systems can be understood more completely.

6.1.2 Effect of Reducing Agent

The role of the reducing agent in the EP is to reduce the metal element from the metal salts through a redox reaction. If there is no reducing agent, no redox reaction can take place in the bath; therefore no deposit is produced. So, the reducing agent is a necessary constituent in the bath as well. Many of the studies of the effects of reducing agent on the EP deposition have been reported in the literature. There is no doubt that the influence of reducing agent in EP is both obvious and enormous.

The effects of reducing agents in a lot of systems from pure metals to quinary alloys have been investigated, such as in pure metals: Pb [5], Cd [64], In [65], Sb [7], Bi [66], Cu [67,68], Ag [9], Ni [11], and Co [12,69]; in binary alloys: Ni–P [15,16], Co–P [17,70], Fe–B [71], Co–B [24,72], and Ni–B [21,22,73]; in ternary alloys: Ni–Sn–P [38], Co–W–P [31], Ni–Co–B [74], Ni–Fe–B [43], Fe–W–B [47], and Fe–Mo–B [48]; in quaternary alloys: Ni–Cu–Sn–P [54], Ni–Fe–La–B [75]; and in quinary alloys: Ni–Co–Fe–La–P [76]. The experimental curve shapes of the effects of reducing agents on plating rates all belong to the class I, no one to the class II. However, there is still a little difference: those are all mountain shapes for the ternary alloy systems and are all rising edges in quaternary and quinary systems, but are mountain and rising edge shapes in the pure metals and binary alloy systems. Only two samples of the shapes on plating rate versus reducing agent in every kind of system are illustrated, as shown in Figs 6.14–6.17.

How do we explain the above experimental data? As the concentration of reductant increases, the electrons in the solution released from reducing agent also increase, and thereby the reducing power of the bath increases. In other words, the quantities of reduced neutral metal atoms and neutral metalloid atoms from

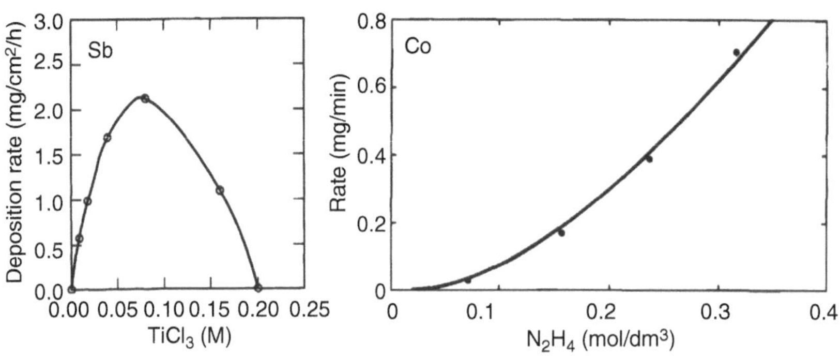

FIGURE 6.14 **Deposition rates of EP pure metals Sb [7] and Co [12] versus reducing agents.**

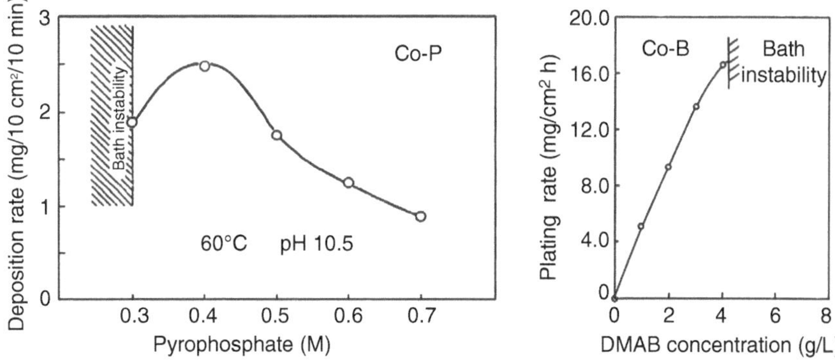

FIGURE 6.15 Deposition rates of EP binary alloy systems Co–P [70] and Co–B [72] versus reducing agents.

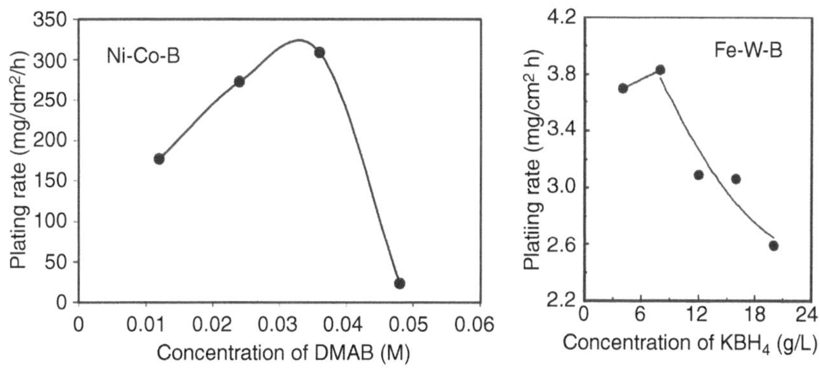

FIGURE 6.16 Deposition rates of EP ternary alloy systems Ni–Co–B [74] and Fe–W–B [47] versus reducing agents.

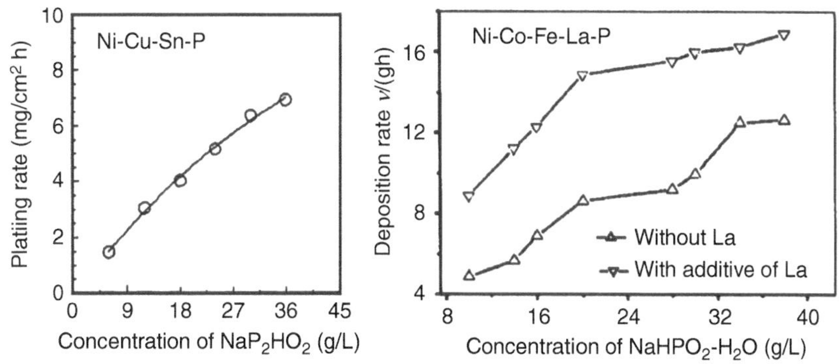

FIGURE 6.17 Deposition rates of EP quaternary and quinary alloy systems Ni–Cu–Sn–P [54] and Ni–Co–Fe–La–P [76] versus reducing agents.

the reducing agent increase. Therefore, the deposition rate increases. From the reaction mechanism of electroless codeposition, this can be seen evidently. We have taken Fe–B as an example. The reaction mechanism of electroless codeposition for Fe–B can be written as [71]

$$BH_4^- + 4OH^- \rightarrow B(OH)_4^- + 2H_2 + 4e \tag{6.2}$$

$$BH_4^- + 8OH^- \rightarrow B(OH)_4^- + 4H_2O + 8e \tag{6.3}$$

$$Fe^{2+} + 2e \rightarrow Fe \downarrow \tag{6.4}$$

$$B(OH)_4^- + 3e \rightarrow B \downarrow + 4OH^- \tag{6.5}$$

These equations show evidently that with increase of reducing agent Na(or K)BH_4 the releasing electrons increase, leading to enhancement of the neutral metal Fe and metalloid B atoms. So, it really confirms the statement.

However, if the reducing agent is present in excess, the stability of the bath deteriorates, and it even decomposes, so that the reduced neutral metal atoms decrease, which makes the deposition rate decrease. That is why there is a maximum in the relationship between deposition rate and reducing agent concentration.

Two points should be made. The first one is about the reducing agent being present in excess. Many experiments indicate that the stability of the bath deteriorates; it can even decompose under such conditions. Our group found that, when making the EP Ni–B [21], precipitates produced in the bath when the KBH_4 concentration went up to about 2 g/L, making the bath unstable and causing the plating rate to decrease. The suitable range of KBH_4 concentration is 0.5–2.0 g/L. We also found such phenomena in the EP of Co–P [17], Fe–B [71], Ni–Sn–P [38], Fe–W–B [47], Fe–Mo–B [48], Co–W–P [31], and Ni–Cu–Sn–P [53,54]. Pearlstein and Weightman found in the EP Co–B [72] that deposition rate increased essentially linearly with DMAB concentration to a maximum at about 4 g/L. When the DMAB concentration was higher than this value, catalytic particles were formed in the solution, which tended to settle to the beaker bottom and resulted in rapid bath decomposition. They even indicated the bath instability evidently in the plot of plating rate versus DMAB (right plot in Fig. 5.15). Narayanana et al. in EP Ni–Co–B ternary alloy deposits [74] found that the plating rate increased to a maximum at about 0.036 M of DMAB; however, when DMAB is used at concentrations higher than 0.036 M, the bath tends to destabilize and at 0.048 M the decomposition of the bath is quite evident. Yang et al. in the EP Ni–Fe–La–B [75] found that the deposition rate increased essentially linearly with DMAB concentration; however, the brightness of deposits declines, even the coating is blistered and peeled off, and in addition a black precipitate occurs in the bottom of beaker when the DMAB concentration is up to 40 mmol/L. When the DMAB concentration was added to 45 mmol/L, the stability of bath solution decreased, which was starting cloudy and decomposition.

From the above experimental data, it can be easy to explain that if the phenomenon of the bath instability just occurs around at the maximum plating rate, then only the rising side but no mountain shape on the plot of plating rate versus reducing agent appears. So that the statement mentioned in Section 6.1.1 that there is not any substantial difference between the mountain shape and the shape of only rising side can be evidently seen, thereby classifying them as one class is proved once again.

The second point is about the phenomenon of lower concentrations of the reducing agent. Nearly all of the experimental data indicate that if the concentration of reducing agent is added up to certain amount, then the deposits can be obtained normally; otherwise no deposits will appear. We pointed out this in our papers about EP Ni–Sn–P [38], Ni–Cu–Sn–P [53,54], Co–P [17], and Co–W–P [31]. Takano et al. [70] not only pointed out such a case, but also indicated this in the plot of plating rate versus pyrophosphate ion concentration, as shown in the right-hand part of Fig. 6.15. The reason is that the bath solution is instable and therefore cannot produce deposits normally if the concentration of reducing agent is too low.

6.1.3 The Effect of Complexing Agent

The complexing agent or chelating agent is an important component of an EP bath solution. The role of adding complexing agent into EP bath is due to its buffering action because it forms a complex with metal ions in solution, thus stabilizing the bath. Otherwise, the free metal ions formed would precipitate rapidly causing the bath to be ineffective. In fact, the metal complex controls the number of free ions that are available to take part in the deposition reaction. The principal functions of complexing agents in the EP plating bath are as follows: the first is to form a metal complex and control the concentration of free metal ions, and the second is to hinder the tendency for the precipitation of basic metal salts. In addition, the complexing agents can play a buffer role that prevents pH value rapid change.

The selection and concentration of complexing agents in EP are important because they are closely related to the plating rate of deposits. In the following, the experimental facts and explanation for the effects of complexing agents on the plating rate will be described, and then how to understand the facts will be discussed.

Like metal salts and reducing agents, the effects of complexing agents in a lot of systems from pure metals to quinary alloys have been investigated, such as in pure metals: Pb [5], Cd [64], Sb [7], In [65,77], Bi [66], Cu [68], Au [78], Ni [11], and Co [69]; in binary alloys: Ni–P [14], Co–P [17], Ni–B [21,73], Fe–B [71], Co–B [24,72]; in ternary alloys: Ni–Sn–P [38], Fe–W–B [47], Fe–Mo–B [48], Ni–W–P [26,27,79], Co–W–P [31]; in quaternary alloys: Ni–Cu–Sn–P [53,54], Ni–Fe–B–La [75]; and in quinary Ni–Co–Fe–La–P alloys [76].

The experimental curve shapes of the effects of complexing agents on plating rates are illustrated in Figs 6.18–6.21. Figure 6.18 shows the shapes of plating rate versus complexing agents for pure metals Au and Ni, which belong to class I. Figures 6.19 and 6.20 show binary and ternary alloy systems; one can

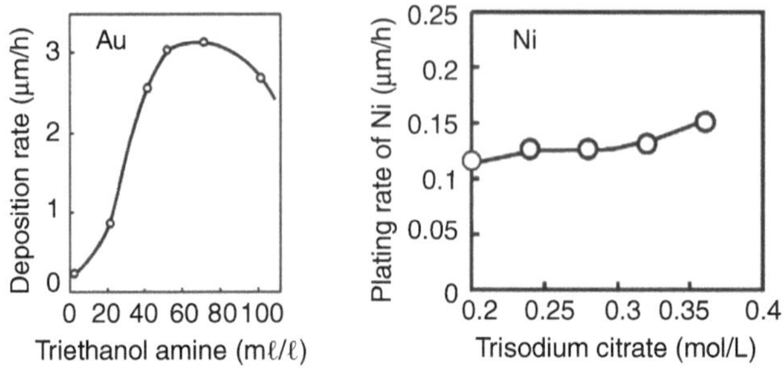

FIGURE 6.18 Deposition rates of EP pure metals Au [78] and Ni [11] versus complexing agents.

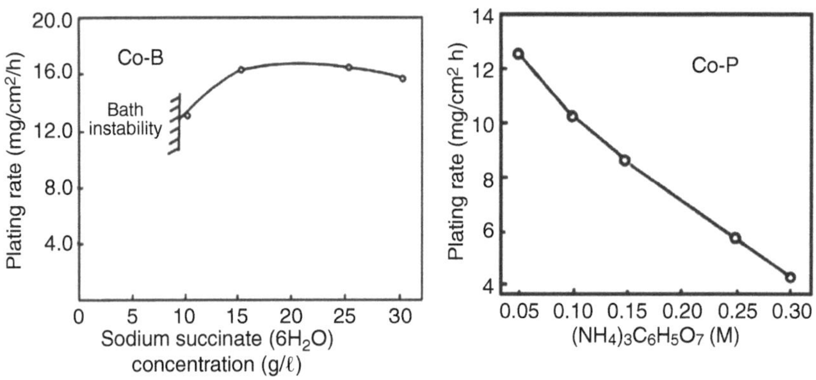

FIGURE 6.19 Deposition rates of EP binary alloy systems Co–B [72] and Co–P [17] versus complexing agents.

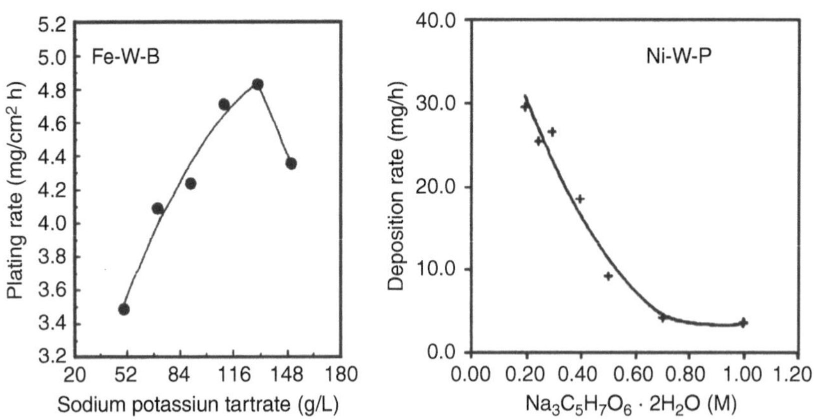

FIGURE 6.20 Deposition rates of EP ternary alloy systems Fe–W–B [47] and Ni–W–P [26] versus complexing agents.

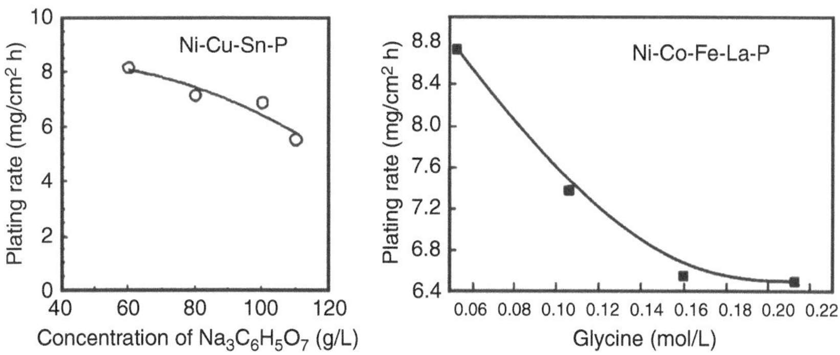

FIGURE 6.21 Deposition rates of EP quaternary and quinary alloy systems Ni–Cu–Sn–P [54] and Ni–Co–Fe–La–P [76] versus complexing agents.

see that the shapes belong to classes I and II. However, the shapes for quaternary and quinary alloy systems are only the class II. Not all of the experimental results are illustrated but just the typical ones as shown in Figs 6.18–6.21 for the sake of brevity. For example, the shapes of Cu [68] and Co [69] also belong to class I; those for Ni–B [80] and Fe–B [71] belong to class I as well. The shapes for Fe–Mo–B [48] and Co–W–P [31] belong to class I, and those for Ni–W–P by other two groups [27,79] all belong to class II as well. However, the experimental conflict phenomenon is not only in one system but rather more, as will appear in several systems and will be illustrated in the following.

The key problem is still how to explain the data. Usually, the original authors have not explained their experimental results in any depth, and no one has explained the data in whole. From the above results, the curve shapes of the plating rate versus complexing agent can be also divided into two classes, classes I and II.

First, the shape of class I will be discussed. In the bath solution, Ni or other metal ions are bound to a well-defined number of radical ions, forming a complex. Under such conditions the bath solution tends to be stable. Otherwise, the free metal ions form precipitate rapidly with other chemicals in the solution, causing the bath to be ineffective and no deposits can be obtained. As the concentration of complexing agent increases, the complexation strengthens, and the tendency to form a precipitate will be reduced. When certain amount of complexing agent is added, a deposit is formed. In such a way, the plating rate increases with the concentration of complexing agent because of the increasing concentration of metals ions released from the complex. When the complexing agent is added at the amount at which the so-called stable complex forms, when all of the free ions are just complexed, the plating rate will approach its maximum value because of the maximum amount of free metal ions released from the complex. After that point, as the concentration of the complexing agent increases over complexation appears, reducing the amount of metal ions released from the complex and thereby decreasing the plating rate. That is why the maximum appears in the plot of plating rate versus complexing agent. If the bath becomes unstable just after maximum, then only the shape of rising side appears, but no falling side occurs. Otherwise, a mountain sharp is formed.

Just mentioned above, only adding the complexing agent at a certain amount will form a deposit. That is because the bath solution only approaches the stability at one point. Many experiments have confirmed this point. For example, Pearlstein and Weightman found for EP Co–B deposition [72] that when no succinate was present, spongy cobalt was formed within the solution accompanied by rapid bath decomposition, and virtually no deposition was produced on the palladium-activated copper specimen. Satisfactory bath stability was provided by the presence of 10 g/L or more of the succinate in the bath. The plot of the Co–B system in the left of Fig. 6.19 shows this case clearly. Such a case appears in nearly all systems in class I.

Now the shape of class II will be discussed. It is obvious that if before reaching a stable complex, some systems always produce a precipitate, then a coating will not be able to form. The deposits can only be formed when the stable complex is formed. In such a case, no maximum occurs, just as the decline in the plating rate as the concentration of the complexing agent. In fact, this is just the case as the author's group found in the EP Ni–Sn–P experiment [38]. Our experiment showed that deposition did not proceed because of the precipitate forming on heating the bath if the concentration of complexing agent $Na_3C_6H_5O_7$ was less than 60 g/L. When the concentration was higher than that value, the deposition rate decreased gradually because excessive complexing agent successively decreased the release of metal ions from the complex. When the concentration of $Na_3C_6H_5O_7$ was higher than 100 g/L, on heating the bath became colloidal, making it ineffective, so no deposit could be formed. Therefore, no maximum but only falling side occurs in this Ni–Sn–P system.

Two points should be stressed. The first is the problem of effects of various complexing agents on the plating rates in the same system done by the same authors. Figure 6.22 is a typical example [80]. For the same EP Cu, the shapes of plating rate versus complexing agents are very different. For triethanolamine

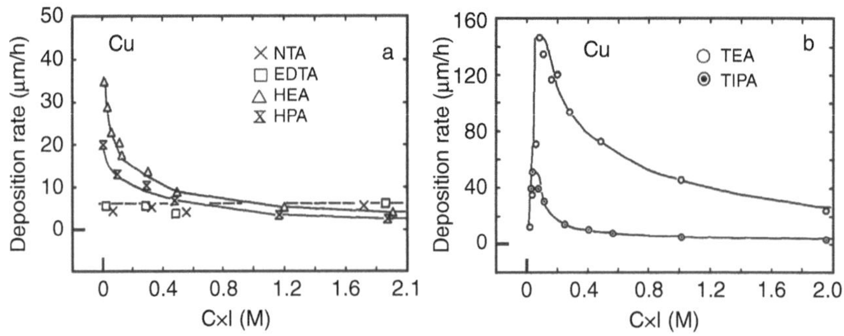

FIGURE 6.22 **Deposition rates of EP Cu versus various complexing agents [80].**

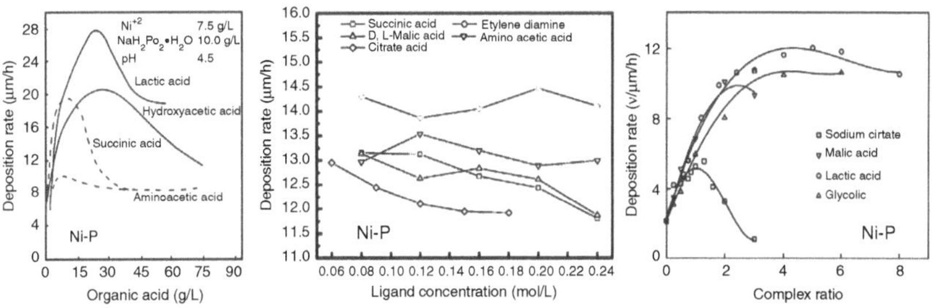

FIGURE 6.23 Deposition rates of EP Ni–P alloy system versus various complexing agents [81–83].

(TEA) and triisopropanolamine (TIPA), the shapes are mountains, but are only falling sides for nitrilotriacetic acid (NTA), ethylenediaminetetraacetic acid (EDTA), N,N,N′,N′-tetrakis (2-hydroxyethyl) ethylenediamine (HEA), and N,N,N′,N′,-tetrakis(2-hydroxypropyl) ethylenediamine (HPA), respectively. Why? The original authors did not explain it, and no one else has tried since. No doubt that there is a need to answer why.

Another problem is experimental conflict phenomenon for the shape of plating rate versus complexing agent obtained by different group authors in the same EP system, as can be seen in a typical plot (Fig. 6.23). The left-hand plot was obtained by de Minjer and Brenner in 1960 [81], the middle plot by Cui et al. in 2005 [82], and right by Jin et al. in 2010 [83]. It can be seen the obvious conflict results from the figure. For example, for succinic acid, a mountain shape was obtained by Minjer and Brenner, but a falling side shape by Cui et al. For citric acid, a mountain shape was obtained by Jin et al., but just a falling side obtained by Cui et al. The authors did not give any reasons for this difference, and did not compare their results with those published many years ago. Are the results by the authors of one group wrong? Or are the other reasons? Obviously, there is a need for further research and explanation.

The last issue is about the relationship between the plating rate and stability constant. A stability constant (formation constant, binding constant) is the equilibrium constant for the formation of a complex in the solution. It is a measure of the strength of the interaction between the reagents that come together to form the complex. There are two main kinds of complex: compounds formed by the interaction of a metal ion with a ligand, and supramolecular complexes. The complex in an EP bath solution is the former kind of complex. It seems that there is a contentious issue about this problem. Because some authors explicitly declared that "electroless nickel complexes with low value (weak) stability constants will show higher deposition rates and yield deposits with lower phosphorus content than baths with numerically larger (strong) stability constants" [84]; and "In general, the complexing agents having smaller stability constant to nickel ion resulted in higher deposition rates ..." [85]. However, these authors did not prove it using theory or experimental results. Here, it is necessary to clarify this point. Using the experimental data published in 2005 [82], a plot can be drawn

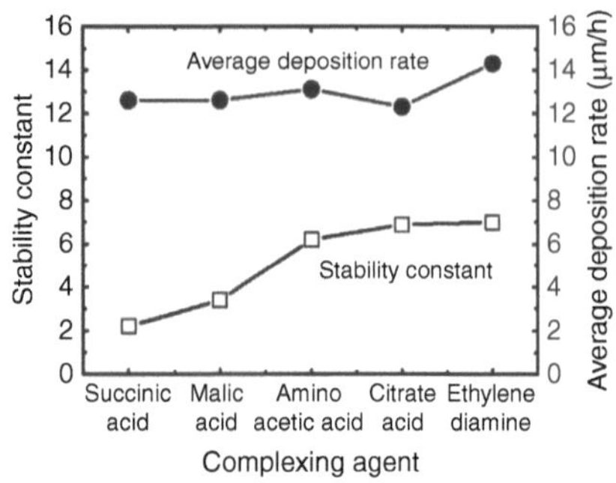

FIGURE 6.24 Stability constant and deposition rates for EP Ni–P alloy system versus various complexing agents.

(Fig. 6.24). It can be seen from the figure that there is no close relationship between stability constant and deposition rates for the EP Ni–P alloy system. For example, the stability constants of succinic acid and ethylene diamine are 2.2 and 7.0, and the corresponding plating rates are 12.6 and 14.3 μm/h, respectively. The fact shows evidently that the higher the stability constant, the higher the plating rate, which is just contrary to the above argument. The plating rate is influenced by many factors, and attempting using a sole stability constant to mainly decide it will be taking a big risk. The experimental results have confirmed that this statement is not correct at least for the most common EP Ni–P alloys.

6.1.4 Effect of Stabilizer

It is just as the name suggests that a stabilizer can make bath solution stable, working for a long time. A stabilizer can also be called an inhibitor. It has been known that without any stabilizers a bath solution can be used for long periods of time. However, if no stabilizers are added, a bath solution can decompose suddenly at any time without signs of warning because of the increase of the evolving hydrogen gas and the precipitation of fine particles in the solution. Therefore, EP usually adds a certain amount of stabilizer.

It is also well known that some stabilizers can accelerate the rate of plating. Such a stabilizer can increase the reducing efficiency of the reducing agent, thereby increasing the plating rate.

A lot of substances can be used as stabilizers, such as lead, tin, arsenic, molybdenum, cadmium or thallium ions, thiourea, etc. Usually, only trace amount of stabilizers is used, and some slightly more amount of stabilizers will cause the plating to stop. So, the amount used in a bath is very critical.

We will discuss the effects of stabilizers on plating rate according to the kinds of stabilizers studied experimentally.

6.1.4.1 Thiourea

Thiourea (Tu) is a convenient and cheaper stabilizer. In addition, thiourea is not toxic. Therefore, Tu is a widely used stabilizer. Early in 1957, de Minjer and

Brenner [86] described the effect of thiourea and other sulfur-containing compounds on the rate of the electroless deposition process for acid nickel. They found that the plot of plating rate versus concentration of Tu was a mountain shape, that is, belonging to class I. Since then, many papers have concentrated on studying the effects of Tu on plating rate from pure metals to alloy systems, such as Cu [87,88] and Ni–P [89–95]. Figures 6.25–6.27 illustrate the typical

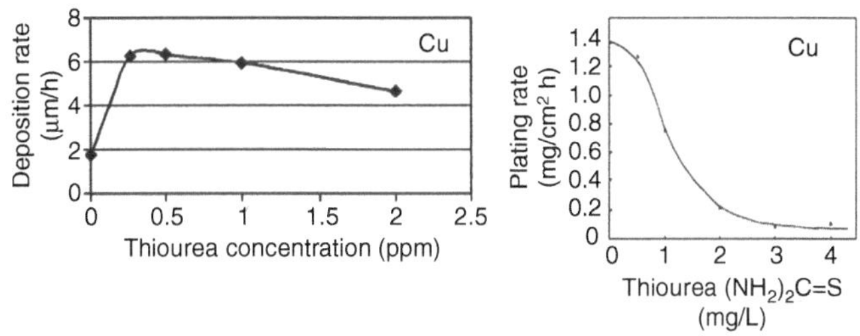

FIGURE 6.25 Plating rates for EP metal Cu versus thiourea. Left cited from [87], right from [88].

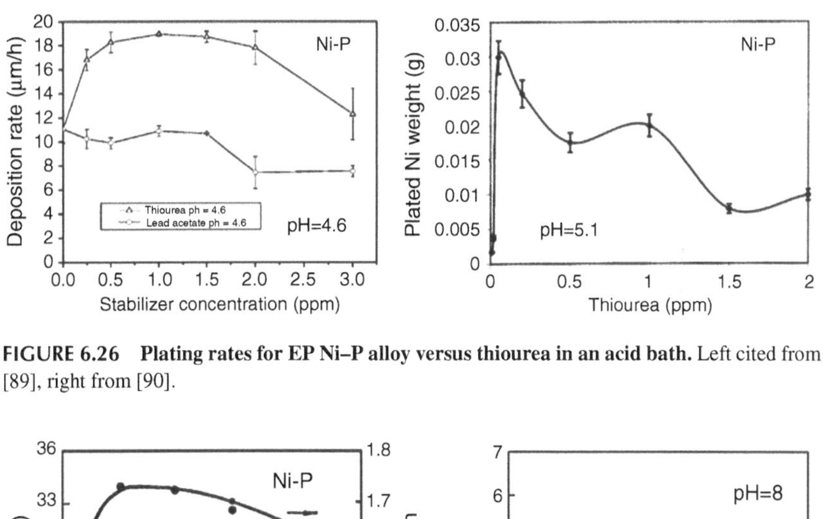

FIGURE 6.26 Plating rates for EP Ni–P alloy versus thiourea in an acid bath. Left cited from [89], right from [90].

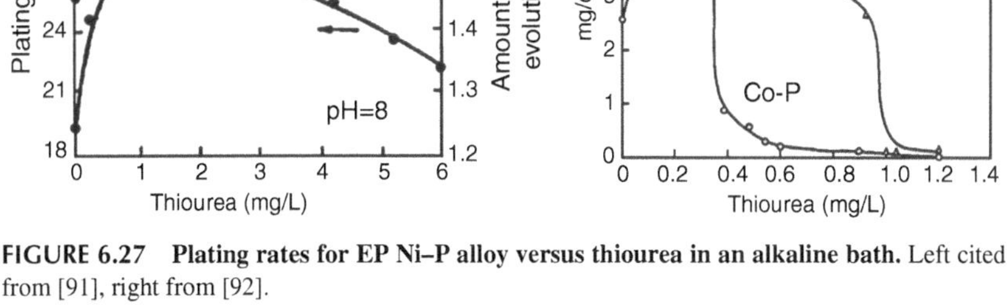

FIGURE 6.27 Plating rates for EP Ni–P alloy versus thiourea in an alkaline bath. Left cited from [91], right from [92].

plots of plating rate versus thiourea. It can be seen that the shapes are both classes I and II. In addition, two maxima were observed in [90] (right plot in Fig. 6.26). The original authors declared that this appears to be the first time a bimodal distribution was recorded for stabilizer concentration. Experimental data either in EP Cu or in Ni–P systems indicate that large amounts of added thiourea will hinder the reaction and stop the growth of the coating.

We first discuss reasons why the class I shape appears. An electroless plating bath is not stable thermodynamically [96]. The bath becomes stable when stabilizers are added to the solution so that the bath cannot decompose rapidly. Adding a small amount of Tu to the plating solution improves the catalytic activity of the metal surface for the oxidation of hypophosphite, causing an increase in the rate of electroless deposition rate. In fact, Li and Kohl [87] found that the color of the deposits changed from black in the absence of Tu in the solution to semibright at 0.5 ppm (small amount) Tu. When excess thiourea was added, more of the Tu adsorbed onto the substrate metal surface. This meant that the number of catalytic sites on the metal surface was reduced due to the formation of a thin film of adsorbed thiourea on the top of the catalytic surface, causing the decrease in the deposition rate. The experiment [87] showed that when the Tu concentration was more than 1.0 ppm (excess amount), the copper deposit became brittle and appeared black in color, similarly to when there was no Tu. This explains the maximum plating rate observed. A group [91] has measured the change curves of the concentration of thiourea with the oxidation current of hypophosphite and stabilizing potential, respectively, and they believed that Tu makes the stable potential negative shift (absolute value increases) or positive shift through increasing or decreasing the oxidation potential of hypophosphite by Tu, then resulting in acceleration or inhibition of the reaction of chemical deposition of nickel. This makes the mountain shape curve that is observed.

As for the two maxima that appeared on the plot of plating rate versus thiourea (right in Fig. 4.26), the original authors [90] explained it as follows. "With continuous addition of thiourea, more thiourea would adsorb onto the substrate surface and would further decrease the plating rate until all available thiourea-binding sites are filled and the plating rate reaches a minimum, as illustrated in part b of Fig. 6.28. When further amounts of thiourea were added, the plating rate increased again because no more thiourea could adsorb onto the substrate surface (all available binding sites have been filled) to decrease the plating rate, and the catalytic process would be predominant again to increase the plating rate, as illustrated in part c of Fig. 6.28. As additional thiourea was added, more formamidine disulfide was produced, and when that dimer had been generated in sufficient amount, its ability to oxidize $H_2PO_3^-$ to $H_3PO_3^-$ inhibited the reduction of Ni^{2+} by $H_2PO_3^-$ thus decreasing the Ni deposition rate. So, with the competing effect of thiourea, the second maximum plating rate at 1 ppm thiourea resulted and the rate continued to decrease with more thiourea added as illustrated in part d of Fig. 6.28". These authors also mentioned why this is the first time they have discovered this effect because "Perhaps, the levels of stabilizer concentrations

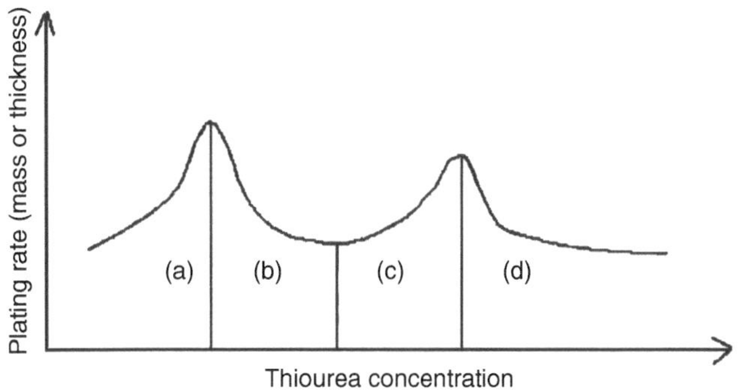

FIGURE 6.28 The dependence of plating rate on thiourea. (a) Rate increased with thiourea where the catalytic process is predominant; (b) Rate decreased where the adsorption of thiourea onto the surface is predominant; (c) Rate increased again where the catalytic process was predominant and no available binding site thiourea; (d) Rate decreased due to the competition between the dimer and Ni(II) for hypophosphite [90].

studied in previous reports could have been very high, which might have prevented the observation of the bimodal distribution recorded in this work." "From comparing the two cases in the Fig. 6.26, a conflict phenomenon can be found. In these two cases, the operating conditions (pH and temperature) are nearly the same, and the composition for nickel hypophosphite bath are basically similar, especially the range of thiourea concentration from 0 to 2 ppm used in [90] is included in the range of 0 to 3 ppm used by another group [89], the shape obtained by [89] is only one maximum without bimodal distribution but two maxima were observed from [90]". The paper of Lin et al. [89] was published in 2002, one year earlier than that by Xu et al. [90], but Xu et al. did not make such comparison. It is no doubt that this experimental conflict requires further explanation.

Now we turn to elucidate the shape of class II appearing in EP Cu and Ni–P. For EP Cu, the original authors explained the reasons for the falling off the plating rate as that "The effect of stabilization is attributed to the strong complexing ability of thiourea with Cu(I) and the surface adsorption of thiourea on copper. The surface adsorption is the main reason for the decrease in plating rate." [88] However, Li and Kohl believed that "A small amount of thiourea, or its derivatives, in the electroless copper solution improves the catalytic activity of the copper surface for the oxidation of hypophosphite, resulting in a higher electroless deposition rate." [87] It is obvious that further study is necessary to reconcile these different data sets and work out the underlying mechanisms.

Kivel and Sallo honestly said that "The reason for the failure of thiourea to enhance the rate of deposition of nickel from alkaline media is not known." [92] However, they believed that "thiourea does not appear to form a monolayer on nickel when adsorption occurs from alkaline media. This monolayer formation seems related to rate enhancement and its absence may explain the failure of thiourea to enhance the rate of alkaline nickel deposition." [92] Such explanation cannot elucidate the very different phenomenon between EP Co–P and

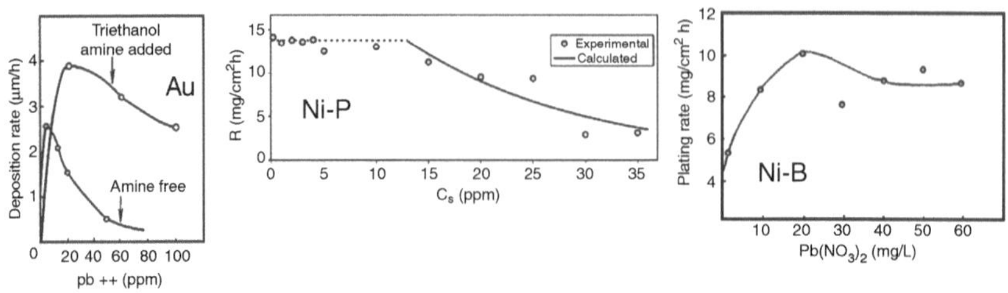

FIGURE 6.29 The dependence of plating rate on Pb^{2+} in EP Au [10]; in EP Ni–P [97], *R* is plating rate, *Cs* is the concentration of Pb^{2+}; and in Ni–B [21].

Ni–P. Especially, after a paper published more than 30 years later, another paper referred to the effect of thiourea on plating rate in similar alkaline electroless Ni–P deposition published [91], in which a class I shape was shown. This obvious conflict over experimental results needs further research to resolve.

6.1.4.2 Pb^{2+}

Pb^{2+} is one of the widely used stabilizers for electroless nickel deposition although it is of great concern in the environment. There are fewer reports of the effects of Pb^{2+} on plating rate than there are for thiourea. Saito et al. [78] in 1988 reported such effect in EP Au; Yin et al. [97] in 2004, Hamdy et al. [98] in 2008, Zhang et al. [99] in 1999, and Lin and Hwang [89] in 2002 reported that in EP Ni–P system, and our group in 1988 reported [21] that in EP Ni–B system. Figure 6.29 shows typical graphs for plating rate versus Pb^{2+} concentration. It can be seen from the figure that the shapes in EP Au and Ni–B systems are class I, but that the Ni–P system belongs to the class II.

When the added concentration of Pb^{2+} is low, the stability of the bath solution increases, enhancing the plating rate. However, the adsorption of Pb^{2+} on the catalytic surface also increases, and adsorbed Pb^{2+} ions occupy more of the surface sites, and thus, hinder the desorption process. Once the surface density of adsorbed Pb^{2+} ions reaches a certain level, the deposition process could be completely inhibited. That is why the maximum plating rate appears. So the plots of plating rate versus Pb^{2+} concentration in EP Au and Ni–B systems can be easily understood.

Considering the shape of the class II in EP Ni–P system, Yin et al. performed a detailed theoretical research in 2004 [97] to describe it, which will be discussed in Section 12.3.3.1 "Effect of Pb^{2+} stabilizer and calculation of plating rate in EN process." From their theoretical model, these authors believe that the stabilizing effect of the lead ion is attributed to its participation as the neutral atom in the lattice of the Ni–P film at the plating frontier, which results in the ascension of Fermi energy. Thus, the oxidation of hypophosphite at the plating frontier is retarded, which in turn slows down the deposition of Ni that needs electrons.

It should be pointed out that in fact the experimental data on the plating rate versus Pb^{2+} concentration in Ni–P deposition reported by the four groups

[89, 97–99] generally belong to class II. However, there is a large difference between their downward trends for the curves of plating rate versus Pb^{2+} plots. The results from Lin and Hwang (Fig. 6.26 left) [89] and from Yin et al. (Fig. 6.29, middle) show that an increase in the concentration of lead acetate from 0 to 1.5 and from 0 to 10 ppm, respectively, the deposition rates are basically maintained at a constant level, and then decrease rather quickly as further enhancing the lead ion. But, Hamdy et al. [Fig. 8 in 98] and Zhang et al. [Fig. 9 in 99] found that there is no constant phase; the decline of the deposition rates starts at zero concentration of lead acetate immediately. Moreover, Hamdy et al. found that the deposition rate is dramatically decreased, especially between 2 mg/L and 5 mg/L lead acetate. At 5 mg/L the nickel deposition rate flattens out and approaches zero. No one proposed explanation exists for such different experimental phenomenon as yet.

6.1.4.3 Other Stabilizers

The reports for the effects of other stabilizers on plating rate are rather few. We will just illustrate ammonia, maleic acid, and thallium acetate as examples, as shown in Fig. 6.30. From the figure one can see that all the shapes belong to class I, although the shape of Ni–P is basically a horizontal straight line.

How to understand these experimental results? Experiment has found that when the adding amount of ammonia was 40 mL/L, the bath solution became turbid and a silver layer was deposited simultaneously on the liquid level after deposition of 20 min. That means the bath is not stable but decomposes. With increasing the amount of ammonia added, the bath trend is to stable, causing the plating rate to be enhanced. The plating rate reaches a maximum at 80 mL/L ammonia. When further increasing ammonia, silver nitrate dissolved in ammonia and generated $Ag(NH_3)_2OH$. $Ag(NH_3)_2OH$ resolved $Ag(NH_3)^{2+}$, afterward $Ag(NH_3)^{2+}$ resolved Ag^+ and $NH_3{\bullet}H_2O$, resulting in the plating rate being decreased. Therefore, the maximum on the plot of the plating rate versus ammonia appeared.

The increase in plating rate with thallium acetate concentration in EP Ni–B is just due to the better stability of the coating bath. In fact, Anik et al. [73] have

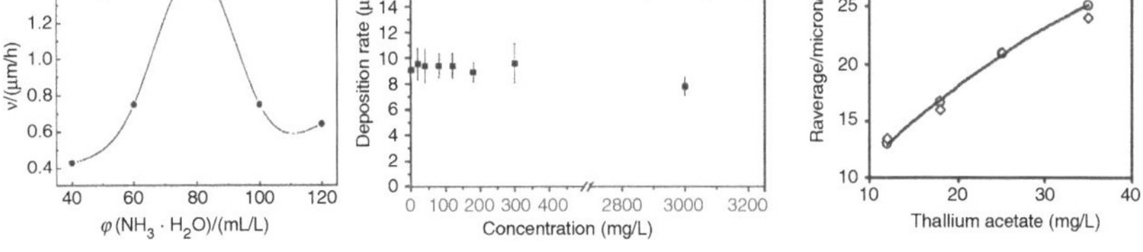

FIGURE 6.30 The dependence of plating rate on ammonia in EP Ag (left) [9], maleic acid in Ni–P (middle) [89], and thallium acetate in Ni–B (right) [73].

found that a threefold increase in the thallium acetate concentration causes an almost twofold increase in the coating rate.

As for the effect of maleic acid on the deposition rate, experimental results observed only a slight change of about 2.45 μm/h in the deposition rates over a very wide concentration range of 0–3000 mg/L, as shown in the middle plot of Fig. 6.30. However, Lin and Hwang [89] did not give an appropriate explanation to it. In addition, Rudnik and Gorgosz [100] in 2007 also measured the effect of maleic acid on plating rate in EP Co–P alloy deposits, finding the plot belonged to the class I (Fig. 4 in [100], not shown here). That is to say, their results are very different from those obtained by Lin and Hwang [89]. Therefore, there is a problem left to be solved.

Besides ammonia, maleic acid, and thallium acetate, the most common stabilizer for electroless copper baths is a steady stream of air bubbled through the solution [101]. In 2003, Hanna et al. measured the effect of several additives on bath stability, studied in electroless Cu baths with and without aeration [102]. Table 6.1 demonstrates the effect of different additives on bath stability as measured by the decomposition time of electroless copper solutions. It can clearly be seen from Table 6.1 that the additive-free baths decompose instantaneously after few minutes (10 min in tartrate bath and 6 min in EDTA bath). However, aeration increases the decomposition time sharply to about 20 times that of baths without aeration. The reason may be attributed to the reoxidation

TABLE 6.1 Decomposition Time (Min) of Electroless Baths (at 50°C for Tartrate Baths and at 75°C for EDTA Baths) [102]

	Tartrate bath, pH 12.5		EDTA bath, pH 13	
Additives	Without aeration	With aeration	Without aeration	With aeration
Additive-free bath	10	170	6	140
+30 mg pyridine	20	180	20	180
+60 mg pyridine*	40	280	30	230
+2 mg/L cytosine	40	240	20	200
+5 mg/L cytosine*	50	300	30	240
+2 mg/L BT	50	200	30	195
+5 mg/L BT*	60	325	35	260
+1 mg/L 2MBT	50	250	40	240
+2 mg/L 2MBT*	95	390	80	320
+0.2 mg/L thiourea*	100	295	120	270
+0.5 mg/L thiourea	150	780	170	725

Optimum concentration.

of cuprous ion formed during the reduction reaction by formaldehyde at higher pH values according to Eq. (6.6):

$$2Cu^{+} + 1/2O_2 + H_2O \rightarrow 2Cu^{2+} + 2OH^{-} \quad (6.6)$$

It should be clearly remembered that the time for decomposition reported in Table 6.1 may vary depending on the type of additives used. For example, although addition of thiourea of 0.5 mg/L increases markedly the decomposition time from 170 min to 780 min, but, the high amount of thiourea has a detrimental effect on the plating rate as mentioned previously.

6.1.5 Effect of Accelerating Agent

In contrast to the stabilizer, an accelerator can enhance the plating rate within a given concentration region. All of the agents added to the bath solution that can accelerate the reaction can be called accelerators or enhancers. Studying accelerators has both theoretical and practical impacts. However, not much research has been carried out to date. The research that has been carried out concentrated on EP Ni–P and Cu systems.

In 1972, Schoenberg reported the change in plating rate of an electroless Cu solution with varying amounts of 8-hydroxy-7-iodo-5-quinoline sulfonic acid (HIOSA) and 2-iodo-3-pyridinol (IP) [103], as shown in the left-hand plot of Fig. 6.31. The plating rates are plotted as the differences (in μm/h) between an additive-free bath and a bath containing a given amount of HIOSA in the plot. It is evident that HIOSA can accelerate the Cu deposition because it enhances the plating rate by as much as 50%. The authors believed that the reason for the accelerative effect of HIQSA in appropriate quantity may be explained by visualizing it adsorbed through an iodide atom on portions of the catalytic surface. The complexing part of the molecule then extends into the bulk of the solution where it is able to form a bond to nearby chelated copper ions. This would tend to accelerate the plating by making it easier for the Cu^{2+}−ligand bond to be broken. However, if a large amount of HIQSA was added, the entire catalytic surface could be poisoned; hence the plating rate will reduce, and the deposition

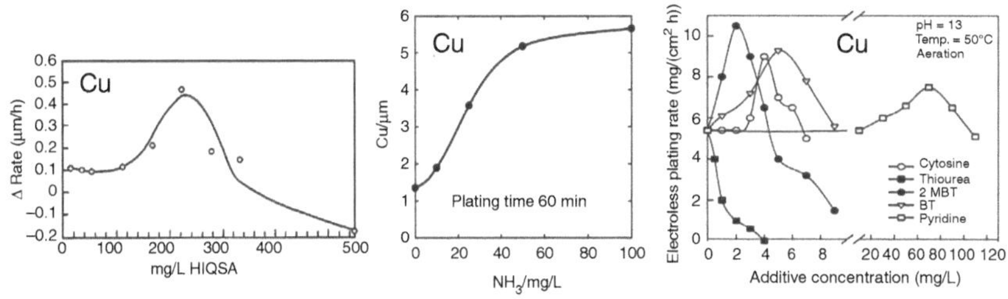

FIGURE 6.31 The dependence of plating rate on accelerant in EP Cu. (Left) HIOSA [103], (middle) NH_3 [104], (right) Cytosine et al. [102].

finally stop. In fact, these authors found that the addition of very large quantities of HIQSA results in the formation of a precipitate, presumably containing copper and HIQSA.

Vaškelis et al. [104] in 2007 reported the accelerating effect of ammonia on electroless copper deposition in alkaline formaldehyde-containing solutions. From the middle plot of Fig. 6.31, it can be seen that the accelerating effect of ammonia is seen beginning from 10 mg/L to 20 mg/L, and at 50–100 mg/L of ammonia a constant copper deposition rate of 5 μm in 1 h is reached. Additions of small amounts of ammonia (1–3 mM) effectively accelerate the autocatalytic reduction of Cu(II) by formaldehyde in EDTA solutions, giving a Cu deposition rate of 20–30°C, a twofold to fourfold increase. The accelerating action of ammonia is explained by assuming that the copper surface of high catalytic activity forms at cathodic reduction from a mixed Cu(II) complex with EDTA and ammonia, and the liberated ammonia molecules participate again in the mixed complex formation at the copper surface.

In 2003, Hanna et al. [102] reported the accelerating effects of five additives in EP Cu. From the right-hand plot in Fig. 6.31, it is evident that the addition of 2-mercaptobenzothiozole (2MBT), benzotriazole (BT), and cytosine in small amounts (<5 mg/L) increases the plating rate significantly as the concentration increases and eventually reaches a maximum, beyond which the rate decreases steadily. However, the plating rate is not affected by pyridine addition up to about 30 mg/L, above which it increases to a maximum at about 60 mg/L followed by a slight decrease. Thiourea has an inhibiting effect on the plating rate and the rate decreases sharply by the addition of very small amounts (0.2 mg/L), so it is not an accelerator under our current criteria. The initial increase of the plating rate by the addition of 2MBT, BT, cytosine, and pyridine may be due to the presence of delocalized k electrons bonds in their structure which enhances the plating rate as a function of pH. This phenomenon is attributed to the greater adsorption ability at low concentrations of these compounds on the active copper surface compared with that of EDTA complexes, which in turn facilitates the copper reduction and accelerates the deposition rate. On the other hand, the decrease in deposition rate may be related to the excess coverage on the copper surface (by these compounds), which makes the reduction of copper ions difficult or even impossible. This is a so-called adsorption phenomenon of overcrowding as suggested by Hung [88] and Nuzzi [105]. It is worth noting that in 2013 a group from China [106] used this statement of adsorption phenomenon of overcrowding to explain the fact that the Ni–P deposition rate decreased gradually from its maximum value as the concentration of the pH stabilizers (ammonium acetate) increased (see Fig. 6 in [106]). In an EDTA bath, the maximum increase of plating rate is increased from 5.4 mg/(cm^2•h) to 10.5 mg/(cm^2•h) or by the addition of 2 mg/L of 2MBT. However, the original authors [102] did not give any explanation for the inhibiting effect on the plating rate of thiourea in EP–Cu (Fig. 6.31 right), except they only described their experimental facts. We have

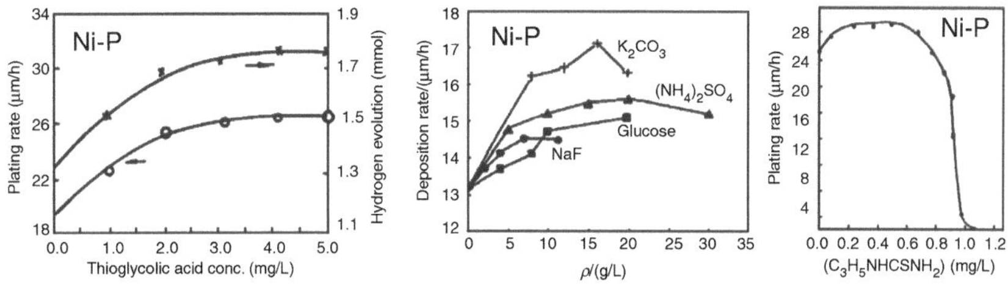

FIGURE 6.32 The dependence of plating rate on accelerant in EP Ni–P alloy. (Left) trioglycolic acid [109], (middle) glucose, NaF, K_2CO_3, and $(NH_4)_2SO_4$ [112], (right) phenylthiourea [108].

mentioned the conflict experimental fact and relative explanations about thiourea in the EP Cu (Fig. 6.25), and readers could see there again.

Li and Kohl [107] in 2003 also reported a copper deposit obtained in a citrate-based electroless copper solution with 0.5 ppm formamidine disulfide (fd). Their results show that fd accelerates the deposition process in the citrate-based electroless copper plating solution. These authors believed that fd accelerates both half reactions: the oxidation of hypophosphite and the reduction of copper [II].

Further research has been carried out on the effects of accelerating agents on the plating rate in Ni–P alloy systems, such as phenylthiourea [108], thioglycolic acid [109], organic acid A_1 [110], mercapto-benzo-thiosole (MBT) [111], glucose, NaF, K_2CO_3, and $(NH_4)_2SO_4$ [112], and on composite accelerators: citrate acid + succinic acid, citrate acid + succinic acid + ammonium chloride, malic acid + aminoacetic acid and malic acid + aminoacetic acid + ammonium chloride [113]. Figure 6.32 shows some typical plots. It is clear that all the curves belong to the I class, just as they do in EP Cu (Fig. 6.31). Usually, the rate of electroless nickel plating with hypophosphite as a reducing agent is below 20 μm/h. From Fig. 6.32, the maximum plating rate can reach to 25–28 μm/h, so this is the basis of the claim for super high-speed electroless nickel plating that appeared in some papers [109].

How to explain the shapes? We just use the left plot in Fig. 6.32 to illustrate it. The plating rate of Ni–P increases from 18.7 μm/h without thioglycolic acid (TA) to 26.4 μm/h with the addition of 4 mg/L of TA. The experimental results also indicated that hydrogen evolution is increased from 1.32 mmol without TA to 1.77 mmol with 4 mg/L. This means that Ni–P alloy deposition and hydrogen evolution and the reduction of Ni^{2+}, H^+, and $H_2PO_2^-$ are associated simultaneously by TA. In order to explain how TA accelerates the reaction, the authors [109] measured the activation energy of the EN plating reaction. They found that trace quantities of TA can decrease the activation energy of the EN plating reaction from 68.5 kJ with no TA to 57.4 kJ with 3 mg/L TA. The decrease of activation energy must accelerate the plating reaction, thereby enhancing the plating rate. Next, they measured the anodic and cathodic polarization curves. The data show that TA affects only the anodic oxidation of $H_2PO_2^-$, but not the cathodic reduction of Ni^{2+}. That is to say that TA is an anodic depolarizer,

TABLE 6.2 Influence of Composite Inorganic Accelerants on Deposition Rate in EP Ni–P Alloy [113]

Basic bath solution and operation conditions			
Nickel sulfate 30 g/L, sodium hypophosphite 34 g/L, lactic acid 10 mL/L, Sodium acetate 20 g/L, K1 1.5 mg/L, pH 5.3, temperature (80 ± 1)°C.			
Composite accelerant		Concentration ρ (g/L)	Plating rate (μm/h)
1	Nothing	0	13.2
2	NaF Glucose	3.5 5.0	15.5
3	NaF K_2CO_3	3.5 6.0	17.2
4	$(NH4)_2SO4$ Glucose	7.5 5.0	15.2
5	$(NH4)_2SO4$ K_2CO_3	7.5 6.0	17.5

increasing the reducing ability of the $H_2PO_2^-$ and the concentration of atomic hydrogen. Therefore, it can accelerate the reduction of Ni^{2+}, H^+, and $H_2PO_2^-$, thereby enhancing the plating rate.

Now we turn to the composite accelerators in the EP Ni–P alloy. Only a very few reports on this topic have been published [112,113]. The studied composite accelerant is composited by two or three simple accelerants. The effects of composite accelerants on plating rate are listed in Tables 6.2 and 6.3. From Table 6.2, the plating rate is only 13.2 μm/h with no accelerant, but increases to 15.2–17.5 μm/h when adding an inorganic composite accelerant to the bath. The data in Table 6.3 show the plating rate of EP Ni–P alloy up to 22.226–25.93 μm/h with a composite accelerant in the bath solution. The accelerate action is very obvious in both cases. Unfortunately, three and four composite accelerants will make microcracks be produced in the coating. There is no explanation of why the composite accelerators can accelerate deposition of Ni–P. Although the author of [113] measured the polarization curve of every single accelerant of the four accelerators, they found that the four accelerators are all the anodic depolarizers, promoting the reducing ability of $H_2PO_2^-$ and thereby causing the enhancement of the plating rate. But does the composite accelerator still be like the single accelerator? This should be confirmed further.

6.1.6 The Effect of Surfactants

The investigation of surfactant was originally aimed at improving activity of substrate surface, so as to obtain a good-quality surface with a high plating

TABLE 6.3 Influence of Composite Accelerants on Deposition Rate of EP Ni–P Alloy [113]

Basic bath solution and operation conditions							
Nickel sulfate 0.095 mol/L, sodium hypophosphite 0.264 mol/L, stabilizer trace, surfactant little, pH 4.8, temperature 90°C, bath loading 0.914 dm^2/L.							
Composite accelerant		Concentration (mol/L)	Plating rate (μm/h)	Macroscopic surface quality	Microscopic surface quality	Stability	P wt% in Ni–P coating
1	Citrate acid Succinic Acid	0.0476 0.169	25.930	Bright	Dense	Stable	4.043
2	Citrate acid Succinic acid Ammonium chloride	0.0476 0.169 0.374	22.755	Bright	Dense	Stable	5.169
3	Malic acid Aminoacetic acid	0.0746 0.2	22.226	Bright	Microcracks	Stable	5.189
4	Malic acid Aminoacetic acid Ammonium chloride	0.0746 0.2 0.374	25.314	Bright	Microcracks	Stable	5.398

rate in electroless deposition. A surfactant is a compound that contains both hydrophilic and hydrophobic segments. The hydrophobic portion of the surfactant is typically a long hydrocarbon chain, referred to as a tail of the molecule. The hydrophilic group often includes anions and cations such as sulfate, phosphate, and tertiary ammonium. According to the nature of their hydrophilic head group, surfactants can be categorized into four groups. The first is an anionic surfactant, which carries a negative charge. This group includes sulfonate salts, alcohol, sulfates, alkyl benzene, phosphoric esters, and carboxylic acid salts. The second is a cationic surfactant, which bears a positive charge. This group includes polyamides and their salt, quaternary ammonium salts, and amine oxides, salt of long chain amines. The third group is nonionic surfactants that carry no charge. The examples of the group are polyethylenated alkyl phenols, alcohol ethoxylates, and alkanolamides. Finally, the fourth is zwitterionic surfactants also known as amphoteric surfactants, which may exhibit either negative or positive charges depending on factors such as the pH of the solution. Long-chain amino acid, betains, and sulfobetain are the examples of this type.

When added to water or solvents, surfactants can enhance the stability of a suspension by increasing the wettability and the surface charge of suspended particles. In addition, they can lessen the surface tension of the systems for the following purposes: wetting, emulsifying, dispersing, foaming, scouring, or lubricity. This meant that the study of surfactants received much attention in recent years.

Ten years ago, a group from Singapore pointed out that "still, no quantitative and systematically qualitative data on the effect of added surfactants on the deposition rates and the corrosion resistance of the electroless Ni–P deposits are available in the literature" [114]. It should be pointed out that the situation has greatly changed today. In 2012, Nwosu et al. [115] reviewed the influence of surfactant incorporation during electroless nickel plating.

The influences of surfactants in many systems have been investigated, such as in Cu [116,117] and Ni–P alloy [118–121], especially in Ni–P composite deposition systems: Ni–P + SiC [122], Ni–P + Al_2O_3 [123,124], Ni–P + ZrO_2 [125,126], Ni–P + PTFE [127–130], Ni–P + diamond [131], and in the ternary alloy systems: Ni–Co–P [132], Ni–Cu–P + PTFE [133].

In order to see the experimental results clearly, the effects of surfactants on the plating rate in four kinds of EP systems (Cu, Ni–P, Ni–P composite system, and ternary alloy system) are illustrated. Two or three typical plots of the effects of surfactants on the plating rate in every system are shown as Figs 6.33–6.36. From the plots one can see that they all belong to class I in EP Cu. In EP Ni–P alloys, the plots are either class I or II. The situation in the Ni–P composite deposition systems is similar to that in the EP Ni–P alloys. EP Ni–P + nano ZrO_2 (middle plot in Fig. 6.35) is evidently class I plot. From the plot in the right of Fig. 6.35, it can be recognized that the plating rates all decreased as those of the three surfactant concentrations increased. This means that they are class II curves. Not only the shape of plating rate versus surfactant concentration is different, which can be divided into classes I and II, but also the value of the plating rate for different surfactants could be very different even in the same EP system deposition. We will only cite EP Ni–P + SiC (left in Fig. 6.35) as an example to illustrate this point. The plating rate is about 15.5 μm/h when adding sodium dodecylbenzenesulfonate (SDBS) but up to about 27.5 μm/h when cooperating the composite surfactant SDBS+ polyethylene glycol (6000) (PEG (6000)), increasing the plating rate by a factor of up to 80%.

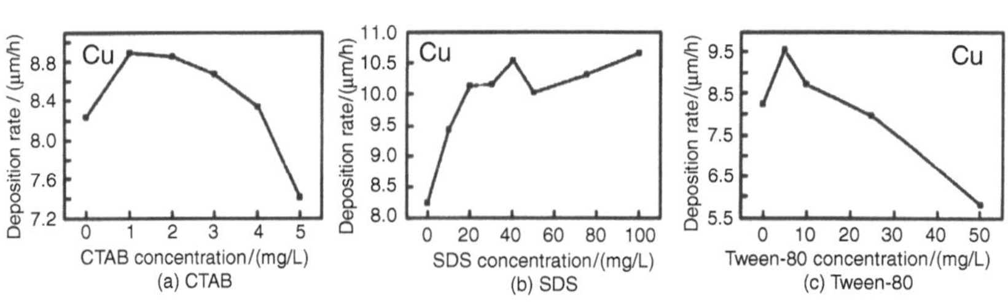

FIGURE 6.33 The dependence of plating rate on surfactant in EP Cu [116]. (Left) cetyltrimethyl ammonium bromide (CTAB), (middle) sodium dodecyl sulfate (SDS), (right) Tween-80.

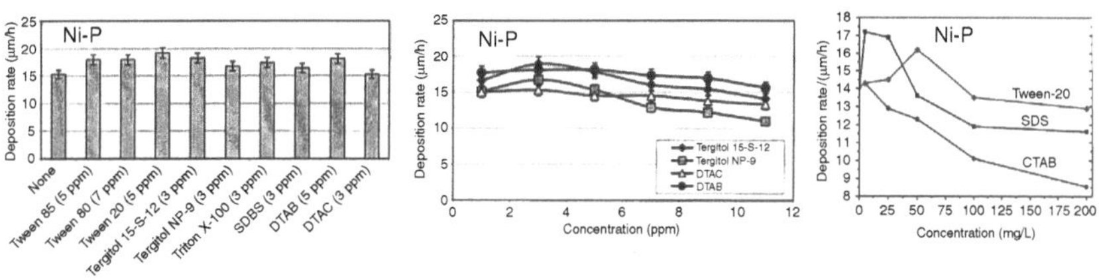

FIGURE 6.34 The dependence of plating rate on surfactants in EP Ni–P alloy. (Left) nine surfactants (two cationic surfactants: dodecyltrimethylammonium bromide (DTAB) and dodecyltrimethylammonium chloride (DTAC); six nonionic surfactants: Triton X-100 and Tergitol 15-S-12, Tergitol NP-9, Tween 20, Tween 80, and Tween 85; and one anionic surfactant: sodium dodecylbenzenesulfonate (SDBS)) [114], (middle) four surfactants (tergitol15-S-12, tergitol NP-9, DTAC and DTAB) [114], (right) three surfactants (sodium dodecyl sulfonate which is also called SDS, CTAB, tween-20) [119].

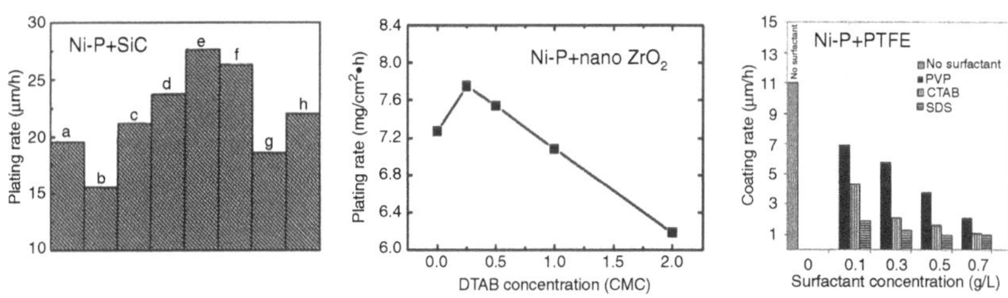

FIGURE 6.35 The dependence of plating rate on surfactants (left) in EP Ni–P + SiC alloy: eight surfactants ((a) SDBS, (b) SDS, (c) polyethylene glycol (6000) (PEG (6000)), (d) Tween-80, (e) SDBS+ PEG (6000), (f) SDS+ PEG (6000), (g) SDBS + Tween-80, and (h) SDS + Tween-80) [122], (middle) in Ni–P + nano ZrO_2 alloy: one surfactant (DTAB) [126], and (right) in Ni–P + PTFE alloy: three surfactants (poly-vinylpyrrolidone (PVP), CTAB, SDS,) [127].

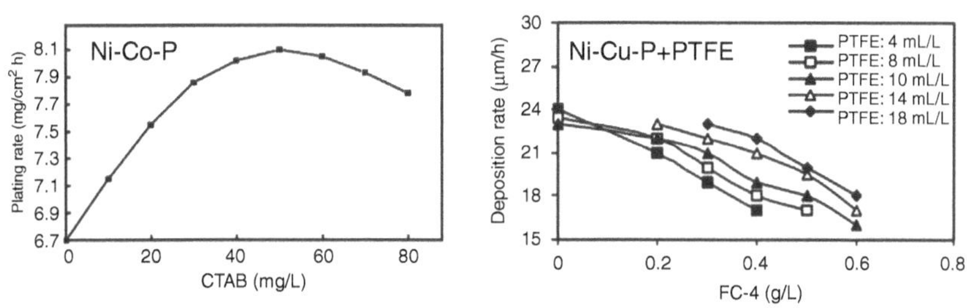

FIGURE 6.36 The dependence of plating rate on surfactants (left) in EP Ni–Co–P alloy: one surfactant (CTAB) [132], and (right) in Ni–Cu–P + PTFE alloy: one surfactant (sodium dodecyl sulfonate (fluorocaruor surfactant FC-4) [133].

We first would like to understand why the deposition rate increases initially and then drops upon a further increase in the surfactant concentration. Just consider the EP Ni–P + ZrO_2. When adding a certain amount of surfactant into the EP bath, several factors will promote deposition and enhance the plating rate. First, the surfactant reduces the contact angle, increasing the wettability of the surfaces to be coated thereby increasing deposition efficiently. In fact, Zielińska et al. found the deposited ZrO_2 from 8.11 wt% for the basic bath up to 12.44 wt% for the bath with 0.50 × critical micelle concentration (cmc) [126]. Second, additional surfactants can be absorbed by the substrate, reducing the interfacial tensions between the generated hydrogen bubbles and the as-deposited substrate or those between the bubble and the plating solution, and can enhance the removal of the hydrogen gas bubbles more quickly and thus form pit-free alloy deposits [134]. The quick removal of hydrogen gas may shift the deposition reactions to favoring the reduction of Ni^{2+} ions. Hence, the addition of surfactants can thus increase the deposition rate. In fact, the measurement from Singapore's group [114] shows that the diminishment of pitting observed on the electroless Ni–P deposits (Fig. 4 in the original paper) is due to the quick removal of hydrogen gas. The adsorption of surfactants on both nonmetallic and metallic substrates to form stable linear cylindrical and half-cylindrical aggregates was first demonstrated by Manne et al. [135] in 1994, who showed direct AFM images for CTAB segregated on graphite. Since then, more reports in this subject have been published [136–146], and in 2009 a group from Canada published a review on surface aggregation of surfactants at electrode surfaces [147]. In order to see the morphology of segregates more clearly, we collect three microscopic images of them as shown in Fig. 6.37. The surfactant can also enhance the electrostatic adsorption of suspended solids on a substrate by increasing their net positive charge. Thus, during the plating process, inert particles adsorb some ions on the surface and are incorporated into the metal matrix by the reduction of some of those adsorbed ions. Cationic surfactant molecules can play the role of the metal ions on the particle surface and undergo reduction reaction increasing the possibility for the particle being embedded in the plating layer. These factors result in an increase in the plating rate.

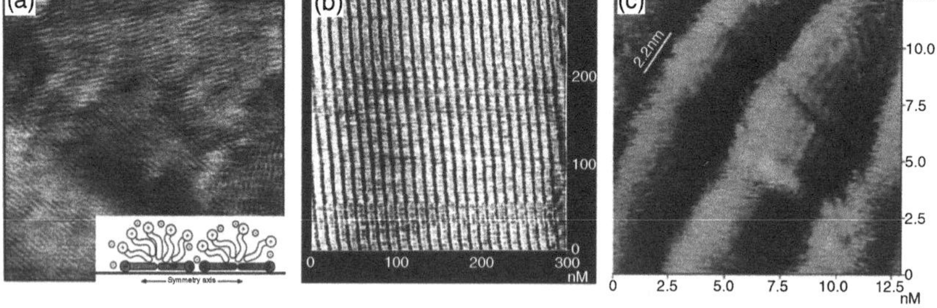

FIGURE 6.37 Microscopic images of surfactants segregate on the nonmetallic and metallic substrates. (a) AFM image of CTAB on graphite, the inset in the bottom is the proposed model of the adsorbate structure [135]; (b) AFM image of showing the coexistence of hemicylinders in the coexistence of SDS and dodecanol at the interface between graphite and aqueous solutions [138]; and high-resolution STM image of hemicylindrical aggregates of N-dodecyl-N,N-dimethyl-3-ammonio-1-propanesulfonate (DDAPS), a model zwitterionic surfactant, at a Au (111) surface [145].

Why does the deposition rate fall dramatically with increasing DTAB concentration? When too large an amount of surfactant is added into the solution, excess surfactant molecules are adsorbed on the surface of substrate, and shield a proportion of the catalytic active sites, affecting the self-catalytic reduction reaction smoothly. Therefore, the amount deposited onto the surface of substrate reduces, resulting in the plating rate decrease.

Now we turn to the class II of the shape of plating rate versus surfactants (the plots in the right of Figs 6.35 and 6.36). There are two facts. One is that the deposition rate falls dramatically with increasing the surfactant concentration. The second is that for a given surfactant concentration, the ionic surfactants show lower deposition rates than nonionic surfactants. The first phenomenon could be due to the fact that when surfactants are added to the solution, the surfactant particles are more likely to cover the cathode surface and so suppress the diffusion of the Ni^{2+} ions toward the interface and as a result hinder codeposition. In fact, Ger and Hwang [130] found from their measurement in EP Ni–P + PTFE system that as surface coverage on the cathode increased with increased surfactant concentration, it would therefore produce significant barriers for the Ni–P codeposition, that is, the deposition rate is decreased dramatically. As for the later phenomenon, the probable reason could be due to the electrostatic charge borne on the ionic surfactants that inhibit the reduction of Ni^{2+}, hence reducing the Ni content as a catalyst in the coating, resulting in the plating rate for it to be lower than that for the nonionic surfactant.

It must be said that the above explanation of the effects of added surfactants to the plating solutions on the deposition rates is qualitative because of the complicated nature of the deposition reactions and adsorption of surfactants to the as-deposited substrates. This point is just as pointed out by a group from Singapore [114] that "because of the incomplete understanding of the detailed mechanisms of the reaction mechanism of electroless deposition, it is difficult to draw any conclusive remark on the beneficial or detrimental effects of the charged surfactants on the deposition rate." Therefore, more work is needed to understand this problem thoroughly.

6.2 EFFECTS OF OPERATING CONDITIONS

6.2.1 Effect of pH Value

The pH value is an important influence parameter in EP. People have paid a great deal of attention to the research of pH value in the bath solution from the very beginning of the discovery of EN until today. The effects of pH value have been reported for many systems, including, in pure metals: Pb [5], Sb [7], Sn [148], In [65,77], Cd [64], Bi [66], Cu [67,102,149–152], Ag [9], Au [153], Ni [11], and Co [12,69,154]; in binary alloys: Ni–P [15,89,98,99,108, 155,156], Co–P [17,70], Ni–B [73,157], Fe–B [71], and Co–B [23,24]; in ternary alloys: Ni–Sn–P [38], Ni–Co–P [158], Ni–Co–B [74,159], Ni–W–P [27,160],

Co–W–P [31,160], Ni–Cu–P [161,162], Ni–Fe–B [42,43], Fe–W–B [47], Fe–Mo–B [48]; in quaternary alloys: Ni–Cu–Sn–P [53], Ni–Fe–La–B [75]; and in EP composite alloys: Ni–P + SiC [163], Ni–P + YSZ [164], Ni–B + Al_2O_3 [165], Ni–B + graphite [166], Cu–P + SiO_2 [167], and Ni–Cu–P + PTFE [133] have been investigated.

Only some typical plot shapes of the plating rate versus pH value are illustrated here, as shown in Figs 6.38–6.40. It can be seen evidently that all of them belong to class I. Of all of the studied EP systems, most have plot shapes that are the same as these, and most of them have mountain shapes. However, there are one or two exceptions, examples of which will be discussed later. It should be said that this is a feature of the effect of pH on the plating rate in EP.

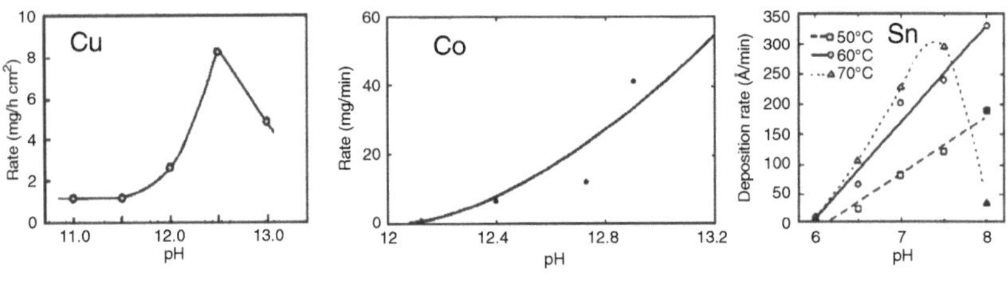

FIGURE 6.38 The dependence of plating rate on pH value. (Left) in EP Cu [150], (middle) in Co [12], and (right) in EP Sn [148].

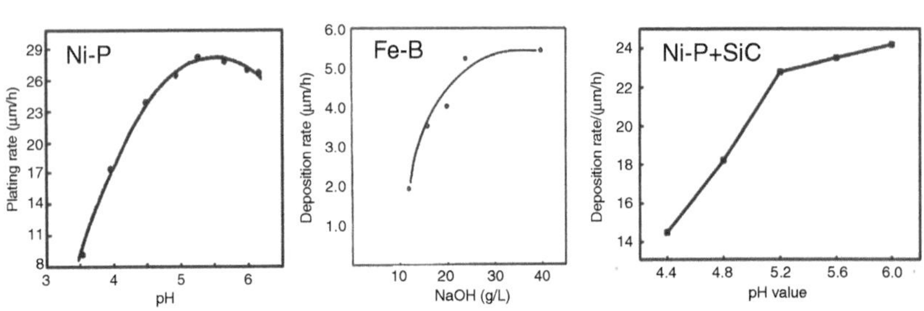

FIGURE 6.39 The dependence of plating rate on pH value. (Left) in EP NiP alloy [15], (middle) in Fe-B alloy [71], and (right) in EP Ni-P+SiC alloy [163].

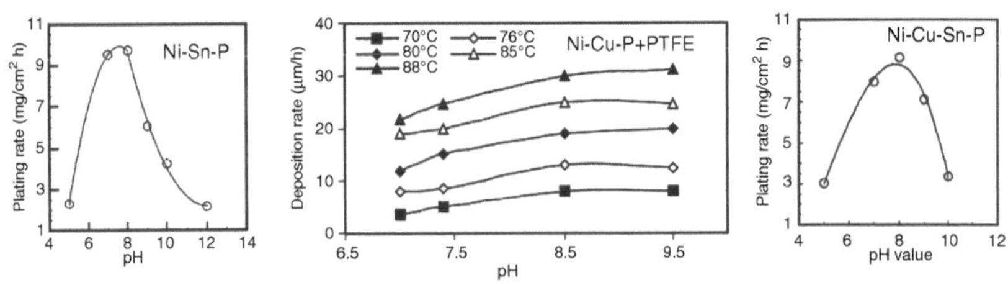

FIGURE 6.40 The dependence of plating rate on pH value. (Left) in EP Ni-Sn-P alloy [38], (middle) in Ni-Cu-P+PTFE alloy [133], and (right) in EP Ni-Cu-Sn-P alloy [53].

From the nature of electroless reaction, the previous phenomenon can be easily understood. We just discuss the mountain shapes more in detail. At low pH values, the H^+ ion concentration in the solution is high, and for the bath using NaH_2PO_2 as a reductant, the reaction equation for deposition of metal Ni atoms can be presented as

$$Ni^{2+} + 2H \rightarrow Ni + 2H^+ \quad (6.7)$$

or the overall reaction equation as

$$Ni^{2+} + 2H_2PO_2^- + 2H_2O \rightarrow Ni + H_2 + 4H^+ + 2HPO_3^{2-} \quad (6.8)$$

One can see from Eq. (6.7) or (6.8) that the more H^+ ions in the solution, the less the deposition of Ni is favored. When the pH is less than a certain value, no deposit of Ni forms at all. Many experiments confirm this point. For example, our group found that the Co–P deposits cannot be obtained when the pH value is less than 7 [17], and Anjana et al. [168] in 1994 confirmed that with Co–P, no deposition was found at pH 8.0. As the pH is raised, the reaction is shifted to the right of the equation. The H^+ ions decrease and the deposition of Ni increases; therefore, the deposition rate increases. After the pH reaches a certain value, for strongly alkaline solutions, two factors appear and become stronger and stronger as the pH is raised continuously. One is the concentration of free nickel ions becomes low [169], and another is the resulting precipitation of basic salts in the bath consumes some nickel ions. Both of them cause the plating rate to decrease. That is why the deposition rate presents a maximum at a particular pH value.

When DMAB was used as reductant in the bath solution, the total reaction equation is

$$(CH_3)_2NHBH_3 + 3Ni^{2+} + 3H_2O \rightarrow 3Ni + H_3BO_3 + (CH_3)_2H_2N^+ + 5H^+ \quad (6.9)$$

It is obvious that the situation is similar to that where NaH_2PO_2 was used. In addition, DMAB is easy to dissolve before pH < 5; thus the plating rate is also increased with pH value. As for using $NaBH_4$ or KBH_4 used as reductant, the reaction equation is different from that for the mentioned two cases:

$$4Co^{2+} + BH_4^- + 8OH^- \rightarrow 4Co + H_2BO_3^- + 5H_2C \quad (6.10)$$

The equation indicates that the more OH^- ions are present, the more metal will be deposited. Therefore, as the pH value increases, that is, increasing the OH^- concentration, the deposition rate also increases. The hydrolysis of $NaBH_4$ in water tends to increase the solution's alkalinity, so if this is limited the coatings cannot be obtained.

After the plating rate reaches its maximum, if the pH value keeps increasing, the concentration of free metal ions becomes lower and lower. Fewer and fewer metal ions become available for reduction in the reaction with NaH_2PO_2

and DMAB as reductant in the bath solution. In addition, the basic salt that is produced precipitates out, causing the solution to thicken, and thus the plating rate decreases. When using $NaBH_4$ as reductant, although the reduction of metal ions becomes rather easy as the solution's alkalinity increases (6.10); however, the plating rate still decreases due to the metal ions gradually decreasing and more and more basic salt produced. In the EP Ni–Sn–P [38] and Ni–Cu–Sn–P [53], the experiments by our group indicated that when the pH is higher than 12, the deposit could still form but the deposition rate is very small because of the formation of a colloidal solution. In 2008, Hamdy et al. [98] were working on EP Ni–P alloy systems. They found that the plating rate reaches a maximum at pH 9, but at pH 10, the solution turns turbid due to precipitation of basic nickel salt or complex, after then the bath solution cannot work normally. It is obvious that if such a phenomenon just appears at around the maximum value, then in the plot of plating rate versus concentration no falling side occurs and only the rising edge can exist.

Now we turn to the exception of one or two examples mentioned above. In 1970, Okinaka [153] found that the plot shapes of the plating rate versus pH value (KOH) in an EP Au system have both a falling side and rising side when KBH_4 or DMAB are used as the reductant, respectively, as shown in the left and middle plots in Fig. 6.41. Okinaka believed that the falling side produced by KBH_4 reductant is contributed from the hydrolysis reaction of KBH_4. The hydrolysis reactions are known to be acid catalyzed, and therefore they proceed more rapidly at lower pH values. So, the plating rate of gold increases with decreasing KOH concentration. However, to prevent spontaneous decomposition, the KOH concentration of the bath must be kept greater than 0.1 mol/L. When DMAB is used in place of borohydride, Okinaka assumed that the actual reducing species in the DMAB bath is also BH_3OH^- and that this species is produced by the direct reaction of DMAB and OH^- ions:

$$(CH_3)_2NH \bullet BH_3 + OH^- \rightarrow (CH_3)_2NH + BH_3OH^- \qquad (6.11)$$

Accordingly, he believed that an opposite dependence appears; that is, the plating rate increases with KOH concentration. It is obvious that such an interpretation does not enable people to truly understand the nature of production for

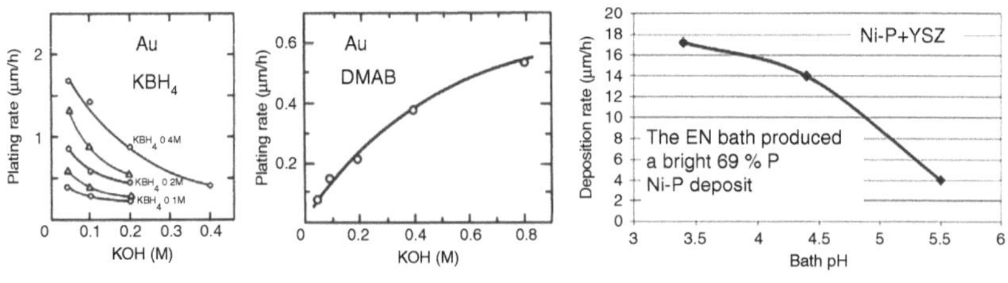

FIGURE 6.41 The dependence of plating rate on pH value. (Left and middle) in EP Au [153], and (right) in EP Ni-P+YSZ composite alloy [164].

the falling side. In particular, EP Ni–B [73], Co–B [23], and Fe–B [71] should be compared, although the reductants used in these deposition reactions are either $NaBH_4$ or KBH_4, but the obtained plot shapes of the plating rate versus pH value all belong to class I, which conflicts directly with that obtained by Okinaka [153]. Accordingly, the explanation of the plot shape in the left of Fig. 6.41 needs to be considered further.

Another issue is the plot shape in EP Ni–P + YSZ [164], which belongs to class II. This was explained by its original authors Baba et al. [164] as follows: "The plating condition – that is the area of the substrate to the bath volume is depleted from its optimum level after certain period of time due to evaporation of solution. On top of that, the addition of ceramic powder to the bath might have disturbed the system and caused this observation. Using larger bath volumes could minimize the effects of temperature variation and evaporation of the solution." These reasons may be difficult to convince people completely, and hence further study is needed.

6.2.2 Effect of Plating Temperature

Temperature is the most important parameter to affect the EN deposition rate. All of the EP reactions are carried out at an appropriate temperature. The effects of temperature on plating rate have been investigated in a lot of systems, including pure metals: Cd [64], In [65], Bi [66], Cu [102,170], Ag [9], Au [10], Ni [11,171], and Co [12,69]; binary alloys: Ni–P [16,98,99,156,172], Co–P [17,70], Ni–B [21,73], Fe–B [71], Co–B [24,72], Cu–P + SiO_2 [167], and Ni–B + SiC [173]; ternary alloys: Co–W–P [31], Fe–W–B [47], Fe–Mo–B [48], and Ni–Fe–B [42]; quaternary and quinary alloys: Ni–Cu–Sn–P [53], Ni–Fe–La–B [75], and Ni–Co–Fe–La–P [76]. All of the shapes for the plating rate versus temperature in these systems are the class I: the mountain shape appears in the Cu, Ag, Ni–B, Fe–B, Ni–B + SiC, Fe–W–B, and Fe–Mo–B systems, and the rising edge shape appears in other systems. No class II shapes have been found. Figure 6.42 shows some typical plots.

All the above experiments show that the electroless plating needs to react at a certain temperature; otherwise, the coating cannot form. The reason is that if temperature is too low the complex metal ions usually cannot be released

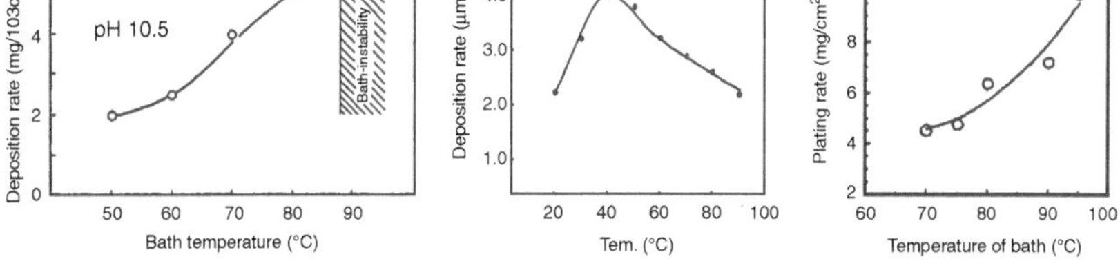

FIGURE 6.42 The dependence of plating rate on temperature. Left in Co–P [70], middle in Fe–B [71], and right in Ni–Cu–Sn–P alloy [53].

from the complexing group of ions because of a lack of energy. Several groups have measured the apparent activation energy of the EP bath which can be obtained from Arrhenius plots (the ln(plating rate) against 1/T relationship), for example, 73.7 kJ/mol in EP Au [10], about 23 kJ/mol in EP Co [12], 26.8 kJ/mol and 68.2 kJ/mol in citrate bath with pH = 9 and pH = 5, respectively, in EP Ni–P alloys [174], 61.9 kJ/mol and 76.6 kJ/mol in ammonium acetate bath with pH = 5.5 and pH = 4.6, respectively, in EP Ni–P alloys [172], and 40 kJ/mol in EP Ni–B alloys [175]. That is to say that some form of energy must be provided in order for the reaction to proceed. Of course, the most effective and convenient forms are heat energy provided by heating. Thereby, the electroless plating must be started at a certain temperature.

Once the EP reaction has started, two reasons can be given to explain the experimental phenomenon that the plating rate increases with increase of temperature. The first is that the concentration of metal ions released from the complexed group of ions increases with increasing temperature, thereby the plating rate increases. The second is mentioned above that the plating rate $\propto \exp(-E_a/RT)$ has been confirmed by both theoretical model and experimental facts [16,84,176]. It is obvious that as the temperature increases, the plating rate must be enhanced.

The plating rate will reach up to a maximum value at certain temperature. After then, the tendency of auto decomposition for the bath solution and precipitation from the solution would increase, which causes its stability to worsen as temperature increases continuously thereby results in the plating rate decreases with temperature. In addition, it is logical that the quality of deposited coatings deteriorates when the temperature is high enough. Many experiments have recorded such a phenomenon. For example, Hanna et al. [102] reported in EP Cu that above 50°C, the tartrate additive-free bath decomposes rapidly and the rate of deposition decreases sharply. In 2008, Hamdy et al. [98] reported that when the temperature exceeded 90°C, the EP Ni–P coating appears dull. Lin et al. reported in EP Ni–B + SiC [173] that the bath solution automatically decomposes and its color becomes black when temperature reaches around 75°C. Yang et al. in EP Ni–Fe–La–B [75] reported that when the temperature reaches 70°C, the bath solution is extremely unstable, appearing cloudy; the coating surface becomes black and brittle, and there is a lot of pitting, peeling, and shedding.

If the bath solution decomposes automatically when the plating rate approaches a given value, and precipitation from the solution becomes such that no deposit can form, then of course only the rising edge occurs, and no mountain shape can be found.

From the previous discussion, it is easy to see that the shape of the plating rate versus temperature is not the class II type.

6.2.3 Effect of Plating Time

There are fewer reports of the dependence of plating rate on plating time than on metal salt, reductant, complexing agent, pH value, and plating temperature.

Such reports appear in EP metal and alloy systems: Cu [87], Ag [177], Au [178], Pd [179], Co–P [180], Ni–P [156,181], Ni–B [21,182], Fe–B [71], Ni–P + SiO_2 [167], Ni–P + Al_2O_3 [98], Ni–Co–B [74], Ni–Co–P [157], Ni–Cu–P [161], Fe–W–B [47], Fe–Mo–B [48], and Co–W–P [31].

It is obvious that as the plating time increases, the concentration of reactants decreases gradually because the formation of the coating must consume the metal and nonmetal ions. In addition, the formation of some products such as hydrated hydroxides or basic salts also affects the speed of the reaction, causing the plating rate to be reduced. For the nonautocatalytic processes, for example, Co and Fe alloys, the plating rate will decrease due to the poor catalytic performance after the coatings has formed. So, the shapes of plating rate versus plating time plots are generally class II, as shown in Fig. 6.43.

But this is not always the case. In some cases, the shapes of plating rate versus plating time are class I, as shown in Fig. 6.44. How can these experimental phenomenon be explained? The original authors explained the shape of the EP Co–P alloy curve [180] as follows "the increase in rate during the first 10 min is due to the very high catalytic activity of palladium formed during activation of the copper surface. After 10 min, the decrease in rate may be due to the deposition of cobalt, which has a low catalytic activity compared with palladium." This explanation may be somewhat unsatisfactory because the very high catalytic activity of palladium formed during activation of the copper surface is also present in other EP systems. For example, in a system in which the substrate was also

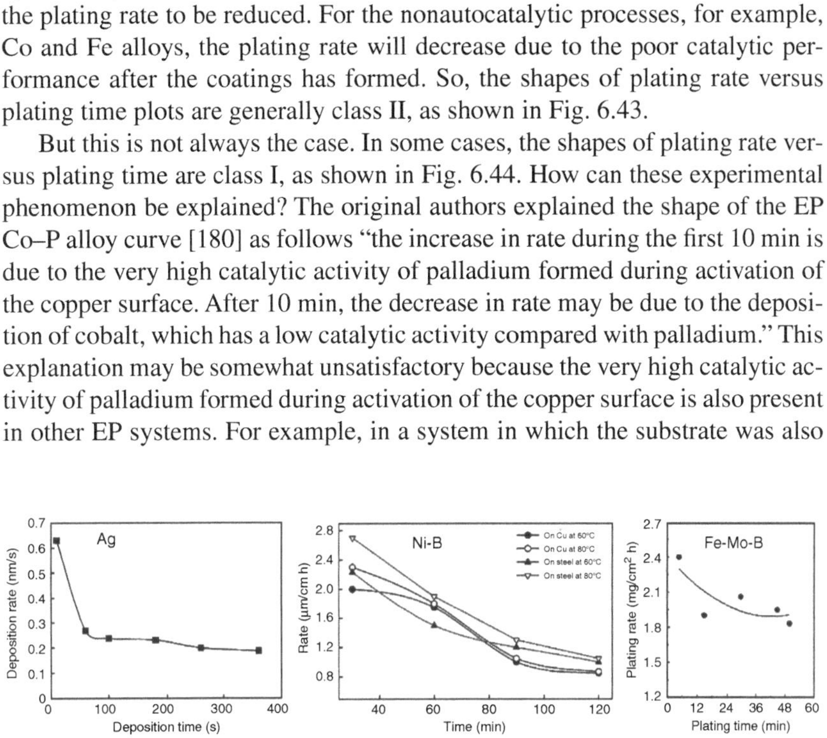

FIGURE 6.43 The dependence of plating rate on plating time. Left in EP Ag [179], middle in EP Ni–B [182], and right in EP Fe–Mo–B alloy [48].

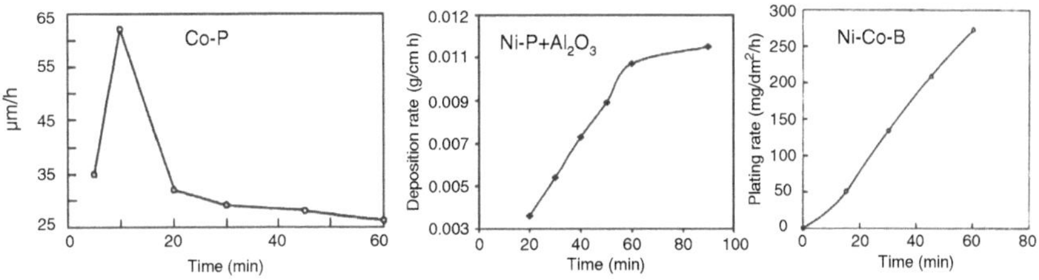

FIGURE 6.44 Deposition rate of EP Co-P [181], Ni–P + Al_2O_3 [98] and Ni–Co–B [74] versus plating time.

Cu that was pretreated by a mixed solution of $SnCl_2$ and $PdCl_2$ before coating in our Co–W–P deposition, the shape obtained was class II, not class I (original Fig. 8 in [31], not shown here). As for the plot of EP Ni–B (middle in Fig. 6.43), the original authors Hamid et al. [98] did not try to explain it. Looking at the plot on the right in Fig. 6.43, the original authors Narayanana [74] just said that "the deposited Ni–Co–B alloy deposit continuously provides a catalytic surface for subsequent coating formation." However, for the EP Ni–Co–P alloy studied by the same group [157] their explanation is that "This is due to the build-up of orthophosphite ions following oxidation of sodium hypophosphite." More work is obviously needed to unite these explanations?

6.3 EFFECTS OF OTHER TECHNOLOGICAL PARAMETERS

In addition to the above effects of bath components and operating conditions, other technological conditions such as stirring (gas flow, mechanical and magnetic stirrings, and ultrasonic radiation), light illumination (ultraviolet light and laser), pressure, and magnetic field all affect the reaction of the bath solution and so the plating rate. Fewer reports are found in the literature than for the factors discussed previously. Even so, the breadth of the topic means that will just cover some selected parameters.

6.3.1 Effect of Stirring

6.3.1.1 Effect of Mechanical Stirring

Very few papers have been published on the effect of mechanical stirring on the plating rate. In 2011, Bulasara et al. [183] investigated the variation of average Ni plating rate with average $NiSO_4$ solution concentration and stirrer speed. They used two loading ratios (defined as substrate surface area per unit volume of plating solution): 196 cm^2/L and 393 cm^2/L. The left plot in Fig. 6.45

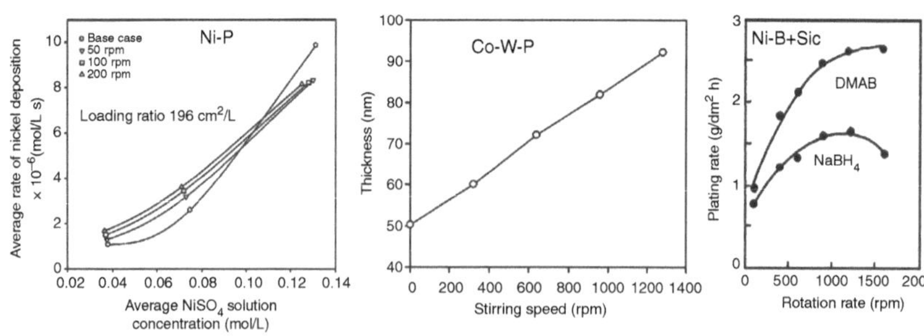

FIGURE 6.45 The left shows variation of average Ni–P plating rate with average $NiSO_4$ solution concentration and stirrer speed, loading ratio = 196 cm^2/L [183]. The middle is coating thickness versus. stirring speed in EP Co–W–P [184]. And the right is the plating rate of Ni–B + SiC versus mechanical rotation rate [185].

just shows the results for loading ratio 196 cm^2/L, but similar results have been obtained for the loading ratio 393 cm^2/L. It can be seen that the values of average nickel deposition rate increased with stirring speed for the initial solution concentrations of 0.04 and 0.08 mol/L and were higher than those of the base case (0 rpm). The stirred plating baths provided about 20–56% excess nickel plating rate when compared with the baths without stirring. However, this was not the case with the higher initial solution concentration (0.16 mol/L) where the values of the average nickel deposition with stirring were lower than those of the base case (without stirring). The authors believe that the 20–56% excess nickel plating rate is caused by mass transfer enhancement with stirring. Their measurements indicate that stirred electroless hypophosphite baths do not function efficiently at higher stirrer speeds and higher solution concentrations. This is attributed to the enhanced nucleation under these conditions in the solution, which contributes significantly to inefficient metal plating.

In 2008, Kim et al. [184] studied the effect of stirring rate on the plating rate of electrolessly deposited Co–W–P thin films. The results are shown in the middle of Fig. 6.45. The deposition rate increases dramatically with solution agitation. The thickness of the film deposited from stagnant solution is 50 nm, while that of the film deposited under a solution agitation of 1280 rpm is 92 nm. It is well known that solution agitation greatly enhances the deposition process due to increased transportation of the ionic species from the bulk of the solution to the electrode surface. Therefore, the deposition rate and the thickness of the films increase with increase in solution agitation.

In 1989, Matsuoka and Hotta [185] measured the deposition rate in composite coatings Ni–P + SiC as a function of mechanical rotation rate. Two bath solutions were used, one contains $NaBH_4$ as reductant, and the other has DMAB. From the results shown in the right of Fig. 6.45, it is clear that as the mechanical rotation rate increases from 0 to 1200 rpm the plating rates increase from 0.78 $g/dm^2{\bullet}h$ to 1.6 $g/dm^2{\bullet}h$ and from 1.0 $g/dm^2{\bullet}h$ to 2.6 $g/dm^2{\bullet}h$ corresponding to the $NaBH_4$ and DMAB baths, respectively. The authors do not suggest an explanation for these enhancements.

6.3.1.2 *Effect of Ultrasound Stirring*

More reports for the effect of ultrasonic radiation on the plating rate have been published, which mainly concentrate on the EP Ni–P alloy systems. Early in 1955, Rich [186] reported the influence of ultrasonic radiation on EN deposition. For the alkaline hypophosphite bath solution, the plating rate was five times greater with 20 kHz ultrasonic radiation compared with ordinary chemical plating. In 1978, Mallory investigated the effect of ultrasound (49 kHz, 30 W) on plating rate in EP Ni–P deposits [176] and showed that ultrasound increased the deposition rate in four different chelating (lactic acid, glycolic acid, glycine, and citric acid + lactic acid) systems. However, there was no effect when using citric acid as a complexing agent. The experimental data for effect of ultrasound on plating rate of various EP alloy deposits are shown in Table 6.4.

TABLE 6.4 Effect of Ultrasound on Plating Rate for EP Alloy Deposits

EP alloys	Chelate (g/L)	Ultrasonic conditions	pH	Temperature (°C)	Deposition rate (μm/h)		Reference (year)
					With ultrasound	No ultrasound	
Ni–P	1, Citric acid 0.2 mol/L 2, Lactic acid 0.3 mol/L 3, Glycolic acid 0.3 mol/L 4, Glycine 0.3 mol/L 5, Citric acid 0.07 mol/L + Lactic acid 0.3 mol/L	49 kHz, 30 W	4.8	87	5 42 34 20 23	5 26 20 14 11	1 (1978)
Ni–P	Citrate sodium 0.1 M	25 kHz, 150 W	4 6 9	80	1.35 6.96 16.9	1.35 2.36 6.8	2 (1986)
Ni–P	Formic acid	Not stated	5.5	80	16.9	16.9	3 (1993)
Ni–P	Sodium citrate 20 g/L	33 Hz,1 W/cm^2	4.8	78–82	11.54	8.37	4 (1997)
Ni–P	Sodium citrate 100 g/L	25 kHz, 400 W	8–9	93	45.0–50.0	15.0	5 (1998)
Ni–P	Sodium citrate 50 g/L	28 kHz 45 kHz 100 kHz	7	50	3.8 4.7 2.4	1.6	6 (2000)

Ni–P	Sodium citrate 30 g/L	80 kHz	100 W 150 W 200 W 250 W	8.1	70	9. 052 10. 93 10. 01 6. 84		7 (2007)
		Unspecified		8.1	50 60 70 75	4.37 8.07 12.09 14.15		
		40 kHz	150–160 200–210 240–250	8.1	Unspecified	7.71 7.1 6.33		
		80 kHz	150–160 200–210 240–250	8.1	Unspecified	10.93 10.01 6.8		
Ni–P	Sodium citrate 25 g/L	20 kHz, 42 W/cm^2		5.2	50 70 90	4.83 14.67 15.89	0.78 0.89 8.67	8 (2012)

(Continued)

TABLE 6.4 Effect of Ultrasound on Plating Rate for EP Alloy Deposits *(cont.)*

EP alloys	Chelate (g/L)	Ultrasonic conditions		pH	Temperature (°C)	Deposition rate (μm/h)		Reference (year)
						With ultrasound	No ultrasound	
Co–P	Sodium citrate 0.2 mol/L	28 kHz	100 W/m^2 300 W/m^2 500 W/m^2	Unspecified	80	2.5* 3.6* 4.2*	1.0*	9 (1993)
	Sodium citrate 0.2 mol/L	28 kHz, 500 W/m^2		7 7.5 8 8.5 9	80	0.05 1.89 5.56 2.22 1.11	0.78 2.0 1.67 0.94 0.67	
Co–W–P	Potassium sodium tartrate 0.5 mol/L	30 W 60 W 90 W 120 W 150 W		10.5	85	20 $g/m^2 \bullet h$ 28 $g/m^2 \bullet h$ 32 $g/m^2 \bullet h$ 38 $g/m^2 \bullet h$ 46 $g/m^2 \bullet h$	15 $g/m^2 \bullet h$	10 (2010)

1 Mallory GO. Trans Inst Met Finish 1978;56:81.
2 Matsuoka M, Hayashi T. Metal Finish 1986;84(3):27.
3 El-Mallah AT, Abbas MH, Nassar MM, Aly IH, Younan MM, Gad MR. Met Finish 1993;91:47.
4 Yang L-X, Hou W-T, Wu Y-S. Trans Inst Met Finish 1997;75:131.
5 Vasudevan R, Narayanan S, Karthik PR. Trans Ind Inst Met Finish 1998;51:445.
6 Kobayashi, Chiba A, Minami N. Ultrasonics 2000;38:678.
7 Yang P, Gong L, Gu B, Sun H. Surf Technol 2007;36(6):61.
8 Cobley AJ, Saez V. Circ World 2012;38(1):12.
9 Matsuda H, Takano O. Trans IMF 1993;71(4):136.
10 Yu Y, Wei G, Hu X, Ge H, Yu Z. Surf Coat Technol 2010;204:2669.
Normalized deposition rate.

In 1986, Matsuoka and Hayashi [187] reported the effect of $TlNO_3$ concentration on the plating rate of Ni–P alloys with or without ultrasound (25 kHz, 150 W). The influence was different depending on the bath pH and the type of complexing agents employed. In the citrate bath with pH 5–9, the deposition rate of Ni–P alloys increased by a factor of two or three compared with the same bath without ultrasonic radiation; however, in the succinate bath with pH 4 to 5.5, the accelerating effect by ultrasonic was not significant. These data are also shown in Table 6.4. It is obvious that the results in the citrate bath obtained by Matsuoka and Hayashi conflict with those obtained by Mallory [176], because Mallory found that ultrasound had no effect when using citric acid as a complexing agent. The authors argued that the reason that ultrasonic radiation enhanced the plating rate was due to the enhancement of mass transfer, whenever the mass transfer was the rate-determining step. The increase in deposition rate seemed to be due to a substantial reduction in thickness on the diffusion layer, which facilitates the mass transfer of the reactants.

In 2000, Kobayashi et al. [188] reported the effect of ultrasound on EP Ni–P alloys. They found that the values of deposition rates were 1.6, 3.8, 4.7, and 2.4 μm/h present in the citrate bath with no radiation, 28, 45, and 100 kHz of ultrasonic frequency at pH = 7.0, 323 K, and 40 g/dm^3 of $NiSO_4{\bullet}6H_2O$. That is to say, the deposition rate with 45 kHz of ultrasound was three times that without ultrasound. The authors believed that the imposition of ultrasound accelerated the diffusion of the Ni^{2+}–citrate complex, and the electroless deposition reaction of nickel is controlled by a mass transfer process whether ultrasound is present or not.

In 2002, Park et al. [189] reported that the amount deposited was increased by 15% using an ultrasound frequency of 40 kHz compared with using 0 Hz frequency. However, at a higher frequency of 68 kHz, the deposit was similar to that at 0 Hz. Yang et al. in 2007 [190] also studied the effect of ultrasonic power on deposition rate in EP Ni–P alloys. When the ultrasonic powers are 100, 150, 200, and 250 W, the corresponding plating rates are 9.0517, 10.927, 10.012, and 6.842 μm/h, respectively. High ultrasonic power (250 W) causes the plating rate to decrease. These authors also found that when the ultrasonic power and other conditions are the same, the plating rate with 80 kHz is 1.5 times than that with 40 kHz. It is obvious that this result is the opposite to that obtained by Park et al. [189]. These authors did not pay attention to such conflicting phenomena in their paper.

In 2012, Cobley and Saez [191] measured the effect of ultrasound (20 kHz) on the plating rate for an electroless Ni–P bath at different operating temperatures, as shown in the left of Fig. 6.46. It can be seen from the figure that when plating occurs in an ultrasonic field the deposition rate is always significantly higher than when no ultrasound stirring is used. This is particularly noticeable at 70°C where the use of ultrasound not only produced plating rates which were much higher than the equivalent rates under silent conditions but, in addition, they were at least 1.5 times higher than

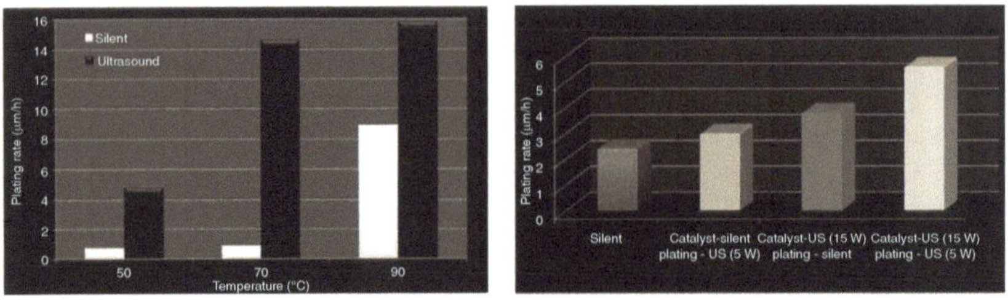

FIGURE 6.46 Left: effect of ultrasound (20 kHz) on the plating rate for an electroless Ni–P bath at different operating temperatures determined using "weight gain" on plated test coupons [191], and right: ultrasound influence on the activation step before electroless coating in EP Cu metal [192].

those obtained without ultrasound at 90°C. This is a very important result in terms of low energy plating and suggests that ultrasound could enable a significant reduction in operating temperature. It is particularly interesting to note the results about the use of ultrasound in both steps (activation and plating) obtained by Touyeras et al. [192] in 2003, which are shown in the right of Fig. 6.46. The best conditions found in the plating rates optimization consist of a very strong ultrasonic agitation (15 W) during the activation step and the lowest acoustic power (5 W) during the first 5 min of plating in EP Cu metal. Plating rates reach up to a value of 5.6 µm/h. This value can be considered important for copper electroless plating processes. The authors believe that ultrasound seems to have a local stirring effect that allows the palladium complexes to dissolve. However, from the results they think that ultrasound may influence other mechanisms involved in the copper deposition, which may be hydrogen degassing, local thermal effects, etc.

In 2007, Abyaneh et al. [193] measured the plating rate of Ni–P deposition versus the concentration of nickel chloride for a bath containing sodium hypophosphite at a concentration of 0.19 mol/L and sodium acetate at 0.04 mol/L (a) in the absence and (b) in the presence of ultrasonic irradiation, as shown in the left of Fig. 6.47. The result shows a 75% increase in the plating rate at nickel

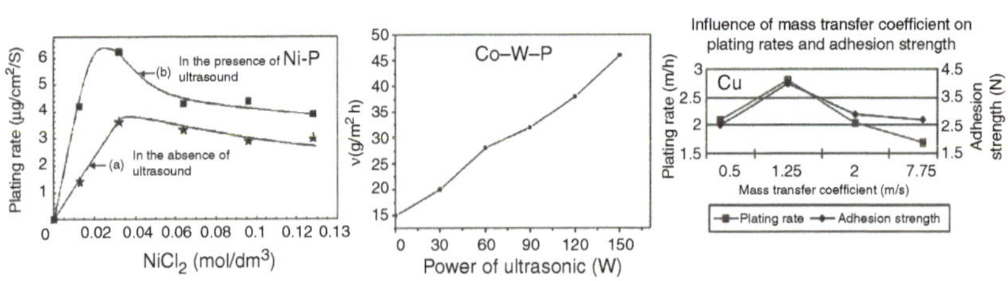

FIGURE 6.47 Effect of ultrasound on the plating rate. Left: compared EP Ni–P (a) in the absence and (b) in the presence of ultrasonic radiation [193]. Middle: effect of power ultrasonic on the plating rate of Co–W–P alloy [160]. Right: influence of mass transfer coefficient on plating rates and adhesion strength in EP Cu [196].

chloride and sodium hypophosphite concentrations of 0.03 and 0.19 mol/L, respectively. In 2010, Yu et al. [160] reported the effects of ultrasound on the deposition speed in an EP Co–W–P alloy, as shown in the middle of Fig. 6.47. Evidently, the deposition speed increases dramatically in the presence of ultrasound. This was due to the ultrasound agitating the solution, which accelerated the deposition process by increasing the transportation rate of ionic species from bulk solution to the electrode surface, as was pointed out by Kim [184]. In 1993, Matsuda and Takano [194] also obtained similar results with EP Co–P alloys (see Fig. 3 in the original paper). For the bath solution (cobalt sulfate 10 mol/m^3, sodium hypophosphite 200 mol/m^3, sodium citrate 200 mol/m^3, and boric acid 500 mol/m^3), they found that the normalized deposition rate increases from 1 to 4.2 when the ultrasonic intensity increased from 0 to 500 W/m^2. In 2005, Touyeras et al. [195] measured the influence of mass transfer coefficient on plating rates and adhesion strength in EP Cu, as shown in the right of Fig. 6.47. They believed that mass transfer is the deciding factor for the influence of ultrasound in EP deposition because their measured plating rate and practical adhesion versus mass transfer coefficient k_d shows a similar dependence for 500 kHz.

Several conclusions can be drawn from these experimental facts and discussion.

The first is that all experiments indicate an increase in plating rate when the bath is stirred (mechanical, especially ultrasonic stirring). Abyaneh et al. [193] even said that all researchers agree on this.

Second, the increase in the plating rate due to the ultrasonic radiation depends upon the composition of the bath solution and the ultrasonic power and frequency. Abyaneh et al. [193] pointed out that only a 15% increase in the plating rate was achieved with an ultrasonic irradiation of 40 kHz obtained by Park et al. [189], but a 75% increase in the plating rate was obtained by themselves. In fact, the complexing agents used by these two groups are different, one is citrate acid [189], and the other is acetic acid [193] though the rest of their baths are nearly the same. But, Abyaneh et al. did not do any analysis of this phenomenon, and did not conduct experimental investigation to investigate the reasons further.

Finally, the explanation for enhancing the plating rate caused by the application of ultrasound is different, no unified view. In 1978, Mallory used the well-known cavitation phenomenon to explain his experimental data [176]. He believed that the cavitation achieved by ultrasound can irradiate the energy to the bath solution system in the form of heat. When the absorption of energy (temperature increase) is highest at the solution–metal (substrate) interface, the exponential term of the plating rate becomes larger as the rate increases. Several authors attribute this phenomenon to mass transfer [187,188,194,195], but convection alone has a minimal effect on the rate of deposition [160]. Other authors believe that the increase in plating rate is because of an increase in the rate of nucleation of crystals. Electroanalytical methods have shown that the rate of nucleation of crystals was significantly increased when ultrasonic irradiation was employed during the electrocrystallization of copper [196,197]. More in-depth research is, therefore, needed to elucidate the effects of ultrasound on the kinetics of electroless deposition. Touyeras et al. [195] also made a similar point when they said: "This influence of ultrasound during the plating step is more difficult to explain."

6.3.2 Effect of Magnetic Field

Only few studies on the effects of magnetic fields on the electroless deposition process have been published, in metals: Ag [198], Co [199]; in binary and ternary alloys: Ni–Co [200], Ni–P [201], Co–Fe [202], and Co–W–P [160]. First, we will illustrate the experimental results, and then we take a look at how to understand them.

Figures 6.48 and 6.49 indicate the effects of magnetic intensity on the plating rate or coating thickness in five EP systems. From these experimental data,

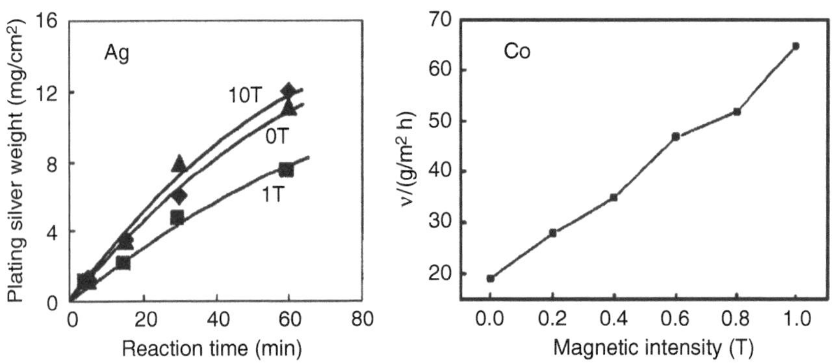

FIGURE 6.48 Effect of magnetic intensity on the plating rate. Left: in EP Ag [198]. Right: in EP Co [199].

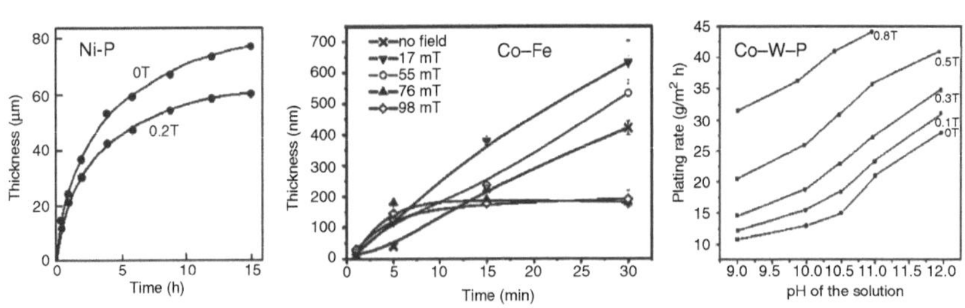

FIGURE 6.49 Effect of magnetic intensity on the plating rate. Left: in EP Ni–P [201]. Middle in EP Co–Fe [202]. Right: in EP Co–W–P [160].

two different effects can be seen. One is that the plating rate increases with increasing magnetic field intensity in the EP Co metal [199] and Co–W–P alloy [160] systems. In the EP Co, the deposition speed was about 20 $g/m^2 \cdot h$ without a magnetic field. However, increasing the magnetic field intensity up to 1 T yielded a 65 $g/m^2 \cdot h$ deposition speed.

Other experimental data showed that the plating rate first increased and then decreased with increasing applied magnetic field intensity, or just the opposite. Plating silver weight decreased quickly as magnetic field increased to 1 T [198]. However, when the magnetic field increased up to 10 T, plating silver weight increased over that of 0 T. The plating weight of Ag is 3.4 mg/cm^2 in 0 T, 2.1 mg/cm^2 in 1 T, and 3.5 mg/cm^2 in 10 T, respectively. In the case of EP Ni–P alloy, deposition rate with no magnetic field was 18.4 μm/h, and in a magnetic field it was 15.2 μm/h [201]. For a 1 h deposition time, the deposition rate was suppressed by 12%, in 12 h it was inhibited by 26%. For the EP Co–Fe alloy, a low value of magnetic field intensity (17 mT) favors the deposition process by almost doubling the deposition rate, from 11.2 nm/min without field to 21.3 nm/min at a field of 17 mT [202]. However, as magnetic field increases (up to 76 mT), the deposition rate tends to decrease.

It is evident that the experimental data are revealing conflicting phenomena, and the effects of magnetic field on the plating rate vary a great deal for reasons that we do not yet understand.

Most researchers agree that a magnetic field can accelerate the EP reaction [200–202]. This is because the Lorentz forces from the magnetohydrodynamic (MHD) act when a magnetic field acts on the bath solution. The Lorentz forces act on the electrolyte/substrate interface, inducing additional electron convection in the solution. This additional convection term might reduce the thickness of the steady-state depletion layer and accelerate the reaction kinetics. Therefore, the deposition rate enhances in the presence of low magnetic field.

In addition, besides the Lorentz forces, when a magnetic field acts on magnetic ions, another magnetic force, that is, field gradient force ($F_{\nabla c} = \chi_m B^2 \nabla c/2\ \mu_0$, χ_m is the molar magnetic susceptibility, c is the concentration, and μ_0 is the permeability of vacuum) arises. The field gradient force would induce a convection effect during the deposition that would affect mass transfer and in turn result in an increase in the reaction speed. However, no one has as yet explained the conflicting experimental results.

6.3.3 Effect of Bath Loading

Bath loading is defined as the ratio of the surface of the working piece to the volume of bath solution. It should be pointed out that although the effect of bath loading on the plating rate has been studied since the 1970s [203], very few reports have been published until now.

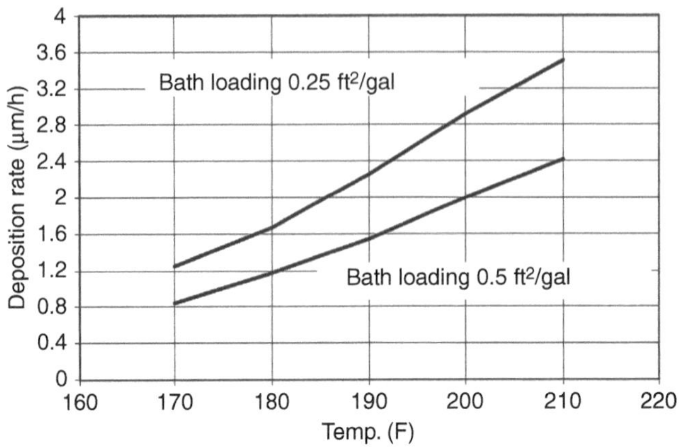

FIGURE 6.50 Effect of bath loading and temperature on the plating rate in EP Ni–P alloys [203]. 0.25 ft^2/gal = 0.5 dm^2/L, 0.5 ft^2/gal = 1.0 dm^2/L.

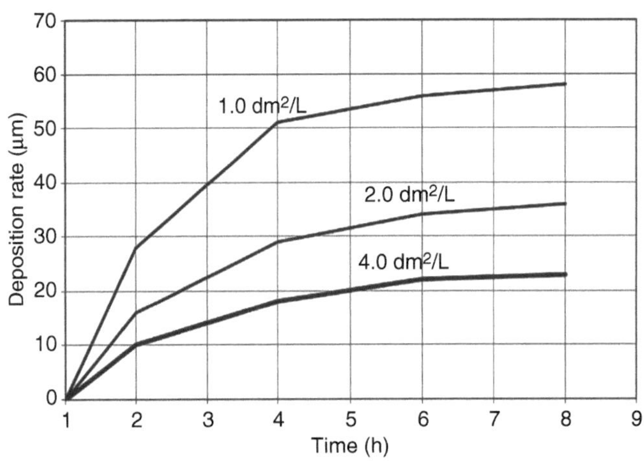

FIGURE 6.51 Effect of bath loading and time on the plating rate in EP Ni–P alloys [204].

The effect of bath loading and temperature on the plating rate in EP Ni–P alloys is shown in Fig. 6.50 [203]. In 1983, Grunwald [204] measured the plating rate versus bath loading and time in EP Ni–P alloys, as shown in Fig. 6.51. These two plots indicate that the plating rate decreases with increasing bath loading. From Fig. 6.51, the plating rate is 10.8 μm/h when bath loading is 1.0 dm^2/L, it decreases down to 6.4 μm/h at a bath loading of 2.0 dm^2/L, and it is only 4.0 μm/h at a bath loading of 4.0 dm^2/L. Commercial baths operate in a bath loading range of 0.1–1.0 dm^2/L.

Our group measured the effect of bath loading on the plating rate in EP Ni–B alloys [21]. The results are shown in Fig. 6.52. It is obvious that the plating rate decreases as the bath loading increases.

It is interesting to note that Sharma et al. [205] in 2005 reported the variation in coating rate for the Ni–P + ZrO_2 + Al_2O_3 + Al_3Zr composite coating for different bath loadings, as shown in Fig. 6.53. This figure shows that the rate of

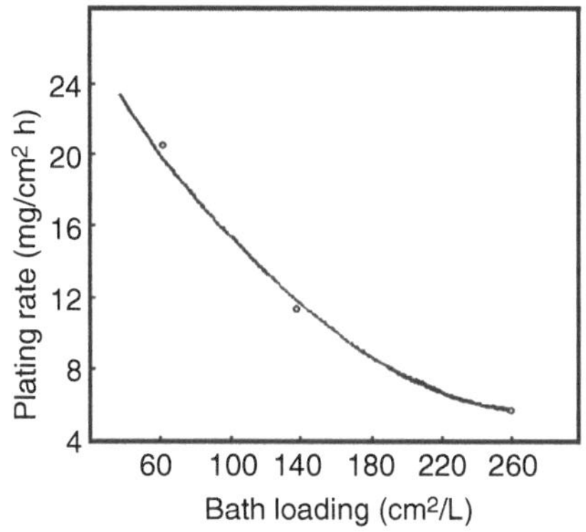

FIGURE 6.52 **Effect of bath loading on the plating rate in EP Ni–B alloys [21].**

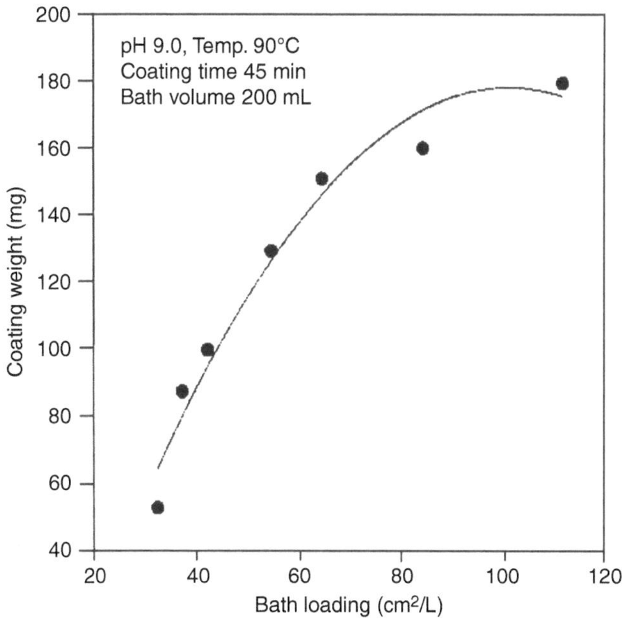

FIGURE 6.53 **Variation of coating weight with bath loading factor in electroless composite Ni–P+ ZrO_2 + Al_2O_3 + Al_3Zr) coating [205].**

deposition increases with an increase in bath loading, which is just opposite to the Ni–P and Ni–B systems reported previously. The reason may be that as the bath loading increases, more catalytic surface is exposed to electroless deposition, which in turn creates more nucleation sites. Of course, more studies are needed to understand and explain these conflicting experimental observations.

6.4 SUMMARY

From the previous description and discussion, the following conclusions can be drawn.

First, research of the effects of various factors on the plating rate has been quite extensive and in-depth. It should be noted that much progress has been made, resulting in a good insight into this subject.

The second is too many parameters, including bath composition (metal salts, reducing agent, complexing agent, stabilizer, accelerating agent, and surfactants), operating conditions (bath pH value, plating temperature, plating time), and other technological parameters (stirring and magnetic field), affect the electroless plating process. This indicates that the electroless plating process is so complicated that its exact mechanism is still some way from being fully understood.

Third, it can be seen from the experimental results reviewed here that there is much conflict over the effect of various parameters on the plating rate in many EP alloy systems. The facts indicate that not only the experimental conflict phenomenon has not been solved, but also much experimental data are not understood correctly. In a word, the theory is not satisfactory, and we can even say that the theory is far behind experiments.

In the analysis and discussion, the author of this book has boldly stated several conclusions. Whether these are correct or not will only be judged by the readers.

Therefore, the final conclusion is that there is still a lot of work needed to be done, not only in experimentally verifying the results, but also more theoretical research is necessary to deepen our understanding of the electroless plating process.

REFERENCES

[1] de Minjer CH, Brenner A. Plating 1957;44:1297.
[2] Mallory GO, Hajdu JB. Electroless plating: fundamentals and applications. Orlando, FL: AESFS; 1990.
[3] Wolfgang R. Electroless nickel plating. Metals Park, OH: ASM International; 1991.
[4] Zhang B, Hu W, Wang L, Shu X, Xie H. Electroplat Pollu Cont 1999;19(5):15.
[5] Senda A, Nakagawa T, Takano Y, Kasanami T. Jpn J Surf Finish 1992;43(4):324.
[6] Molenaar A, de Bakker JWG. J Electrochem Soc 1989;136(2):378.
[7] Senda A, Takano Y, Nakagawa T. Jpn J Surf Finish 1992;43(6):589.
[8] daRosa CP, Maboudian R, Iglesia E. J Electrochem Soc 2008;155(6):E70.

[9] He Y, Liu J, Wang Z, Wang Y, He Z. Electroplat Finish 2010;29(9):22.
[10] Iacovangelo CD. J Electrochem Soc 1991;138(4):976.
[11] Nakao S, Kim D-H, Obata K, Inazawa S, Majima M, Koyama K, Tani Y. Surf Coat Technol 2003;169–170:132.
[12] Djokić SS. J Electrochem Soc 1997;144(7):2358.
[13] Pearlstein F, Weightmen RF. Plating 1969;56(10):1158.
[14] Xie H. Master thesis, Study on effects of alloying elements on preparation, structure and property of ED and MA amorphous alloys. Changsha: Hunan University; July 1995.
[15] Han K, Wu Y, Zhang M, Wang J. Trans IMF 1996;4(39):91.
[16] Małecki A, Micek-Ilnicka A. Surf Coat Technol 2000;123:72.
[17] Zhang B, Chen B, Ye G, Wang L. Electroplat Pollu Cont 1997;17(6):19.
[18] Shi K, Huang Y, Liao Z-J, Zu E-F, Zhang L-B, Wang L. Mater Protection 2005;38(6):27.
[19] Huang G-F, Huang W-Q, Wang L, Zou B-S, Wang Q-L, Zhang J-H. Int J Electrochem Sci 2008;3:145.
[20] Akimoto Y, Noshino Y, Sato M, Utsunomiya T. J Metal Finish Soc Jpn 1974;25(12):666.
[21] Hu W, Zhang B, Zhang H. Plat. Finish 1988;10(6):10.
[22] Shrivastava PB, Venkatramani N, Rohatgi VK, Totalani MK, Mital CK. Metal Finish 1985;83(2):65.
[23] Chang YH, Lin CC, Hung MP, Chin TS. J Electrochem Soc 1986;133(5):985.
[24] Jiang D, Zhang C, Liao S, Zhang B. Electroplat Finish 2007;26(2):5.
[25] Huang G-f, Huang WE, Wang L, Zhu Y, Wang Q, Zhang J. J Hunan Univ (Natural Sci) 2007;34(1):52.
[26] Zhang B, Hu W, Qu X, Zhang Q, Zhang H, Tan Z. Trans IMF 1996;74(2):69.
[27] Guo Z. Electroplt Poll Cont 1993;13(3):11.
[28] Lu G, Giovanni Z. J Electrochem Soc 2003;150(11):C777.
[29] Szczygieł B, Turkiewicz A. Appl Surf Sci 2009;255:8414.
[30] Younan MM, Shoeib M. Galvanotechnik 2002;4:932.
[31] Zhang B, Yi G, Chen B. J Surf Ind (Chinese Taiwan) 1998;69:31.
[32] Shacham-Diamand Y, Sverdlov Y, Petrov N. J Electrochem Soc 2001;48(3):C162.
[33] Dulal SMSI, Kim TH, Rhee H, Sung JY, Kim C-K. J Alloys Comp 2009;467:370.
[34] Tarozaitė R, Sukackienė Z, Sudavičius A, Juškėnas R, Selskis A, Jagminienė A, Norkus E. Mater Chem Phys 2009;117:117.
[35] Chen H, Zhong Xi. J Guangxi Uni (Nat Sci Ed) 2005;30(4):306.
[36] Koiwa I, Usuda M, Yamada K, Osaka T. J Electrochem Soc 1988;135(3):718.
[37] Lee S, Liang H. P&SF 1991;78(9):82.
[38] Xie H, Zhang B, Yang Q. Trans IMF 1999;77(3):99.
[39] Rudnik E, Mucha M. Surf Eng 2011;27(9):683.
[40] Depew JR, Speliotis DE. P&SF 1976;54(6):705.
[41] Ma J, Wu Y, Hang Y, Wang W, Qiao Y, Zheng Y, Liu Y, Huang X, Li G, Zhang L. Metallic Funct Mater 2004;11(1):4.
[42] Schemeckenbecher AF. Plating 1971;58(12):905.
[43] Kim D, Matsuda H, Aoki K, Takano O. P&SF 1996;83(2):78.
[44] Yazima H, Numata H, Ohno I. Jpn J Surf Finish 1997;48(1):62.
[45] Dadvand N, Jarjoura G, Kipouros GJ. J Mater Sci: Mater Electron 2008;19:51.
[46] Wang L, Zhao L, Huang G-f, Yuan X, Zhang B, Zhang J. Surf Coat Technol 2000;126:272.
[47] Yi G, Wang L, Ouyang Y, Mao L, Guo Y, Zhang B, Xie H. Metal Finish 1996;94(11):31.
[48] Yi G, Zhang B, Wang L, Ouyang Y, Xie H. J Electrochem Soc 2013;160(9):D403.
[49] Wang L, Zhao L, Hu W, Zhang B, Shu X, Shun X. Metal Finish 2001;99(6):92.

[50] Osaka T, Arai K, Masubuchi N. J Metal Finish Soc Jpn 1988;39(12):778.
[51] Hu W, Wu L, Wang L, Zhang B. Mater Protec 1994;27(5):10.
[52] Shimachi H, Ozawa S, Tamura K, Osaka T. J Electrochem Soc 1994;141(6):1471.
[53] Xie H, Zhang B, Xu X, Liao S. J Changsha Commun Uni 1998;14(3):24.
[54] Xie H, Zhang B, Xu X. Metal Finish 1999;97(10):35.
[55] Wang L, Ouyang Y, Liao S, Zhang B, Hu W. Mater Protec 1997;30(1):4.
[56] Wang L, Zhao L, Ouyang Y, Liao S, Zhang B. Rare Metals 2000;19(4):261.
[57] Wang S, Zhang Q. Rare Metal Mater Eng 2007;36(12):2224.
[58] Guo H, Liu H, Li W, Xie G. Electroplat Finish 2010;29(7):20.
[59] Takano O, Matsuda H. Metal Finish 1986;84(6):60.
[60] Osaka T, Kasa N. J Metal Finish Soc Jpn 1981;32(6):309.
[61] Rohan JF, Ahern BM, Reynolds K, Crowley St, Healy DA, Rhen FMF, Roy S. Electrochim Acta 2009;54:1851.
[62] Yokoshima T, Kaneko D, Akahori M, Nam H-S, Osaka T. J Electroanal Chem 2000;491:197.
[63] Kim D-H, Aoki K, Takano O. Jpn J Surf Finish 1995;46(1):81.
[64] Senda A, Takano Y, Nakagawa T. Jpn J Surf Finish 1992;43(10):967.
[65] Senda A, Takano Y, Nakagawa T. Jpn J Surf Finish 1992;43(7):694.
[66] Senda A, Takano Y, Nakagawa T. Jpn J Surf Finish 1993;44(2):145.
[67] Honma H, Kobayashi T. J Electrochem Soc 1994;141(3):730.
[68] Koyano H, Kato M, Takenouchi H. J Electrochem Soc 1992;139(11):3112.
[69] Takano O, Shigeta T, Ishibash S. J Metal Finish Soc Jpn 1967;18(8):299.
[70] Takano O, Matsuda N. J Metal Finish Soc Jpn 1980;31(1):28.
[71] Hu W, Zhang B. Trans IMF 1993;71(1):30.
[72] Pearlstein F, Weightman RF. J Electrochem Soc 1984 1023;121(8).
[73] Anik M, Körpe E, Şen E. Surf Coat Technol 2008;202:1718.
[74] Narayanana TSNS, Stephan A, Guruskanthan S. Surf Coat Technol 2004;179:56.
[75] Yang L, Xuan T, Ju Z, Zhang L. Electroplat Poll Contr 2006;26(4):19.
[76] Qin X, Huang Y, Du Z, Yue Y. Chin J Rare Metals 2007;31(5):631.
[77] Nishikawa H, Minami S. J Metal Finish Soc Jpn 1982;33(3):114.
[78] Saito M, Takami H, Sato M, Kiso M, US Patent 4,792,469. Electroless gold plating solution, December 20, 1988.
[79] Pearlstein F, Weightman RF. Electroless deposition of nickel and cobalt based alloys. AD837160 August 9, 1968.
[80] Kondo K, Ishikawa J, Takenaka O, Matsubara T, Irie M. J Electrochem Soc 1991;138(12):3629.
[81] de Minjer C, Brenner A. Plating 1960;44:63.
[82] Cui G, Li N, Li D, Chi M. J Electrochem Soc 2005;152(10):C669.
[83] Jin Y, Yu H, Yang D, Sun D. Rare Metals 2010;29(4):401.
[84] Mallory OG, Hajdu JB. Electroless plating: fundamentals and applications. Norwich, NY: Noyes Publications/William Andrew Publishing, LLC; 1990, 71.
[85] Saito T, Sato E, Matsuoka M, Iwakura C. J Appl Electrochem 1998;28(5):559.
[86] Li Jun, Kohl PA. J Electrochem Soc 2002;149(12):C631.
[87] Hung A. J Electrochem Soc 1985;132(5):1047.
[88] Changsong D, Yiyong W, Dianlong W, Xinguo H. Electropla Pollu Cont 1997;17(4):10.
[89] Lin KL, Hwang J-W. Mater Chem Phys 2002;76:204.
[90] Xu H, Brito J, Sadik OA. J Electrochem Soc 2003;150(11):C816.
[91] Han K, Fan J. J Appl Sci 1997;15(2):229.
[92] Kivel J, Sallo JS. J Electrochem Soc 1965;112(12):1201.
[93] Cheong WJ, Luan BL, Shoesmith DW. Appl Surf Sci 2004;229:282.

[94] Han K, Jing Li F. Metal Finish 1997;95(2):73.
[95] Baskaran I, Sankara Narayanan TSN, Stephen A. Mater Chem Phys 2006;99:117.
[96] Nareus H. Plating 1967;51:380.
[97] Yin X, Hong L, Chen B-H. J Phys Chem B 2004;108:10919.
[98] Hamdy AS, Shoeib MA, Hady H, Salam FA. J Appl Electrochem 2008;38:385.
[99] Zhang S, De Baets J, Vereeken M, Vervaet A, Van Calster A. J Electrochem Soc 1999;146(8).
[100] Rudnik E, Gorgosz J. Surf Coat Technol 2007;201:6953.
[101] Innes WP, Grunwald JJ, Ottavio ED, Toller WH, Carmicheal C. Plating 1969;56:51.
[102] Hanna F, Hamid ZA, Aal AA. Mater Lett 2003;58:104.
[103] Schoenberg LN. J Electrochem Soc 1972;119(11):1491.
[104] Vaškelis A, Jačiauskienė J, Stalnionienė I, Norkus E. J Electroanal Chem 2007;600:6.
[105] Nuzzi FJ. P&SF 1983;70:51.
[106] Hu B, Sun R, Yu G, Liu L, Xie Z, He X, Zhang X. Surf Coat Technol 2013;228:84.
[107] Li J, Kohl PA. J Electrochem Soc 2003;150(8):C558.
[108] Han K, Wu Y, Zhang M, Wang J. Trans IMF 1996;74(3):91.
[109] Fang JL, Lu Y, Han KP. P&SF 1997;84(9):91.
[110] Guan S, Zhang Q, Wu J, Hu R. Electroplat Pollu Cont 2000;20(3):24.
[111] Ambata R, Zhou W. Surf Coat Technol 2004;179:124.
[112] An M, Yang Z, Xu X, Zhang J, Tu Z. Chin J Nonfer Met 2007;9(1):155.
[113] Xia C. Mater Prote 1996;29(1):4.
[114] Chen B-H, Hong L, Ma Y, Ko T-M. Ind Eng Chem Res 2002;41:2668.
[115] Nwosu N, Davidson A, Hindle C, Barker M. Ind Eng Chem Res 2012;51:5635.
[116] Zhu Y, Kong X, Jiang Q, Wang W. Chi Surf Eng 2012;25(1):76.
[117] Gan X, Zhou K, Hu W, Zhang D. Surf Coat Technol 2012;206:3405.
[118] Sudagara J, Liana JS, Jianga Q, Jianga ZH, Li GY, Elansezhian R. Prog Org Coat 2012;74:788.
[119] Zhou F, Li X, Gao X. Chi J Inorg Chem 2009;25(9):1584.
[120] Liu S, Jiang T, Li J. Chi Surfac Deterg Cosm 2009;39(6):390.
[121] Liu S, Jiang T, Zou Y. Chem Eng (China) 2007;135(9):37.
[122] Huang W, Xue Y. Mater Eng 2010;No.1:23.
[123] Zhou Q, Shao Z, He C, Shao Z, Cai Q, Gao W. J Chi Soc Corr Protec 2007;27(1):27.
[124] Afroukhteh S, Dehghaniann C, Emamy M. Prog Nat Sci Mater Inter 2012;22(4):318.
[125] Feng C, Wang Q. J Nanchang Hangkong Uni (Nat Sci) 2011;25(4):86.
[126] Zielińska K, Stankiewicz A, Szczygieł I. J Coll Interf Sci 2012;377:362.
[127] Mafi IR, Dehghanian C. Appl Surf Sci 2011;257:8653.
[128] Nishira M, Takano O. Jpn J Surf Finish 1991;42(8):839.
[129] Zhang H, Liu J, Du C. J Mater Prot 2009;40(5):36.
[130] Ger M-D, Hwang BJ. Mater Chem Phys 2002;76:38.
[131] Xiang Y, Zhang J, Jin C. Mater Eng (China) 2000;No. 4:22.
[132] Jiang T, Liu S, Zhou H. China Surf Deter Cosm 2007;37(4):219.
[133] Zhao Q, Liu Y. Surf Coat Technol 2005;200:2510.
[134] Karuppusamy K, Anantharam R. Metal Finish 1992;90(5):15.
[135] Manne S, Cleveland JP, Gaub HE, Stucky GD, Hansma PK. Langmuir 1994;10:4409.
[136] Manne S, Gaub HE. Science 1995;270(5241):1480.
[137] Wanless EJ, Ducker WA. J Phys Chem 1996;100:3207.
[138] Wanless EJ, Davey TW, Ducker WA. Langmuir 1997;13:4223.
[139] Atkin R, Craig VSJ, Wanless EJ, Biggs S. J Phys Chem B 2003;107(13):2978.
[140] Hu K, Bard AJ. Langmuir 1997;13:5418.
[141] Jaschke M, Butt H-J, Gaub HE, Manne S. Langmuir 1997;13:1381.

[142] Lamont RE, Ducker WA. J Am Chem Soc 1998;120:7602.
[143] Burgess I, Jeffrey CA, Cai X, Szymanski G, Galus Z, Lipkowski J. Langmuir 1999;15:2607.
[144] Tang Z, Wang E. J Electroanal Chem 2001;496:82.
[145] Xu S, Chen M, Cholewa E, Szymanski G, Lipkowski J. Langmuir 2007;23:6937.
[146] Notley SM. J Phys Chem B 2009;113(42):13895.
[147] Chen M, Burgess I, Lipkowski J. Surf Sci 2009;603:1878.
[148] Senda A, Morita K, Takano Y. Jpn J Surf Finish 1993;44(11):966.
[149] Schoenberg LN. J Electrochem Soc 1971;118(10):1571.
[150] Duffy J, Pearson L, Paunovic M. J Electrochem Soc 1983;130(4):876.
[151] Li J, Kohl PA. P&SF 2004;91(2):2.
[152] Anik T, Touhami ME, Himm K, Schireen S, Belkhmima RA, Abouchane M, Cissé M. Int J Electrochem Sci 2012;7:2009.
[153] Okinaka Y. Plating 1970;57:914.
[154] Paunovic M, Nguyen T, Mukherjee R, Sambucetti C, Romankiw LT. J Electrochem Soc 1995;142(5):1495.
[155] Bu J, Huang G. Surf Technol 2006;35(4):48.
[156] Bi H, Kou KC, Rider AE, Ostrikov K, Wu HW, Wang ZC. Appl Surf Sci 2009;255:6888.
[157] Evans WT, Schlesinger M. J Electrochem Soc 1994;141(1):78.
[158] Narayanana TSNS, Selvakumarb S, Stephen A. Surf Coat Technol 2003;172:298.
[159] Aoki K, Takano O. P& SF 1990;77(3):48.
[160] Yu Y, Wei G, Hu X, Ge H, Yu Z. Surf Coat Technol 2010;204:2669.
[161] Cherkaoui M, Srhiri A, Chassaing E. P&SF 1992;79(11):68.
[162] Ashassi-Sorkhabi H, Dolati H, Parvini-Ahmadi N, Manzoori J. Appl Surf Sci 2002;185:155.
[163] Gou Y, Huang W, Zeng R, Zhu Y. Trans Nonferrous Met Soc China 2010;20:s674.
[164] Baba NB, Waugh W, Davidson AM. World Acad Sci Eng Technol 2009;49:715.
[165] Wu F, Chu S. Surf Technol 1994;23(4):154.
[166] Nishira M, Yamagishi K, Takano O. Jpn J Surf Finish 1993;44(11):961.
[167] Xue M, Li X, Ma M. Chem Ind Times 2001;15(9):18.
[168] Anjana L, Selvam M, Srinivasan KN, Vasudevan T. Metal Finish 1994;92(8):47.
[169] Abrantes LM, Correia JP. J Electrochem Soc 1994;141(9):2356.
[170] Oita M, Matsuoka M, Iwakura C. Electrochem Acta 1997;42(9):1435.
[171] Cheng SL, Peng HC. J Electrochem Soc 2010;157(2):D81.
[172] Tzeng G-S. J Appl Electrochem 1996;26:969.
[173] Lin Z, Zhu Y, Huang W. Surf Technol 2005;34(4):45.
[174] Matsuoka M, Hayashi T. Metal Finish 1986;84(3):27.
[175] Ohno I, Wakabayashi O, Haruyama S. J Electrochem Soc 1985;132(10):2323.
[176] Mallory GO. Trans Inst Met Finish 1978;56:81.
[177] Tong H, Zhu L, Li M, Wang C. Electrochim Acta 2003;48:2473.
[178] Sullivan AM, Kohl PA. J Electrochem Soc 1995;142(7):2250.
[179] Seshimo M, Ozawa M, Sone M, Sakurai M, Kameyama H. J Memb Sci 2008;324:181.
[180] Anjana L, Selvan M, Srinivasan KN, Vasudevan T. Metal Finish 1994;92(8):47.
[181] John S, Shanmugam NV, Shenoi BA. Metal Finish 1982;80(4):47.
[182] Hamid ZA, Hassan HB, Attyia AM. Surf Coat Technol 2010;205:2348.
[183] Bulasara VK, Thakuria H, Uppaluri R, Purkait MK. J Mater Proc Technol 2011;211:1488.
[184] Kim TH, Dulal SMSI, Park CH, Chae H, Kim C-K. Surf Coat Technol 2008;202:4861.
[185] Matsuoka M, Hotta T. J Surf Finish Soc Jpn 1989;40(7):831.
[186] Rich SR. Plating 1955;42:1407.
[187] Matsuoka M, Hayashi T. Metal Finish 1986;84(3):27.

[188] Kobayashi K, Chiba A, Minami N. Ultrasonics 2000;38:678.
[189] Park YS, Kim TH, Lee MH, Kwon SC. Sur Coat Technol 2002;153:245.
[190] Yang P, Gong L, Gu B, Sun H. Surf Technol 2007;36(6):61.
[191] Cobley AJ, Saez V. Circ World 2012;38(1):12.
[192] Touyeras F, Hihn JY, Delalande S, Viennet R, Doche ML. Ultrasonics Sonochem 2003;10:363.
[193] Abyaneh MY, Sterritt A, Mason TJ. J Electrochem Soc 2007;154(9):D467.
[194] Matsuda H, Takano O. Trans Inst Metal Finish 1993;71(4):138.
[195] Touyeras F, Hihn JY, Bourgoin X, Jacques B, Hallez L, Branger V. Ultrasonics Sonochem 2005;12:13.
[196] Abyaneh MY, Fleischmann M. Electrochim Acta 1982;27:1513.
[197] Abyaneh MY. J Electroanal Chem 2006;586:196.
[198] Yonemochi S-I, Iwasaka T, Aogaki R. Jpn J Surf Finish 1999;50(11):1015.
[199] Yu Y, Li W, Lou J, Ge H, Sun L, Jiang L, Wei G. Rare Metals 2012;31(2):125.
[200] Lee C-C, Chou T-C. Ind Eng Chem Res 1998;37:1815.
[201] Chiba A, Ogawa T. Jpn J Surf Finish 1989;40:568.
[202] Borioli C, Franz S, Cavallotti PL, Cantoni M, Bertacco R. J Electrochem Soc 2010;157(8):D437.
[203] MFPP-Process Guide, 11, Sept.–Oct. 1975, p. 144–52.
[204] Grunwald P. Galvanotechnik 1983;74:1286.
[205] Sharma SB, Agarwala RC, Agarwala V, Satyanarayana KG. Metall Mater Trans B 2005;36B:23.

Chapter 7

Green Electroless Plating

Chapter Outline

7.1 WHAT IS GREEN ELECTROLESS PLATING?

Standard electroless plating uses some hazardous substances that can harm both human health and the environment. They are usually found in the reductant, complexing agent, stabilizer, accelerating agent, or brightening agent used in the bath solution. In EN plating, ethylene diamine tetraacetic acid (EDTA) is currently the most widely complexing agent. Although EDTA is low cost, it forms stable heavy metal complexes that are highly mobile in the environment, and also increases the total nitrogen content of waste waters. How to break EDTA down efficiently in water, either chemically or biochemically, is a difficult technological problem that remains to be solved. Because of this, European countries have prohibited the discharge of aqueous effluent containing EDTA [1]. Another problem constituent is lead. Pb^{2+} is a commonly used stabilizer in EN. However, lead is toxic to humans, causing hematological damage, brain damage, anemia, and kidney malfunction [2]. Therefore, the World Health Organization (WHO) recommends that Pb concentrations be below 10 μg/L in natural water systems [3]. The Restriction of Hazardous Substances (RoHS) directive [4] and the End-of-Life Vehicle (ELV) directive [5] specify the limit for Pb in an electroless nickel deposit to be less than 1.0 mg/L [6]. In conventional electroless plating (EP), Cd^{2+} (as cadmium sulfate and cadmium acetate) is usually used as a brightener [7,8]. But, Cd is also a highly toxic element. In addition, it is well known that electroless deposits from conventional cadmium-brightened systems would almost certainly contain in excess of 0.01 wt% of cadmium. The RoHS directive specifies the limit for Cd up to 0.01% by weight [4].

Amorphous and Nano Alloys Electroless Depositions. http://dx.doi.org/10.1016/B978-0-12-802685-4.00007-8

In the traditional EP Cu, formaldehyde was usually used acting as a reductant and EDTA as a complexing agent [9–11], especially in all commercial electroless copper solutions. Deposition of high purity copper in through-hole plating (THP) uses formaldehyde as the reducing agent [10,12]. Alternatively, HF solution can directly deposit Cu onto Si through a redox reaction with silicon without the need for other reducing agents [13]. In addition to the risk from EDTA, both formaldehyde and HF are environmentally hazardous due to their corrosiveness and high toxicity (formaldehyde is a known carcinogen).

Traditionally, the EP Au bath solution must contain cyanide in the form of gold cyanide, potassium cyanide, etc. [14–17]. In the conventional EP Ag deposition, cyanide was also used sometimes [18]. The toxicity of cyanide is well known.

It is evident from this that normal EP is harmful to both human health and the ecological environment. EP has helped various industry progress, benefiting humankind. However, EP is also one of the culprits that has caused ecological damage. That is why a new path should be opened. A low-cost, no hazardous substance, and environmentally friendly chemical route is expected to be further developed to meet increasingly stringent environmental regulations. That is green electroless plating. Therefore, the definition of green EP could be an EP using eco-agents, having a higher deposition rate (e.g., 18–25 μm/h in EN), not expensive and having a stable solution for longer recycling use (e.g., 10–12 cycles of EN).

In the following sections, green EN EP Cu, Ag, and Au will be discussed to see how green EP is progressing.

7.2 GREEN ELECTROLESS PLATING OF EN

In principle, as long as the bath solution does not use Pb^{2+}, EDTA, and Cd^{2+} as reductant, complexing agent, and brightener, respectively, the EN can be called green chemical plating. In fact, even in the early studies, there are a significant number of reports about meeting green standards with EN [Ni–P:19–23, Ni–B:24,25]. At the time, the world environmental status was not so serious and restrictive regulations for the protection of the environment have not been worked out. The first EU RoHS directive was promulgated in 2002, and in China RoHS was promulgated in 2006. Of course, these articles did not specify that their EN was environmentally friendly. Only recently, especially after those restrictive regulations were promulgated have the published papers which concern environment friendly obviously increased.

Many papers on the stabilizers in lead- and cadmium-free EN processes for EU RoHS and End-of-Life Vehicle Directives have been published [26–36]. Some of these will be discussed more in detail in the remainder of this chapter.

In 2004, Cheong et al. [28] investigated the effects of two nontoxic electroless nickel bath stabilizers: thiourea and maleic acid. They found that the bath stability increases significantly with either thiourea or maleic acid. For thiourea, the bath life increased from 1.3 ± 0.5 to >264 min as the stabilizer concentration was increased from 0.5 mg/L to 2.0 mg/L. For the maleic acid stabilizer, the bath life increased from 0.7 ± 0.1 min to 101 ± 9 min over the studied stabilizer concentration range of 0–3000 mg/L. These authors also measured the effect of thiourea and maleic acid on the deposition rate of the electroless nickel system (see the left plot in Fig. 6.26). An enhancement of the deposition rate was observed for thiourea, which reached a maximum of 15.4 μm/h at 2.2 mg/L thiourea. However, deposition rates in the presence of maleic acid stabilizer showed no significant change.

In 2006, Baskaran et al. [29] reported the effect of two nontoxic stabilizers, namely thiourea and succinic acid, on the formation and characteristics of electroless Ni–P deposits obtained from an acidic hypophosphite-reduced electroless nickel bath. To compare these stabilizers with one in normal use, the effect of the toxic stabilizer lead acetate was also studied. The basic bath composition and operating conditions are: nickel sulfate hexahydrate 21.2 g/L, sodium hypophosphite monohydrate 24 g/L, lactic acid 28 g/L, propionic acid 2.2 g/L, pH 4.50, temperature 90 ±1°C. The stabilizers are succinic acid 12.0 g/L, thiourea 0.8 ppm, and lead acetate 1.0 ppm. The results show that thiourea accelerates the rate of deposition up to 0.8 ppm and starts to inhibit it at a concentration of 1 ppm. Similarly, succinic acid increases the plating rate up to 12 g/L and exhibits an inhibiting effect at higher concentrations of the order of 15 g/L. However, lead acetate inhibits the rate of deposition even at a concentration of 0.5 ppm and the extent of inhibition is increased when the concentration is higher than 1 ppm. The plating rates for Ni–P without additive, Ni–P with thiourea, succinic acid, and lead acetate are 15 ±1, 25 ±1, 19 ±1, and 13 ± 1 μm/h, respectively, which also show that thiourea and succinic acid accelerate the deposition, but lead acetate reduces the plating rate.

The stabilizing effects of three stabilizers KI, benzotriazole, and organic heterocyclic compounds A containing sulfur on EP Ni–P were investigated by Liu et al. in 2006 [30]. The results show that all three stabilizers improve greatly the stability of the bath, as shown in Fig. 7.1. The stability of the bath solution

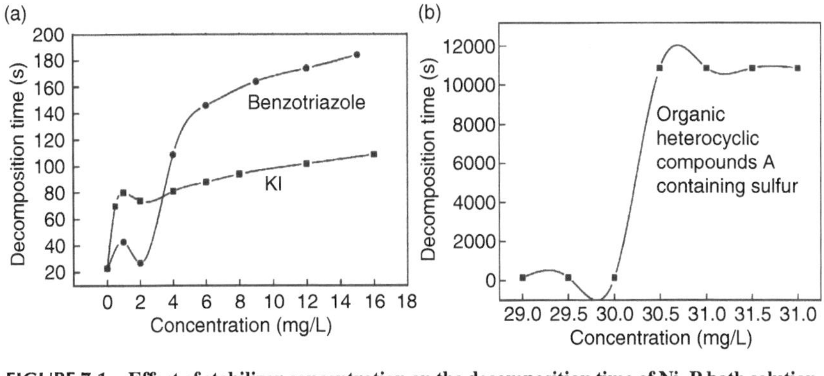

FIGURE 7.1 Effect of stabilizer concentration on the decomposition time of Ni–P bath solution. (a) Stabilizers KI and benzotriazole, (b) organic heterocyclic compounds A containing sulfur [30].

increases from 23 s to 109 s as the concentration of KI increases from 0 to 16 mg/L. When the concentration of benzotriazole increases from 0 to 15 mg/L, the stability of bath solution increases from 27 s to 186 s. As for the stability of A, the stability of bath is only 156 s at a concentration of 30 mg/L, but when its concentration is larger than 30.5 mg/L, the stability of bath solution exceeds 3 h. Generally speaking, the two stabilizers KI and A accelerate the plating rate but benzotriazole reduces it (see Fig. 3 in original paper [30]). Stabilizer A increases the plating rate from about 13.7 μm/h (the plating rate of Ni–P without stabilizer) to in the range 14–16 μm/h as its concentration increases from 0 to 45 mg/L. Further increasing its concentration up to 50 mg/L basically stops the deposition as the plating rate is only 2.9 μm under these conditions. For KI, the plating rate is enhanced from 13.7 μm/h up to around 16.0 μm/h when its concentration increases from 0 to 16 mg/L. However, the stabilizer benzotriazole evidently decreases the plating rate. The plating rate is 9.7 μm/h at a concentration of 12 mg/L, and only 0.8 μm/h at a concentration of 15 mg/L.

In 2008, Dong et al. [33] studied the effects of four stabilizers ($Na_2S_2O_3$, S1, $CuSO_4$, and benzotriazole) in EP Ni–P deposition. All of the four stabilizers increase the both stability. The stabilizer stability is 25, 630, 1200, 850, and 800 s for no stabilizer, $Na_2S_2O_3$ (2 mg/L), S1 (10 mg/L), $CuSO_4$ (2 mg/L), and benzotriazole (2 mg/L), respectively. The amount in parentheses is the concentration of stabilizer added. The plating rates are from 20 μm/h to 28 μm/h, 15 μm/h to 20 μm/h, and 15 μm/h to 20 μm/h when the concentration of $Na_2S_2O_3$ ranges from 0.5 mg/L to 2.5 mg/L, S1 from 5 mg/L to 25 mg/L, and $CuSO_4$ from 0.5 mg/L to 2.5 mg/L, respectively. But the plating rate is less than 15 μm/h when the concentration of benzotriazole ranges from 0.5 mg/L to 2.0 mg/L, which is comparable with the results obtained by Liu [30].

In 2009, a group from Singapore [34] reported the influence of a model phosphine stabilizer – triphenylphosphine-3,3′,3″-trisulfonic acid trisodium salt (TPPTS) in EP Ni–P deposition. They found that the plating rate is basically around 13 μm/h when the concentration of TPPTS ranges from 10^{-9} mol/L to 10^{-4} mol/L, which is below the value (15 μm/h) obtained when no stabilizer is employed in the same plating solution. This means that the phosphine stabilizers suppress the plating rate of Ni–P. As mentioned earlier, the sulfur-containing stabilizers promote the nickel deposition rate, so the way that kinetics of EN are affected by the phosphine stabilizers must be different than that for sulfur-containing stabilizers. It seems that TPPTS does not form a complex with the Ni^{2+} ion, and it is likely that this phosphine is an EN stabilizer of cathodic type.

The effects of five stabilizers TU, KI, KIO_3, an organic heterocyclic compound A containing sulfur (A), and $CuSO_4$ in EN were reported in 2009 [35]. The authors measured the stability of these stabilizers, and used the stability coefficient (SC) of bath solution rather than the decomposition time to indicate stability. The definition of the stability coefficient for a stabilizer in EN bath solution under normal operating conditions is the ratio of the plating amount of Ni in the coating to the amount Ni in the bath solution. So, the stability coefficient

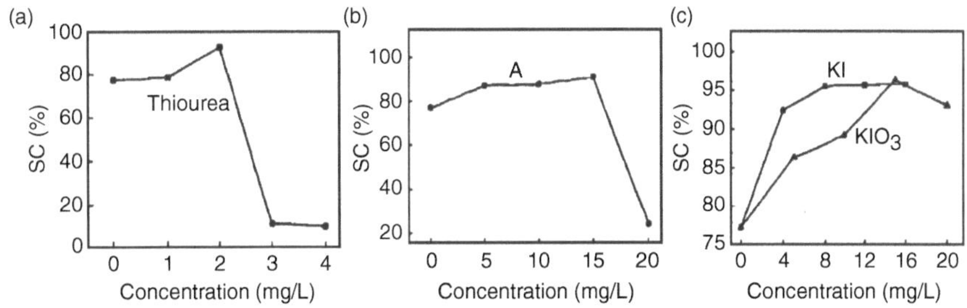

FIGURE 7.2 Effect of stabilizer concentration on the stability coefficient of Ni–P bath solution. (a) Stabilizer thiourea, (b) $CusO_4$, and (c) KI and KIO_3 [35].

SC = $(M_1/M_0) \times 100\%$, where M_1 is the content Ni in the coating, and M_0 is the amount Ni in the bath solution. Figure 7.2 shows the relationship between SC and the concentration of stabilizer for four stabilizers in EN deposition: (a) thiourea, (b) A, (c) KI and KIO_3. Stabilizer $CuSO_4$ is not shown in the figure. It can be seen that all SCs increase when thiourea, A, KI, or KIO_3 are added. The SC increases from 77% up to 92% as the concentration of thiourea increases from 0 to 2 mg/L. However, deposition stops when the concentration is over 3 mg/L. A is an organic heterocyclic compound containing sulfur. When this is added to the bath, the SC improves quickly. The SC increases up to 96% when the concentration of A reaches 15 mg/L. After then, the bath solution is poisoned and the reaction stops. The scope for using A is relatively large, and more than that of thiourea. After adding KI and KIO_3 respectively, the SC increases gradually, reaching a maximum value of 96% when the concentrations of KI and KIO_3 reach 8 and 15 mg/L, respectively. $CuSO_4$ does not contribute to the stability of bath solution, and the SC basically remains around 77%, fluctuating by less than 3%. The authors believe that adding too much sulfuric acid may cause the bath solution to decompose, thereby only a little effect to the SC. It should be pointed out that this result is different from that obtained by Dong et al. [33], where adding CuSO4 causes the decomposition time to increase from 25 s (without adding $CuSO_4$) to 800 s. Is the effect just due to the different methods of measuring stability (stability coefficient vs. decomposition time) or some other reason? Further work is needed to reconcile these conflicting results.

These five stabilizers have, overall, little effect on the plating rate. Thiourea and A are similar: the plating rate with added thiourea in the EN bath fluctuates around 12.9 μm/h which is the same as that without stabilizer, and that for added A is a little less than 12.9 μm/h. But adding A can work up to 15 mg/L, adding thiourea is only worked to 3 mg/L. The measured results for the other three stabilizers are shown in Fig. 7.3. The measured concentration range of stabilizer is from 0 to 20 mg/L. For all of them, the plating rate increases and decreases. The maximum value is not more than 14 μm/h, which is only a little higher than the EP plating rate without stabilizer (12.9 μm/h). The plating rates for adding KIO_3 and $CuSO_4$ reached their minimum at a concentration

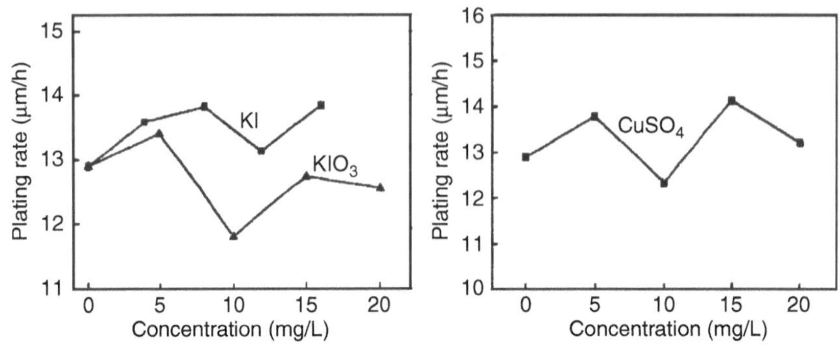

FIGURE 7.3 Effect of stabilizer concentration on the plating rate in Ni–P bath solution. Left shows the stabilizers of KI and KIO_3, right shows $CusO_4$ [35].

of 10 mg/L. The minimum values are about 12.4 and 11.8 μm/h for KIO_3 and $CuSO_4$, respectively. But the minimum value for KI is reached at 12 mg/L, and it is still a little higher (about 13.2 μm/h) than 12.9 μm/h. Compared to the plot in the right of Fig. 7.3 to the aforementioned results of the plating rate for stabilizer $CuSO_4$ obtained by Dong et al. [33] (see their original Figure 3), it can be seen that the trends are different. There the plating rate just decreases in the concentration range studied from 0.5 mg/L to 2.5 mg/L of $CuSO_4$. Here, the plating rate is increased at that concentration range. This experimental conflict needs to be resolved by further research.

It is interesting to consider the paper by Seifzadeh and Rajabalizadeh published in 2013 [36]. These authors developed an environmentally friendly method for electroless Ni–P plating on magnesium alloy. The two conventional methods, zinc immersion and direct electroless nickel plating for EN plating on magnesium alloys, are toxic. The zinc immersion procedure is not only complicated, but cyanide must be used in the copper plating step. Although the direct EN plating procedure has the advantages of simplicity and suitability for magnesium alloys with high aluminum content, it usually uses hexavalent chromium compounds and HF acid during the etching and conversion coating treatments, respectively. It is obvious that all of these materials increase the danger of the operation and are not friendly to the environment [37,38]. As an alternative, the authors first treated the substrate in a cerium–lanthanum–permanganate (CLP) conversion coating bath. Then, the electroless Ni–P deposition was carried out on the conversion coating from a sulfate solution. The bath composition and operation parameters for CLP conversion treatment are: $Ce(NO)_3$ 0.4 g/L, $La(NO)_3$ 0.4 g/L, $KMnO_4$ 25 g/L, immersion for 15 min at room temperature. The bath composition and operation parameters for electroless Ni–P deposition are: $NiSO_4 \cdot 6H_2O$ 15 g/L, $Na_2H_2PO_2 \cdot H_2O$ 14 g/L, $NaC_2H_3O_2$ 13 g/L, HF(40% V/V) 12 mL/L, NH_4HF_2 8 g/L, thiourea 1 ppm, pH 6.4, plating time 3 h plating, temperature 65°C. The results show that a uniform and amorphous nickel coating with high phosphorus content was successfully deposited on the AZ61 alloy after the cerium–lanthanum–permanganate conversion coating treatment. The authors believe that the CLP treatment can be considered a green alternative to the dangerous chromium and HF treatment. Compared with the bare and CLP-treated substrates, the Ni–P deposits showed more nobler corrosion potential and also lower corrosion current density. In addition, the results illustrated that the Ni–P coating on the AZ61 magnesium alloy exhibited high-corrosion resistance in an accelerated corrosion test. This could be due to its fine grain structure and the low porosity of the coating.

In 2007, Zhang et al. [39] used a different phosphate–manganese conversion film as the pretreated layer to deposit the electroless Ni–P alloy on the AZ91D magnesium alloy substrate. The pretreatment composition and operation conditions are: $C_2H_4O_2$ 20 mL/L, CH_3CH_2OH 50 mL, HNO_3 (80% V/V) 5 mL, pretreatment time 3 min. The results show that a Ni–P coating with fine and dense structure was obtained on the AZ91D magnesium alloy, which shows better corrosion resistance than the Ni–P with chromium oxide plus HF as pretreatment.

Very few papers considering lead and cadmium-free brighteners have been published [40–42]. Here, we will just discuss the work by Wang et al. in 2007 [41]. This group studied the effect of single component of organic metal salts and organic compounds, and combined brighteners on deposits on bright electroless nickel plating. Their results show that the brightness effect of single-component brighteners is very poor. But, a brightening effect for electroless nickel deposits can be achieved by the use of a poly component brightener made up by mixing sulphur-containing organic compound and other components. The results are shown in Table 7.1. Thus, the possibility of achieving lead-free and cadmium-free bright electroless nickel deposits brightener was ascertained. Experimental results of the full spectrum of direct reading inductively coupled plasma emission spectrometry instrument (ICP) test of deposits show that the lead and cadmium in the deposit are all up to the standard of European Union, provided the lead content in the main salt is well controlled.

TABLE 7.1 Action of Combined Brighteners on Deposit Brightness

Brightener	Concentration (mg/L)	Brightening effects
Organic compound containing sulfur + $Ce_2(SO_4)_3 \cdot 8H_2O$	1 + 15	Plating coating semibright
2-Naphthol + $Ce_2(SO_4)_3 \cdot 8H_2O$	15 + 15	Whitish and flooding coating
Organic compound containing sulfur + cystine	1 + 0.5	Uniform brighter coating
Organic compound containing sulfur + $CuSO_4 \cdot 5H_2O$	1 + 15	Plating coating semibright
Multicomponent brightener	20 mL/L	Bright coating, good effect

7.3 GREEN ELECTROLESS PLATING ON Cu

EP copper is widely used in many branches of industry, including electromagnetic interference shielding of electronic components, through-hole plating in printed circuit boards, decorative plating of household utensils, conductive traces in electronic interconnection devices, and integrated circuit manufacturing [43,44]. It was pointed out at the beginning of this chapter that traditional electroless copper-plating solutions often use formaldehyde as the reducing agent and EDTA as the ligand in the electroless copper plating technique due to their low cost and the high quality of the deposit. However they are environmental hazards; hence other reductant and complexing agent needed to be developed to replace HCHO (formaldehyde) and EDTA to meet the requirements of the green electroless plating.

Today, many reducing agents have been developed as substitutes for formaldehyde, such as hypophosphite [45–65], glyoxylic acid [66–77], borane–dimethylamine complex (DMAB) [78–82], sodium bisulfate ($NaHSO_3$), sodium thiosulfate pentahydrate ($Na_2S_2O_3 \bullet 5H_2O$) [83], Co(II) [84–87], Fe(II) [88,89], and saccharides [90].

7.3.1 Hypophosphite

First, we will describe the hypophosphite alternative to formaldehyde. Researchers have found that using sodium hypophosphite as the reducing agent instead of formaldehyde is especially attractive because of its low pH, low cost, and relative safety [45,48,55]. However, new problems have appeared by using hypophosphite, and the main difficulty results from the fact that the oxidation of hypophosphite is not catalyzed by copper. Because of this little or no plating on a pure copper surface can be obtained. This meant that additive(s) to overcome this problem were actively researched. For example, adding nickel ions (or other metal ions) to the bath, resulting in a very small amount of codeposited nickel in the copper deposit can catalyze the oxidation of the reducing agent. The nickel then serves to catalyze the oxidation of hypophosphite enabling continuous copper deposition [50,55].

Formally using the hypophosphite as reducing agent in electroless plating copper to substitute conventionally used formaldehyde can be traced back to 1980. Kukans et al. were granted a US patent "electroless copper composition solution using a hypophosphite reducing agent" in that year [45]. After checking earlier patents, they said in their patent: "the bath composition examples invariably employ formaldehyde-type reducing agents for the copper formulations and, in contrast, hypophosphites for the nickel formulations. There is no suggestion in the patent art that the hypophosphite of the nickel baths could be substituted for formaldehyde in copper baths." Today, many papers have focused on the substitution of formaldehyde with hypophosphite in EP Cu [45–65].

In 1980, Goldstein et al. [46] used a bath solution of EP Cu made up as follows: $CuCl_2 \bullet 2H_2O$ 0.024 mol/L, $CoCl_2 \bullet 6H_2O$, 0.005–0.01 mol/L, $NiSO_4 \bullet 6H_2O$

0.002 mol/L, KNa-Tartrate, 0.037 mol/L, NaOH, 0.156 mol/L, $NaH_2PO_2{\bullet}H_2O$, 0.20 mol/L, time 10 min, temperature 60°C. The results show that when a bath formulation without any nickel or cobalt autocatalysis promoter with immersion times of 10, 30, and 60 min, the Cu deposit thickness builds to about 0.38 μm and then terminates. When a small amount of cobalt ion is added to the bath formula, the deposits are pink, indicating good conductivity, and adherent to the substrate. No termination of deposit occurs, and the deposition rate increases linearly with immersion time. In addition, higher cobalt ion levels appear to accelerate the plating rate. Alternatively, when using nickel ions instead of cobalt, and varying the nickel ion levels, higher levels do not appear to dramatically accelerate the plating rate, in contrast to that observed with cobalt.

In 1997, Cheng et al. [48] reported electroless copper plating using hypophosphite as the reducing agent. The bath composition and operating conditions for EP Cu are: cupric sulfate 10 g/L, citrate 25 g/L, hypophosphite 35 g/L, nickel sulfate 1.6 g/L, boric acid 35 g/L, pH 10, and temperature 70°C. They found that the maximum plating rate is about 7 μm/h. However, the authors believed that compared with a formaldehyde-type bath, the hypophosphite-type bath should be operated at high temperature, and Ni^{2+} ions must be added to catalyze the reaction. In addition, EDTA was used for stability.

In 2003, Li and Kohl [53] investigated the deposition characteristics of accelerated nonformaldehyde. The electroless copper plating solution contained copper sulfate 0.04 mol/L, nickel sulfate 0.00095–0.0057 mol/L, sodium hypophosphite 0.17 mol/L, sodium citrate (as the complexing agent) 0.051 mol/L, *N*-(2-hydroxyethyl)ethylenediaminetriacetic acid trisodium salt hydrate (HEDTA) 0.015 mol/L, boric acid 0.485 mol/L, polyethylene glycol 200 ppm, and formamidine disulfide (fd) (as the accelerator) 0.5 ppm. The pH was adjusted using NaOH or H_2SO_4 to a final value of 9.0–9.3. The temperature was held at 72.5°C. Sodium citrate is the complexing agent for chelating the copper and nickel ions, and HEDTA was added to maintain the stability of fd. The experiments show that a copper deposit with a pink tint was obtained in a citrate-based electroless copper solution with 0.5 ppm fd. The average deposition rate for 30 min plating is 7.15 μm/h, which is higher than that in the HEDTA-based electroless copper solution using thiourea or fd as accelerators (5.78–6.36 mm/h). At the Cu^{2+}/Ni^{2+} molar ratio of 42, the citrate-based electroless copper solution had a low deposition rate in the absence of fd (1.27 μm/h for 30 min) and the copper deposit was dark once the palladium catalyst on the substrate surface was covered with copper. This indicates that fd accelerates the deposition process in the citrate-based electroless copper plating solution. Formamidine disulfide accelerates both half reactions, the oxidation of hypophosphite and the reduction of copper (II). As for the effect of nickel ions, the authors found that in the absence of nickel ions, a deposition rate of 2.91 μm/h was obtained in the solution with 0.5 ppm fd. Figure 7.4 shows the effects of $NiSO_4$ concentration on the plating rate and resistivity of the Cu coating. The deposition rate of the electroless plating increased with the initial addition of nickel ions and was constant with

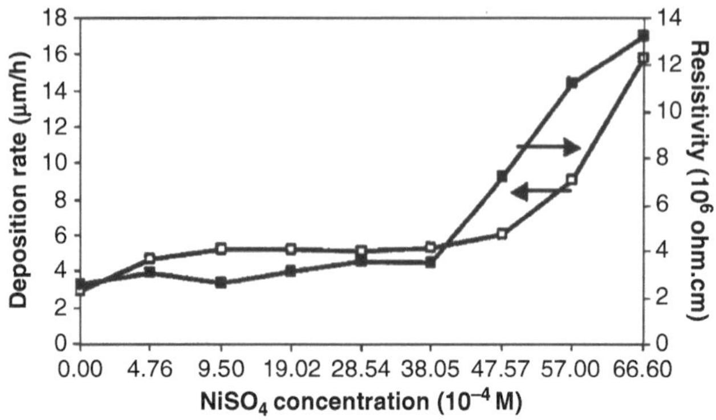

FIGURE 7.4 **Effect of the addition of the concentration of nickel ions in the electroless copper plating solution on the deposition rate (60 min plating) and resistivity [53].**

nickel ion concentrations ranging from 0.000476 mol/L to 0.0038 mol/L. After this point, the deposition rate increased dramatically as the concentration of nickel ions increased further. At 0.00666 mol/L nickel (II), the deposition rate is up to about 16.0 μm/h. The resistivity and plating rate are almost parallel to increased nickel concentration.

In 2006, Touir et al. [56] reported electroless deposition of copper-rich Cu–Ni–P alloys in acidic solution using hypophosphite as the reducing agent. The authors believe that no bath composition has been proposed before for the electroless deposition of copper from acid hypophosphite solutions, except for some with very low copper content [91]. Moreover, very low plating rates were obtained at pH 5.3 [91]. Composition of the bath solutions is: $CuSO_4•5H_2O$ 0–1.6 × 10^{-3} mol/L, $NiSO_4•7H_2O$ 0.10–0.085 mol/L, $Na_3C_6H_5O_7•2H_2O$ 0–0.30 mol/L, $NH_4CH_3CO_2$ 0–1.0 mol/L, $NaH_2PO_2•H_2O$ 0.28 mol/L, pH 5 and *T* 60–80°C. The Cu content in the obtained Cu–Ni–P deposits is up to 97 wt%. The phosphorus content of the alloy decreased from 13% to 1% with increasing Cu content. The plating rate ranges from 5 μm/h to 1 μm/h, and decreases when the copper sulfate concentration in the solution is increased. In addition, it increases with increasing pH or temperature, but the influence is less pronounced than in alkaline solutions. Anik et al. in 2012 [63] studied the effect of pH solution on the electroless deposition of copper-rich Cu–Ni–P alloys, using hypophosphite as reducing agent. The bath solution they used is nearly the same as that used by Touir et al. [56]. Figure 7.5 shows the plating rate vs. pH of the bath. The plating rate increases remarkably from 2 μm/h to 6.5 μm/h when the pH of the solution increases from 5 to 9; this behavior is probably due to the effect of pH on the anodic reaction.

A group compared electroless copper plating processes using sodium hypophosphite and formaldehyde as reductants. In order to compare, in 2008, Yang et al. [60] used two different bath solutions. The bath solution containing sodium hypophosphite is: $CuSO_4•5H_2O$ 10 g/L, $NaH_2PO_2•H_2O$ 30 g/L,

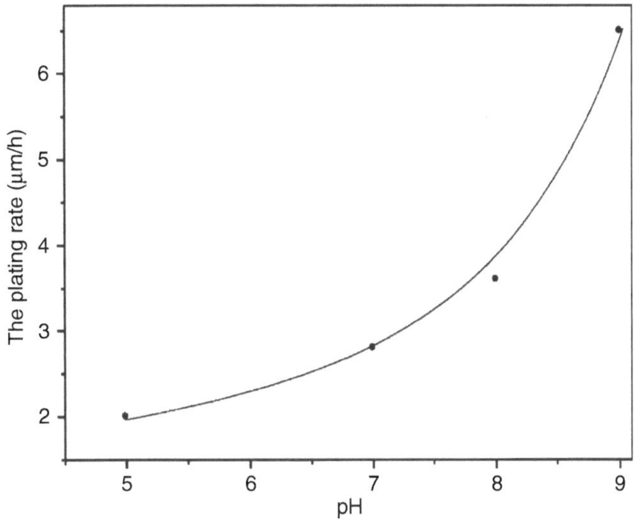

FIGURE 7.5 **The effect of pH on the plating rate (bath: Cu^{2+} 0.0032M, Ni^{2+} 0.081 mol/L, cit 0.3 mol/L, NH_4Cl 1 mol/L, NaH_2PO_2 0.28 mol/L and T = 78°C) [63].**

$Na_3C_6H_5O_7$•$2H_2O$ 23.5 g/L, 2,2′-dipyridyl 15 mg/L, $NiSO_4$•$6H_2O$ 2 g/L, pH 9, T 70°C, and plating time 30 min. The composition and operation conditions for formaldehyde bath are: $CuSO_4$•$5H_2O$ 10 g/L, HCHO (37%) 15 mL/L, EDTA•2Na 40 g/L, 2,2′-dipyridyl 25 mg/L, pH 12.5, T 70°C, and plating time 30 min. The results are shown in Table 7.2. The electrolyte solution that used sodium hypophosphite as a reducing agent is more stable than that in which using formaldehyde was the reducing agent. Also, the electroless plating rate of the sodium hypophosphite solution is higher than that of the formaldehyde bath. However, the physical properties (conductivity, tensile strength, and ductibility) of deposits obtained through using formaldehyde are better than that from the sodium hypophosphite solution. In addition, the copper content in formaldehyde's deposit is nearly 100% (wt), whereas the deposit for sodium hypophosphite bath contains 93.9% (wt) Cu and 6.1% (wt) Ni. From this comparison, it is entirely feasible to replace formaldehyde with hypophosphite as a reductant in copper electroless plating.

7.3.2 Glyoxylic Acid

A number of reports for replacing formaldehyde with glyoxylic acid as reductant have been published [66–77]. They cannot all be discussed here, so only a selection will be presented. Early in 1991, Honma et al. [66] investigated electroless copper plating using glyoxylic acid as a reducing agent. However, the authors still used EDTA in their bath. The bath solution and operating conditions are: $CuSO_4$•$5H_2O$ 0.03 mol/L, glyoxylic acid 0.20 mol/L, EDTA 0.25 mol/L, 2,2′-bipyridyl 10 mg/L, pH was adjusted to 12.5 with sodium

TABLE 7.2 Effect of Recycling of Bath Solution on the Plating Rate and Coating Appearance

	Sodium hypophosphite bath		Formaldehyde bath	
Number of recycling	*Plating rate ($\mu m/h$)*	*Coating appearance*	*Plating rate ($\mu m/h$)*	*Coating appearance*
1	3.96	Brightness	2.73	Brightness
2	3.78	Brightness	2.52	Brightness
3	3.90	Brightness	2.40	Brightness
4	3.96	Brightness	Bath solution decomposition	
5	3.66	Semibrightness		
6	4.86	Matt		
7	5.88	Dark, black		
Physical properties of coating	Sodium hypophosphite bath		Formaldehyde bath	
Resistivity ($\mu\Omega \cdot cm$)	10.59		3.06	
Tensile strength (N/mm^2)	143.9		167.8	
Elongation (%)	21.3		27.0	

hydroxide, and temperature 60°C. Figure 7.6 shows the effect of ① glyoxylic acid concentration, ② pH value, and ③ bath temperature on the plating rate. It can be seen that the plating rate increases with glyoxylic acid concentration, pH value, and bath temperature, respectively. Plating rates are considerably faster and bath stability was superior to formaldehyde baths under standard operating conditions. Moreover, it was confirmed that deposition uniformity of whole wall coverage is superior to that of the formaldehyde bath. Glyoxylic acid is a nonvolatile chemical that showed good reducing power in electroless copper

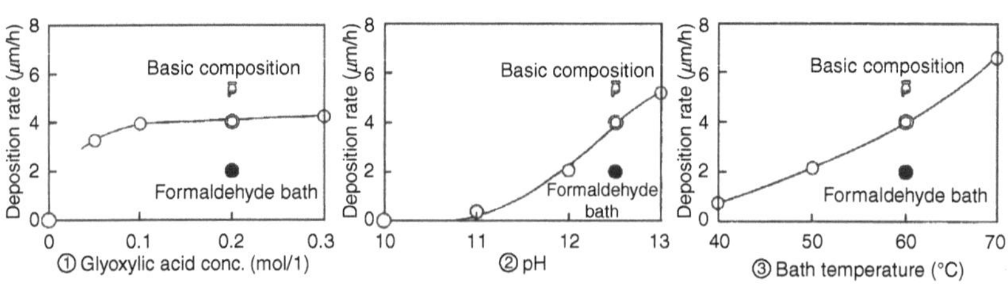

FIGURE 7.6 Relation between plating condition and deposition rate in EP Cu. ① Glyoxylic acid concentration, ② pH value, ③ bath temperature [66].

plating, and can therefore replace formaldehyde, alleviating the environmental problems associated with the latter.

In 2000, Shacham-Diamand [68] found that EP copper from solutions that contain glyoxylic acid as reducing agent formed a conformal thin film compatible with very large-scale integrated technology for integrated circuit and microelectromechanical system applications. Glyoxylic acid is an alternative reducing agent that can replace formaldehyde for basic copper sulfate based solutions without compromising thin film properties and process capabilities. The resistivity of thin copper film, thicker than 300 nm, deposited from glyoxylic-based solution was 2.3–2.4 μΩ•cm as deposited, and 1.9–2.0 μΩ•cm after a 200°C anneal for 30 min. Electroless copper was deposited from solution with glyoxylic acid into via contact of 0.2 μm diameter and up to 4:1 aspect ratio. The deposit was smooth and conformal.

In 2008, Wu and Sha [74] compared electroless copper deposits that were plated on epoxy boards in solutions containing either formaldehyde or glyoxylic acid as reducing agent at different concentrations and operating temperatures. They found that deposits with smoother surface can be acquired by replacing formaldehyde with glyoxylic acid as reducing agent in an electroless copper plating bath operated at 45°C; there is no significant amount of residual stress on the surface of the deposits from the solutions with either of the reducing agents; glyoxylic acid has great potential to be the substitute for formaldehyde in electroless copper plating processes.

7.3.3 DMAB

Early in 1989, Jagannathan et al. [78] used substituted DMAB for the traditional formaldehyde as reductant in EP Cu in their US Patent. The preferred plating bath composition consists of the following: copper sulfate 4 g/L, EDTA (preferably a disodium salt) 20 g/L, triethanolamine 50 mL/L, DMAB 4 g/L, sodium cyanide 1.6–2.0 μL/L, 1,10-phenanthroline 22 μg/L. The preferable pH is 8.7. The resulting bath is operated at 60°C. The plating rate is between 2 μm/h and 3 μm/h on epoxy substrates activated with a Pd/Sn colloid. It is observed that the quality of the copper is comparable to that achieved with conventional copper–formaldehyde bath. Unfortunately, sodium cyanide was used in the bath, which is toxic. Therefore, this approach is not a real green electroless plating.

In 2009, Lu [79] reported electroless deposition of Cu on poly(ethylene terephthalate) (PET) fabric modified with 3-mercaptopropyltriethoxysilane. The bath solution and operating conditions are: $CuCl_2•2H_2O$ 9 g/L, H_3BO_3 6 g/L, $EDTA•2Na•2H_2O$ 18 g/L, DMAB 7 g/L, pH 7 temperature 60°C, and plating time 4 h. Figure 7.7 shows the SEM (a) and EDX spectrum (b) of the modified fabric sample. It can be seen from the figure that the deposits became more compact and smoother on modified fabric. The color of the deposits also had a corresponding improvement, changing from dark brown to copper-bright. A bright appearance of the deposits usually indicates better mechanical and

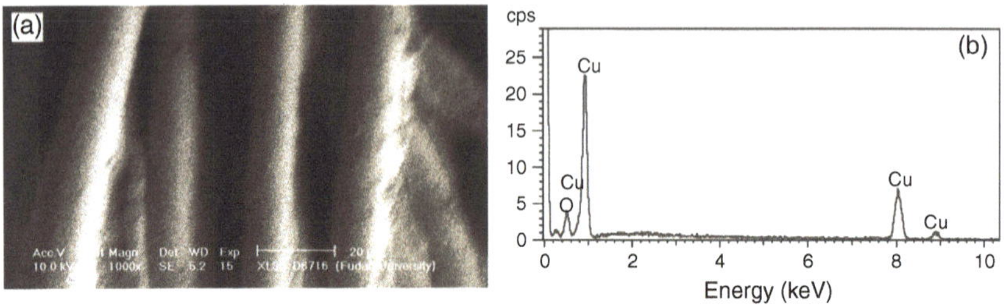

FIGURE 7.7 SEM photos (a) and EDX spectrum (b) of copper plating PET fabrics after ultrasonic washing and pretreatment [79].

electrical properties. The coating is composed of Cu, and other possible ingredients, such as B, N, are invisible. In addition, copper plating on modified fabric has good adherence stability and high electrical conductivity before and after ultrasonic washing, while copper coating on fabric without modification is easily destroyed during the washing process, which leads to the textile changing from conductor to dielectric.

In 2010, Plana et al. [80] studied the kinetics of the electroless deposition of copper on gold, using DMAB as a reducing agent. The bath solution and operating conditions are: copper sulfate (Cu resource) 0.030 mol/L, reducing agent DMAB 0.065 (or 0.033) mol/L, triethanolamine (added to control the pH) 0.3 mol/L, 1,5,8,12-tetraazadodecane (employed as a complexing agent for the copper (II) ions to prevent the spontaneous formation of copper hydroxide) 0.045 mol/L, pH 11.6, and temperature 55.0 ± 0.5°C. The copper deposition rate in the electroless bath is 50 nm/min. The measured kinetic data show that the oxidation and reduction processes are not additively coupled for the specific case of the electroless deposition of copper by DMAB. In this case, the fact that the rate of DMAB oxidation differs on Cu and Au and the overall rates decrease when the two half-reactions are separated means that the mixed potential theory does not hold for this system.

In 2012, Kulyk et al. [81] reported copper electroless plating in weakly alkaline electrolytes using DMAB as a reducing agent for metallization on polymer films. EP Cu deposition was performed in solutions containing $CuSO_4{\bullet}5H_2O$ 0.03 mol/L, EDTA 0.035 mol/L, DMAB 0.1 mol/L, and triethanolamine (TEA) 0.38 mol/L. The pH was adjusted by the addition of small amounts of tetramethylammonium hydroxide and phosphoric acid. Typical photographs of the deposited films show that copper films deposited in low pH solutions (7.5 and 8) are shiny and uniform; films of similar thickness but deposited at higher pH levels (8.3–9) have a rough surface and often have large defects (on millimeter scale). The authors believe the defects are due to the high deposition rate, the short initiation time, and the high rate of hydrogen evolution during the higher-pH ELP process. Low pH (7.5 and 8) copper films show the strongest adhesion (the T-peel adhesion strength = 2.5 N/cm, see Fig. 9 in [81]) to the polymer substrate (PI) substrate, comparable to the highest values (T-peel adhesion strength = 7 N/cm) obtained so far on polymer substrates with plasma-modified surfaces [92]. Thin films deposited at pH = 7.5 and thick films deposited at pH = 9 show the highest conductivity. The dependence of the observed growth rates of copper films on the pH of the deposition solutions shows that the plating rate basically is linearly enhanced from 0.105 μm/h at 7.5 pH to 1.85 μm/h at 9.0 pH. It is obvious that such plating rate is rather low, as acknowledged by the authors, who said that "the only drawback for this deposition condition is the very low growth rate."

7.3.4 Sodium Bisulfate

Only one paper on using sodium bisulfate as reductant to replace formaldehyde in EP Cu has been found. In 2009, Chen et al. [83] reported the electroless deposition of a copper sulfide coating on polyacrylonitrile (PAN) with a chelating agent of triethanolamine. However, they obtained a CuS coating, not pure Cu on the PAN film. The aim is to improve the EMI shielding effectiveness. The composition and operating conditions of the electroless copper sulfide bath are: $CuSO_4 \cdot 5H_2O$ 0.60 mol/L, $NaHSO_3$ 0.30 mol/L, $Na_2S_2O_3 \cdot 5H_2O$ 0.60 mol/L, TEA 0.10–0.6 mol/L, pH 1–2, temperature 75°C, plating time 60 min. The formation of copper sulfide can be represented by the following equations [3]:

$$2SO_2O_3^{2-} \rightarrow S_4O_6^{2-} + 2e^- \tag{7.1}$$

$$S_2O_3^{2-} + H^+ \rightarrow S + HSO_3^- \tag{7.2}$$

$$S + 2e^- \rightarrow S^{2-} \tag{7.3}$$

$$Cu^{2+} + S^{2-} \rightarrow CuS \downarrow \tag{7.4}$$

The average thickness of the CuS layer increased from 76 nm to 247 nm as the TEA concentration increased from 0.00 M to 0.60 M. Therefore, the EMI SE of the composites increased from 10–12 dB to 25–27 dB.

7.3.5 Co^{2+} and Fe^{2+}

Using a Co^{2+} complex as a reducing agent instead of formaldehyde in EP Cu has been investigated by a group from Lithuania [84–87]. In 1999, Vaškelis and Norkus reported the first autocatalytic processes of Cu^{2+} reduction by Co^{2+} complexes [84]. The electroless copper plating solution contained: Cu^{2+} 0.05 mol/L, Co^{2+} 0.15 mol/L, ethylenediamine 0.4, 0.6, or 1.2 mol/L, pH 5.5–11, temperature 50 and 70°C, and loading 1.4 dm^2/L. The pH value of the solutions was adjusted by using an acid with the same anion as in Cu^{2+} and Co^{2+} salts. The results show that the electroless Cu deposition rate reaches ca. 6 μm/h (50–70°C) in fairly stable plating solutions. The Cu deposit has a good quality. The authors found by experimentation that the Co^{2+} complex reducing agent can be regenerated by electrochemical reduction of the Co^{3+} species formed in the plating process.

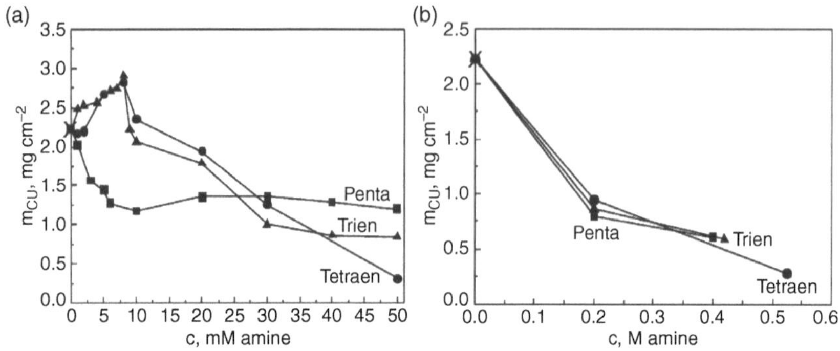

FIGURE 7.8 Dependence of the copper deposition rate on the concentration of amines added. (a) Lower concentrations (mmol range) and (b) higher investigated amine concentrations. Solution composition: $CuCl_2$ 0.05 mol/L, $CoCl_2$ 0.15 mol/L, diethylenetriamine (Dien) 0.6 mol/L. Solution pH 6.3, deposition time 15 min, temperature 20°C. X initial solution; ▲ triethylenetetramine (Trien), ● tetraethylenepentamine (Tetraen), ■ pentaethylenehexamine (Penta) [87].

In 2012, this group published a paper of a comparative study of electroless copper deposition using different Co^{2+}–amine complex compounds as reducing agents [87]. They claimed that tetraethylenepentamine (Tetraen) and triethylenetetramine (Trien) were applied as ligands in electroless copper plating systems using Co^{2+}–amine complexes as reducing agents for the first time. The solutions contained: $CuCl_2$ 0.05 mol/L, $CoCl_2$ 0.15 mol/L, amine 0.4 mol/L. Solution pH was adjusted by adding HCl. The solutions were purged of oxygen in the air by using Ar. The experiments were carried out at 20 ± 1°C. Figure 7.8 shows the dependence of the copper deposition rate on the concentration of amines added. At lower concentrations (mmol range) (Fig. 7.8a), pentaethylenehexamine (Penta) has a retarding effect and diminishes the plating rate at its concentration 10 mmol/L. In contrast, the same amount of tetraethylenepentamine (Tetraen) or triethylenetetramine (Trien) has an accelerating effect and increases the plating rate from 2.25 mg/cm^2 to about 3 mg/cm^2. A further increase in the concentration of the amines mentioned results in a decrease of the copper plating rate. Thus, small amounts of Tetraen or Trien (up to 10 mmol/L) have an accelerating effect, which merges into retardation at higher investigated amine concentrations, whereas Penta exhibits a retarding effect in all concentration ranges investigated (Fig. 7.8b). It can be also seen from (b) that higher concentrations (0.2–0.5 mol/L) of Penta, Tetraen, and Trien have a more expressed retarding effect (when compared with lower concentrations). In the last case, the retarding behavior of all three amines is practically the same or very similar.

Because Co is an expensive metal, a Japanese group examined electroless copper deposition using Fe^{2+} compound as a reducing agent [88,89]. In 2002, Sone et al. [88] used the following bath composition and plating conditions. Trisidium citrate dehydrate 0.1 mol/L, ethylenediamine 1.2 mol/L, $CuSO_4 \cdot 5H_2O$ 0.1 mol/L, $FeSO_4 \cdot 7H_2O$) 0.03 mol/L, 2′2-bipyridyl 100 ppm, Cl–(NaCl)

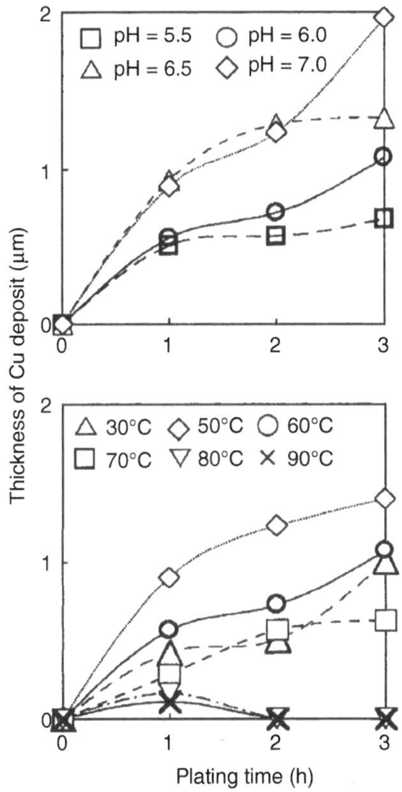

FIGURE 7.9 Effect of plating conditions on deposition rate [88].

1.2 mol/L, pH (adjusted by H_2SO_4) 6.0, bath temperature 60°C, and atmosphere N_2. Deposition rates versus changing pH and bath temperature are shown in Fig. 7.9. Although the deposition rate increased with an increase in pH, Cu fine particles were generated and the stability of the bath decreased at pHs greater than 6.5. As the bath temperature increased, the deposition rate increased up to 50°C, but particles formed in the bath. The authors found that the optimal bath conditions were a pH of 6.0 and a temperature 60°C. Comparing the surface of samples obtained by the conventional method using formaldehyde and that using a Co^{2+} complex as a reducing agent, the surface of the sample obtained from this study showed polycrystalline structure with many block-like particles. By adding 100 ppm of polyethylene glycol to the bath, this block-like structure was reduced and the surface became smooth. Deposition thickness of about 1.7 μm was obtained after 3 h of plating under optimal conditions. In 2004, these authors studied further electroless copper deposition using Fe^{2+} compound as a reducing agent. The authors found that particle formation in the bath could be avoided by adding a source of chloride ion. Adding polyethylene glycol (PEG) into the bath improves the copper deposition rate and makes the surface smooth.

Under optimal conditions, over 1 μm of deposit with a smooth surface was obtained after 3 h of plating, which contained no iron impurities. The electrical resistivity of the copper film is about 3–4 μΩ•cm, the same as that of electroplated copper films. However, the lifetime of this bath was not very long, so further improvement is required to put this method into practical use.

7.3.6 Saccharide

In 2009, Poole et al. claimed in their patent [90] that they used reducing agents selected from a group of fructose esters to substituting for formaldehyde. Their invention hopes to provide improved electroless copper compositions that are stable, provide uniform copper deposits, and are environmentally friendly. As mentioned above, hypophosphite and DMAB have been suggested as a replacement for formaldehyde. However, the plating rates of baths containing hypophosphites are generally too slow. As for DMAB, it tends to be significantly more expensive and has associated health and safety issues, the resultant borates have adverse effects on crops on release to the environment and DMAB is toxic. So, these authors used fructose as a reducing agent. The electroless bath composition included at least copper chloride (source of copper ions) 7 g/L, fructose (reducing agent) 47 g/L, and 5,5-dimethylhydantoin (chelating agent) 64 g/L. In addition, electroless copper compositions included one or more conventional complexing or chelating agents in conventional amounts. The temperature of the compositions was maintained at 55°C and a pH of 13.2 during electroless copper deposition. Deposition time was 20 min. Both conductive and nonconductive substrates can be used in the EP Cu. All of the electroless copper compositions were stable. There was no observable Cu_2O precipitate. All of the copper deposits appeared salmon pink which indicated that the copper deposits were suitable for industrial applications. No dark brown patchy deposits were observed. In addition to composition stability and good copper deposition, typically copper plating rate may range from 1.5 μm/h to 9 μm/h, which was acceptable for industrial applications.

7.3.7 Green Ligand for EP Cu

As mentioned above, although the environmentally toxic reductant formaldehyde has been replaced by various green reducing agents the non green complexing agents EDTA was still used in some bath solutions. So, investigations were carried out to find greener alternatives to EDTA. Early in 1991, Jagannathan et al. [93] reported tetra-aza ligand systems as complexing agents for electroless deposition of copper. An electroless copper plating bath comprises copper sulfate 4–15 g/L, DMAB 3–10 g/L, tetra-aza ligand (1,5,8,12 tetraazadodecane = 1,2 Bis (3-aminopropylamino) ethane) 40–200 mmol/L, buffer (triethanolamine) 25–100 mL/L, pH of the bath 7–12, temperature 45–70°C. Surfactants can facilitate hydrogen solution, so sodium lauryl sulfate (10–50 mg/L) is added to the bath. Another additive, 2,2 bipyridine, is sometimes used to ensure long-term stability and to achieve desirable metallurgy such as brightness, ductility, and resistivity. Therefore, 30–600 mg/L of 2,2 bipyridine is also added to the bath. Plating rates of 1–4 μm/h have been observed with various compositions in the aforementioned pH range. The measured resistivity ranges between 1.9 μΩ•cm and 2.4 μΩ•cm, which is comparable to those obtained with the conventional formaldehyde process. The bath can be used for metal deposition at lower pH and for providing an opportunity to use additive processing for metallization in the presence of polyimide, positive photoresists, and other alkali-sensitive materials.

From 2005 to 2012, a group from Lithuania published several papers [94–97], in which they proposed four kinds of environmentally friendly chemicals as alternatives to EDTA ligand for electroless copper plating solutions. The first kind contains xylitol, D-mannitol, and D-sorbitol [94]. The second and third are trisodium acetate and saccharose, respectively [95, 96]. The fourth includes D-, L-, and DL-tartrate [97]. These authors claimed that these ligrands are all "environment-friendly ligands for alkaline electroless copper plating systems." However, it is a pity that they used formaldehyde as a reducing agent in all of these alkaline EP Cu bath solutions. The plating rates for all of the four kinds of ligands are rather low. For the three ligands of the first kind the plating rate only reaches a maximum value of 2–2.5 μm/h in a very narrow range of pH (13–13.5) (see Figs 2 and 4 in [94]). The rate of copper deposition in citrate-containing electroless copper solutions under optimal operating conditions (pH 12.75) reaches about 3 μm/h (Fig. 4 in [95]). When using saccharose, the plating rate is 2 μm/h at pH 12.75 (Fig. 3a in [96]). As for the fourth kind of ligands used in EP Cu, the highest plating rate (ca. 3 μm/h) was obtained from the solutions containing the racemic mixture of DL-tartrate at pH 12.5 (Fig. 3 1c and 2c in [97]).

7.4 GREEN ELECTROLESS PLATING Ag

In Chapter 2, it was indicated that nearly all commercial silver plating is conducted in formulations that contain cyanide and formaldehyde. Cyanide silver plating processes have many advantages, which are that cyanide is an excellent complexing agent; cyanide silver plating solutions produce a pleasing pure white, excellent adhesion, fully bright deposit, as demanded by customers; and such processes are easy to chemically control, and have a broad range of concentrations that will yield successful deposits. However, the toxicity of cyanide and formaldehyde is well known. So research has been carried out on the possibilities of producing a green EP Ag without using cyanide or formaldehyde [84, 98–104].

In 1994, Mandich et al. [98] was granted a US Patent. In the patent, the authors proposed an electroless silver plating solution comprising a silver(I) complex, a thiosulfate salt, and a sulfite salt. They claim that this electroless silver plating solution uses a novel reducing agent combination of thiosulfate and sulfite that is far superior to those of conventional silver plating solutions containing formaldehyde, reducing sugars, borohydride, hydrazine, and other reducing agents. The electroless silver plating solution contains 3 g/L of silver as a silver (I) complex, a solution of 200 g/L sodium thiosulfate, and 20 g/L of sodium sulfite. No strong chelating agent was used. The pH was adjusted with potassium carbonate solution to pH 7.5 and the solution heated to 65°C. The plating rate of silver was 6.2 μm/h. The appearance of deposit Ag was dull grey. The bath showed no plate-out or other instability.

In 1999, Vaškelis and Norkus [84] reported an autocatalytic process of silver (I) reduction by cobalt(II) complexes. The main plating solution contained $AgNO_3$ 0.04 mol/L, $CoSO_4$ 0.10 mol/L, $(NH_4)_2SO_4$ 2 mol/L, and NH_3 concentration 4 mol/L. The pH of the solutions was changed by manipulating the molar ratio NH_4^+:NH_3. Electroless coatings were deposited onto glass plates (2 ×2.5 cm) at 20 or 50°C. The plating rate increases with the solution pH in the range 9–10, and after reaching values over 1 μm/h it decreases rather sharply at a higher solution alkalinity. Comparing EP Cu using the Co^{2+} as reductant, its plating rate is up to ca. 6 μm/h (see Section 7.3.5), so the plating rate of Ag from this bath solution is fairly low. The authors believed that the fall of the silver plating rate at a higher solution pH could be related to the stability of Co^{2+} complexes – the possibility of their hydrolysis.

Using vitamin B2 in the synthesis of Ag, nanospheres, nanowires, and nanorods were reported by Nadagouda and Varma in 2008 [102]. They claimed that this is the first report of a green chemistry approach using vitamin B2 as a reducing agent in the synthesis of Ag without using any harmful reducing agents, such as sodium borohydride or hydroxylamine hydrochloride and any special capping or dispersing agent. Why using vitamin B2 as a reducing agent as well as a capping agent? The reasons were due to its high-water solubility, biodegradability, and low toxicity compared with other reducing agents. The bath solution consisted of silver nitrite, vitamin B2, acetic acid, ethylene glycol, *N*-methylpyrrolidinone (NMP), and sucrose as a capping agent. Because the reduction potential of vitamin B2 is −0.3 V versus a saturated calomel electrode (SCE) [105], it is then sufficient to reduce silver (0.80 V versus SCE). The average particle size of Ag nanoparticles was found to be 6.1 ± 0.1, 5.9 ± 0.1, and 6.1 ± 0.1 nm in ethylene glycol, acetic acid, and NMP, respectively.

In 2013, Kim et al. [104] published a paper on environmentally friendly electroless plating for Ag/TiO_2-coated core-shell magnetic particles using ultrasonic treatment without toxic additives. The bath solution for plating Ag layer on the magnetic particles (TiO_2-coated Fe) consisted of silver nitrate, glucose, sodium citrate, potassium hydroxide, and ammonium hydroxide. Figure 7.10 shows a schematic illustration of the procedure used to prepare Fe/TiO_2/Ag core-shell composite microspheres. A TiO_2 layer was used as a surface modification layer that bound to silver ions and promoted silver metal deposition. A continuous and compact silver layer could be formed with a high degree of

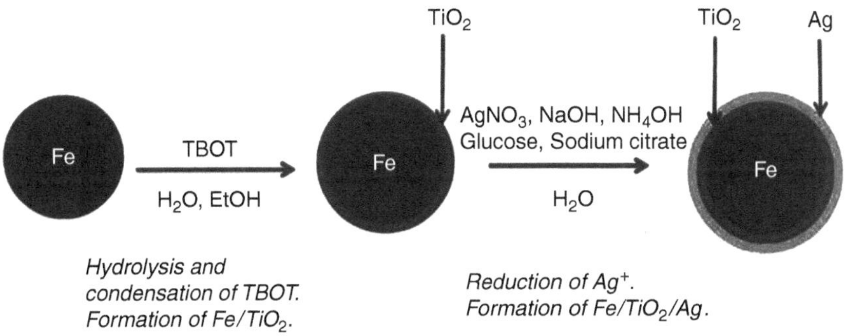

FIGURE 7.10 Schematic illustration of the procedure used to prepare Fe/TiO_2/Ag core-shell composite microspheres via TiO_2-assisted electroless silver plating. TBOT (titanium *n*-butoxide) [104].

morphological control by introducing ultrasonication and adjusting the ammonium hydroxide concentration.

7.5 GREEN ELECTROLESS PLATING Au

Practical electroless gold plating bath still mostly utilizes Au(I) cyanide complex as the source of gold and KCN as the stabilizer, which releases free cyanide ions during plating. The released free cyanide is highly toxic. Therefore, cyanide-free type plating bath is now needed from the view point of safety and environmental concern. Key components of electroless gold plating bath are Au complex (gold source) and reducing agent. So, the real cyanide-free type electroless Au plating bath must contain some compound other than cyanide Au complex (gold source) and must be an environmentally-friendly reducing agent. How to select a noncyanide Au complex? Such Au(I) complex ligand candidates must be able to replace conventional cyanide ion by that has been used as the strongest ligand for Au(I) ion so far. When making the choice, one should consider compounds that strike a suitable balance between reducing power of the reagents and moderate stability of the noncyanide gold complex. In such a way, practically stable noncyanide electroless Au plating can really meet rather severe requirements for high-performance electronics manufacturing applications.

Many papers on green EP Au have been published to date [106–118].

Green EP Au can be traced back to 1967 [106]. Luce in his US Patent proposed a noncyanide electroless Au plating bath, which consists of $HAuCl_4 \cdot 3H_2O$ 0.025 mol/L, sodium sulphite 1.6 mol/L, ammonium chloride 1.5 mol/L, cyclohexylamine 1.3 mol/L, hydrazine 1.8 mol/L, pH 7–12, temperature 85°C. The plating rate is 3.81 μm/h.

In 1995, Inoue et al. [110] reported stable noncyanide electroless Au plating. After experimentation, the authors obtained an improved noncyanide electroless Au bath formulation: sodium tetrachloroaurate 0.005–0.025 mol/L, sodium thiosulfate 0.04–0.2 mol/L, sodium sulfite 0.2–0.6 mol/L, sodium tetraborate 0.066–0.13 mol/L, thiourea 0.0013–0.013 mol/L, hydroquinone 0.0018–0.018 mol/L, pH 7.5–8.0, temperature 60–80°C. The bath load is the Au surface area of 50 cm^2/L. The bath life without load is greater than 100 h. The bath life with load means that the improved bath stands over three turns of plating operation with replenishment and can be used for at least 2 months or 6 months period after a fresh startup under a mass manufacturing condition or small experimental condition, respectively. The definition of bath life is that the time duration until gold particle precipitates appears in operating bath. Remarkable features of the bath formulation are much lower content of thiourea, lower value of optimum pH, lower bath temperature, and longer bath life without and with load. Average stable gold deposition rate is around 0.9 μm/h.

Hu et al. in 2008 [114] developed a simple and environmentally-friendly electroless plating solution of chloroauric acid ($HAuCl_4$) and hydrogen peroxide (H_2O_2) for depositing gold film onto (3-aminopropyl)-trimethoxysilane (APTMS)-coated glass surface. The plating bath is an aqueous solution containing 10 mL of 0.01% $HAuCl_4$ and 500 μL of 30% H_2O_2, which was agitated continuously to ensure the formation of a uniform gold film. The plating was performed at room temperature. The reaction can be completed within 8–10 min. The thickness of gold film measured by AFM is approximately 45 nm corresponding to the plating time of 8 min. The overall reaction equation is

$$2AuCl_4^- + 3H_2O_2 = 2Au + 3O_2 + 8Cl^- + 6H^+ \tag{7.5}$$

The standard redox potential of the overall reaction is +0.307 V, which shows that H_2O_2 is thermodynamically capable of reducing $AuCl_4^-$ to Au.

Gold nanoparticles (AuNPs) for conducting microbeads were prepared using the green electroless plating method by Tokonami et al. in 2009 [116]. The AuNPs conducting microbeads were performed by three steps, as shown in Fig. 7.11. The first step was to prepare AuNPs with green EP. The bath solution was that a 30 mL aliquot of 1 wt% $HAuCl_4$ was added to 170 mL of 0.26 wt% sodium citrate and stirred at 80°C for 20 min. TEM measurements indicated that the AuNP produced had a mean diameter of 31.1 ± 2.4 nm. The second one was to deposit AuNP on plastic beads, which was performed as follows. The microbead was added to 22.5 mL of the colloidal gold dispersion containing the binder molecule, and the mixture was stirred at room temperature for 1 day. The additional amounts of the beads and thiol molecules are listed in Table I in reference [116]. The third step was AuNP growth. The operation process was: AuNP-deposited beads (10 mg) were added to a 25 mL aqueous solution containing 50% ethanol, 18 mM of poly(vinyl alcohol), 0.18 of mM sodium as corbate as a reducing agent, and 0.19 mmol/L of chloroauric acid. The mixture was stirred at 50°C for 1 h. The electrical resistance of the AuNP-fixed single acrylic resin beads (3.8 MΩ) was successfully improved (0.5 Ω) by the AuNP-growth method; this is comparable to the resistance of commercially available beads

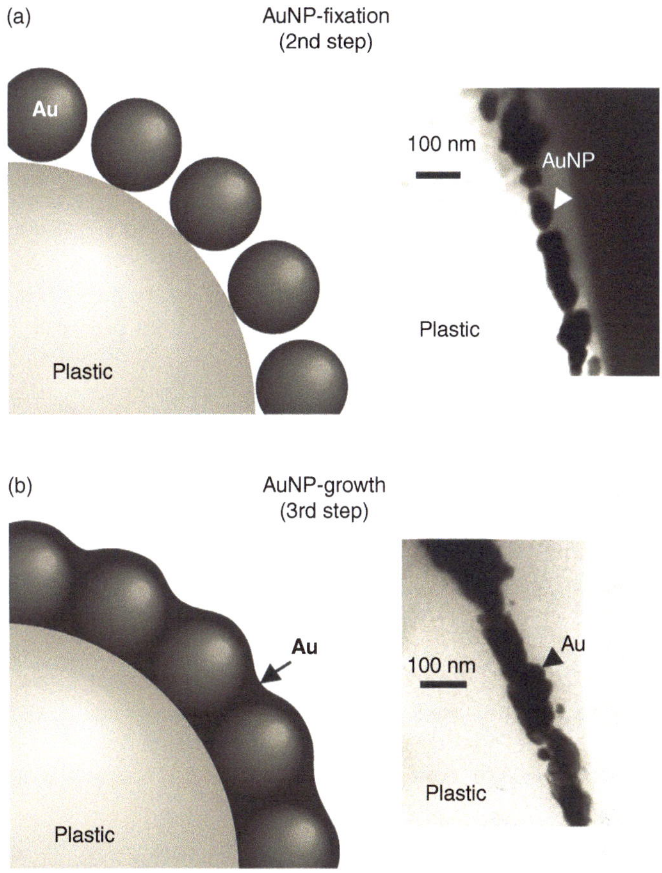

FIGURE 7.11 Schematic illustration for the preparation procedure [(a) second step, AuNP-fixation, (b) third step, AuNP-growth] of the conducting microbead. The right images are the corresponding TEM images for second and third steps, respectively [116].

(0.7 Ω). The prepared microbeads were successfully applied to anisotropic conductive films (ACFs), and a 1.9 Ω resistance was obtained. This suggests that the beads are suitable for application to the ACF. In 2011, these authors [117] studied such microbeads further. They were determined to show that the AuNP-plated microbead not only had good electrical properties, but also had high thermal and physical stabilities. These are attributed to the adhesion between the AuNPs and the microbead via the binder. Therefore, the presence of the binder leads to the formation of a superior metallic thin layer on the microbead with a high flexibility compared with that formed by conventional methods. This AuNP-plating method facilitates the realization of low environmental emissions since non green chemicals such as cyanide, chromic acid, and strong alkalis and acids, which are usually used in conventional electroless plating, are not required in the process. In addition, lesser amount of Au is required because of the use of AuNPs, leading to a low-cost manufacturing.

7.6 SUMMARY

After description and discussion of the green electroless plating more in detail, some conclusions can be drawn.

Until today the green EP has been researched broadly and deeply. Green electroless plating not only with Ni–P alloys, but also Cu, Ag, and Au have all been all investigated to produce a process that does not seriously harm human health and the environment. However, there are still areas that need investigation for the green EP of other metal and alloy systems.

Hazardous substances, such as Pb^{2+}, EDTA, Cd^{2+}, formaldehyde, cyanide used as reductant, complexing agent, brightener, or metal source, respectively, in the traditional EP have been all substituted by chemicals that are nontoxic and not harmful to human health and ecological environment in the green EP.

Not all chemicals used in the green EP approaches are all real green substances, sometimes non green chemicals such as EDTA are still used in bath solution by some authors, but they still put their chemical plating method called green chemical plating. In the above discussion, we have repeatedly pointed out this point.

It should be said that the aim of the real-green EP eco-agent (the used chemicals are nontoxic and not harmful to human health or ecological environment), higher deposition rate, not expensive, and stable solution for longer recycling use has not been reached completely up today. More research is needed to achieve this end.

REFERENCES

[1] Dietz KH, Fine Lines in High Yield (Part CLXI): some health and effluent disposal considerations in PWB fabrication. (Tech Talk), CircuiTree Mag., No. 2, 2009.
[2] Ensafi AA, Shiraz AZ. J Hazard Mater 2008;150:554.
[3] Solsona F. Guidelines for Drinking Water Quality Standards in Developing Countries (OPS/CEPIS/PUB/02.77). Lima: Pan American Health Organization, Pan American Sanitary Bureau, Regional Office of the World Health Organization; 2002.
[4] Anonymous, RoHS Regulations-Government Guidance Notes, in, Department for Business, Innovation & Skills, Eco-design and Product Regulation Unit, Environmental & Technical Regulation Directorate, London, UK, 2010.
[5] Anonymous, Council Directive 2000/53/EC on End of Life Vehicles, in, Environment Directorate, European Commission, Brussels, Belgium, 2003.
[6] Rahman IMM, Furusho Y, Begum ZA, Izatt N, Bruening R, Sabarudin A, Hasegawa H. Microchem J 2011;98:103.
[7] Sun H, Feng L, Ma H, Lou H. J Shandong Uni Architec Eng 2005;20(1):82.
[8] Wang H, Chen H, Sun A, Xu J. Surf Technol 2008;37(3):14.
[9] Weber CJ, Pickering HW, Weil KG. J Electrochem Soc 1997;144:2364.
[10] Hanna F, Hamid AZ, Aal AA. Mater Lett 2004;58(1–2):104.

[11] Vaskelis A, Jačiauskienė J, Stalnioniene I, Norkus E. J Electroanal Chem 2007;600(1):6.
[12] Deckert CA, P&SF 1995;82(2):48; 1995;82(3):58.
[13] daRosa CP, Maboudian R, Iglesia E. J Electrochem Soc 2008;155(6):E70.
[14] Brookshire RR, US Patent 2,976,181 Method of gold plating by chemical reduction, March 21, 1961.
[15] Okinaka Y. Plating 1970;57(9):914.
[16] Ali HO, Christie IRA. Gold Bull 1984;17(4):118.
[17] Ko JW, Koo HC, Kim DW, Seo SM, Kang TJ, Kwon Y, Yoon JL, Cheon JH, Kim YH, Kim JJ, Park YJ. J Electrochem Soc 2010;157(1):D46.
[18] Pearlstein F, Wightman RF. Plating 1971;58(10):1014.
[19] Brenner A, Riddell GE. J Res Bur Stand 1946;37:31.
[20] Schwartz M. Proc AES 1960;47:176.
[21] Brenner A, Riddell G. Surf Technol 1980;10:81.
[22] Lambert MR, Duquette DJ. Thin Solid Films 1989;177:207.
[23] Zhang B, Xie H. Mater Sci Eng 2000;A281:286.
[24] Mallory GO. Plating 1971;58:319.
[25] Di Giampaolo AR, Ordofiez JG, Gugliemacci JM, Lira J. Surf Coat Technol 1997;89:127.
[26] Zhou H, Hu Y, Liu J, Li S. Electroplat Pollu Cont 1999;19(1):22.
[27] Lin K-L, Hwang J-W. Mater Chem Phys 2002;76:204.
[28] Cheong WJ, Luana BL, Shoesmith DW. Appl Surf Sci 2004;229:282.
[29] Baskaran I, Sankara Narayanan TSN, Stephen A. Mater Chem Phys 2006;99:117.
[30] Liu H, Li N, Bi S. Electroplat Pollu Cont 2006;26(2):20.
[31] Liu H, Li N, Bi S, Zhang D. Electroplat Finish 2008;27(3):19.
[32] Chen Z, Liu Q, Wang W, Kong Y, Dong R, Mitsuzaki N. Mater Protec 2008;41(4):33.
[33] Dong K, Li D, Wu G. Surf Technol 2008;37(3):12.
[34] Wang K, Hong L, Liu Z-L. Ind Eng Chem Res 2009;48:1727.
[35] Zheng Z, Li N, Li D. Electroplat Pollu 2009;29(1):30.
[36] Seifzadeh D, Rajabalizadeh Z. Surf Coat Technol 2013;218:119.
[37] Gray JE, Luan B. J Alloys Compd 2002;336:88.
[38] Ambat R, Zhou W. Surf Coat Technol 2004;179:124.
[39] Zhang WX, He JG, Jiang ZH, Jiang Q, Lian JS. Surf Coat Technol 2007;201:4594.
[40] Sun K, Zhang Y. Electroplat Pollu Cont 1998;18(3):18.
[41] Wang D, Su R, Dai C. Electroplat Finish 2007;26(1):41.
[42] Xue L, Yin R, He Y, Sun J, Zhao H, Yu Z. Plat Finish 2007;29(2):27.
[43] Deckert CA. P&SF 1995;82(2):48.
[44] Kou SC, Hung A. P&SF 2003;90(3):44.
[45] Kukans PE, Grunwald JJ, Ferrier DR, Sawoska D A, US Patent 4,209,331 Electroless copper composition solution using a hypophosphite reducing agent, June 24, 1980.
[46] Goldstein R, Kukanskis PE, Grunwald JJ, US Patent 4,265,943 Method and composition for continuous electroless copper deposition using a hypophosphite reducing agent in the presence of cobalt or nickel ions, May 5, 1980.
[47] Kukanskis PE, Grunwald JJ, Ferrier DR, Sawoska DA, US Patent 4,279,948 Electroless copper deposition solution using a hypophosphite reducing agent, June 21, 1981.
[48] Cheng DH, Xu WY, Zhang ZY, Yiao ZH. Metal Finish 1997;95(1):34.
[49] Lee C-Y, Huang T-H, US Patent 6,046,107 Electroless copper deposition solution using a hypophosphite reducing agent, April 4, 2000.
[50] Honma H, Fujinami T, Terashima Y, Hayashi S, Shimizu S, US Patent 6,193,789 Bl Electroless copper plating solution and method for electroless copper plating, February 27, 2001.

[51] Li J, Kohl PA. J Electrochem Soc 2002;149(12):C631.
[52] Honma H, Fujinami Y, Terashima Y, Hayashi S, Shimizu S, EP 0 909 838 B1, Electroless copper plating solution and method for electroless copper plating, November 13, 2002.
[53] Li J, Kohl PA. J Electrochem Soc 2003;150(8):C558.
[54] Grunwald J, US Patent 6,524,490 Bl Method for electroless copper deposition using a hypophosphite reducing agent, February 25, 2003.
[55] Li J, Hayden H, Kohl PA. Electrochim Acta 2004;49:1789.
[56] Touir R, Larhzil H, Ebntouhami M, Cherkaoui M, Chassaing E. J Appl Electrochem 2006;36.
[57] Yang F, Yang B, Lu B, Huang L, Xu S, Zhou S. Acta Phys Chim Sin 2006;22(11):1317.
[58] Gan X, Wu Y, Liu L, Shen B, Hu W. Surf Coat Technol 2007;201:7018.
[59] Gan X, Wu Y, Liu L, Hu W. J Appl Electrochem 2007;37(8):899.
[60] Yang F, Yang B, Huang L, Xu S, Yao G, Zhou S. Plat Finish 2008;30(8):12.
[61] Zhao J, Li Z, Wang S, Yin L. Electroplat Poll Cont 2009;29(2):35.
[62] Liu C, Huo X, Li Y. Electroplat Poll Cont 2009;29(6):34.
[63] Anik T, Touhami ME, Himm K, Schireen S, Belkhmima RA, Abouchane M, Cissé M. Int J Electrochem Sci 2012;7:2009.
[64] Gan X, Zhou K, Hu W, Zhang D. Surf Coat Technol 2012;206:3405.
[65] Yuan XL, Gao J, Yang ZF, Wang ZX, Wang ZL. Surf Eng 2012;28(5):377.
[66] Honma H, Komatsu M, Fujinami T. Jpn J Surf Finish 1991;42(9):913.
[67] Honma H, Kobayashi T. J Electrochem Soc 1994;141(3):730.
[68] Shacham-Diamand YY. Electrochem Solid State Lett 2000;3(6):279.
[69] Tang D, Wei Z, Shao Y, Sun J, Lee Y, You S, Li J. Chin J Nonferr Metals 2003;13(5):1252.
[70] Yabe A, Sekiguchi J, EP 1 681 372 A1 Electroless copper plating solution and method for electroless copper plating, July 19, 2006.
[71] Wang J, Yang Z, Xie H, Jin Z, Hu N. Surf Technol 2006;35(3):19.
[72] Wu L, Yang F, Hang L, Sun S, Zhou S. Electrochemistry 2005;11(4):402.
[73] Shen D, Yang F, Wu H. Electrochemistry 2007;13(1):67.
[74] Wu X, Sha W. Synth React Inorg, Metal-Orga, Nano-Metal Chem 2008;38:292.
[75] Liu R, Li D, Yang Z. Fine Chem Intermedilates 2009;39(3):68.
[76] Mishra KG, Paramguru RK. African J Pure Appl Chem 2010;4(6):87.
[77] Wang L, Sun L, Li J. BioResources 2011;6(3):3493.
[78] Jagannathan R, Krishnan M, Wandy GP, US Patent 4,818,286 Electroless copper plating bath, April 4, 1989.
[79] Lu Y. Appl Surf Sci 2009;255:8430.
[80] Plana D, Campbell AI, Patole SN, Shul G, Dryfe RAW. Langmuir 2010;26(12):10334.
[81] Kulyk N, Cherevko S, Chung C. Electrochim Acta 2012;59:179.
[82] Liao Y, Zhang S, Dryfe R. Particuology 2012;10:487.
[83] Chen Y-H, Huang C-Y, Lai F-D, Roan M-L, Chen K-N, Yeh J-T. Thin Solid Films 2009;517:4984.
[84] Vaškelis A, Norkus E. Electrochim Acta 1999;44:3667.
[85] Vaškelis A, Jačiauskienė J, Jagminienė A, Norkus E. Solid State Sci 2002;4:1299.
[86] Jagminien A, Vaškelis A, Stankevièienė I, Tamašiūnait LT. Chemija 2005;16(2):33.
[87] Norkus E, Stankevičienė I, Tamašiūnaitė LT, Prušinskas K. Chemija 2012;23(1):1.
[88] Sone M, Kobayakawa K, Saito M, Sato Y. Jpn J Surf Finish 2002;53(8):559.
[89] Sone M, Kobayakawa K, Saito M, Sato Y. Electrochim Acta 2004;49:233.
[90] Poole MA, Cobley AJ, Singh A, Hirst DV, US Patent 7,611,569 Electroless copper compositions, November 3, 2009.

[91] Hung A, Chen K-M. J Electrochem Soc 1989;136:72.
[92] Wang WC, Kang ET, Neoh KG. Appl Surf Sci 2002;199:52.
[93] Jagannathan R, Knarr RF, Krishnan M, Wandy GP, US Patent 5,059,243 Tetra aza ligand systems as complexing agents for electroless deposition of copper, October 22, 1991.
[94] Norkus E, Vaškelis A, Jačiauskienė J, Vaičiunienė J, Gaidamauskas E, Macalady DL. J Appl Electrochem 2005;35:41.
[95] Norkus E, Kepenienė V, Vaškelis A, Jačiauskienė J, Stalnionienė I, Stalnionis G. Chemija 2006;17(4):20.
[96] Norkus E, Prušinskas K, Vaškelis A, Jačiauskienė J, Stalnionienė I, Macalady DL. Carbohyd Res 2007;342:71.
[97] Norkus E, Kepenienė V, Stalnionienė I. Chemija 2012;23(3):155.
[98] Mandich NV, Krulik GA, Singh R, US Patent 5,322,553 Electroless silver plating composition, June 21, 1994.
[99] Vaškelis A, Jagminienė A, Juškėnas R, Matulionis E, Norkus E. Surf Coat Technol 1996;82:165.
[100] Norkus E, Vaškelis A, Jagminienė A, Tamašiūnait LT. J Appl Electrochem 2001;31:1061.
[101] Kozlov AS, Palanisamy T, Narasimhan D, US Patent 6,387,542 Electroless silver plating, May 14, 2002.
[102] Nadagouda MN, Varma RS. J Nanomater 2008. Article ID 782358, 8 pages.
[103] Heuck FCA, Staufer U. J Microelectromech Syst 2011;20(2):451.
[104] Kim S-D, Choe W-G, Jeong J-R, (2013). Ultrason Sonochem, 20(6), 1456.
[105] Massey V. Biochem Soc Trans 2000;28(4):283.
[106] Luce BM, US Patent 3,300,328 Electroless plating of gold, January 24, 1967.
[107] Ushio J, Miyazawa O, Yokono H, Tomizawa Y, US Patent 4,880,464 Electroless gold plating solution, November 14, 1989.
[108] Iacovangelo CD. J Electrochem Soc 1991;138(4):976.
[109] Krulik GA, Mandich NV, Singh R, US Patent 5,318,521 Plating rate improvement for electroless silver and gold plating, January 7, 1994.
[110] Inoue T, Ando S, Okudaira H, Ushio J, Tomizawa A, Takehara H, Shimazaki T, Yamamoto H, Yokono H, In: IEEE Proc. electronic components and technology conference, 45th, 21–24 May 1995, p. 1059–67.
[111] Cai Jiqing. Electroplat Poll Cont 1997;17(3):14.
[112] Sato J, Kato M, Otani H, Homma T, Okinaka Y, Osaka T, Yoshioka O. J Electrochem Soc 2002;149(3):C168.
[113] Sanada M, US Patent 7,384,452 B2 Non-cyanide electroless gold plating solution and process for electroless gold plating, June 10, 2008.
[114] Hu J, Li W, Chen J, Zhang X, Zhao X. Surf Coat Technol 2008;202:2922.
[115] Aiba A, Kawamura K, Takahashi H, US Patent 7,396,394 Electroless gold plating solution, July 8, 2008.
[116] Tokonami S, Yamamoto Y, Mizutani Y, Ota I, Shiigi H, Nagaoka T. J Electrochem Soc 2009;156(12):D558.
[117] Tokonami S, Shirai S, Ota I, Shibutani N, Yamamoto Y, Shiigi H, Nagaoka T. J Electrochem Soc 2011;158(12):D689.
[118] Tang X, Wang J, Wang C, Shen B. Surf Coat Technol 2011;206:1382.

Part III

Composition, Microscopic Structure, and Surface Morphology of Electroless Deposits

After the description and discussion of the preparation technology of electroless plating deposits that is detailed in the previous chapters, we turn to discuss the EP deposits themselves. The first subject is their compositions and microscopic structure. The composition of the coating, and of course other external factors such as temperature, determine the structure and the various properties of the coating. Therefore, it is vital to understand the composition and structure of EP coatings. In addition, by collecting the SEM micrographs of electroless alloy depositions together and discussing them, people may understand the nature of electroless plating deposition more clearly and deeply. Most of the published papers about EP almost always contained a few SEM or TEM images. Some papers on EP even used SEM of their investigated coatings predominantly. However, this issue was not especially a point of focus in any book's chapter or review papers. That is why we use a chapter to discuss it. This is another feature of the book.

Chapter 8

Composition and Microstructure

Chapter Outline

Amorphous and Nano Alloys Electroless Depositions. http://dx.doi.org/10.1016/B978-0-12-802685-4.00008-X

Studying the compositions of EN coatings can be traced back to the 1950s. In early 1956, Gutzeit and Mapp reported the composition and structure of "Kanigen" coating in their paper "Kanigen: Chemical nickel plating" [1]. The bath composition and operating conditions were nickel ions 0.07 mol/L, hypophosphite ions 0.230 mol/L, an initial pH 4.5 to 5, and temperature above 90°C. Coatings obtained in the acid solutions are primarily metallic nickel containing 10.8–17.4 at% phosphorus. X-ray and electron diffraction measurements indicate that Kanigen coatings have the structure of an amorphous, solid substance with liquid-like disorder of the atoms. In addition, the deposition of different thicknesses (0.0254–0.254 mm) of Kanigen's study has no effect on the amorphous structure of the coating. After 1 year, in 1957, Goldenstein et al. [2] reported from X-ray and electron diffraction results that as-plated electroless Ni–P deposits containing 12.5–17.4 at% P are amorphous.

Since then, the measurements of composition and structure of EN films have continued to the present. In fact, in addition to the EN coating, such measurements for other EP binary, ternary, and multicomponent alloy deposits, and EP composite alloy coatings have also been performed. Therefore, the measurements of composition and structure for the EP alloy films can be found in a lot of papers [3–47,56–69,70–72,84–117].

8.1 COMPOSITION AND MICROSTRUCTURES OF EP ALLOY DEPOSITS

It is well known that the atoms in a crystal solid are arranged in order. Therefore, the obvious structural feature in a crystal is long-range structural periodicity. When the X-ray photons strike a crystalline plan, for example, {111}, narrow well-defined diffraction peaks are formed. If its crystallinity is perfect, then the diffraction peak must become a line. The electron diffraction pattern of a crystal is composed of several sharp diffraction rings.

As for amorphous solids, the arrangement of its atoms is short-range order but long-range disorder. In other words, for the amorphous solid, long-range structural periodicity is absent because of the randomness in the orientation of the semicrystals, and its atoms are arranged in a continuous liquid-like random packing without abrupt structural discontinuities. So, its X-ray diffraction would consist of only a few broad overlapping peaks. For a perfect amorphous solid, its electron diffraction pattern would be a single diffuse halo.

That is why, the scientists always used X-ray diffraction and TEM to measure the EP structures.

Metal science and materials science tell us that the crystal structure of metal alloys is controlled by their composition. Of course, temperature also affects their structures. Here, in the EP deposits, the situation will be similar. Moreover, this issue can be confirmed by the facts presented in the following subsections. The structure and composition of EP deposits are closely related; therefore, we discuss the two subjects together.

8.1.1 Ni–P Alloy Deposits

8.1.1.1 Studies of Composition and Structure Before 1980

EN is the main EP. Moreover, the properties of a deposit are closely related to its composition and structure. Lots of measurements of the composition and structure in EN coatings have been reported since the discovery of EN [3–47]. Even today, such investigations can be found in some papers [38–47].

In the research into the composition and structure of EN deposits, it is very common in the literature to classify the deposit on the basis of its phosphorous content [14,21,23,24,28,30–32,35–38,42], that is, the EN deposits can be classified as low (1–5 wt% P = 1.88–9.07 at% P), medium (6–8 wt% P = 10.79–14.15 at%), and high (9 wt% P and more = 15.78 at% and more).

As mentioned in the beginning of this chapter, in some early work using the Kanigen electroless nickel bath, Goldenstein et al. [2] used X-ray diffraction techniques to show that the as-deposited EN coatings have an amorphous structure with a "liquid-like" disorder of the atoms, and that the phosphorus is randomly distributed throughout the coating. These coatings contained 12.5–17.4 at% phosphorus and were plated from an acid bath at 90°C.

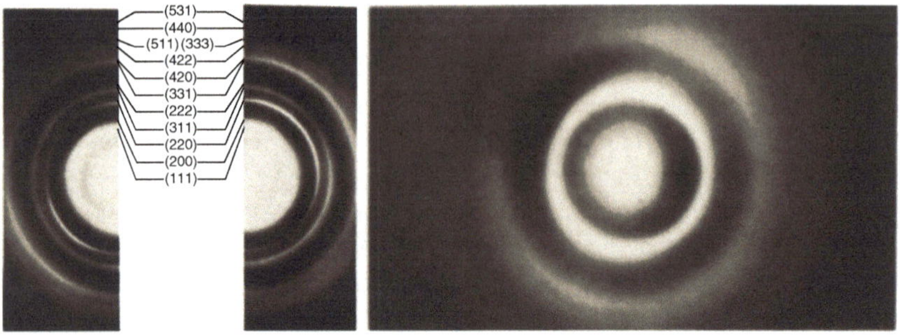

FIGURE 8.1 Electron transmission diffraction patterns of as-plated EN alloys from the alkaline-chloride bath. Left: P content 9.1 at%, indexed according to a fcc structure. Right: P content 15.13 at% [4].

However, Graham et al. [4] found that the EN coatings contained from 8.37 at% to 16.43 at% phosphorus with plating from alkaline baths (pH 8.2–8.4 at 90°C) they were crystalline. In addition, they found that no significant differences were observed between the X-ray diffraction patterns of alloys deposited from the alkaline–chloride bath and the patterns of deposits from the acid-sulfate bath. Figure 8.1 shows the TEM diffraction patterns of P contents of 9.1 and 15.13 at%. Though the crystallinity of the as-deposited coating was decreased with increase of phosphorus content, the image of P 15.13 at% sample showed that its structure was still not an amorphous state. They concluded that the structure of the as-deposited coating was a supersaturated solid solution of phosphorus in a microcrystalline nickel matrix. It was estimated from the Scherrer formula and TEM micrograph that the crystals were of the order of 10 μm. Using electron diffraction, Marton and Schlesinger [7] in 1968 found an amorphous structure for as-deposited nickel coatings. They observed that this structure existed independently of the size of the nucleated nickel islands. They used an acid bath of pH 5.3 at 25°C. Figure 8.2 shows typical diffraction pattern of as-deposited Ni–P islands, and fresh thin Ni–P films (P content of 9.1 at%). It is obvious that this result is very different from the above result obtained by Graham et al., where the diffraction pattern indicates a crystal structure.

In 1970, Cargill III [8] used X-ray analysis to study the structure of noncrystalline Ni–P alloys in detail. The samples were prepared by electroplating, but his analysis is very useful, so we still cite some of its observations here. He measured the structures of Ni–P with the contents of 73.8–88.8 Ni at% (26.2–11.2 P at%). Figure 8.3 shows total interference functions $I(K)$ for five samples (73.8–81.4 Ni at%). It was found that the first peak in this function is much sharper than the subsequent peaks and the second peak has a shoulder on the large k side. This feature is common to the interference functions for all five amorphous metallic alloys. It is well known that liquid has a typical amorphous structure. In order to compare the two, we inset $I(K)$ of liquid [9] in the upper right of the figure. The main difference between liquid and amorphous state in $I(K)$ is that

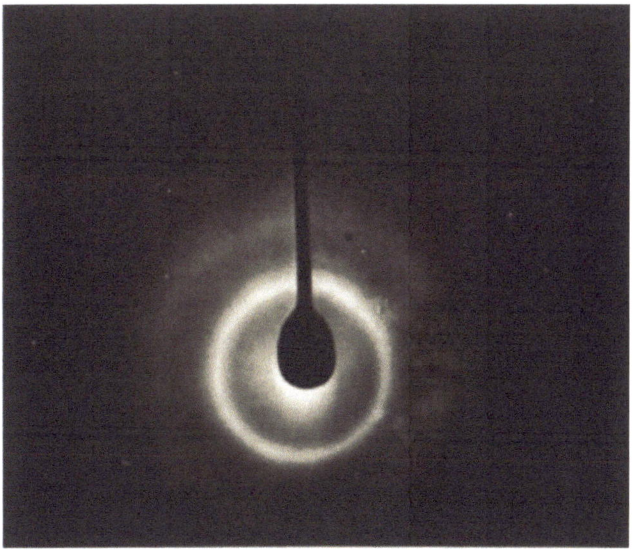

FIGURE 8.2 **Typical diffraction pattern of as-deposited Ni–P islands, and fresh thin Ni–P films (P content of 9.1 at%) [7].**

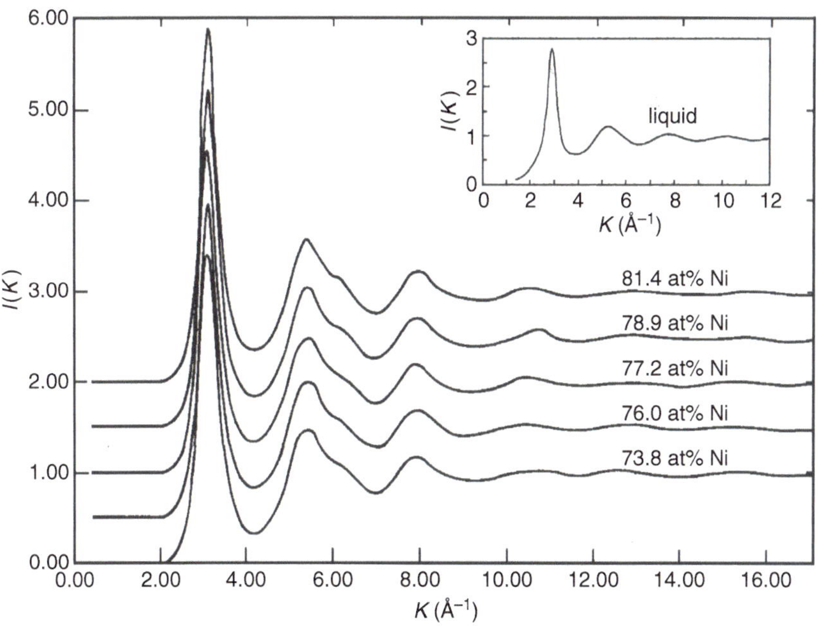

FIGURE 8.3 **Total interference functions *I*(*K*) for five EP Ni–P samples, with displaced zeros.** The image inset in the upper right is *I*(*K*) of liquid, which has been added by the present author.

the second peak has no shoulder on the large *K* side for liquid, but it has a shoulder on the large *K* side for amorphous state. The alloy of 88.8 at% Ni yielded a diffraction pattern of three broad peaks on which were superimposed four sharp peaks at positions corresponding approximately to the (111), (222), (333), and (444) reflections in crystalline nickel. The first peak of the 84.8 at% Ni sample decreased in intensity by 6% when the diffracting sample was tilted with respect to the plane of the incident and diffracted beams, indicating an anisotropic contribution of the type seen in the 88.8 at% Ni sample. Therefore, Cargill did not put these two samples (11.2 and 15.2 P at%) into the group of amorphous alloys but believed that they belonged to crystalline phase. Moreover, Cargill III measured the minimum average microcrystal grain sizes D of the amorphous Ni–P alloys and found that they were between 5.7 and 13.3 nm. In a study by Chow et al. in 1972 [10], nickel coatings were plated in alkaline baths of pH 7.5–10.5. TEM analysis showed that the deposits had diffraction patterns consisting of broad but discrete nickel rings that became narrower as the pH increased. The proposed structure was a supersaturated solid solution of phosphorus in a very fine polycrystalline nickel matrix. The investigation of Randin et al. in 1967 [6] showed that differential thermal analysis of coatings plated from an acid bath of pH 4–5 with a phosphorus content of 6.43–19.6 at% also indicated a supersaturated solid solution of phosphorus in nickel. It was suggested that this is a metastable state, with the equilibrium state being nickel plus Ni_3P.

All of these early works on the compositions, especially on the structures of EN, are not consistent, but may even conflict. Some authors believed from their experimental results that the EN deposits had a supersaturated solid solution of phosphorus in nickel. In contrast, some other investigators examined the TEM and X-ray for electroless Ni–P deposits and concluded that the deposits had an amorphous structure. These conflicting results are thought to be due to the unstable structure of the deposited alloy, and the fact that detailed structures of electroless Ni–P deposits are not well understood. Therefore, more basic studies are required to determine whether electroless nickel deposits are microcrystalline supersaturated solid solution or amorphous.

8.1.1.2 Researches of Composition and Structure from the 1980s till Today

It may be that many reports published until the 1970s did not solve the problem of structure of EN as mentioned above; many investigators still pay particular attention to this issue even now [12–45]. Even very recently, some authors were interested in this problem [41–45]. Of course, not all papers have the composition and structure of EN as their main point. Although they may touch on the composition and structure of EN, this issue does not form the primary content of these papers. In order to see this issue clearly, typical experimental results for the composition and structure of EN alloy deposits published during the period from the 1980s to 2013 are listed in Table 8.1. The key point is how to identify and classify the structures of low, middle, and high phosphorus ENs.

TABLE 8.1 Composition and Structure of EN Alloy Deposits

No.	Composition, at% P	Structure	X-ray pattern	TEM diffraction pattern	Ref. (Year)
1	13–15 16–18 19–21	Microcrystalline Amorphous and microcrystalline Amorphous		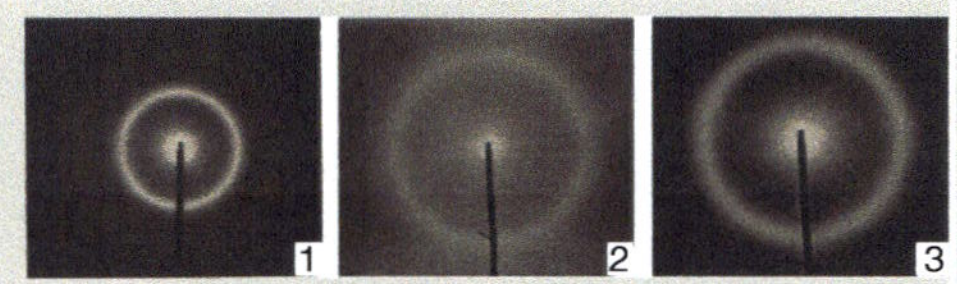	13 (1986)
2	1.88–5.54 11.65–14.15 18.2–19	Crystalline Amorphous Amorphous			14 (1988)
3	13.19 16.62 17.51	Phosphorus-supersaturated nickel solid solution with grain size 4 to 5 nm	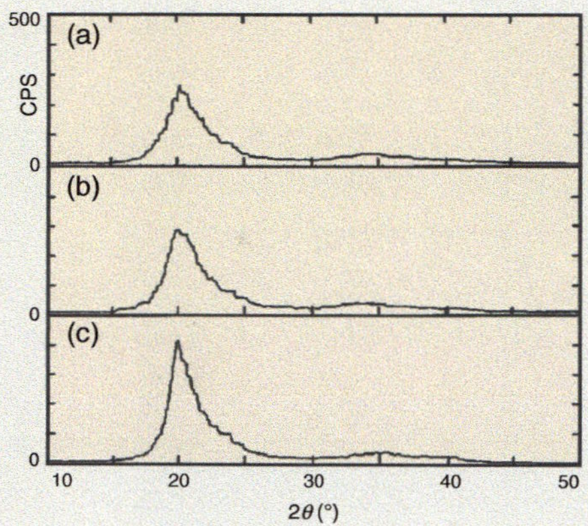	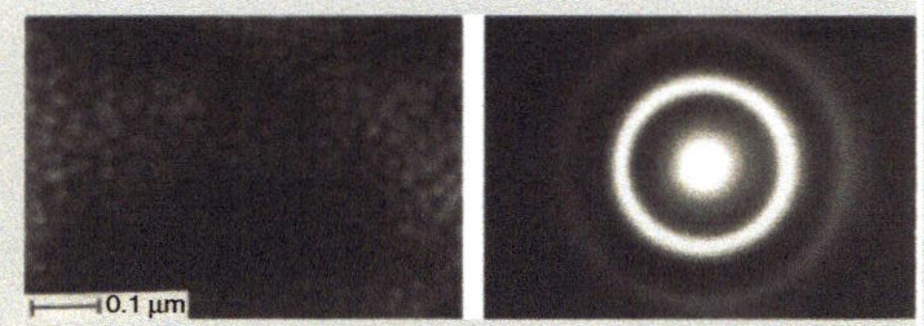	15 (1988)

(*Continued*)

TABLE 8.1 Composition and Structure of EN Alloy Deposits (*cont.*)

No.	Composition, at% P	Structure	X-ray pattern	TEM diffraction pattern	Ref. (Year)
4	10.79	Amorphous	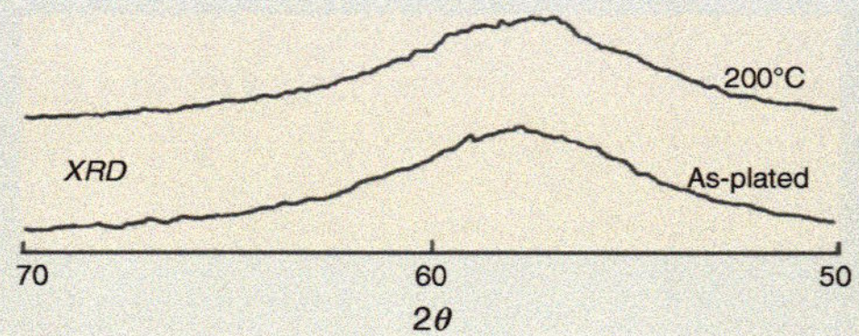	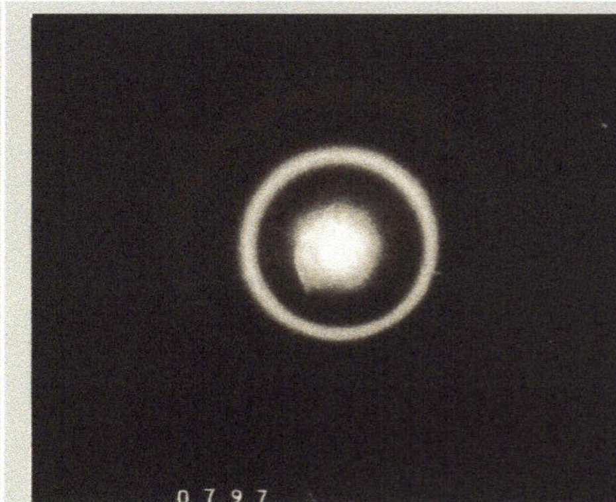	17 (1989)
5	11.3 17.5 19.8 21.3 23.0	fcc Ni–P nano crystals Amorphous Amorphous Amorphous Amorphous	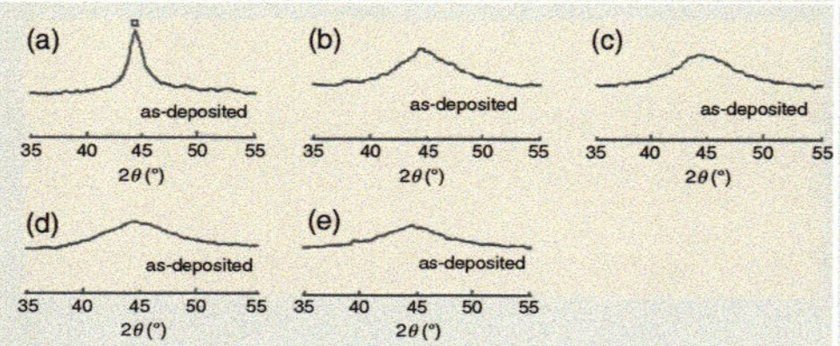	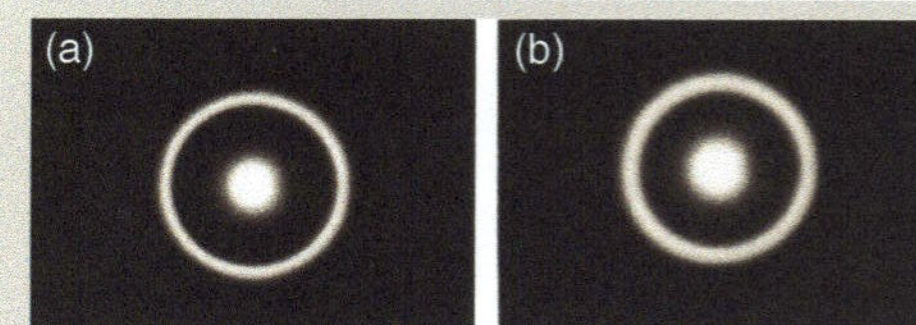	19 (1990)

6	7.85 8.72; 11.47; 15.78	Microcrystalline Amorphous	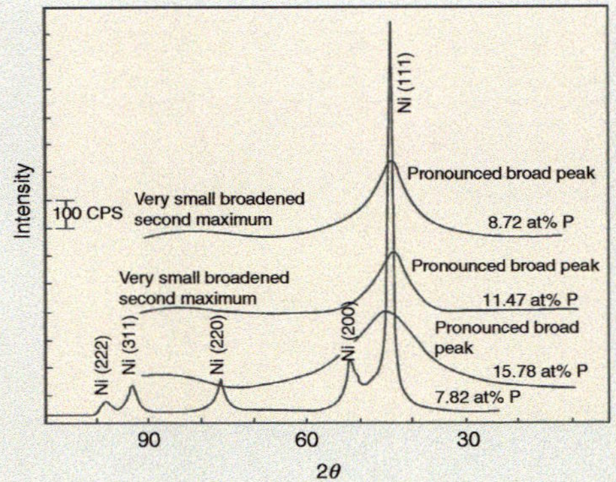		20 (1993)
7	Low P, 7.32 Middle P 9.1–14.15 High P, 4.60>14.15	Crystalline Semi-amorphous structure Amorphous		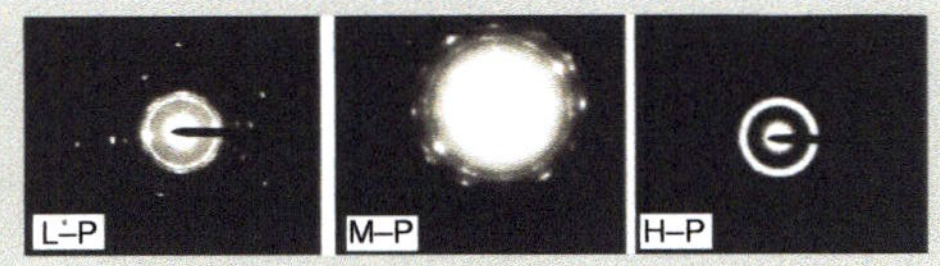	21 (1994)
8	7.93–15.94	Amorphous and microcrystalline			22 (1996)
9	6, 4.60 5, 8.03 1–4, 12.53– 28.71	Crystalline Microcrystalline Amorphous	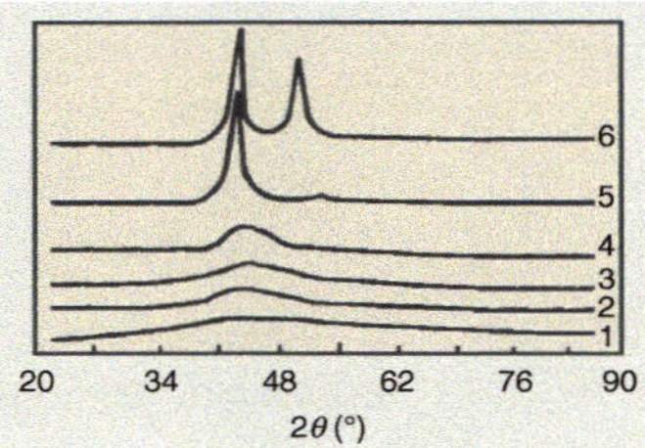		24 (1998)

(Continued)

TABLE 8.1 Composition and Structure of EN Alloy Deposits (*cont.*)

No.	Composition, at% P	Structure	X-ray pattern	TEM diffraction pattern	Ref. (Year)
10	9.1–10.79	A semi-amorphous structure			27 (2000)
11	≤10.8 ≥17.4	fcc Ni–P nano crystals An amorphous Ni–P phase	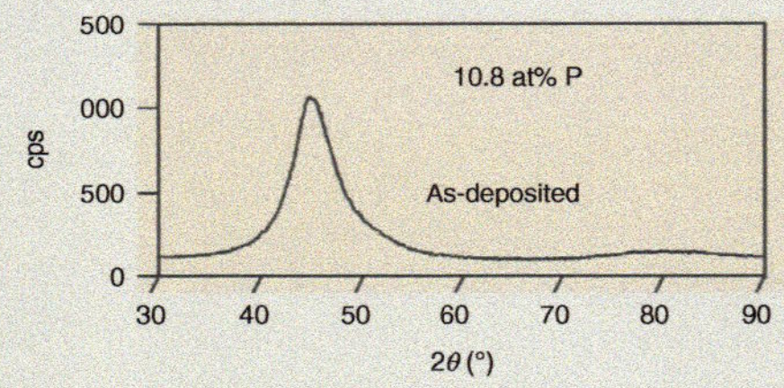	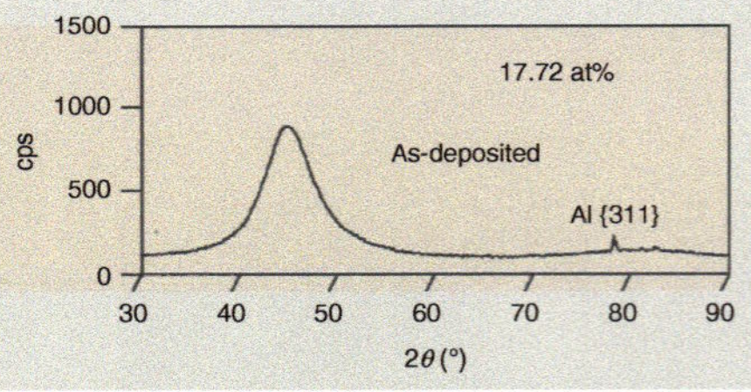	28 (2000)
12	13.6 and 19.3	Fully amorphous	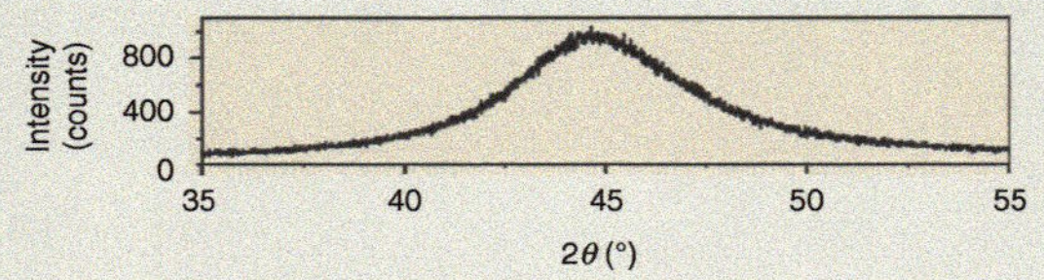		29 (2001)
13	M-P, 10.79–15.78	Amorphous	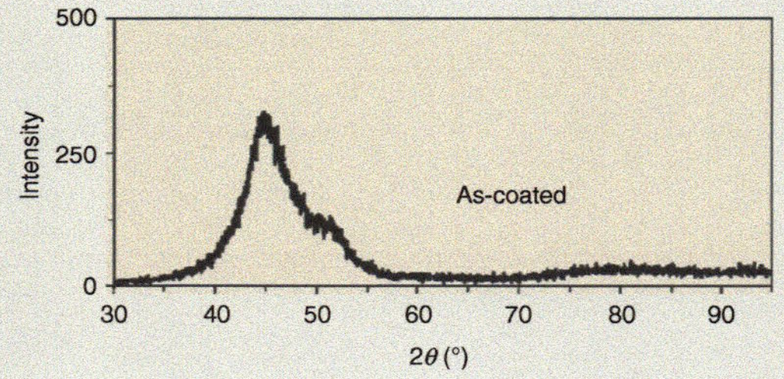		30 (2001)

14	5.54–9.1 9.1–14.15	Amorphous and microcrystalline	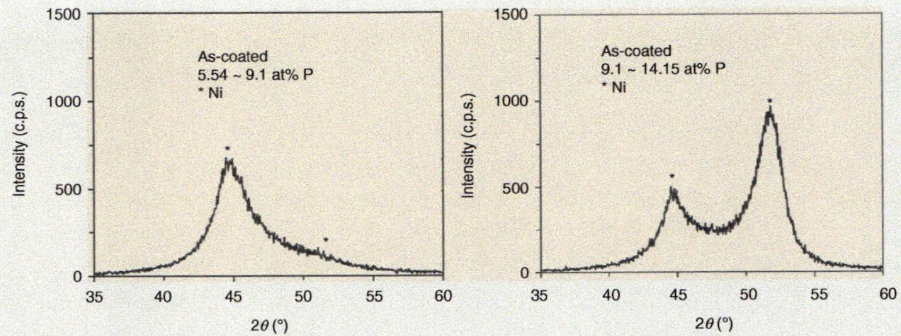	31 (2002)
15	17.4 23.55	Amorphous	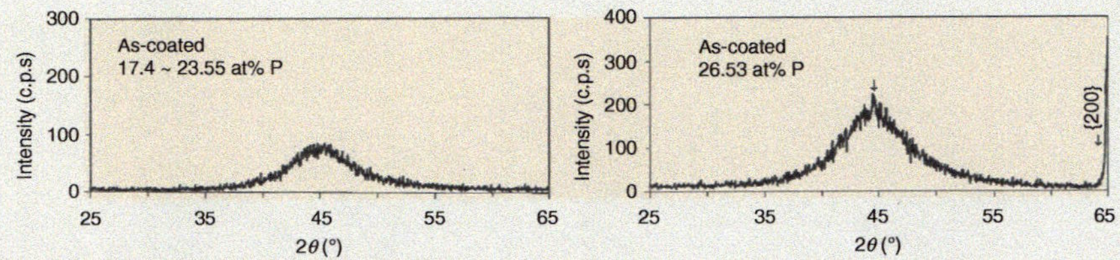	32 (2002)
16	5.54 15.3 23.55	Definite crystallinity Amorphous Amorphous	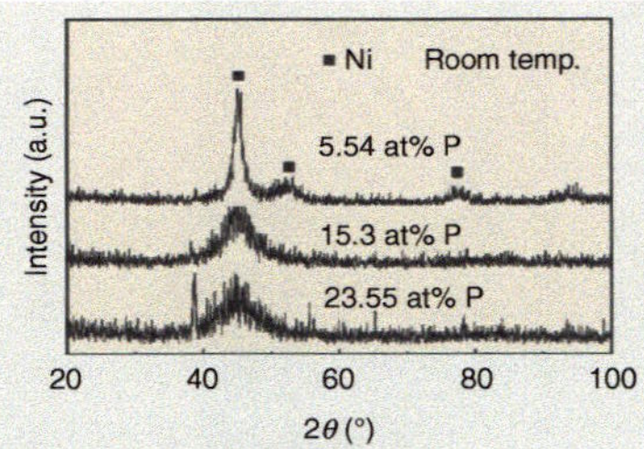	33 (2002)

(Continued)

TABLE 8.1 Composition and Structure of EN Alloy Deposits (*cont.*)

No.	Composition, at% P	Structure	X-ray pattern	TEM diffraction pattern	Ref. (Year)
17	15.78 20.54 26.53	Partially amorphous Fully amorphous Fully amorphous	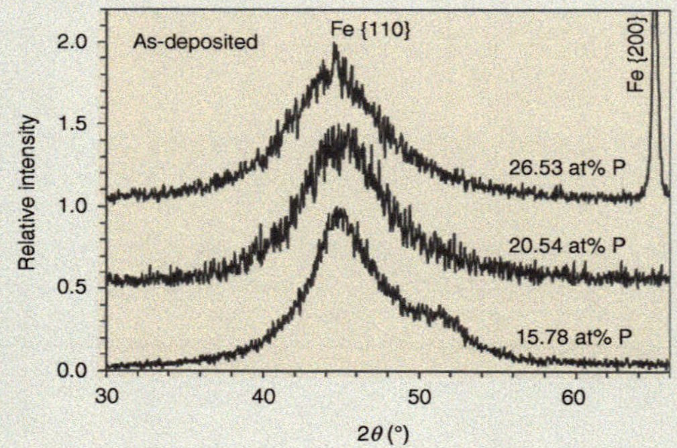		34 (2003)
18	5.54–9.1 9.1–14.15 10.79–15.78 17.4–23.55	Amorphous and microcrystalline Amorphous and microcrystalline Amorphous and microcrystalline Fully amorphous			36 (2003)
19	6.39 13.33 20.74	Nanocrystalline (grain size ~13 nm) Amorphous and microcrystalline Fully amorphous			37 (2004)

20	6 8 12 20	fcc Ni crystallites fcc Ni crystallites fcc Ni crystallites Amorphous	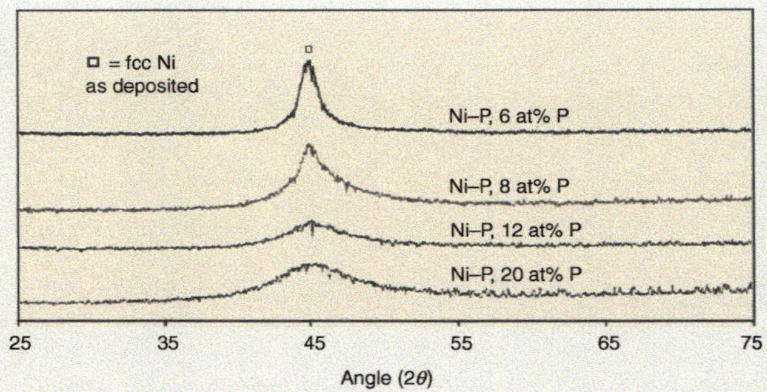	38 (2005)
21	12–14 18–20	fcc Ni crystallites Amorphous*		39 (2006)
22	9.36	Amorphous and crystallized phase	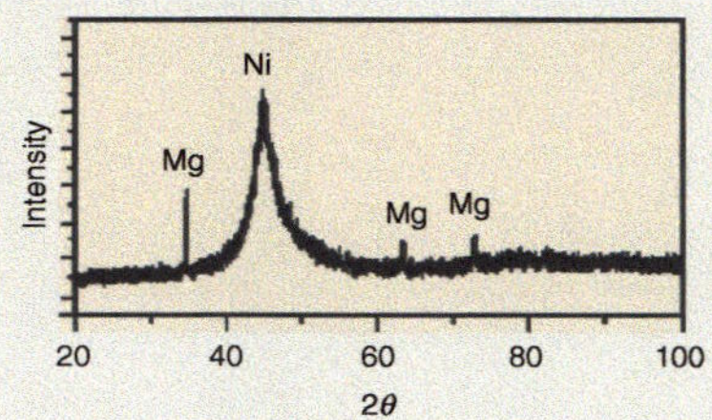	41 (2008)

(Continued)

TABLE 8.1 Composition and Structure of EN Alloy Deposits (*cont.*)

No.	Composition, at% P	Structure	X-ray pattern	TEM diffraction pattern	Ref. (Year)
23	15.78	Amorphous	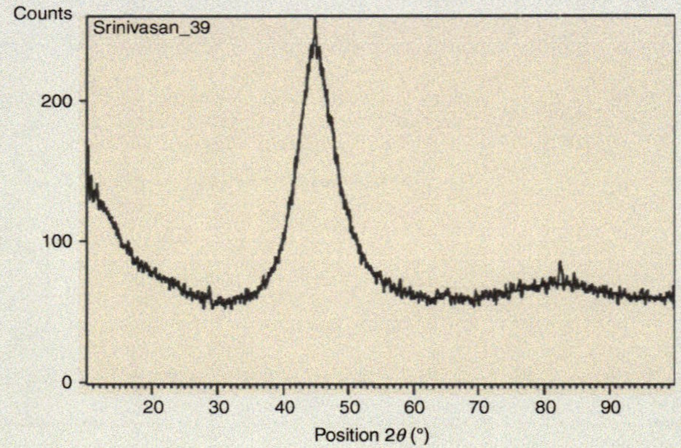		42 (2009)
24	17.4	Amorphous			43 (2011)
25	22.84	Amorphous	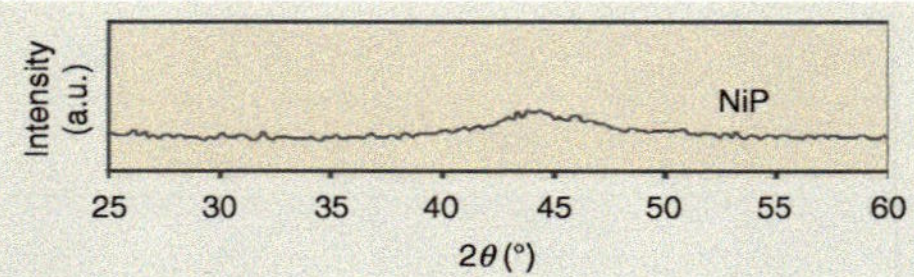		44 (2012)

26	22	Amorphous and microcrystalline			45 (2013)
27	14.1	Amorphous	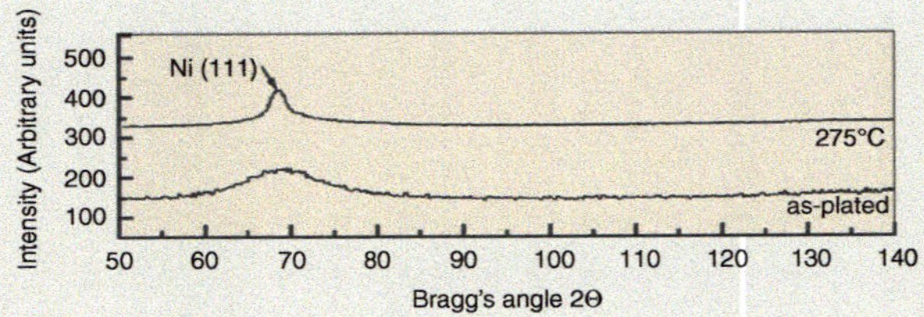		46 (2013)

**The lowest phosphorous content required to obtain an amorphous Ni–P film is about 16 at%.*

Before analyzing and discussing the data in Table 8.1, it should be noted that some authors have pointed out conflicting experimental results regarding the composition and structure of EN, which appeared in the early research. For example, in 1988, Park and Lee [15] said, "These conflicting results are thought to be due to the unstable structure of the deposited alloy, and the detailed structures of electroless Ni–P deposits are not well understood." That is why so many authors are still paying attention to study the structure and composition of EN from the 1980s till today.

From Table 8.1, several points can be made, as follows.

First, the common aim of most studies is concentrated on the characterization and separation of low phosphorus, medium phosphorus, and high phosphorus EN deposits.

Second, it can be said with certainty that the structures are crystalline nickel (a supersaturated solid solution of phosphorus in nickel) for the EN deposits of low P content, there are amorphous and microcrystalline for the EN deposits of medium P contents, and there is amorphous phase for the EN deposits of high P contents.

Third, there are no strict and definite boundaries for dividing between low phosphorus and medium phosphorus categories, and between middle phosphorus and high phosphorus categories. In other words, there are no common and consistent conclusions for such separation of low P, middle P, and high P of EN deposits. Of course, we have noted that some authors have made proposals regarding such a separation. For example, in 1997, Parkinson [23] said, "The industry normally identifies electroless nickel coatings according to their phosphorus content, for example, low phosphorus 3.72–9.1 at% P, medium phosphorus 10.79–15.78 at% P, and high phosphorus 17.4–22.07 at% P." In 2000, Apachitei et al. [28] said, "Their structure changed from a nanocrystalline supersaturated solid solution of phosphorus in nickel ($P \leq 10.79$ at%) to an amorphous nickel phosphorus phase ($P \geq 17.4$ at%)." In 2011, Qin et al. [43] also said, "XRD studies have shown that deposits with more than 12.5 at% are amorphous."

Fourth, some authors conducted their experiments definitely according to the division of low, middle, and high phosphorus. Such studies are certainly welcome. However, as can be seen from the examples cited here, their conclusions obtained are not consistent. In 1994, Martyak [21] concluded that low phosphorus deposits were crystalline. Medium phosphorus deposits were more continuous than low phosphorus coatings and exhibited a semiamorphous structure. High phosphorus deposits were continuous and amorphous. In 2000, Martyak et al. [27] concluded that low concentrations of phosphorus in nickel result in a microcrystalline coating, whereas medium levels of phosphorus yield coatings that contain both crystalline and amorphous components. High concentrations of occluded phosphorus produce EN deposits that are completely amorphous. In 2001, Keong et al. [32] found that as-coated low phosphorus deposits are either high crystalline or consist of microcrystalline nickel. As-coated medium phosphorus deposits are either fully amorphous or mixtures of microcrystalline nickel and amorphous phases. High phosphorus as-coated deposits have demonstrated that the deposits could be fully amorphous, and mixtures of microcrystalline nickel and amorphous.

Fifth, there is still obvious conflict over the phenomena described in Table 8.1. For example, Nos. 1 and 2 indicate that the structures of EN deposits with 13–15 at% P and 13.19–17.51 at% are microcrystalline or phosphorus-supersaturated nickel solid solution with grain size 4–5 nm, but No. 3 shows that the structure of the EN with 11.65–14.15 at% is amorphous. The content of P is only 10.79 at% in No. 4, but the structure is amorphous. No. 7 shows that EN deposits with 8.72, 11.47, and 15.78 at% P are all amorphous. Also No. 12 shows the structure is fully amorphous, but the P content is only 13.6 at%.

Sixth, most of the author(s) did not analyze and list the reasons why the structures of their coatings are just amorphous or a mixture of amorphous and microcrystalline structure or crystalline. There are two exceptions to this. Park and Lee [15] (No. 3 in Table 8.1) gave three reasons to explain the structures of their prepared coatings with P contents of 13.19, 16.62, and 17.51 at%. They believed that the structures of these coatings were a phosphorus-supersaturated nickel solid solution with a grain size of 4–5 nm rather than amorphous phase. The first reason is that the first-peak intensities in three interference functions are all higher than 4. The values are considered to be too high for amorphous materials, since the first peaks in interference function of amorphous materials are, in general, known to have values less than four. However, the authors did not mention the fact in the same original Fig. 3 that the second peak has a split shoulder on the large k side in three interference functions, which is the typical feature of an amorphous phase [8]. Of course, the authors did not give any explanation for that. The interested reader may consult the original paper for the other two reasons in detail. The paper by Rajam et al. [20] gives a different result. They concluded that with the exception of 7.85 at% P, all the deposits (8.72, 11.47, and 15.78 at% P) are amorphous (No. 6 in Table 8.1). They listed three reasons for this. The first one is that the diffractogram shows a pronounced broad peak and a very small broadened second maximum, which is characteristic of the diffraction pattern of amorphous materials (for other two reasons, please see the original paper). Comparing the XRD they obtained (No. 6 in Table 8.1) with that obtained by Park et al. (No. 3 in Table 8.1), it is easy to see that the two XRD patterns are basically the same. But, Park and Lee classified their deposits (13.19, 16.62, and 17.51 at% P) as crystalline (phosphorus-supersaturated nickel solid solution) with grain size 4–5 nm, while Rajam et al. classified their coatings (8.72, 11.47, and 15.78 at% P) as amorphous phases. Moreover, Rajam et al. did not cite the paper by Park and Lee in their paper, even though their paper was published 5 years later than that by Park and Lee. The reasons from these two groups are not unified and it seems that they have difficulty in convincing readers.

Seventh, despite the problems mentioned above, the composition and structure of low, medium, and high P EN coatings may be classified as follows.

L–P	<9.1 at% P	Crystalline with nanosize grains
M–P	9.1–15 at% P	Nano size crystalline embedded in amorphous matrix
H–P	≥15 at% P	Amorphous phase

This is the common result of the work of many authors over the last 60 years.

8.1.1.3 *Effects of Several Parameters on Composition and Structure of EN*

A considerable part of many articles has been devoted to the influences of several parameters on the composition and structure of EN, such as concentration of metal salt ions, pH value, temperature, etc. These studies were basically published in recent years.

In 2000, Małecki and Micek-Ilnicka [26] investigated the influence of nickel ion concentration on the weight percentage of phosphorus in Ni–P deposition, as shown in the left of Fig. 8.4. It can be seen that the P content in the deposits decreases as the nickel ion concentration increases. The authors did not give any explanation of their experimental results. In 2002, the author's group [35] also measured the composition and structure of EN when studying the formation range of its amorphous phases. The results shown in the right of Fig. 8.4 are the similar to those obtained by Małecki and Micek-Ilnicka. It is interesting to note that one is an acid bath [26], another is an alkaline bath [35], but the trend of less P content in the deposits is similar to the increase in the concentration of Ni ions. In fact, this experimental phenomenon is not difficult to understand. As the concentration of nickel ions increases, more Ni metal is reduced to Ni^{2+}. However, due to the other parameters in bath solution are fixed, the $H_2PO_2^-$ must be decreased because it gradually consumers for the growing Ni–P coating,

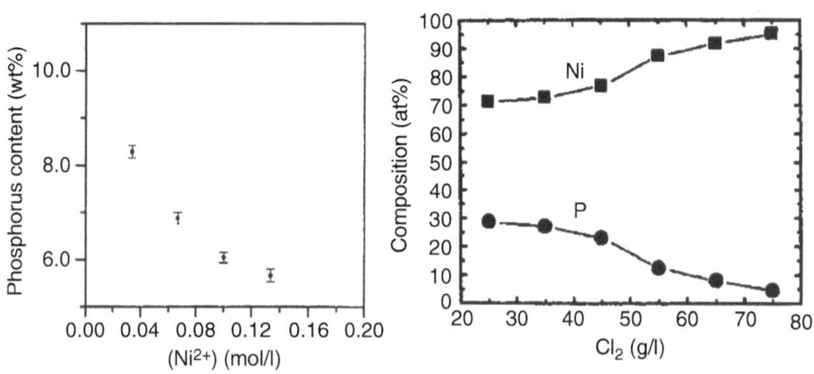

FIGURE 8.4 Effect of nickel ion concentration on the content of P in Ni–P deposition. Left: $[H_2PO_2^-]$ = 0.28 mol/L, $[CH_3COOH]$ = 0.5 mol/L, pH 4.8, temperature 360 K [26]. Right: $NiCl_2•6H_2O$ 55 g/l; $NaH_2PO_2•H_2O$ 18 g/l; $Na_3C_6H_5O_7•2H_2O$ 80 g/l; pH 9; T 90°C [35].

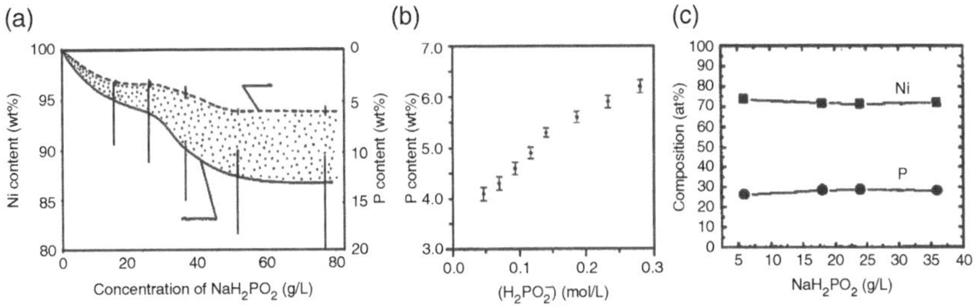

FIGURE 8.5 (a) Nickel and phosphorus contents in deposits from plating bath of various NaH_2PO_2 conc. [pH 10.0, 55°C] [11]. (b) The influence of hypophosphite ion concentration on the weight percentage of phosphorus in Ni–P deposition ([Ni^{2+}] = 0.1 mol/L, [CH_3COOH] = 0.5 mol/L, pH 4.8, temperature 360 K) [26]. Dependence of the compositions of Ni–P deposits on the concentration of sodium hypophosphite ($NiCl_2$•$6H_2O$ 55 g/l; NaH_2PO_2•H_2O 18 g/l; $Na_3C_6H_5O_7$•$2H_2O$ 80 g/l; pH 9; T 90°C) [35].

thereby the P element reduced from $H_2PO_2^-$ decreases, leading to the P content in the castings reduction.

Experimental data show that the concentration of hypophosphite ion impacts the composition of EN deposits. In 1984, Sugita and Ueno [11], in 2000 Małecki and Micek-Ilnicka [26], and in 2002 the author's group [35] performed such measurements. The results are shown in Fig. 8.5a, b, and c, respectively. Data show that the phosphorus component increases with increasing concentration of NaH_2PO_2 in these three cases. In plot c, the P contents are 26.25, 28.61, 28.84, and 28.31 at% when the concentrations of NaH_2PO_2 are 6, 18, 24, and 36 g/L, respectively, which means their enhancement is very small. The former two sets of authors gave no explanations of their experimental results. As the concentrations of all other parameters in bath solution are not changed, the reduced phosphorus elements enhance when the concentration of NaH_2PO_2 increases. Therefore, the P contents in the deposits increase with increasing reductant agent. From this, the experimental results can be easily understood.

Several sets of authors investigated the effects of pH value on the compositions of EN coatings [11,33,38,42,46,47]. All the data obtained from these authors indicate that the P content decreases more or less linearly with an increase of pH of the plating solution. Figure 1 in [11], Fig. 1 in [33] (Fig. 8.7b), Table 2 in [42], Table 3 in [46], and Table 1 in [47] show such trends, and interested readers can view the original articles. In 2005, Antonelli et al. [38] measured P content vs. pH for Ni–P films and compared to data from another nine groups, as shown in Fig. 8.6. The figure clearly shows that the P content in EN deposits decreases nearly linearly with an increase of pH value. This trend can easily be understood in the light of the proposed mechanisms for electroless Ni–P deposition from a hypophosphite bath. As all of the mechanisms of EP depositions will be discussed in detail in Chapter 10, here we will just use some mechanisms to explain this experimental trend.

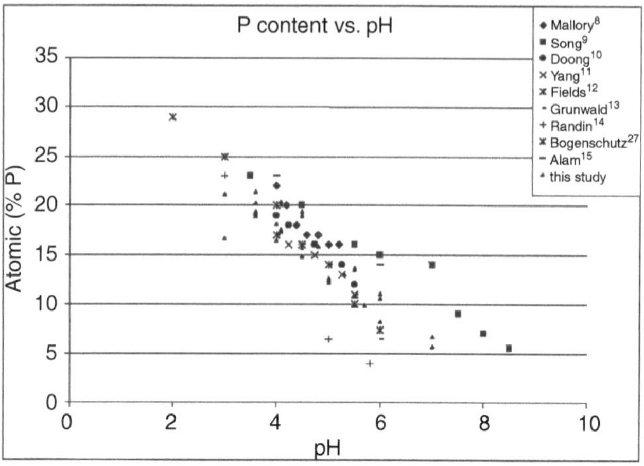

FIGURE 8.6 P content vs pH for Ni–P films from this study compared to data from other groups [38]. 8, Mallory GO. Electroless plating fundamentals and applications. In: Mallory GO, Hajdu JB, editors. New York: Noyes Publications/William Andrew Publishing; 1990. p. 62–63. 9, Song JY, Yu J Thin Solid Films 2002;415:167. 10, Doong JC, Duh JG. Surf Coat Technol 1993;58:19. 11, Yang TY, Personal communication, (1992). 12, Fields W, Duncan R, Zickraff JR. Electroless plating. Metals Park, OH: Publication of ASM Committee on EN-Plating; 1984. 13, Grünwald P. Galvanotechnik 1983;74:1286. 14, Randin JP, Hintermann HE. J Electrochem Soc 1970;160:117. 15, Alam MO, Chan YC, Hung KC. Microelectron Reliab 2002;42:1065. 27, Bogenshütz AF, Jostan JL, Mussinger W. Galvanotechnik 1969;60:507.

Each of these mechanisms suggests a reaction for the incorporation of P in the film. From the atomic hydrogen and hydride transfer mechanisms proposed by Brenner and Riddell in 1946 [48] and by Hersch and Lukes in 1955–1956 [49], the reaction for the incorporation of P in the film is

$$H_2PO_2^- + H \rightarrow P\downarrow + OH^- + H_2O \tag{8.1}$$

From the hydroxyl ion coordination mechanism of Salvago and Cavallotti [50] in 1972, the reaction equation for producing P element in the EN deposit is

$$Ni_{cat} + H_2PO_2^- \rightarrow P\downarrow + NiOH^+ + OH^- \tag{8.2}$$

And from the electrochemical mechanism by Brenner and Riddell [51] in 1947, the reaction equation for producing P element is

$$H_2PO_2^- + H^+ + e \rightarrow P\downarrow + 2H_2O \tag{8.3}$$

In Eqs (8.1) and (8.2), OH^- ions are a product of the reaction that produces P. When the pH increases, the OH^- ion concentration rises, which leads to the reaction swinging in the opposite direction, therefore decreasing the amount of P deposited. In Eq. (8.3), H^+ is one of the reactants. As the pH value increases, the H^+ ion concentration decreases, which decreases the amount of P in the products. That is to say, these mechanisms explain this phenomenon clearly.

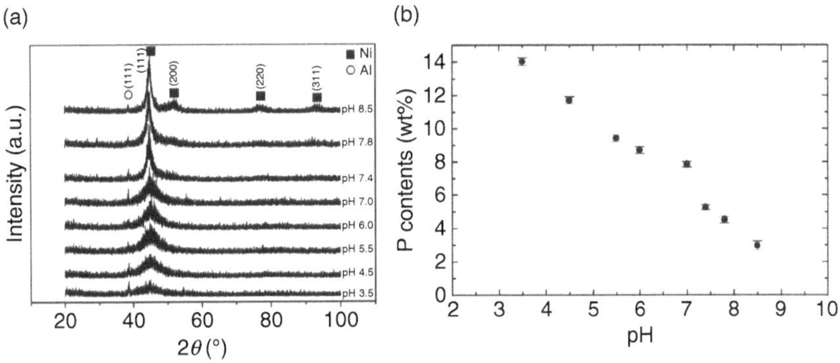

FIGURE 8.7 (a) XRD patterns and (b) variations of the P content of the as-prepared EN films plated under different pH of the plating solution [33].

In 2002, Song and Yu [33] also measured XRD patterns of as-deposited EN coatings plated under different plating solution pHs. The results are shown in Fig. 8.7a and b. It can be seen that the Ni (111) peak became sharper with increasing pH (i.e., decreasing P content in the film). From the reports of the literature, these authors are sure that EN films with less than 9.1 at% P have a nanocrystalline structure with an average grain size of approximately 2–15 nm, and films with more than approximately 12.5 at% P show an amorphous phase. Within the intermediate concentration range, nanocrystallites were reported to be embedded in an amorphous matrix. Moreover, from Fig. 8.7a the authors believed that only the 8.5 pH (5.54 at% P) film showed definite crystallinity in the as-prepared condition.

Research into the influence of temperature on composition and structure of EN coatings is rather few. In 2009, Srinivasan and John [42] studied the effect of bath temperature on the phosphorous content at a pH of 5, as shown in Fig. 8.8. It is obvious from the figure that the phosphorous content increases in the deposit as the temperature increases. In 2011, Qin [43] also investigated the effect of temperature on chemical compositions of Ni–P deposits (please see Fig. 3 in [43]). The results showed that the percent of phosphorus in the deposit increased slightly with increasing temperature of reaction. Such an experimental phenomenon should not be difficult to understand. It may be that with an increase of temperature of reaction, deposition rate of phosphorus increased more rapidly than that of nickel. Consequently, the proportion of phosphorus in the deposit increased.

8.1.1.4 Cross-Sectional Compositional and Structural Profile of the EN Deposits

Several groups reported the cross-sectional compositional varying profile of the EN deposits [4,17,25,29,40]. In 1965, Graham et al. [4] studied typical electron probe microanalysis (EPMA) traverses of as-plated EN deposits, as shown in

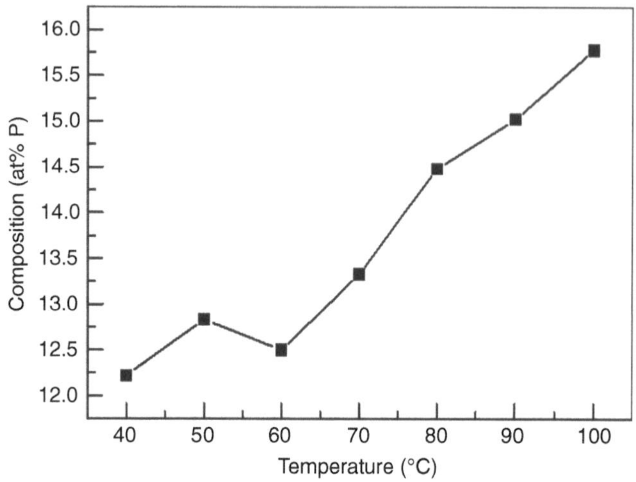

FIGURE 8.8 The phosphorous content of the as-prepared EN films as varying temperature of bath solution [42].

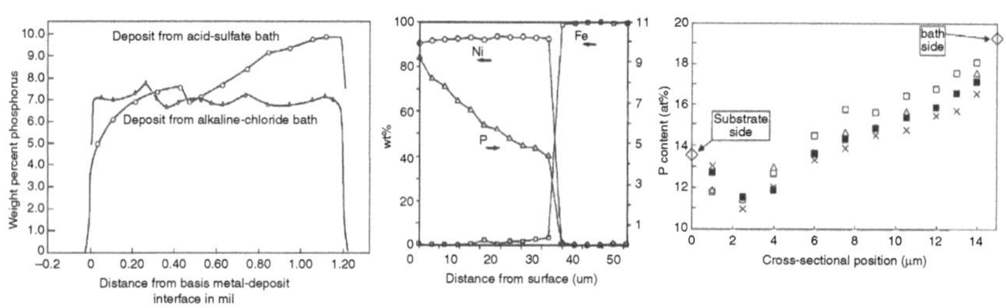

FIGURE 8.9 Chemical compositions of traverse profile of the as-deposited EN deposits [4,17,29]. 1 mil = 2.54 μm.

the left of Fig. 8.9. In 1989, Lin and Lai [17] reported the elemental profile of the as-deposited electroless Ni–P measured by EDS, as indicated in the middle of Fig. 8.9. In 2001, Révész et al. [29] investigated the cross-sectional compositional profile of the EN deposit as determined by EPMA, as shown in the right of Fig. 8.9. It is evident that the P content is decreased from the surface of deposits to the substrate side. This is the common trend for the results by these three groups. The bath solutions used by Graham et al. and Lin and Lai were acid bath, and that used by Révész et al. was not indicated. It is evident in the left of Fig. 8.9 that the P content in the deposit is basically constant for the deposit from alkaline bath. The same group obtained very different results from acid and alkaline baths. Also, in the middle of Fig. 8.9, the nickel content is constant though the P content is decreased.

Using glow discharge optical spectroscopy (GDOS), Staia et al. [25] in 1999 determined the depth profile of an as-deposited EN coating. The results show that the phosphorous content of 14.85 at% is constant throughout the coating thickness. The nickel content is also constant throughout the coating thickness, which is similar to that obtained by Lin and Lai (middle of Fig. 8.9). These authors believed that this phenomenon resulted from a rigorous control of pH during the deposition process. It is interesting to note that Révész et al. said definitely [29], "We are not aware of a similar study reported on a composition variation through the thickness of electroless plated Ni–P alloys." Obviously, this is not the case.

Lin and Lai [17] not only measured the variation of phosphorus in the traverses of as-plated EN deposits at room temperature, but also measured such variation in the heat-treated deposits. The data show the same gradient trend for the heat-treated deposits for heat treatment temperatures up to 800°C. However, for higher temperatures such as 700°C and 800°C, the cross section microstructures show two phases and these are accompanied by a random distribution of phosphorus across the deposit.

As to the origin of this composition graduation, two groups [4,29] attributed it to the change in pH. The pH cannot be kept at a constant value during plating, and the phosphorus content will increase if there were a rapid decrease in the pH of the bath. It is obvious that such observations cannot explain the conflict between the P content decreases and the phosphorous content being constant throughout the coating thickness. More in-depth studies are needed to explain these conflicting experimental facts. Révész et al. [29] also pointed out in their paper that "However, to elucidate these mechanisms would require a more elaborate study,"

In 1947, Brenner and Riddell [51] not only proposed the electrochemical mechanism for EN, but they also measured the cross-sectional microstructures of EN deposits, as shown in Fig. 8.10. Of course, this is the first picture of microstructure of EP coatings in the world. Figure 8.10a is the cross-sectional microstructures of as-plated EN film, which has laminations running parallel to the base metal, and within each layer there is a columnar structure perpendicular to the base metal. After heating at 400°C for 1/2 h, the width trends narrowed. When annealing at 800°C for 1/2 h, the details of the layered structure

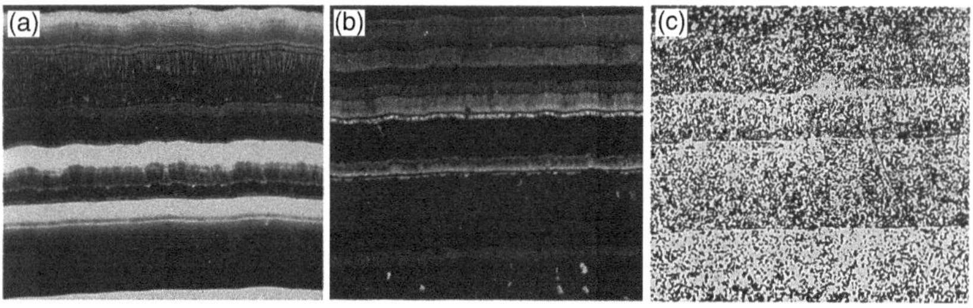

FIGURE 8.10 Cross-sectional microstructures of electroless nickel deposits. (a) From an acid solution showing laminations and columnar structure, 500×. (b) After annealing at 400°C for 1/2 h, 500×. (c) After annealing at 800°C for 1/2h, showing apparent precipitation, 500× [51].

disappear, but precipitation occurs. Then in 1965 Graham et al. [4] not only measured the cross-sectional compositional varying profile of the EN deposit, but also measured its cross-sectional structure profile. They obtained similar cross-sectional structure profiles to those of Brenner and Riddell. They measured the lamellae and found a lamellar thickness of approximately 0.51 μm for the as-plated deposit from an alkaline–chloride bath. The authors have found further that after heat treatments at temperatures as high as 400°C the microstructures of the EN alloys have the same lamellar appearance as the as-plated alloys, even though the supersaturated solid solution has transformed to the equilibrium phases of nickel and Ni_3P. However, heat treatments at 600°C for 120 min result in the disappearance of the lamellar structure. The two equilibrium phases are evident in the micrograph, but the structure is too fine to be clearly resolved.

It is interesting that in 2007 Balaraju et al. [40] not only measured the cross-sectional microstructures but the compositions of traverse profile of the as-deposited Ni–P, Ni–Sn–P, Ni–W–P, and quaternary Ni–Sn–W–P deposits. Figure 8.11a–d shows the microstructure of the electroless nickel alloys cross-sectioned along the thickness direction. All the cross-sections reveal a lamellar/banded structure due to the variation in the composition of the deposit, which

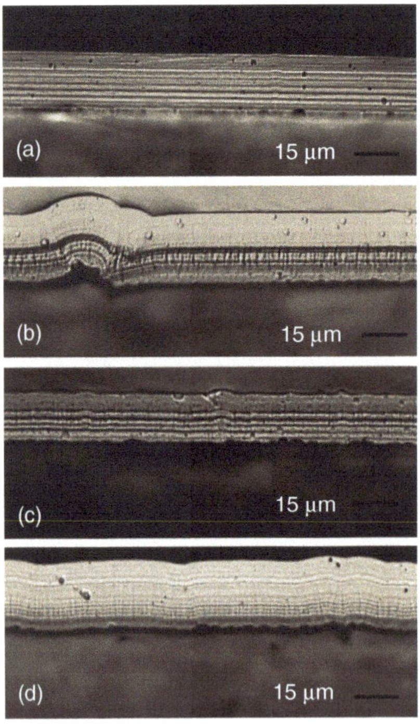

FIGURE 8.11 Optical micrographs of cross-sections of as-plated electroless nickel polyalloy coatings. (a) Ni–P; (b) Ni–Sn–P; (c) Ni–W–P; (d) Ni–Sn–W–P [40].

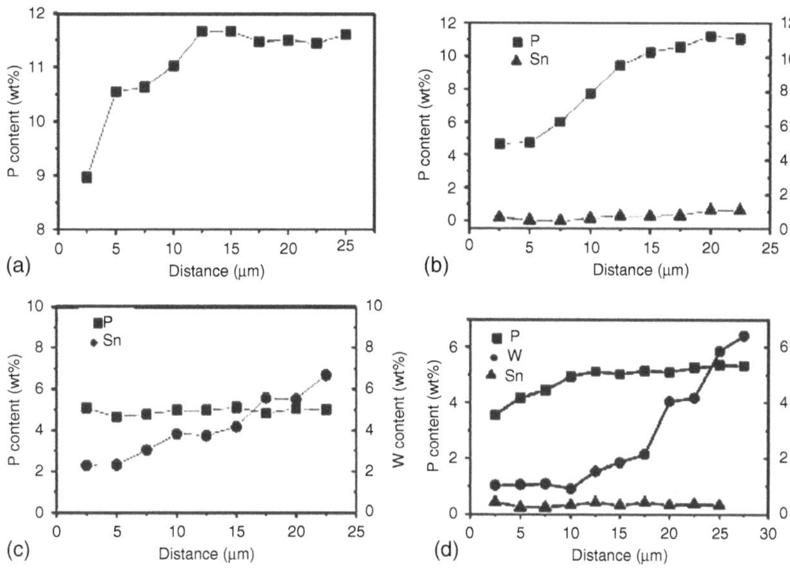

FIGURE 8.12 Composition across the lamellar structure of as-plated coatings. (a) Ni–P; (b) Ni–Sn–P; (c) Ni–W–P; (d) Ni–Sn–W–P [40].

is similar to that shown in Fig. 8.10. The concentration profiling of P, Sn, and W along the thickness direction of the coating was carried out using EDAX, and is shown in Fig. 8.12. In the binary Ni–P alloy (Fig. 8.12a), the observed concentration variation of P from interior to exterior was between 9 and 12 wt%. The P-concentration variation in ternary and quaternary coatings was as follows (Fig. 8.12b–d): 5–11 wt% in Ni–Sn–P, around 5 wt% (essentially constant) in Ni–W–P, 4–5 wt% in Ni–n–P alloys. Sn content in both ternary Ni–Sn–P and quaternary Ni–W–Sn–P alloys was almost constant. However, there is a remarkable concentration variation of W in both ternary Ni–W–P and quaternary Ni–W–Sn–P alloys deposits (1–7 wt%).

To explain these data, these authors on the one hand adopted the above similar considerations to illustrate the lamellar or banded structure in the deposits, that is, "the formation of laminates is due to the periodical fluctuations in the pH of plating solution adjacent to the deposit surface." On the other hand, they believed that "the lamellar or banded structure in the deposits can be ascribed to the variation in the compositions of binary, ternary and quaternary electroless nickel alloys with thickness from the interface of the deposit/substrate to the surface of the deposit." Which is the deciding factor? If this is ascribed to the variation in the compositions, then how does one use the constant or near constant contents of P in Ni–W–P and Ni–W–Sn–P alloy deposits and of Sn in both Ni–Sn–P and Ni–W–Sn–P alloy coatings to explain the data? Considering the above discussion and to the later discussion on the similar experimental phenomena in Ni–B coating, more work is needed to understand this conflict.

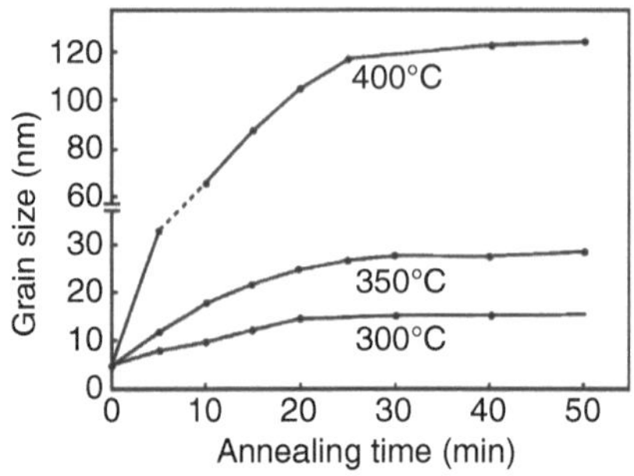

FIGURE 8.13 **Average grain size as a function of annealing time for electroless Ni-2.8 at% P deposits annealed at various temperatures [52].**

8.1.1.5 Grain Size of the EN Deposits

As research progressed on the nanoelectroless plating, measurements of nano grain size have appeared [27,36,52–55]. In 1998, Zhang et al. [52] reported average grain size as a function of annealing time for electroless Ni-2.8 at% P deposits annealed at various temperatures. The grain size of the as-plated deposit is about 5–6 nm. As the annealing time and temperature are increased, the grain size increases, as shown in Fig. 8.13. At 400°C, Ni-2.8 at% P deposits are very unstable; their grain size increases rapidly in the initial stage of annealing and reaches equilibrium due to grain boundary pinning by the free surface after 60 min annealing.

In 2000, Martyak and Drake [27] investigated the variation of grain size of EN deposit with P content of 9.1–10.79 at% as annealing temperature. They found that changes in particle size (4.8 nm) did not occur until the annealing temperature exceeded 300°C for 1 h. However, the size increased to 24.6 and 61.9 nm when the annealing temperature was up to 400°C and 500°C for 1 h, respectively. In addition, these authors found that the amorphous component decreased from 22.4% for the as-plated deposit to 19.6%, 14.9%, and 9.7% annealed at 100°C, 200°C, and 300°C, respectively. No amorphous component was detected above 300°C. In 2003, Guo et al. [36] obtained similar results using four EN deposits: Ni–P1 (5.54–9.1 at%), Ni–P2 (9.1–14.15 at%), Ni–P3 (10.79–15.78 at%) and Ni–P4 (17.4–23.55 at%). As shown in No. 18 of Table 8.1, in the as-deposited condition, coatings Ni–P1, Ni–P2, and Ni–P3 are a mixture of amorphous and microcrystalline materials, whereas the plating Ni–P4 is fully amorphous. Figure 8.14 shows the results of grain size in four EN films. Before the heating temperature less than 400°C, the grain size of the four deposits almost does not change. But after continuous heating the grain size of all the films increases sharply when the temperature reaches above 400°C. Moreover, the crystalline nickel formed is largely randomly orientated.

In 2002, Lu and Giovanni [53] studied effect of P content on the grain size of EN films. When the P content is 5, 13, and 22 at% in the films, the grain size is 2.719, 0.874, and 0.722 nm, respectively. In other words, grain size gradually

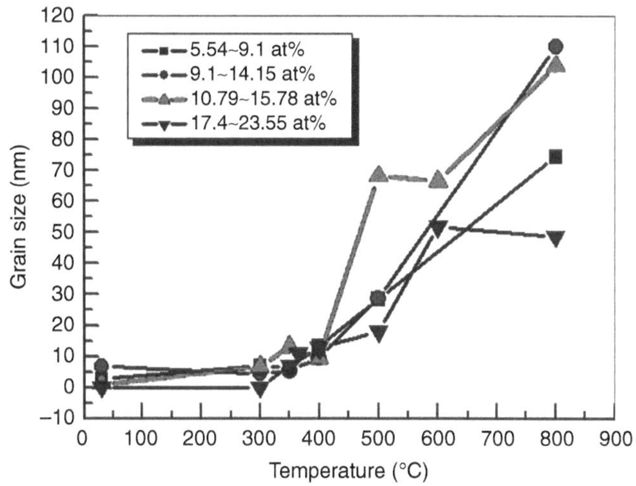

FIGURE 8.14 Grain size of four EN films calculated from different nickel reflections vs. temperature [36].

decreases with increasing P content, and perhaps a transition to an amorphous structure.

In 2000, Farber et al. [54] investigated the microstructures of electroless plated and thermally aged nanocrystalline Ni-3.6 at% phosphorus layers using a tomographic atom probe (TAP). They found for the first time a continuous P-segregation in the grain boundaries of the nanocrystalline structure after heat treatments at 250 and 400°C. After the deposit had been heated up to 400°C, they observed directly a continuous, about 0.8-nm-thick P-segregation zone along the grain boundaries with a concentration of about 14 at% P. Assuming the existence of a metastable equilibrium and using a simple mass balance calculation, these authors derived an equation of the relation between the size and P content in EN films:

$$d(\mathrm{CP}) = \delta/\{1-[(C_{\mathrm{P}}-C_{\mathrm{gb}})/(C_{\mathrm{g}}-C_{\mathrm{gb}})]^{1/3}\} \quad (8.4)$$

where d is the grain size, C_{P} the P content, δ the width of the grain boundary zone, C_{g} the P content within the grains, and C_{gb} the maximum P concentration in Ni grain boundaries. Using experimental data, that is, δ = 0.8 nm, C_{g} = 1 at%, and C_{gb} = 15 at%, Farber et al. obtained the curve shown in Fig. 8.15. With this curve it may be possible to predict grain sizes of other Ni–P alloys within the thermal stability region. It can be seen in Fig. 8.15 that the prediction of Eq. (8.4) is in pretty good agreement with the experimental data, although scatter spots are shown in the figure. Moreover, both experimental data and theoretical results predict that the grain size in the as-deposited condition was found to decrease with the increase of P content. The papers cited in Fig. 8.15 by Farber et al. are not indicated here, the interested reader is referred to the

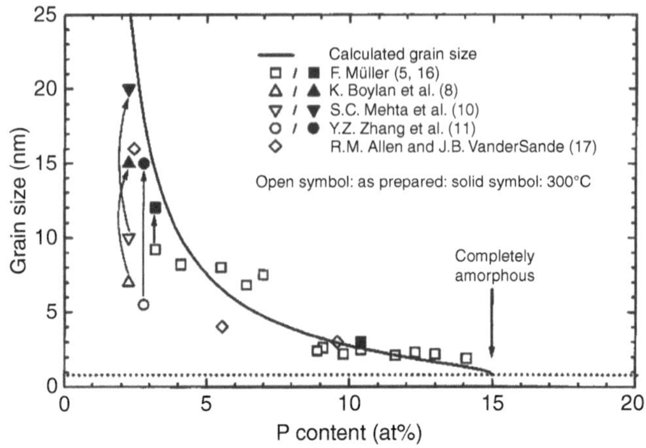

FIGURE 8.15 Relationship between grain size and P content. The solid line represents the predicted relation based on Eq. (8.4). The experimentally determined grain sizes of Ni–P alloys after preparation and after heat treatment at low temperatures before Ni_3P precipitation is observed are depicted as open and solid symbols, respectively [54].

original papers. In 2005, Gao et al. [55] reconsidered such a problem. They used C_{gb} as 19 at% and C_g as 1 at% in their study, and obtained a similar curve "F-A19" to that in Fig. 8.15, which they called "F-A15." Of course, these authors also added their own experimental data of the size of nano-particle grains in EN alloys into the similar figure, that is, Fig. 2 in their paper.

8.1.2 Other EP Binary Deposits

In addition to the composition and structure of binary Ni–P alloy films, those of other binary alloy deposits such as Ni–B, Co–P, Co–B, and Fe–B deposits have been also investigated.

8.1.2.1 Ni–B Binary Alloy Films

A large number of papers on composition and structure of Ni–B binary EP alloys have been published [56–69]. First, the effect of reductant in an EP Ni–B bath on the composition and structure of the deposited film is considered. In 1975, Watanabe and Tanabe [58] reported the effects of $NaBH_4$. Figure 8.16a shows their results of relation between quantity of added reducing agent in plating bath and concentrations of boron in the deposited film. The increase in the reducing agent concentration increases the concentration of boron in the deposited Ni–B film. When the reductant concentration is 400 mg/L or above, the B content is basically not changed (>40 at%), and the morphology of the deposited Ni–B film is an amorphous state. When the boron concentration is lower than 40.12 at%, the deposited film is crystalline but in the case of the saturated concentration (40.12.at%), the structure became amorphous.

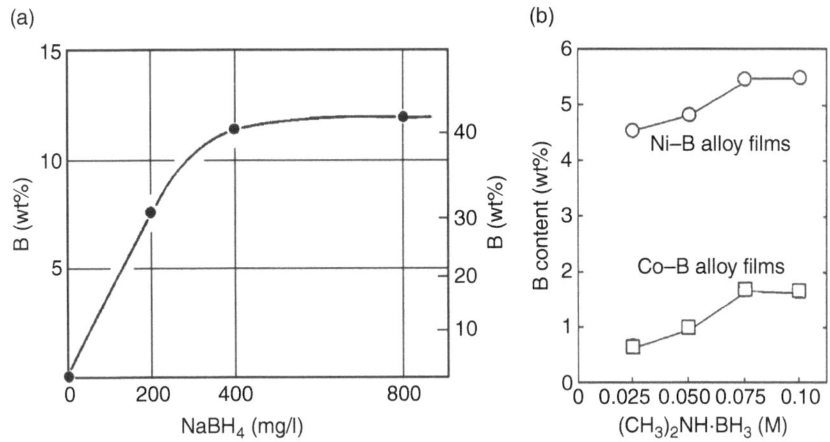

FIGURE 8.16 Relation between quantity of added reducing agent in plating bath and concentrations of boron in the deposited film. (a) Reducing agent $NaBH_4$ [58]. (b) Reducing agent DMAB [61].

In 1981, Suda et al. [59] investigated the influence of DMAB as a reducing agent on the composition and structure of Ni–B coatings. When the concentration of DMAB is 0.5, 1, 3, 6, 9, and 12 kg/m^3, they found that the B content in deposit is 4.0, 5.2, 8.5, 12.7, 15.0, and 16.2 at%, respectively. Figure 8.17 shows the microstructure of the Ni–B alloy film with different B contents, which were deposited by using different concentrations of reducing agent DMAB. Even in the alloy film with the highest boron concentration, the coating is not a homogenous

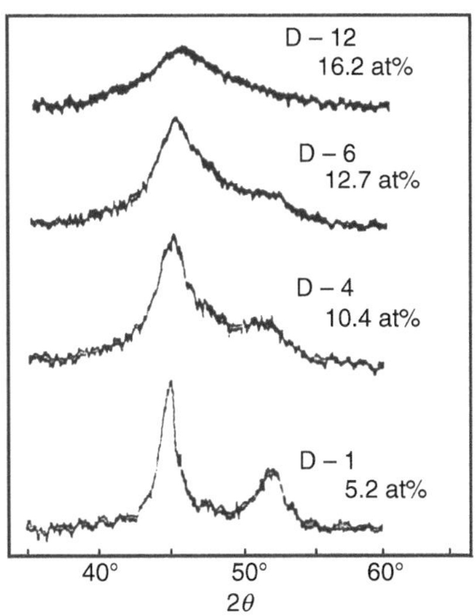

FIGURE 8.17 X-ray diffraction patterns of various Ni–B alloy films (59).

amorphous phase and contains a very small amount of quite fine crystals. These authors called this structure the "amorphous-like structure." In 1998, Saito et al. [61] also studied the effect of DMAB concentration on the boron content of Ni–B alloy films. They had similar results as shown in the upper curve of Fig. 8.16b. The B content increased as the amount of boron source increased. However, it reached a limiting value at concentrations higher than 0.075 mol/L. It is not difficult to understand these experimental phenomena: The reason is that when all the parameters are fixed in the bath solution, the B element reduced increases as enhance of DMAB, thereby the B content in the deposit raises.

Second, we take a look on the effect of complexing agent. In 1998, Saito et al. [61] reported effect of four complexing agents (sodium tartrate, sodium citrate, sodium malonate, and succinic acid) on the boron content of Ni–B alloy films, as shown in Fig. 8.18. For the three complexing agents (sodium tartrate, sodium citrate, and sodium malonate), the B content increases with that of the complexing agent. However, the B content decreases slightly as the succinic acid concentration increases. The original authors did not give any explanation of these findings. More work is needed to elucidate these data.

Third, we will mention the effect of stabilizer on the composition and structure of EP Ni–B films. In 1973, Gorbunova et al. [57] reported the effect of two kinds of stabilizers ($TlNO_3$ and 2-mereaptobenzothiazole + $PbC1_2$) on the composition and structure of EP Ni–B alloys. A, B, and C indicate the specimens obtained by using stabilizer $TlNO_3$, none, and 2-mereaptobenzothiazole + $PbC1_2$, respectively, but all other parameters of the bath are kept the same. The boron compositions are 19.6 (A), 24.7 (B), and 27.1 (C) at%, respectively. Moreover, the as-plated nickel–boron alloys containing 19.6 at% B consist of two phases: a solid solution of boron and hydrogen in face-centered cubic β-nickel and an amorphous phase. The samples with boron compositions 23.7 at% or 27.1 at% B

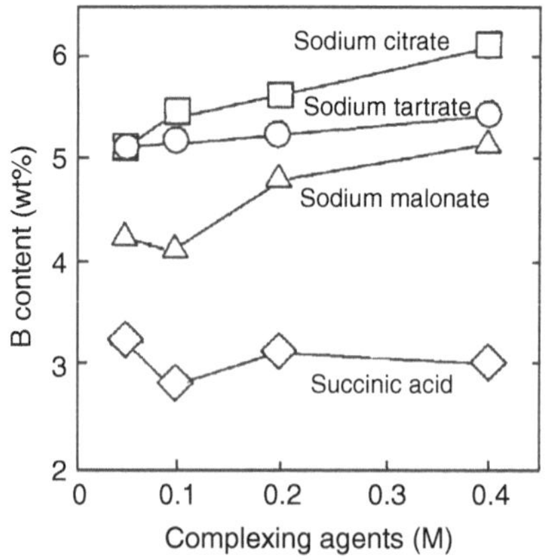

FIGURE 8.18 Effect of complexing agent on deposition rate and boron content of Ni–B alloy films. ○$Na_2C_4H_4O_6{\bullet}2H_2O$; □$Na_3C_6H_5O_7{\bullet}2H_2O$; △$Na_2C_3H_2O_4{\bullet}H_2O$; ◇$C_4H_6O_4$ [61].

are single-phase alloys of amorphous structure. Stabilizer $TlNO_3$ suppresses the reduction of B element, causing the lower B content in the deposit. In 2002, Delaunois and Lienard [62] also investigated the effects of stabilizers Tl^+ ($TlNO_3$) and $Pb2^+$ ($PbWO_4$) on EP Ni–B coatings. The added amounts are 0.11 g/L $TlNO_3$ and 0.02 g/L $PbWO_4$, respectively. The amounts of other chemicals are different in these two baths. The results show that a ternary Ni–B–Tl deposit is obtained with 3.99 wt% B and 6.90 wt% Tl when adding $TlNO_3$, but still Ni–B coating is obtained with 5.75 wt% B and only 1.25 wt% Pb when adding $PbWO_4$. The Ni–B–Tl deposit is almost amorphous with an extremely fine Ni crystallized structure near the nanometer size. The Ni–B–(Pb) structure is extremely finely crystallized with a Ni crystal structure near the nanometer size, which is almost amorphous, as it is the case for as-deposited Ni–B–Tl layer.

Fourth, pH value in the bath solution affects the composition and structure of EP Ni–B films. Mallory in 1971 [56] researched the effect of pH value. He found that the B contents are 19.55, 6.27, and 1.2 at% when the pH values are 5.1–6.0, 6.9–7.6, and 9.0–9.5, respectively. That is to say, the increase in the pH value causes a decrease in the B content in the deposit. This situation is similar to that in the EN described above. However, this is very different from the work by Saito et al. in 1998 [61]. Figure 8.19 shows the effect of pH on boron content of Ni–B alloy films (the upper curve). It is evident that the B content of Ni–B alloys was little changed with pH and remained nearly constant at around 5.2 wt%. This is very different from that obtained by Mallory.

Fifth, the amorphous nature of the Ni–B deposits with certain B content has been mentioned in the preceding paragraphs. Here discussions will be continued further using very recent published data. In 2005, Krishnaveni et al. [65] reported electroless Ni–B coatings and confirmed that the structure of Ni-27.4

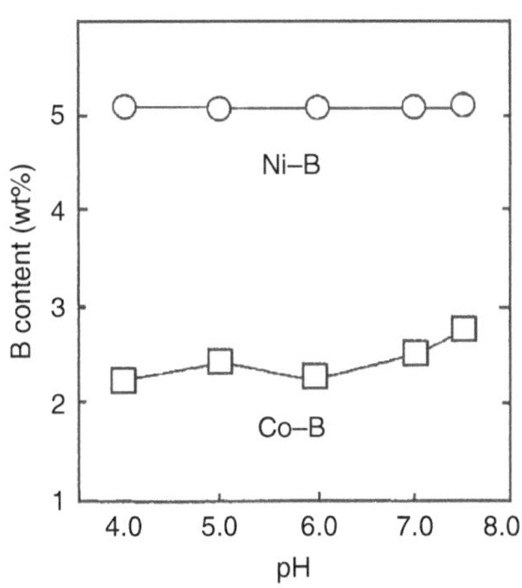

FIGURE 8.19 Effect of pH on boron content of Ni-B and Co–B alloy films. ○Ni–B alloy film; □Co-B alloy film [61].

at% B deposit was amorphous. In 2012, Wang et al. [69] believed that when the B content is 26.6 at% or higher, the as-deposited Ni–B coating is completely amorphous. Also in 2012, Vitry et al. [68] prepared the EP Ni–B films with 5 wt% (~22 at%) boron, approximately 1 wt% lead, the balance being nickel (94 wt%). Their experiments show the amorphous or nanocrystalline character of the sample, but further analysis is needed to ascertain the crystallinity. The obtained average grain size is very small (close to 1 nm), which means that the material is not fully amorphous and presents short-range order but it is not fully nanocrystalline either. In 2004, Sankara Narayanan and Seshadri [63] fabricated a sample of EP Ni–B deposit containing 93.36 wt% nickel, 6.4 wt% (~27 at%) boron, and 0.24 wt% thallium. XRD patterns and TEM images of the electroless Ni–B coating in the as-plated condition show a typical characteristic of amorphous structure (see Figs 2 and 3 in [63]). It is more interesting to note that these authors' TEM analysis was carried out at different regions and indicates the presence of microcracks, which have spread throughout the film. Their presence is more pronounced at the edges, as shown in the left picture of Fig. 8.20. Moreover, their analysis performed at the cracks/boundary regions reveals the presence of nanometer-sized crystals, which are concentrated in these regions. The dark field image confirms the presence of very small crystallites (white spots), which are nanocrystalline in size indicated in the right image of Fig. 8.20. Therefore, these authors conclude that EP Ni–B films in the as-plated condition consist of two phases; the major part is amorphous whereas a small portion of it is nanocrystalline.

The last point is the composition and structure of cross section in EP Ni–B deposits. Figure 8.21 shows the microstructures of EP Ni–B deposits obtained by Gorbunova et al. [57] in 1973 and by Rajagopal [60] in 1983, respectively. They all indicate a column-lamellar structure, which is similar to that of electroless Ni–P alloys, as shown in Figs 8.10 and 8.11. These two sets of authors believed that this phenomenon seems to arise from the periodic change in the ratio

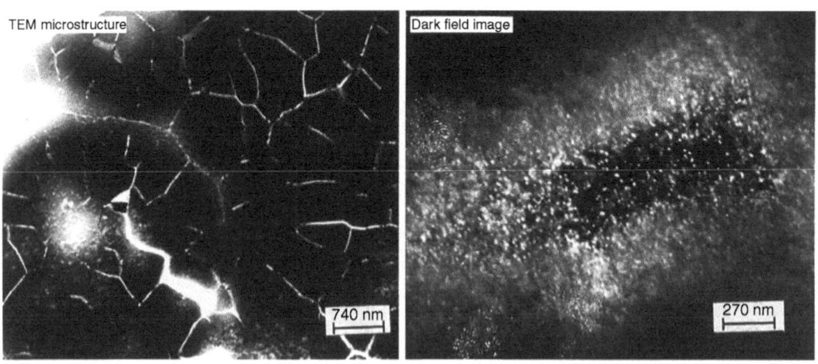

FIGURE 8.20 TEM microstructure of electroless Ni–B coating in the as-plated condition. Dark field image of electroless Ni–B coating in the as-plated condition at the crack/boundary region [63].

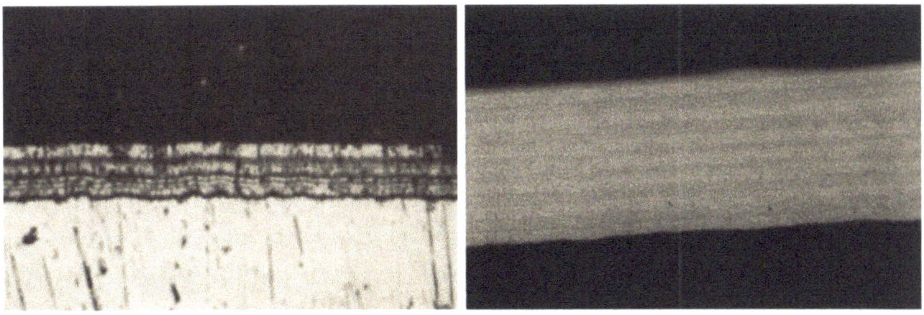

FIGURE 8.21 Cross section microstructure of electroless nickel-boron plate. Left: magnification ×225 [57]; right: ×400 [60].

of nickel to boron caused by pH changes near the substrate. Moreover, at 400°C heating, the structure remained the same, except that the laminations became broad and pronounced. The laminations could not be seen after 800°C heating and the microstructure revealed a typical grain pattern. This also is similar to that of the EN deposits mentioned in the preceding paragraphs. In 2005, Rao et al. [66] also reported cross section micrograph of the Ni–B deposit, as shown in the left of Fig. 8.22. It can be seen that there exist morphological variations in the film between the outer part and the inner part. The inner part with a thickness of ca. 3 μm shows a columnar morphology, while the outer part of about 7 μm shows a pronounced granular morphology. It is obvious that this morphology is different from that shown in Figs 8.10 and 8.11. The authors generally explained that the morphological variations may have resulted from the changes of growth mode during the deposition. Vitry et al. [68] in 2012 reported in their Fig. 2 that the thickness of the deposit is 16.5 μm, and the cross-section observation reveals their all columnar morphology. The authors stated that this is due

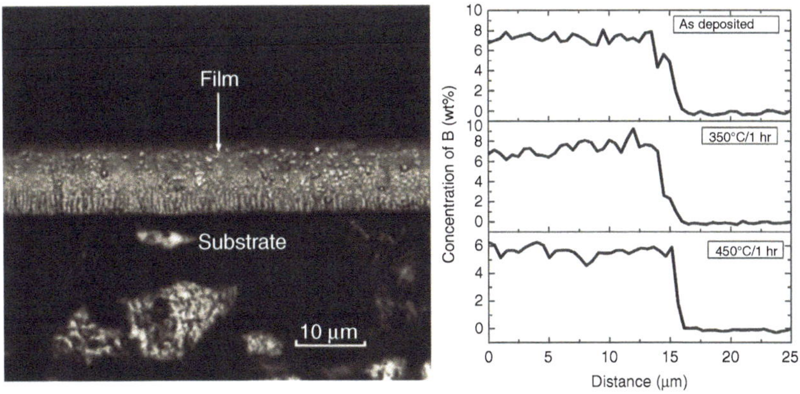

FIGURE 8.22 Cross section micrograph of the Ni–B deposit (left) [66]. EPMA composition profiles of B along the coating cross section of as-deposited, 350 and 450°C heat-treated samples [67].

to the growth mechanism of the deposit. This is again a conflicting experimental phenomenon.

In 2011, Pal et al. [67] measured the composition profiles of B along the coating cross section of as-deposited, 350 and 450°C heat treated Ni–B samples using EPMA, as shown in the right of Fig. 8.22. It is obvious that B content in the coating is nearly constant throughout the coating thickness, but is substantially decreased with increasing heat treatment temperatures. This result is similar to that obtained by Lin and Lai [17]. These authors did not give any explanation.

8.1.2.2 Co–P and Co–B Binary Alloy Films

Generally speaking, there are not many publications on the composition and structure of EP Co–P or Co–B [61,70–72]. The author's group in 1997 investigated the composition and structure of EP Co–P alloy films [71]. They found that the P content in the deposit increases with increase of complexing agent ammonium citrate, as shown in Table 8.2. What is the reason? Our experiments also show that the plating rate of alloy deposit always reduces from its stating point (Fig. 3 in [71]), which means that the so-called stable complexation reaction is just reached, making the plating rate maximum. If increasing the concentration of complexing agent again it speeds up the complexation excessively, then Co ions released from the excess of complexing agent gradually decrease, and the deposited Co atoms reduce, thereby the Co content in the coatings decreases, causing the increase of P content. XRD patterns show that the structure of Co–P coatings with different P contents is a solid solution of P atoms in the Co lattice. The height of the diffraction spectrum peak reduces with increasing the P content but the width of it rises, which means the grain size decreases.

Takano and Matsuda in 1980 [70] reported the relationship between the P content of the EP Co–P deposited film and solution pH, as indicated in the left of Fig. 8.23. As the pH value increases, the P content in the coatings decreases. Tarozaitė et al. in 2009 [72] studied the P content in the Co–P films as changing the pH value with and without additive H_3BO_3, as indicated in the right of Fig. 8.23. The amount of phosphorus incorporated into cobalt films

TABLE 8.2 Effects of the Concentration of Complexing Agent on the Composition in EP Co–P Alloys [71]

Sample	I	II	III	IV
Ammonium citrate, mol/L	0.1	0.15	0.25	0.3
Co, at%	98.66	98.46	98.04	97.19
P, at%	1.34	1.54	1.96	2.81

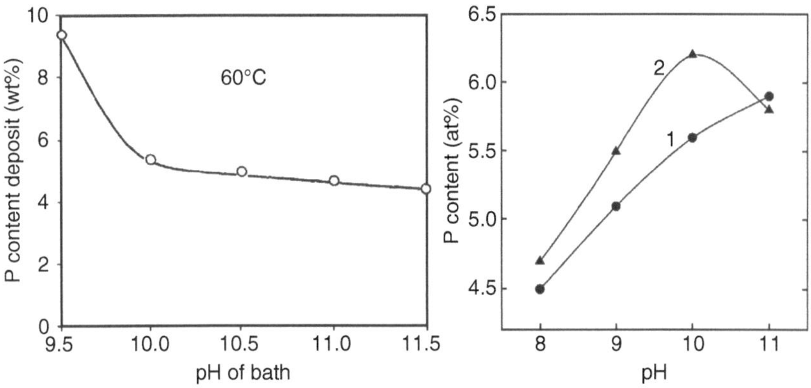

FIGURE 8.23 Relation between P content of the deposited Co–P film and solution pH. Left: composition of solutions (mol/L): $CoSO_4$—0.1, NaH_2PO_2—0.5, $Na_2P_2O_7$—0.4, $(NH_4)_2SO_4$—0.5 [70]. Right: composition of solutions (mol/L): $CoSO_4$—0.1, glycine—0.6, NaH_2PO_2—0.5; 1—without additive H_3BO_3, 2—with 0.3 mol/L H_3BO_3 [72].

obtained from solutions without addition of H_3BO_3 increases with increase in solution alkalinity in all pH ranges investigated (the right in Fig. 8.23, curve 1), whereas when the H_3BO_3 additive was added to solutions, the maximum content of phosphorus was determined at around pH 10 (the right in Fig. 8.23, curve 2). The authors did not give an explanation of their data. It is obvious that these results (the left and right plots in Fig. 8.23) are not similar, but very different and even contradictory. Tarozaitė et al. also reported the dependence of Co–P phosphorus content in the films on concentration of succinic acid (Fig. 2 in [72]). The P content increases as the buffering additive succinic acid enhances, which is from 5.8 to 6.3 at% as the succinic acid increased from 0 to 0.5 mol/L.

In 1974, Pearstein and Weightman [73] measured the composition of EP Co–B deposits from acid bath (cobalt sulfate 25 g/L, DMAB 4 g/L, sodium citrate 25 g/L, pH 5, 70°C) is 8.95 at% B. However, there was C 0.97 wt%, and N 0.05 wt% in the deposit, indicating the presence of organic or organometallic compounds in the deposits. In 1998, Saito et al. [61] also measured the effect of DMAB concentration on the boron content of Co–B alloy films prepared in an acid bath (cobalt sulfate 0.10 mol/L, g/L, DMAB 0.025–0.10 mol/L, sodium citrate 0.10 mol/L, sodium malonate 0.10 mol/L, pH 6, 60°C, plating time 1 h). The results are shown in Fig. 8.16b (the lower curve). The B content increases as the amount of boron (DMAB) increases. However, it reaches a limiting value at concentrations higher than 0.075M. The reason for enhancing the B content by increasing the level of DMAB in Co–B deposits is just the same as that for Ni–B deposits, mentioned previously, so it will not be repeated here. In 1986, Chang et al. [74] found that the composition of the Co–B films is independent of pH, BH_4^- concentration, and temperature. It is obvious that this result (using

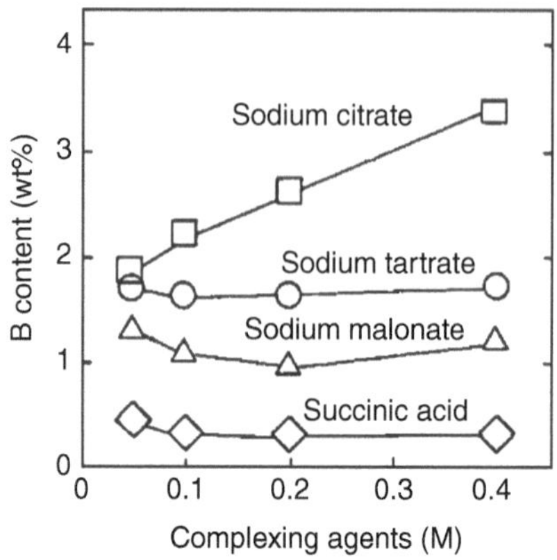

FIGURE 8.24 Effect of complexing agent on boron content of Co–B alloy films. ○$Na_2C_4H_4O_6$•$2H_2O$; □$Na_3C_6H_5O_7$•$2H_2O$; △$Na_2C_3H_2O_4$•H_2O; ◇$C_4H_6O_4$ [61].

$NaBH_4$ as reductant) is opposite to that from Sato et al. (using DMAB as reductant), and the reasons for this need to be elucidated further.

As for the effect of complexing agent on B content in Co–B deposits, in 1998 Saito et al. [61] reported the effect of four complexing agents (sodium tartrate, sodium citrate, sodium malonate, and succinic acid). The results are illustrated in Fig. 8.24. For the three complexing agents (sodium tartrate, sodium malonate, and succinic acid) the B content is not much changed basically, even a slight reduction with an increase in the complexing agent. However, the B content increases lineally with enhancing the complexing agent sodium citrate. These situations are different from those in Ni–B deposits, as shown in Fig. 8.18. Again, the original authors did not give any explanation of this. More work is needed to resolve this conflict.

The effect of pH value on the composition of EP Co–B films was also reported by Saito et al. [61]. The data is indicated in Fig. 8.19 too (the lower curve). When the pH increases, the B content in the coatings increases, which is different from that observed in Ni–B deposit. It is obvious that more data is needed to elucidate this situation clearly.

In 1998, Chang et al. [74] found that the structure of as-deposited Co–B films is independent of pH and solution concentration of Co^{2+} and BH_4^- ions. Jiang et al. in 2007 [75] reported similar results because the results of their X-ray diffraction analysis show that as-plated Co–B alloy is amorphous. In addition, the structure of Co–B alloy films is not fundamentally changed across the ranges studied for all five parameters (concentrations of $CoSO_4$, $NaC_4H_4O_6$, $NaBH_4$ and NaOH, and bath temperature), revealing still an amorphous material. Figure 8.25 shows the XRD patterns of the effects of three parameters (concentrations of $CoSO_4$, $NaC_4H_4O_6$, $NaBH_4$). The structure of Co–B alloys containing different B contents was also reported by Saito et al. [61]. Their measurements indicate that Co–B films with high boron content, for example,

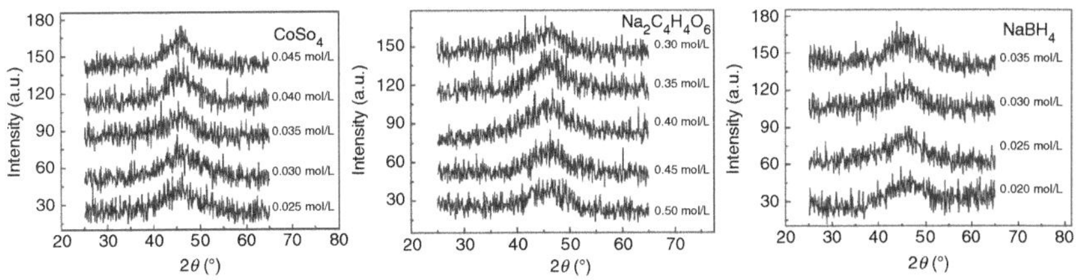

FIGURE 8.25 Effect of concentrations of $CoSO_4$, $NaC_4H_4O_6$, and $NaBH_4$ on Co–B film structure [75]. The composition and structure of EP Fe–B alloys were described and analyzed in Chapter 5, so this is not repeated here.

25 at% B are characterized by the amorphous structure as indicated by the broadening of the diffraction line. The XRD diffraction lines corresponding to Co {002} were recognized with decreasing boron content to values less than 5.2 at%, which indicated that structural change from amorphous to crystalline may occur depending on boron content. A fine crystal structure was detected for as-plated Co–B films with boron content of 2.66 at%.

8.1.3 Binary Alloy-Based ECP Deposits

Only a few papers have been published on the composition and structure of binary alloy based ECP deposits [76–83]. Discussion of composition and structure in binary alloy-based ECP deposits will be able to ascertain whether the intentionally added insoluble particles can impact and in what way they would affect the composition and structure of plain binary EP alloy films. In 1998, Leóna et al. [76] reported phosphorus content and the percentage BN(h), and the structure in the coating with addition of BN(h). They have found from XRD analyses that the electroless ECP Ni–P coatings with phosphorous content between 9.07 at% and 10.79 at% have a microcrystalline structure (Fig. 3A in [76]), while the coatings with a phosphorous content of approximately 17.3 at% show an amorphous structure (Fig. 3B in [76]). Both structures have been seen in the above for similar phosphorous contents in the Ni–P coatings. The well-defined peaks corresponding to the presence of BN(h) are observed in the XRD diffraction patterns of ECP Ni–P films. The data illustrates that BN(h) particles have no big impact on the composition and structure of Ni–P coatings. In 2005, Gao et al. [77] investigated the composition and structure of nanometer-sized Al_2O_3 composite coatings prepared by electroless deposition. The average composition analyzed by EDS and chemical method showed that the content of Al_2O_3 particles was 2.3 wt%. The phosphorus content of Ni–P + Al_2O_3 composite coatings was 10 wt%, which was slightly lower than that in Ni–P coatings (11 wt%) because of the presence of Al_2O_3 particles. The TEM bright-field (BF) image selected area diffraction patterns of the as-deposited electroless Ni–P + Al_2O_3 coating showed that the nanometer-sized Al_2O_3 particles were distributed homogeneously in the amorphous matrix of composite coating. Analysis of the XRD spectra indicated that the existence of nanometer-sized Al_2O_3 particles did not change the structure of the matrix, and both coatings as-deposited were amorphous.

Balaraju et al. in 2006 [78] studied the structure of electroless Ni–P + Si_3N_4, Ni–P + CeO_2, and Ni–P + TiO_2 composite coatings. Chemical analysis of the plain electroless Ni–P coating shows that it contains 89 wt% Ni and 11 wt% P, whereas with the incorporation of Si_3N_4, CeO_2, and TiO_2 particles in the Ni–P coating, the nickel and phosphorus contents vary slightly. The nickel and phosphorus contents of the composite coatings are as follows: 10.1 wt% phosphorus and 81.79 wt% nickel for electroless Ni–P + Si_3N_4 coating; 10.18 wt% phosphorus and 82.38 wt% nickel for electroless Ni–P + CeO_2 coating; 10.40 wt% phosphorus and 84.18 wt% nickel for Ni–P + TiO_2 coating. XRD pattern, TEM, selected area electron diffraction pattern, and energy-dispersive X-ray analysis reveal that the incorporation of Si_3N_4, CeO_2, and TiO_2 particles in the Ni–P matrix does not have any influence on the structure. This incorporation of intentionally insoluble particles in the Ni–P matrix does not have any influence on the structure and may change only slightly with the P composition, as has also been confirmed by other authors. For example, in 2005, Wu et al. [79] investigated Ni–P, Ni–P + SiC, Ni–P + PTFE, and Ni–P + PTFE + SiC, and found that the as-deposited coatings had amorphous Ni–P structure (Fig. 8.26).

In 2006, Novakovic et al. [80] studied electroless Ni–P + TiO_2 composite coatings. The P composition reduces from 12.5 wt% in Ni–P deposit to 11.4 wt% in Ni–P + TiO_2 composite coatings when the contents of TiO_2 in the bath are ≤ 10 g/L. The TiO_2 particles embed into the amorphous nickel matrix. In 2013, Allahkaram et al. [81] reported the ECP of Ni–P + nano ZnO. They found that

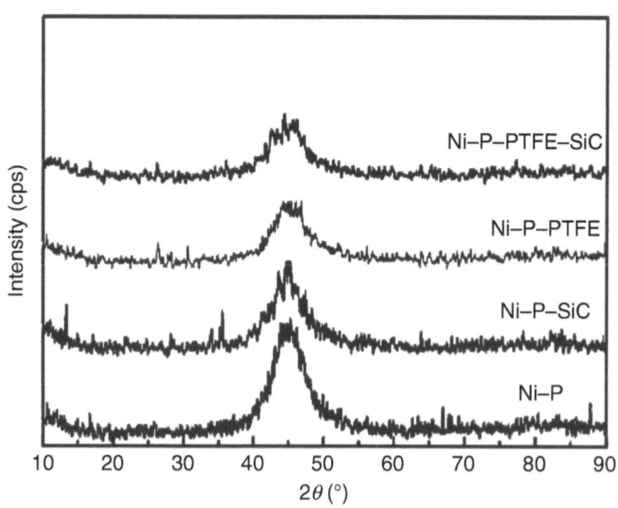

FIGURE 8.26 **XRD patterns of as-deposited of Ni–P and Ni–P composites [79].**

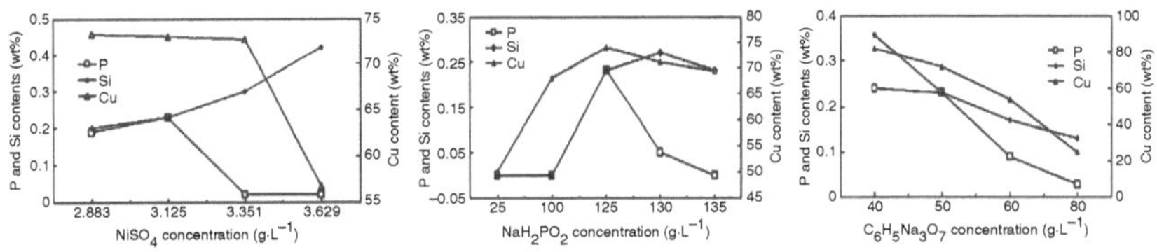

FIGURE 8.27 Effect of $NiSO_4$ (left), NaH_2PO_2 (middle), and $Na_3C_6H_5O_7$ (right) concentrations on the coating composition [83].

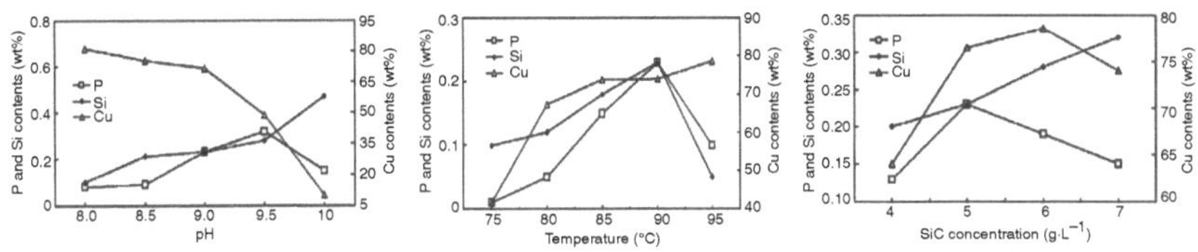

FIGURE 8.28 Effect of pH value (left), temperature (middle), and SiC concentration (right) on the coating composition [83].

the XRD patterns of both composite and pure Ni–P coatings show an amorphous phase, the amount of phosphorous in the composite coating (8.9 wt%) in comparison to Ni–P (9.4 wt%) is only a small reduction. However, Lee in 2012 [82] found a large decrease in P content in Ni–P + TiO_2 ECP coatings, from 20.03 wt% in the Ni–P coating to 8.35 wt% in the Ni–P + nano TiO_2 composite coating, when 15 g/L TiO_2 was added to the plating solution. Similar results have been obtained from electroless Ni–P + nano SiO_2 composite coatings. Some explanation is needed to understand these divergent data, as was pointed out by Allahkaram et al. [81] in 2013, "The reason for this phenomenon is unidentified."

Faraji et al. in 2011 [83] measured the variation of compositions of ECP Ni–P + micrometer (4~10 μm in diameter) SiC coatings. They measured the effects of six parameters (nickel sulfate, sodium hypophosphite, sodium citrate, pH value, temperature, and silicon carbide particles) on the compositions of the films, as shown in Figs 8.27 and 8.28. Here we just mention the variation of P content in the deposit. It is evident from the figures that the P content increases to a maximum value and then decreases as all of the five parameters (nickel sulfate, sodium hypophosphite, pH value, temperature, and silicon carbide particles) increased. Only for sodium citrate, as its concentration raises, P content (also Cu and Si) decreases. Sodium citrate indicates a strong suppressing effect on the composition in the deposits. The reasons for this need to be understood further.

8.2 COMPOSITION AND MICROSTRUCTURES OF EP TERNARY AND MULTICOMPONENT ALLOY DEPOSITS

Many papers on the composition and structure in EP ternary and quaternary alloy deposits have been published [44,46,84–117]. Most of them were published after 2000, with only a few papers being published before 1990. Of all the alloy systems, the main focuses were on Ni–W–P [44,84–89] and Ni–Cu–P [47,90–93]. As for the dependent factors, most authors considered the metal salt, reductant, and pH value.

8.2.1 Effects of Metal Salts on Composition and Structure in Ternary and Quaternary Alloy Coatings

Many groups investigated the effects of metal salts on composition and structure in ternary and quaternary alloy coatings [44,47,72,86,88–90,92–94,97,100,101, 103–113], so much of the experimental data can be subdivided into three categories. These are known as the regular, confusion, and contradiction types. Most of the experimental data belong to the first category, in which the content of corresponding metal in the deposits will be increased, but the contents of other metal and metalloid elements decrease when the concentration of the adding metal salt in the bath increases. That is why this category is called regular type. Figure 8.29 shows three typical plots of compositions versus metal salts for the Ni–Co–P (left) [94], Ni–Mo–P (middle) [108], and Ni–Cu–Sn–P (right) [107] alloy deposits in the regular type. It is evident that, when the adding amount of $CoSO_4$/[$CoSO_4$ + $NiSO_4$] enhances, the Co content really increases, but the Ni and P contents decrease for the Co–Ni–P alloy deposit. The same situation appears in the Ni–Mo–P ternary alloy films. As for the quaternary alloy deposit, the Ni content increases, but all of the Cu, Sn, and P contents decrease with rising $NiCl_2$ in the bath. All the experimental data obtained from Ni–Mo–P [108,110], Ni–W–P [86,89], Ni–Cu–P [47,92,93], Co–W–P [101], Ni–Co–P [94], Ni–Co–B [97], and Ni–Cu–B [103] systems reach the same conclusion. Here, they will not be described in order to save space; interested readers can refer to the original papers. Such experimental data are not difficult to understand. While adding metal ions without change of other conditions, the released metal element enhances the metal atoms, thereby the content of such metal increases. The contents of other metals and metalloids thus decrease. Such an interpretation also adapted to the quaternary alloy films.

The second category is called the confusion type, which means that the variation of the contents of the metal and metalloid components in the deposit

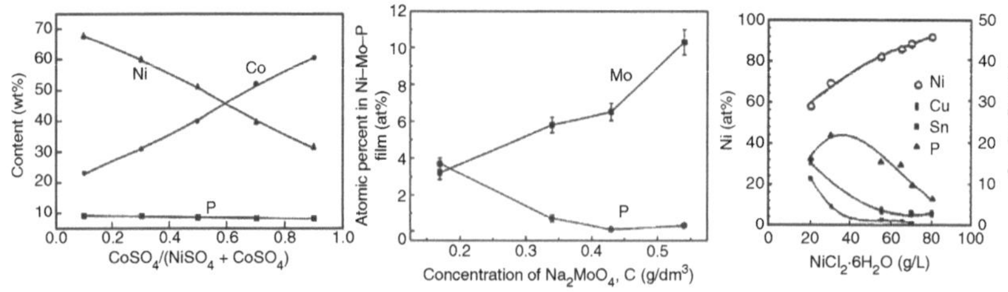

FIGURE 8.29 The chemical compositions of ternary and quaternary deposits versus metal salt concentrations. Left: Ni–Co–P [94], middle: Ni–Mo–P [109], right: Ni–Cu–Sn–P [107].

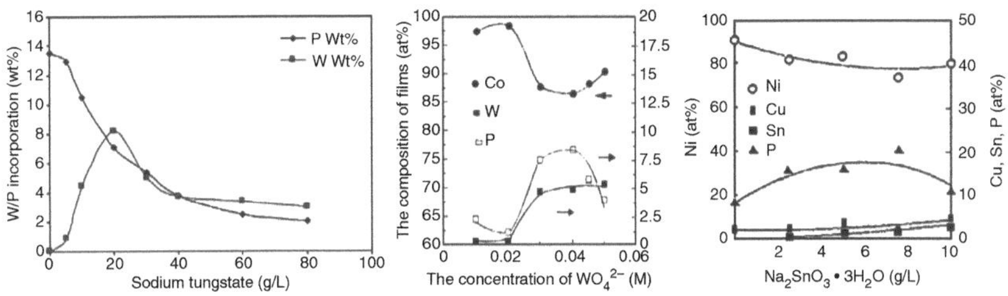

FIGURE 8.30 The chemical compositions of ternary and quaternary deposits versus metal salt concentrations. Left: Ni–W–P [44], middle: Co–W–P [101], right: Ni–Cu–Sn–P [107].

is not ordered but confused when the concentration of metal salts in the bath solution is increased. Figure 8.30 shows three typical plots of this type. In the Ni–W–P deposit [44] (the left of Fig. 8.30), the W content first enhances, reaching a maximum at 20 g/L of sodium tungstate, and then decreases as the concentration of sodium tungstate increases further. The P content always decreases. The original authors did not give any explanation for such a confused phenomenon; they did not say why the W content reaches the maximum value at 20 g/L of sodium tungstate. In the middle plot of Co–W–P in the figure, although the author's group in the original paper [101] has noted that as the concentration of sodium tungstate increases, the W content also increases, and the P content in the coating also increases before the concentration of sodium tungstate reaches 0.04 M, which means that it promotes P codeposition. Therefore, the Co content first reduces, before the concentration of sodium tungstate is less than 0.04 M, and then rises after that concentration. That is to say, the content of Co varies inversely with the changes in W and P. However, we did not explain the confusion phenomena of appearing minimum and maximum values of Co and P contents [101]. A similar confusing phenomenon also occurs with the Ni–Cu–Sn–P deposit [107] because the maximum value of the P content appears in the deposit as increasing the concentration of sodium stannate (right in Fig. 8.30). It is obvious that these confusing phenomena need to be studied further.

Because some experimental results on the effects of metal salt concentration on composition are contradictory, the third category is called contradiction type. Figure 8.31 shows two samples. Yokoshima et al. in 2000 [112] and Azizi et al. in 2010 [113] used the same Co–Ni–Fe–B alloy deposits. The left and right plots all show that as the $NiSO_4$ concentration increases, the Ni content increases and the Co content decreases linearly in the deposited films. However, the Fe content remains constant in the left plot, but changes from 20.25% to 28.56 wt% as the molar ratio of $NiSO_4/(NiSO_4 + CoSO_4)$ varies from 0 to 1 in the right plot. It is obvious that these two data about Fe contents are contradictory. The later authors [113] did not note such conflict; hence they did not give any explanation for it.

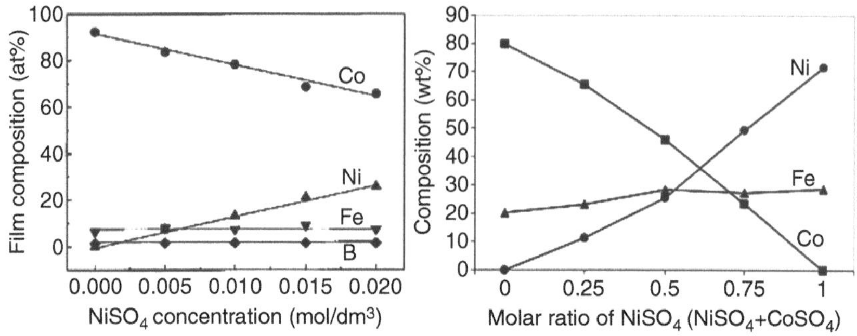

FIGURE 8.31 The chemical compositions of quaternary deposits versus metal salt concentrations. Left: Co–Ni–Fe–B [112], right: Co–Ni–Fe–B [113].

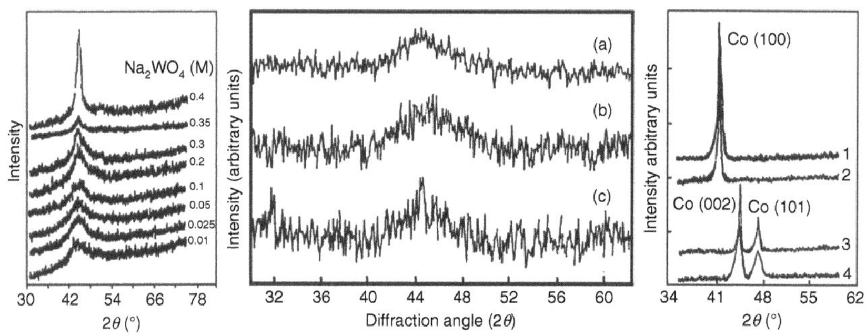

FIGURE 8.32 The XRD patterns of ternary deposits with various metal salt concentrations. Left: Ni–W–P [86], middle: Ni–Co–P [94], right: Co–W–P [101].

For the effects of metal salts on structure of ternary and quaternary deposits, the nature of structure simply depends on the compositions of the deposits. Figure 8.32 shows three samples in ternary alloy deposits. The author's group in 1996 [86] reported the XDR patterns of Ni–W–P deposits with the Na_2WO_4 concentration ranging from 0.01 to 0.4 M (left plot in the figure). The diffraction patterns for the amorphous alloys are observed for the deposits within the sodium tungstate concentration range from 0.01 M to 0.35 M, but an obvious diffraction peak corresponds to a crystalline phase at the concentration of sodium tungstate of 0.4 M. Therefore, the formation range of amorphous structure for the deposits ranges from the 0.01 M to 0.35 M of sodium tungstate. The measured P content in the coating (Fig. 1 in [86]) indicates that the P content of deposits is higher (larger than 5.5 at%) for the concentration of sodium tungstate from 0.01 M to 0.35 M, but is lower (1–4 at%) for that of 0.4 M. Phosphorus is known to facilitate the formation of amorphous structures; thereby the structures produced by the sodium tungstate concentration range from 0.01 M to 0.35 M are amorphous.

The middle plot in Fig. 8.32 shows the XRD patterns of Ni–Co–P films [94]. The (a), (b), and (c) correspond to the metallic ratios $CoSO_4$/($CoSO_4$ + $NiSO_4$) = 03, 05, and 0.7. The figure clearly shows the amorphous nature of the deposits. The compositions are shown in Fig. 8.28 (the left plot), which are Ni: 60, 51.4, 39.7 wt%; Co: 31, 40, 52.2 wt%; and P: 9, 8.6, 8.1 wt%, respectively. Because the P content is relatively large (from 8.1 wt% to 9.0 wt% for samples (a), (b), and (c)), nucleation of nickel/cobalt phase is prevented, and this has resulted in an amorphous structure.

The right plot in Fig. 8.32 was obtained by the author's group for Co–W–P deposits [101]. The compositions of the four samples are 1, $Co_{98.5}W_{0.4}P_{1.1}$, 2, $Co_{97.5}W_{0.3}P_{2.2}$, 3, $Co_{88.5}W_{5.7}P_{5.8}$, and 4, $Co_{86.7}W_{4.9}P_{8.4}$. The XRD patterns of four samples show obvious crystalline characteristic and no amorphous nature, which is a solid solution of W and P atoms in the Co lattice. W and P atoms replaced the Co atoms in the Co hcp structure. Why are they all crystalline? The metalloid element P content is not large, and the largest is only 8.4 at% P, but this is enough to promote the formation of crystalline coating. On the other hand, because the radius of a W atom is much larger than that of a Co atom, rather large microstress occurs leading to a broadened diffraction peak. Moreover, W and P as impurity atoms during the deposition process will lead to increased nucleation centers, which promotes the grain refinement thereby making the diffraction lines shorter and wider. In fact, the diffraction lines in the right plot of Fig. 8.32 are just like that. The peaks of the diffraction lines become lower and broader, especially for the samples 3 and 4. It means that the above explanation of the diffraction patterns of EP Co–W–P deposits is reasonable.

8.2.2 Effects of Reductant on Composition and Structure in Ternary and Quaternary Alloy Coatings

Relative to the metal salts, there are fewer research articles on the effects of reductant on ternary and quaternary alloy deposits. The effect of reductant on composition in ternary and quaternary deposits can be subdivided into two classes. One is that as the NaH_2PO_2 concentrations increase, the P content increases evidently, one or two metal contents thereby decrease and the other metal basically varies a little or increases slightly. The three samples in Fig. 8.33 illustrate such a trend in composition variation clearly. The explanation to this phenomenon is similar to that for the binary Ni–P deposits mentioned before.

Another class is shown in Fig. 8.34. The left plot for EP Ni–W–B coatings using DMAB as reductant was obtained by Drovosekov et al. in 2005 [98]. An increase in DMAB concentration from 1.2 g/L to 2.5 g/L results in a decrease in the boron concentration in the alloy from 2.1 wt% to 1.8 wt%. These results differ from the usual trend when boron content in Ni–B deposits increases with an increase in the concentration of a reducing agent at the chemical reduction of nickel ions with DMAB or boron hydride. For example, it is just opposite to the results for Ni–B films in Fig. 8.16. The middle plot in Fig. 8.34 for EP Ni–Fe–P–B

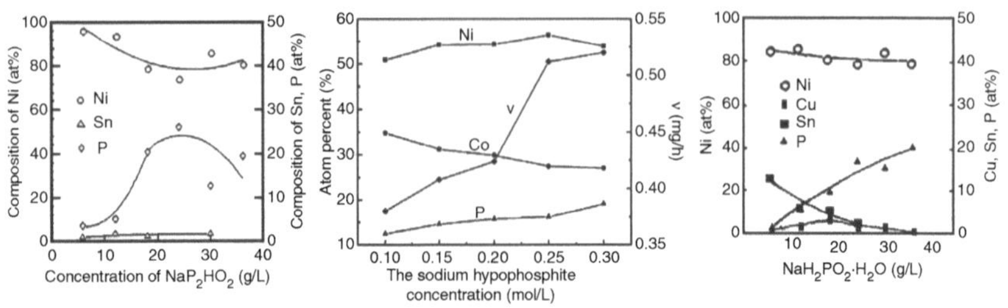

FIGURE 8.33 The chemical compositions of ternary and quaternary deposits versus NaH_2PO_2 concentrations. Left: Ni–Sn–P [111]; middle: Ni–Co–P [95]; right: Ni–Cu–Sn–P [107].

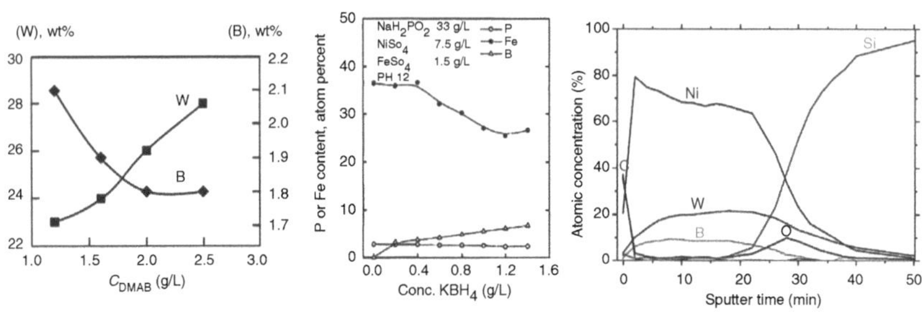

FIGURE 8.34 The chemical compositions of ternary and quaternary deposits versus DMAB [98] and KBH_4 [104] concentrations. Left: Ni–W–B [98]; middle: Ni–Fe–P–B [104]. Right: compositions versus section distant in Ni–W–B films [114].

alloy deposits was obtained by the author's group in 2000 [104]. The boron content increases from zero to about 6.7 at% with increasing KBH_4 from zero to 1.4 g/L. Evidently, the trend of variation for these two sets of data is in conflict when there is an increase in the concentration of reductant. As for the fact that the B content increases, and Fe and P decreases with increasing KBH_4 in the Ni–Fe–P–B alloy deposits, it would be easily understood according to the explanation for Ni–B deposits mentioned before. As for why the B content decreases with increasing DMAB concentration in Ni–W–B deposits, the reasons given by Drovosekov et al. [98] themselves are as follows. The change in the alloy composition with an increase in the DMAB concentration in the solution is seemingly determined by the rate ratio of the competitive reactions of the nickel and tungsten ion reduction and DMAB oxidation, which in turn depends on the chemical composition of the alloy, that is, its surface catalytic activity. The acceleration of the tungsten ion reduction with an increase in the DMAB concentration results in the enrichment of the alloy in tungsten, which lowers the surface catalytic activity with respect to the reduction of boron and its inclusion in the alloy deposit. The problem is the reason why the tungsten ion reduction would be accelerated with an increase in the DMAB concentration. The authors did not give any experimental data or reasons to explain it. In addition, they did not give

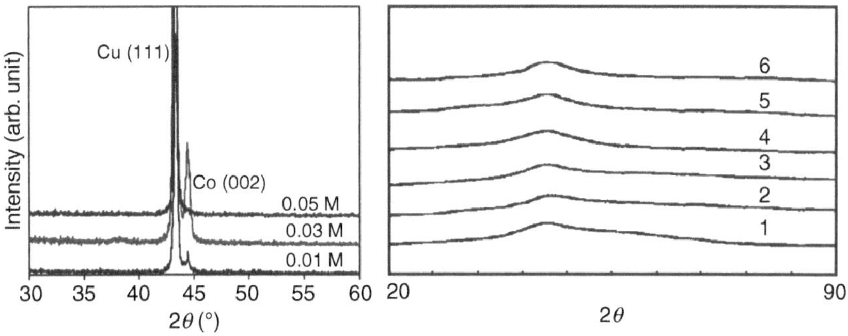

FIGURE 8.35 The XRD patterns of ternary and quaternary deposits prepared by different concentrations of reductant agents. The reductant agents are all using NaH_2PO_2. Left: Co–W–P deposits [115]; right: Ni–Cu–Sn–P deposits, the concentrations of NaH_2PO_2 are 1. 6; 2. 12; 3. 18; 4. 24; 5. 30; and 6. 36 g/L, respectively [107].

other supporting materials to confirm their view. However, the work by Duhin et al. in 2009 [114] on the same EP Ni–W–B deposits used DMAB as a reducing agent (the right plot in Fig. 8.34), and showed that the variation of the B and W content in the whole section of film is not opposite but similar. Of course, the work by Duhin et al. does not measure directly the compositions versus DMAB concentration. In a word, these problems need more work to be understood.

Only a small number of papers refer to the effects of reductant agents on the structures of ternary and quaternary deposits [105,107,115]. Figure 8.35 shows two samples of Co–W–P [115] and Ni–Cu–Sn–P [107]. It can be seen from the left plot in the figure that the films deposited from baths with hypophosphite ion concentration less than 0.05 mol/L are crystalline, whereas the film deposited from a solution containing 0.05 mol/L hypophosphite ion is amorphous. This is simply because of the presence of higher amount of phosphorus (>12 at%) in the film. The εCo (002) peak on the XRD spectrum of the films deposited from the least amount of hypophosphite ion containing solution is very small compared to the same peaks of other crystalline films. This might be due to the effect of higher tungsten (<10 at%) content in the film. The combined amount of phosphorus and tungsten in other crystalline films is less than 12 at%.

The XRD patterns of the right plot in the figure indicate the structures of all six Ni–Cu–Sn–P films are amorphous nature (Fig. 6.35) [107]. Compared to the compositions of these six deposits (the right plot in Fig. 8.33), the P contents are 1.66, 5.93, 9.67, 17.79, 14.43, 20.51 at%, respectively. From the discussion on the structure of Ni–P above, the amorphous nature of the last four samples could be easy to understand. However, the P contents in the former two samples are only 1.66 and 5.93 at%; why are they still of an amorphous nature? That is because they contain 1.28, 1.64 at% Sn and 12.30, 6.98 at% Cu, respectively. The combined actions of these elements promote their amorphous structures. We will analyze and discuss it in Chapter 11 in more detail.

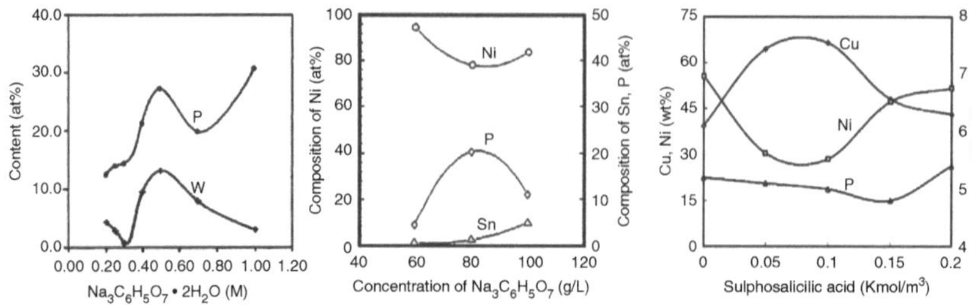

FIGURE 8.36 The chemical compositions of ternary deposits versus complexing agent concentrations. Left: Ni–W–P $Na_3C_6H_5O_7$ [86]; middle: Ni–Sn–P $Na_3C_6H_5O_7$ [105]; right: Ni–Cu–P sulphosalicilic acid [93].

8.2.3 Effects of Complexing Agents on Composition and Structure in Ternary and Quaternary Alloy Coatings

It should be said that there are few publications on the research of effects of complexing agents on composition and structure in ternary and quaternary alloy coatings [84–86,93,105,111,115]. Figure 8.36 shows three samples for compositions versus complexing agents. The left-hand one for Ni–W–P deposits and the middle plot for Ni–Sn–P films were obtained by the author's group in 1996 and 1999, respectively [86,105]. The right-hand plot of Ni–Cu–P coatings in the figure was measured by Ashassi-Sorkhabi et al. in 2002 [93]. From the figure, the common feature for these three deposits is that the concentrations of nonbase metals (W, Sn, and Cu) rise as the complexing agent concentration increases up to a certain value. The metalloid element P in the Ni–W–P and Ni–Sn–P coatings increases, but that in Ni–Cu–P films basically does not vary. This experimental data can be understood because at the very beginning, the plating solution may be in a stable complex state, the amount of based metal (Ni) in the deposits decreases. Therefore, the deposition of another metal (W, Sn, and Cu) and metalloid P will be enhanced. However, after the concentrations of complexing agents increase up to certain values, the Ni content rises, that of W and Cu decrease and that of P either enhances or decreases. This confusing fact is not clear, and more work is needed.

In 1994, Kim et al. [111] also measured the effect of sodium citrate on composition in Ne–Fe–P films. Their Fig. 3(b) in [111] shows that the Fe and P contents decrease as the concentration of sodium citrate increases from 0.05 to 0.25 mol/L. This means that Ni content in the coating increases with the range of sodium citrate. The authors did not explain this fact.

Three papers have reported the structures of Ni–W–P [86], Ni–Sn–P [105], and Co–W–P [115] deposits using various concentrations of complexing agents. All the complexing agents use sodium citrate. Figure 8.37 shows the XRD patterns for the three alloy films. The left plot of Fig. 8.37 was obtained by the author's group [86]. As can be seen, the Ni–W–P deposits obtained in a bath containing 0.2, 0.25 mpl/L, and larger than 0.6 mol/L sodium citrate are crystalline, but the deposits obtained in bath solutions containing 0.3–0.5 mol/L are amorphous.

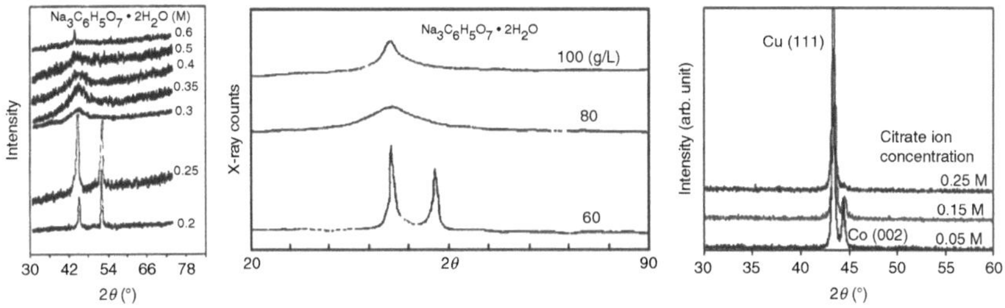

FIGURE 8.37 The XRD patterns of ternary deposits using different complexing agent concentrations. Left: Ni–W–P, $Na_3C_6H_5O_7$ [86]; middle: Ni–Sn–P, $Na_3C_6H_5O_7$ [105]; right: Co–W–P, $Na_3C_6H_5O_7$ [115].

Therefore, the formation of amorphous structure in the deposits ranges from 0.3 to 0.5 mol/L of sodium citrate. The measured P and W contents (see Fig. 2 in [86]) are 14, 15, 27.2, 22.5 at% and 3.2, 0, 13.2, 10.8 at% when the concentrations of $Na_3C_6H_5O_7$ are 0.2, 0.3, 0.5, 0.6, respectively. It is evident that when the concentrations of $Na_3C_6H_5O_7$ are less than 0.3 mol/L or larger than 0.5 mol/L, the W contents are rather large, but the P contents are not large enough, thereby the structures of the respective coatings are crystalline, but not amorphous.

As for the middle plot in Fig. 8.37 [105], which is also obtained by the author's group, one can see that the deposit prepared by 60 g/L sodium citrate is crystalline, whereas the other two samples prepared with 80 and 100 g/L sodium citrate are amorphous. The crystalline nature for the former results from its low P content (4.5 at%) and only 0.7 at% Sn. The amorphous nature for the last two samples is caused by their high P content (11.1–20 at%) and a certain value of tin (1.2–4.8 at%).

The right plot in Fig. 8.37 is obtained by Dula et al. in 2009 [115]. It suggests that the films deposited from electrolytes of lower citrate ion concentrations are crystalline, whereas the film deposited from 0.25 mol/L citrate ion containing solution is amorphous. This is because of higher phosphorus content in the film plated from high citrate-containing bath. The measured compositions by these authors are that Co content decreases from 92 to 88 at% with an increase in citrate ion concentration in the solution from 0.05 to 0.25 mol/L. On the other hand, phosphorus in the films increases from 3 to 9 at%. There is little effect of citrate ions on the tungsten content in the films, and this is around 2.6 at%. The maximum P content is only 9 at%, and still there is 2.6 at% W in the deposit. So, it seems that the amorphous nature of the deposit prepared by 0.25 mol/L citrate ion containing solution is not easy to explain.

8.2.4 Effects of pH Value on Composition and Structure in Ternary and Quaternary Alloy Coatings

Reports of the effect of pH value on composition have been published for Ni–W–P [84,85], Ni–Co–P [46,95], Ni–Cu–P [90], Ni–Mo–P [109], Ni–Fe–P

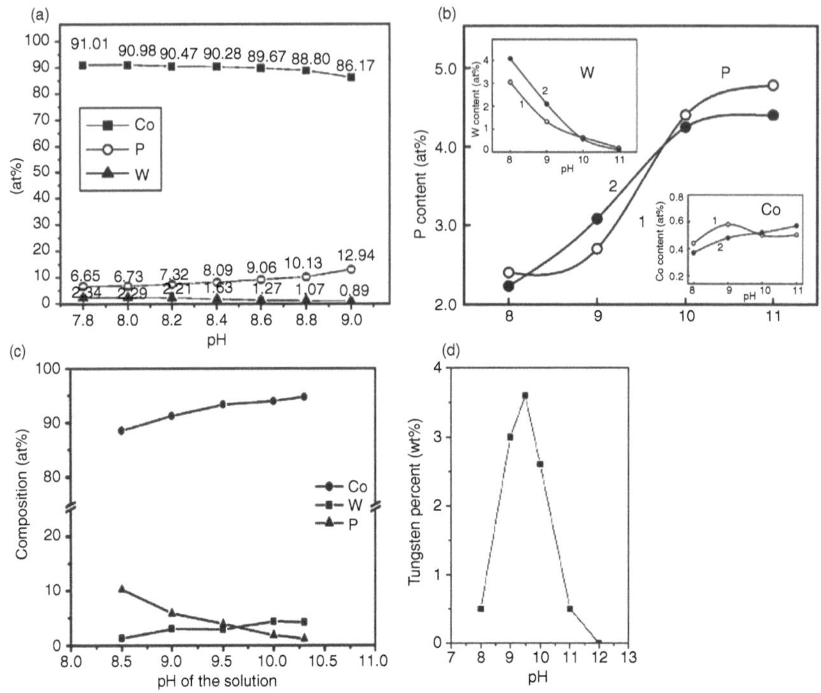

FIGURE 8.38 The chemical compositions of EP ternary Co–W–P deposits versus pH value. (a) [116]; (b) [72]; (c) [99]; (d) [96].

[111], and Co–W–P [72,99,116] alloy deposits. Figure 8.38 shows the compositions versus pH values for the four EP Co–W–P deposits by different author groups [72,96,99,116]. In Fig. 8.38(a), obtained by Wu et al. in 2008 [116], one can see that the P content increases, but those of Co and W decrease with increase in the pH. In Fig. 8.38(b) showing data of Tarozaitė et al. in 2009 [72], P and Co increase but W decreases. In Fig. 8.38(c) by Kim et al. in 2008 [99], P decreases (from 10 to 1 at%) but Co and W increase. Figure 8.38d was obtained by Abdel Aal in 2008 [96]. A maximum of W appeared. It can be seen that the change in composition with increase of pH is very different in each of the four cases. The results by Wu et al. and those by Kim et al. directly oppose one another. The data by Abdel Aal et al. are opposite to those from all other three cases. Are these four sets of experimental data all really consistent with the nature of things? Which set of experimental data reflects the nature of things? How do these authors explain their own experimental results?

Tarozaitė et al. [72] did not give any explanation of their data (Fig. 8.38b). But the other two sets of authors did explain their experimental results from the viewpoint of electrochemical reactions or mechanism involved in electroless

Co–W–P plating. Kim et al. [99] used DMAB and NaH_2PO_2 as reductant agents, and proposed the following reaction equations:

$$(CH_3)_2HN{\cdot}BH_3 + 3HOH + OH^- \rightarrow (CH_3)_2H_2N^+ + B(OH)_4^- + 5H^+ + 6e^- \tag{8.5}$$

$$H_2PO_2^- + OH^- \rightarrow H_2PO_3^- + 1/2H_2 + e^- \tag{8.6}$$

$$Co^{2+} + 2e^- \rightarrow Co \tag{8.7}$$

$$H_2PO_2^- + e^- \rightarrow P + 2OH^- \tag{8.8}$$

They believe that at higher pH values, reaction in Eq. (8.6) is facilitated to proceed forward decreasing the concentration of the hypophosphite ions in the solution. Consequently, the amount of phosphorus in the deposit decreases. As the phosphorus content in the film decreases, the amounts of cobalt and tungsten in the film increase.

Wu et al. in 2008 [116] only used NaH_2PO_2 as reductant agents in the Co–W–P plating, so there is no Eq. (8.5), but Eqs (8.6)–(8.8) are basically the same. Their basic starting point is that the simultaneous decrease of Co and W contents with increasing pH is attributed to the OH^--related kinetics [117] stated in the preceding section and to a codeposition phenomenon. Then they use Eqs (8.6) and (8.8) to explain the variation of P content. As the Co deposition was retarded by the OH^--related kinetics, more electrons were available to react with $H2PO_2^-$, and hence, more elemental P was precipitated. The P content hence increases with decreasing Co content in electroless Co–W–P films in the chosen pH range. Another set of author group (Abdel Aal et al. [96]) also explained their results: increase in initial pH value from 8 to 9.5 tends to increase the W content from 2.0 to 3.6 wt%. But, further increase in pH value up to 12 lowers the W and the P contents. They believed that this behavior can be attributed to the idea that an increase in pH enhances Co^{2+} reduction and $H_2PO_2^-$ oxidation kinetics, but does not affect $H_2PO_2^-$ reduction to P. However, beyond pH 9.5, the bath destabilizes and decomposes within a short span of time. Notably, it is difficult to operate the plating bath at a pH higher than 9.5 without using a suitable stabilizer.

It is obvious that each explanation cannot be used to elucidate other facts, that is, no unified explanation can depict these very different and even conflicting sets of data. It should be noted that these four papers were not published in the same year (three [96,99,116] in 2008, one in 2009 [115]), but the paper published later did not compare their data with those published before. Once again, there is a conflict of experimental phenomena, and further work is needed.

Very few papers have reported the structures of Co–W–P [99,116] deposits using various pH values. The results are shown in Fig. 8.39. The left plot in Fig. 8.39 depicts the XRD patterns of Co–W–P films plated at a pH in a range of 7.6–9.0 obtained by Wu et al. in 2008 [116]. The three most intense

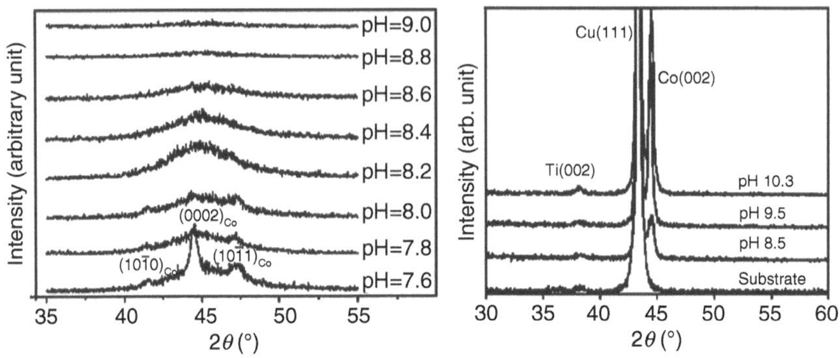

FIGURE 8.39 The XRD patterns of ternary deposits with different pH values. Left: Co–W–P [116]; right: Co–W–P [99].

peaks corresponding to hexagonal-phase Co (10$\bar{1}$0) at $2\theta = 41.68°$, (0002) at $2\theta = 44.76°$, and (101$\bar{1}$) at $2\theta = 47.57°$ indicate that the polycrystalline structure constitutes a Co–W–P layer deposited at low pH. The gradual diminution of these characteristic features with increasing pH indicates the emergence of nanocrystalline, and eventually amorphous, structure in Co–W–P films. The right plot in Fig. 8.39 indicates the XRD spectra of the substrate and the Co–W–P films deposited at various pH values (8.5 to 10.3) obtained by Kim et al. in 2008 [99]. In these spectra, strong peaks at $2\theta = 43.5°$ correspond to substrate copper with preferred orientation of {111}. The small peaks at $2\theta = 38.4°$ are related to the titanium seed layer. The spectra of all the films deposited at various pH values show peaks at $2\theta = 44.5°$, which indicate the presence of hexagonal cobalt (εCo) with a preferred orientation of {002}. The intensity of the εCo {002} peak increases with increase in the pH of the bath, so the average grain sizes of the crystalline films can be calculated using the Debye–Scherrer equation, which increases from 8 to 15 nm with increase in the pH of the solution.

8.2.5 Effects of Temperature on Composition and Structure in Ternary and Quaternary Alloy Coatings

Only a few papers on the effects of temperature on composition and structure in ternary and quaternary alloy coatings have been published [98,99,109]. In 2010, Liu et al. [109] published the results for Ni–Mo–P films, and in 2008 Kim et al. [99] reported data for the Co–W–P deposits. The results are shown in the left and right plots of Fig. 8.40, respectively. For the Ni–Mo–P films, the Mo content increases and the P content apparently reduces with increasing bath temperature. However, the content of P increased when the bath temperature was increased up to 363 K. The authors believe that the rise in P content may be due to the dominant reaction of the P reduction compared with the reduction of Mo.

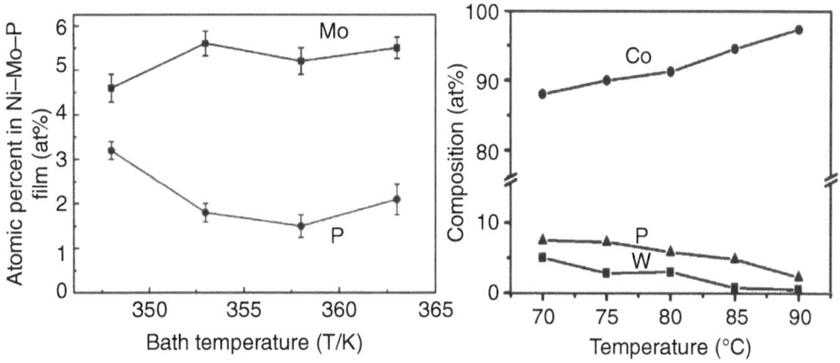

FIGURE 8.40 The chemical compositions of ternary deposits versus temperature. Left: Ni–Mo–P [109]; right: Co–W–P [99].

As for the Co–W–P deposits, the cobalt content increases from 88 to 97 at%, but both tungsten and phosphorus contents decrease (tungsten from 5 to less than 1 at% and phosphorus from 7 to 2 at%) with increase of the bath temperature from 70 to 90°C. What are the reasons? It is well known that the mobility of the ionic species increases with increase in solution temperature, but the cobalt ion is less bulky than the tungstate and hypophosphite ions; therefore, comparatively more cobalt ions reach the electrode surface with increase in solution temperature, which is possibly the reason for the increase in cobalt content in the films.

Figure 8.41 shows the XRD spectra of the Co–W–P films deposited from baths of various temperatures [99]. The spectra suggest that all the films deposited

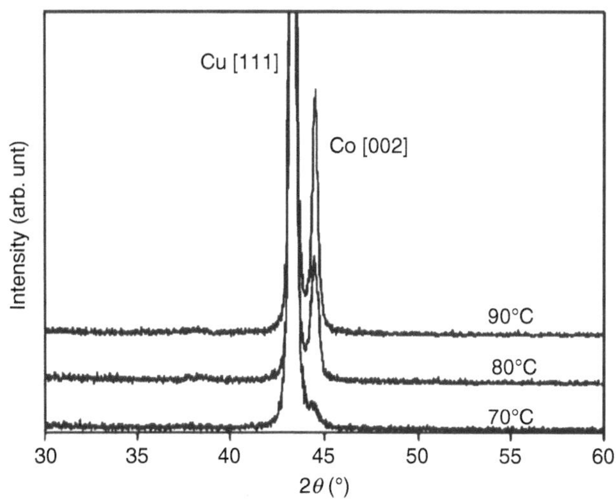

FIGURE 8.41 The XRD patterns of Co–W–P deposits with different temperatures [99].

in the temperature range used in this study are crystalline in nature with direction of {002} of εCo. The crystalline nature of the deposits is due to the P contents being less than 7 at%. The intensity of the εCo {002} peak of the film deposited at 70°C is very low. However, the εCo {002} peak height gradually increases with increase in the deposition temperature. From the figure, the average grain size of the films also increases linearly from 5 to 18 nm with temperature from 70 to 90°C.

8.2.6 Influence of Ultrasound on Composition and Structure of EN Deposits

In Chapter 6, the effect of ultrasonic radiation on plating rate in binary and ternary deposits has been discussed more in detail. Because ultrasound affects the plating rate, it must also impact the composition and structure of the EP deposits. Early in 1978, Mallory [118] studied the compositions of Ni–P deposits with ultrasound. He also compared the composition for the EP Ni–P alloy deposits with and without ultrasound, as shown in Table 8.3. The table indicates when the coordinating atoms in the complexing agent are only oxygen, the phosphorus content is lower in the deposits plated with ultrasound. The lower phosphorus content of the ultrasonically produced deposits can be attributed to higher temperature at the solution metal interface due to absorption of the ultrasonic energy at the interface. Figure 8.42 indicates the XRD patterns of as-plated Ni–P deposits without ultrasound (A, sample 5n, $P = 17.4$ wt%), and with ultrasound (B, sample 5u, $P = 15.1$ wt%). The patterns show a low ordered structure with crystallite diameters measuring 1.7 and 1.9 nm, respectively. Both deposits show the strong {111} refection of electroless Ni–P deposits. However, the most important factor in distinguishing between the two deposits is the presence

TABLE 8.3 Comparison of Composition for the EP Ni–P Alloy Deposits With and Without Ultrasound [118]

	P content of deposits, at%	
Bath composition, chelating agent*	Ultrasonic irradiation (49 kHz, 30 W)	Normal or no ultrasound
1, Citric acid 0.2 mol/L	22.5	23.9
2, Lactic acid 0.3 mol/L	12.2	14.2
3, Glycolic acid 0.3 mol/L,	13.8	16.1
4, Glycine 0.3 mol/L	10.1	8.9
5, Citric acid 0.07 mol/L + Lactic acid 0.3 mol/L	15.1	17.4

Except complexing agent, all other bath composition and operation conditions are the same for the five cases: Ni^{2+} 0.1 mol/L, $NaH_2PO_2•H_2O$ 0.28 mol/L, Thiourea 0.6 ppm, pH (adjusted with KOH) 4–8, temperature 87°C.

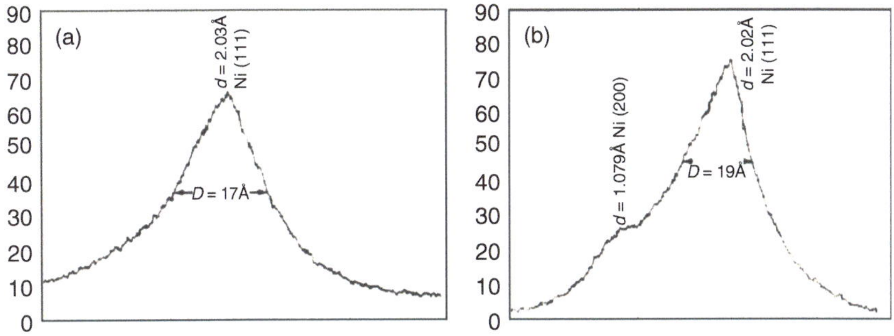

FIGURE 8.42 The XRD patterns of EP as-plated Ni–P deposits. (a) Sample 5n ($P = 17.4$ wt%) without ultrasound. (b) Sample 5u ($P = 15.1$ wt%) with ultrasound [118].

of the {200} reflection in the diffraction pattern of deposit 5u in the as-plated condition. The {200} refraction is absent in the diffraction pattern of deposit 5n.

Yang et al. [119] also found that the phosphorus content in the as-plated Ni–P films decreased from 10.49 to 8.39 at% in the presence of ultrasound (33 Hz, 1 M/cm^2). They believed that this decrease may be attributed to alteration of the pH near the activated surface. During deposition without ultrasound, the local concentration of H^+ would be high, which promotes the formation of phosphorus. However, if in the presence of ultrasound, the H^+ concentration near the activated surface of the substrate were reduced, the higher pH near the surface would result in a decrease of phosphorus content in the deposit formed. In 2012, Cobley and Saez [120] reported again the effect of ultrasound on the phosphorus content of deposits plated from an electroless nickel bath at different operating temperatures. The bath solution was operated at 50, 70, and 90°C, and the ultrasound was 20 kHz with a power density of 42 W/cm^2. As can be seen in Fig. 8.43, the P content decreases in the presence of ultrasound at all three temperatures. These authors believe that ultrasonic cavitations cause the phenomenon.

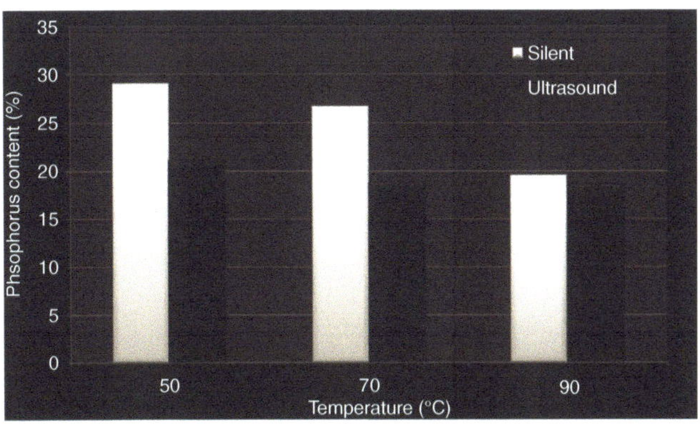

FIGURE 8.43 Effect of ultrasound on the phosphorus content of deposits plated from an electroless nickel bath at different operating temperatures [120].

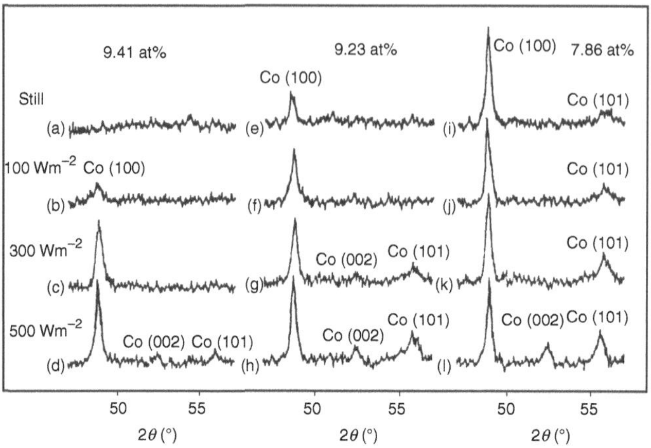

FIGURE 8.44 Effect of ultrasonic radiation on the microstructure of Co–P films deposited from different cobalt sulfate concentration solution [121].

In 1993, Matsuda and Takano [121] reported the effect of ultrasonic radiation on the phosphorus content of electroless deposited Co–P films. The ultrasound used was 28 kHz with ultrasonic intensities of 100, 300, and 500 W/m^2. The P content also decreased in the presence of ultrasound, as in the Ni–P deposits mentioned previously. For example, the phosphorus content of one deposit reduces from 9.41 to 6.96 at% with the intensity of 500 W/m^2. Figure 8.44 shows the XRD patterns of three Co–P samples with variation of composition and ultrasonic intensity. It can be seen that even a low intensity radiation of 100 W/m^2 produces a remarkable change from an amorphous structure to a crystalline structure. The peak height of the Co {100} plane rises and the half width of the peaks decreases as the intensity of ultrasonic radiation increases. With a higher intensity of 500 W/m^2 peaks in Co {002} and Co {101} are also observed.

Research into the effect of ultrasonic irradiation on structure and properties of electroless Co–Fe–B–Ce alloy coating was carried out in 2004 [122]. Xuan et al. found that the compositions of Co–Fe–B–Ce alloy deposits are Co 93.85, Fe 3.87, B 2.11, and Ce 0.17 at% without ultrasound, and 96.05, 2.71, 1.04, and 0.20 at% with ultrasound. The reasons are that the ultrasonic cavitations produce more H^+, OH^-, and H_2O_2, consuming some amounts of reductant agent KBH_4, thereby reducing the B content that can codeposit with Co and Ce. Moreover, the radius of Ce is much larger than those of Co and B, causing a large distortion energy, thus retarding the reduction and deposition of Fe atoms. That is why the Fe content decreases in the presence of ultrasound. As for structure, the results show that the as-plated Co–Fe–B–Ce deposits are microcrystalline structure without ultrasonic radiation, but becomes crystalline and has Co{100}, Co {002} and Co {101} diffraction peaks in the presence of

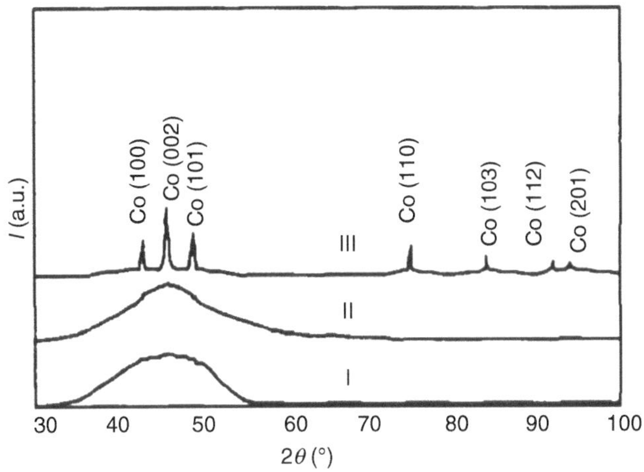

FIGURE 8.45 XRD patterns of electroless Co–Fe–B–Ce alloy coatings. I—Co-Fe-B (without ultrasound) II—Co-Fe-B-Ce (without ultrasound) β—Co-Fe-B-Ce (with ultrasound) [122].

ultrasonic radiation, as shown in Fig. 8.45. What are the reasons? Ultrasound promotes the increase of Co and Ce content and decreases the B content in the EP Co–Fe–B–Ce deposits, reducing their amorphous nature; as a result the deposit becomes crystalline in structure.

8.3 CRYSTALLIZATION OF EP AMORPHOUS ALLOYS

From the structures of EP alloys described in detail in the previous section, it can be seen that the as-deposited EP alloy coatings are usually either fully amorphous or mixtures of microcrystalline (and/or nanocrystalline) and amorphous phases. Only in a few cases, for example, very low phosphorus EN deposits are the structures of as-deposited alloys either highly crystalline or consist of microcrystalline (and/or nanocrystalline) phases. It is well known that the amorphous phase is a metastable state; it will change into stable phases when heated to a given temperature. This is the crystallization transformation of amorphous alloys. Just because of the EP alloy deposits basically containing some amorphous components (some data to be discussed in the next sections), chemical plating coatings thus are not stable, therefore almost all of them have a crystallization transformation problem. The study of crystallization of EP alloys not only reveals their nature more detail but also there is such a need in practical application. When using an amorphous EP alloy coating at high temperatures, it is necessary to know the temperature range within which the amorphous state will not change. If the coating transforms to the crystalline state, the excellent performance of amorphous phase would be reduced, or could even disappear. That is why much research has focused on the crystallization transformation of EP amorphous alloy deposits.

The crystallization of EP alloys has been investigated by many different methods: differential scanning calorimetry (DSC), X-ray and electron diffraction, small-angle X-ray scattering, transmission electron microscopy, magneto thermal methods, and by the change of alloy properties such as microhardness and magnetic properties during crystallization. Generally speaking, the crystallization process of EP coatings is not a single-step process, but takes place in a series of stages involving one or more intermediate crystalline phases.

The crystallization transformation can be subdivided into four aspects for discussion, which are crystallization process and products, crystallization kinetics and transformation volume, heating rate and crystallization temperature, and crystallization activation energy.

A point should be made first, that is the experimental data of transformation products, crystallization temperature, and crystallization activation energy obtained by different author groups usually are not consistent with each other, but are different. This situation is similar to the composition and structure of EP deposits mentioned and discussed above. In order to enable the readers to pay sufficient attention to this issue, the original comments from some author groups are cited here. In 1996, Kumar and Nair [22] said: "Nevertheless, there appears to be conflicting opinions regarding both structural aspects and crystallization process of these deposits." Sridhar and Udaya Bhat said in 2013 [45]: "Though the literature has references for formation of various Ni–P compounds, there is no unanimity on the type of the compound." In 2005, Gao et al. [55] said: "Different results were reported regarding the microstructures in the as-deposited condition and the stable phases after heat treatments."

8.3.1 Crystallization Process and Products of EP Alloys Deposits

Many author groups have investigated the crystallization process and products of EP alloys deposits [14,19,30,32,47,77,94,102,103,108,123–125]. These investigations concentrate on EN and EP Ni–P-based alloy coatings.

8.3.1.1 Crystallization Process and Products of EN Alloys Films

In 1988, Mai et al. [14] reported the structural changes induced by heating in EN alloy coatings at three phosphorus compositions: 1.88–5.54, 11.65–14.15, and 18.2–19 at%, which are typical values in the crystallization process of EN alloy coatings. Figure 8.46 shows the XRD patterns of these three alloy coatings annealed at 485, 385, and 485°C, respectively. The crystallization process and products of the three films can be summarized by the following equations.

Low phosphorus composition (1.88–5.54 at% P)

$$\begin{aligned}\text{Ni–P crystalline} &\xrightarrow{535^{\circ}\text{C},1\text{h}} \text{single-phase nickel, a texture}\\ &\xrightarrow{425^{\circ}\text{C},1\text{h}} \text{Ni}_3\text{P phase was detected}\end{aligned} \tag{8.9}$$

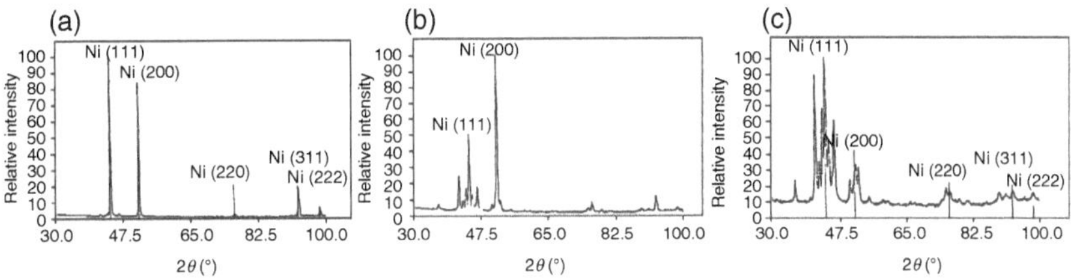

FIGURE 8.46 XRD patterns obtained from EN coatings: (a) coating with 1.88–5.54 at% P annealed at 485°C; (b) coating with 11.65–14.15 at% P annealed at 485°C; (c) coating with 18.2–19 at% P annealed at 385°C. The indicated peaks are those of the nickel phase. The additional peaks are those of b.c.t, nickel phosphide [122].

Medium phosphorus composition (11.65–14.15 at% P)

$$\begin{aligned}\text{Amorphous phase} &\xrightarrow{350^\circ\text{C},1\text{h}} \text{f.c.c. Ni} \\ &\xrightarrow{400^\circ\text{C},1\text{h}} \text{Ni}_3\text{P} \\ &\xrightarrow{800^\circ\text{C},1\text{h}} \text{Texture strength increases with temperature to } 800^\circ\text{C}\end{aligned} \tag{8.10}$$

High phosphorus composition (18.2–19 at% P)

$$\begin{aligned}\text{Amorphous phase} &\xrightarrow{385^\circ\text{C},1\text{h}} \text{Ni} + \text{Ni}_3\text{P} \\ &\xrightarrow{425^\circ\text{C},1\text{h}} \text{A duplex texture}\end{aligned} \tag{8.11}$$

Hur et al. [19] in 1990 studied the crystallization and products of EN (P from 11.3 to 23.0 at%) deposits in detail. The P content (at%) of each deposit is, A 11.3, B 17.5, C 19.8, D 21.3, and E 23.0. Heat treatment of the deposits was carried out under argon atmosphere in a tube furnace at 200–450°C for 2 h. Their results are as follows.

$$\text{Deposits A: Ni–P (solid solution)} \xrightarrow{1} \text{Ni} + \text{Am.}' \xrightarrow{1'} \text{Ni} + \text{Ni}_3\text{P} \tag{8.12}$$

$$\begin{aligned}&\text{Deposits B: Ni–P (Am.)} \xrightarrow{2} \text{Ni0} + \text{Am.}' \xrightarrow{2'} \\ &\text{Ni0} + (\text{Ni}' + \text{Ni}_3(\text{P},\text{Ni})) \xrightarrow{2''} \text{Ni0} + (\text{Ni} + \text{Ni}_3\text{P})\end{aligned} \tag{8.13}$$

$$\begin{aligned}&\text{Deposits C: Ni–P (Am.)} \xrightarrow{3} \text{Ni0} + \text{Am.}' \xrightarrow{3'} \\ &\text{Ni0} + (\text{Ni}' + \text{Ni}_3(\text{P},\text{Ni})) \xrightarrow{3''} \text{Ni0} + (\text{Ni} + \text{Ni}_3\text{P})\end{aligned} \tag{8.14}$$

$$\text{Deposits D: Ni–P (Am.)} \xrightarrow{4} \text{Ni}_3(\text{P},\text{Ni}) \xrightarrow{4'} \text{Ni} + \text{Ni}_3\text{P} \tag{8.15}$$

$$\text{Deposits E: Ni–P (Am.)} \xrightarrow{5} \text{Ni}_5(\text{P},\text{Ni})_2 \xrightarrow{5'} \text{Ni} + \text{Ni}_3\text{P} \tag{8.16}$$

where Ni0 is precrystallized nickel, Am′ phosphorus-enriched amorphous, Ni′ nickel rejected from Ni_3(P, Ni) due to its limited solubility into Ni_3P, Ni_3(P, Ni) and $Ni_5(P, Ni)_2$ are off-stoichiometric phases in which P and Ni atoms share phosphorus sites. It is obvious that the P content in the sample C is the same as the one in the above high P sample (both are around 19 at%), but the crystallization process and products in both cases are very different. These authors noted that the eutectic crystallization, direct crystallization of amorphous into a mixture of Ni and Ni_3P phase, did not take place. The eutectic composition of Ni–P system is at 19 P at% [126].

In 2002, Keong et al. [32] reported the crystallization kinetics and phase transformation behavior of EN deposits with high phosphorus content. Figure 8.47 and Table 8.4 show the XRD patterns and crystallization process and products of samples A and B in as-deposited condition and after heating to the end temperatures of 300–800°C at 20°C/min. Therefore, the sequence of phase transformation for these two samples is

Sample A: amorphous phase $\rightarrow$ coexistence of stable Ni_3P, and metastable Ni_2P, $Ni_{12}P_5$ and unidentified phases $\rightarrow$ stable Ni_3P phase. (8.17)

Sample B: amorphous phase $\rightarrow$ coexistence of stable Ni_3P, fcc nickel, and metastable Ni_2P, $Ni_{12}P_5$ and unidentified phases $\rightarrow$ stable Ni_3P and fcc nickel phases. (8.18)

The results for the transformation products are different from those obtained by Hur et al. [19] above.

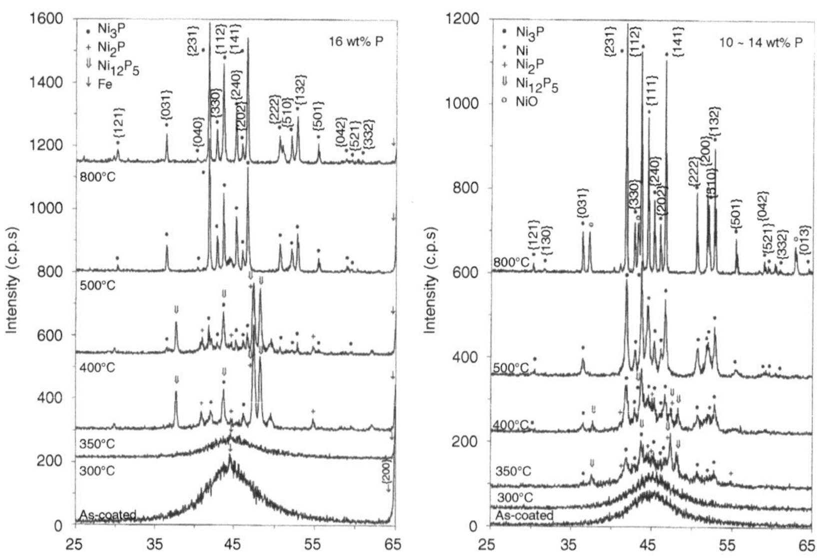

FIGURE 8.47 XRD profiles of the EN deposit with 26.5 at% P (sample A, left) and 17.4~23.6 at% P (sample B, right) in as-deposited condition and after heating to the end temperatures of 300–800°C at 20°C/min. For the sake of clarity the XRD profiles were shifted arbitrarily on the vertical scale [32].

TABLE 8.4 Comparison of Phase Compositions of Samples A and B After Heating to Different End Temperatures at a Constant Heating Rate of 20°C/min [32]

Temperature (°C)	Sample A, 26.5 at% P	Sample B, 17.4–23.6 at% P
As-deposited	Am.	Am.
300	Am.	Am.
350	Am.+Ni_2P+$Ni_{12}P_5$+Ni_3P	Am. +Ni_2P+$Ni_{12}P_5$+Ni+Ni_3P
400	Am.+ Ni_2P+$Ni_{12}P_5$+Ni_3P	Am. +Ni_2P+$Ni_{12}P_5$+Ni+Ni_3P
500	Am.+ Ni_3P	Am. +Ni+Ni_3P
800	Ni_3P	NiO+Ni+Ni_3P

Am.: amorphous phase

From Fig. 8.48, one can see clearly that the component of amorphous phase in these two samples A and B decreases gradually during heating process, and becomes zero at around 800°C. That is to say, the amorphous phase in the EP films is not abruptly but gradually transformed to a metastable or stable phases. In fact, this is a typical case, and the transformation of amorphous

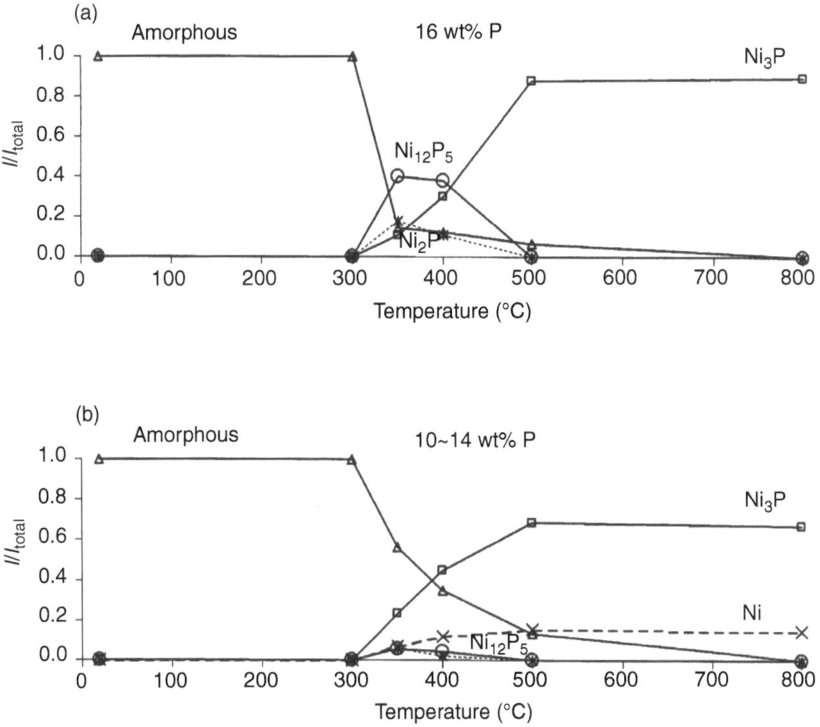

FIGURE 8.48 Variation of the integrated intensity ratio (I/I_{total}) over heating temperature in (a) sample A and (b) sample B [32].

phase in other EP deposits is also like this. For example, Martyak and Drake [27] in 2000 measured the contents of amorphous phases in the EN deposits of as-plated and annealed at various temperatures. The amorphous content (%) is 22.4, 19.6, 14.9, 9.7, and zero in the coatings of as-plated and annealed at 100, 200, 300, and 400°C.

The last point regarding the crystallization of EN deposits we would like to discuss is the work by Balaraju et al. [40] in 2007. They reported the phase compositions obtained from XRD data for EN alloy coating Ni-19.45 at% P in as-plated and heat treated at about peak transformation temperatures. The results are as follows.

$$\begin{aligned}\text{Amorphous phase} &\xrightarrow{\text{RT.}} \mathrm{Ni}\,(1.2\ \mathrm{nm}) + \mathrm{Ni_{12}P_5} \\ &\xrightarrow{355^\circ\mathrm{C,1h}} \mathrm{Ni}\,(42.9\ \mathrm{nm}) + \mathrm{Ni_3P} + \mathrm{Ni_5P_2}, \mathrm{Ni_7P_3} \\ &\xrightarrow{420^\circ\mathrm{C,1h}} \mathrm{Ni}(42.9\ \mathrm{nm}) + \mathrm{Ni_3P}\end{aligned} \quad (8.19)$$

One can see that the results are different with the above data.

From the discussion in the preceding sections, the following conclusions can be drawn. The first is that some metastable phases produce during the heating process. The second is that the emergence of the first crystallization phase depends on the temperature and the P content. The third is that the data on the starting crystallization phase and the metastable phases from different author groups usually are different, even though the P content and temperature are basically the same. The last one is that the stable phases are Ni and Ni_3P; moreover, they are basically the same from various groups.

8.3.1.2 Crystallization Process and Products of Other EP Binary Alloys Films

Early in 1973, Gorbunova et al. [57] reported the structure of the as-plated and annealed Ni–B plates. They studied the crystallization of three compositions of EP Ni–B deposits: Ni-19.6 at% B, Ni-24.7 at% B and Ni-27.1 at% B. The as-plated Ni–B alloys containing 16.9 at% B consist of two phases: a solid solution of boron and hydrogen in face-centered cubic β-nickel and an amorphous phase. The samples with 24.7 and 27.1 at% B are single-phase alloys of amorphous structure. The crystallization process and products of three samples are given by the following equations.

Ni-19.6 at% B

$$\text{Amorphous phase} + \text{f.c.c.Ni}\,(\mathrm{B,H}) \xrightarrow{150\sim700^\circ\mathrm{C,1h}} \mathrm{Ni} + \mathrm{Ni_3B} \quad (8.20)$$

Ni-24.7 at% B

$$\begin{aligned}\text{Amorphous phase} &\xrightarrow{150^\circ\mathrm{C,1h}} \text{Amorphous} \xrightarrow{200\sim250^\circ\mathrm{C,1h}} \mathrm{Ni} + \mathrm{Ni_3B} \\ &\xrightarrow{300\sim550^\circ\mathrm{C,1h}} \mathrm{N} + \mathrm{Ni_3B} + \mathrm{Ni_2B} \xrightarrow{600\sim700^\circ\mathrm{C,1h}} \mathrm{Ni_3B} + \mathrm{Ni_2B}\end{aligned} \quad (8.21)$$

Ni-27.1 at% B

$$\text{Amorphous phase} \xrightarrow{200^\circ\text{C},1\text{h}} \text{Amorphous} \xrightarrow{250\sim300^\circ\text{C},1\text{h}} \text{Ni} + \text{Ni}_3\text{B} \xrightarrow{350\sim400^\circ\text{C},1\text{h}} \text{Ni} + \text{Ni}_3\text{B} + \text{Ni}_2\text{B} \xrightarrow{450\sim700^\circ\text{C},1\text{h}} \text{Ni}_3\text{B} + \text{Ni}_2\text{B} \quad (8.22)$$

It can be seen that the amorphous phase will remain at 150 and 200°C for the Ni-24.7 at% B and Ni-27.1 at% B samples, respectively, which means that the higher B content in the Ni–B amorphous film, the higher crystallization temperature. This problem will be discussed more in detail below.

In 2004 Sankara Narayanan and Seshadri [63] investigated the crystallization process of EP Ni-27 at% B film. Their results are below.

$$\text{Amorphous phase (major)} + \text{nanocrystalline (small portion)} \xrightarrow{325^\circ\text{C},1\text{h}} \text{Ni} + \text{Ni}_3\text{B} \xrightarrow{450^\circ\text{C},1\text{h}} \text{Ni} + \text{Ni}_2\text{B} \xrightarrow{600^\circ\text{C},1\text{h}} \text{Ni} + \text{Ni}_2\text{B} \quad (8.23)$$

Comparing with the above results of Ni-27.1 at% B obtained by Gorbunova et al., it can be seen that both transformation products are different though both the compositions are the same.

Pal et al. [67] in 2011 also reported the crystallization of Ni-25.74 at% B deposit. The results are:

$$\text{Amorphous phase (major)} + \text{crystalline Ni (small amount)} \xrightarrow{300^\circ\text{C},1\text{h}} \text{Ni} + \text{Ni}_3\text{B} \xrightarrow{350^\circ\text{C},1\text{h}} \text{Ni} + \text{Ni}_3\text{B} \xrightarrow{450^\circ\text{C},1\text{h}} \text{Ni} + \text{Ni}_3\text{B} + \text{Ni}_2\text{B} \quad (8.24)$$

The results indicated that the XRD pattern of 350°C/1 h heat-treated sample shows a pattern similar to the 300°C/1 h sample. The only difference observed is in the intensity of the Ni peak which is much higher in the case of 350°C/1 h, while the residual amorphous hump still persists at this temperature though its integrated intensity has decreased. Moreover, heat treatment at 450°C for 1 h shows that the amorphous hump is smaller.

It is interesting to note that in 1998 Saito et al. [61] compared the crystallization of Co–B and Ni–B. Figure 8.49 compares the XRD patterns of electroless Co–B and Ni–B alloy films before and after heat treatment. The crystallization process and products are as given in the following equations.

Co-25.82 at% B

$$\text{Amorphous phase} \xrightarrow{500^\circ\text{C},1\text{h}} \text{hcp Co} + \text{Co}_2\text{B} + \text{Co}_3\text{B} + \text{Co}_4\text{B} \quad (8.25)$$

Ni-24.34 at% B

$$\text{Amorphous phase} \xrightarrow{300^\circ\text{C},1\text{h}} \text{Ni} + \text{Ni}_3\text{B} \xrightarrow{>300^\circ\text{C},1\text{h}} \text{no appreciable change in X-ray diffraction pattern.} \quad (8.26)$$

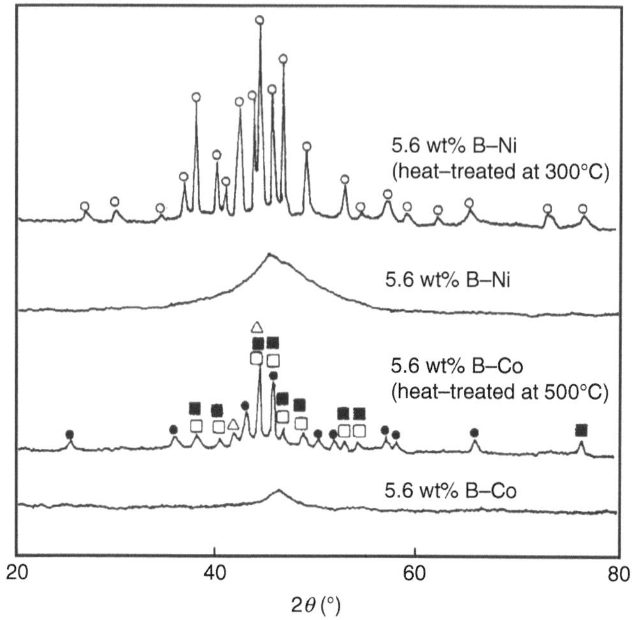

FIGURE 8.49 XRD patterns of electroless Ni-B and Co–B alloy films before and after heat treatment. Key: (○) Ni_3B; (●) Co_2B; (□) Co_3B; (■) Co_4B; (△) h.c.p. Co [61].

It is evident that the crystallization temperature of Co–B alloys is higher than that of Ni–B alloys for the same boron content. These authors believed that the precipitation of nickel boride at lower temperature compared to cobalt boride is consistent with the fact that the absolute value of the formation enthalpy (ΔH) of nickel borides is greater than that of cobalt borides.

8.3.1.3 Crystallization Process and Products of EP Ternary Alloys Coatings

Many papers on the crystallization process and products of EP ternary alloys coatings have been published [39,44,46,47,87,91,92,94,97,102,103,108, 127–130]. Among them, much researches of the crystallization process are in EP Ni–W–P alloy coatings papers [39,44,87,127,128]. Tien et al. [87] in 2004 reported the crystallization process and products on EP Ni-14.5 P + 3.5 W (at%) deposit, and comparing these from Ni-14.9 at% P deposit. Figure 8.50 illustrates the XRD patterns of a Ni-14.5 P-3.5 W (at%) deposit annealed at various temperatures for 4 h: (a) as-deposited; (b) 350; (c) 380; (d) 400; (e) 450; (f) 500; (g) 550°C. The crystallization process and products can be shown as follows.

$$\begin{aligned} &\text{Ni–W–P amorphous} \xrightarrow{350^\circ\text{C},4\text{h}} \text{No change (still amorphous structure)} \\ &\xrightarrow{380^\circ\text{C},4\text{h}} Ni_3P + Ni \text{ (a mixture of amorphous and the crystallization phase)} \\ &\xrightarrow{550^\circ\text{C},4\text{h}} Ni_3P + Ni \text{ (amorphous phase decreased)} \\ &\xrightarrow{500^\circ\text{C},550^\circ\text{C},4\text{h}} Ni_3P + Ni - W \text{ (both continue to grow sharper)} \end{aligned} \tag{8.27}$$

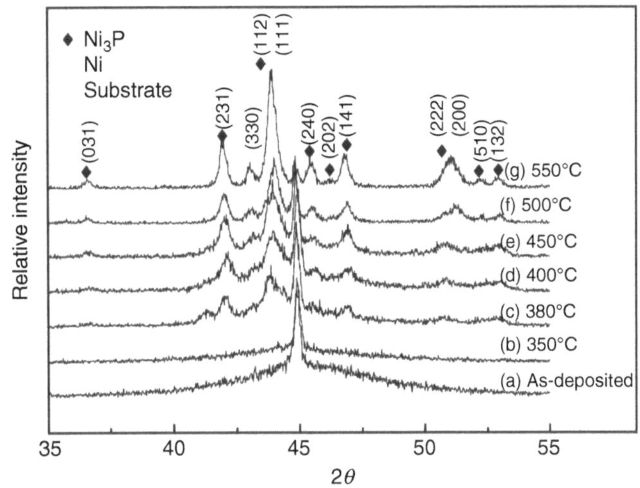

FIGURE 8.50 XRD spectra of Ni–14.5 P–3.5 W (at%) deposit annealed at various temperatures for 4 h. (a) As-deposited; (b) 350; (c) 380; (d) 400; (e) 450; (f) 500; (g) 550°C [87].

The authors found that tungsten peaks cannot be observed in the heat-treated Ni–P–W coating, indicating that tungsten had dissolved completely into the nickel matrix. In contrast, both amorphous and crystalline structures coexisted in the Ni–P–W alloy coating. The proportion of the amorphous structure decreased with increasing temperature. Moreover, they found that after annealing at 350°C for 4 h, the Ni–P coating gave rise to precipitation of a Ni_3P phase, and no metastable phases were observed. However, the Ni–W–P deposit shows no change, still amorphous structure, which means that W raises the crystallization temperature of Ni–W–P deposit, or W suppresses the nucleation of the new phases in Ni–P films. Antonelli et al. [39] also found evidently such phenomenon in their experiment of Ni–W–P deposit. But they found tetragonal Ni_3P peaks began to appear at 350°C for the Ni–W–P film with 18 atom % P and 2.7 atom % W.

In 2012, Balaraju et al. [44] reported phase transformation behavior of amorphous/nanocrystalline Ni–P alloys containing various W and P contents. Seven Ni–W–P alloy films were studied, which are:

NiWP-5	Ni-12.9 P + 0.94 W (wt%)
NiWP-10	Ni-10.5 P + 4.6 W (wt%)
NiWP-20	Ni-7 P + 8.3 W (wt%)
NiWP-30	Ni-5.4 P + 5.1 W (wt%)
NiWP-40	Ni-3.9 P + 3.9 W (wt%)
NiWP-60	Ni-2.5 P + 3.5 W (wt%)
NiWP-80	Ni-2.1 P + 3.1 W (wt%)

All the deposits were subjected to isothermal heat treatment for 1 h at the respective exothermic peak temperatures. Figure 8.51 shows the XRD patterns

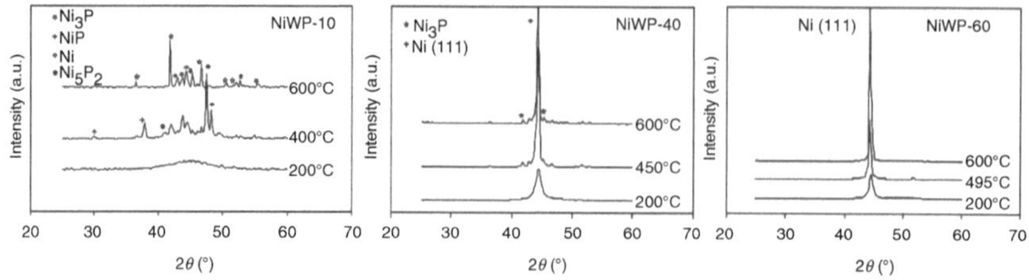

FIGURE 8.51 XRD patterns of NiWP-10 (left), NiWP-40 (middle), and NiWP-60 (right) deposits at various heat treatment temperatures [44].

of the three alloy deposits. The crystallization process, products, and relative explanation of these deposits can be shown as follows.

NiWP-5

$$\begin{aligned}\text{Ni–W–P amorphous} &\xrightarrow{200^\circ\text{C}, 1\,\text{h}} \text{No change}\\ &\xrightarrow{370^\circ\text{C}, 1\,\text{h}} \text{NiP} + \text{Ni}_5\text{P}_2 \text{ (metastable phases)}\\ &\xrightarrow{415^\circ\text{C}, 600^\circ\text{C}, 1\,\text{h}} \text{Ni}_3\text{P} + \text{Ni (both are stable crystalline phases)}\\ &\xrightarrow{500^\circ\text{C}, 550^\circ\text{C}} \text{Ni}_3\text{P} + \text{Ni-W (both continue to grow sharper)}\end{aligned} \tag{8.28}$$

NiWP-10

For the NiWP-10 deposit, both stable and metastable phases could be observed even after heat treatment at 400°C for 1 h and transformation to stable phase occurs only upon heat treatment at 600°C (left in Fig. 8.51).

NiWP-20

In NiWP-20 deposit metastable phases are present (with very low intensity peak profiles) for heat-treated deposits at temperatures, except those treated at 600°C. However, heat treatment at 600°C leads to the formation of only stable Ni {111} and Ni_3P peaks.

NiWP-30 and NiWP-40 (middle in Fig. 8.51)

Low-intensity Ni_3P peak along with very high-intensity Ni {111} peak is seen for NiWP-30 and NiWP-40 deposits heat treated at and above 415°C.

NiWP-60 (right in Fig. 8.51) and **NiWP-80**

In contrast, NiWP-60 and NiWP-80 deposits (having similar amounts of W: 4 wt% and P: 2.2 wt%) reveal only a single Ni (111) peak at all heat treatment temperatures.

The diffraction patterns of all EP ternary Ni–W–P alloy deposits, in both as-plated and heat-treated conditions, exhibit no characteristic peak of "W"; instead a shift in the Ni {111} peak is observed due to the codeposition of W in these deposits.

It is evident that these data are different from the above Ni–W–P deposit.

In addition, Koiwa et al. [129] in 1986 and [108] in 1987 reported the crystallization of Ni–Mo–P deposits, Zhang et al. [130] in 2003 published the crystallization for the amorphous state of Ni–Mo–P coating. The sample they used for heating treatment 1 h is ≥8 wt% P, 2–5 wt% Mo. The crystallization process and products are as fellows.

$$
\begin{aligned}
\text{Ni–Mo–P amorphous} &\xrightarrow{250\sim300^\circ\text{C},1\text{h}} \text{No change (Low temperature structural relaxation stage)} \\
&\xrightarrow{350\sim400^\circ\text{C},1\text{h}} Ni_5P_2, Ni_3P, Ni, MoNiP, Ni_3Mo, MoNi_4 \ \text{(metastable phases)} \\
&\xrightarrow{500^\circ\text{C},1\text{h}} Ni_5P_4 \ \text{appear (crystalline phases proliferation stage)} \\
&\xrightarrow{700^\circ\text{C},1\text{h}} \text{Completion of crystallization process (the peak increase over} \\
&\qquad \text{and the shape of cells disappeared)}
\end{aligned}
\tag{8.29}
$$

The data is different from those published by Koiwa et al. [108], because the results of these authors indicate that for the Ni-12.2 P-1.3 W wt% sample, a broad peak corresponding to the Ni {111} surface is observed after 330°C heat treatment, and for the Ni-5.3 P-6.4 W wt% sample, a broad peak corresponding to the Ni {111} surface was observed after 355°C heat treatment.

In 2013, Ahmad et al. [46] published their crystallization data for Ni–Co–P deposits. Figure 8.52 shows their results. The deposits were annealed at different temperatures: that is, 275, 300, 400, 500, 600, and 700°C, respectively. According to Fig. 8.52, the crystallization and products can be illustrated in the following schema.

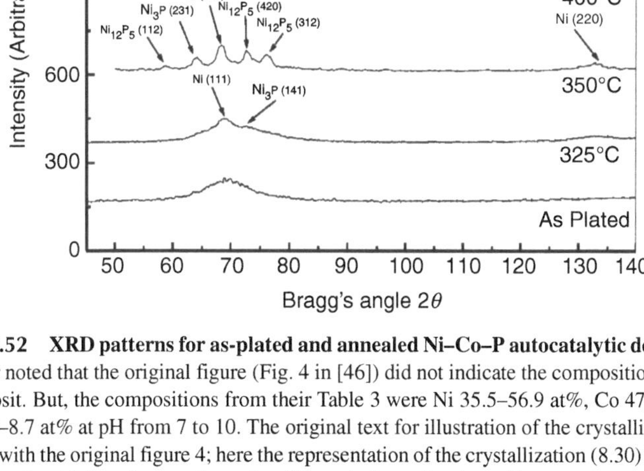

FIGURE 8.52 XRD patterns for as-plated and annealed Ni–Co–P autocatalytic deposits [46]. The author noted that the original figure (Fig. 4 in [46]) did not indicate the composition of the Ni–Co–P deposit. But, the compositions from their Table 3 were Ni 35.5–56.9 at%, Co 47.5–34.4 at% and P 17.0–8.7 at% at pH from 7 to 10. The original text for illustration of the crystallization is not consisting with the original figure 4; here the representation of the crystallization (8.30) is according to the figure. In addition, these authors also did not indicate the time of heat treatment.

$$\begin{aligned}\text{Ni–Co–P amorphous} &\xrightarrow{300^\circ\text{C}} \text{No change (structural relaxation)} \xrightarrow{325^\circ\text{C}} Ni_3P+Ni\\ &\xrightarrow{350^\circ\text{C}} Ni_{12}P_5+Ni_3P+Ni \xrightarrow{400^\circ\text{C}} Ni_{12}P_5+Ni_3P+Ni+Co\\ &\xrightarrow{500^\circ\text{C}} Ni_{12}P_5+Ni_5P_2+Ni_3P+Ni+Co+Co_2P\\ &\xrightarrow{700^\circ\text{C}} \text{The same as treatment at } 500^\circ\text{C}\end{aligned} \tag{8.30}$$

These data are different from those obtained by Narayanana et al. [94] in 2003. For example, Narayanana et al. found that the XRD pattern of Ni–31.0Co–9.0P (wt%) deposit annealed at 300°C for 1 h reveals the presence of hexagonal Ni_5P_2 phase, but the above Ahmad et al. data indicate that the Ni_5P_2 phase only appears after annealing at 500°C. Moreover, results of Narayanana et al. show that transformation of metastable to stable phase (Ni + Ni_3P) after annealing at >400°C for 1 h, but such a situation did not occur ever over 700°C in the experiments of Ahmad et al. [46].

In 1995, the author's group investigated the crystallization of EP amorphous Ni–Cu–B [103] and Ni–B–P [102] alloys. The sample $Ni_{72.9}Cu_{5.1}B_{22}$ (at%) was annealed at various temperatures for 1 h. The crystallization process as well as transition process of metastable crystalline phase of the amorphous alloy can be represented as follows:

$$\begin{aligned}\text{Ni–Cu–B amorphous} &\xrightarrow{275^\circ\text{C,1h}} Ni+Ni_3B+\text{amorphous}\\ &\xrightarrow{425^\circ\text{C,1h}} Ni+Ni_3B+Ni_2B \xrightarrow{525^\circ\text{C,1h}} Ni+Ni_3B\text{ (Stable phase)}\end{aligned} \tag{8.31}$$

Compared with the above Ni–B deposits, the starting crystallization temperature of Ni–Cu–B alloy films shows no increase. The as-plated amorphous alloy $Ni_{80.7}B_{15.2}P_{4.1}$ (at%) was annealed at 520°C for different times: (a) as-plated, (b) 20 S, (c) 2 min, (d) 10 min, (e) 30 min, and (f) 180 min. Figure 8.53 shows the XRD patterns for as-plated and annealed Ni–B–P amorphous deposits. The characteristic of crystallization is very different from those for Ni–P or Ni–B alloys, and can be described as follows:

$$\begin{aligned}&\text{Amorphous} \rightarrow Ni\,(fcc)+Ni_3P\,(\text{tetragonal})\\ &\text{Amorphous} \rightarrow Ni_3B\,(\text{orthogonal})\\ &\text{Amorphous} \rightarrow Ni_2P\,(\text{orthgogonal})\end{aligned} \tag{8.32}$$

From the above discussion on EP ternary alloy deposits, some remarks can be drawn as follows. The first is that some metastable phases are produced during the heating process. The second is that the metastable phases are related to the ternary alloy systems. The third is that appearance of the starting crystallization phase depends on the temperature and the P content. The fourth is that even for the same system, for example, Ni–W–P alloy deposits, the data on the starting crystallization phase and the metastable phases from different author groups are usually different. Therefore, considering the situation of the above-discussed Ni–P and other binary alloy coatings, further work is still needed to study these problems.

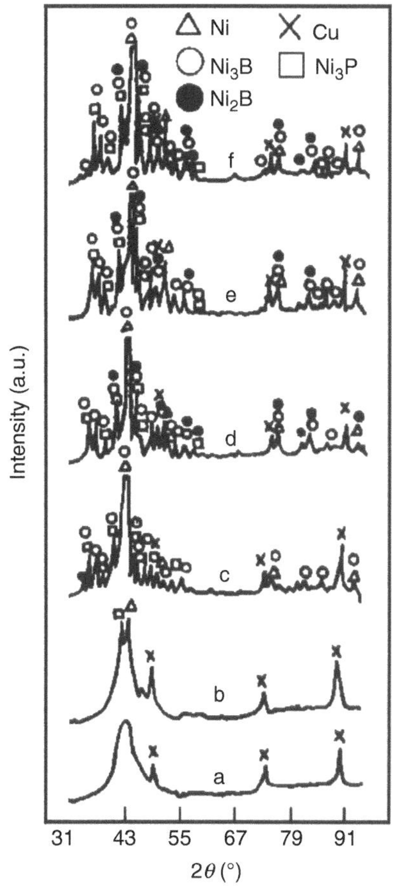

FIGURE 8.53 XRD patterns for as-plated and annealed Ni–B–P amorphous deposits. (a) as-plated, (b) 520°C for 20 S, (c) for 2 min, (d) for 10 min, (e) for 30 min, and (f) for 180 min [102].

8.3.1.4 Crystallization Process and Products of EP Quaternary Alloys Coatings

Only few papers on the crystallization process and products of EP quaternary alloys coatings have been published until now. Balaraju et al. [40] in 2007 reported the structure and phase transformation behavior of electroless Ni–P alloys containing tin and tungsten. It is interesting to note that these authors not only investigated the structure and phase transformation behavior of Ni–W–Sn–P, but also reported Ni–P, Ni–W–P, and Ni–Sn–P coatings as well. Thereby, one can make a comparison of the crystallization process and products of these four kinds of coatings easily and directly. The compositions are, Ni-11.3 P (wt%), Ni-10.3 P-0.4 Sn (wt%), Ni-5.1 P-7.8 W (wt%), and Ni–6.4P-Trace Sn-6.9 W (wt%). The coatings were annealed for 1 h at respective transformation temperatures. The phase compositions obtained from XRD data for electroless Ni–P alloy coatings in as-plated and heat-treated at about peak transformation temperatures are as given. Calculated grain sizes (nm) using the Debye–Scherrer method are given in parentheses.

Ni-11.3 P (wt%)

$$\begin{aligned} \text{Ni–P} &\xrightarrow{\text{RT}} \text{Ni}\,(1.2\text{ nm}) + \text{Ni}_{12}\text{P}_5 \\ &\xrightarrow{355°\text{C}, 1\text{ h}} \text{Ni}\,(42.9\text{ nm}) + \text{Ni}_3\text{P} + \text{Ni}_5\text{P}_2 + \text{Ni}_7\text{P}_3 \\ &\xrightarrow{420°\text{C}, 1\text{ h}} \text{Ni}\,(42.9) + \text{Ni}_3\text{P}\,(\text{stable phase}) \end{aligned} \tag{8.33}$$

Ni-10.3 P-0.4 Sn (wt%)

$$\begin{aligned} \text{Ni–Sn–P} &\xrightarrow{\text{RT}} \text{Ni}\,(1.2\text{ nm}) + \text{Ni}_{12}\text{P}_5 \xrightarrow{360°\text{C}, 1\text{ h}} \text{Ni}\,(29.3\text{ nm}) \\ &\quad + \text{NiP} + \text{Ni}_2\text{P} + \text{Ni}_5\text{P}_2 + \text{Ni}_5\text{P}_4 \\ &\xrightarrow{420°\text{C}, 1\text{ h}} \text{Ni}\,(29.3\text{ nm}) + \text{NiP} + \text{Ni}_2\text{P} + \text{Ni}_5\text{P}_2 + \text{Ni}_5\text{P}_4 \\ &\xrightarrow{550°\text{C}, 1\text{ h}} \text{Ni}\,(30.3\text{ nm}) + \text{Ni}_3\text{P} + \text{Ni}_5\text{P}_2 \xrightarrow{800°\text{C}, 1\text{ h}} \text{Ni} + \text{Ni}_3\text{P}\,(\text{stable phase}) \end{aligned} \tag{8.34}$$

Ni-5.1 P-7.8 W (wt%)

$$\begin{aligned} \text{Ni–W–P} &\xrightarrow{\text{RT}} \text{Ni}\,(1.2\text{ nm}) + \text{Ni}_{12}\text{P}_5 \xrightarrow{305°\text{C}, 1\text{ h}} \text{Ni}\,(8.3\text{ nm}) + \text{Ni}_{12}\text{P}_5 \\ &\xrightarrow{400°\text{C}, 1\text{ h}} \text{Ni}\,(14.0\text{ nm}) + \text{NiP}_2 + \text{Ni}_5\text{P}_2 + \text{Ni}_5\text{P}_4 + \text{Ni}_3\text{P} \\ &\xrightarrow{430°\text{C}, 1\text{ h}} \text{Ni}\,(17.9) + \text{Ni}_3\text{P} \xrightarrow{530°\text{C}, 1\text{ h}} \text{Ni}\,(27.2) + \text{Ni}_3\text{P}\,(\text{stable phase}) \end{aligned} \tag{8.35}$$

Ni-6.4 P-Trace Sn-6.9 W (wt%)

$$\begin{aligned} \text{Ni–W– Sn–P} &\xrightarrow{\text{RT}} \text{Ni}\,(1.9\text{ nm}) \xrightarrow{300°\text{C}, 1\text{ h}} \text{Ni}\,(8.2\text{ nm}) + \text{Ni–P} + \text{Ni}_5\text{P}_4 \\ &\xrightarrow{405°\text{C}, 1\text{ h}} \text{Ni}\,(15.0\text{ nm}) + \text{NiP}_2 + \text{Ni}_5\text{P}_4 + \text{Ni}_3\text{P} \\ &\xrightarrow{440°\text{C}, 1\text{ h}} \text{Ni}\,(18.9) + \text{Ni}_3\text{P} \xrightarrow{550°\text{C}, 1\text{ h}} \text{Ni}\,(29.2) + \text{Ni}_3\text{P}\,(\text{stable phase}) \end{aligned} \tag{8.36}$$

From the above crystallization process and products for the four alloy deposits, one can easily draw the following conclusions. (1) Adding one or more alloying elements (W, Sn) in the electroless Ni–P matrix affects the phase transformation behavior. (2) The starting phase transformation temperatures are 355–360°C for Ni–P and Ni–Sn–P, and 300–305°C for Ni–W–P and Ni–W–Sn–P, which is closely related to their P contents. The phase transformation temperatures increase with decreasing phosphorus content in deposits. This issue will be discussed further later. (3) Metastable phases NiP, Ni_2P, NiP_2, Ni_5P_2, Ni_5P_4, Ni_7P_3, and $Ni_{12}P_5$ are present in the annealed deposits. Among them, NiP_2 is very rare in related literature. (4) The temperatures of formation of Ni_3P are different, 355°C for Ni–P, 400–405°C for Ni–W–P and Ni–W–Sn–P, and 550°C for Ni–Sn–P coating. (5) Complete transformation of metastable phases to stable Ni and Ni_3P phases in 1 h annealed samples is also very different. It occurred

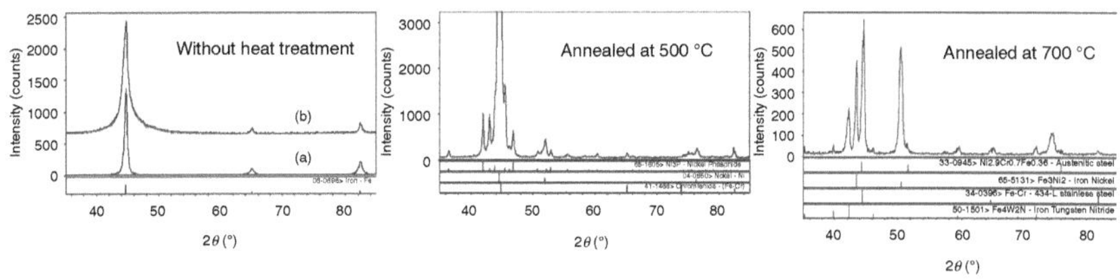

FIGURE 8.54 XRD patterns of the Ni–W–Cr–P alloy coatings. (Left plot) substrate (carbon steel) and (b) as-deposited Ni–W–Cr–P alloy coatings without heat treatment. (middle plot) Ni–W–Cr–P alloy coatings annealed at 400°C for 1 h. (Right plot) Ni–W–Cr–P alloy coatings annealed at 700°C for 1 h [131].

at 355°C for Ni–P, 400–405°C for Ni–W–P and Ni–W–Sn–P, and even up to 890°C for Ni–Sn–P coating. (6) These authors' experimental data show that codeposited Sn and W alloying elements in the Ni–P matrix had inhibited the Ni_3P formation around 350°C, which can be seen in the unalloyed Ni–P coating. It should be said that these authors did not give explanation and reasons for further understanding these phenomena. Welcome to further study these problems.

In 2008, Zhang et al. [131] reported the effects of annealing temperature on the crystal structure and properties of electroless deposited Ni–W–Cr–P alloy coatings. Figure 8.54 shows the XRD patterns of (a) substrate (carbon steel) and (b) as-deposited Ni–W–Cr–P alloy coatings without heat treatment (left plot). The middle plot in Fig. 8.54 shows the XRD pattern of the Ni–W–Cr–P alloy coatings annealed at 500°C for 1 h. The right-hand plot in the figure illustrates the XRD pattern of the Ni–W–Cr–P alloy coatings annealed at 700°C for 1 h. The composition of Ni–W–Cr–P alloy coatings is Ni (86.6–87.1), W (4.1–6.2), Cr (2.3–3.2), and P (4.6–5.6) (see Table 2 in [131]*). The crystallization process and products are as follows.

$$
\begin{aligned}
&\text{Ni–W–Cr–P amorphous} \xrightarrow{300^\circ\text{C},1\text{h}} \text{Ni} + \text{Ni}_3\text{P} + \text{amorphous} \\
&\xrightarrow{400^\circ\text{C},1\text{h}} \text{Ni} + \text{Ni}_3\text{P} + \text{Fe-Cr} \xrightarrow{500^\circ\text{C},1\text{h}} \text{Ni}_3\text{P} + \text{Ni}_{2.9}\text{Fe}_{0.36}\text{Cr}_{0.7} + \text{Fe-Cr} \\
&\xrightarrow{600^\circ\text{C},1\text{h}} \text{Ni}_3\text{P} + \text{Ni}_{17}\text{W}_3 + \text{Cr}_{17}\text{Ni}_3 + \text{Fe-Cr} + \text{Ni}_{2.9}\text{Fe}_{0.36}\text{Cr}_{0.7} \qquad (8.37) \\
&\xrightarrow{700^\circ\text{C},1\text{h}} \text{Fe}_3\text{Ni}_2 + \text{Fe-Cr} + \text{Fe}_4\text{W}_2\text{N} + \text{Ni}_{2.9}\text{Fe}_{0.36}\text{Cr}_{0.7} \\
&\xrightarrow{800^\circ\text{C},1\text{h}} \text{Ni}_{17}\text{W}_3 + \text{Ni-Cr-Fe}
\end{aligned}
$$

It is evident that comparing the above Ni–W–Sn–P, a lot of new metastable phases appear. Thinking of the process in (8.37), people feel suspicious about three questions, at the very least there is a need to study further in order to understand them more clearly. The first is that why does the $Ni_{17}W_3$ phase appear annealed at 600 and 800°C for 1 h, but not at 700°C for 1 h? The second is that why Ni_3P phase appears annealed at 300–600°C for 1 h, but not at 700 and 800°C for 1 h? Where is the phosphorus?

*The original authors indicate the composition in %, but not shown in wt% or at%.

8.3.2 Crystallization Temperature and Activation Energy of EP Alloy Deposits

At the beginning of the section, we analyzed the reasons for crystallization of EP alloy coatings. From the viewpoint of atomic movements, the crystallization process can be subdivided into different stages. The first is local atomic movements by thermal activation. Usually, this occurs at low temperature. The atoms cannot move long distance because of insufficient thermal activation force. Such local atomic movements could happen even at room temperature. At this point, structural relaxation occurs in the plain alloy materials [132]. The main feature in the term is the elimination of internal stress. However, reports on structural relaxation of EP alloy deposits are basically absent from the literature. After a certain annealing temperature, enough thermal activation force can promote the long-distance movement of atoms; then decomposition of original amorphous or nanostructure metastable phases starts and metastable or even stable phases form.

In the crystallization process, the process itself and products have been discussed at the start of this section. Two other important issues are crystallization temperature and activation energy. The temperature at which the crystallization begins is crystallization temperature. Usually, the crystallization temperature is not a single point, but a range of temperatures. Figure 8.55 shows the DSC curves of the crystallization transformation processes in amorphous Ni–P alloys schematically: (a) hypoeutectic alloys, (b) eutectic alloys, and (c) hypereutectic alloys [133,134]. The key temperatures T_{xp}, T_x, and T_{xr} depend on the heating rate and actually denote temperature intervals. Here T_x is the crystallization temperature, above which the sample is fully crystalline. T_{xp} is related to the precipitation of Ni crystallites (precrystallization temperature), involving a change of the amorphous matrix and an increase in its P content. At the recrystallization temperature, T_{xr}, an off-stoichiometric (Ni-rich) Ni_5P_2 phase decomposes into final equilibrium phases: crystalline Ni and Ni_3P. Complete crystallization, that is, the equilibrium state is reached only after annealing at elevated temperatures for several hours, as has been shown and discussed above more in detail. The inset plot in part (c) of Fig. 8.55 shows the real measured DSC curves of Ni-8 at% P deposit at the heating rates of 20,10, 5, and 2°C/min. If there are several crystallization temperature peaks, only the first peak temperature is adopted as crystallization temperature later.

The driving force of the atomic motion is thermal energy. The activation energy is used to describe the required energy of thermal activation leading to atomic movement. The activation energy can be defined as the minimum energy required by the reactants to undergo a reaction. Here, in the crystallization process, the activation energy is the atomic minimum energy required to surmount the interatomic potential barrier.

Early in 1889, Arrhenius introduced the term "activation energy" [135]. Based on the previous work of the Dutch physical chemist Jacobus van't Hoff

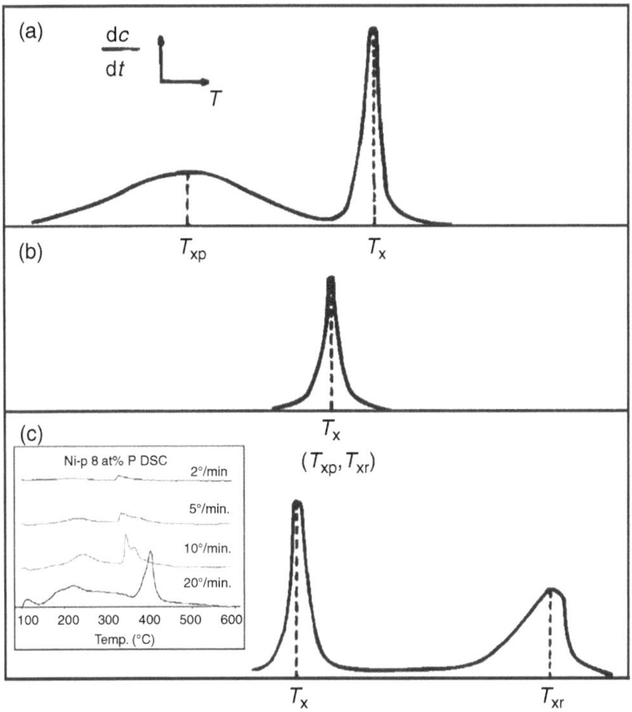

FIGURE 8.55 Differential scanning calorimetry (DSC) curves of the crystallization transformation processes in amorphous Ni–P alloys shown schematically. (a) Hypoeutectic alloys, (b) eutectic alloys, and (c) hypereutectic alloys [133,134].

in 1884, Arrhenius showed that the relationship between temperature and the rate constant for a reaction obeyed the following equation.

$$k = A\exp(-E_a/RT) \tag{8.38}$$

where k is the rate constant for the reaction, A is a proportionality constant that varies from one reaction to another (often called the frequency factor), E_a is the activation energy for the reaction, R is the ideal gas constant in joules per mole kelvin, and T is the temperature in kelvin. This is the famous Arrhenius equation. Furthermore, plotting the experimental results of ln$k$ against $1/T$ (the so-called Arrhenius plot) yields a straight line in many cases, and from the slope of this line one can easily calculate the activation energy E_a. It can be seen from Eq. (8.38) that the larger this ratio of E_a/RT, the smaller the rate. This means that high temperature and low activation energy mean larger rate constants, and thus faster reactions.

There are two practical methods to determine the activation energy. One of these is the Kissinger approach [136]. For the non-isothermal crystallization processes, the Kissinger relation is frequently used, and is expressed as

$$\ln(v/T^2) = -E_a/RT + \text{constant} \tag{8.39}$$

where v is the heating rate (K/min), T is a characteristic temperature, E_a is the activation energy, and R is the universal gas constant (8.314 J/K-mole). By plotting $\ln(v/T^2)$ versus $1/T$, one can derive the value of E_a from the slope of straight line plotted.

Another method was proposed by Duswalt [137]. He used the onset temperatures obtained at different heating rates to determine the activation energy E_a for the crystallization process of EP alloy deposits. The equation is

$$d(\ln v)/d(1/T) = -E_a/R \tag{8.39}$$

where R is the gas constant, v is the DSC heating rate, and T is the onset of crystallization temperature (in K).

These two methods are the ones usually used for determining the activation energy. Besides them, there are other isothermal methods for determining the activation energy, which are not used so often. Interested readers can refer to the relevant literature, for example, in Ref. [138].

It must be pointed out that the activation energy of the same EP alloy deposit determined by these two methods is usually not the same. In the following, we will list the specific data. The reason for such a difference is that the experimental thermograms obtained by DSC of major crystallization process are not always reproducible at the region between the crystallization onset and end temperatures [34].

Table 8.5 lists the reported crystallization temperatures and activation energies of EP alloy deposits.

8.3.2.1 Crystallization Temperature of EP Alloy Deposits

It should be said from Table 8.5 that many data for the crystallization temperature have been reported. Analyzing these data of crystallization temperature, some conclusions could be drawn as follows.

The first is that for the EP binary alloy coatings of Ni–P, Ni–B, and Fe–B, the phase transformation temperatures are closely related to the content of the metalloid elements P or B. Usually, the crystallization temperature increases with increasing the content of P or B. For example, No. 7 in Table 8.5, the crystallization temperature of Ni–P deposits increases from 200 to 300°C, 320 to 350°C, 355 to 335°C while increasing the P content from 17.5, 19.8, 21.3 to 23.0 at%, respectively. All the heating rates are 10°C/min. In no. 31 in Table 8.5, the crystallization temperature of Ni–B deposits increases from 302°C, 304°C to 305°C as increasing the B content from 19.6, 24.7 to 27.1 at%, respectively. All the heating rates are 10°C/min. Similar results are found for Fe–B deposits.

TABLE 8.5 Crystallization Temperatures and Activation Energies of EP Alloy Deposits

No.	Alloy system	Activation energy (kJ/mol)	Method	Avrami exponent n	Crystallization temperature °C (onset or first peak)	Literature (year)
1 2	Ni[a]	285.1 291				1 (1962) 2 (1976)
3 1	Co[a]	280.5 273.9				3 (1951) 1 (1862)
4 5	Fe[a]	281.4 240				4 (1960) 5 (1963)
6	Ni-19 P (at%)	227.6				6 (1985)
7	Ni-17.5 P (at%)				200–300 (at 10°C/min)	7 (1990)
	Ni-19.8 P (at%)				320–350 (at 10°C/min)	7
	Ni-21.3 P (at%)				355 (at 10°C/min)	7
	Ni-23.0 P (at%)				335 (at 10°C/min)	7
8	Ni-11.9 P (wt%)	280±12	Kissinger[b]	2.3	351 (at 40°C/min)	8 (1995)
	Ni-12.2 P+1.7 Cu (wt%)	224±7	Kissinger	1.6	371 (at 40°C/min)	8
	Ni-11 P+1.1 Sn (wt%)	408±38	Kissinger	2.2	388 (at 40°C/min)	8
	Ni-11.7 P+1.4 Sb (wt%)	560+36	Kissinger	2.5	396 (at 40°C/min)	8
9	Ni-2.8 P (at%)	180±3.5	Unspecified			9 (1998)
10	Ni-19.3 P (at%)	228	Kissinger			10 (2001)
11	Ni-10.79–15.78 P (at%)	239±5	Kissinger		330–360 (at 5–50°C/min)	11 (2001)
12	Ni-5.54–9.1 P (at%)	322 ± 54	Kissinger		386–408 (at 5–50°C/min)	12 (2002)
	Ni-9.1–14.2 P (at%)	230 ± 6	Kissinger		332–362 (at 5–50°C/min)	12

(Continued)

TABLE 8.5 Crystallization Temperatures and Activation Energies of EP Alloy Deposits (*cont.*)

No.	Alloy system	Activation energy (kJ/mol)	Method	Avrami exponent n	Crystallization temperature °C (onset or first peak)	Literature (year)
13	Ni-26.53 P (at%)	210±17	Kissinger		325–354 (at 5–50°C/min)	13 (2002)
	Ni-17.4–23.55 P (at%)	225±20	Kissinger		346–366 (at 5–50°C/min)	13
14	Ni-26.5 P (at%)	210 ± 17	Kissinger	1.51±0.01		14 (2003)
	Ni 20.5 P (at%)	225 ± 20	Kissinger	1.57±0.08		14
	Ni-15.8 P (at%)	239 ±5	Kissinger	2.6±0.1		14
15	Ni-20 P (at%)	220 (seed Cu)	Kissinger		~350 (at 20°C/min)	15 (2005)
	Ni-12 P (at%)	210 (seed Cu)	Kissinger		350 (at 20°C/min)	15
	Ni-8 P (at%)	240 (seed Cu)	Kissinger		390 (at 20°C/min)	15
	Ni-6 P (at%)	280 (seed Cu)	Kissinger		400 (at 20°C/min)	15
16	Ni-11 P (wt%)	274	Kissinger	3.09	349 (at 20°C/min)	16 (2005)
	Ni-10 P+2.3 Al_2O_3 (wt%)	163	Kissinger	2.68	340 (at 20°C/min)	16
17	Ni-11 P (wt%)	234.39	Duswalt[c]		318.4–339.2 (at 3–15°C/min)	17 (2006)
	Ni-11 P+8.10 Si_3N_4 (wt%)	237.75	Duswalt		319.7–340.2 (at 3–15°C/min)	17
18	Ni-12 P (at%)	210	Kissinger		350 (at 20°C/min)	18 (2006)
	Ni-20 P (at%)	220	Kissinger		350 (at 20°C/min)	18
	Ni-14 P+3.6 W (at%)	206	Kissinger		420 (at 20°C/min)	18
	Ni-18 P+2.7 W (at%)	270	Kissinger		400 (at 20°C/min)	18

19	Ni-11.3 P (wt%)	222	Kissinger		360 (at 40°C/min)	19 (2007)
	Ni-10.3 P+0.4 Sn (wt%)	258	Kissinger		365 (at 40°C/min)	19
	Ni-5.1 P+7.8 W (wt%)	82	Kissinger		320 (at 40°C/min)	19
	Ni-6.4 P+Trace Sn+6.9 W (wt%)	192	Kissinger		307 (at 40°C/min)	19
20	Ni-12 P (wt%)	174	Duswalt		332 (at 20°C/min)	20 (2007)
	Ni-8.5 P+2.5 W (wt%)	198	Duswalt		336 (at 20°C/min)	20
	Ni-5.2 P+12–14 W (wt%)	243	Duswalt		338 (at 20°C/min)	20
21	Ni-3.7 P (wt%)	280	Kissinger		402–412 (at 10–40°C/min)	21 (2008)
	Ni-3.4 P+Si_3N_4 (wt%)	261	Kissinger		391–417 (at 10–40°C/min)	21
22	Ni-13.5 P (wt%)	200±3	Kissinger			22 (2012)
	Ni-12.9 P+0.94 W (wt%)	272	Kissinger			22
	Ni-10.5 P+4.6 W (wt%)	295	Kissinger			22
	Ni-7 P+8.3 W (wt%)	222	Kissinger			22
	Ni-5.4 P+5.1 W (wt%)	290	Kissinger			22
	Ni-3.9 P+3.9 W (wt%)	355	Kissinger			22
	Ni-2.5 P+3.5 W (wt%)	327	Kissinger			22
	Ni-2.1 P+3.1 W (wt%)	356	Kissinger			22

(Continued)

TABLE 8.5 Crystallization Temperatures and Activation Energies of EP Alloy Deposits (*cont.*)

No.	Alloy system	Activation energy (kJ/mol)	Method	Avrami exponent n	Crystallization temperature °C (onset or first peak)	Literature (year)
23	Ni-4.9 P (wt%)	269.2	Kissinger		395	23 (1987)
	Ni-11.9 P (wt%)	196.78	Kissinger		315	23
	Ni-12.2 P+1.3 Mo (wt%)	239.9	Kissinger		355	23
	Ni-5.3 P+6.4 Mo (wt%)	269.63	Kissinger		377	23
24	Ni-11.6 P+21.66 Cu (wt%)	229.1	Kissinger		374 (at 20°C/min)	24 (1999)
25	Ni-20 P+23 Co (at%) Ni-15 P+42 Co (at%) Ni-11 P+53 Co (at%) Ni-14 P+62 Co (at%) Ni-11 P+74 Co (at%)	42.901 46.865 48.685 65.994 82.291	Kissinger Kissinger Kissinger Kissinger Kissinger			25 (2009) 25 25 25 25
26	Co–P	60.2	Arrhenius plot[d]			26 (2008)
27	Co–P	48.1	Arrhenius plot			27 (2010)
28	Co–P	46.8	Arrhenius plot			28 (2012)
29	Co–B	33	Arrhenius plot			29 (2008)
30	Co–B	94.5 LT(15–25°C) 10.9 HT(30–45°C)	Arrhenius plot			30 (2014)
31	Ni-19.6 B (at%)				302 (at 10°C/min)	31 (1973)
	Ni-24.7 B (at%)				304 (at 10°C/min)	31
	Ni-27.1 B (at%)				305 (at 10°C/min)	31

32	Ni-22 B (at%)	93	Kissinger		300 (at 10°C/min)	32 (1995)
	Ni-22 B+3.8 Cu (at%)	144	Kissinger		313 (at 10°C/min)	32
	Ni-22 B+5.1 Cu (at%)	114	Kissinger		313 (at 10°C/min)	32
	Ni-22 B+7.0 Cu (at%)	109	Kissinger		315 (at 10°C/min)	32
	Ni-22 B+7.3 Cu (at%)	99	Kissinger		315 (at 10°C/min)	32
	Ni-22 B+8.0 Cu (at%)	98	Kissinger		310 (at 10°C/min)	32
33	Ni-15.2 P+4.1 B (at%)				260 (at 20°C/min)	33 (1995)
34	Ni-27.1 B (at%)				306 (at 20°C/min)	34 (2004)
35	Fe-18 B (at%)	184.5	Kissinger		447–495 (at 10–80°C/min)	35 (1993)
	Fe-24 B (at%)	146.7	Kissinger		440–500 (at 10–80°C/min)	35
	Fe-27.4 B (at%)	129.1	Kissinger		436–503 (at 10–80°C/min)	35
36	Fe-18.9 Mo-6.4 B (at%)				472 (at 10°C/min)	36 (1997)
	Fe-16.3 Mo-10.1 B (at%)				475 (at 10°C/min)	36
	Fe-14.0 Mo-14.9 B (at%)				495 (at 10°C/min)	36
	Fe-13.3 Mo-23.3 B (at%)				498 (at 10°C/min)	36
	Fe-11.3 Mo-30.0 B (at%)				497 (at 10°C/min)	36
37	Fe-17.4 Sn-12.1 B (at%)	175	Kissinger		446 (at 12.5°C/min)	37 (1999)
	Fe-26.9 Sn-27.0 B (at%)	300	Kissinger		456 (at 12.5°C/min)	37

1 Hirano KI, Agarwala R, Averbach B, Cohen M. J Appl Phys, 1962;33(10):3049.
2 Meier K, Mehrer H, Lessmann E, Schole W, Phys Stat Sol (b), 1976 ;78:689.
3 Nix FC, Jabot, Jr FE. Phys Rev 1951;82(1):72.
4 Borg RJ, Birchcnal1 CE. Trans AIME, 1960;980:218.

(Continued)

TABLE 8.5 Crystallization Temperatures and Activation Energies of EP Alloy Deposits (*cont.*)

5 Grahama D, Tomlina DH. Philos Mag 19638(93):1581.
6 Mahoney MW, Dynes PJ. Scripta Met 1985;19:539.
7 Hur KH, Jeong JH, Lee DN. J Mater Sci 1990;25:2573.
8 Krasteva N, Armyanov S, Georgieva J, Avramova N, Fotty V. J Electron Mater, 1995;24:941.
9 Zhang YZ, Wu YY, Yao M. J Mater Sci Lett 1998;17:37.
10 Révész Á, Lendvai J, Lóránth J, Pádár J, Bakonyi I. J Electrochem Soc, 2001;148(11):C715.
11 Keong KG, Sha W, Malinov S. Acta Metall Sin (Eng Lett) 2001;14:419.
12 Keong KG, Sha W, Malinov S. J Mater Sci 2002;37:4445.
13 Keong KG, Sha W, Malinov S. J Alloys Compd 2002;334:192.
14 Keong KG, Sha W, Malinov S. J Non-Cryst Solids 2003;324:230.
15 Antonelli SB, Allen TL, Johnson DC, Dubin VM. J Electrochem Soc 2005;152(9):J120.
16 Jiaqiang G, Yating W, Lei L, Wenbin H. Mater Lett 2005;59(2/3):391.
17 Balaraju JN, Sankara TSN, Narayanan, Seshadri SK. Mater Res Bull 2006;41:847.
18 Antonelli SB, Allen TL, Johnson DC, Dubin VM. J Electrochem Soc 2006;153(6):J46.
19 Balaraju JN, Millath Jahan S, Jain A, Rajam KS. J Alloys Comp 2007;436:319.
20 Palaniapp M, Seshadri SK. Mater Sci Eng A 2007;460–461:638.
21 Balaraju JN, Rajam KS. J Alloys Comp 2008;459:311.
22 Balaraju JN, Kalavati, Manikandanath NT, William Grips VK. Surf Coat Technol 2012;206:2682.
23 Koiwa I, Yamada K, Usuda M, Osaka T. J Metal Finish Soc Jpn J Surf Finish Soc Jpn 1987;38(10):488.
24 Chen C-J, Lin K-L. J Electrochem Soc 1999;146(1):137.
25 Liu WL, Hsieh SH, Chen WJ, Hsu YC. Appl Surf Sci 2009;255:3880.
26 Eom KS, Cho KW, Kwon HS. J Power Sources 2008;180(1):484.
27 Zhang X, Zhao J, Cheng F, Liang J, Tao Z, Chen J. Intern J Hydro Ener 2010;35(15):8363.
28 Oh TH, Kwon S. Intern J Hydro Ener 2012;37(21):15925.
29 Dai H, Liang Y, Wang P, Cheng H. J Power Sources 2008;177:17.
30 Muir SS, Chen Z, Wood BJ, Wang L, (Max) Lu GQ, Yao X. Intern J Hydro Ener 2014;39:414.
31 Gorbunova KM, Ivanov MV, Moiseev VP. J Electrochem Soc 1973;120:613.
32 Hu W, Wu L, Wang L, Zhang B, Guan H. Physica B 1995;212:195.
33 Zhang H, Zhang B, Tan Z, Lei Q. Mater Mach Eng 1995;12(1):19.
34 Sankara Narayanan TS N, Seshadri SK. J Alloys Comp 2004;365:197.
35 Hu W, Zhang B. Trans Inst Metal Finish 1993;71(1):30.
36 Wang L, Zhang B, Yi K, Ouyang Y, Hu W. J Alloys Comp 1997;255(12).
37 Wang L, Zhao L, Zhang B, Hu W, Shu X, Sheng X, Fang Z. J Alloys Comp 1999;287:234.

[a]*Self-diffusion activation energy.*
[b]*Kissinger HE. Anal Chem 1957;29:1702.*
[c]*Duswalt AA. Thermochem Acta 1974;8:57.*
[d]*Arrhenius plot [135].*

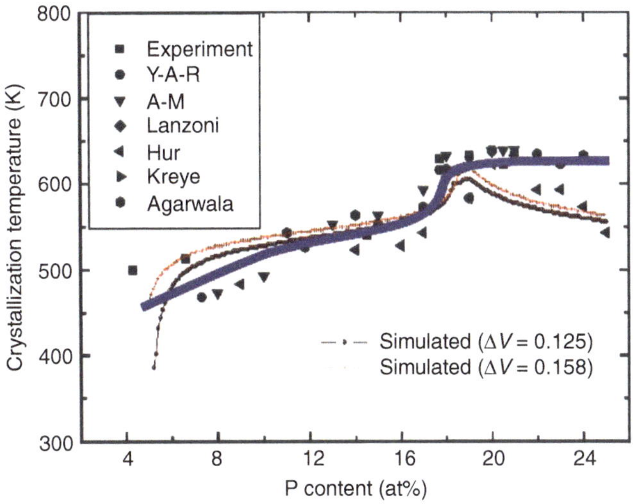

FIGURE 8.56 Crystallization temperatures of Ni–P alloys with various P contents [55]. The papers cited by these authors are given in [55] to which interested readers can refer. They will not be repeated here.

From No. 35 in Table 8.5, it can be seen that the crystallization temperature of Fe–B deposits increases from 447 to 495°C, 440–500°C to 436–503°C as increasing the B content from 18, 24 to 27.4 at%, respectively. The heating rates all are 10–80°C/min.

It is interesting to note that Gao et al. [55] in 2005 collected the data of crystallization temperatures obtained at various heating rates by other investigators, coupled with their own data, and drew a figure of crystallization temperature versus P content, as shown in Fig. 8.56. They simulated crystallization temperature, and compared them with the experimental data in the figure. The two curves with peak point are the Gao et al.'s simulated results. The "experiment" in the figure is their experimental data. The blue curve in the figure was added by the author, because it may look like that according to the shown experimental points. Because the eutectic composition of Ni–P system is at 19 P at% [126], therefore the data in Fig. 8.56 shows evidently that for the hypereutectic alloys, most of the experimental data for crystallization temperatures are not decreased, but nearly maintain constant, so the data do not support the occurrence of peak point at the eutectic alloy. It seems that the appearance of the peak point comes from the simulated model by Gao et al. [55]. Moreover, these authors have noted, "results by other investigators were obtained at various heating rates, but the simulation results were at 0.3 K/s. Amorphous Ni–P alloy turns to crystal with different excess volumes for various P contents. So the simulated results should be corrected in future works." However, they believed, "the simulated result is acceptable to some extent."

However, there are opposite results, for example, the data in no. 12. When the heating rates are 5–50°C/min, the crystallization temperature of Ni–P

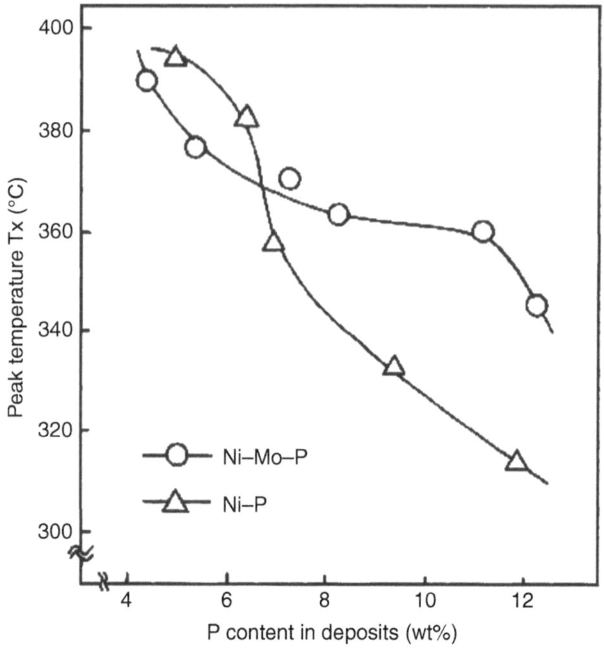

FIGURE 8.57 **Effect of phosphorus content on peak temperature of sharp exothermic peaks for the Ni–P and Ni–Mo–P alloy films measured at a heating rate of 10°C/min [108].**

deposits decreases from 386–408°C to 332–362°C as the P content increases from 5.54–9.1 at% to 9.1–14.2 at%, respectively. Koiwa et al. [108] also found that the crystallization temperature in Ni–P and Ni–Mo–P alloy deposits decreases with increased P content, as shown in Fig. 8.57. As yet there is no reasonable explanation of such conflicting experimental data.

The second is that the crystallization temperature of Ni–P and Fe-B alloy deposits increases with the heating rate. Nos. 11~13, 17, and 21 in Table 8.5 of Ni–P alloy all show this relationship.

For example, the crystallization temperature of Ni–P deposits in No. 11 of Table 8.5 raises from 330, 340, 344, 349, 354, 356 to 360°C with enhancing the heating rate from 5, 10, 15, 20, 30, 40 to 50°C/min, respectively. Again, the crystallization temperature of Fe-18 at% B alloy coating in No. 35 of Table 8.5 increases from 447, 462, 478 to 495°C with increased heating rate from 10, 20, 40 to 80°C/min, respectively. For other two deposits of Fe-24 B (at%) Fe-27.4 B (at%) (No. 35 in Table 8.5), their crystallization temperature is also increased with increased heating rate. Why does this happen? All of the groups that reported such phenomenon either in Ni–P or in Fe–B alloy deposits do not give any explanation of this phenomenon. This problem needs to be studied further.

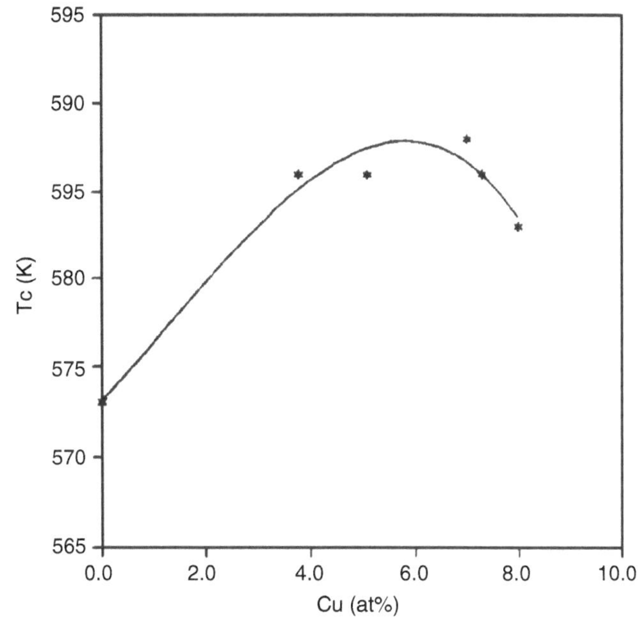

FIGURE 8.58 **The dependence of the first peak crystallization temperature on the copper content [103].**

The third is the effects of additive element on the crystallization temperature. From the data for No. 32 in Table 8.5, one can see that addition of Cu into Ni–B deposit, the crystallization temperature enhances as increase of Cu, as shown in Fig. 8.58. Antonelli et al. in 2006 (No.18 in Table 8.5) found that the added W of electroless Ni–P films does significantly increase crystallization temperatures in Ni–P films, which is 500°C (at 20°C/min) for Ni-18 at% P-2.7 at% W compared 350°C (at 20°C/min) for Ni-20 at% P. This shows that tungsten has stabilized the amorphous state. This occurs because the W inhibits the nucleation of the phase in Ni–P films, because W has limited ability to suppress face-centered cubic Ni nucleation due to being soluble in Ni. Krasteva et al in 1995 (No. 8 in Table 8.4) investigated the effects of Cu, Sn, or Sb in Ni–P deposits. Their results show that the addition of Cu, Sn, or Sb to Ni–P amorphous alloys increases the heat of crystallization. The indications are that the inclusion of Cu, Sn, or Sb in EN alloy increases its thermal stability.

8.3.2.2 Crystallization Activation Energy of EP Alloy Deposits

The measured crystallization activation energy of EP alloy deposits is also listed in Table 8.5. In order to compare, the activation energies of self-diffusion for the pure metals Ni, Co, and Fe are also given in the table. The following conclusions can be drawn from the data.

Firstly, the crystallization activation energies in an EP amorphous alloy with near similar compositions measured by various author(s) are usually different. The reason is the measurement techniques are not always the same, and the samples of different author(s) are not really the same.

Secondly, the activation energies in EP Ni–P, Ni–B, Co–P, Co–B, and Fe–B alloys are all less than the self-diffusion energies in pure metals when the metalloids P and B are added to the Ni, Co, and Fe, respectively. The radii of the metalloids P and B are largely less than those of metals Ni, Co, and Fe. According to the theory of diffusion, smaller sized atoms in the EP amorphous alloys make atomic diffusion easier, and therefore reduce the activation energies.

Thirdly, the sample with lower phosphorus or boron content has higher crystallization activation energy. The data of Nos. 12, 13, and 15 in Table 8.5 for Ni–P coatings and No. 35 in the table for Fe–B deposits all confirm this point. However, there is an opposite situation, that is, No. 18 in the table for Ni–P deposits. The activation energy is 220 kJ/mol for Ni-20 at% P, but 210 kJ/mol for Ni-12 at% P. The sample with higher phosphorus has higher activation energy, which contradicts the above facts. It should be said that no explanation of these conflicting data has as yet been given.

Fourthly, most of the additive metal elements increase the activation energy in the Ni–P films with similar P content, that is, improving the amorphous stabilization. Nos. 18 and 22 in Table 8.5 for Ni–W–P, No. 23 for Ni–Mo–P, No. 8 for Ni–Sn–P and Ni–Sb–P, and No. 19 for Ni–Sn–P all confirm that this is true. These experimental data should not be difficult to understand. The radii of these atoms are all much larger than that of Ni and P. Addition of these metals into Ni–P coating enhances the cohesive strength of atomic interaction, and more force is needed to move these atoms over long distances, thereby increasing the activation energy.

Fifthly, the activation energies in the EP Ni–Co–P system (no. 25 in Table 8.5) increase from 42.901 to 82.291 kJ/mol as the Co content enhances from 23 to 74 at%. In addition, the values of the activation energies are much less than those in the systems mentioned in the fourth point such as in Ni–W–P, Ni–Mo–P, Ni–Sn–P, and Ni–Sb–P alloys. The original authors Liu et al. (Ref. 25 in Table 8.5) did not give explanations for these facts, but they said that, "In order to understand the results in this work basing on the first principle a further study is needed in the future."

The final point is that the activation energy in Ni–Cu–B alloy deposits versus the Cu content is rather special, as shown in Fig. 8.59. The activation energy increases at first and then decreases as the Cu content increases. The author's group did not give explanation to this phenomenon in the original paper [103]. Further work should be done to understand the data clearly.

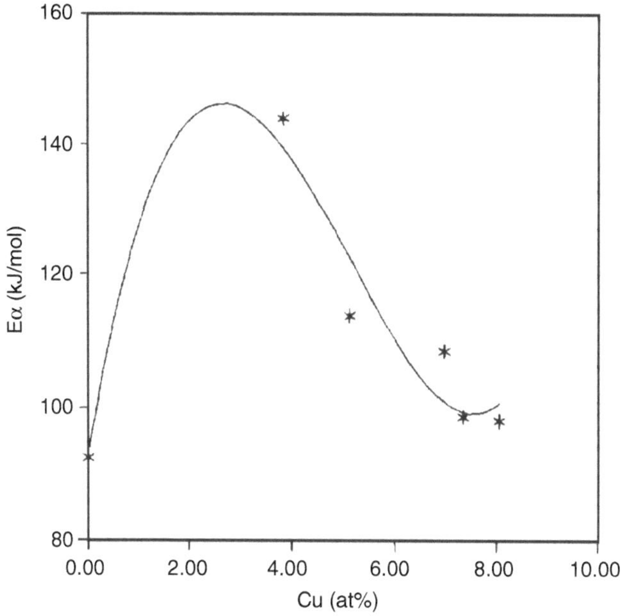

FIGURE 8.59 **The dependence of the activation energy for the first peak crystallization temperature on the copper content [103].**

8.3.3 Crystallization Transformation Kinetics of EP Alloy Deposits

Crystallization kinetics studies the dynamic process of a solid phase transformation, that is, the time (t) dependence of the transformed volume fraction (X) or the degree of transformation. Early in the 1930s, Johnson, Mehl, Avrami, and Kolmogorov (JMAK) [139–141] proposed the so-called classical solid phase transformation theory, which is used very often in describing the kinetics of nucleation-and-growth-type crystallization reactions. However, due to the constraints of the classical JMAK model and thus its limited applicability [142], the much more general modular kinetic model, which comprises unconstrained combinations of nucleation, growth, and impingement modes, was developed and applied also for kinetic analysis of crystallization reactions [143–147]. According to the JMAK theory, the transformed volume fraction, X, at constant temperature is given by

$$X(t) = 1 - \exp(-kt^n) \tag{8.41}$$

where $k = k(t)$ is a temperature-dependent rate constant and n is the growth exponent. n is usually called the Avrami exponent, because it was first proposed by Avrami. According to Avrami, exponent n is closely related to the growth shape of new phase. When $3 \le n \le 4$, the new phase is 3D growth; $2 \le n \le 3$, is layered growth; and $1 \leq n \leq 2$, is linear growth. By means of the JMAK equation for isothermal transformation

$$\ln \ln[1/(1-X)] = \ln K + n \ln(t-\tau) \tag{8.42}$$

where τ is the incubation time, n can be obtained by plotting $\ln \ln[1/(1 - X)]$ versus $\ln(t - \tau)$. There are two types of difficulty in using this traditionally isothermal method to determine the Avrami exponent. On one hand, its value depends strictly on the incubation time, which cannot be given exactly. On the other hand, isothermal heating is more difficult to perform than nonisothermal anneal by differential scanning calorimetry (DSC) or differential temperature analysis (DTA). Therefore, it is necessary to develop a method to determine the n value under nonisothermal conditions. We have done it. The author's group has proposed a method of multiscanning rate for the measurement of Avrami exponent for an amorphous alloy [148]. The equation is

$$n = -\frac{\ln[-\ln(1-X_1) - \ln \ln(1-X_2)}{\ln v_1 - \ln v_2} \tag{8.43}$$

So long as we choose several different heating rates, v_1, v_2, ⋯ and measure the crystallization fractions transformed, X_1, X_2, ⋯ at the same temperature, the Avrami exponent can be calculated using the equation.

Only a very small number of papers report applying the crystallization kinetics theory to analyze the crystallization kinetics of EP alloy coatings. Keong et al. [34] in 2003 measured $X(t)$ at different annealed temperatures using DSC experimental data for the EN deposit, and then compared them to the results from the modified JMAK theoretical equation (the original equation is based on the isothermal but not nonisothermal transformation conditions). Figure 8.60 shows their measured and calculated from the JMAK equation results of the crystallization fractions transformed heating at 5–50°C/min in EN deposits. The solid lines are the experimental data, and the dashed lines are the theoretical results from the JMAK equation. The P contents in sample A and B are 26.53 at% and 15.78 at% P, respectively. It can be seen that the overall fittings of X plots for Sample A (Fig. 8.60a) are satisfactory, though not all of them are well fitted. On the other hand, the fittings between the simulated and experimental X plots for Sample B are not satisfactory as shown in Fig. 8.60b. The XRD pattern for the as-plated sample A indicates that A is fully amorphous, but B is partially amorphous. These authors therefore believed that those unsatisfactory results are likely caused by the partially amorphous nature of the as-deposited condition for sample B (see Fig. 3 in [34]).

As for the n, their measured values are 1.51 and 2.6 for samples A and B, respectively. According to the above-mentioned classical phase transformation theory, the major crystallization processes of samples A are considered to be linear growth with decreasing nucleation rate, and the similar process of sample B is regarded as layered growth with decreasing nucleation rate.

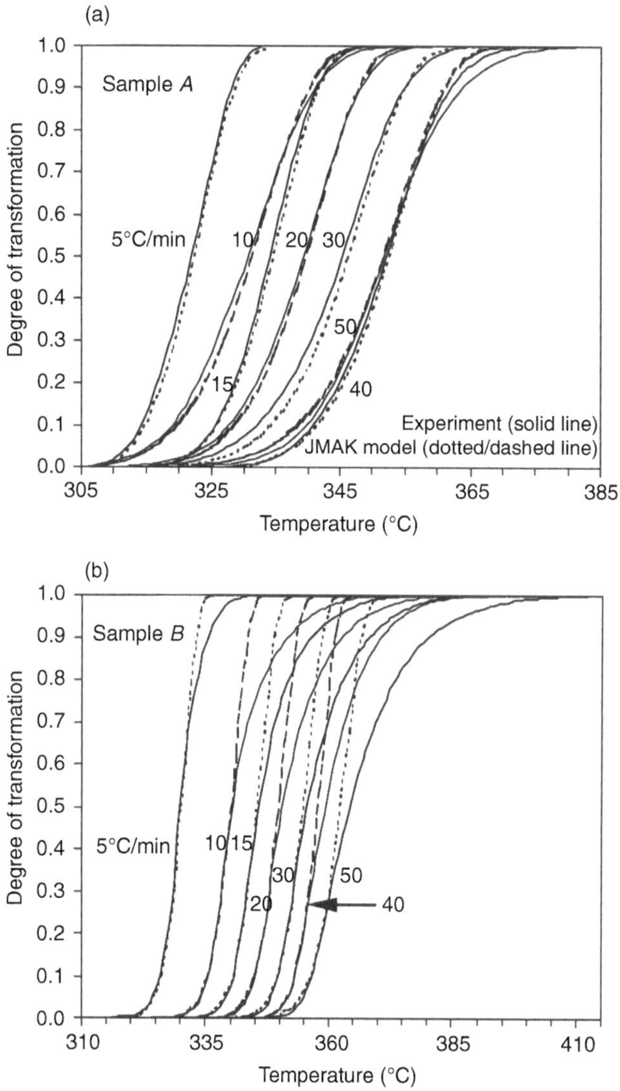

FIGURE 8.60 Degree of transformation, X, for samples A, and B (heating at 5~50°C/min) calculated from the JMAK model and experimental DSC data [34].

From Table 8.5, it can be seen that the reported n values in EP systems are very few, so it is difficult to discuss the effect of varying parameters further, for example, the effect of P or B content, or the impact of adding metal elements, etc. There are very few papers on crystallization kinetics and measurements of the $X(t)$ and n in the EP alloy deposits, and more work in this area would be very informative.

8.4 SUMMARY

Now, we can draw some overall conclusions from the discussion so far.

As of 2015, the composition and structure of nearly all kinds of EP alloy coatings including binary, ternary, and quaternary alloy films have been investigated. Studies of the compositions and structures of EP alloy systems, especially the study of the effects of various parameters such as metal salts, reductant, complexing agent, pH, and temperature on the compositions and structures of EP alloy systems, mean that we now have a good understanding of this subject (i.e., compositions and structures of EP alloy deposits).

However, less research has been undertaken on the compositions and structures of EP Co-based alloy systems, especially Fe-based alloy systems. More work in this field would be welcome.

Many investigations have been carried out on the crystallization transformation of EP amorphous and nano materials, giving us a pretty good understanding of this subject. Today, a lot of data on the crystallization process and products, crystallization temperature and activation energy of EP amorphous and nano materials have been published. This situation could not have been imagined in the half century after the discovery of EN in 1947 by Brenner, as most of these data were published in the last 10 years.

The research on the crystallization kinetics of EP alloy materials, even in EN alloy films, was very few. Studies on the crystallization process of EP Fe–B based alloy systems are also rare. Further research would help to strengthen this field also.

Many experimental facts have not yet been explained. This is particularly the case for the experiments that have conflicting phenomena, as mentioned in many times in this chapter. The phenomenon that the theory is behind experiment meets once again. The explanation of the composition and structure and crystallization behavior of the amorphous and nano deposits needs to be strengthened.

REFERENCES

[1] Gutzeit G, Mapp ET. Anti-Corr Meth Mater 1956;3(10):331.
[2] Goldenstein AW, Rostoker W, Sehossberger F, Gutzeit G. J Electrochem Soc 1957;104:104.
[3] Graham AH, Lindsay RW, Read HJ. J Electrochem Soc 1962;109:1200.
[4] Graham AH, Lindsay RW, Read HJ. J Electrochem Soc 1965;112:401.
[5] Albert PA, Kovac Z, Lilienthal HR, McGuire TR, Nakamura YJ. Appl Phys 1967;38:1258.
[6] Randin JP, Maire PA, Saurer E, Hintermann HE. J Electrochem Soc 1967;114(12):442.
[7] Marton JP, Schlesinger M. J Electrochem Soc 1968;115(1):16.
[8] Cargill GS III. Appl Phys 1970;41(1):12.
[9] Balder NC, North DM, Wagner CNJ. Phys Rev 1969;177(1):47.
[10] Chow SL, Hedgecock NE, Schlesinger M, Rezek J. J Electrochem Soc 1972;119:1614.
[11] Sugita K, Ueno N. J Electrochem Soc 1984;131(1):111.
[12] Matsuoka M, Hayashi T. Metal Finish 1986;84(3).

[13] Tyagi SVS, Tandon VK, Ray S. Bull Mater Sci 1986;8(3):433.
[14] Mai QX, Daniels RD, Harpalani HB. Thin Solid Films 1988;166:235.
[15] Park SH, Lee DN. J Mater Sci 1988;23(5):1643.
[16] Agarwala RC, Ray S. Z Metall 1988;79:472.
[17] Lin K-L, Lai P-J. J Electrochem Soc 1989;136(12):3803.
[18] Lambert MR, Duquette DJ. Thin Solid Films 1989;177:207.
[19] Hur K-H, Jeong J-H, Lee DN. J Mater Sci 1990;25:2573.
[20] Rajam KS, Rajagopal I, Rajagopalan SR, Visvalathan B. Mater Chem Phys 1993;33:289.
[21] Martyak NM. Chem Mater 1994;6:1667.
[22] Kumar PS, Nair PK. J Mater Process Technol 1996;56:511.
[23] Parkinson R. Properties and Applications of Electroless Nickel Deposits, Nickel Development Institute, Technical Series No. 10081 1997.
[24] Zhang B, Xie H. Rare Metal Mater Eng 1998;27(2):79.
[25] Staia MH, Puchi ES, Castro G, Ramirez FO, Lewis DB. Thin Solid Films 1999; 355–356:472.
[26] Małecki A, Micek-Ilnicka A. Surf Coat Technol 2000;123:72.
[27] Martyak NM, Drake K. J Alloys Comp 2000;312:30.
[28] Apachitei I, Duszczyk J. Surf Coat Technol 2000;132:89.
[29] Révész Á, Lendvai J, Lóránth J, Pádár J, Bakonyi I. J Electrochem Soc 2001;148(11):C715.
[30] Keong KG, Sha W, Malinov S. Acta Metall Sin (Eng Lett) 2001;14:419.
[31] Keong KG, Sha W, Malinov S. J Mater Sci 2002;37:4445.
[32] Keong KG, Sha W, Malinov S. J Alloys Compd 2002;334:192.
[33] Song JY, Yu Jin. Thin Solid Films 2002;415:167.
[34] Keong KG, Sha W, Malinov S. J Non-Cryst Solids 2003;324:230.
[35] Xie H, Zhang B. J Mater Proc Technol 2002;124(1/2):8.
[36] Guo Z, Keong KG, Sha W. J Alloys Comp 2003;358:112.
[37] Gao Y, Zheng ZJ, Zhu M, Luo CP. Mater Sci Eng A 2004;381:98.
[38] Antonelli SB, Allen TL, Johnson DC, Dubin VM. J Electrochem Soc 2005;152(9):J120.
[39] Antonelli SB, Allen TL, Johnson DC, Dubin VM. J Electrochem Soc 2006;153(6):J46.
[40] Balaraju JN, Millath Jahan S, Jain A, Rajam KS. J Alloys Comp 2007;436:319.
[41] He YD, Fu HF, Li XG, Gao W. Scr Mater 2008;58(6):504.
[42] Srinivasan KN, John S. J Alloys Comp 2009;486:447.
[43] Qin W. J Coat Technol Res 2011;8(1):135.
[44] Balaraju JN, Kalavati Manikandanath NT, William Grips VK. Surf Coat Technol 2012;206:2682.
[45] Sridhar N, Udaya Bhat K. J Mater 2013;V 2013. Article ID 985763, 7 pages.
[46] Ahmad A, Ghauri KM, Ali L, Noshahi AA, Farr JPG, Pak. J. Eng Appl Sci 2013;12:20.
[47] Yu H-S, Luo S-F, Wang Y-R. Surf Coat Technol 2001;148:143.
[48] Brenner A, Riddell G. J Res Natl Bur Stand 1946;37:31.
[49] Hersch P. Trans Inst Met Finish 1955-1956;33:417.
[50] Salvago G, Cavallotti P. Plating 1972;59:665.
[51] Brenner A, Riddell GE. J Res Natl Bur Stand 1947;39:385.
[52] Zhang YZ, Wu YY, Yao M. J Mater Sci Lett 1998;17:37.
[53] Lu Guojin, Giovanni Z. Electrochim Acta 2002;47:2969.
[54] Farber B, Cadel E, Menand A, Schmitz G, Kirchheim R. Acta Mater 2000;48:789.
[55] Gao J, Wu Y, Liu L, Shen B, Hu W. Mater Lett 2005;59:1665.
[56] Mallory GO. Plating 1971;58:319.
[57] Gorbunova KM, Ivanov MV, Moissev VP. J Electrochem Soc 1973;120:613.

[58] Watanabe T, Tanabe Y. J Japan Inst Metals 1975;39(8):831.
[59] Suda H, Watanabe T, Misaki Y, Tanabe Y. J Japan Inst Metals 1981;45(2):117.
[60] Rajagopal I. Bull Mater Sci 1983;(3 & 4):323.
[61] Saito T, Sato E, Matsuoka M, Iwakura C. J Appl Electrochem 1998;28(5):559.
[62] Delaunois F, Lienard P. Surf Coat Technol 2002;160:239.
[63] Sankara Narayanan TSN, Seshadri SK. J Alloys Comp 2004;365:197.
[64] Shi Z, Wang D, Ding Z. Appl Surf Sci 2004;221:62.
[65] Krishnaveni K, Sankara Narayanan TSN, Seshadri SK. Surf Coat Technol 2005;190:115.
[66] Rao Q, Bi G, Lu Q, Wang H, Fan X. Appl Surf Sci 2005;240:28.
[67] Pal S, Verma N, Jayaram V, Biswas SK, Riddle Y. Mater Sci Eng A 2011;528:8269.
[68] Vitry V, Kanta A-F, Dille J, Delaunois F. Surf Coat Technol 2012;206:3444.
[69] Wang ZC, Jia F, Yu L, Qi ZB, Tang Y, Song G-L. Surf Coat Technol 2012;206:3676.
[70] Takano O, Matsuda N. J Metal Finish Soc Jpn 1980;31(1):28.
[71] Zhang B, Chen B, Yi G, Wang L. Electropla Pollu Cont 1997;17(6):20.
[72] Tarozaitė R, Sukackienė Z, Sudavičius A, Juškėnas R, Selskis A, Jagminienė A, Norkus E. Mater Chem Phys 2009;117:117.
[73] Pearstein F, Weightman F. J Electronchem Soc 1974;121(4):1023.
[74] Chang YH, Lin CC, Hung MP, Chin TS. J Electrochem Soc 1986;133(5):985.
[75] Jiang D, Zhang C, Liao S, Zhang B. Electroplat Finish 2007;26(2):5.
[76] Leóna OA, Staia MH, Hintermann HE. Surf Coat Technol 1998;108–109:461.
[77] Gao J, Wu Y, Liu L, Hu W. Mater Lett 2005;59(2/3):391.
[78] Balaraju JN, Sankara Narayanan TSN, Seshadri SK. Mater Res Bull 2006;41:847.
[79] Wu Y, Liu L, Shen B, Hu W. J Mater Sci 2005;40(18):5057.
[80] Novakovic J, Vassiliou P, Samara K, Argyropoulos Th. Surf Coat Technol 2006;201(3/4):895.
[81] Allahkaram SR, Faezi Alivand R, Bakhsh MS, Iranian J. Mater Sci Eng 2013;10(1):10.
[82] Lee C-K. Int J Electrochem Sci 2012;7:8487.
[83] Faraji S, Rahim AA, Mohamed N, Sipaut CS. Int J Miner Metall Mater 2011;18(5):615.
[84] Pearlstein F, Weightman RF. Electroless deposition of nickel and cobalt based alloys, AD837160 August 9, 1968.
[85] Aoki K, Takano O. P& SF 1990;77(3):48.
[86] Zhang B, Hu W, Qu X, Zhang Q, Zhang H, Tan Z. Trans IMF 1996;74(2):69.
[87] Tien S-K, Duh J-G, Chen Y-I. Surf Coat Technol 2004;177–178:532.
[88] Palaniapp M, Seshadri SK. Mater Sci Eng A 2007;460–461:638.
[89] Balaraju JN, Kalavati, Rajam KS. J Alloys Compd 2009;486:468.
[90] Cherkaoui M, Srhiri A, Chassaing E. P&SF 1992;79(11):68.
[91] Krasteva N, Armyanov S, Georgieva J, Avramova N, Fotty V. J Electron Mater 1995;24:941.
[92] Chen C-J, Lin K-L. J Electrochem Soc 1999;146(1):137.
[93] Ashassi-Sorkhabi H, Dolati H, Parvini-Ahmadi N, Manzoori J. Appl Surf Sci 2002;185:155.
[94] Narayanana TSNS, Selvakumarb S, Stephen A. Surf Coat Technol 2003;172:298.
[95] Pang J, Li Q, Wang W, Xu X, Zhai J. Surf Coat Technol 2011;205:4237.
[96] Abdel Aal A, Barakat H, Abdel Hamid Z. Surf Coat Technol 2008;202:4591.
[97] Sankara Narayanan TSN, Stephan A, Guruskanthan S. Surf Coat Technol 2004;179:56.
[98] Drovosekov AB, Ivanov MV, Krutskikh VM, Lubnin EN, Polukarov YuM. Protect Metals 2005;4:55.
[99] Kim TH, Dulal SMSI, Chang HP, Chae H, Kim CK. Surf Coat Technol 2008;202:4861.
[100] Shacham-Diamand Y, Sverdlov Y, Petrov N. J Electrochem Soc 2001;148(3):C162.
[101] Zhang B, Ye G, Chen B. Surf Treat (Chinese Taiwan) 1998;69:31.

[102] Zhang H, Zhang B, Tan Z, Lei Q. Mater Mech Eng 1995;19(1):19.
[103] Hu W, Wu L, Wang L, Zhang B, Guan H. Physica B 1995;212:195.
[104] Wang L, Zhao L, Huang G, Yuan X, Zhang B, Zhang J. Surf Coat Technol 2000;126:272.
[105] Xie H, Zhang B, Yang Q. Trans IMF 1999;77(3):99.
[106] Xie H, Zhang B, Xu X, Liao S. J Changsha Commun Uni 1998;14(3):24.
[107] Zhang B, Xie H, Xu X. Metal Finish 1999;97(10):35.
[108] Koiwa I, Yamada K, Usuda M, Osaka T. J Metal Finish Soc Jpn 1987;38(10):488.
[109] Liu D, Yang Z, Zhang C. Mater Sci Eng B 2010;166:67.
[110] Osaka T, Arai K, Masubuchi N. J Metal Finish Soc Jpn 1988;39(120):778.
[111] Kim DH, Matsuda H, Aoki K, Takano O. Jpn J Surf Finish 1994;45:202.
[112] Yokoshima T, Kaneko D, Akahori M, Nam H-S, Osaka T. J Electroanal Chem 2000;491:197.
[113] Azizi A, Mohammadi M, Sadrnezhaad SK. International conference on materials heat treatment (ICMH 2010), Islamic Azad University, Majlesi Branch, May, 11–14, 2010. Isfahan, Iran.
[114] Duhin A, Sverdlov Y, Feldman Y, Shacham-Diamand Y. Electrochim Acta 2009;54:6036.
[115] Dula SMSI, Ho KT, Rhee H, Yong SJ, Kim C-K. J Alloys Comp 2009;467:370.
[116] Wu WC, Hsieh T-E, Pan H-C. J Electrochem Soc 2008;155(5):D369.
[117] Petrov N, Sverdlov Y, Shacham-Diamand Y. J Electrochem Soc 2002;149:C187.
[118] Mallory GO. Trans Inst Met Finish 1978;56:81.
[119] Yang L-X, Hou W-T, Wu Y-S. Trans IMF 1997;75:131.
[120] Cobley AJ, Saez V. Circ World 2012;38(1):12.
[121] Matsuda H, Takano O. Trans IMF 1993;71(4):138.
[122] Xuan T, Zhang L, Huang Q. Nonferrous Metals 2004;56(4):27.
[123] Hu W, Zhang B. Trans IMF 1993;71(1):30.
[124] Wang L, Zhang B, Yi K, Ouyang Y, Hu W. J Alloys Comp 1997;255(12):231.
[125] Wang L, Zhao L, Zhang B, Hu W, Shu X, Sheng X, Fang Z. J Alloys Comp 1999;287:234.
[126] In: Baker H, Okamoto H, editors. Binary alloy phase diagrams, in Alloy phase diagrams, vol.3. Materials Park. OH: ASM International; 1992. p. P2*313.
[127] Tsai Y-Y, Wu F-B, Chen Y-I, Peng P-J, Duh J-G, Tsai S-Y. Surf Coat Technol 2001;146–147:502.
[128] Palaniappa M, Seshadri SK. Mater Sci Eng A 2007;460–461:638.
[129] Koiwa I, Nishikawa M, Yamada K, Osaka T. Bull Chem Soc Jap 1986;59:133.
[130] Zhang Y, Fang Y-K, Duan J-G. Heat Treat Metals 2003;28(3):28.
[131] Zhang L, Jin Y, Peng B, Zhang Y, Wang X, Yang Q, Yu J. Appl Surf Sci 2008;255:1686.
[132] Martin JW, Doherty RD, Cantor B. Stability of microstructure in metallic systems (Cambridge solid state science series), 2nd ed. Cambridge University Press, 1997.
[133] Bakonyi I, Cziraki A, Nagy I, Hosso M. Z Metallkd 1986;77:425.
[134] Krasteva N, Fotty V, Armyanov S. J Electrochem Soc 1994;141(10):2864.
[135] Arrhenius S. Recherches sur la conductivité galvanique des électrolytes, doctoral dissertation, Stockholm, Royal Publishing House, PA. Norstedt & Söner, 1884, p 89; Arrhenius S. Z Phys Chem 1889;4:226.
[136] Kissinger HE. Anal Chem 1957;29:1702.
[137] Duswalt AA. Thermochim Acta 1974;8:57.
[138] Hu W, Wang L, Wu L, Zhang B, Guan H. Physica B 1994;203:147.
[139] Kolmogorov AN. Izv Akad Nauk SSSR M 1937;3:355.
[140] Johnson WA, Mehl. RF. Trans AIME 1939;135:416.
[141] Avrami MJ. Chem Phys 1939;7(12):1103. 9 (1941) 177.
[142] Mittemeijer EJ. Fundamentals of materials science. Heidelberg: Springer Verlag; 2010.

[143] Mittemeijer EJ. J Mater Sci 1992;27(15):3977.
[144] Liu F, Sommer F, Mittemeijer EJ. J Mater Sci 2004;39(5):1621.
[145] Liu F, Sommer F, Bos C, Mittemeijer EJ. Int Mater Rev 2007;52(4):193.
[146] Nitsche H, Sommer F, Mittemeijer EJ. J Non-Cryst Solids 2005;351(49–51):3760.
[147] Cardenas-Leal JL, Vazquez J, Lopez-Alemany PL, Villares P, Jimenez-Garay R. J Alloy Comp 2009;471(1–2):44.
[148] Tan Z, Zhang B, Zhang H. J Mater Sci 1992;27:4021.

Chapter 9

Surface Morphologies

Chapter Outline

Amorphous and Nano Alloys Electroless Depositions. http://dx.doi.org/10.1016/B978-0-12-802685-4.00009-1

Studying the surface morphology of EP alloy films is very important. Surface morphology is the most important part of the microstructure of EP deposits. If the surface morphology of EP deposits is known, its microstructure is basically known. Next, features of the surface morphology, such as uniformity, roughness, density, holes, and defects, are closely related to the preparation technique and conditions. From the surface morphology, we can conversely revise the original technology conditions to obtain a better deposit. Also, many applications such as metal internal conjunctions, hard drives in the IT industry, corrosion, and other surface chemical reactions are all related to the surface conditions of the EP coating, and they all have certain requirements on the surface condition. Fourth, the incorporated particles in electroless composite plating (ECP) and the nanoparticles in nano EP can be clearly identified and the size of the nanoparticles can be measured directly from the surface morphology. Last, a better understanding of the nucleation and growth in the initial stages of electroless deposition is essential in order to understand how its morphology can be controlled. This last point will be discussed in more detail in the next chapter.

Just because of development of a variety of electron microscopy, surface morphology study was able to implement. The most frequently used electron microscopes in studying the surface morphology are the scanning electron microscope (SEM), transmission microscope (TEM), and atomic force microscopy (AFM).

As we mentioned at the very beginning of Part III, one, two, or more SEM photographs usually appeared in an EP article. However, in the author's opinion, three aspects of problems in surface morphology of EP alloy deposits still exist. The first is that the surface morphology photographs are scattered among a large number of articles, there is as yet no review paper or book collecting and analyzing these morphology data. That is why we are using this chapter to analyze and discuss them specially. As we know that the morphology of the amorphous alloys prepared by splat quenching is a featureless microstructure (it will be illustrated clearly using a figure later) and many of the electroless alloy deposits are amorphous, it may mistakenly be believed that the morphology of the deposits is not complex and not worthy of study. This may be the second problem. The final issue is that systematic investigation of the surface morphology of deposits almost is very rare. Even for the usual Ni–P deposits, only one report by the author and Taglauer [1] can be found. Therefore, the content of that paper will act as a skeleton in the development of our understanding of the surface morphologies of EP alloy coatings. After this, the data of the surface morphologies of EP alloy coatings collected from a large number of papers are analyzed and discussed according to their classified names.

It should be pointed out that some pictures of surface morphology in EP films have been presented before. Particularly in Chapter 5, many SEM, TEM, and AFM images of various EP nanomaterials were presented for the sake of completeness in the analysis and discussion there. So, we will not repeat them in this chapter.

9.1 SKELETON UNDERSTANDING OF SURFACE MORPHOLOGIES OF THE EP ALLOY COATINGS

In 2005, the author's and Taglauer's group jointly published a paper titled, "Influence of alloying elements on the surface morphology of electroless deposits for Ni–P based alloy systems" [1]. Taglauer and the author planned this article together; the author's group prepared the samples and measured the compositions, X-ray diffraction (XRD) patterns, and corrosion rates; and SEM morphologies were recorded in the Max-Planck-Institut für Plasmaphysik (IPP) by Lindig, Taglauer, and the author. To the best of our knowledge, this is the only paper that systematically investigated the surface morphology of deposits. This paper systematically reported the amorphous and crystalline deposits of Ni–P, Ni–Sn–P, and Ni–Sn–Cu–P alloy systems at magnifications of 500×, 1000×, 5000×, 10,000×, and 20,000×. The first finding was that the morphology only can be displayed clearly with magnifications of 5000× for most of these deposits; some of them need up to 10,000×. It is very difficult to see any features with magnifications below 1000×. The features look like cauliflower, some like arranged oranges with nearly uniform size, or like arranged potatoes. Our results also show that the size of these features decreases with increasing contents of P, Sn + P, and Sn + Cu + P in the deposits. A formula relating the size and content could be worked out, which is an inverse proportional relationship. The SEM pictures of these deposits indicate that the deposits are really dense, thus causing them to have superior properties.

9.1.1 What Magnification Can See the Morphology Clearly?

In order to decide what magnification can see the surface morphology clearly, magnifications of 500×, 1000×, 5000×, 10,000×, and 20,000× are used to observe the surface morphology of the amorphous and crystalline deposits of Ni–P, Ni–Sn–P, and Ni–Sn–Cu–P alloy systems. Figure 9.1 shows SEM morphology of Ni-23.1P at% electroless deposit at different magnifications: (a) 500×; (b) 1000×; (c) 5000×; and (d) 10,000×. It can be seen that the features cannot be recognized if the magnification is less than 5000×, but can be seen clearly with magnifications higher than 5000×. It is worth noting that as discussed in Chapter 8 Ni-23.1P at% electroless deposit is a definite amorphous alloy.

Figure 9.2 shows the SEM morphology of Ni-0.5Sn-10.7P at% electroless deposit with different magnifications. (a) 500×; (b) 1000×; (c) 5000×; (d) 10,000×. It is evident that the features can only be seen with magnifications higher than 5000×, similar to the case of Ni–P binary deposits. In addition, though only 0.5 Sn at% is cooperated into Ni–P, the surface morphology is finer than that of the Ni–P deposit.

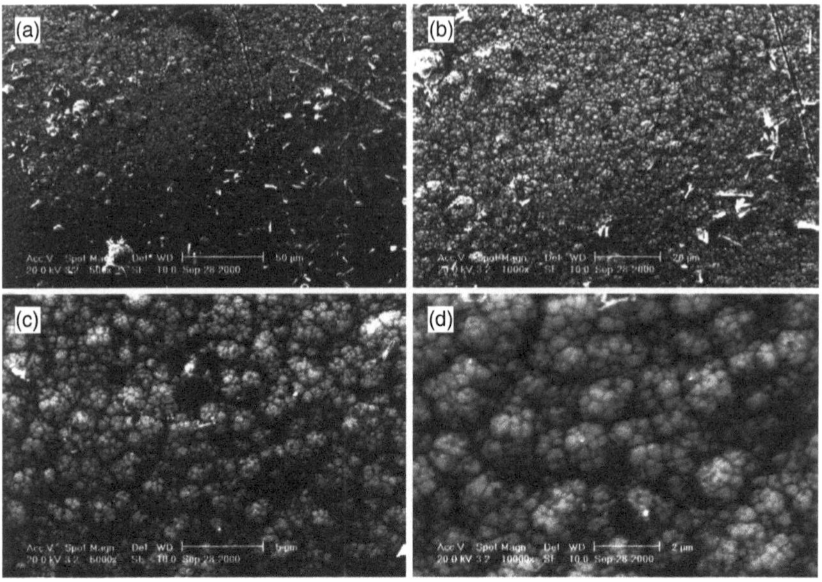

FIGURE 9.1 SEM morphology of Ni-23.1P at% electroless deposit with different magnifications. (a) 500×; (b) 1000×; (c) 5000×; (d) 10,000× [1].

FIGURE 9.2 SEM morphology of Ni-0.5Sn-10.7P at% electroless deposit with different magnifications. (a) 500×; (b) 1000×; (c) 5000×; (d) 10,000× [1].

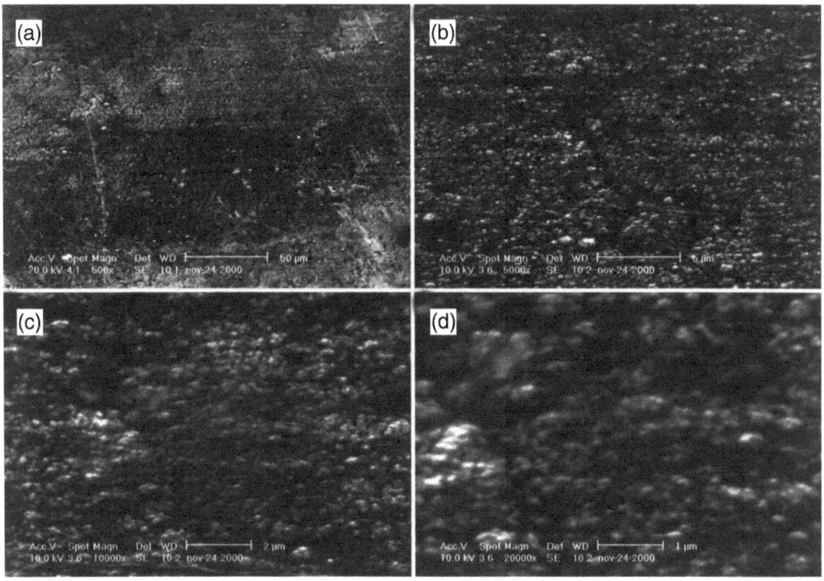

FIGURE 9.3 **SEM morphology of Ni-4.7Sn-5.1Cu-21.9P at% electroless deposit with different magnifications. (a) 500×; (b) 5000×; (c) 10,000×; (d) 20,000× [1].**

Figure 9.3 shows the SEM morphology of Ni-4.7Sn-5.1Cu-21.9P at% electroless deposit with different magnifications. (a) 500×; (b) 5000×; (c) 10,000×; (d) 20,000×. It can be seen that the features are only clearly resolved at a magnification of 10,000×, and it is difficult to see the features even at 5000×. This is because the surface morphology is further refined by the cooperation of Sn and Cu in Ni–P deposits together.

These facts show that the magnification of SEM images must be high enough for their features can be seen; otherwise they are difficult to distinguish. First, it was found that the morphology only can be seen clearly at magnifications of 10,000×. It is very difficult to see any features with magnifications below 1000×. That is why some reported photographs in the literature are not clear because the magnification used was not high enough.

The reports for the effects of magnification on clarity are sparse. Engin Ayturka and Ma in 2009 [2] used two magnifications 500× and 3000× to investigate the effect of agitation on the deposition morphology of Pd and Ag. The results (Fig. 6 in [2]) indicate that most of the SEM images are not very clear at low magnification (500×), but basically are clear at high (3000×) magnification. Their results are similar to the above data. In 2006, Sankara Narayanan et al. [3] reported the surface morphology of electroless Ni–P coatings with (a) high P; (b) medium P; and (c) low P, as shown in Fig. 9.4. It is evident that the features from these three images are not clear, because of the low magnification of 1000×.

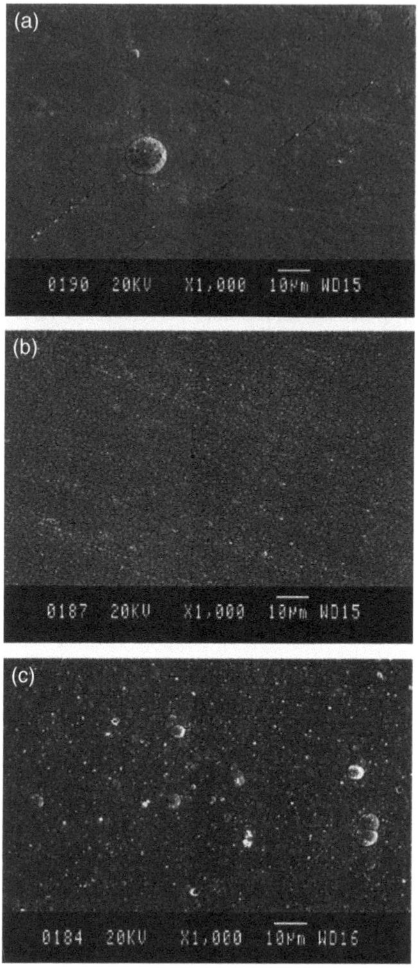

FIGURE 9.4 **Surface morphology of electroless Ni–P coatings: (a) Ni–high P coating; (b) Ni–medium P coating; and (c) Ni–low P coating [3].**

9.1.2 What Are the Morphological Features for Ni–P Based Alloy Deposits?

From Fig. 9.1, one can see that the picture resembles bundles of daisy or looks like cauliflower for the binary Ni–P alloy films. For the ternary alloy Ni–Sn–P coatings, Fig. 9.2 shows that the particles look like arranged oranges or potatoes with nearly uniform size. As for the morphology of quaternary Ni–Sn–Co–P deposits, their features as shown in Fig. 9.3 are not very obvious, maybe like a Scottish jam.

9.1.3 What is the Influence of Alloying Elements on the Surface Morphology of Ni–P Based Alloy Deposits?

In order to understand the influence of alloying elements on the surface morphology in Ni–P based alloy deposits, changes in P, Sn + P, and Sn + Cu + P levels are studied. In binary Ni–P alloy system, five different phosphorus levels (8.0, 12.5, 23.1, 27.4, and 28.7 at%) were prepared. Figure 9.5 shows their images with a common magnification of 10,000×. The figures show that the size of particles decreases as the content of P in the samples increases.

Figure 9.6 shows SEM morphologies for the four ternary Ni–Sn–P deposits: (a) Ni-8.9P, (b) Ni-0.5Sn-10.7P, (c) Ni-1.2Sn-20.3P; (d) Ni-3.0Sn-17.5P in at% with a common magnification of 10,000× for comparison purposes. Again, it can be seen that the particle size decreases with increasing content of P + Sn. It is interesting

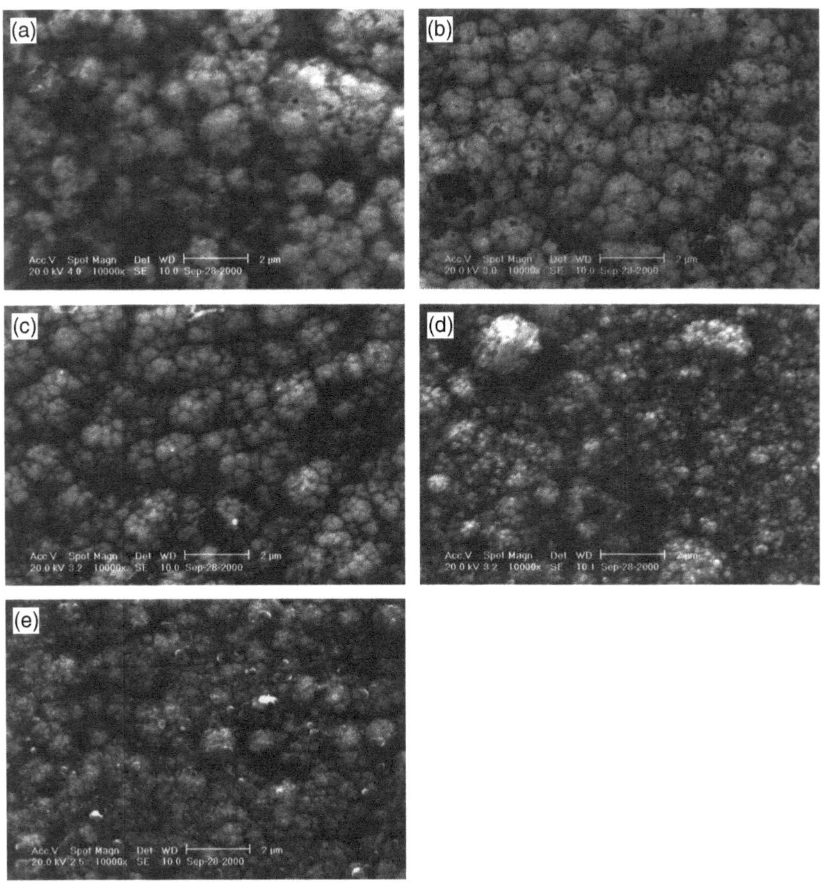

FIGURE 9.5 Comparison of the SEM morphologies with common magnification of 10,000× for the five binary Ni–P deposits: (a) Ni-8.0P; (b) Ni-12.5P; (c) Ni-23.1P; (d) Ni-27.4P; (e) Ni-28.7P at% [1].

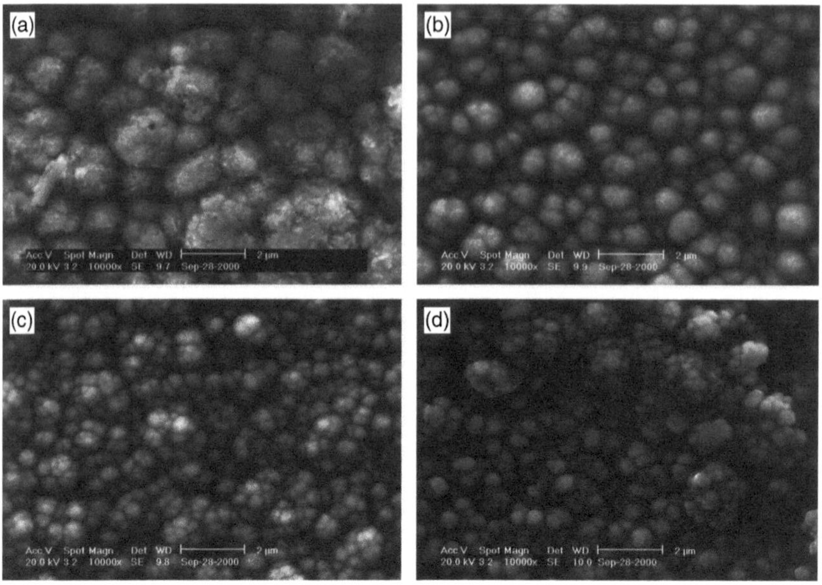

FIGURE 9.6 Comparison of the SEM morphologies with common magnification of 10,000× for the four ternary Ni–Sn–P deposits: (a) Ni-0Sn-8.9P, (b) Ni-0.5Sn-10.7P, (c) Ni-1.2Sn-20.3P; (d) Ni-3.0Sn-17.5P in at% [1].

to note that the size of particles is comparable when the contents of Sn + P in the samples are nearly the same (c, Ni-1.2Sn-20.3P and d, Ni-3.0Sn-17.5P in at%).

Figure 9.7 compares the SEM photographs for the five samples (Ni-0.3Sn-3.0Cu-9.6P; Ni-1.0Sn-0Cu-14.5P; Ni-1.3Sn-2.5Cu-15.3P; Ni-4.7Sn-5.1Cu-21.9P; and Ni-11.4Sn-15.1Cu16.1P in at%) with a common magnification of 10,000×. Generally speaking, the situation is similar to Ni–P and Ni–Sn–P alloy deposits, that is, the greater the content of Sn + Cu + P, the smaller the particle size. However, there is an exception for the samples of b and c because the particles of c are somewhat larger than those of b while the content of Sn + Cu + P of c is somewhat larger than that of Sn + P of b.

9.1.4 Is there a Quantitative Relationship Between the Particle Size and Alloy Composition?

Why is it that the higher the contents of P, Sn + P, and Sn + Cu + P, the lower of the size of the particles? During deposition, the P atoms released from the complex atomic group become the nucleation sites of the depositing alloy in the electroless Ni–P deposit. The greater the number of released P atoms, that is, the greater the content of P in the deposit, the larger the number of nucleation sites, therefore the smaller the size of particles. This explanation is similar for the Ni–Sn–P and Ni–Sn–Cu–P deposits. So, our results can be easily understood from the point of view of the theory of phase transformation.

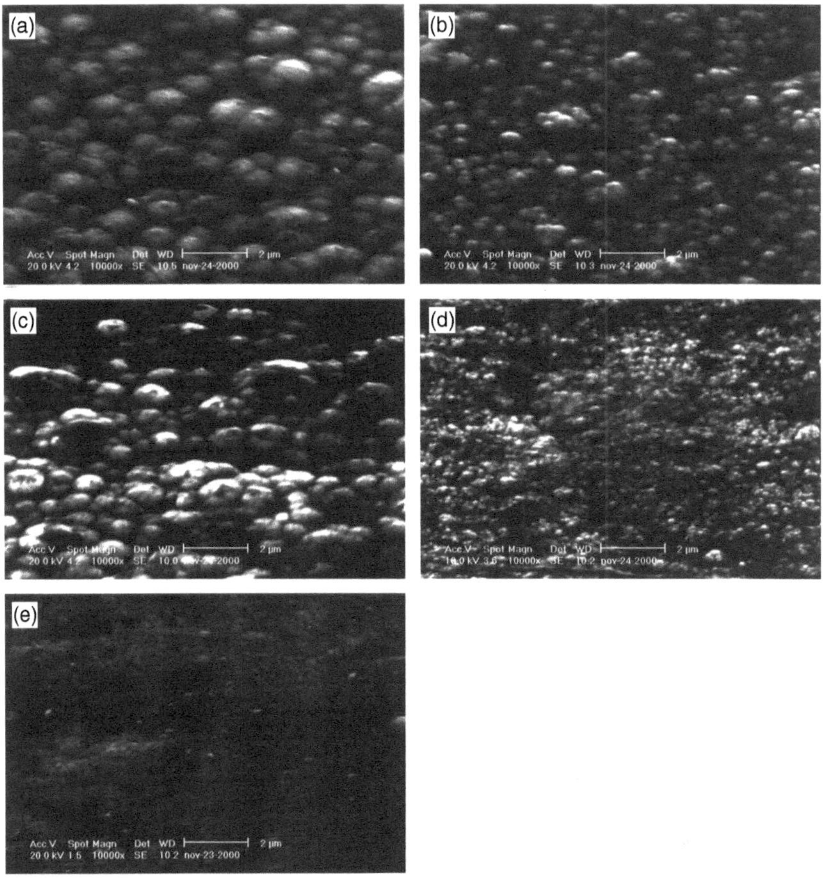

FIGURE 9.7 Comparison of the SEM morphologies with common magnification of 10,000× for the five binary Ni-Sn–Cu–P deposits: (a) Ni-0.3Sn-3.0Cu-9.6P; (b) Ni-1.0Sn-0Cu-14.5P; (c) Ni-1.3Sn-2.5Cu-15.3P; (d) Ni-4.7Sn-5.1Cu-21.9P; (e) Ni-11.4Sn-15.1Cu16.1P in at% [1].

Not only can the experimental data be explained, but an approximate equation between the size of particles and the content of samples can also be worked out from the results given above. For these three Ni–P based alloy deposits, the equation is [1]

$$r = A/x \tag{9.1}$$

where r is the average particle radius in micrometers, x is the content in atomic percentage of P, Sn + P and Sn + Cu + P for Ni–P, Ni–Sn–P, and Ni–Sn–Cu–P deposits, respectively. A is a constant, which is 0.04 for Ni–P, 0.045 for Ni–Sn–P, and 0.055 for Ni–Sn–Cu–P. This is a very simple relationship, which has not been found in the literature to date.

9.1.5 Should the Surface Morphology of Electroless Amorphous Coatings Be a Distinctive Pattern or Featureless?

It is well known that the as-quenched alloy exhibits a featureless, homogeneous microstructure typical of an amorphous state, without any detectable crystalline inclusions. Figure 9.8 shows SEM, TEM, and XRD images of the ribbons of amorphous alloys obtained by the melting quench method from three sets of groups. (a) and (b) in the figure are the SEM image and XRD pattern of the Cu–Ni–Sn–P amorphous alloys prepared by splat quenching by the author's group in 2005 [1] and 1992 [4]. In order to illustrate the issue more clearly, the magnification used was up to 50,000×. The SEM micrograph is featureless even at such a large magnification. The XRD pattern is typical of an amorphous state.

Figure 9.8c and d was obtained by Babilas and Nowosielski in 2010 [5]. Figure 9.8c shows TEM and electron diffraction pattern of the as-cast glassy $Fe_{72}B_{20}Si_4Nb_4$ ribbon with a thickness of 0.03 mm, which was manufactured by the "chill-block melt spinning" (CBMS) technique. TEM reveals only a changing of contrast and no appreciable contrast corresponding to a crystalline phase is seen. The electron diffraction pattern consists only of broad diffraction halo rings. Figure 9.8d shows the X-ray diffraction patterns of $Fe_{72}B_{20}Si_4Nb_4$ glassy ribbons in as-cast state with thickness from 0.03 to 0.20 mm, which shows the broad diffraction halo characteristic of an amorphous structure.

Figure 9.8e and f is from Révész et al. in 2010 [6]. The SEM images were taken of a polished cross-section of the as-quenched amorphous $Mg_{65}Ni_{20}Cu_5Y_{10}$ ribbon. They said that "the as-quenched alloy exhibits a featureless, homogeneous microstructure typical of an amorphous state, without any detectable crystalline inclusions." Figure 9.8f shows XRD patterns of the as-quenched amorphous alloy and its disks deformed by high pressure torsion. *N* denotes the number of total revolutions. The XRD pattern of the as-quenched ribbon consists of a broad symmetric halo that indicates a typical homogeneous amorphous phase.

In a word, all these experimental data from three sets of groups definitely confirm that the surface morphology of amorphous alloys manufactured by melting quenching is featureless even at a magnification of 50,000×.

However, except for the three samples Ni-8P, Ni-8.9P, and Ni-0.3Sn-3.0Cu-9.6P in at%, all 10 samples studied here are amorphous in structure (see Fig. 8 in [1]). The structure for sample Ni-0.5Sn-10.7P at% has not been determined. The results above show that all these electroless amorphous Ni–P based alloy deposits have morphological features. That is to say that the surface morphology of EP amorphous Ni–P based alloy deposits is very different from those of the amorphous alloys by splat quenching.

Is there anything wrong? Should the morphology of amorphous alloys manufactured by melt quenching have special features, and that by electroless plating be featureless? No, the morphologies of these two kinds of alloy systems have no problems, they are all reflected in the objective facts. Then how do we

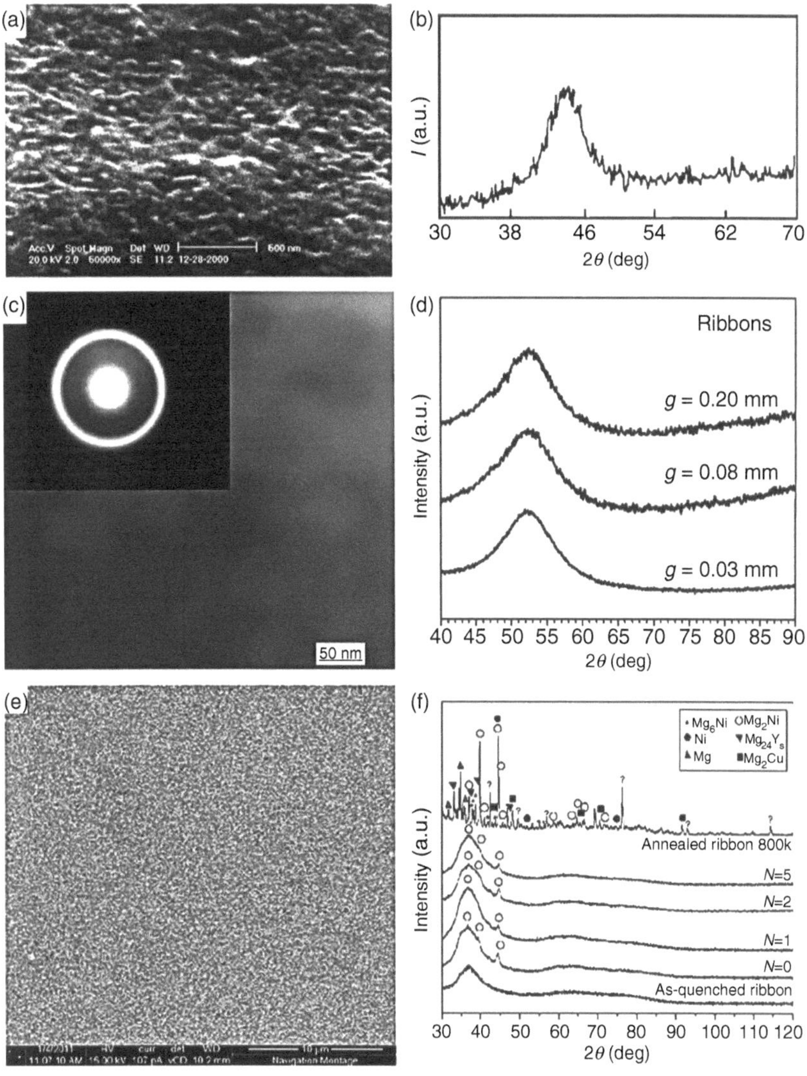

FIGURE 9.8 SEM morphology with 50,000×: (a) [1] and XRD pattern (b) [4] of the amorphous $Cu_{77}Ni_8Sn_8P_7$ (wt%) alloy by splat quenching. (c) [5] TEM and electron diffraction pattern of the as-cast glassy $Fe_{72}B_{20}Si_4Nb_4$ ribbon with a thickness of 0.03 mm, which was manufactured by the "chill-block melt spinning" (CBMS) technique. (d) [5] X-ray diffraction patterns of $Fe_{72}B_{20}Si_4Nb_4$ glassy ribbons in as-cast state with thickness from 0.03 to 0.20 mm. (e) [6] SEM images taken on the polished cross-section of the as-quenched amorphous $Mg_{65}Ni_{20}Cu_5Y_{10}$ ribbon. (f) [6] XRD patterns of an as-quenched amorphous $Mg_{65}Ni_{20}Cu_5Y_{10}$ alloy and its disks deformed by high pressure torsion. N denotes the number of total revolutions.

explain the difference? In the electroless plating, the deposit is not formed in a very short time (seconds) but rather longer, usually about 1 h. First, the released neutral atoms form many amorphous nuclei, and then these nuclei gradually grow. Therefore, the granular feature of the amorphous deposit could be seen in the SEM images. However, the amorphous alloys prepared by splat quenching are formed very fast so that the amorphous granulation cannot develop, so there is no feature in their SEM pictures.

9.2 THE EFFECT OF ALLOYING ELEMENTS ON SEM

The number of alloying elements affects the surface morphology of an EP film significantly. The following sections will discuss them according to their type; EP pure metals, binary, ternary, and multicomponent alloys.

In order to make the content more compact, usually just one or two typical images are given. The selection of the so-called typical image is according to the viewpoint of the author. Of course, if several groups published similar pictures, we will specify this.

Moreover, the surface morphologies of EP nano 3D to 0D materials have been presented before, so they will not be repeated here.

9.2.1 The Surface Morphology of EP Pure Metals

First, the surface morphology of EP pure metals Cu, Ag, Au, etc., will be described and analyzed.

Many groups have published pictures of the surface morphology of EP Cu in recent decades [7–25]. Figure 9.9a shows the SEM picture of Cu film by Aina Hung, obtained in 1986 [7]. The electroless copper was deposited from electroless copper bath for 1 h. The bath contained 0.055 mol/L $CuSO_4$, 0.24 mol/L sodium potassium tartrate, and 0.47 mol/L paraformaldehyde and NaOH to give a pH of 12.50. The presence of guanidine hydrochloride in the bath (up to 15 mg/L) causes the surface of deposited copper to contain smaller spherical

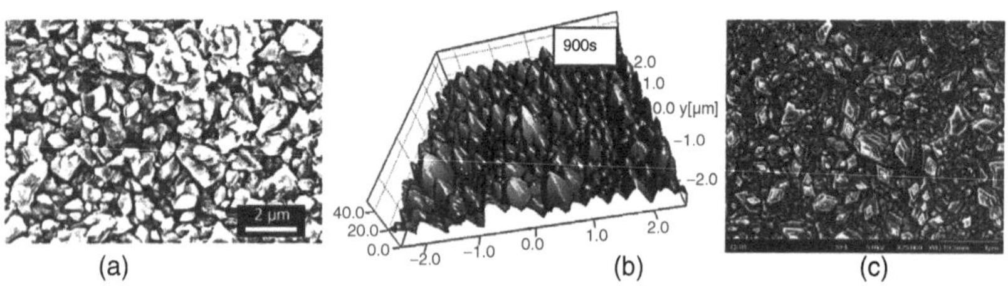

FIGURE 9.9 Surface morphology of EP Cu. (a) SEM picture of copper film electroless copper from electroless copper bath with presence of 5 mg/L guanidine hydrochloride in the bath [7]. After 10 min, the film coverage seemed to be complete. (b) AFM micrograph of the Cu film after deposition for 900 s [24]. (c) Diamond pyramid structure in similar size of electroless Cu deposit on stainless steel substrate [21].

grains and become porous, that is, the surface area for deposition does increase to some extent. The middle plot of Fig. 9.9 shows the 5 μm × 5 μm AFM micrograph of the Cu film after deposition for 900 s. In 2011, Scudieroa et al. [24] used 10 mmol/L $CuSO_4$ to deposit a Cu film on a Si substrate for different deposition times of 1, 30, 60, 300, and 900 s. The experiments found that even for relatively brief immersion times, coverage is fairly uniform even though the average particle size is very small (see Fig. 2 in [24]). As the deposition time increases, the particle diameter rapidly grows and large clusters appear. Figure 9.9b shows the longest plating time and largest particles. After deposition for 30 min, the average roughness (R_a) was estimated to be about 15 nm and the mean size of the Cu particle about 100 nm. At this point, the film appears granular and relatively dense. Even though a similar surface morphology of Cu film to that in Fig. 9.9a was obtained by most of the groups [7,9,11,12,15–23], some other features of surface morphologies still occurred. For example, Fig. 9.9c shows the special feature obtained by Wu and Sha [21], which is a diamond pyramid structure in similar size. They obtained this surface morphology of electroless copper deposit on a stainless-steel substrate. Moreover, they also found the same with the deposits on epoxy board substrates, these diamond pyramid structures grow mainly upward.

There are also a lot of articles published that investigate the surface morphology of electroless plating Ag [26–35]. The typical surface morphology of EP Ag deposited on various substrates is spherical particles, bead-like spots, or a nodular structure, as shown in Fig. 9.10. Figure 9.10a was obtained by Kortenaar et al. in 2001 [29]. This is an SEM image of a silicon wafer that has been dipped in a solution containing 0.25% HF and 5.9 mmol/L $AgNO_3$ for 10 s. It is evident that Ag particles are spherical and monodisperse with an average size of around 15 nm. In 2004, Brevnov et al. [31] carried out electroless deposition of silver on films containing 99.5% aluminum and 0.5% copper. They found that the electroless Ag deposition proceeds via the formation of spherical particles of silver randomly distributed on the surface. Both the particle density and average particle diameter increase with the deposition time. The average diameter of particles

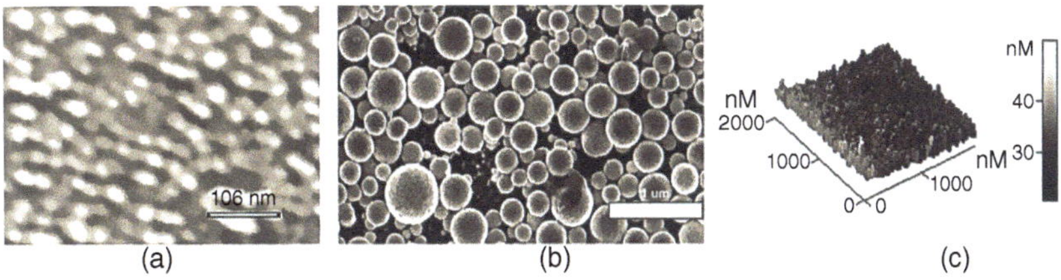

FIGURE 9.10 Surface morphology of EP Ag. (a) SEM micrograph of a silicon wafer after dipping into a solution containing 0.25% HF and 5.9 mM AgNO3 for 10 s [29]. (b) SEM micrograph collected after electroless deposition of silver for 2 h [31]. (c) AFM image of the silicon surface after the preparation of Ag seed layer. The preparation was performed in a solution of 0.005 mol/L $AgNO_3$ + 0.06 mol/L HF for 5 s [30].

increases from 50 nm after 9 min of deposition to 180 nm after 120 min, just as shown in Fig. 9.10b. Their experiment shows that electroless deposition results in a distribution of particle diameters, indicating that the silver phase formation is a continuous process (e.g., new nanoparticles are formed while old particles increase in diameter).

In 2003, Tong et al. [30] reported the surface morphology of EP Ag on a Si(100) substrate. Figure 9.10c is the AFM image of the silicon surface after the preparation of an Ag seed layer. The preparation was performed in a solution of 0.005 mol/L $AgNO_3$ + 0.06 mol/L HF for 5 s. Their experiment showed that a thin silver layer is deposited on the surface of the silicon wafer. Bead-like spots are observed. The distribution of the spots is very dense and even. The roughness of the Ag layer is 2.1 nm. In addition, in 2010, Gu et al. [35] fabricated nanoporous Ag films on copper alloy substrates by the galvanic replacement reaction from the choline chloride-based deep eutectic solvents (DES) containing AgCl. These authors measured the surface morphology of as-deposited Ag films from 0.1 mol/L AgCl DES solution at RT for 5 min deposition. They found that some nodular structures with the size of about 10 nm are formed on the larger Ag grains, thus making the faceted Ag grains lose their regular geometries. More importantly, several nanometric voids also exist around the Ag grains even after the deposition time of 5 min. Thus, the films obtained from 0.1 mol/L AgCl DES solution at RT can also be called porous Ag.

The study of surface morphology in EP Au films is still rather more [36–43]. It should be said that the features of surface morphology in EP Au alloy deposits are rather more dispersed. Figure 9.11 shows the possible representative features. In 1998, Hou et al. [38] reported the surface morphology of self-assembled monolayers of electroless gold on different substrates. Figure 9.11a shows the surface of a film of electroless gold supported on a glass microscope slide. The roughness of the surface is R_z = 80 nm, which is much greater than that of evaporated gold (R_z = 20 nm). After the sample was annealed at 250°C for 3 h, the gold grains increased in size. In addition, the packing of gold grains became more uniform and the surface roughness (R_z = 40 nm) was notably reduced. The average thickness of the films of electroless gold deposited on glass sub-

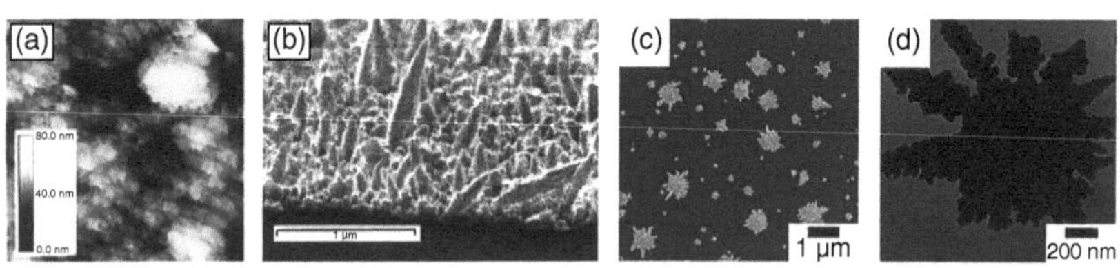

FIGURE 9.11 Surface morphology of EP Au. (a) STM image of as-plated electroless Au on glass microscope slides. The size of image is 1.0 μm × 1.0 μm [38]. (b) SEM micrograph of sample plated for 30 min in 1 mM Au bath [39]. (c) TEM and (d) SEM micrographs gold nanoflowers formed under DMDPB Langmuir monolayers at 35°C with the subphase concentration of 1×10^{-4} mol/L. The reaction times are 30 min (c, d) [43].

strates is about 100 nm. In 2002, Magagnin et al. [39] investigated the surface morphology of EP Au films. The films were deposited for 5 s on Si(111) and Ge(100) and (111), respectively. The formation of small clusters in the range of 10–30 nm, distributed uniformly across the surface, was observed for silicon substrates. The film thickness was estimated to be about 50 nm. The rms roughness grows rapidly for gold on silicon, from a range of 2 nm after 60 s to about 18 nm after 10 min of plating. Figure 9.11c shows the cross sections after fracture of the samples. Gold nanodendrites formed under dimethyldipalmitylammonium bromide (DMDPB) monolayers were investigated by Ding et al. in 2011 [43]. When the copper grids were covered on the surfaces of the Langmuir monolayers of DMDPB formed at the air/$HAuCl_4$ aqueous solution interfaces, nanostructures composed of dendrites appeared. After 2 min, flowerlike nanostructures appeared. From the electron diffraction (ED) pattern, they found that the film is fcc Au crystalline, indicating that the nanostructures are polycrystalline. Ten minutes later, the dendrites grew. After 30 min, the dendrites grew further. It can be seen from a well-developed dendrite shown in Fig. 9.11d that the dendrite is made up of a trunk and two side branches that are in the same plane and perpendicular to the trunk. Figure 9.11c shows the Au nanoflowers that are the same as in (d) at low magnification.

9.2.2 The Surface Morphology of EP Binary Alloy Films

Many papers have reported the surface morphology of EP Ni–B alloy coatings [44–54]. Some of the SEM pictures cannot be seen clearly, such as Figs. 1 to 4 in [44], Fig. 1 in [45], Fig. 1 in [49], etc., which may result from the magnification not being large enough. The surface morphologies in most of EP Ni–B coatings are a "cauliflower"-like structure or a columnar, nodular structure. Using SEM and XRD equipment in 2011, Bulbul [54] reported the effects of deposition parameters on the crystallographic orientation of electroless Ni–B coatings. SEM analysis revealed that the Ni–B coatings developed six types: pea-like, maize-like, primary nodular, blackberry-like or grape-like, broccoli-like, and cauliflower-like morphological structures depending on the deposition parameters, as shown in Fig. 9.12. XRD results also showed that these structures exhibited different levels of amorphous character. They found from the XRD results that the sixth type was the most amorphous of these morphologies. The tendency for the coating to become amorphous increased together with the growth of the Ni–B phase, and this was an indicator of transformation from grainy structure to nodular. In addition, the concentration of $NaBH_4$ had the greatest effect on the morphological and crystallographic development of electroless Ni–B coatings. The composition and conditions of the six samples in Fig. 9.12 are given in Table 9.1.

From the above work by Bulbul, two points can be made. The first is that the surface morphology of EP Ni–B films is very similar to that for Ni–P deposits described in Section 9.1. Various interesting surface morphologies can

FIGURE 9.12 Six surface morphology of Ni–B alloy coatings prepared by different bath composition and operation conditions, which are illustrate in Table 9.1 [54].

be obtained in these two cases. The second is that the features can only be seen clearly at magnifications from 5000× to 10,000× in Ni–B coatings (from Fig. 9.12), which is also similar to the case for Ni–P binary deposits.

Several groups have reported the surface morphologies of EP Co–P deposits [55–61]. Figure 9.13a shows the typical surface morphology of as-plated Co–P film, which was deposited on Cu/Si substrate. At least four groups reported such surface morphology [55,57,58,61]. No special name was given to the morphology by these authors. However, they stated that these micrographs show

TABLE 9.1 Bath Composition and Operation Conditions for Ni–B Samples in Fig. 9.12

Run no	Temperature (°C)	$NiCl_2$ (g/L)	$NaBH_4$ (g/L)	Plating time (min)
2	80	20	0.6	60
3	80	40	1.2	90
5	95	40	0.6	30
6	95	10	1.2	60
8	88	10	0.6	90
9	88	20	1.2	30
Constant parameters	Ethylenediamine and sodium hydroxide:90 g/L (the same value for both); lead nitrate: 0.0145 g/L; pH ≈ 13.5			

relatively large crystals separated by crevices. From the XRD pattern, Tarozaitė et al. [61] concluded that the crystallites of the coating have crystallographic planes (100) and (001) parallel to the surface, that is, the closest packed planes of Co–P grow on the closest packed plane (111) of Cu. The pattern analyzed by the grazing incidence showed that the outer or topmost layers of Co–P coating do not contain crystallites with crystallographic plane (001) parallel to the surface. Apart from the typical morphology, a sharply outlined wave-like structure with a dense network of shallow pits (Fig. 9.13b) or globular three-dimensional structures were also detected [59,60].

It is interesting to note that early in 1966, Miksic et al. [56] investigated the electron micrographs for Co–P films with increasing phosphorus concentration

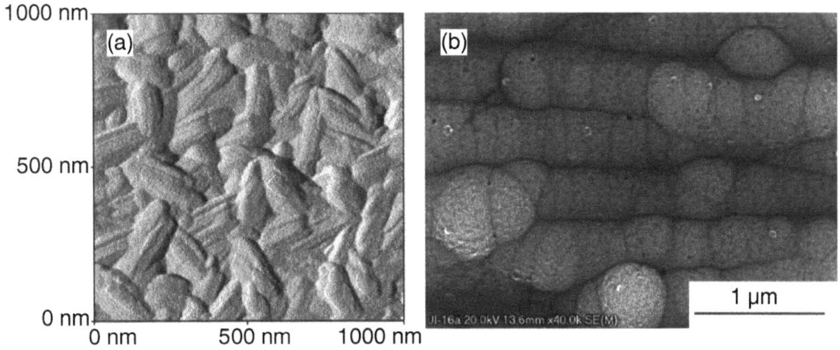

FIGURE 9.13 Surface morphologies for as-plated Co-P films. (a) AFM images of Co–P film deposited on Cu/Si substrate in solution containing (mol/L): $CoSO_4$ 0.1, glycine 0.6, NaH_2PO_2 0.5, succinic acid 0.3, pH 10. Thickness of the film: 0.2 μm [61]. (b) Morphology of the as-plated Co–P deposit [60].

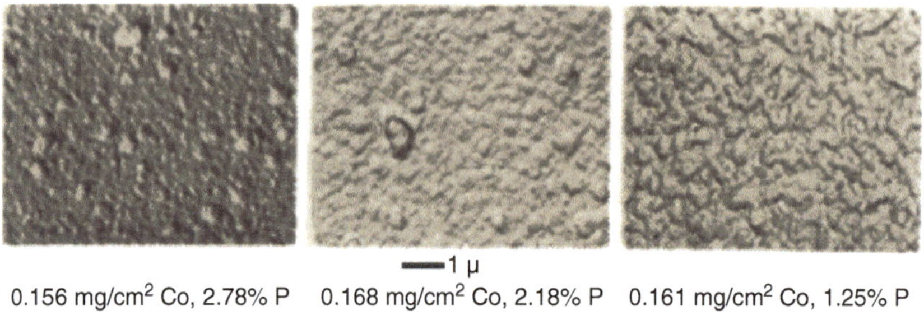

FIGURE 9.14 Electron micrographs for cobalt-phosphorus films with increasing phosphorus concentration at constant cobalt weight. Magnification 10,000× [56].

at constant cobalt weight, as shown in Fig. 9.14. It can be seen that the aggregate size as shown by electron micrographs decreases with increasing phosphorus concentration at constant cobalt weight. These micrographs also show a crevice-type surface structure at a phosphorus concentration below approximately 2% and a non-crevice-type surface structure at a phosphorus concentration above 2%. Such research, that is, surface morphology with change of P content in the binary alloy films is rarely seen in the literature. In addition, only when the magnification is raised up to 10,000× can these features be seen clearly.

9.2.3 The Surface Morphology of EP Multicomponent Alloy Films

9.2.3.1 *The Surface Morphology of EP Ternary Alloy Films*

A lot of works referring to the surface morphology of EP ternary alloy films have been published [62–84]. Here, we will only choose some representative works to describe and discuss, rather than survey all the literature.

In 2008, Abdel Aal et al. [71] reported nanocrystalline, soft, ferromagnetic Ni–Co–P thin films on Al alloy by low-temperature electroless deposition. The deposit obtained at optimum conditions (pH 9.5 and deposition time 15 min) was compact and smooth with nodular grains as shown in Fig. 9.15. The observed randomness in distribution of spherical nano particles with size in the range 150–250 nm is due to the partially amorphous structure in Ni–Co–P film grown in the perpendicular direction to the substrate surface. It can be seen in Fig. 9.15 that incorporating Co into the Ni–P deposit means that the features only can be see clearly at a magnification of 20,000×. If these results are compared to the SEM images of Ni–P films mentioned in Section 9.1, the size of the particles decreases with incorporation of Co into Ni–P deposit because the features of Ni–P films can be seen clearly at magnifications of less than 10,000×.

If W is incorporated into Ni–P films, how does the surface morphology change? Figure 9.16 indicates such a change, as obtained by Wu et al. in 2003 [65]. The features of SEM in EP binary Ni–P and ternary Ni–W–P coatings are basically the same, all revealing nodular characteristics. Moreover, the size of

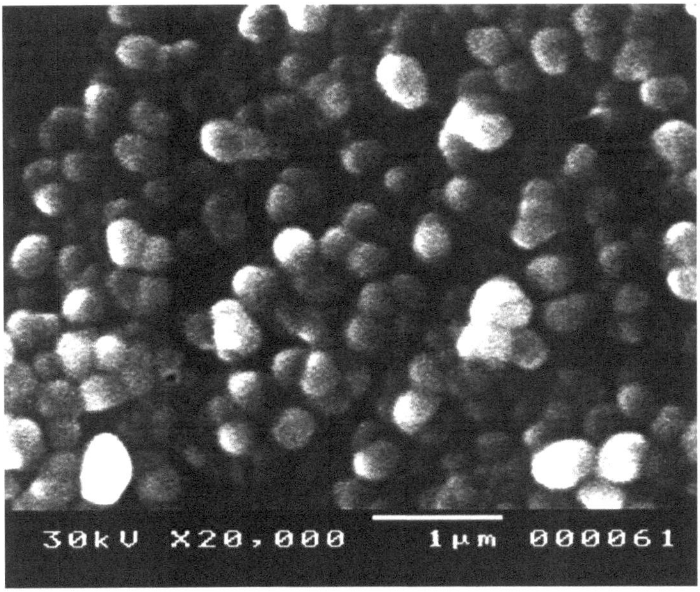

FIGURE 9.15 SEM images of Ni–Co–P films (pH 9.5 and deposition time 15 min). Magnification 20,000× [71].

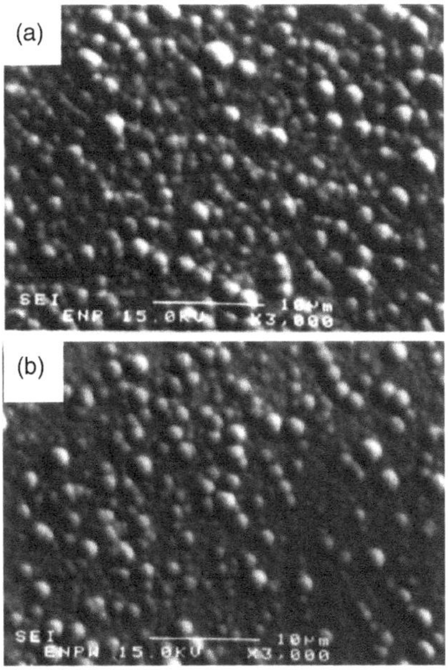

FIGURE 9.16 Scanning electron images of (a) electroless Ni–P, and (b) electroless Ni–W–P coatings [65].

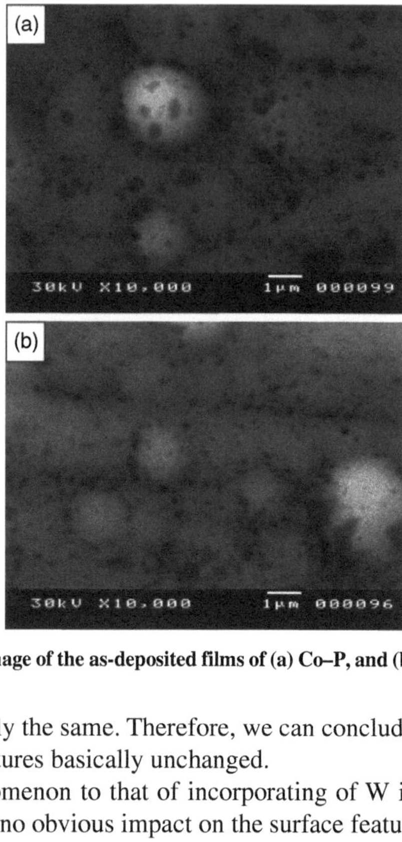

FIGURE 9.17 **SEM image of the as-deposited films of (a) Co–P, and (b) Co–P–3.6 wt%W [70].**

the particles is nearly the same. Therefore, we can conclude that adding W into Ni–P leaves the features basically unchanged.

A similar phenomenon to that of incorporating of W into Ni–P films, that is, having basically no obvious impact on the surface features, also appeared in Co–W–P [70,74] deposits. In 2008, Abdel Aala et al. [70] reported the surface morphologies of Co–P and Co–W–P coatings, as shown in Fig. 9.17. There was no significance change in surface morphology between Co–P and Co–W–P deposits. They all have a nodular structure, and the size of particles is basically the same. In addition, the images show that both Co–P and Co–W–P coatings have an amorphous structure. In 2009, Dulal et al. [74] obtained similar results (Fig. 10 in [74]. Please note: the features are not very clear because the magnification is not large enough). They found that the crystalline Co–W–P film has small spherical crystallites with diameters less than 40 nm.

However, in the paper by Einati et al. in 2005 [67], a contradictory phenomenon appeared. These authors studied EP Co–B and Co–W–B deposits. Figure 9.18 shows the HRSEM (high-resolution scanning electron microscope) pictures of as-plated $Co_{96}B_{2.2}$ (a) and $Co_{85.6}W_{8.7}B_{5.1}$ (b) films. It can be seen that $Co_{85.6}W_{8.7}B_{5.1}$ contains the largest amount of boron and therefore has a different morphology. There are three different particle sizes that vary from 125, 85 to 20 nm. $Co_{96}B_{2.2}$ (a) is a very uniform and dense layer; the concentration of holes in it is smaller than in $Co_{85.6}W_{8.7}B_{5.1}$. Particles of as-deposited $Co_{96}B_{2.2}$ (a) are about 10–13 and 20 nm. That is to say, W incorporation into Co–B film

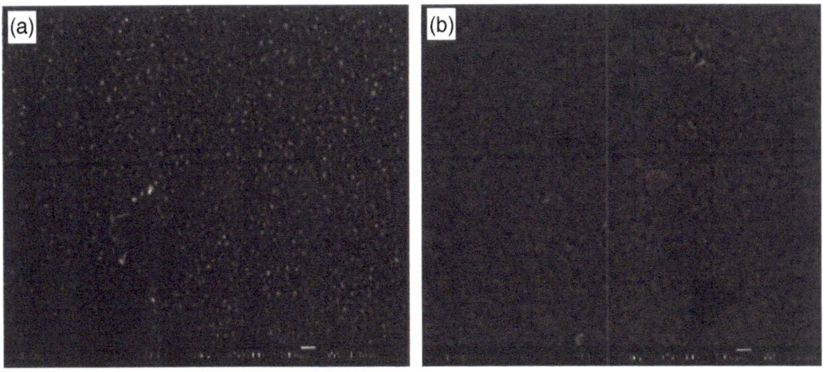

FIGURE 9.18 HRSEM pictures of as-plated $Co_{96}B_{2.2}$ (a) and $Co_{85.6}W_{8.7}B_{5.1}$ (b) films [67].

increases the diameters of the particles, which is just opposite to what was seen in the Co–W–P films.

A few groups investigated the effect of incorporation of Mo on the surface morphology in Ni–P [81,82] Co–P [77], and Co–B [76]. In 1996, Lo et al. [81] measured the surface morphology at different contents of Mo in EP Ni–Mo–P deposits. The compositions are (wt%): (a) Mo 0, P 11.80; (b) Mo 2.21, P 13.43; (c) Mo 11.08, P 3.34; (d) Mo 14.37, P 2.56. They found that the particle size of the electroless Ni–Mo–P decreases with increasing Mo-complex concentration. The crystal size and the crystallinity increase with Mo content, as shown in Fig. 9.19.

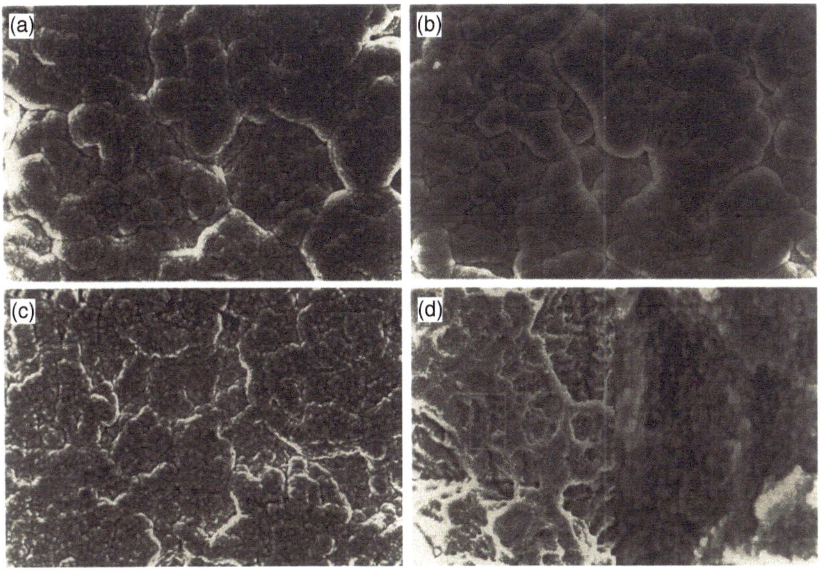

FIGURE 9.19 SEM micrographs of electroless Ni–Mo–P electrodes: (a) Mo 0, P 11.80; (b) Mo 2.21, P 13.43; (c) Mo 11.08, P 3.34; (d) Mo 14.37, P 2.56 [81].

In 2011, Rudnik and Mucha [77] studied the influence of the concentration of Mo(IV) compounds on the morphology of EP Co–Mo–P alloy from alkaline citrate bath. Figure 9.20 indicates the effect of concentration of Mo(VI) species on morphology of as-plated Co–P and Co–Mo–P deposits. The compositions of the deposits are (at%): (a) Mo 0, P 6.63; (b) Mo 0, P 8.04; (c) Mo 0.54, P 8.16; (d) Mo 1.56, P 7.66; (e) Mo 6.95, P 1.62; and (f) Mo 4.03, P 0.53. The Co–P film (a) consisted of star-like grain characteristics for hexagonal cobalt and cobalt-rich deposits, which are very different from those described above. This morphology evolved gradually via needle and nodular to the fine-grained structure with increased Mo(VI) concentration. Deposit (f) seemed to be flocculent at the grain boundaries. It suggests incorporation of nonmetallic occlusions. Significant changes in the surface morphology confirm the adsorption of molybdate anions. These enhance the nucleation stage and deposition of the metal as a fine grained layer. Therefore, the structure of the deposits evolved from ramified coarse grains with clearly visible grain boundaries

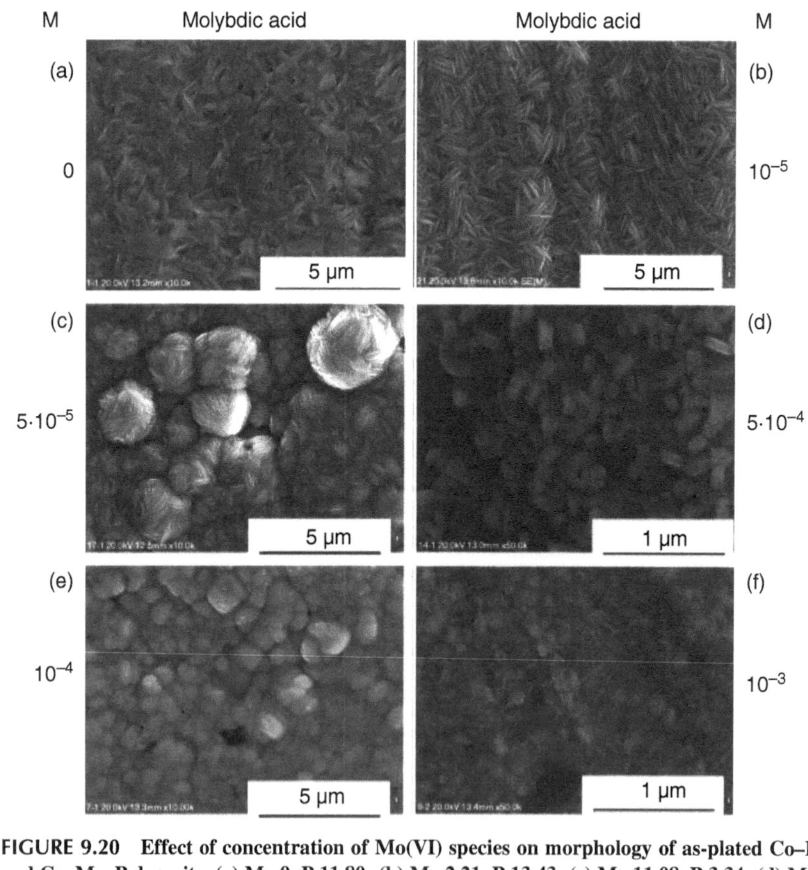

FIGURE 9.20 Effect of concentration of Mo(VI) species on morphology of as-plated Co–P and Co–Mo–P deposits: (a) Mo 0, P 11.80; (b) Mo 2.21, P 13.43; (c) Mo 11.08, P 3.34; (d) Mo 14.37, P 2.56; (e) Mo 6.95, P 1.62; and (f) Mo 4.03, P 0.53 [77].

to fine-grained deposits. However, it is just as pointed out by these authors that "more detailed studies are necessary". Dai et al. in 2010 [76] reported the morphology of Co–W–B (76.3 wt% Co, 17.7 wt% Mo, and 6.0 wt% B). The feature of the morphology is typical nodular or spherical particles, which is very dense (Fig. 1(c) in [76]). However, these authors did not measure the size of the particles.

Several groups have reported the effect of incorporation of Cu into Ni–P deposits on surface morphology [72,73,83]. Only the author's group reported SEM images of EP Ni–Cu–B alloy films [84]. For the Ni–Cu–P deposits, all these groups reported the uniform spherical nodular surface structure. Balaraju et al. [83] compared two SEM micrographs of as-plated ternary Ni–Cu–P coatings at magnifications 1000× and 5000×. The features at 1000× are not clear, but very clear at 5000× (Fig. 3 in [83]). Moreover, they found that the number of nodules increases with the Cu content in the Ni–Cu–P deposits, so the surface of higher Cu content is not smooth. In 2009, Guo et al. [72] investigated the surface morphology of the Ni–Cu–P deposits plated on the polyester fabric by SEM and then compared with the Ni–P deposits, as shown in Fig. 9.21. The surface morphology of the latter exhibits the coarse nodular structure (Fig. 9.21a). Copper deposition plays a major role in reducing or inhibiting the activation of additional nodule sites. The refinement in the microstructure of the nodularity, with smooth and bright Ni–Cu–P deposits, can be clearly seen in Fig. 9.21b. The reason is that introduction of copper ions activates natural nucleation sites, which results in an increase in the concentration and thus the production of more nuclei. In other words, the introduction of copper into the deposit seems to suppress the growth of the nodules by inhibiting further growth. This is probably the main reason for the attainment of bright, smooth Ni–Cu–P alloy deposits. It is noted that the magnification is rather high, at 30,000×. The author's group investigated the surface morphology of $Ni_{78-x}Cu_xB_{22}$. $x = 0.0$, 3.8, 5.1, 7.0, 7.3, and 8.0 at% [84]. We have found that the resultant plating deposits are uniform and adhered to the substrate. They range from smooth and bright to mat and semibright. The SEM micrographs of these alloys (Fig. 9.22) reveal

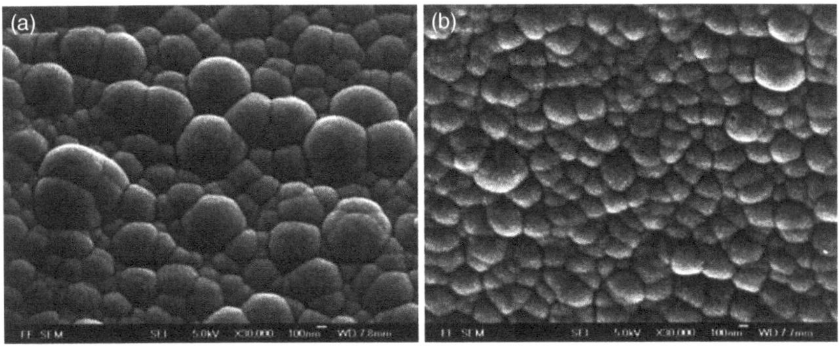

FIGURE 9.21 **SEM micrographs of Ni–P (a) and Ni–Cu–P (b) deposits with magnifications of 30,000× [72].**

FIGURE 9.22 Microstructure of electroless plating $Ni_{78-x}Cu_xB_{22}$ alloys: (a) x = 0; (b) x = 3.8; (c) x = 5.1; (d) x = 7.0; (e) x = 7.3; and (f) x = 8.0 at%. Magnification form 1,500X to 5,000X. [84]

that the deposits are porous and composed of granulas with different shapes and sizes, and the single large granula changes to a cluster of small granulas with increased Cu content. That is to say that adding Cu may suppress the growth of the granulas, which is consistent with that in Ni–Cu–P mentioned above.

The last metallic element we would like to discuss is Sn. In Section 9.1, the results from the author's group indicated that with incorporation of Sn into Ni–P films, the particles size decreases with increasing content of Sn + P. What results have been obtained by other groups? In 2010, Zou et al. [75] investigated the effect of Sn addition on the surface morphology of electroless Ni–P–Sn ternary deposits. Figure 9.23 shows the SEM micrographs of the as-plated EP

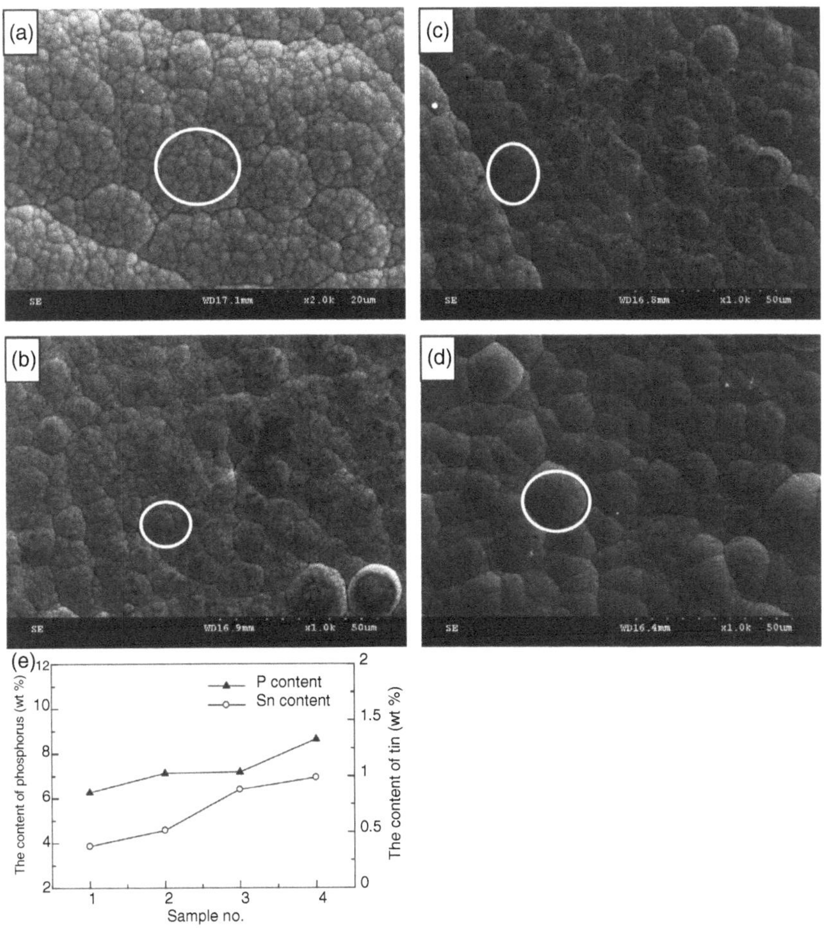

FIGURE 9.23 SEM micrographs of the as-plated EP Ni-Sn-P coatings, (a) sample 1, (b) sample 2, (c) sample 3 and (d) sample 4. (e) The contents of P and Sn of four samples in EP Ni–P-Sn deposits. [74]

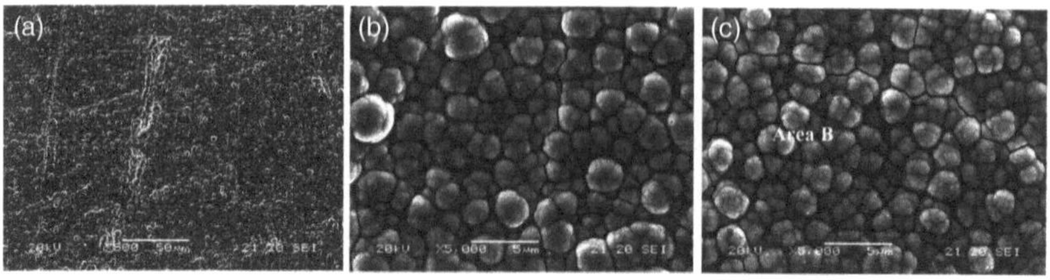

FIGURE 9.24 SEM images of electroless plated low Sn (a and b) and high Sn (c) Ni–Sn–P alloys [85].

Ni–Sn–P coatings of (a) sample 1, (b) sample 2, (c) sample 3, and (d) sample 4. Figure 9.23e shows the contents of P and Sn of four samples in EP Ni–P–Sn deposits. It is seen that the surface morphology of all these Ni–P–Sn deposits showed a similar spherical nodular structure and one big nodule that included many small nodules. As the Sn + P content increased, the boundary between the small nodules in one big nodule gradually disappeared, as shown in Fig. 9.23a–d and marked by a circle. In other words, the size of particles increases with increase of the content of Sn + P, which contradicts the results of the author's group. However, the data obtained by Georgieva et al. in 2005 [85] are consistent with those of the author's group. Figure 9.24 shows their results. (a) (500×) and (b) (5000×) are the SEM images of electroless plated low Sn content (Sn 3.2 at%, Ni 75.4 at%, P 21.4 at%) Ni–Sn–P alloys, and (c) (5000×) is the SEM image of electroless plated high Sn (average over the coating Sn 62.8 at%, Ni 34.6 at%, P 2.6 at%) Sn–Ni–P alloy. It is seen that the morphologies of both low and high Sn samples are the typical globular morphology for electroless nickel coatings. These authors did not measure the size of the particles. However, it can be seen from the figure that the size of particles in (c) is slightly less than that in (b). Therefore, with increase of Sn + P, the size of particles decreases, which is in agreement with that obtained by the author's group. Figure 9.24a and b are taken from the same low tin Ni–Sn–P alloy sample, but the magnification is different, (a) is 500× and (b) is 5000×. It is obvious that the features cannot be seen clearly at 500×, which confirms again that these features can only be seen clearly at high magnifications.

9.2.3.2 The Surface Morphology of EP Quaternary and Quinary Alloy Films

Many papers have been published that study the surface morphology of quaternary alloy films [86–96]. However, there is only one paper for the morphology in quinary alloy deposit Ni–P–Fe–Zn–Cu [97]. Three papers by one group investigated the morphology of Ni–W–Cu–P [88,90,91]. Two groups reported the morphology in Co–W–P–B [94,96]. Each one is reported by one group, respectively in another six systems: Ni–Sn–Cu–P [85], Ni–W–Sn–P [93], Co–Ni–W–P [89], Co–Ni–Fe–B [86], Co–Ni–Cu–P [87], Co–Ni–Ce–B [92], and Co–Re–Ni–P [95].

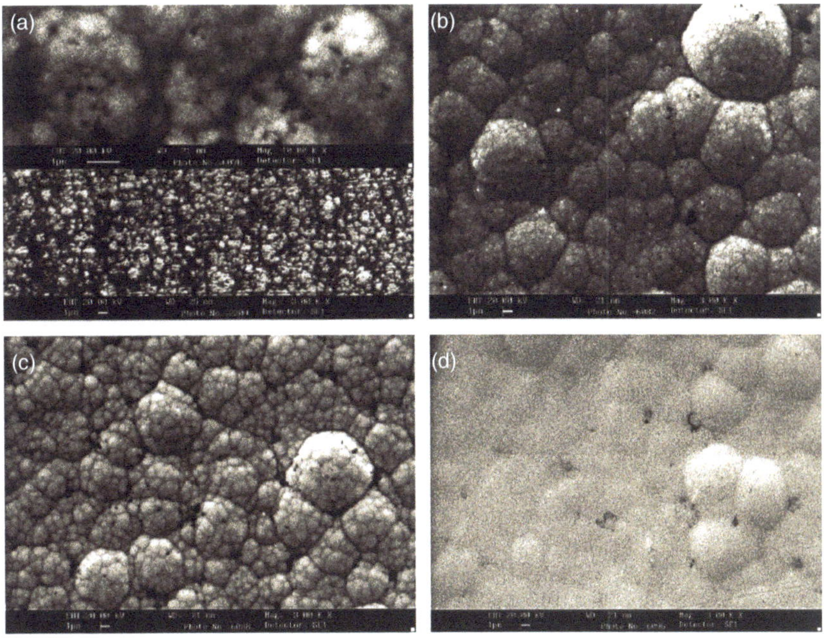

FIGURE 9.25 **SEM micrographs of the as-plated electroless nickel coatings (3000×) (a) Ni–P, (b) Ni–Cu–P, (c) Ni–W–P, and (d) Ni–W–Cu–P. The upper part in (a) is 10,000× [88].**

In 2005, Balaraju et al. [88] reported the SEM micrographs of Ni–W–Cu–P films, and compared them with those for Ni–P, Ni–Cu–P and Ni–W–P deposits, as shown in Fig. 9.25. It can be seen that the surface morphology of Ni–P deposit showed the spherical nodular structure as shown in (a). Because the features are very indistinct at 3000×, the authors increased the magnification to 10,000×, as shown in the upper part of (a). Now the features can be seen very clearly. The ternary Ni–Cu–P and Ni–W–P deposits exhibited a coarse nodular structure as shown in (b) and (c). Refinement in microstructure and disappearance of the nodularity is clearly seen in (d). In other words, the smooth deposit surface is seen for quaternary Ni–W–Cu–P deposit. The reason may be that Cu codeposition has played a major role in reducing/inhibiting the activation of additional nodule sites. These authors obtained the same results in their other two papers: ([90], Fig. 4) and ([91], Fig. 2).

In 2007, Shacham-Diamand et al. [94] and in 2009 Koo Hyo-Chol et al. [96] investigated the morphology of Co–W–P–B alloy deposits. The SEM images of EP Co–W–P–B films are shown in Fig. 9.26. Figure 9.26a was obtained by Shacham-Diamand et al., and (b) by Hyo-Chol et al. The magnification in (a) is 100,000×. The features in (a) can be seen clearly, and are obvious nodular shapes. The features in (b) are very fuzzy, which may be because the magnification is not high enough. The original authors did not indicate the magnification in their images.

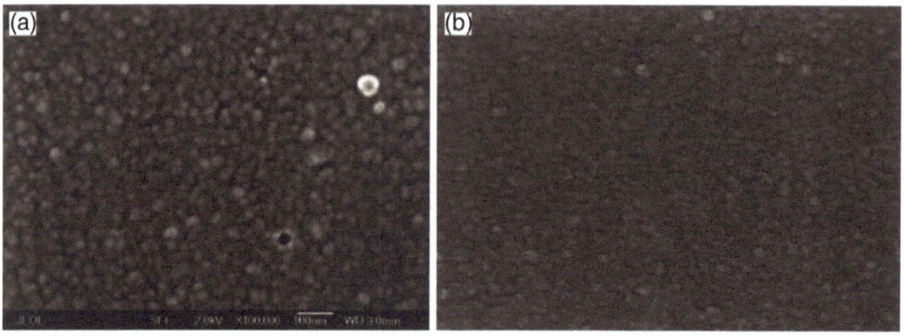

FIGURE 9.26 **SEM images of electroless deposited Co–W–P–B films on thin Cu layer. (a) [94] and (b) [96].**

In 2009, Ayoub and Ibrahim reported the surface morphology of quinary alloy Ni–P–Fe–Zn–Cu deposits [97]. The composition of EP Ni–P–Fe–Zn deposits is 5.3 wt% P, 2.23 wt% Fe, and 6.99 wt% Zn. Incorporation of copper sulfate resulted in the deposition of Ni–P–Fe–Zn–Cu increased zinc (8.77 wt%) and reduced phosphorous (4.16 wt%) and iron (1.53 wt%). The content of Cu is 10.86 wt%. From these results, it could be concluded that Cu has inhibited the codeposition of P and Fe in the EP Ni–P–Fe–Zn–Cu coatings.

Figure 9.27 shows the surface morphologies of the Ni–P–Fe–Zn deposits (a) and that in the Ni–P–Fe–Zn–Cu deposit (b). It can be seen that the morphology in Ni–P–Fe–Zn alloy film is a coarse nodular structure, and pores are also visible and appear as dark regions. However, the morphology in the Ni–P–Fe–Zn–Cu deposit shows smooth surface with fine deposit due to the presence of isotropous copper resulting in hemispherical islands made of larger crystals and compact grain boundary deposits. Therefore, the effects of the incorporation of copper element on the morphology of the Ni–P–Fe–Zn deposit are obvious, and are similar as the effects of Cu incorporated into ternary alloy deposits as mentioned above.

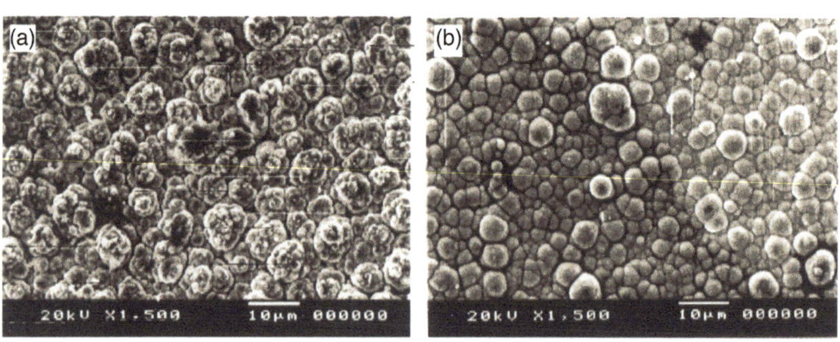

FIGURE 9.27 **The surface morphologies of the as-deposited Ni–P–Fe–Zn (a) and Ni–P–Fe–Zn–Cu (b) films [97].**

9.3 SURFACE MORPHOLOGY OF ECP ALLOY DEPOSITS

Introducing intentionally insoluble or sparingly soluble particulate matter into conventional electroless plating impacts the surface morphology of alloy deposits. This means the properties of deposits will be affected. Here, we only describe and discuss the morphology of plain ECP, in which the size of the added particles is not less than 1 μm. Nano ECP will not be discussed here, since Chapter 4 covered nanomaterials.

Considerable work has been carried out on the surface morphology of plain ECP metal and alloy coatings [98–130]. The matrixes in ECP systems studied include pure metal such as Ag, Cu, and Co, binary alloys such as Ni–P, Ni–B, Co–P, and Cu-P, and ternary alloys such as Ni–W–P, Ni–Cu–P, and Ni–Zn–P. Most of them consider one or two insoluble particles being added into Ni–P based alloy films. As for the particles, many kinds of them have been included, including oxides (Al_2O_3, TiO_2, ZrO_2, and Fe_3O_4), carbides (SiC, B_4C, WC, carbon, and diamond), sulfides (MoS_2 and WS_2), nitrides (BN and Si_3N_4), and other particles such as PTFE and PVP. The following sections cover just some of these.

In 1998, Apachitei et al. [98] investigated the codeposition of some hard particles (SiC, Al_2O_3, and B) within an electroless Ni–P matrix. The average size of SiC, Al_2O_3-I, Al_2O_3-S, Saffil δ-Al_2O_3, and B particles is 3.2, 2.8, 3.4, 7.2, and 1 μm, respectively. Using the same type of electroless bath, the three types of hard particles – SiC, Al_2O_3, and B – were successfully codeposited in the Ni-(8.22 wt%) P matrix. To prevent bath decomposition during plating, the ratio particle/bath volume was kept constant for all types of hard materials. Figure 9.28 indicates the surface morphology of the Ni–P and Ni–P + X coating. From this figure, a uniform distribution of particles on coating surfaces can be observed. Moreover, a lamellar Ni–P grain (nodular) orientation with different grain sizes is distinguished on the Ni–P deposit (Fig. 9.28a). This is a consequence of the preferential growth of Ni–P grains on substrate defects, which resulted from the metallurgical working process of hot rolling. This occurred because the substrates used for plating electroless Ni–P coatings were an aluminum 6063-T6 alloy which had experienced such metallurgical processing. The size and grain orientation are modified when the particles are incorporated throughout the deposit (Fig. 9.28b–f). But, the surface morphologies are very different from that of Ni–P deposits. In addition, these morphologies are related to their original shapes of insoluble particles. In order to see this point more clearly, the SEM micrographs of particles used as reinforcement in electroless Ni–P coatings: (b) SiC, (c) Al_2O_3-I, (d) Al_2O_3-S, (e) Saffil α-Al_2O_3, and (f) B were inset into the left lower part of every corresponding ECP image, respectively. The most typical samples are Al_2O_3-S and Saffil α-Al_2O_3, and their shapes are spherical morphology and cylindrical fiber. The resulting ECP morphologies are rather uniform and dense nodular (it is not very clear maybe because the magnification is not high enough) and like bamboo leaves,

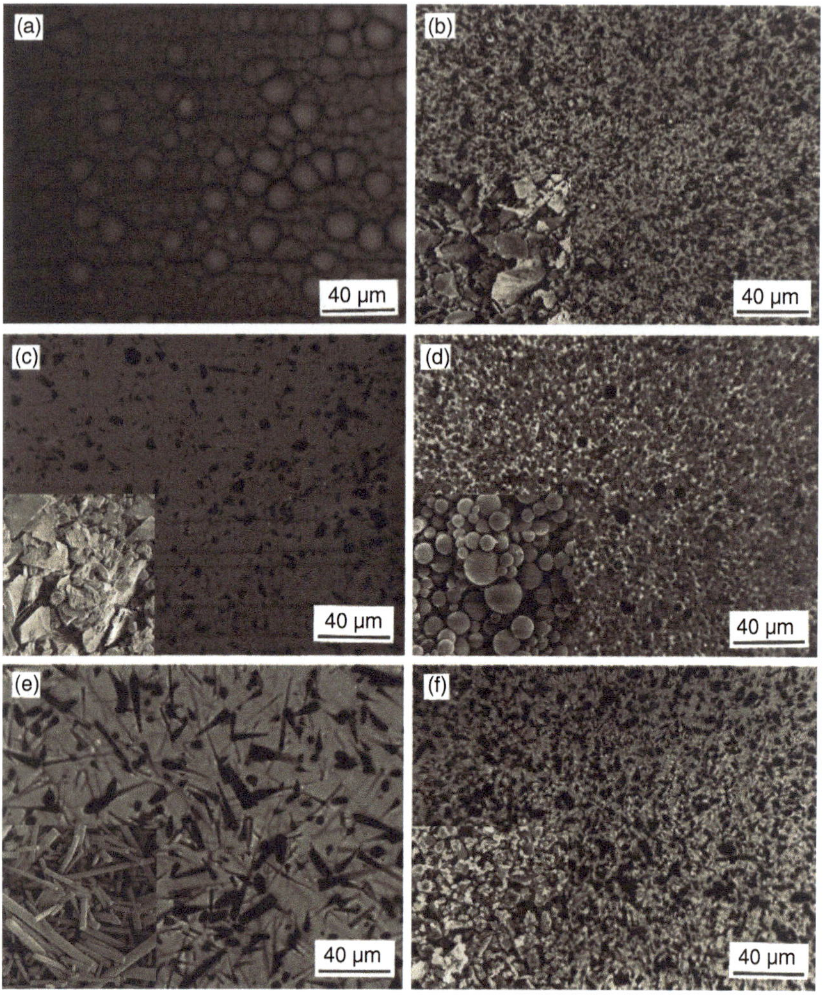

FIGURE 9.28 **Surface morphology of as-deposited coatings: (a) Ni–P, (b) Ni–P + SiC, (c) Ni–P + Al_2O_3-I, (d) Ni–P+ Al_2O_3-S, (e) Ni–P + Saffil α- Al_2O_3, and (f) Ni–P + B [98].**

respectively. It evidently shows that these morphologies are really related to their original shapes of insoluble particles. These authors believed that only in the Ni–P + Al_2O_3-I coating (Fig. 9.28c), the lamellar grain orientation is still present and this might be the result of a low particle concentration in the deposit (9.7 vol%).

In 2002, Abdel Hamid et al. [102] investigated electroless Ni–P composite deposits incorporating ZrO_2, TiO_2, and Al_2O_3 on 6061 Al alloy from acidic bath. The ceramic particles including ZrO_2, TiO_2, and Al_2O_3 0–12 g/L with grain size from 7 to 10 μm were added to the bath. Figure 9.29 shows the microstructure of the distribution of the different particles on the surface of composite coatings.

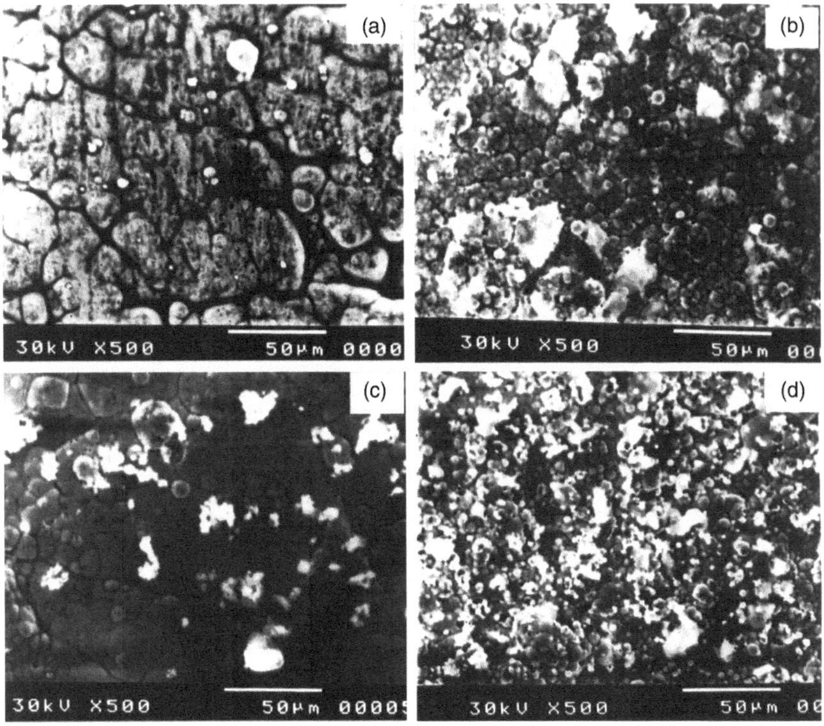

FIGURE 9.29 The microstructure of the distribution of the different reinforcement particles on the surface of composite coatings. (a) Ni –P, (b) Ni–P + TiO_2, (c) Ni –P + ZrO_2, and (d) Ni–P + Al_2O_3 [102].

The figure shows a very high volume percent of Al_2O_3 particles in the deposit if compared with another two reinforced particles. However, the surface morphologies of the ECP coatings are very different from that of Ni–P alloy. Of course, they are also different from each other. Although the surface morphologies for Ni–P + TiO_2 and Ni–P + Al_2O_3 all have similar nodular shapes the size of their particles is less than that in Ni–P deposits.

The surface morphology of the Ni–P, Ni–P + ZrO_2, Ni–W–P, and Ni–W–P + ZrO_2 coatings was reported by Szczygieł et al. in 2008 [115]. The size of the ZrO_2 particles they used is up to 5 μm with a specific surface of 4.753 m^2 g/L. Figure 8.30a–d shows the scanning electron microscopy of Ni–P, Ni–P + ZrO_2, Ni–W–P, and Ni–W–P + ZrO_2 coatings, which indicates considerable differences in the appearance of their surface morphology. The Ni–P coating (Fig. 9.30a) is fine grained. Most of the grains are from 2 to 5 μm in size. The coating containing second phase particles in the form of ZrO_2 powder (Fig. 9.30b) looks completely different. It is made up of spheroids up to 20 μm in size. The three-component Ni–W–P coating shows a similar spheroidal morphology but with much smaller grains: mostly 5–10 μm in diameter (Fig. 9.30c).

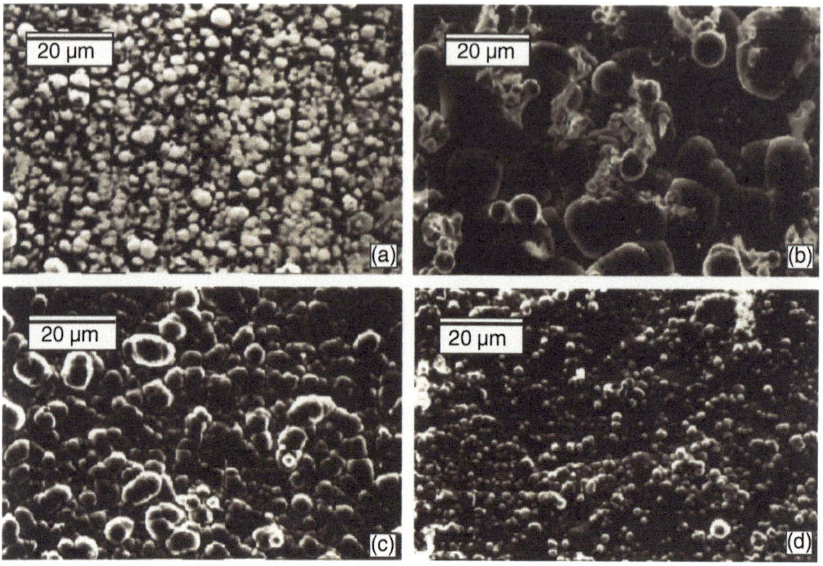

FIGURE 9.30 **SEM surface morphology of electroless nickel coatings (complexing agent: glycine; pH = 6.0, T = 75°C) (magnification 1250×): (a) Ni–P; (b) Ni–P + ZrO_2; (c) Ni–W–P; (d) Ni–W–P + ZrO_2 [115].**

Under the microscope the Ni–W–P + ZrO_2 composite alloy coating seems to be the smoothest of all the examined coatings (Fig. 9.30d). But on its surface one can observe numerous poorly compacted grains about 1 μm in diameter. Comparing the SEM image (Fig. 9.30b) to the similar morphology of Ni–P + ZrO_2 obtained by Abdel Hamid et al. [102] (Fig. 9.29c), the difference can be clearly seen. These authors measured the sizes of crystallites in multicomponent nickel coatings, and they were 1.4596, 1.6723, 1.6328, and 1.9434 nm for the Ni–P, Ni–P + ZrO_2, Ni–W–P, and Ni–W–P + ZrO_2 coatings, respectively. With incorporation of ZrO_2 particles in Ni–P and Ni–W–P coatings, the sizes of crystallites in Ni–P + ZrO_2 and Ni–W–P + ZrO_2 coatings increase but only enhance by 14.6% and 19%, respectively.

In 2011, Mohammadi and Ghorbani [125] investigated the morphology of electroless nickel composite coatings with PTFE and/or MoS_2 particles. The sizes of PTFE and MoS_2 particles they used are 1.0 μm and 1.5 μm. Figure 9.31 illustrates the scanning electron micrographs of the EN and EN composite coatings, which are very different from each other. As indicated in Fig. 9.31a, the EN coating has a typical spherical nodular structure. Figure 9.31b and c shows that the PTFE and MoS_2 particles are well dispersed in the EN + PTFE and EN + MoS_2 coatings, respectively. As can be seen, the surface of the EN + PTFE coating is more homogenous and defect-free in comparison with EN + MoS_2 coating, which contains some surface defects, cited by the black arrows in Fig. 9.31c. Fig. 9.31d shows the surface morphology in the EN + PTFE + MoS_2

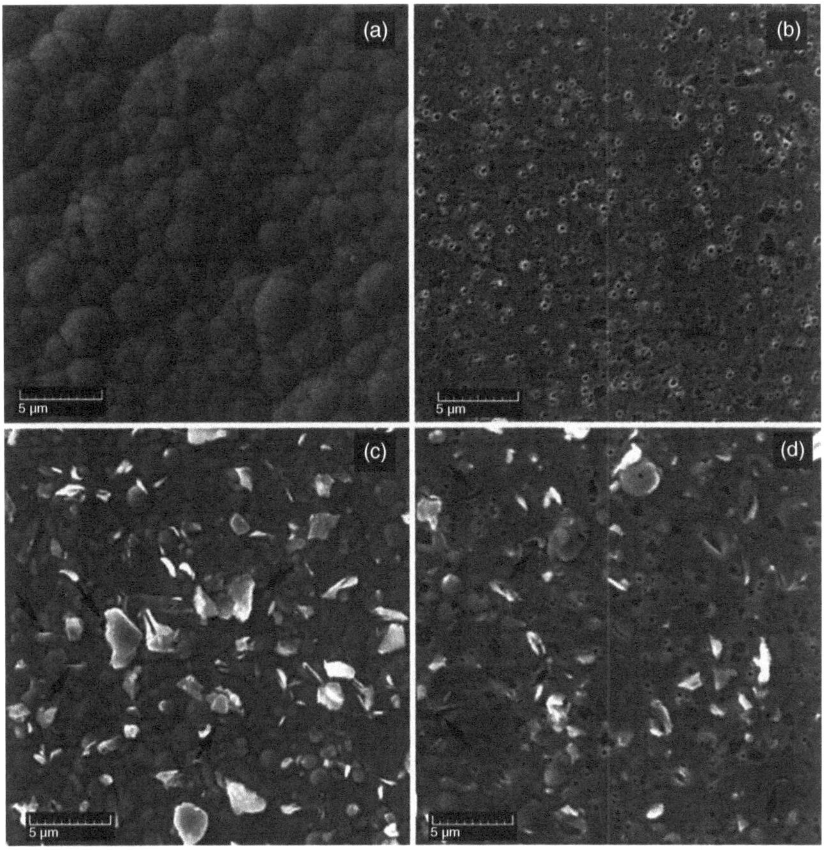

FIGURE 9.31 SEM micrographs of the surface of the coatings: (a) EN; (b) EN + PTFE; (c) EN + MoS_2; (d) EN + PTFE + MoS_2 [125].

coating, which is simultaneous codeposition of the PTFE and MoS_2 particles into the EN deposit. It can be seen that the EN + PTFE + MoS_2 coating is not completely continuous and contains some cracks and porosities as in the case of the EN + MoS_2 coating. It is obvious that the codeposition of these particles, separately and simultaneously, changes the EN morphology because of change in the coating compactness and also its surface energy. These authors also measured the microscopic cross-sectional images of the EN and EN + PTFE + MoS_2 coatings (Fig. 2 in [125]) which indicate that these coatings form a uniform deposition with identical thickness. The authors believed that this is one of the special properties of the electroless depositing process.

In order to compare the morphology of the same ECP coating obtained by different groups, we illustrate SEM images of incorporation of SiC into Ni–P and Cu–P as a sample, as shown in Fig. 9.32. Figure 9.32a is the SEM image of EN [119], which indicates a typical spherical nodular structure. Other SEM

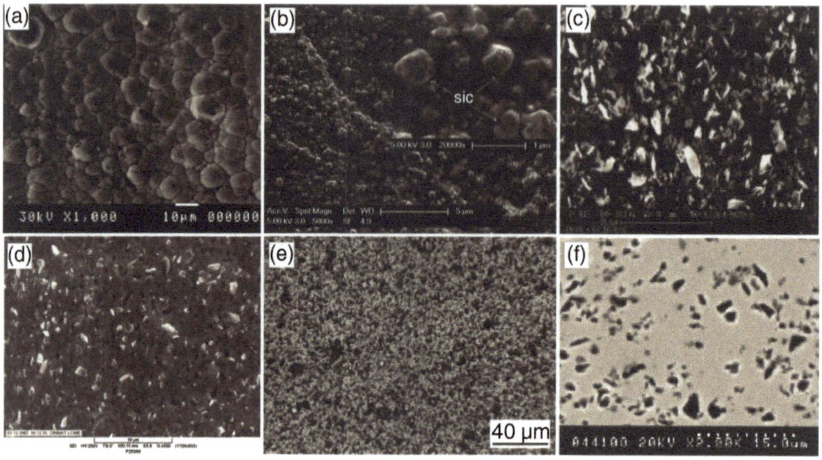

FIGURE 9.32 SEM images of the coatings: (a) EN [119]; (b) Cu–P + SiC [124]; (c) EN + SiC [103]; (d) EN + SiC [108]; (e) EN + SiC [98]; and (f) EN + SiC [112].

images of ECP films fall into five groups: (b) Cu–P + SiC from Faraji et al. [124], (c) EN + SiC from Huang et al. [103], (d) EN + SiC from Malfatti et al. [108], (e) EN + SiC from Apachitei et al. [98], and (f) EN + SiC from Wu et al. [112], respectively. It is evident that these morphologies are different. Only the morphology of (b) has evident nodular structure (the magnification is 20,000×, so it is very clear). Image (e) is just the same as Fig. 9.28b, and as mentioned above, its morphology may be dense nodular (it is not clear; this may be caused by its magnification not being high enough). Other three SEM images are somewhat like leaves of some trees or bamboo leaves. Why they are so different? Why the same ECP coatings (c to f all are Ni–P + SiC films) have so different surface morphology? It is maybe the case that only further work can answer this problem.

9.4 EFFECTS OF VARIOUS PARAMETERS ON SEM

Much attention was paid to the effects of various parameters on surface morphology. We will describe and discuss them successively in the following sections. Only one or two samples will be illustrated for every influencing parameter in order to just reveal the characteristics of the surface morphology.

9.4.1 Effects of the Concentration of Metal Salts on SEM

Some groups reported the effect of the concentration of metal salts in the plating bath on surface morphology. In 2002, Younan and Shoeib [131] produced the SEM photomicrographs of deposited Ni–W–P alloy by changing the concentration of sodium tungstate in the bath. Three groups reported the SEM morphologies of deposited Ni–Cu–P alloy by changing the concentration of copper sulfate in bath [132–134]. In 2010, Wei et al. [135] studied the SEM

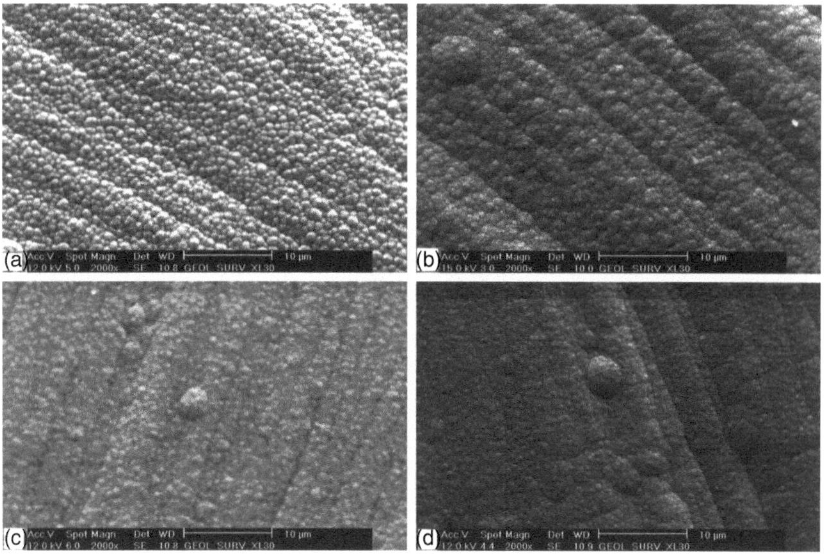

FIGURE 9.33 SEM photomicrographs of deposited Ni–W–P alloy, under constant conditions, from the basic bath with: (left top) 0, (left down) 0.06, (right top) 0.12, and (right down) 0.15 mol/L of sodium tungstate [131].

morphologies of deposited Co–Pt–W–P alloy by changing the concentration of chloroplatinate in the bath.

Figure 9.33 shows the SEM photomicrographs of deposited Ni–W–P alloy, under constant conditions, from the basic bath with different concentrations of sodium tungstate: (a) 0, (b) 0.06, (c) 0.12, and (d) 0.15 mol/L. From the figure, it can be seen that the surface structure of Ni–P deposit showed the spherical nodular structure and coarse grain size. The Ni–W–P deposits still have spherical nodular structure; however, the grains are compact, fine, and have a more uniform crystalline structure, as shown in Fig. 9.33b–d as compared with that of Ni–P alloy. The authors measured the composition of deposits with changing the concentration of sodium tungstate (Fig. 1 in [131]). The contents of W and P (wt%) are from 0 to about 9.5 and from about 6.9 to 3.5 as increase of sodium tungstate from 0 to 0.15 mol/L, respectively. That is to say that with increase of sodium tungstate, that is, with increase of content of W, the size of particles in SEM images decreases. The surface morphology of Ni–W–P deposits containing more than 8 wt% tungsten had a very fine and more compact crystal as shown in Fig. 9.33c and d.

In 2004, Younan and Abo El-Enine [133] investigated the SEM photomicrographs of Ni–Cu–P alloy deposited with various concentrations of copper sulfate (0.000, 0.002, 0.004, 0.006, and 0.010 mol/L) in the plating bath. Their resulting SEM photomicrographs are shown in Fig 7 of [133]. The alloy Cu content increased linearly up to 18.2 wt% with increasing Cu^{2+} concentration up to

0.010 mol/L in the plating bath. The surface topography of Ni–P deposit showed the spherical nodular structure with very fine and compact grains. After adding copper sulfate into the bath, the electroless Ni–Cu–P deposits still have spherical structure. But, the grains are accumulated and compacted to give an apparent large particle size as compared with that of Ni–P alloy. Moreover, with increasing the alloy copper content, the grain boundary of the accumulated particles and its grooves decreases. It is a contradiction to the results in Ni–W–P deposits mentioned above because the size of particles in SEM images decreases with increase of the content of W. In 2005, Balaraju et al. [134] obtained similar results, that is, the size of particles of SEM images in Ni–Cu–P deposits increases with increasing Cu content (see Fig. 3 in [134]). However, Jiang and Wu in 2000 [132] obtained a different result from these two groups. They also changed the concentrations of copper sulfate (a 0; b 0.05; c 0.1; d 0.2; e 0.4; and f 0.6 g/L) in the plating bath, and then checked the features of the SEM photos in the Ni–Cu–P deposits. The measured contents of Cu and P (wt%) in the coatings are 0, 2.54, 4.28, 6.49, 9.27, 11.58 and 9.25, 8.07, 7.75, 6.02, 5.46, 3.77 as the concentrations of copper sulfate (g/L) in the plating bath are 0, 0.05, 0.1, 0.2, 0.4, and 0.6, respectively. In other words, with increase of copper sulfate in the bath, Cu content in the deposits increases but P decreases. Figure 9.34 shows the results of SEM images of the Ni–Cu–P coatings. The surface topography of Ni–P deposit still illustrates the conventional spherical nodular structure. With addition of copper sulfate in the bath, the electroless Ni–Cu–P deposits still have spherical structure. But, the grains are smaller than that in Ni–P deposit first, which means that codeposition of a small amount of Cu in the plating layer can play the role of grain refinement. Then, after the amount of adding copper sulfate is greater than 0.1 g/L, the grains gradually increase, the small particles agglomerate and form

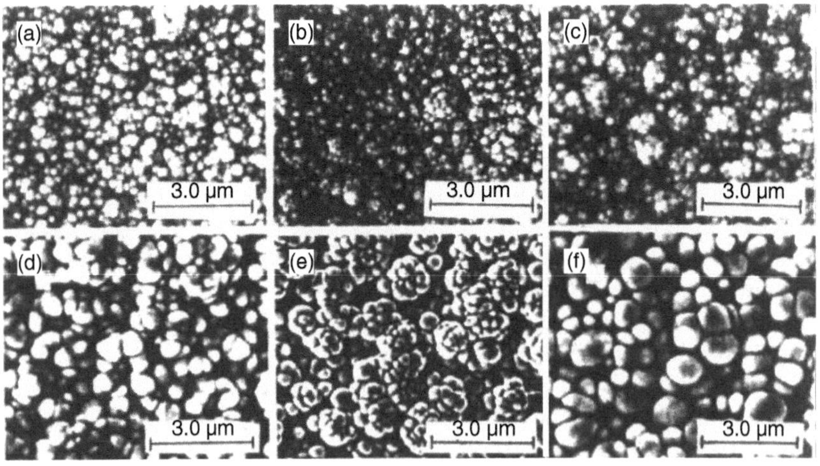

FIGURE 9.34 SEM photos of Ni–Cu–P alloy coatings. The concentrations of copper sulfate in the plating bath (g/L) are (a) 0; (b) 0.05; (c) 0.1; (d) 0.2; (e) 0.4; and (f) 0.6 [132].

larger particles, and the surface of the grains becomes rough. Thereby, this variation characteristic of grains illustrates its respective feature with that obtained by the above two groups. These authors did not give any explanation to their results. The explanation for this difference is obviously needed.

Another work that shows obvious difference with the above results was reported by Wei et al. in 2010 [135]. They investigated the surface morphology of Co–Pt–W–P alloy deposits by changing the concentration of $[PtCl_4^{2-}]/[Co^{2+}]$ in the plating bath. They studied four different metal salts ratio $[PtCl_4^{2-}]/[Co2+]$ in the plating bath: (a) 1:25; (b) 2:25; (c) 3:25; and (d) 4:25. The results are shown in Fig. 9.35. At low Pt content (Fig. 8.35a), the surface of Co–Pt–W–P is covered with roundish clusters of diameter 100 nm. When Co–Pt–W–P thin films are electroless deposited from the solutions with $[PtCl_4^{2-}]/[Co^{2+}]$ = 2:25, oval cluster structures are observed (Fig. 9.35b). As $[PtCl_4^{2-}]/[Co^{2+}]$ is up to 3.25, the surface of Co–Pt–W–P is covered with hexagonal aggregates with a diameter of 50–100 nm (Fig. 9.35c). At $[PtCl_4^{2-}]/[Co^{2+}]$ = 4.25, the hexagonal aggregates become larger and the surface looks unlevel (Fig. 9.25d). With increase of Pt content (unfortunately the authors did not measure the composition with changing the $[PtCl_4^{2-}]/[Co^{2+}]$), the shapes of the grains in the coatings are greatly changed. The authors did not give any explanation to this fact.

In 2010, Azizi et al. [136] also reported similar phenomenon in EP NiCoFe–B coatings. They changed the concentration of molar ratios of $NiSO_4$/ ($NiSO_4$ + $CoSO_4$) in the coating bath: 0, 0.25, 0.50, 0.75, and 1.0, respectively.

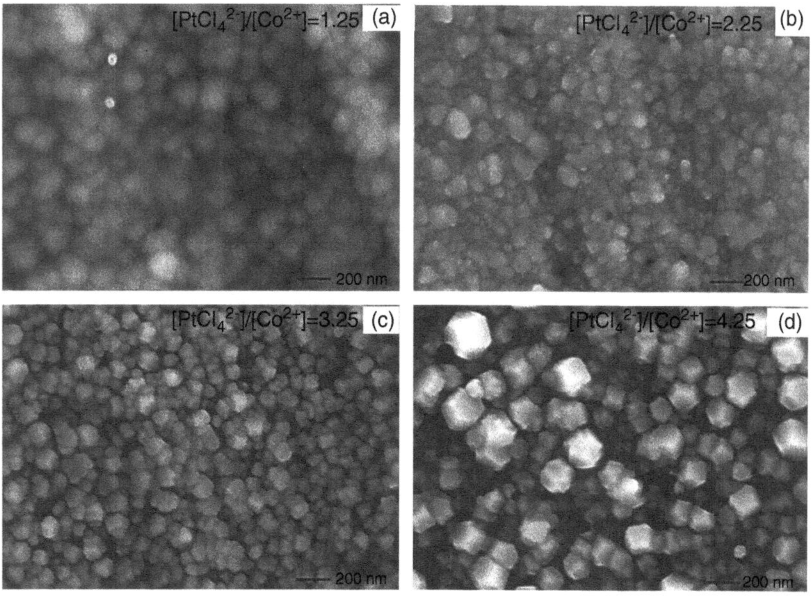

FIGURE 9.35 Surface morphology of electroless deposited Co–Pt–W–P under different metal salts ratio $[PtCl_4^{2-}]/[Co^{2+}]$ in the plating bath: (a) 1:25; (b) 2:25; (c) 3:25; and (d) 4:25 [135].

The corresponding composition of CoNiFe coatings is Ni from 0 to 70 wt%, Co from 80 to 0 wt%, and Fe from 20 to 28 wt% (Fig. 2 in [136]). They found that the CoFe–B coatings obtained from bath with a molar ratio of 0 have globular morphologies (Fig. 2 in [136]). Addition of the nickel to this coating bath ($NiSO_4/(NiSO_4 + CoSO_4)$) molar ratio of 0.25 leads to less particulate size of coating and changes the morphology of coating. This coating has pyramidal-shape morphology. On further increase in nickel content, the coating changes the coating morphology to a mixture of pyramidal shape and cauliflower shape for deposit obtained from the equal content of the nickel and cobalt in the coating bath. $NiSO_4/(NiSO_4 + CoSO_4)$ molar ratios of 0.75 and 1 show depositions with larger morphology than the other investigated coating. These coatings indicate the smooth surface with globular morphology. However, the authors still did not explain the reasons why this occurred.

9.4.2 Effects of Reductant on SEM

Several groups investigated the effect of reductant in the plating bath on surface morphology [137–141]. In 2006, Baskaran et al. [140] reported the morphology of EP Ni–B deposits by varying the reductant $NaBH_4$ concentration (0.2, 0.4, 0.8, and 1.0 g/L) in the plating bath. The measured B contents in the coatings are 0.6, 1.2, 2.6, and 3.2 wt%. Figure 9.36 indicates the SEM images of Ni–B

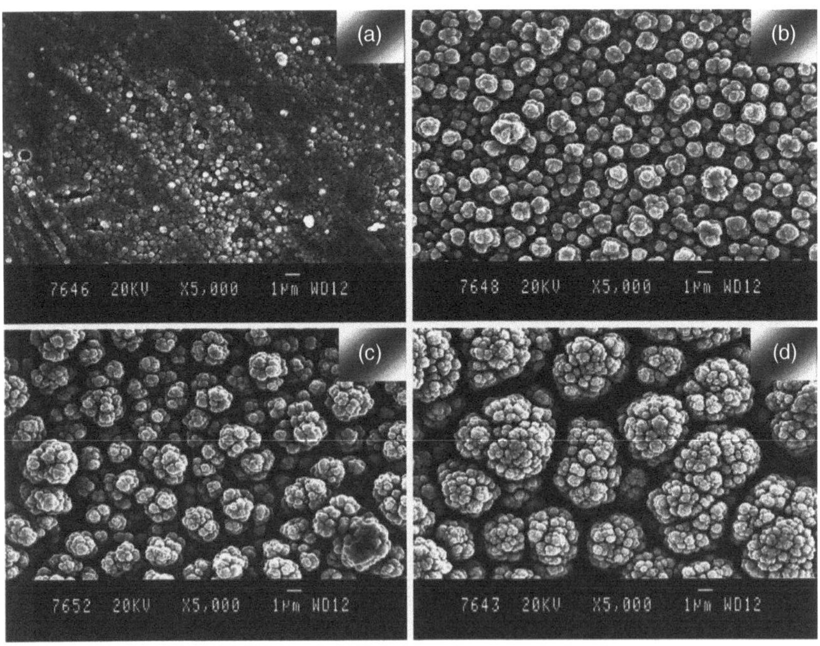

FIGURE 9.36 SEM of electroless Ni–B coatings obtained at different concentrations of $NaBH_4$ (a) 0.2 g/L (NB2), (b) 0.4 g/L (NB4), (c) 0.8 g/L (NB8), and (d) 1.0 g/L (NB10) [140].

coatings at different concentrations of reductant $NaBH_4$. The coatings are uniform and they consist of agglomerates of nickel that are randomly distributed. The morphology resembles a typical cauliflower-type feature. Comparing the four morphology images of the Ni–B coatings, it can be seen that the size of the nodules increases with increase in concentration of $NaBH_4$. These nodules combined to form a granular-type structure, which is quite evident in Fig. 9.36d. This granular structure makes the electroless Ni–B coatings naturally lubricious and enables it to achieve a higher wear resistance by reducing the surface contact [4,5]. De Robertis et al. in 2008 also studied the effect of reductant on morphology of EP deposits [141]. The NaH_2PO_2 concentrations in the baths they used are: (a) 0.10 mol/L, (b) 0.20 mol/L, (c) 0.30 mol/L, and (d) 0.40 mol/L. The EDS semiquantitative analysis provided the alloys composition: $Pd_{93.9}P_{6.1}$, $Pd_{88.1}P_{11.9}$, $Pd_{86.1}P_{13.9}$, and $Pd_{85.7}P_{14.3}$. Despite the increase in the phosphorus content with the increase in NaH_2PO_2 concentration in the plating bath, the distribution of codeposited phosphorus in the alloy is uniform. The growth type and morphology of the deposits are not significantly altered by the use of different NaH_2PO_2 concentrations in the plating bath. As shown in Fig. 9.37, a 3D growth of quasispherical nuclei that coalesce laterally occurs forming compact layers; the development of new layers is clearly seen in Fig. 9.37c and d. Pd–P films prepared from solutions where the reducing agent concentration is lower than 0.20 mol/L (Fig. 9.37 A) present cracks. The cracks appeared are likely due to the stress caused by the trapped hydrogen gas. Raising the NaH_2PO_2 concentration, the

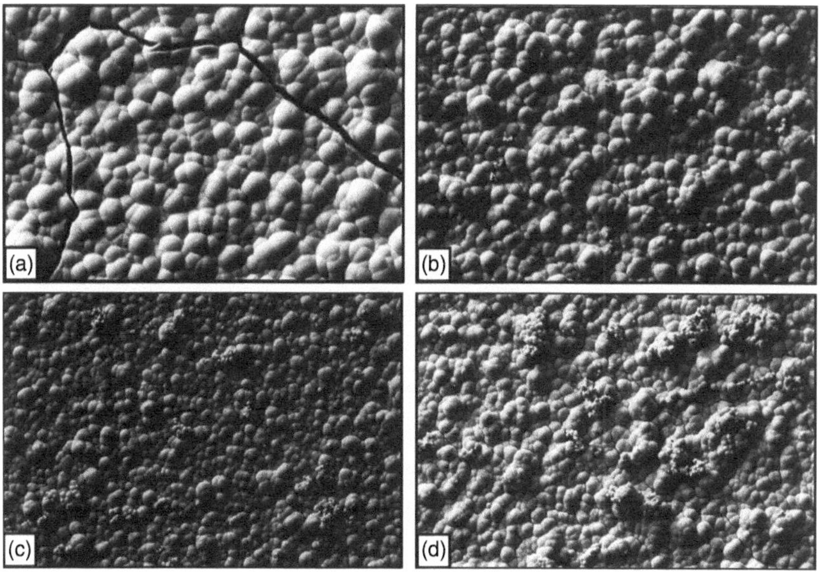

FIGURE 9.37 SEM of Pd–P electroless films over carbon steel 1045, deposition time of 120 min and NaH_2PO_2 concentration in the bath of: (a) 0.10 mol/L, (b) 0.20 mol/L, (c) 0.30 mol/L, amd (d) 0.40 mol/L [141].

deposit cracking is less noticeable (Fig. 9.37b to d) and the alloy displays better adherence to the substrate; this feature can be related to the difficulty for the cracks to propagate in amorphous material in the absence of long-range order.

9.4.3 Effects of Complexing Agents on SEM

Only a few papers about the effect of complexing agent on SEM of the EP coatings have been published [142–145]. In 2012, Babanejhad et al. investigated the effect of complexing agent on the morphology of electroless deposited Ni–P alloy [145]. They changed the lactic acid concentrations of 0 g/L, 14 g/L, and 28 g/L in the plating bath. The corresponding contents of P in the deposits are 0, around 8 and 14 wt% (Fig. 5 in [145]). The substrates used were (20 × 20 × 4 mm) St37 steel sheets. Figure 9.38 illustrates the SEM (a, b, and c) and AFM (d, e, and f) images of the EN deposits. Figure 9.38a shows SEM micrograph of electroless Ni–P deposit obtained from deposition bath without lactic acid. Although small and bright grain centers as initial Ni–P deposit can be found on the surface, the main microstructure of St37 substrate is almost visible. It is obvious that this SEM image is different from the convenient nodular structure in EN deposits. SEM micrograph of electroless Ni–P deposit with 14 g/L lactic acid has been shown in Fig. 9.38b. Main microstructures of substrate St37 disappeared completely and a rough layer of electroless Ni–P deposit covered the entire surface. Increasing lactic acid concentration to 28 g/L changed the surface morphology of deposits from a rough layer to a smooth layer as seen from Fig. 9.38c. AFM images of deposits with different concentrations of lactic acid are shown in Fig. 9.38d–f. Figure 9.38d shows the initial Ni–P growing

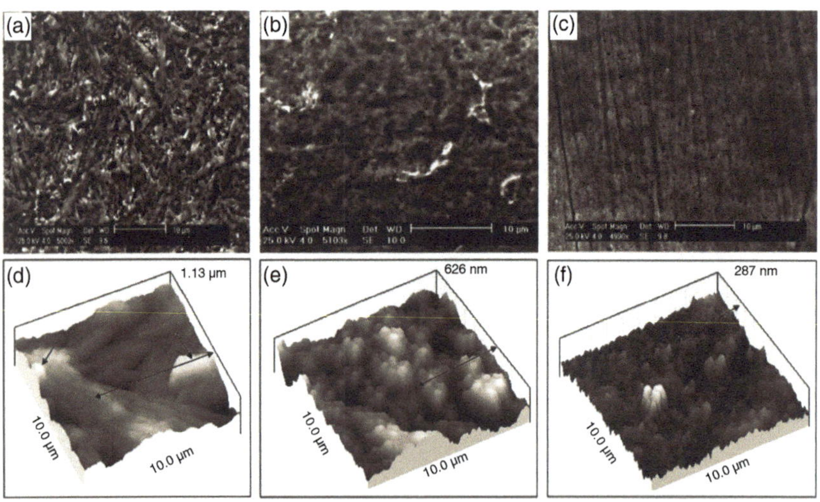

FIGURE 9.38 SEM micrographs and AFM images of electroless Ni–P deposit obtained from deposition bath without lactic acid (a and d), with 14 g/L lactic acid (b and e) and with 28 g/L lactic acid (c and f) [145].

centers in a coniform structure on the surface (arrows) with no concentration of lactic acid. Average surface roughness is evaluated as ~600 nm with standard deviation of 200 nm. As seen from Fig. 9.38e, the surface has been completely covered with several high regions as coniform structure and low regions as crater structure. Initial Ni–P deposits in coniform structure seem to be consisting of some agglomerated nodules. Average surface roughness of deposit was evaluated as 280 nm with a standard deviation of 150 nm. Figure 9.38f shows the AFM image of deposit with 28 g/L lactic acid concentration. It shows a nodular-like structure with a few coniform regions. Average surface roughness of deposit was evaluated as 100 nm with a standard deviation of 100 nm. With increase of lactic acid concentration from 0 g/l to 28 g/L, coniform structures changed to nodular structures. Huang and Cui in 2007 [143] also investigated effect of complexing agent on the morphology of EP Ni–P alloy, but they obtained a very different morphology from the above results. They selected aminoacetic as a main complex in the deposition bath. From the obtained SEM images of the deposits (Fig. 3 in [143]), they observed spherical clusters on the surface when there is no concentration of aminoacetic in the bath. That is the usual surface morphology in EN deposits. With the concentration of aminoacetic increasing in the deposition bath, the deposited films more and more smooth. The surface was very smooth, no spherical clusters appeared, and a mirror bright deposition was obtained at a consecration of aminoacetic up to 20 g/L.

Effect of complexing agent on electroless copper plating on the surface of Al_2O_3 ceramic was reported by Wang et al. in 2010 [144]. The complexing agents they used are EDTA, $NaKC_4H_4O_6$, and a compound of the two agents. The results are shown in Fig. 9.39. The surface features of the deposits are basically nodular shapes although it is not so clear because the magnification may not be large enough. When EDTA is used as a sole complexing agent, the surface roughness of the plating layer is 0.39 μm, agglomerated copper particles are formed and uniformity of plating layer is poorer; when $NaKC_4H_4O_6$ is used as a sole complexing agent, the surface roughness is 0.46 μm, and some impurity particles with a diameter of 2–5 μm are observed; and when EDTA and $NaKC_4H_4O_6$ are used as a compound complexing agent, the surface roughness is 0.35 μm, and thus obtained copper coating has a compact microstructure, a uniform and dense copper particle distribution, as well as a smooth and clean surface. As for the

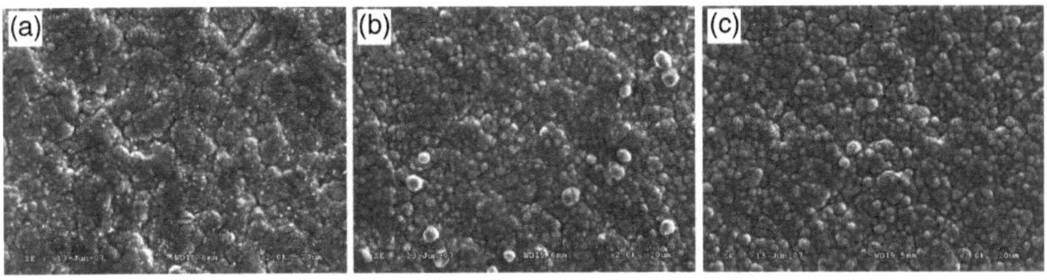

FIGURE 9.39 SEM micrographs 2,000X of electroless Cu deposit obtained from deposition bath with complexing agent: (a) EDTA, (b) sodium-potassium tartrate and (c) EDTA+ sodium-potassium tartrate [144].

color of the deposits, these authors found that the Cu coating layer was dark red, bright surface, smooth, but little small black dots appear when EDTA was used solely. When $NaKC_4H_4O_6$ was used as a sole complexing agent, the coating surface was pink, bright surface; there are also a few small black dots and impurity particles. When the compound complexing agent was used, the coating surface was pink, bright surface and smooth, clean with no other impurities.

9.4.4 Effects of Stabilizers on SEM

Several papers on effect of stabilizers on surface morphology in EP coatings have been published [7,60,146–151]. The deposits include Cu, Ag, Ni–P, Ni–B, and Co–P, and the stabilizers are thiourea (Tu), maleic acid (MA), guanidine hydrochloride, benzotriazole (BTA), 5-aminotetrazole (ATRA), 3-amino-1,2,4-triazole, Pb^{2+}, and saccharin (SAC). We will select some examples to explain and describe as follows.

In 2008, Hyo-Chol Koo and Jae Jeong Kim [149] reported the effects of three stabilizing agents (BTA, ATRA, and 3-amino-1,2,4-triazole) on film morphology in EP Ag. An optimum additive concentration was obtained through comparing film roughness. So, they used stabilizers at concentrations of 0.67 mmol/L BTA, 3.9 mmol/L ATRA, and 0.3 mmol/L of 3-amino-1,2,4-triazole. Figure 9.40 illustrates surface morphologies of Ag films with stabilizing agents. The addition

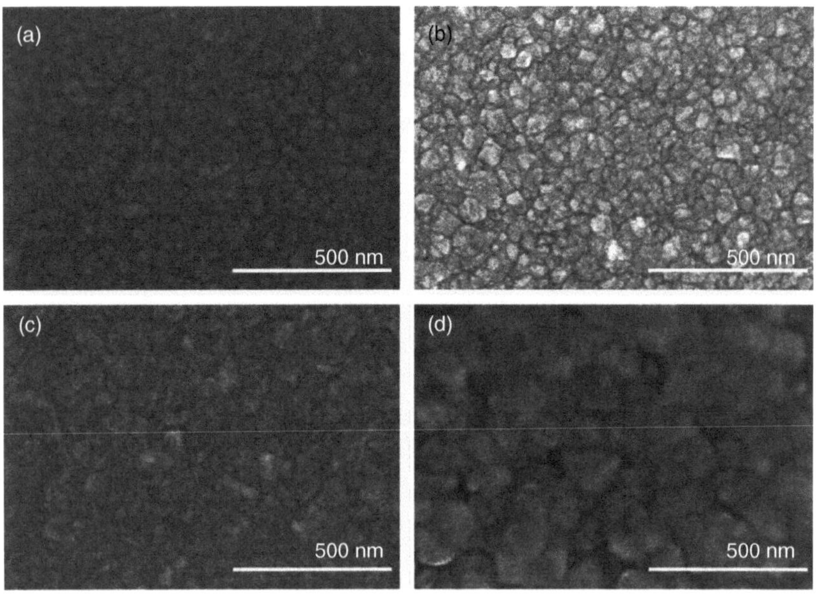

FIGURE 9.40 Surface morphologies of electroless deposited Ag films with various stabilizers after 10 min deposition: (a) without additives, (b) with 0.67 mmol/L of BTA, (c) with 3.9 mmol/L of ATRA, and (d) with 0.3 mmol/L of 3-amino-1,2,4-triazole [149].

of ATRA and 3-amino-1,2,4-triazole induced smooth surface morphologies, which showed no significant difference compared with the nonadditive case, while the addition of BTA made distinct grains.

In 2004, Cheong et al. [147] investigated the effects of two electroless nickel bath stabilizers (thiourea and maleic acid) on morphology of Ni–P deposits. The concentrations of Tu are 0, 0.3, 0.9, 1.3, 15, and 2.0 mg/L. Figure 9.41 shows the corresponding SEM pictures. The horizontal ridges in the SEM micrographs outline the underlying copper substrate morphology. The surface produced with no stabilizer was very smooth (Fig. 9.41a), but started to become nodular with 0.3 mg/L thiourea (Fig. 9.41b). The nodular structure ("orange peel" shape) became more evident with a 1.3 mg/L addition of thiourea (Fig. 9.41d), and small pits developed on the surface. This nodular structure completely evolved into a cauliflower-like structure when the thiourea concentration was increased to 2.0 mg/L (Fig. 9.41f). The transition between nodular and cauliflower-like structures is apparent at 1.5 mg/L (Fig. 9.41e). They studied the SEM morphology at various concentrations of MA: 40, 80, 180, 300, 2000, and 3000 mg/L. The surface features are different from those obtained by Tu (Fig. 4 in [147]). When adding 40 and 80 mg/L of MA into an EN deposit, ripple-like (or say wave-like) structures were formed on the surface. A surface-smoothing effect was observed at 180 and 300 mg/L MA. Small nodules started to form at a stabilizer concentration of 2000 mg/L, and became more evident at 3000 mg/L MA. The authors also measured the grain size distribution using an image analysis program (Image-Pro Plus). The changes in the grain size distribution with the addition of the stabilizer are shown in original Fig. 8 in [147]. It can be seen that the addition of 0.9 mg/L thiourea produces more, smaller grains compared with the deposit produced from the solution with no stabilizer. Moreover,

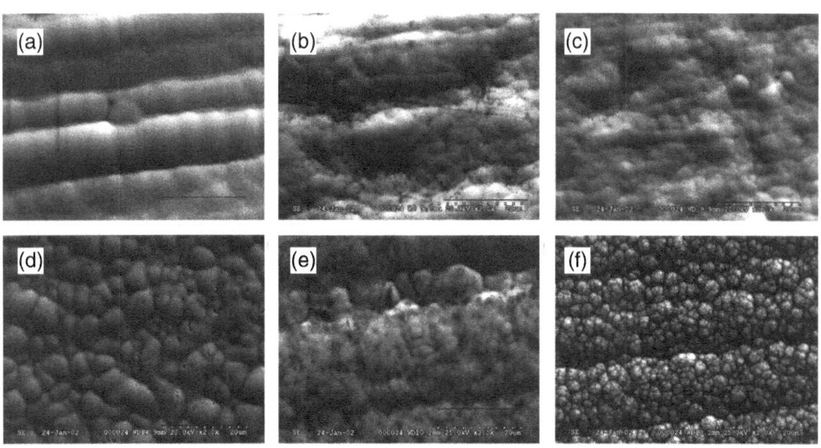

FIGURE 9.41 SEM micrographs (2000×) of the electroless Ni–P deposits from the plating solution with different concentrations of thiourea stabilizer. The black lines in the SEM micrographs indicate 20 mm [147].

a reduction in the number of larger grains is observed. With the addition of 40, 180, and 3000 mg/L thiourea, the percentage of smaller grains of ~1–2 nm size increased significantly from 40 to 72%.

These authors have explained their above experimental facts. As it is known that during the electrochemical deposition process, new grains will grow if the size of an initial cluster exceeds the critical dimension N_c. The larger the N_c, the more favorable it is for the growth of a single crystal from a previously nucleated seed grain. According to Tian et al. [151], the critical dimension N_c for a 2D-like nucleus can be described as follows:

$$N_c = \frac{bs\varepsilon}{(Ze\eta)^2} \tag{9.2}$$

where s, ε, Z, η, and b are the area occupied by one atom on the surface of the nucleus, the edge energy, the effective electron number, the overpotential, and a constant, respectively. Since the added stabilizers adsorb to the catalytic-free nickel surface, the rate of the charge transfer reaction will be decreased. Thus, the open circuit potential during the deposition and deposition potential obtained from an anodic and a cathodic half cell reactions tend to increase, respectively. From Eq. (9.2), the increase in the overpotential decreases N_c. This will enhance the nucleation process, resulting in an increasing number of smaller sized grains. As the concentration increases, the number of small sized grains is increased. In such a way, the obtained above data can be understood clearly.

The influence of MA or succinic acid (SA) on surface morphology in the EP Co–P coatings was reported by Rudnik and Gorgosz in 2007 [60]. The effects of the various concentrations of the stabilizers on the morphology of the Co–P deposits are shown in Fig. 9.42. The surface produced without additives was characterized by a sharply outlined wave-like structure with a dense net of shallow pits (Fig. 9.42a). In the presence of MA this morphology evolved gradually into the nodular structure (Fig. 9.42b,c). At higher MA concentrations the metal was deposited in the form spherical grains with diameters of about 0.25–0.5 μm (Fig. 9.42 d,e). A spherical structure was formed on the surface upon the influence of SA (Fig. 9.42 f). In a word, as the concentration of MA increased, the wave-like structure evolved into a nodular one.

Rudnik and Gorgosz [60] used the similar reasoning to the previous section to explain their experimental data. They believed that the significant changes in the surface morphology resulted from the MA adsorption. During the electrochemical deposition the increase in the electrode over potential leads to the decreasing dimensions of the metal grains because the adding MA adsorbs to the catalytic nickel surface. This enhances the nucleation process and increases number of the grains with smaller sizes. As the concentration of MA increased the wave-like structure evolved into a nodular one.

The effect of Tu and saccharin (SAC) on the surface morphology of an electroless Ni–B plating layer on the AZ91D magnesium alloy was investigated by

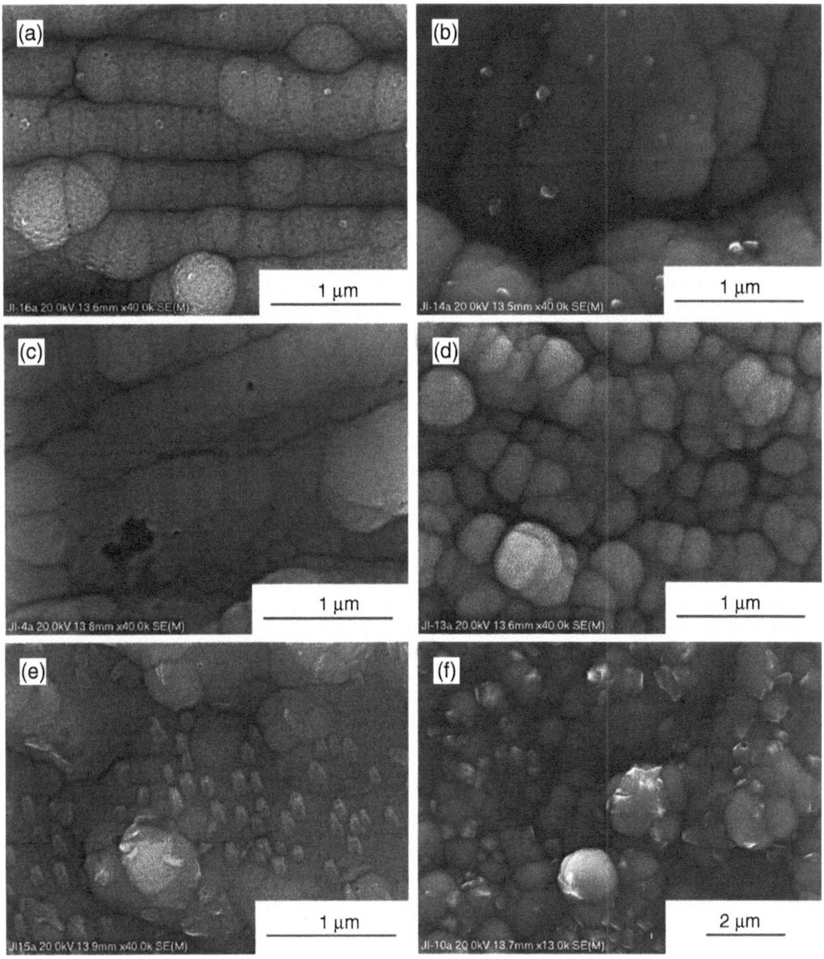

FIGURE 9.42 The morphology of the as-plated Co–P deposits: (a) no additives; (b) 1 × 10^{-3} mol/L MA; (c) 5 × 10^{-3} mol/L MA; (d) 7.5 ×10^{-3} mol/L MA; (e) 1 × 10^{-2} mol/L MA; (f) 7.5 × 10^{-3} mol/L SA [60].

Wang et al. in 2012 [150]. Figure 9.43 illustrates the surface morphologies of the Ni–B plating coatings from the baths without and with Tu and SAC stabilizers. Without additives, the plating particles formed on the alloy surface are not uniform with size ranging from 1 to 10 μm. The spatial gaps between particles indicate incomplete coverage of the plating coatings on AZ91D. The authors found that hydrogen was generated when the sample was dipped in the plating solution. Black sediments were observed at the bottom of the plating bath after plating, and the obtained coating had a dark black color. These experimental facts suggest an unsuccessful plating on the magnesium alloy surface. On the contrary, uniform and compact plating particles with smaller size of 2–4 μm

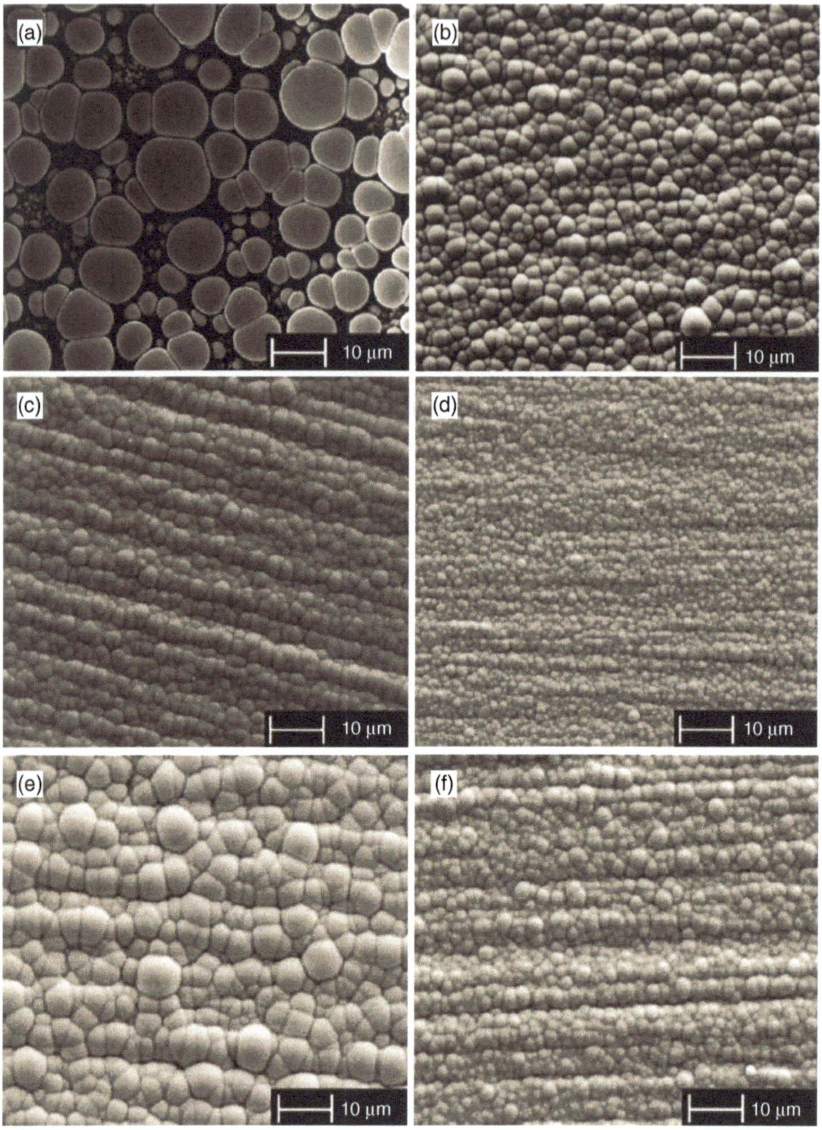

FIGURE 9.43 **SEM images of electroless Ni–B coating prepared in the plating bath with and without stabilizers: a. without additives; b. 0.5 mg/L thiourea + 2 g/L saccharin; with various concentrations of: c. 0.5 mg/L thiourea; d. 2.0 mg/L thiourea; e. 1.0 g/L saccharin; f. 4.0 g/L saccharin [150].**

can be obtained with small amounts of the two stabilizers in the plating bath. These authors believe that the difference in surface morphology as shown in Fig. 9.43a and b should be related to the influence of the two additives on the growth of Ni–B particles and their inhibition effect on hydrogen evolution during the plating process. Tu and SAC slowed down the coating particle growth in the prolonged period. Figure 9.43c to f shows the individual influences of Tu and SAC on the surface morphology of the Ni–B plating. Slightly different surface morphologies can be seen with different concentrations of the two stabilizers. The surface becomes nodular at 0.5 mg/L Tu and evolves into a cauliflower structure with smaller plating particles at 2.0 mg/L. This phenomenon is consistent with the findings by Cheong et al. [147] in Ni–P plating baths mentioned above. The increase in SAC concentration from 1.0 to 4.0 g/L significantly refines the nodular plating particle sizes. At 4.0 g/L SAC, the nodules coalesce together, which could enhance the compactness of the deposits. The grain refinement effect of Tu is more evident than SAC. This means the Tu may more effectively retard the Ni–B coating particle growth than SAC during plating.

9.4.5 Effects of Surfactants on SEM

As mentioned in Section 6.1.6 of Chapter 6, surfactants are surface active agents that lower the surface tension of a liquid, allowing it to spread easily, and lower the interfacial tension between two liquids. So as surfactants can enhance the stability of a suspension by increasing the wettability and the surface charge of suspended particles, it is easy to imagine that surfactants must influence the surface morphology of EP coatings. Many articles on the effects of surfactants on surface morphology in deposits have been published [152–160].

Silver nanostructures with controlled morphologies have been prepared on copper and aluminum substrates by a surfactant-assisted colloidal chemical method in 2011 [157]. Zhou et al. selected two surfactants: poly(vinyl pyrrolidone) (PVP), a polymeric surfactant and cetyltrimethylammonium bromide (CTAB), a cationic surfactant. Figure 9.44 and Table 9.2 show their results of morphology in the Ag coatings. Three distinctive conclusions can be obtained from the figure and the table. The first is that the morphologies are very distinct on the Cu substrate and Al substrate, respectively, even though the same surfactant (0.1 g PVP) was used and other conditions were identical. As is indicated in Fig. 9.44a, the typical SEM image is Ag microplates about 100 nm in thickness on copper substrate. Some microplates exhibit multilateral morphology, but others show irregular shape. However, the Ag microstructures fabricated on aluminum substrate (sample B) are formed by numerous hexagonal prismatic Ag microrods with diameter of 100–150 nm and length of 1–2 μm (the set picture is low-magnification SEM images of sample B). These authors further confirm that the products consisted of a large quantity of uniform microrods by using TEM. Why under the same reaction conditions, the Ag nanostructures formed on an aluminum substrate were microrods, instead of microplates formed on copper substrate? According to the authors, Al is more active than Cu, it is easier to reduce Ag atom from the $AgNO_3$ solutions. Ag seeds formed around the aluminum substrate are much more than that around the copper substrate and have much more time to further growth and self-organizing superstructures. This phenomenon implies that the control

FIGURE 9.44 SEM images of the Ag microstructures obtained with 0.1 g PVP and 5 mL $AgNO_3$ (0.02 M) at 50°C for 8 h (a) on a copper substrate; (b) on an aluminum substrate; inset in (b) is low magnification SEM images of sample B; (c) high-magnification SEM images of the as-synthesized Ag microarchitectures on an aluminum substrate for 8 h at 50°C with 0 g CTAB, 5 mL $AgNO_3$ (0.02 M); (d) SEM images of the sample D on an aluminum substrate obtained with 0.1 g PVP, 5 mL $AgNO_3$ (0.04), 2 mL HCl (2 M), for 24 h; and (e) the samples prepared without any surfactants [157].

TABLE 9.2 The Morphologies of the Ag Microstructures Obtained by Different Synthetic Conditions

Sample	Substrate	Surfactant	Morphology
a	Cu	PVP (0.1 g PVP)	Microplates
b	Al	PVP (0.1 g PVP)	Assembly of microrods
c	Al	CTAB (0.1 g)	Assembly of nanobelts
d	Al	PVP (0.1 g PVP, 2 mL HCl (2 M))	Assembly of nanosheets
e	Al	No any surfactants	Irregular morphology

of the nucleation process and growth rate of Ag particles may be the key step to affecting the morphology of Ag nanostructures. In a word, the aluminum substrate helps Ag seeds self-assemble into superstructures. The second is that using the same substrate (Al), Ag assembly of nanobelts was obtained by using surfactant CTAB (Fig. 9.44c), but assembly of microsheets by using PVP (0.1 g) and 2 mL of HCl (2 mol/L) (Fig. 9.44d). The widths of nanobelts are in the range from 30 to 50 nm, and the diameters of microspheres are between 10 and 15 μm. Why does the surfactant play a crucial role in the morphology of Ag nanostructures on Al substrate? According to these authors, CTAB and PVP are cationic and polymeric surfactants, respectively, and they could adsorb on the different crystal facets of Ag nanoparticles at initial generation stages. Moreover, they could guide the different anisotropic crystal growth because the absorption intensity of surfactants was different. Cationic surfactant CTAB with positive charges had strong selectivity adsorption on the special crystal facets caused by electrostatic action and would be particularly conductive to the formation of nanobelt structures. As the reaction continued, Ag nanobelts were fabricated with the help of CTAB, which can control the growth rates of various faces of silver. Ag nanobelts can self-assemble into a spherical joint structure. As for the formation of nanosheets, it is necessary for the bath to consist of besides 0.1 g PVP, 2 mL of HCl (2 mol/L), and 0.04 M $AgNO_3$. The experiments by these authors found that if only 2 mL of HCl (2 mol/L) was added in the reaction system (0.02 mol/L $AgNO_3$, 0.1 g PVP), a curdy white mass of AgCl precipitated in the solutions as soon as hydrochloric acid was added. After 24 h, some irregular particles were obtained on the Al substrate without any superstructures. The reason is that Ag(I) ions had been almost entirely consumed by the formation of AgCl, there were not enough Ag nanoparticles to form the self-assembled structures on the Al substrate. If $AgNO_3$ was up to 0.04 mol/L, the shape of the samples changed remarkably. After 8 h, a self-assembly of nanosheets (sample D) was obtained (Fig. 9.44d). The thickness of nanosheets is in the range between 50 and 100 nm.

Extending the reaction time to 24 h, the small polyhedral blocks disappeared from the product, and a uniform plate-like structure appeared finally (the inset in Fig. 9.44d). The third is that if no surfactant was added and other conditions unchanged, then the products show irregular morphology (Fig. 9.44e).

Using polyethylene glycol (PEG) as a surfactant in the copper deposition solution to improve the physical properties of the copper deposits was reported by Gan in 2012 [158]. They changed the concentration of PEG from 0 to 3.33×10^{-5} mol/L to see the variation of the Cu surface morphology. Figure 9.45 shows the surface morphology of the copper deposits obtained at different PEG concentrations. In the absence of PEG, the Cu deposits exhibited a rough topography and porous microstructure, as shown in Fig. 9.45a. The topography became smoother obviously and the color changed from dark brown to dark red with the addition of PEG from 0.83×10^{-5} mol/L to 3.33×10^{-5} mol/L, as shown in Fig. 9.45b–d. Generally speaking, it is evident from the figure that the particles in the morphology decreased by adding the surfactant PEG in the Cu bath. The growth direction of copper crystal grains all showed vertical to the substrate surface, which resulted in a loose microstructure. The authors did not explain the reasons for the variation of Cu morphology.

In 2002, Chen et al. [152] investigated the effect of nine different surfactants: six nonionic (Tergitol 15-S-12, Tergitol NP-9, Triton X-100, Tween 20, Tween 80, and Tween 85), two cationic (dodecyltrimethylammonium bromide

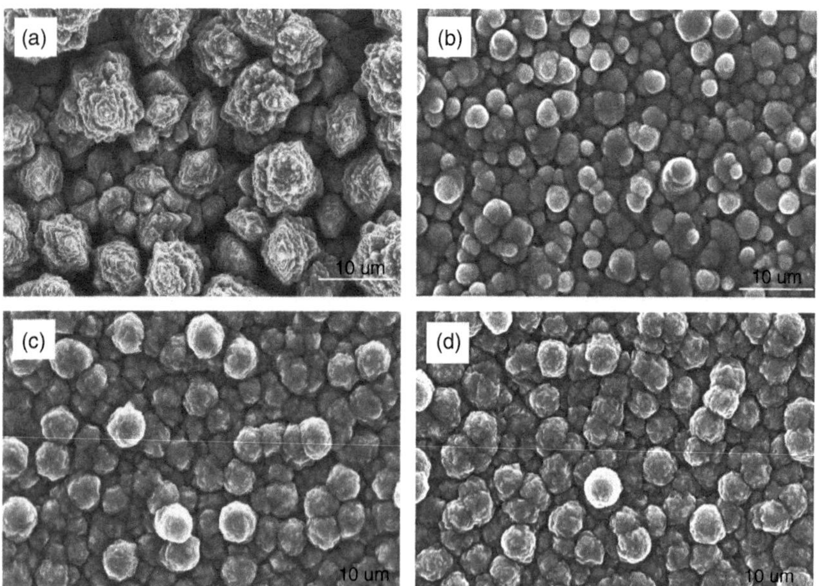

FIGURE 9.45 SEM photographs of the Cu deposits surface obtained from the bath at different PEG concentrations: (a) without PEG; (b) 0.83×10^{-5} mol/L; (c) 1.67×10^{-5} mol/L; and (d) 3.33×10^{-5} mol/L [158].

(DTAB) and dodecyltrimethylammonium chloride (DTAC)) and one anionic (sodium dodecylbenzenesulfonate (SDBS)) on surface morphology in EN coatings. Figure 9.46 shows SEM micrographs of the EN deposits from the plating solution without and with different surfactants. It clearly demonstrates that pits with dimensions of a few micrometers are seen in the sample deposited from the surfactant-free plating solution (Fig. 9.46a). The usual nodular structure seems not to be seen. This is somewhat surprising. The authors did not explain why their results differed from other studies. It can be seen from Fig. 9.46b–g that adding small amounts of nonionic surfactants except Tergitol NP-9 to the plating baths can reduce the formation of micro pitting on the deposits. Only a very few pits were observed in the Ni–P deposit plated from the bath containing Tergitol NP-9. Its presence in the plating solution successfully prevents the formation

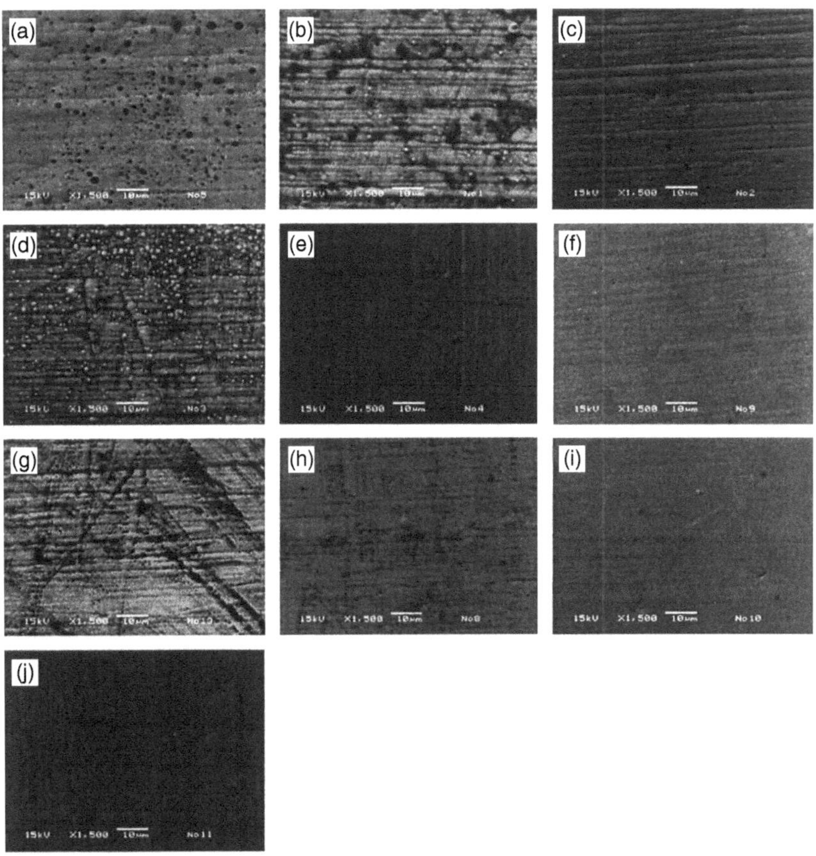

FIGURE 9.46 **SEM micrographs of the EN deposits from the plating solution with different surfactants: (a) no surfactant; (b) Tween 85 (5 ppm); (c) Tween 80 (7 ppm); (d) Tween 20 (5 ppm); (e) Tergitol 15-S-12 (3 ppm); (f)Tergitol NP-9 (3 ppm); (g) Triton X-100 (3 ppm); (h) SDBS (3 ppm); (i) DTAB (5 ppm); (j) DTAC (3 ppm). All have the same magnification 1500× [152].**

of pits and gives a very smooth surface. However, an overdose of Tergitol NP-9, viz., greater than 9 ppm, to the plating solutions, instead, would give the deposits a very rough surface. Adding Tween or Triton X-100 surfactants to the bath solutions produces quite a ragged surface on the EN deposits. Its terrain is like ripples, and the pits exist randomly. If adding a large concentration of Tergitol 15-S-12, greater than ca. 50 ppm, to the plating bath, the morphology of the resulting Ni–P alloys is quite similar to that resulting from adding a large concentration of Tergitol NP-9, that is, a very rough surface. On the contrary, it is evident from Fig. 9.46h–j that the initial presence of ionic surfactants in the plating baths, in general, produces the Ni–P deposits with smoother morphology. Pits are not so obvious, and the terrains are more flat on the resulting deposits.

Several groups also studied the surface morphology of EN deposits by addition of various surfactants [153–156,160]. Generally speaking, the surface morphologies have nodular structures even though various surfactants, for example, SDS, CTAB, PVP, DTAB, SLS, and Brij 30 were added in the EN or ECP Ni–P plating baths. The interested reader can read them. Here, we just introduce the work of Zielińska et al. in 2012 [160]. These authors also used DTAB as surfactant in the Ni–P + ZrO_2 ECP plating bath, thereby we can compare it with the above results. The ZrO_2 particles are below 100 nm in size. Figure 9.47 shows the SEM images of Ni–P+ ZrO_2 coatings: (a) without surfactant and (b) with 2.00× critical micelle concentration (cmc) of DTAB. The obtained composite layers have the spherical nodular structure, and the ZrO_2 grains embed into the Ni–P matrix. It is obvious that this morphology is very different from the above results in Fig. 9.46i. The same surfactant DTAB is used in both cases, of course, the amount of DTAB is different in both cases. Does just the different amount of DTAB or other reasons cause such difference? It seems that more work is needed.

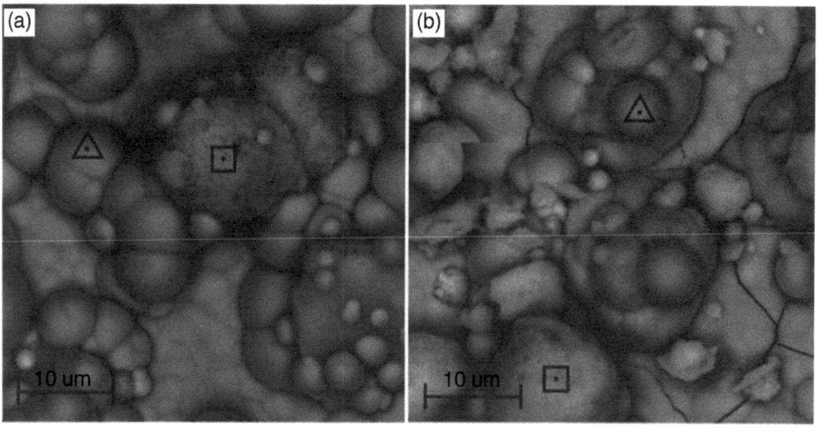

FIGURE 9.47 The scanning electron micrographs of Ni–P + nano-ZrO_2 coatings: (a) without surfactant and (b) with 2.00 × critical micelle concentration (cmc) of DTAB [160].

9.4.6 Effects of pH Values on SEM

Many papers have been published on the effects of the pH value on surface morphology of EP films [131,135,161–173]. We just select some of them to analyze and describe. In 2008, Malel et al. [169] investigated the effect of the pH value on surface morphology of EP Au coatings, as shown in Fig. 9.48. The used substrates were prepared by evaporation of a 200 nm Pd film onto the glass substrate. They find that efficient EP of Au can be observed at pH 1–7. At pH 1, massive EP of small Au nanoparticles can be seen. The deposition is fast and the rate of nuclei formation is high compared to particle growth (Fig. 9.48a). The coverage of the surface is complete and higher than at elevated pH. As the pH increases, Au deposition is less dense, the particles are bigger, and more uniform, which indicates that the rate of nucleation decreased, thereby favoring crystal growth. At pH 7, the formation of large microstructures (of the order of tens of micrometers) with hexagonal symmetry and composed of pyramidal nanoparticles is clearly seen. The average size of the nanoparticles changes from 23 (pH 1) to 32 nm (pH 7). The authors thus conclude that the pH value has a significant effect on the nature of the deposited gold and could be correlated with the kinetics of gold ions reduction by hydroquinone (H_2Q).

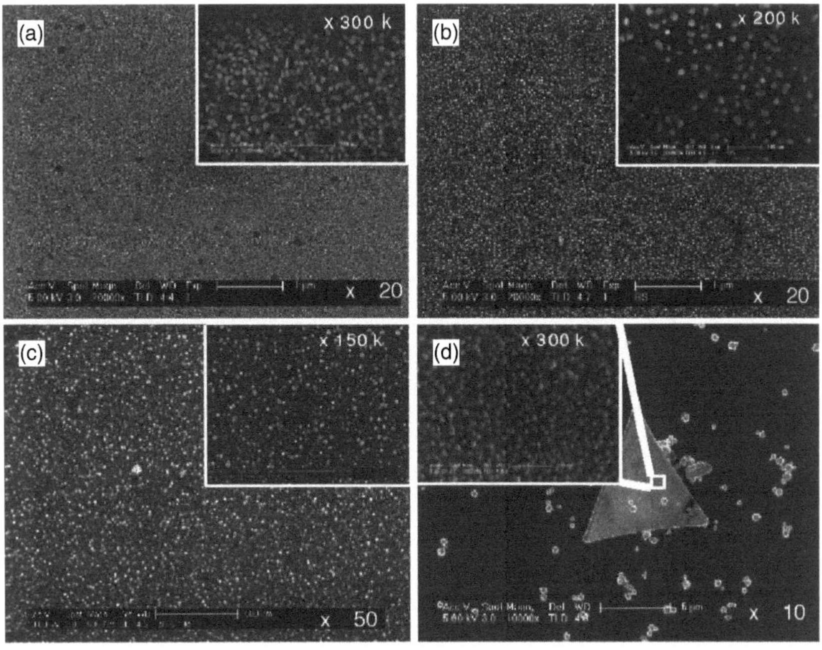

FIGURE 9.48 SEM images of bulk gold ED on Pd (10 s deposition) at different pH: (a) pH 1, (b) pH 3, (c) pH 5, and (d) pH 7. All solutions consisted of 5 mmol/L H_2Q and 2 mM $HAuCl_4$ [169].

In 1983, Cortijo and Schlesinger [161] reported a comparative study of the microstructure of electroless Ni–P thin films grown from an alkaline metallizing bath with different pH values (7.9–12.3). Figure 9.49 presents TEM images of EP EN films. Based on electron transmittivity, the authors judge the films to have grown to about the same thickness each, which estimated them to be about 5000 nm. It can be seen from the figure that the higher the pH value, the larger and more sparsely distributed on the surface the metal islands are. The authors measured the crystallite sizes, which vary from about 100 nm for pH 8 to 200 nm for the pH 12. After then, two groups measured the morphologies of

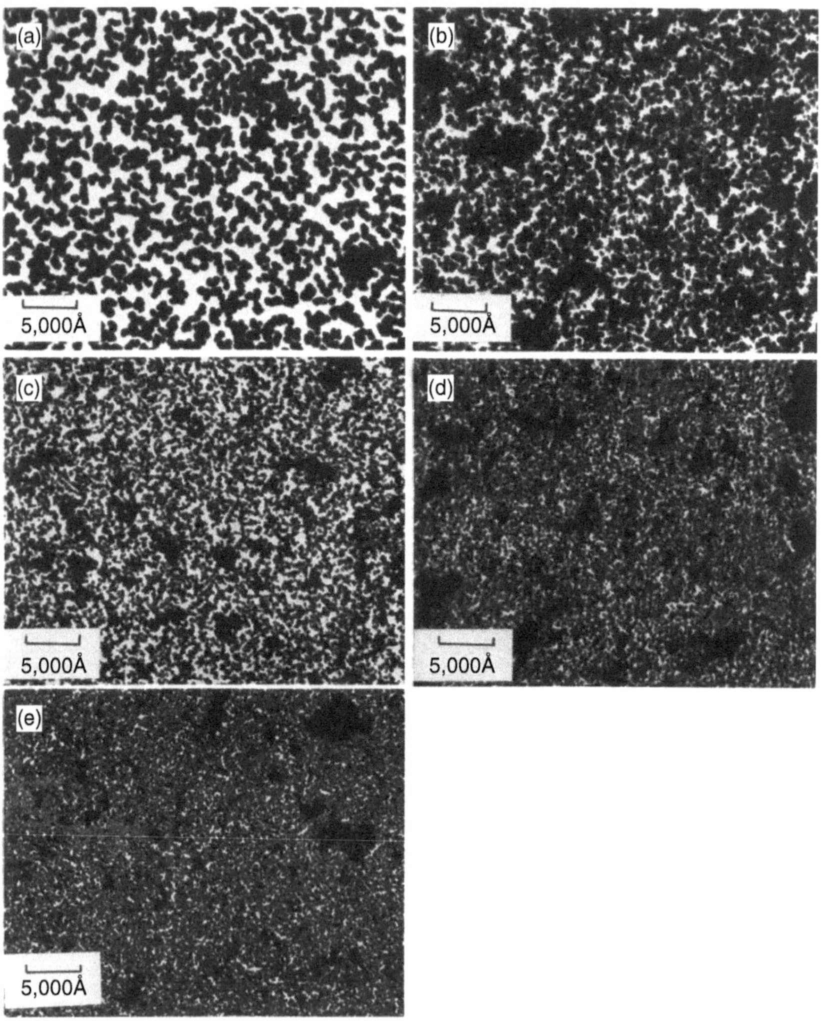

FIGURE 9.49 **TEM images of electroless deposited EN from an alkaline solution grown at different pH values: (a) pH = 12.3, (b) pH = 10.8, (c) pH = 9.9, (d) pH = 9.0, (e) pH = 7.9 [161].**

Ni–P + YSZ and Ni–P + SiC deposits by changing the pH values in bath solutions [170,172]. They found that if the pH value is too low, the sample is not fully covered by the composite coating. As the pH increases to an appropriate value, the composite coatings show a typical spherical nodular structure with good uniformity and dense coverage (Fig. 9 in [170] and Fig. 2 in [172]).

The effects of the pH value on morphologies in Co–P and Co–Ni–P coatings were reported by Nicholson et al. in 1986 [162]. The features and trends with variation of pH values in these two kinds of deposits are basically similar. For the Co–Ni–P bath, grain size decreases with increasing pH, and film continuity improves in the pH range 7.8–8.4, as shown in the pictures of Fig. 9.50. Increasing pH further continues to decrease grain size until, ultimately, very small grains are produced. They also studied the nucleation process of Co–P and Co–Ni–P deposits plated at different pH values (original Figs. 5 and 7 in [162]). SEM pictures of the nucleation process of Co–Ni–P films plated at low pH indicate that at low thickness, the nucleation sites form needle-shaped crystallites that spread to cover the substrate with increased plating time. As the pH is increased, the needle shapes are less predominant. Crystallites begin to have a more circular appearance. However, they found that the nucleation and growth process appears quite different in the Co–P bath. The difference is most pronounced at low pH, and Co–P nucleation sites appear round (original Fig. 7 in [162]). The number of nucleation sites increases until the substrate is covered. Then, the morphology has no difference with that in Co–Ni–P deposits. That is why we said that they are basically but not completely similar for the features and trends with variation of pH values in these two kinds of deposits above.

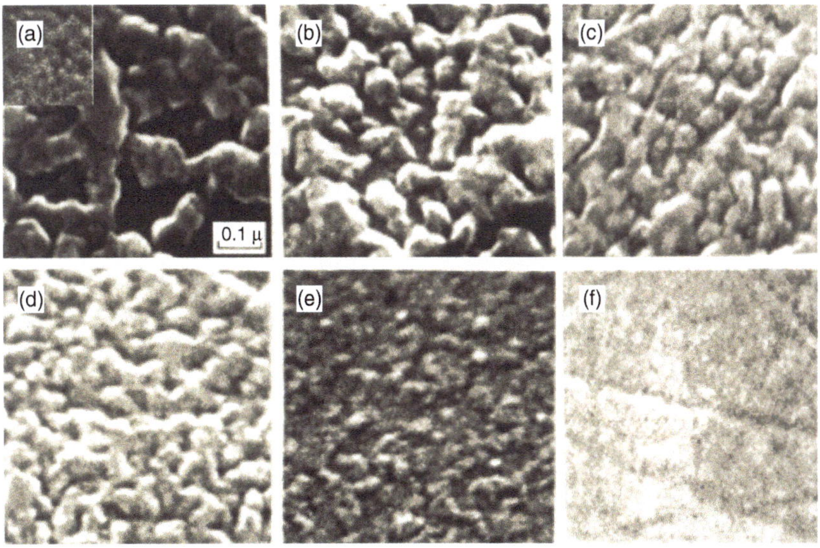

FIGURE 9.50 Electron micrographs of Co–Ni–P films (80 s) (a) pH = 7.8, (b) pH = 8.0, (c) pH = 8.2, (d) pH = 8.4, (e) pH = 9.1, and (f) pH= 9.3 [161].

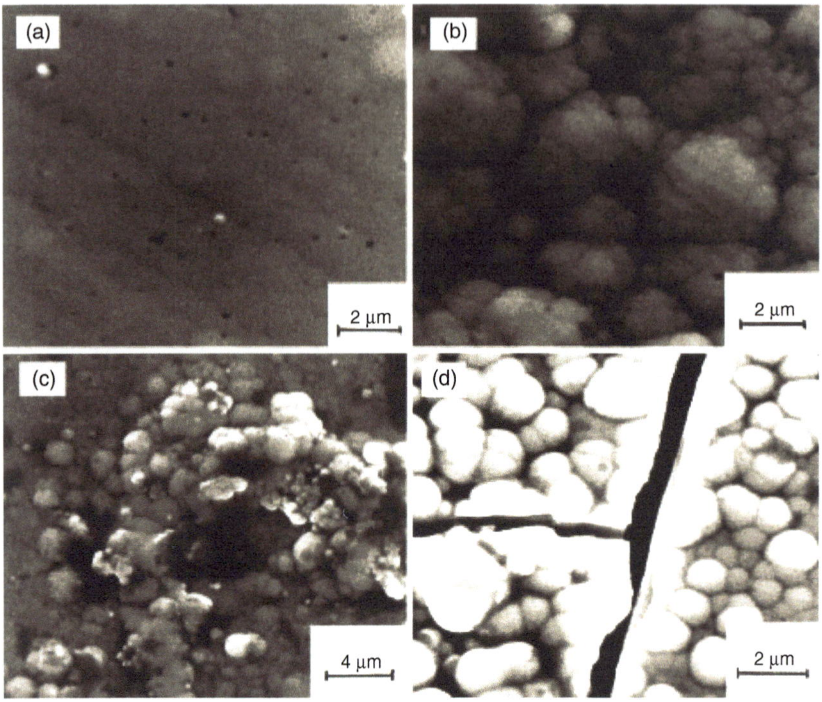

FIGURE 9.51 SEM images of the deposits obtained at different bath pH. (a) pH 8.0, $Ni_{79}Zn_{8.5}P_{13}$; (b) pH 9.5, $Ni7_{6.3}Zn_{12.5}P_{11.2}$; (c) pH 10.0, $Ni7_{9.8}Zn_{13.3}P_7$; and (d) pH 11.2, $Ni_{83.6}Zn_{10}P_5$ [164].

Three groups reported the effect of pH value on surface morphology in Ni–Zn–P deposits [163–165]. Figure 9.51 indicates the SEM images of the Ni–Zn–P films obtained at different bath pH (from 8.0 to 11.2). All the deposits were bright or semi-bright gray. The coating obtained from pH 8.0 is uniform, homogenous, and smooth (Fig. 9.51a). As the pH increases, the coatings are also homogenous and uniform, but show rounded formations (Figs. 9.51b–d). However, at pH 11.2 (Fig. 9.51d), the deposit is cracked and does not adhere to the iron substrate. The cracks in the deposits are generally due to internal stress, which depends on several parameters such as the nature and state of the substrate, the composition of the deposit, and hydrogen penetration during the deposition. The decrease in the phosphorus content in the deposit from 13% at pH 8 to 5% at pH 11.2 is the probable cause of the cracks.

The variation in the SEM pictures of the quaternary Co–Ni–W–P deposits obtained at various bath pH values was reported by Cheng et al. in 2007 [168]. The obvious feature of such morphology is the change from nodular to dendrite shape, as shown in Fig. 9.52. The surface morphology is cellular at pH 7.23 and 8.06, and the cellular grain grows up as pH increases from 7.23 to 8.06. However, when pH is increased to 8.85, the surface morphology is changed from cellular to dendritic. At pH 9.11, the dentritic grains become smaller. These authors said that

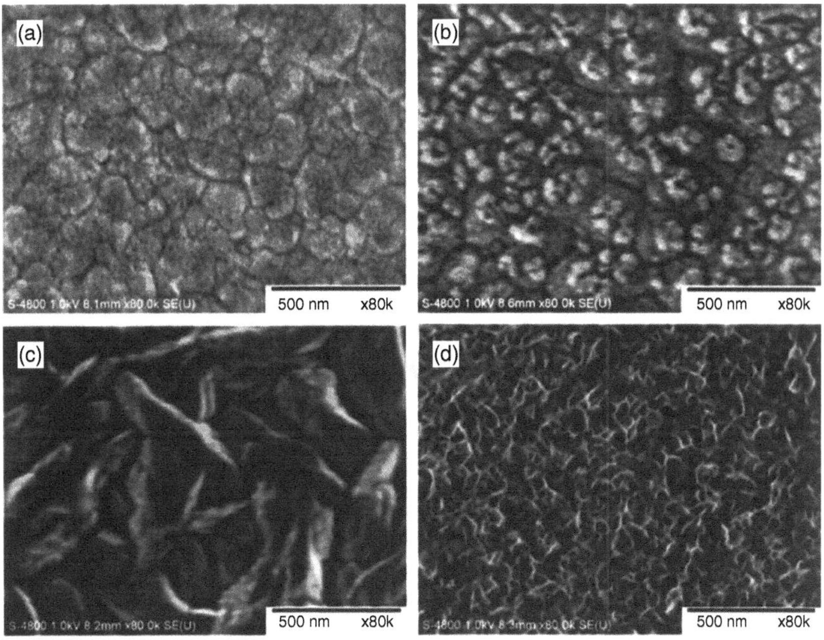

FIGURE 9.52 Dependence of surface morphology of Co–Ni–W–P films on bath pH. (a) pH = 7:23, (b) pH = 8:06, (c) pH = 8:85, (d) pH = 9:11 [168].

mechanism of the growth of the cellular-like grains and the dentritic-like structure surface are complicated and not yet very clear. Such variation of the SEM pictures with bath pH was also reported by Wei et al. in quaternary Co–Pt–W–P deposits in 2010 [135]. They observed hexagonal clusters and irregular-shaped crystallites when the films were deposited from electrolytes of pH 10.0 and 11.0. When the films were deposited from baths at pH 12.0, they found smoother surfaces and nodular clusters. Then, Co–Pt–W–P thin films prepared from the bath of pH 13.0 have spherical nodular structures and the surface looks compact covered and very smooth. All the features can be seen from the original Fig. 5 in [135]. These authors believed that this distinct variation in surface morphology is probably due to the difference of film composition with bath pH value.

9.4.7 Effects of Plating Temperature on SEM

A considerable number of articles described the effects of plating temperature on surface morphology in EP deposits [54,57,153,174–180]. In 1991, Iacovangelo [174] investigated SEM images of Au deposits on gold foil as a function of bath temperature: (a) 55°C, (b) 65°C, (c) 75°C, and (d) 85°C, as shown in Fig. 9.53. The striations shown mirror the original striations in the gold foil substrates. Concurrent with the increase in plating rate was coarsening of the grains and an improvement in uniformity.

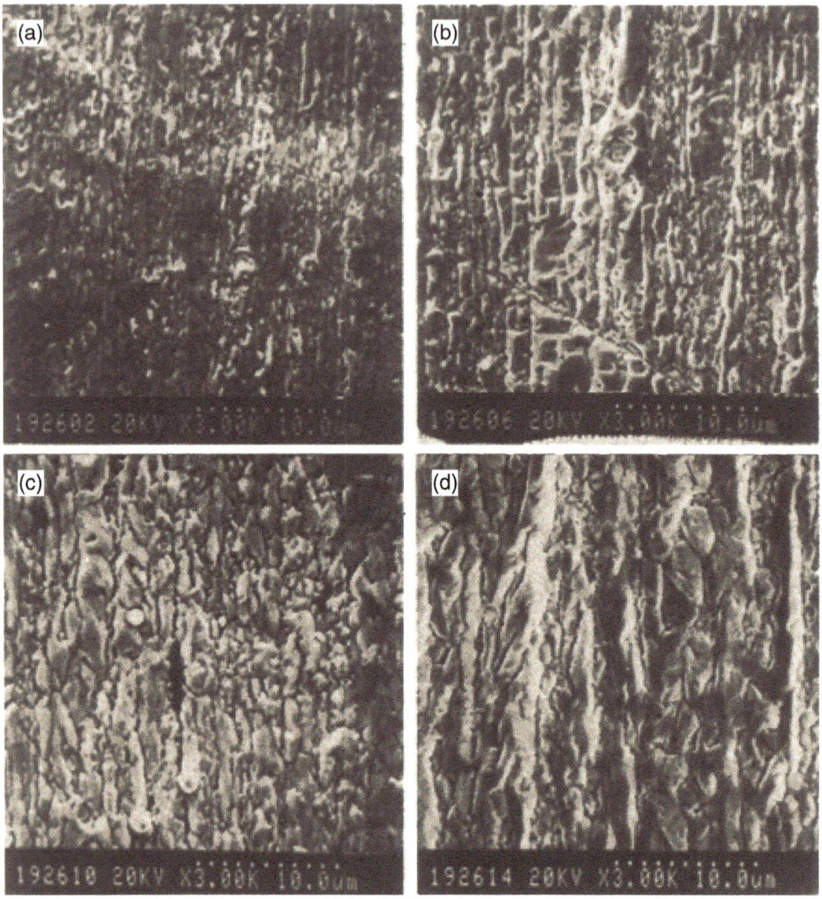

FIGURE 9.53 SEM pictures of Au deposits on gold foil as a function of bath temperature: (a) 55°C, (b) 65°C, (c) 75°C, and (d) 85°C [174].

The morphology versus the temperature in the plating bath for Ni–P coatings was investigated by Valova et al. in 1996 [176]. Figure 9.54 indicates the SEM images of coatings deposited on an aluminum substrate for 6 min: (a) Ni–P 88°C, (b) Ni–P 93°C, (c) Ni–Cu–P 88°C, (d) Ni–Cu–P 93°C. The SEM pictures in both EP Ni–P and Ni–Cu–P alloys display the typical globular feature of EN films. The higher the deposition temperature, the larger the globules' size is. The thicker deposits (at 93°C) are smoother. These authors also studied the surface morphology of the alloy coatings (Fig. 2 in [176]) deposited at the same temperature but on Fe substrates, and founded the similar features. However, the surface morphology follows the topology of the substrate at the initial stages of thin film formation. The former (for Al substrate) depends strongly on preplate substrate preparation. According to the contents in Section 9.1, it is seen that the magnification 1500× in Fig. 9.54 was too low, so that the features could not be seen very clearly. In 1996,

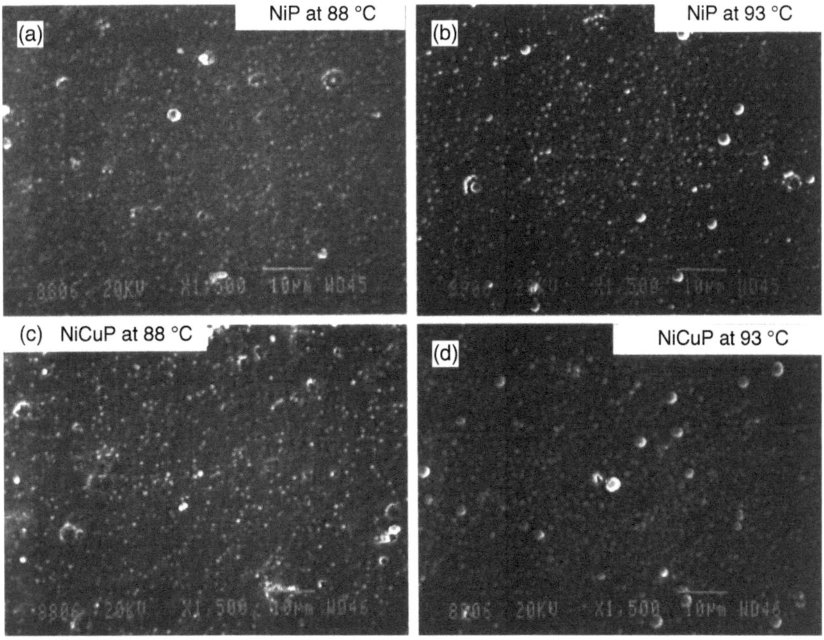

FIGURE 9.54 SEM images of coatings deposited on an Al substrate for 6 mm: (a) Ni–P 88°C, (b) Ni–P 93°C, (c) Ni–Cu–P 88°C, (d) Ni–Cu–P 93°C [176].

Tzeng also studied the effect of temperature on the microstructure of EN deposits from the bath of $NiSO_4$, NaH_2PO_2, and CH_3COONH_4 at pH 5.0. The bath temperatures are 70, 90. and 95°C. They obtained the similar results (Fig. 8 in [177]), not only the features are the typical nodular structure, but also the particles in the film are increased as the temperature increases. It should be noted that the magnification in Fig. 9.54 is not large enough; otherwise the features must be seen more clearly.

Bulbul in 2011 [54] reported the effect of bath temperature on surface morphology of electroless Ni–B coatings, as shown in Fig. 9.55. As seen in Fig. 9.55a, the Ni–B coating produced at temperature of 80°C has a granular structure. Along with an increase in diffusion owing to the temperature being elevated from 80°C to 88°C, the detached grains coalesced. However, no discernible change with respect to grain size was observed during this temperature raise. But, as the temperature was increased to 95°C, the grain size increased and the grains were changed to a nodular structure by forming pairs or triples at different configurations. As a result, the average grain size increased approximately two-fold as a consequence of increasing the temperature from 80°C to 95°C and the morphological structure was transformed to "primary nodular" from a granular structure.

As the bath temperature increases, enormous changes take place in the surface morphology of quaternary Co–Pt–W–P thin films, which was reported by Wei et al. in 2010 [135]. Inhomogeneous surface appears when the Co–Pt–W–P thin films were deposited at 65°C, as seen in Fig. 9.56a. Some crystals gather

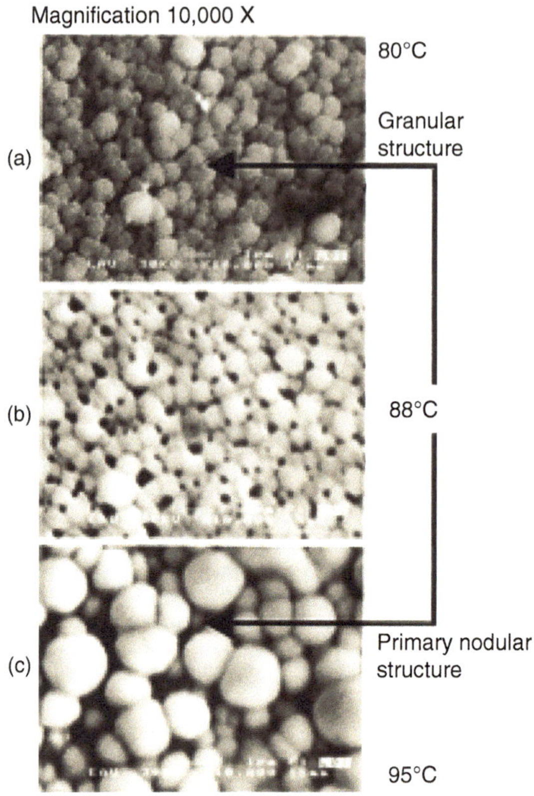

FIGURE 9.55 The effect of bath temperature on morphology in Ni–B deposits obtained at constant $NaBH_4$ amount (0.6 g/L) [54].

together and the surface looks rough. When the bath temperature was 75°C, more crystals gathered and the clusters look bigger (Fig. 9.57b). Then, the surface is covered with the spherical-like clusters when the Co–Pt–W–P thin films were deposited at 80°C and 90°C (Fig. 9.57c and d). It is evident that the surface morphology of EP Co–Pt–W–P deposits becomes more uniform and consistent with increase of the bath temperature. It must be pointed out that the original authors did not give any explanation to their enormous changes of SEM at different temperatures, especially some big particles appear at 75°C.

9.4.8 Effects of Plating Time on SEM

It is not difficult to imagine that the plating time must impact the surface morphology of the deposits. As the plating time increases, the components in bath and the plating rate will change, the nucleation and growth of crystal will vary, and thereby the morphology will change. Many papers discussing deposits have been published about the effect of plating time on surface morphology in EP [181–212].

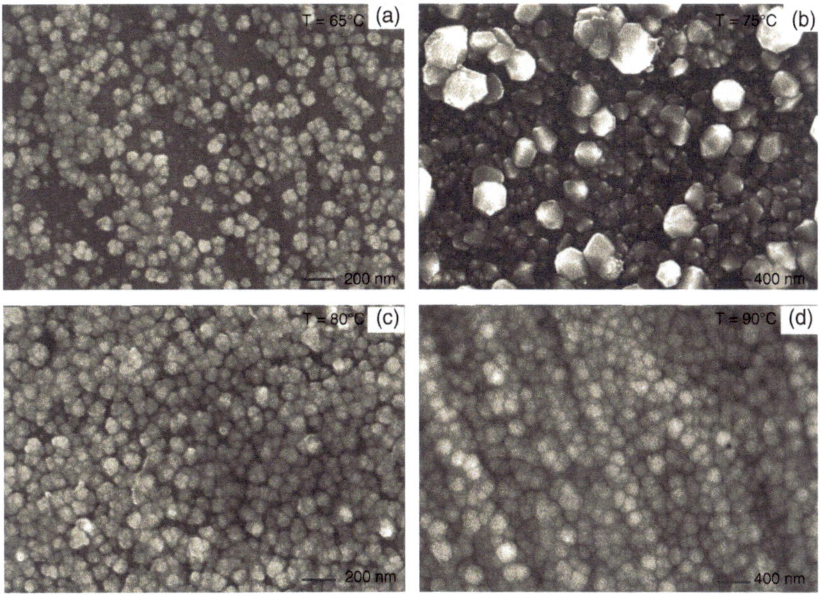

FIGURE 9.56 Surface morphology of electroless deposited Co–Pt–W–P under different bath temperatures: (a) 65°C; (b) 75°C; (c) 80°C; (d) 90°C; bath pH = 13.0; CoSO4 . 0.1 mol/L; $(NH_4)_2PtCl_6$ 0.004 mol/L; Na_2WO_4 0.005 mol/L; NaH_2PO_2 0.4 mol/L; $C_4H_4KNaO_6$ 0.3 mol/L; H_3BO_3 0.3 mol/L. Film thickness: 520 nm [135].

In 1997, Weber et al. [184] investigated the change in surface morphology with plating time for electroless copper. The substrate they used was highly oriented pyrolytic graphite (HOPG). Figure 9.57 shows the sequence of STM images during electroless copper deposition of the same area; the slow scan direction is indicated by arrows. The times for scanning the areas are (a) 0 to 56 s, (b) 56 to ll2 s; (c) 225 to 281 s; (d) 340 to 396 s; (e) 396 to 452 s; (f) 510 to 566 s. Three different crystals are denoted by arrows in every figure. Until Fig. 9.57c, these crystals grew faster than their surroundings, creating a valley in between them. The maximum difference in height is about 9 nm. In Fig. 9.57d, the valley is growing faster than the crystallites. When the scan time is from 396 to 452 s, the valley has nearly the same height as the first two crystals. In the last image (Fig. 9.57f), the previously faster growing crystallites are hardly recognizable anymore, as they are mostly covered by the now faster growing surroundings. These images show that the deposition rate in EP Cu fluctuates locally and with time. Other groups have investigated the same process [189,190,203], but the surface morphology appears to be inconsistent. For example, the results reported by Zhong et al. in 2006 [189] are very similar to those of Weber at al. Figure 9.58 is their morphology pictures for the Cu films deposited for 60, 90, and 120 s. The substrate plated for 60 s merely contains several large Cu grains with an average grain size of 4 μm (Fig. 9.58a). When the plating time

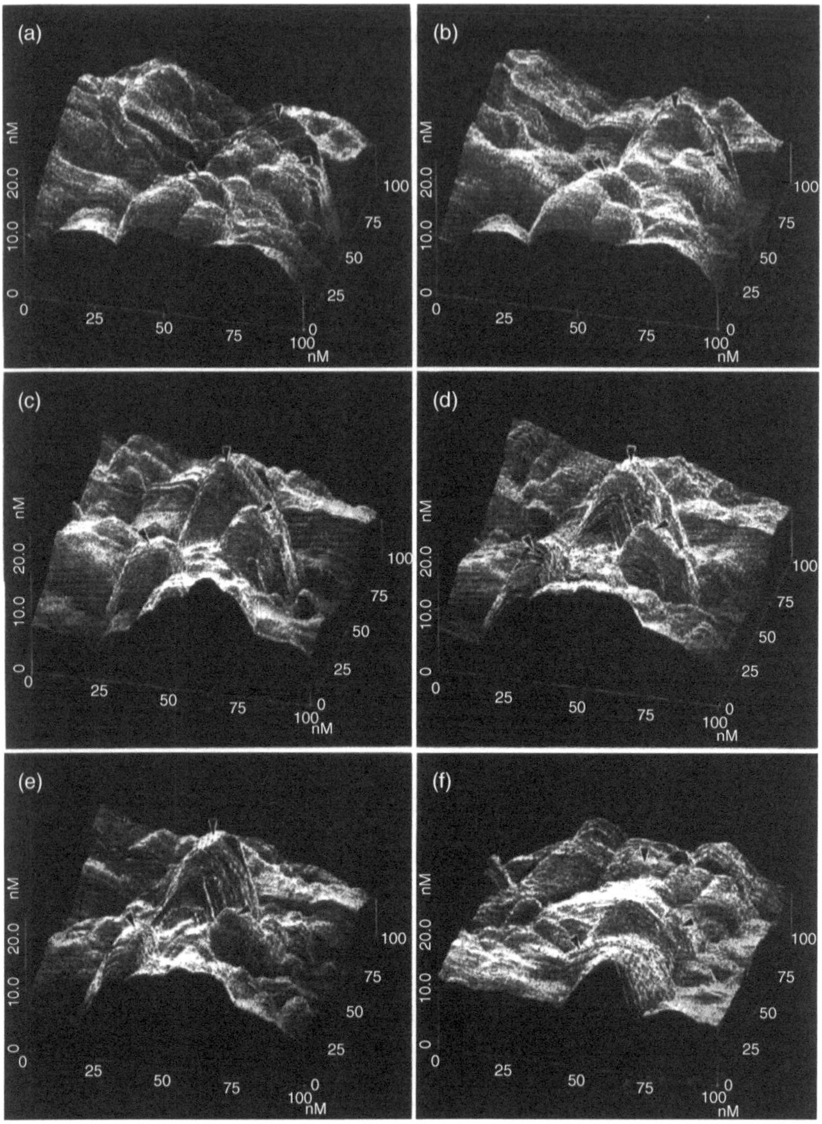

FIGURE 9.57 **Sequence of STM images during electroless copper deposition on the same area; the slow scan direction is indicated by arrows; (a) 0 to 56 s ↓, (b) 56 to 112 s,↑; (c) 225 to 281 s, ↑; (d) 340 to 396 s, ↓; (e) 396 to 452 s, ↑; (f) 510 to 566 s, ↓ [184].**

is increased to 90 s, the number of deposited Cu grains is increased. These Cu grains are more homogeneous and symmetric, but their size reduces to 1.36 μm. As the deposition time is increased to 2 min, an electroless Cu film is produced. The average grain size is reduced to 1.25 μm at that time. Moreover, the surface coverage is still not complete. Some voids are clearly observed on the TiN

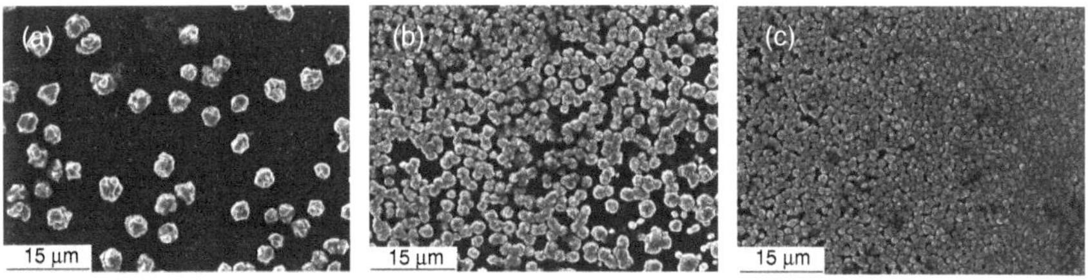

FIGURE 9.58 The morphology of as-deposited Cu films on TiN barriers for different plating times. The plating time is a 60 s, b 90 s, and c 120 s. The solution is 0.045 mol/L $CuSO_4$ + 2 mol/L NH_4F, and the bath temperature is 50°C [189].

surface as revealed in Fig. 9.58c. It is evident that only the plating time is larger than 120 s, the Cu film can be basically formed. But, the Cu film can be seen in Fig. 9.57a and b, the plating time is less than 120 s. Of course, it should be noted that the substrates are different, and the bath solutions are also not completely the same in the two cases.

At least four groups have published their results on the morphology of EP Pd films varying with plating time [183,197,201,204]. Here, we will just describe and discuss the work by Kitiwan and Atong in 2010 [204]. Figure 9.59 indicates the deposited Pd microstructure as a function of plating time: (a) 0 min, (b) 10 min, (c) 20 min, (d) 30 min, (e) 1 h, (f) 2 h, (g) 3 h, (h) 3 h with vacuum application for 15 min. The substrate used in this experiment is the alumina support, which contains the macro pore structure with an average pore size of 0.26 μm and a total porosity of 45% (Fig. 9.59a). After 10 min plating, Pd metal quickly started depositing on the surface of support, and this reduced metal was also found to penetrate into the pore of alumina substrate as shown in Fig. 9.59b. The plating reaction uniformly occurred onto the substrate surface due to the homogeneous distribution of palladium nano-particle seeds from the initial state. From Fig. 9.59b–d, the number of deposited Pd particles increases with deposition time from 0 to 30 min. After plating up to 30 min (Fig. 9.59d), most of the substrate is covered by Pd particles but pinholes on the Pd film still exist. After the deposition

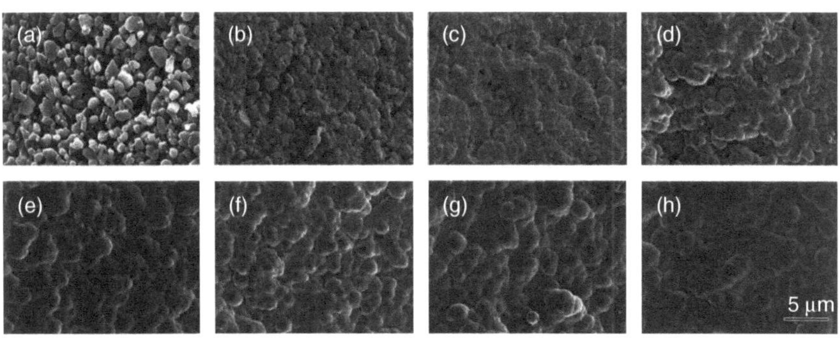

FIGURE 9.59 Deposited Pd microstructure as a function of plating time: (a) 0 min, (b) 10 min, (c) 20 min, (d) 30 min, (e) 1 h, (f) 2 h, (g) 3 h, (h) 3 h with vacuum application for 15 min [204].

for 1 h, the number and size of Pd particles do not change with the extension of deposition time, but the voids existing between Pd grains decrease. The experiments indicate that overall, the plating seems to be completed after the plating time up to about 3 h (Fig. 9.59g). At that time, no defects such as voids or pinholes can be found. Fig. 9.59h illustrates that when vacuum is applied from the inside of the porous alumina tube, the Pd film then becomes more dense packing and smooth surface. Seshimo et al. reported the similar results on the morphology of EP Pd films plated with different times [197,201]. The substrate used in the experiment also is nanoporous γ-alumina. From the six surface SEM images shown in their Fig. 5 [197], it clearly shows the Pd deposition onto the nanoporous γ-alumina took place in two distinct steps. In the first step, from 2 to 4 h, the Pd was deposited into the nanoporous γ-alumina pores and formed a covering over the nanoporous γ-alumina surface. Next, once the Pd completely blanketed the pores, a granulate Pd was deposited over the surface with a progressively larger grain size as the electroless plating reaction proceeded. That is to say, the Pd deposition on the nanoporous γ-alumina changed from a needle-shaped deposition to a spheroidal deposition at some point between 4 and 5 h during the electroless plating reaction. However, the plating time required to cover the pores completely in the work reported by Kitiwan and Atong in 2010 is ca. 1 h, but that reported by Seshimo et al. in 2008 is ca. 3–4 h, thereby it has some difference in both cases.

In 2012, Vitry et al. [211] studied the effect of plating time on surface morphology in EP Ni–B deposits in detail. The substrates chosen by the authors were mild steel (St-37) sheets (1 mm thick). The immersion time in the nickel–boron deposition bath is (a) 15 s, (b) 30 s, (c) 60 s, (d) 90 s, (e) 4 min, (f) 7 min, (g) 10 min, (h) 30 min, and (i) 1 h, respectively. The resulting SEM images are shown in Fig. 9.60. Immersions shorter than 15 s in the plating bath brought no evidence of nickel deposition detectable visually by SEM. However, some nickel is already detected by Energy-Dispersive X-ray Analysis (EDX) (see Table 2 in [211]). After immersion in the bath for 15 s, a few nodules formed on the substrates are found preferentially on the scratches and surface defects of the substrate (Fig. 9.60a). The EDX results indicate that there is nearly 6% of Ni in the top 1 μm. After 30 s, evidence of Ni deposition is easily observed in the form of numerous nodules that can be found on the whole surface of the sample (Fig. 9.60b). At this time, the Ni layer is not yet continuous and the size of the nodules stays between 0.1 and 0.2 μm. After 60 s (Fig. 9.60c), the nodules are bigger and the Ni layer appears nearly continuous. The size of the nodules is bigger after an immersion of 90 s to 4 min (Fig. 9.60d and e), and a leveling of the surface is observed. The small residual amount of Fe detected by EDX after 4 min indicates that the coating is then continuous and nearly densified. The evolution of coating top morphology is similar after 7 min (Fig. 9.60f), but refining of the size of the top nodules is observed after 10 min (Fig. 9.60g) and is followed at up to 30 min by a slight thickening of the columns accompanied by leveling of the coating (Fig. 9.60h). After 1 h, the intercolumnar spaces appear to be nearly completely filled out and the apparent size of the nodules is quite small (Fig. 9.60i). The typical cauliflower-like surface texture is then observed clearly.

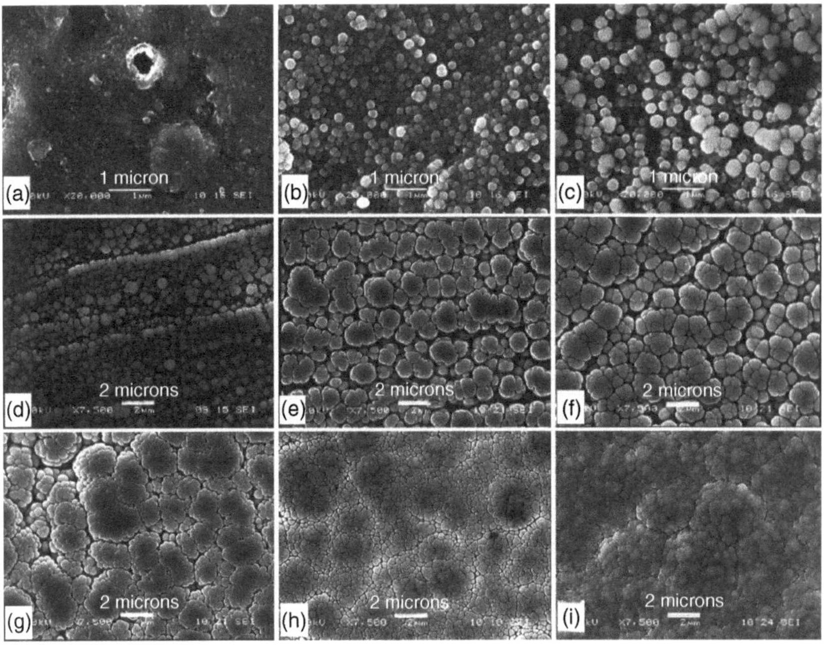

FIGURE 9.60 Surface of mild steel (St-37) samples after an immersion of (a) 15 s, (b) 30 s, (c) 60 s, (d) 90 s, (e) 4 min, (f) 7 min, (g) 10 min, (h) 30 min, and (i) 1 h in the nickel–boron deposition bath [211].

It should be noted again that if the magnification is less than 20,000× in Fig. 9.60a, b, and c, or less than 7500× for other six pictures in the figure, then the features will not be seen clearly. Once again, the appropriate magnification is important to see the features of SEM images in EP alloy deposits clearly.

Now, we compare the surface morphology of EN alloy films deposited with various plating times obtained by three groups [191,194,198]. The first row in Fig. 9.61 was obtained by Liu et al. in 2008 [194]. They investigated microdots of Ni–P alloy deposited on ITO substrates using a nonisothermal deposition (NITD) electroless method. The deposition time is (a) 3, (b) 5, and (c) 10 min. The dots formed a quasiregular array, the mean sizes of these microdots were about 330 ± 75, 411 ± 84, and 981 ± 147 nm, respectively, which increased with the reaction time until a layer of film was formed on the substrate. The number and density of these microdots initially increased then decreased as the reaction time increased. These authors believe the increase of mean size of these dots was owing to the autocatalytic characteristic of Ni dots to catalyze the dot growth in the following plating process. The second row was obtained by Khan et al. in 2007 [191]. Aluminum alloys Al 1050 and Al 6061 were used as substrates. After 1 min of electroless deposition (Fig. 9.61, second row a), the SEM shows a significant change. Nickel deposition starts to develop by nucleation after dissolution of the zincated aluminum substrate, as shown in Fig. 9.61, second

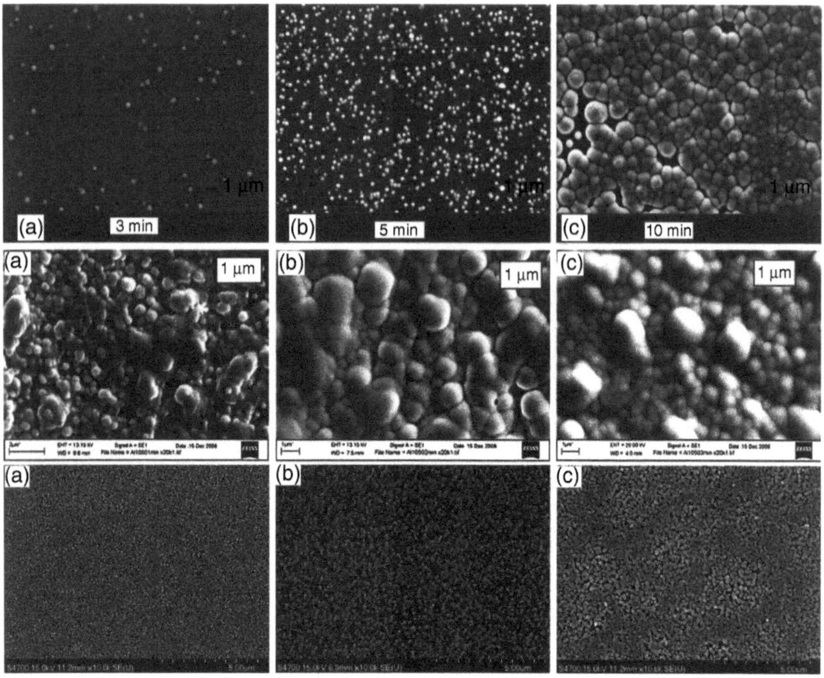

FIGURE 9.61 First row: the Ni–P microdots grew in (a) 3, (b) 5, and (c) 10 min, respectively. The bath temperature is 20°C, the substrate temperature is 170°C, and the width of clearance is 0.05 mm [194]. Second row: SEM pictures of Al 1050 surface after different time of electroless nickel immersion. (a) 1 min, (b) 2 min, and (c) 3 min [191]. Third row: SEM images of surface morphologies of Ni electroless deposition on P-type bare Si (without additional doping) with electrolyte containing 3.5 mol/L of NH_4F at various times. (a) 1 min, (b) 5 min, and (c) 15 min [207].

row b. The growth and coalescence of such nuclei result in a continuous layer of Ni–P on the Al surface (Fig. 9.61, second row c) after 3 min of deposition on Al 1050. The growing Ni–P layer covered over almost all the substrate within 4 min of deposition. The third row was obtained by Lee et al. in 2011 [207]. They used P-type Si as a substrate. As shown in Fig. 9.61, third row a, small grains about 100 nm in size covered almost the entire Si surface after 1 min of deposition time. Interestingly, after 5 min, instead of growth of the small Ni grains, the surface coverage by Ni had decreased. This indicated that the reduction of Ni stopped by termination of the electron supply by Si dissolution in NH_4F. In addition, the dissolution of Ni grains also occurred during the process. After 15 min, however, the Ni layer fully covered the Si substrate and the grain size was obviously bigger than what it had attained at 5 min. As the plating time continuously increased, such a deposition and dissociation of Ni was observed once more. Till the time passed the 45 min, a conformal Ni layer with a smooth surface morphology was obtained and further dissolution was not observed again. These authors believe that this phenomenon of dissociation of Ni results from the dissolution rate is faster than that when the grain size is smaller. From these results, it is evident that as the plating time is enhanced, the growth mode of EN is not the same, deposition needs 10 min to cover the whole substrate in the first experiment, but only needs 1 min in the third experiment. Of course, the obvious phenomenon of dissociation of Ni was only observed in the third experiment, but not observed in the other two experiments, at least the two groups did not mention this issue.

The last study of the effect of the plating time on surface morphology we would like to mention is that of EP Co–Ni–P alloy deposits [192]. The authors used an n-type silicon with (100) surface acting as the substrate for electroless Co–Ni–P film deposition. Figure 9.62 illustrates the SEM surface images and

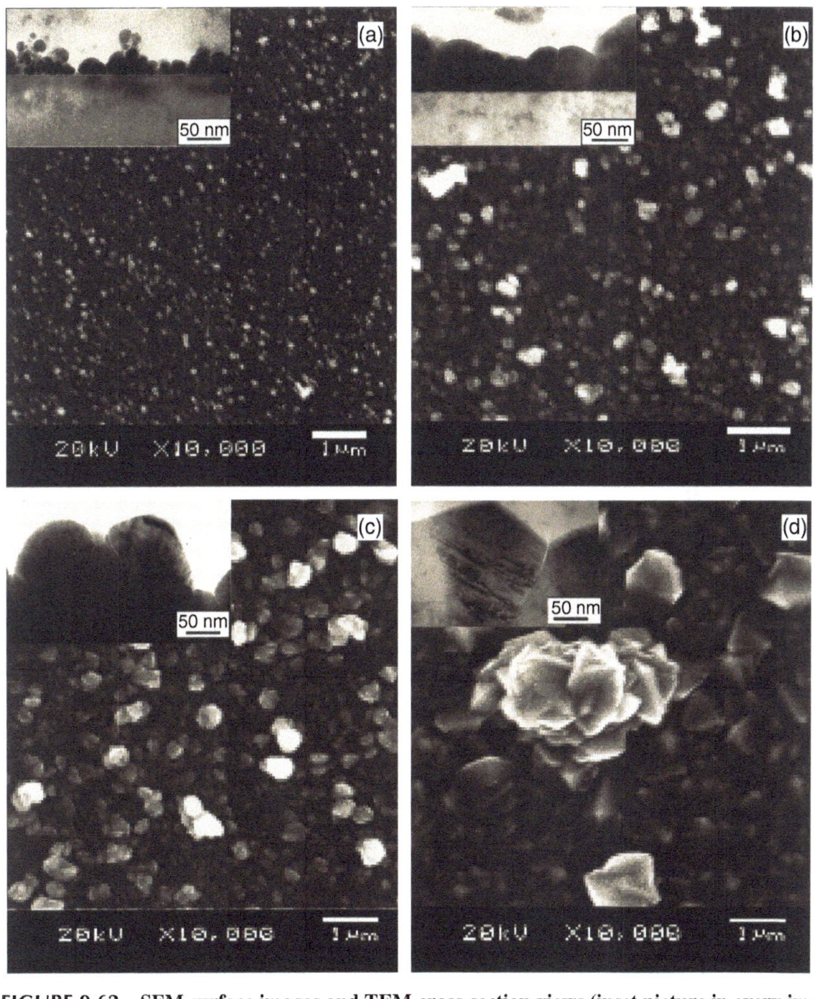

FIGURE 9.62 SEM surface images and TEM cross-section views (inset picture in every image) of electroless Co–Ni–P films deposited in 5 wt% $NiSO_4$–$CoSO_4$ bath for (a) 20 s; (b) 50 s; (c) 1.5 min; and (d) 5 min [192].

TEM cross-section views (inset picture in every image) of electroless Co–Ni–P films deposited in 5 wt% $NiSO_4$–$CoSO_4$ bath for (a) 20 s; (b) 50 s; (c) 1.5 min; and (d) 5 min, respectively. As shown in Fig. 9.62a, the electroless Co–Ni–P alloys with the form of spherical particles were dispersively deposited on the surface of the silicon substrate in 20 s of plating time. The particles grew larger as the plating time increased up to 50 s (Fig. 9.62b). The size of spherical particles increased more and more with the increase of plating time until they touched each other in the direction parallel to the surface of the substrate. Then, the electroless deposits grew only in the direction vertical to the substrate surface and their form turned into polyhedrons. At that time, the plating time is 1.5 min, as seen in Fig. 9.62c. It is clearly seen from Fig. 9.62d that the electroless Co–Ni–P deposits have polyhedral form. This variation of morphology in EP deposits as increasing the plating time is a somewhat special.

9.4.9 The Effects of Heat Treatment on SEM

As discussed in Chapter 8, amorphous and metastable microcrystalline electroless deposits must transform gradually to stable crystalline phases under heating. Therefore, the original as-plated surface morphology will change because of the atomic movement and phase transformation that occurs during heat treatment. Scientists paid much attention to the effect of heat treatment on surface morphology in EP coatings; thereby a lot of papers on this subject have been published [44,213–227]. As before, only some examples will be described and discussed in this section.

It must be noted that studies have concentrated on the effect of heat treatment on morphology of the EN system [213,214,217,220,225–227]. Hur et al. in 1990 [214] investigated such subject more in detail. A 5086 Al sheet, which was previously zincated, was used as a substrate. Figure 9.63 depicts the TEM microstructures and electron diffraction patterns of deposits contained P 11.3 at% in various heating stages. On heating to 300°C grain coarsening occurred (Fig. 9.63a), and the diffraction rings corresponding to fcc nickel crystallites became sharper and the inner ring was separated into (1 1 1) and (2 0 0) rings. The diffraction pattern of Ni_3P phase is, however, not shown (inset in Fig. 9.63a). On heating the deposit further to 350°C, the diffraction spots corresponding to Ni_3P crystallites (tetragonal, a = 0.895 nm, c = 0.439 nm) appeared

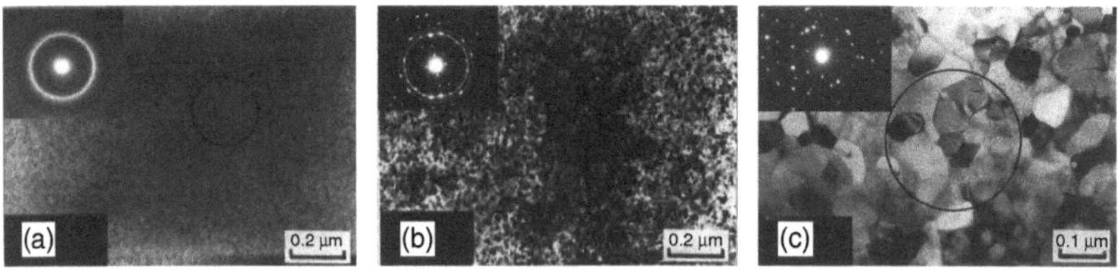

FIGURE 9.63 TEM and electron diffraction patterns (inset pictures) of the electroless Ni-P alloy obtained from solution A in various hot stages. (a) 300°C; (b) 350°C; and (c) 450°C [213]

together with much sharper diffraction rings of nickel (inset in Fig. 9.63b). The fact that the spot pattern appeared indicates the Ni_3P crystallite has already grown up large. The shape of Ni_3P grain is, however, indistinct (Fig. 9.63). All of the results suggest that in the early stage pre-existing fcc nickel grains coarsened, and on further heating, the Ni_3P crystallites nucleated in the phosphorus rich matrix and grew rapidly by encircling nickel grains. On further heating to 450°C nickel grains grew up to about 50 nm and the diffraction ring became spot pattern (Fig. 9.63c and inset in the figure). Refer to the content on the crystallization process and products of this alloy system in Section 8.2.1.1, and the reader will more clearly understand. It is interesting to note that Rabizadeh et al. in 2010 [226] found that heat treatment at different temperatures has not had any significant effect on morphology of the coatings. From the measured SEM morphology images of as-plated and heat-treated EN coating at 200°C (2 h), 400°C (1 h), and 600°C (15 min) in a vacuum environment (see Fig. 1 in [226]), these authors draw this conclusion.

In 2004,Tharamani et al. [219] reported SEM patterns of Cr–P alloy films of as-prepared and after heat treatment (350°C for 5 h under N_2 atmosphere), as shown in Fig. 9.64. The feature of the morphology in as-plated coatings is a flake that is big in the middle and small at both ends. After heat treatment, the shape of the flake has not changed, but the size is increased (Fig. 9.64b), possibly due to formation of new phases. This is in accordance with the observed XRD data because CrP and Cr_3P phases appeared in the as-deposited sample, but Cr_3P and $Cr_{12}P_7$ phases occurred in the sample after heat treatment (see Fig. 2 in [219]). Moreover, these authors observed the uniform distribution of P and Cr in the alloy, respectively.

The effect of heat treatment on morphology in EP Co–W–P coatings was studied by Chang Shou-Yi and Lu Chia-Ling in 2008 [222]. Blanket Si wafers with a SiC layer of 100 nm thickness were used as substrates in their study. Figure 9.65 shows the SEM pictures of the as-deposited layer and the layers after heat treatment at 400, 500, and 600°C, respectively. The experiments found that when the Co–W–P layers deposited with different plating times (0–10 min) at 70°C, Co–W–P particles of 20–30 nm in size first packed, then grew, and then coalesced/adhered to each other to form a continuous film (Fig. 9.65a). The feature of the particles is their nodular structure. As deposited for 10 min, a

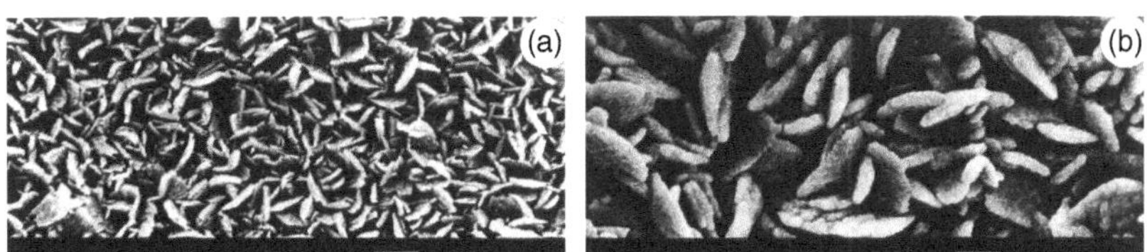

FIGURE 9.64 SEM images of Cr–P alloy: (a) as-prepared; (b) after heat treatment (350°C for 5 h under N_2 atmosphere) [219].

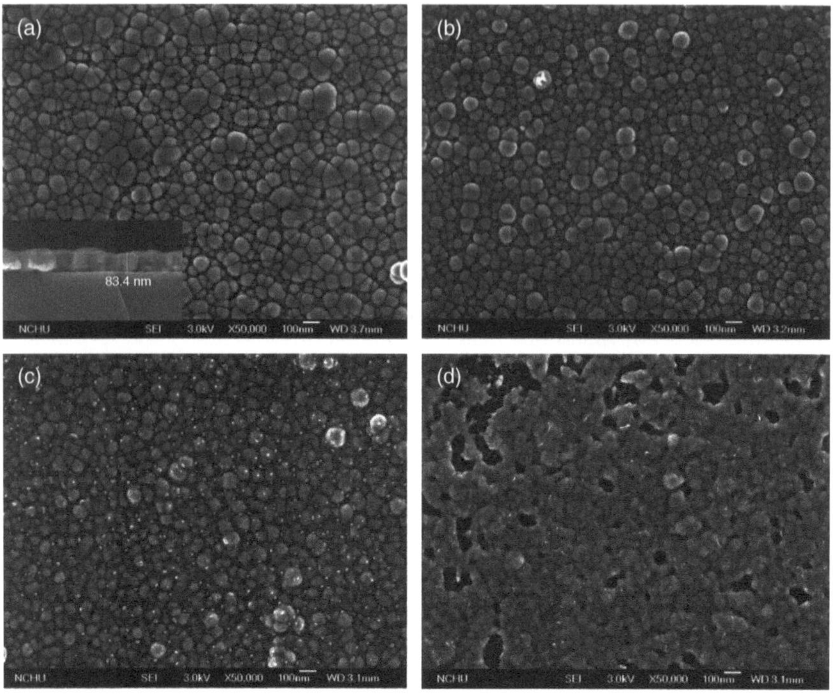

FIGURE 9.65 SEM surface morphologies of EP Co–W–P layer deposited at 70°C for 10 min (a), and Co–W–P layers deposited at 70°C for 10 min after thermal annealing at (b) 400, (c) 500, and (d) 600°C for 30 min. Inset picture in (a) is the cross-sectional microstructure of as-plated sample [219].

continuous and smooth Co–W–P layer with a thickness of ~80 nm was obtained, as seen from the inset picture in Fig. 9.65a. The film then grew vertically, during which small particles disappeared and only large particles were retained with boundaries. From the AFM analyses, the surface roughness of the Co–W–P layer was measured as ~9.3 nm. After the as-plated Co–W–P layer heat annealing at 400°C (Fig. 9.65b), a continuous film resulting from the coalescence of large Co–W–P particles was observed. After thermal annealing at 500°C (Fig. 9.65c), Co–W–P particle boundaries became unclear, and the densification and continuity of the Co–W–P layer were improved, a consequence associated with the elimination of some other defects. Some small precipitates ~10 nm in size were found at the Co–W–P particle boundaries. These precipitates might be the early stage nuclei for the formation of compounds or crystalline phases. These authors pointed out that because they are too small to identify, further examination is still needed. With a further increase in the annealing temperature to 600°C, as shown in Fig. 9.65d, the particle boundaries completely disappeared. However, a morphology very different from that of as-deposited film was observed along with some remaining pores, probably attributable to the formation of some compounds or crystalline phases at high temperatures.

9.5 SUMMARY

After we analyze and discuss the surface morphology in EP metal and alloy coatings in detail, some conclusions can be drawn as follows.

Due to the high level of interest in the surface morphology in EP deposits, a great many papers have been published on the subject, which gives us so much source material, and allows analysis of the topic from various points of view, for example, all kinds of EP deposits, effects of nearly all parameters on morphology.

Till date, nearly all aspects in the surface morphology in EP deposits have been investigated. However, systematic works, analyzing and reviewing works are very rare. Thereby research on such aspect is needed.

As mentioned in the text, the morphology in a similar EP system obtained by different groups may not be the same or similar but different. For example, the surface morphology of EN alloy films deposited with various plating times obtained by three groups [191,194,198] as shown in Fig. 9.61 is just like this. This phenomenon is obviously in need of further study.

The surface topography is strongly dependent on the characteristics of electroless plating solution. It is obvious that that there will be a wide variety of surface morphology structures. In fact, we have seen a lot of surface textures in this chapter. Can they be classified as several catalogues? How can they be classified? No papers in the literature have investigated this problem until now.

Especially, why do so many morphology textures occur? How do we explain them theoretically? No one has answered these questions in his/her papers. Even though for the typical nodular and cauliflower-like surface textures that usually appear in EN, Ni–B, and ternary Ni-based alloy deposits, all authors, including the author of the book, did not really explain why the morphology (obtained by themselves) is just like that. The problem of the theory falling behind the experiments is once again seen here.

REFERENCES

[1] Zhang B, Xie H, Liao S, Lindig S, Taglauer E. Prakt Metallogr 2005;42:9.

[2] Engin Ayturka M, Hua MY. J Memb Sci 2009;330(1–2):233.

[3] Sankara Narayanan TSN, Baskaran I, Krishnaveni K, Parthiban S. Surf Coat Technol 2006;200(11):3438.

[4] Tan Z, Zhang B, Zhang H. J Rare Earths 1992;10(1):355.

[5] Babilas R, Nowosielski R. Arch Mater Sci Eng 2010;44(1):5.

[6] Révész Á, Kis-Téth Á, Varga LK, Schafler E, Bakonyi I, Spassov T. Intern J Hydro Eng 2010;37:5769.

[7] Hung A. J Electrochem Soc 1986;133(7):1350.

[8] Honma H, Kobayashi T. J Electrochem Soc 1994;141(3):730.

[9] Lin Y, Yen S. Appl Surf Sci 2001;178:116.
[10] Hsu H-H, Lin K-H, Lin S-J, Yeh J-W. J Electrochem Soc 2001;148(1):C47.
[11] Tseng W-T, Lo C-H, Lee S-C. J Electrochem Soc 2001;148(5):C327.
[12] Li J, Kohl PA. J Electrochem Soc 2002;149(12):C631.
[13] Guazzone F, Engwall EE, Mardilovich IP, Ma YH. Prepr Pap-Am Chem Soc Div Fuel Chem 2004;49(2):895.
[14] Sone M, Kobayakawa K, Saitou M, Sato Y. Electrochim Acta 2004;49:233.
[15] Li Y, Chen D, Lu Q, Qian X, Zhu Z, Yin J. Appl Surf Sci 2005;241:471.
[16] Zhong S, Yang Z-G, Cai J. J Electrochem Soc 2005;152(3):C143.
[17] Zhong S, Yang Z-G, Cai J, Wang J, Li H-M. J Electrochem Soc 2006;153(6):C405.
[18] Kim Y-S, Shin J, Cho J-H, Ten Eyck GA, Liu D-L, Pimanpang S, Lu T-M, Senkevich JJ, Shin H-S. Surf Coat Technol 2006;200:5760.
[19] Kao C-Y, Chou K-S. Electrochem Solid-State Lett 2007;10(3):D32.
[20] Zhang H-P, Jiang Z-H, Lian J-S, Hou X-F. Trans Nonferrous Met Soc China 2007;17:s736.
[21] Wu X, Sha W. Appl Surf Sci 2008;255:2813.
[22] daRosa CP, Maboudian R, Iglesia E. J Electrochem Soc 2008;155(6):E70.
[23] Pan Z, Hu A, Hang T, Duan Y, Li M, Mao D. J Electrochem Soc 2010;157(1):D74.
[24] Scudieroa L, Fasasi A, Griffiths PR. Appl Surf Sci 2011;257:4422.
[25] Koo H-C, Saha R, Kohl PA. J Electrochem Soc 2011;158(12):D698.
[26] Vaškelis A, Jagminienk A, Juškhas R, Matulionis E, Norkus E. Surf Coat Technol 1996; 82:165.
[27] Xia Y, Venkateswaran N, Qin D, Tien J, Whitesides George M. Langmuir 1998;14:363.
[28] Shacham-Diamand Y, Inberg A, Sverdlov Y, Croitoru N. J Electrochem Soc 2000; 147(9):3345.
[29] ten Kortenaar MV, de Goeij JJM, Kolar ZI, Frens G, Lusse PJ, Zuiddam MR, van der Drift E. J Electrochem Soc 2001;148(1):C28.
[30] Tong H, Zhu L, Li M, Wang C. Electrochim Acta 2003;48:2473.
[31] Brevnov DA, Olson TS, López GP, Atanassov P. J Phys Chem B 2004;108:17531.
[32] Tong H, Kong LB, Wang CM. Thin Solid Films 2006;496:360.
[33] Koo H-C, Kim SY, Cho SK, Kim JJ. J Electrochem Soc 2008;155(9):D558.
[34] Ghaemi M, Makhlooghi F, Adelkhani H, Aghazadeh M, Shiri HM. Int J Electrochem Sci 2010;5:131.
[35] Gu CD, Xu XJ, Tu JP. J Phys Chem C 2010;114:13614.
[36] Okinaka Y, Sard R, Wolowodiuk C, Craft WH, Retajczyk TF. J Electrochem Soc 1974;121(1):56.
[37] Phaner M, Stremsdoerfer G, Wang LY, Martin JR, de Villeneuve C, Porte L. J Electrochem Soc 1991;138(3):874.
[38] Hou Z, Abbott NL, Stroeve P. Langmuir 1998;14:3287.
[39] Magagnin L, Maboudian R, Carraro C. J Phys Chem B 2002;106:401.
[40] Yamada K, Gasparac R, Martin CR. J Electrochem Soc 2004;151(1):E14.
[41] Osaka T, Okinaka Y, Sasano J, Kato M. Sci Technol Adv Mater 2006;7:425.
[42] Ahn W, Taylor B, Dall'Asén AG, Roper DK. Langmuir 2008;24:4174.
[43] Ding H-P, Wang M, Chen L-J, Fan W, Lee Y-I, Qian D-J, Hao J, Liu H-G. Collo Surf A Physicochem Eng Aspects 2011;387:1.
[44] Hedgecock N, Tung P, Schlesinger M. J Electrochem Soc 1975;122(7):866.
[45] Watanabe T, Tanabe Y. Mater Sci Eng 1976;23:97.
[46] Delaunois F, Lienard P. Surf Coat Technol 2002;160:239.
[47] Krishnaveni K, Narayanan TSNS, Seshadri SK. Surf Coat Technol 2005;190:115.

[48] Anik M, Körpe E, Şen E. Surf Coat Technol 2008;202:1718.
[49] Kanta A-F, Poelman M, Vitry V, Delaunois F. J Alloys Comp 2010;505:151.
[50] Hassan HB, Abdel Hamid Z. Intern J Hydro Ener 2011;36:849.
[51] Vitry V, Sens A, Kanta A-F, Delaunois F. Surf Coat Technol 2012;206:3421.
[52] Vitry V, Kanta A-F, Dille J, Delaunois F. Surf Coat Technol 2012;206:3444.
[53] Wang ZC, Jia F, Yu L, Qi ZB, Tang Y, Song G-L. Surf Coat Technol 2012;206:3676.
[54] Bulbul F. Met Mater Int 2011;17(1):67.
[55] Fisher RD, Koopman DE. J Electrochem Soc 1964;111(2):263.
[56] Miksic MG, Travieso R, Arcus A, Wright RH. J Electrochem Soc 1966;113(4):360.
[57] Frieze AS, Sard R, Weil R. J Electrochem Soc 1968;115(6):586.
[58] Tarozaitė R, Kurtinaitienė M, Džiūvė A, Jusys Z. Surf Coat Technol 1999;115:57.
[59] Magagnin L, Sirtori V, Seregni S, Origo A, Cavallotti PL. Electrochim Acta 2005;50:4621.
[60] Rudnik E, Gorgosz J. Surf Coat Technol 2007;201:6953.
[61] Tarozaitė R, Sukackienė Z, Sudavičius A, Juškėnas R, Selskis A, Jagminienė A, Norkus E. Mater Chem Phys 2009;117:117.
[62] Schmeckenbecher AF. J Electrochem Soc 1966;113(8):778.
[63] Takano O, Matsuda H, Yoneda M. J Metal Finish Soc Jpn 1981;32(12):610.
[64] Valova E, Georgiev I, Armyanov S, Delplancke J-L, Tachev D, Tsacheva Ts, Dille J. J Electrochem Soc 2001;148(4):C266.
[65] Wu F-B, Tien S-K, Duh J-G, Wang J-H. Surf Coat Technol 2003;166:60.
[66] Vemagiri JK, Murthy SK, Gunasekaran RA, Coane P, Varahramyan K. Mater Lett 2003; 57:4098.
[67] Einati H, Bogush V, Sverdlov Y, Rosenberg Y, Shacham-Diamand Y. Microelect Eng 2005; 82:623.
[68] Georgieva J, Armyanov S. J Solid State Electrochem 2007;11:869.
[69] Dai HB, Liang Y, Wang P, Yao XD, Rufford T, Lu M, Cheng HM. Intern J Hydro Ener 2008;33:4405.
[70] Abdel Aal A, Barakat H, Abdel Hamid Z. Surf Coat Technol 2008;202:4591.
[71] Abdel Aal A, Shaaban A, Abdel Hamid Z. Appl Surf Sci 2008;254:1966.
[72] Guo RH, Jiang SQ, Yuen CWM, Ng MCF. J Appl Electrochem 2009;39:907.
[73] Abdel Aal A, Shehata Aly M. Appl Surf Sci 2009;255:6652.
[74] Dulal SMSI, Kim TH, Rhee H, Sung JY, Kim C-K. J Alloys Comp 2009;467(370).
[75] Yong Z, YanHai C, Lin C, Wen L. Mater Trans 2010;51(2):277.
[76] Dai HB, Li-Li G, Liang Y, Kang X-D, Wang P. J Power Sour 2010;195:307.
[77] Rudnik E, Mucha M. Surf Eng 2011;27(9):683.
[78] Rakap M, Kalu EE, Özkar S. Intern J Hydro Ener 2011;36:254.
[79] Kumar A, Singh A, Kumar M, Kumar D, Barthwal S. J Mater Sci Mater Electron 2011;22:1495.
[80] Pang J, Li Q, Wang W, Xu X, Zhai J. Surf Coat Technol 2011;205:4237.
[81] Lo YL, Chou SC, Hwang BJ. J Appl Electrochem 1996;26:733.
[82] Lu Z, Gao Y, Zheng Z. China Surf Eng 2005;(No. 2):45.
[83] Balaraju JN, Anandan C, Rajam KS. Appl Surf Sci 2005;250(1–4):88.
[84] Hu W, Wu L, Wanf L, Zhang B, Guan H. Physica B 1995;212:195.
[85] Georgieva J, Kawashima S, Armyanov S, Valova E, Hubin A, Koyama Y, Steenhaut O, Haydu J, Delplancke J-L, Tsacheva Ts. J Electrochem Soc 2005;152(11):C783.
[86] Yokoshima T, Kaneko D, Akahori M, Hyo-Seung N, Osaka T. J Electroanal Chem 2000;491:197.
[87] Tarozaite R, Stalnionis G, Sudavicius A, Kurtinaitiene M. Surf Coat Technol 2002;161:150.

[88] Balaraju JN, Rajam KS. Surf Coat Technol 2005;195:154.
[89] Cheng H, Li D, Wang J, Cheng X, Guan J, Zhang O, Yuan R. Funct Mater 2005;36(9):1331.
[90] Balaraju JN, Millath Jahan S, Anandan C, Rajam KS. Surf Coat Technol 2006;200:4885.
[91] Balaraju JN, Ezhil Selvi V, William Grips VK, Rajam KS. Electrochim Acta 2006;52:1064.
[92] Xuan T-P, Zhang L, Huang Q. Trans Nonferrous Met Soc China 2006;16:363.
[93] Balaraju JN, Millath Jahan S, Jain A, Rajam KS. J Alloys Comp 2007;436:319.
[94] Shacham-Diamand Y, Sverdlov Y, Bogush V, Ofek-Almog R. J Solid State Electrochem 11 2007;929.
[95] Valova E, Armyanov S, Dille J, Van Ingelgem Y, Hubin A, Steenhaut O. J Electrochem Soc 2008;155(6):D449.
[96] Koo H-C, Ki CS, Joong KO, Suh M-W, Im Y, Jeong KJ. J Electrochem Soc 2009;156(7):D236.
[97] Ayoub IE, Ibrahim IS. Isotope Rad Res 2009;4(suppl. 2):1561. 41.
[98] Apachitei I, Duszczyk J, Katgerman L, Overkamp PJB. Scr Mater 1998;38(9):1383.
[99] Leóna OA, Staiaa MH, Hintermann HE. Surf Coat Technol 1998;108–109:461.
[100] Bozzini B, Martini C, Cavallotti PL, Lanzoni E. Wear 1999;225:806.
[101] Reddy VVN, Ramamoorthy B, Nair PK. Wear 2000;239:111.
[102] Abdel Hamid Z, Abou Elkhair MT. Mater Lett 2002;57:720.
[103] Huang YS, Zeng XT, Annergren I, Liu FM. Surf Coat Technol 2003;167:207.
[104] Alirezaei Sh, Monirvaghefi SM, Salehi M, Saatchi A. Surf Coat Technol 2004;184(2–3):170.
[105] Zhao Q, Liu Y, Abel EW. Mater Chem Phys 2004;87:332.
[106] Wu Y, Liu L, Bin S, Hu W. J Mater Sci 2005;40(18):5057.
[107] Zhao Q, Liu Y, Wang C. Appl Surf Sci 2005;252:1620.
[108] Malfatti CF, Ferreira ZJ, Santos CB, Souza BV, Fallavena EP, Vaillant S, Bonino J-P. Corr Sci 2005;47:567.
[109] Balaraju JN, Kalavati, Rajam KS. Surf Coat Technol 2006;200:3933.
[110] Madej M, Ozimina D, Piwoński I. Trib Lett 2006;22(2):135.
[111] Novakovic J, Vassiliou P, Samara KL, Argyropoulos Th. Surf Coat Technol 2006;201:895.
[112] Wu YT, Lei L, Shen B, Hu WB. Surf Coat Technol 2006;201:441.
[113] Dasa CM, Limaye PK, Grover AK, Suri AK. J Alloys Comp 2007;436:328.
[114] Abdel Hamid Z, El Badry SA, Abdel Aal A. Surf Coat Technol 2007;201:5948.
[115] Szczygieł B, Turkiewicz A, Serafińczuk J. Surf Coat Technol 2008;202:1904.
[116] Balaraju JN, Rajam KS. J Alloys Comp 2008;459:311.
[117] Wang L, Zhu X, Zhou J, Ouyang S. J Wuhan Uni Technol Mater (Scc Ed) 2008;23(2):169.
[118] Novakovic J, Vassiliou P. Electrochim Acta 2009;54(9):2499.
[119] Khalifa ORM, Sakr E. The Open Corr J 2009;2:211.
[120] Zuleta AA, Galvis OA, Castaño JG, Echeverría F, Bolivar FJ, Hierro MP, Pérez-Trujillo FJ. Surf Coat Technol 2009;203:3569.
[121] Balaraju JN, Ezhil Selvi V, Rajam KS. Mater Chem Phys 2010;120:546.
[122] Balaraju JN, Kalavati, Rajam KS. Surf Coat Technol 2010;205:575.
[123] Ranganatha S, Venkatesha TV, Vathsala K. Appl Surf Sci 2010;256:7377.
[124] Faraji S, Rahim AA, Mohamed N, Sipaut CS, Raja B. Mater Chem Phys 2011;129:1063.
[125] Mohammadi M, Ghorbani M. J Coat Technol Res 2011;8(4):527.
[126] Faraji S, Rahim AA, Mohamed N, Sipaut CS. Surf Coat Technol 2011;206:1259.
[127] Sivandipoor I, Ashrafizadeh F. Appl Surf Sci 2012;263:314.
[128] He M-F, Hu W-B, Zhong C, Weng J-F, Shen B, Wu Y-t. Trans Nonferrous Met Soc China 2012;22:2586.
[129] Li Z, Wang J, Lu J, Meng J. Appl Surf Sci 2013;264:516.
[130] Bai H, Ma N, Lang J, Jin Y, Zhu C, Ma Y. Mater Des 2013;46:740.

[131] Younan MM, Shoeib M. Galvanotechnik 2002;4:932.
[132] Jiang T-X, Wu H-H. Chin J Appl Chem 2000;17(5):578.
[133] Younan MM, Abo El-Enine SA. Galvanotechnik 2004;12:2876.
[134] Balaraju JN, Anandan C, Rajam KS. Appl Surf Sci 2005;250:88.
[135] Wei G, Jin Y, Yu Y, Ge H, Wang X, Jiang L, Sun L. Thin Solid Films 2010;519:1259.
[136] Azizi A, Mohammadi M, Sadrnezhaad SK. International Conference on Materials Heat Treatment (ICMH 2010), Islamic Azad University, Majlesi Branch, May, 11–14, 2010, Isfahan, Iran.
[137] DePew JR. J Electrochem Soc 1973;120(9):1187.
[138] Cortijo RO, Schlesinger M. J Electrochem Soc 1984;131(12):2800.
[139] Watanabe H, Honma H. J Electrochem Soc 1997;144(2):471.
[140] Baskaran I, Sakthi Kumar R, Sankara Narayanan TSN, Stephen A. Surf Coat Technol 2006;200:6888.
[141] De Robertis E, Abrantes LM, Motheo AJ. Thin Solid Films 2008;516:6266.
[142] Lo YL, Joe HB. J Electrochem Soc 1996;143(7):2158.
[143] Huang YS, Cui FZ. Surf Coat Technol 2007;201:5416.
[144] Wang X-H, Zhou H, Feng P-Z, Sun Z. Electroplat Pollu Cont 2010;30(2):19.
[145] Babanejhad A, Hashemi M, Rahmatallahpur Y, Nozad ShA. Bull Mater Sci 2012;35(4):561.
[146] Chu SZ, Sakairi M, Takahashi H. J Electrochem Soc 2000;147(4):1423.
[147] Cheong WJ, Luan BL, Shoesmith DW. Appl Surf Sci 2004;229:282.
[148] Cheong WJ, Luan BL, Shoesmith DW. Corr Sci 2007;49:1777.
[149] Koo H-C, Jeong KJ. J Electrochem Soc 2008;155(3):D176.
[150] Wang Z-C, Yu L, Jia F, Song G-L. J Electrochem Soc 2012;159(7):D406.
[151] Tian M, Wang J, Kurtz J, Mallouk TE, Chan MHW. Nano Lett 2003;3(7):919.
[152] Chen B-H, Hong L, Ma Y, Ko T-M. Ind Eng Chem Res 2002;41:2668.
[153] Elansezhian R, Ramamoorthy B, Kesavan Nair P. J Mater Proc Technol 2009;209:233.
[154] Liu D, Yan Y, Lee K, Yu J. Mater Corr 2009;60(9):690.
[155] Farzaneh A, Ehteshamzadeh M, Ghorbani M, Vazifeh Mehrabani J. J Coat Technol Res 2010;7(5):547.
[156] Mafi IR, Dehghanian C. Appl Surf Sci 2011;257:8653.
[157] Zhou Y, Sun W, Dong L, Chu Y. Appl Surf Sci 2011;257:10395.
[158] Gan X, Zhou K, Hu W, Zhang D. Surf Coat Technol 2012;206:3405.
[159] Sudagara J, Liana JS, Jianga Q, Jianga ZH, Li GY, Elansezhian R. Prog Org Coat 2012;74:788.
[160] Zielińska K, Stankiewicz A, Szczygieł I. J Coll Inter Sci 2012;377:362.
[161] Cortijo RO, Schlesinger M. J Electrochem Soc 1983;130(12):2341.
[162] Nicholson EL, Khan MR. J Electrochem Soc 1986;133(11):2342.
[163] Schlesinger M, Meng X, Snyder DD. J Electrochem Soc 1991;138(2):406.
[164] Bouanani M, Cherkaoui F, Fratesi R, Roventi G, Barucca G. J Appl Electrochem 1999;29:637.
[165] Veeraraghavan B, Haran B, Kumaraguru SP, Popov B. J Electrochem Soc 2003;150(4):B131.
[166] Murthy SK, Vemagiri JK, Gunasekaran RA, Coane P, Varahramyan K. J Electrochem Soc 2004;151(1):C1.
[167] Liang M-W, Yen H-T, Hsieh T-E. J Elect Mater 2006;35(7):1593.
[168] Cheng H, Xu S, Zou S, Zhang Q, Zheng J, Zhao W, Guan J, Sun Z. IEEE Trans Mag 2007;43(6):2433.
[169] Malel E, Mandler D. J Electrochem Soc 2008;155(6):D459.
[170] Bahiyah Baba N, Waugh W, Davidson AM. World Acad Sci Eng Technol 2009;49:715.
[171] Armini S, Maestre Caro A. J Electrochem Soc 2010;157(1):D74.

[172] Gou Y-N, Huang W-J, Zeng R-C, Zhu Y. Trans Nonferr Met Soc China 2010;20:s674.
[173] Firoozbakht M, Monirvaghefi SM, Niroumand B. J Alloys Comp 2011;509S:S496.
[174] Iacovangelo CD. J Electrochem Soc 1991;138(4):976.
[175] Yang L-J. J Electrochem Soc 1991;138(7):2086.
[176] Valova E, Armyanov S, Delplancke J-L, Winand R, Steenhaut O, Vereecken J. J Electrochem Soc 1996;143(9):2804.
[177] Tzeng G-S. J Appl Electrochem 1996;26:969.
[178] Blackwood DJ, Li YY, Chow GM. J Electrochem Soc 2002;149(3):D27.
[179] Liu WL, Hsieh SH, Tsai TK, Chen WJ. J Electrochem Soc 2004;151(10):C680.
[180] Yang L, Luan B. J Electrochem Soc 2006;152(7):C474.
[181] Chow SL, Hedgecock NE, Schlesinger M, Rezek J. J Electrochem Soc 1972;119(12):1614.
[182] Homma T, Yamazaki T, Osaka T. J Electrochem Soc 1992;139(3):732.
[183] Brandow SL, Dressick WJ, Marrian CRK, Chow G-M, Calvert JM. J Electrochem Soc 1995;142(7):2233.
[184] Weber CJ, Pickering HW, Weil KG. J Electrochem Soc 1997;144(7):2364.
[185] Lee YG, Chou T-C. J Electrochem Soc 2003;150(1):H21.
[186] Jun L, Kohl PA. J Electrochem Soc 2003;150(8):C558.
[187] Tsaia T-K, Chao C-G. Appl Surf Sci 2004;233:180.
[188] Brevnov DA, Olson TS, López GP, Atanassov P. J Phys Chem B 2004;108:17531.
[189] Zhong S, Yang Z-G, Cai J, Wang J, Li H-M. J Electrochem Soc 2006;153(6):C405.
[190] Kim Y-S, Shin J, Cho J-H, Ten Eyck GA, Liu D-L, Pimanpang S, Lu T-M, Senkevich JJ, Shin H-S. Surf Coat Technol 2006;200:5760.
[191] Khan E, Oduoza CF, Pearson T. J Appl Electrochem 2007;37:1375.
[192] Liu WL, Chen WJ, Tsai TK, Hsieh SH, Chang SY. Appl Surf Sci 2007;253(8):3843.
[193] Hung CC, Lin CC, Shih HC. Diam Rel Mater 2008;17:853.
[194] Liu Y-M, Shieh M-J, Sung Y, Ger M-D. J Alloys Comp 2008;453:407.
[195] Koura N, Nagase H, Sato A, Kumakura S, Takeuchi K, Ui K, Tsuda T, Loong CK. J Electrochem Soc 2008;155(2):D155.
[196] Wu WC, Hsieh T-E, Pan H-C. J Electrochem Soc 2008;155(5):D369.
[197] Seshimo M, Ozawa M, Sone M, Sakurai M, Kameyama H. J Memb Sci 2008;324:181.
[198] Chou Y-H, Sung Y, Bai C-Y, Ger M-D. J Electrochem Soc 2008;155(9):D551.
[199] Liu Y-M, Sung Y, Pu N-W, Chou Y-H, Yeh K-C, Ger M-D. Thin Solid Films 2008;517:727.
[200] Elsentriecy HH, Azumi K. J Electrochem Soc 2009;156(2):D70.
[201] Seshimo M, Hirai T, Rahman MM, Ozawa M, Sone M, Sakurai M, Higo Y, Kameyama H. J Memb Sci 2009;342:321.
[202] Cheng SL, Peng HC. J Electrochem Soc 2010;157(2):D81.
[203] Han WK, Hwang GH, Hong SJ, An HH, Yoon CS, Kim JH, Lee MJ, Hong G, Park KS, Kang SG. Mater Chem Phys 2010;123:401.
[204] Kitiwan M, Atong D. J Mater Sci Technol 2010;26(12):1148.
[205] Blackwood DJ, Khoo SM. Solar Ener Mater Solar Cells 2010;94:1201.
[206] Byeon JH, Kim J-W. J Electrochem Soc 2011;158(1):D15.
[207] Lee D, Kim D, Yoo B. J Electrochem Soc 2011;158(8):D490.
[208] Yoshida N, Tasaki T, Chang MT-F, Shibata A, Ishiyama C, Sone M. Microelec Eng 2011;88:2401.
[209] Hao J, Han M-J, Xu Z, Li J, Meng X. Nanoscale Res. Lett. 2011;6:263.
[210] Hoon BJ, Kim Y-W. ACS Appl Mater Interfaces 2011;3:2912.
[211] Vitry V, Sens A, Kanta A-F, Delaunois F. Appl Surf Sci 2012;263:640.
[212] Latha N, Raj V, Selvam M. Bull Mater Sci 2013;36(4):719.

[213] Koiwa I, Usuda M, Osaka T. J Electrochem Soc 1988;135(5):1222.
[214] Hur K-H, Jeong J-H, Nyung LD. J Mater Sci 1990;25:2573.
[215] Osaka T, Homma T, Inoue K. J Electrochem Soc 1991;138(2):538.
[216] Osaka T, Higashikaw T, lizuk A, Takaif M, Kim M. J Electrochom Soc 1998;145(7):2420.
[217] Chu SZ, Sakairi M, Takahashi H, Qiu ZX. J Electrochem Soc 1999;146(2):537.
[218] Kim M, Yokoshima T, Osaka T. J Electrochem Soc 2001;148(11):C753.
[219] Tharamani CN, Shahina Begum N, Mayanna SM. Mater Chem Phys 2004;83:278.
[220] Baskaran I, Sankara Narayanan TSN, Stephen A. Mater Chem Phys 2006;99:117.
[221] Wang S-L, Hong L-L. J Alloys Comp 2007;429:99.
[222] Chang S-Y, Lu C-L. J Electrochem Soc 2008;155(3):D234.
[223] Liu H, Guo R-X, Zong Y, He B-Q, Liu Z. Trans Nonferrous Met Soc China 2010;20:1024.
[224] Wei S, Wu X, Sarililah W. Mater Sci Eng B 2010;168:95.
[225] Rabizadeh T, Allahkaram SR, Zarebidaki A. Mater Des 2010;31:3174.
[226] Li X, Duan Y-x, Zhao Y, Zhu L. Prog Nat Sci Mater Intern 2011;21:392.
[227] Islam M, Shehbaz T. Surf Coat Technol 2011;205:4397.

Part IV

Kinetics, Mechanism, and Theory of Electroless Plating

The content of this part includes the formation process of electroless deposits, their formation mechanisms and a theoretical description of the coating formation.

There was new progress especially in the theory of describing formation, and people usually not pay good attention to them in existing electroless plating books. Here we pay particular attention to them.

Chapter 10

Mechanism of Electroless Plating

Chapter Outline

From the preceding chapters, we can see that electroless plating (EP) produces a wide variety of materials and is full of tricks. Naturally, there is curiosity about what might be controlling the EP process. These are the problems that the reaction mechanism of EP could answer. Once we have a complete mechanism of EP, all of the EP experimental facts could be understood and explained. This is an ideal situation, and is very difficult to reach.

When Brenner and Ridell [1] published their discovery of EN in 1946, they also proposed the first reaction mechanism for the process. In the following 70 years or so, some scientists proposed their own mechanisms of EP. This was because every mechanism can explain some features of EP, but cannot explain others. The so-called universal reaction mechanism of EP has not appeared so far due to the complexity of the process. The deposition reaction of metal ion and the oxidation

Amorphous and Nano Alloys Electroless Depositions. http://dx.doi.org/10.1016/B978-0-12-802685-4.00010-8

reaction of reducing agent are linked with each other. The oxidation reaction of the reducing agent proceeds via various reaction intermediates, and continuously takes place only when the deposited metal surface has catalytic activity for the reaction of the reducing agent. Other components in the bath solution, such as complexing agent, stabilizer, surfactant, and other additives also affect the reactions. Moreover, the technique parameters pH value, plating time, and temperature also affect the EP process. Developing a theory (reaction mechanism) is difficult since it has to reflect and control all these complex factors in it. In other words, things become complicated, the task is too daunting, and it is very difficult to perform. The fact that the reaction mechanism of EP deposits is not completely understood has been also pointed out by other groups, for example, Małecki and Micek-Ilnicka in 2000 said: "until now, the plating mechanism has not been fully explained" [2].

In this chapter, we will first simply describe and discuss the existing mechanisms in EP deposits, and then concentrate on analyzing the problems and the phenomena of contrary experimental results. Recently, there have been quite a few papers published that point out the shortcomings and deficiencies about the existing EP reaction mechanism. Only knowing the situation fully and clearly, can people be aware of their starting point. Then, the direction for further work becomes obvious, whether it is improving the existing mechanism, or developing your own new mechanism of EP deposits.

10.1 THE EXISTING REACTION MECHANISM OF EP DEPOSITS

Since Brenner and Riddell published their first paper in 1946 [1], four principal reaction mechanisms have been proposed to explain electroless nickel deposition. In 1990 Mallory summarized the principal reaction mechanisms proposed [3], and explained the electroless nickel deposition. The four principal mechanisms had all been proposed before the rapid development of period (1980–1999) for the electroless plating. Brenner and Riddell in 1946 [1] proposed the first mechanism of electroless nickel plating, that is, the atomic hydrogen mechanism. The second mechanism is called the hydride transfer mechanism, which was first suggested by Hersch [4] in 1956. The electrochemical mechanism is the third one [5,6], and the fourth is the metal hydroxide mechanism, which was first proposed by Cavallotti and Salvago in 1968 [7].

10.1.1 The So-Called Four Classical Mechanisms

10.1.1.1 Atomic Hydrogen Mechanism

As mentioned above, the atomic hydrogen mechanism was first proposed by Brenner and Riddell [1]. They used the following reaction equations to describe the formation of Ni and H_2:

$$NiCl_2 + NaH_2PO_2 + H_2O \rightarrow Ni\downarrow + 2HCI + NaH_2PO_3 \tag{10.1}$$

or

$$Ni^{2+} + H_2PO_2^- + H_2O \rightarrow Ni^0 \downarrow + 2H^+ + H_2PO_3^- \quad (10.2)$$

and

$$NaH_2PO_2 + H_2O \rightarrow NaH_2PO_3 + H_2\uparrow \quad (10.3)$$

Equation (10.1) represents the production of metal Ni, which results from the metal salt $NiCl_2$ reduced by reductant NaH_2PO_2. In fact, Brenner and Riddell believed that the actual nickel reductant is atomic hydrogen. The atomic hydrogen, generated by the reaction of water with hypophosphite, is absorbed at the catalytic surface. The Ni ions (Ni^{2+}) gained the electrons released from the atomic hydrogen, and were reduced to metallic Ni. This is probably the reason why the later people called it the atomic hydrogen mechanism [3]. Equation (10.2) indicates that the hypophosphite is oxidized by the water, and then hydrogen was produced and released.

Experiments evidently indicate that the deposit is not pure nickel but contains phosphorus, that is, the alloys of nickel and phosphorus when the reductant hypophosphite is used. However, Eqs. (10.1) and (10.2) completely fail to account for the phosphorus component of the alloy.

Gutzeit in 1959 [8] looked at that problem and resolved it. He believed that the formation of elemental phosphorus was the result of reaction between hypophosphite and atomic hydrogen:

$$H_2PO_2^- + H^+ \rightarrow P \downarrow + H_2O + OH^- \quad (10.4)$$

Moreover, the formation of atomic hydrogen may also be the result of the dehydration of the hypophosphite ion:

$$H_2PO_2^- \rightarrow PO_2^- + 2H \quad (10.5)$$

$$PO_2^- + H_2O \rightarrow H_2PO_3^- + H^+ \quad (10.6)$$

From Eq. (10.1), it is easy to see that the reduction of nickel ion is catalyzed by certain metals, including nickel, and as nickel is produced by the reaction, it is therefore autocatalytic. This explains why the reaction, which is rather slow in starting, proceeds with so much vigor after it begins. In addition, OH^- produced in Eq. (10.4) will increase the pH at the interface between coating and bath solution. This pH increase will hence increase the reaction speed in Eq. (10.2), which increases the amount of H^+. The increase of H^+ causes the pH value to decrease, thereby increasing the rate of reaction in Eq. (10.4), and the amount of OH^- decreases. Such cyclic variation of pH value leads to the periodical change of P content in the film, causing the layer structure in electroless plating coatings. That is to say, the layer structure in EN deposits stems from the wave variation of pH value in the bath solution. Therefore, the atomic hydrogen mechanism can help us to understand the layer structure of EN films.

However, the atomic hydrogen mechanism cannot explain the simultaneous reduction of nickel and hydrogen. Brenner found that the use efficiency of hypophosphite is around 37% at the best conditions [9]. In other words, to deposit 1 g of nickel, 5 g of sodium hypophosphite is required [9,10]. The atomic hydrogen mechanism cannot explain such fact as well.

10.1.1.2 The Hydride Transfer Mechanism

The hydride transfer mechanism was first suggested by Hersch in 1955 [4] and then modified by Lukes in 1964 [11]. Lukes believed that the hypophosphite ion $H_2PO_2^-$ acted as the donor of hydride ions (H^-), but did not decompose and release atomic hydrogen H^+. Therefore, Lukes proposed the main reaction equation of the hydride transfer theory was [11,12]

$$H_2PO_2^- + H_2O \rightarrow HPO_3^- + 2H^+ + H^- \quad (10.7)$$

Thus, hydride ions (H^-) on a metal-catalyzed surface can react with available nickel ions to produce a deposit and evolve hydrogen gas:

$$Ni^{2+} + 2H^- \rightarrow Ni\downarrow + H_2\uparrow \quad (10.8)$$

P was incorporated with Ni according to the reaction equation postulated by Gutzeit in 1959 [12,13]:

$$H_2PO_2^- + 2H \rightarrow P\downarrow + 2H_2O \quad (10.9)$$

10.1.1.3 Electrochemical Mechanism

The electrochemical mechanism was originally proposed by Brenner and Riddell in 1947 [14], and was later modified by Machu and El-Gendi in 1959 [15]. The electrochemical mechanism suggests that the deposition of Ni would be an electrolytic process occurring via local cells on a catalytic surface. In this mechanism, the whole EN process can be divided into two parts, namely, the anodic (oxidation) reaction and the cathodic (reduction) reactions. The anodic reaction between hypophosphite and water generates the electrons for the cathodic reactions. The hydrogen gas and elemental phosphorus are byproducts of nickel deposition. Therefore, this is a pure electrochemical reaction. The reaction equations are [16]:

Anodic:

$$H_2PO_2^- + H_2O \rightarrow H_2PO_3^- + 2H^+ + 2e \quad (10.10)$$

Cathodic:

$$Ni^{2+} + 2e \rightarrow Ni\downarrow \quad (10.11)$$

$$2H^{+} + 2e \rightarrow H_2\uparrow \quad (10.12)$$

$$H_2PO_2^{-} + e \rightarrow P + 2OH^{-} \quad (10.13)$$

Hwang and Lin in 1995 [16] investigated the nucleation and growth of electroless Co–P on highly oriented pyrolytic graphite (HOPG) with tapping mode AFM (TMAFM) and concluded the reaction mechanism appears to be an electrochemical mechanism.

10.1.1.4 Metal Hydroxide Mechanism

The metal hydroxide mechanism was first proposed by Cavallotti and Salvago in 1968 [7], and then supported by the calorimetric study of Randin and Hintermann in 1970 [17]. The proposers of the metal hydroxide mechanism believed that hydrolyzed nickel ions were the actual nickel reactant in Ni–P deposition. First, Ni^{2+} reacted with the hydroxyl ions OH^{-} produced by ionization of water at the catalytic nickel surface and produced the hydrolyzed nickel species $Ni(OH)_2$. Then, the hydrolyzed nickel species $Ni(OH)_2$ as the actual nickel reactant produced the Ni, P, and H_2 through the following reactions:

$$Ni(OH)_2 + H_2PO_2^{-} \rightarrow NiOH + H_2PO_3^{-} + H \quad (10.14)$$

$$NiOH + H_2PO_2^{-} \rightarrow Ni\downarrow + H_2PO_3^{-} + H \quad (10.15)$$

$$Ni + H_2PO_2^{-} \rightarrow P\downarrow + NiOH + OH^{-} \quad (10.16)$$

$$NiOH + H_2O \rightarrow Ni(OH)_2 + H \quad (10.17)$$

$$H + H \rightarrow H_2\uparrow \quad (10.18)$$

The metal hydroxide mechanism successfully explains the lamellar structure of the Ni–P deposit, as well as the different behavior of electroless copper, silver, and palladium plating in which the metal ions are reduced by hypophosphite without P codeposition [3].

10.1.2 Mixed Potential Theory

The concept of mixed potential was first introduced by Wagner and Traud [18]. They used the mixed potential concept to explain the corrosion processes in Zn dissolution from a Zn/Hg amalgam. These two authors found that such dissolution was dependent on the amalgam electrochemical potential, but independent of the accompanying H_2 evolution partial reaction. Saito in 1966 [19] and Paunovic in 1968 [20] used the mixed potential concept to describe the EN plating reactions. This mechanism postulated that the potential of a surface undergoing electroless deposition could be regarded as a mixed potential intermediate in value between the potentials of its constituent anodic and cathodic partial reactions. Electroless plating can be considered the superposition of anodic and cathodic reactions at the mixed (deposition) potential. Accordingly, the rates of the anodic reactions are independent of the cathodic reactions occurring simultaneously at the catalytic surface, and the rates of the separate partial reactions depend only on the electrode potential, the mixed potential. Therefore, the mixed potential theory consists of two simple hypotheses: (1) any electrochemical reaction can be divided into two or more partial oxidation and reduction reactions, and (2) there can be no net accumulation of electric charge during an electrochemical reaction. The second hypothesis is a restatement of the law of conservation of charge. It follows that during the reaction of an electrically isolated metal sample, the total rate of oxidation must equal the total of reduction.

The key problem of the mixed potential theory (MPT) is the polarization curves, that is, the Wagner–Traud diagram. The polarization curve is also the current–potential (i–E) curve. The overall reaction of the electroless deposition can be described electrochemically in terms of three current–potential (i–E) curves, as shown schematically in Fig. 10.1. The two full curves represent the two partial anodic and cathodic reactions, one is the $i_c = f(E)$, the current–potential curve for the reduction of M^{z+} ions, the other is $i_a = f(E)$, the current–potential curve for oxidation of the reducing agent (Red). The intersecting points for the polarization curves for anodic and cathodic partial reactions with the potential axis are their respective equilibrium potential values, denoted by E^0_{Red} and E^0_{Mz+}, respectively. The third curve, that is, the dashed curve $i_{total} = f(E)$, shows the current–potential curve for the full electroless solution reaction. From Fig. 10.1, two points on the curve $i_{total} = f(E)$ can be seen. One is its intersecting point with the potential axis. At this intersection the current is zero

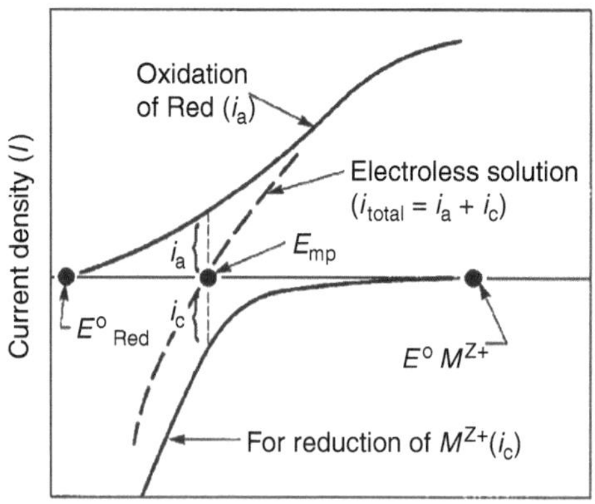

FIGURE 10.1 Current–potential curves (Wagner–Traud diagram) for a generalized electroless deposition reaction. The dashed line indicates the curve for the complete electroless solution. The partial anodic and cathodic currents are represented by i_a and i_c, respectively. *(Adapted from [21].)*

and $i_c = i_a$ for $i_{total} = 0$. The potential in such a case is just the mixed potential, E_{mp}. Therefore, if an electroless system follows MPT, the rate of metal deposition equals the rate of reductant oxidation at E_{mp}. Another is $i_{total} = i_c + i_a$ at any point of the $i_{total} = f(E)$ curve, that is, the total current density, i_{total}, is composed of two components, which is the result of the addition of current densities of the two partial processes.

There is an alternative method of presenting the current–potential curves for electroless metal deposition, that is, the Evans diagram. This was proposed by Evans, who used it to describe the corrosion process of metal. In an Evans diagram, the abscissa represents the log value, as shown in Fig. 10.2. According to the Evans diagram of the mixed potential theory, the current–potential curves for individual processes, $i_c = i_M = f(E)$ and $i_a = i_{Red} = f(E)$, intersect. The features of the coordinates of the intersection in Evans diagram are: (1) the abscissa, the current density of the intersection, is the deposition current density i_{dep} (i.e., log i_{dep}), that is, the rate of electroless deposition in terms of mA/cm^2; and (2) the ordinate, the potential of the intersection, is the mixed potential, E_{mp}.

Using the Faraday's Laws of Electrolysis, the weight of the deposited metal is

$$W = (ItA)/(nF) \quad (10.19)$$

where

W = weight of plated metal in grams
I = current in coulombs per second
t = time in seconds
A = atomic weight of the metal in grams per mole
n = valence of the dissolved metal in solution in equivalents per mole

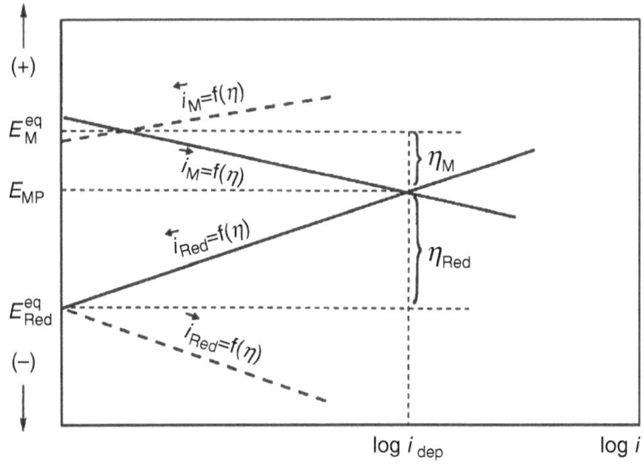

FIGURE 10.2 Evans diagram of current–potential curves for a system with two different simultaneous electrochemical reactions [22].

F = Faraday's constant in coulombs per equivalent $F = 96485.309$ coulombs/equivalent.

It is easy to get the plating rate of metal from Eq. (10.19):

$$R = (600000\, IA)/(nF\rho S) \tag{10.20}$$

where ρ is the density in grams per cubic centimeter and S is the surface area of the plated part in square centimeters. So, for the electroless plating Ni and Cu, the plating rates are

$$R = i \times 1.09\ \text{mg/h}\cdot\text{cm}^2 \tag{10.21}$$

$$R = i \times 1.18\ \text{mg/h}\cdot\text{cm}^2 \tag{10.22}$$

where i is the current density, given in mA/cm^2, respectively. Therefore, once the current density i is known from the current–potential curves diagram (Wagner–Traud diagram or Evans diagram), the plating rate of metal can be easily calculated.

Many research groups have tried to verify the correctness of the mixed potential theory. The first was Paunovic himself. The basis for proposing this theory is that the result of the plating rate he obtained is in agreement with the mixed potential for EP Cu. The bath solution for plating Cu he used contains 0.1 M $CuSO_4 \cdot 5H_2O$, 0.175 M EDTA, 0.05 M CH_2O, NaOH to pH 12.5, and temperature at 24°C (±0.5). He used a gravimetric method (by weighing before and after deposition) to measure the average rate of electroless deposition of Cu, and the result is 1.9 mg/h·cm^2. This value is obtained if the time of deposition is counted from the instant of immersion of the copper plate (substrate) into the solution. If the time of deposition is counted from the instant the mixed potential is reached (about 4 min after immersion of the Cu substrate), the average value of the mixed potential during copper deposition is −0.65 V versus saturated calomel electrode (SCE). Then, he constructed the Evans diagram to determine the plating rate and mixed potential theoretically. He measured the current–potential curves for the reduction of cupric ions in a solution, and that for the oxidation of formaldehyde at the single electrode, respectively, to construct the Evans diagram, as shown in Fig. 10.3. From the Evans diagram, the coordinates of intersection are (1) abscissa, $i = 1.9\ 10^{-3}$A/cm^2, and (2) ordinate, $E = -0.65$ V versus SCE. Therefore, the mixed potential is –0.65 mV, which is in excellent agreement with the experimental result –0.65 mV. From Eq. (10.22), the plating rate of EP Cu is $R = 1.9 \times 1.18$ mg/h·cm^2 = 2.2 mg/h·cm^2. A comparison of these two values, 2.2 (theoretical) and 1.9 (experimental, gravimetric), shows that there is relatively good agreement between the theoretical and the experimental (gravimetric) data. So, we can say that the mixed potential theory is essentially verified for the case of electroless copper deposition.

A number of other groups have reported that their electroless Cu formulations also behaved in a manner consistent with MPT, such as Donahue [23],

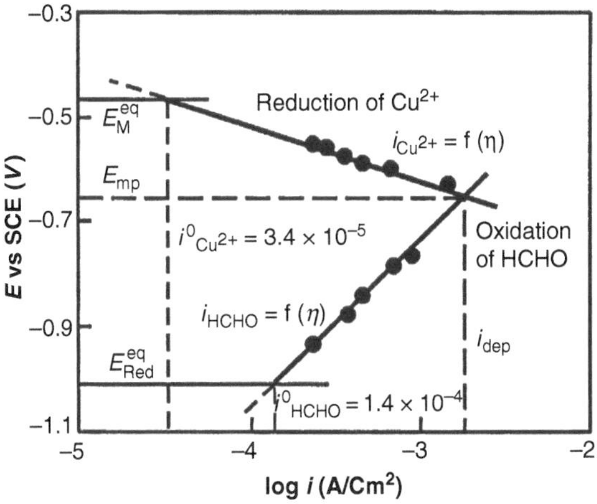

FIGURE 10.3 Current–potential curves for the reduction of Cu^{2+} ions and the oxidation of reducing agent Red, formaldehyde, combined into one graph (an Evans diagram). Solution for the Tafel line for the reduction of Cu^{2+} ions: 0.1 M $CuSO_4$, 0.175 M EDTA, pH 12.50, E^{eq} (Cu/Cu^{2+}) = −0.47 V versus SCE; for the oxidation of formaldehyde: 0.05 M HCHO and 0.075M EDTA, pH 12.50, E^{eq} (HCHO) = −1.0 V versus SCE; temperature 25 ± 0.5°C [20].

Molenaar et al. [24], El-Raghy and Abo-Salama [25], and Hung and Ohno [26]. Here, we will provide more for the work by Hung and Ohno [26] only, because they used hypophosphite as the reducing agent for electroless copper deposition and verified that their results agreed with MPT.

In 1990, Hung and Ohno [26] investigated the hypophosphite-reduced electroless copper deposition by partial anodic and partial cathodic polarization curves at various pH and various concentrations of metal ions and reducing agents. The complete bath polarization curves were also studied. From the results, these authors concluded that although the reaction of hypophosphite-reduced system for EP Cu is very complicated, it still follows the mixed potential theory. However, although they said that the deposition rate is higher at higher pH and higher concentration of hypophosphite, no data was given in the paper to verify this conclusion.

In 1983, Paunovic [27] tested whether the MPT is in agreement with the experimental data in EN. The bath solution of EN he used contained $NiSO_4$ 39.4 g/L, sodium citrate 20 g/L, 85% lactic acid 10 g/L, and DMAB 2 g/L, as the reducing agent. The temperature was held at 40 ± 0.5°C and pH 7.00 (measured at 25°C), adjusted by NH_4OH. Current–potential curves for the partial cathodic reaction (absence of DMAB) and the partial anodic reaction (absence of $NiSO_4$) are shown in Fig. 10.4 in the form of a Wagner–Traud diagram. The potential where $I = 0$ is the mixed potential E_{mp}, and it has a value of −840 mV versus SCE. At the mixed potential E_{mp}, the values of the partial currents i_c and i_a are equal to the deposition rate itself expressed in terms of current. From Fig. 10.4, the deposition rate is 1.38 mA. Because the electrode area used is 0.68 cm², so the deposition

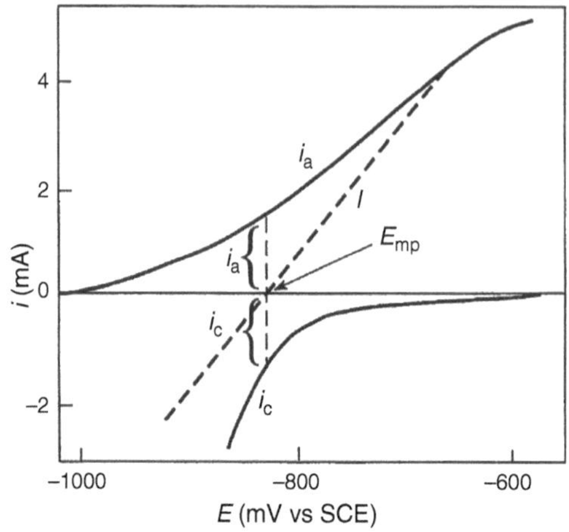

FIGURE 10.4 Wagner–Traud diagram for electroless Ni–B deposition: E_{mp} = −840 mV versus SCE. Electrode area 0.68 cm^2 [27].

rate also can represent as 2.03 mA/cm^2. Using Eq. (10.22), the deposition rate R is $R = i \times 1.09$ $mg/h \cdot cm^2 = 2.2$ $mg/h \cdot cm^2$. Paunovic obtained the same result using the Evans diagram method. The gravimetrically determined rate of deposition is 2.4 $mg/h \cdot cm^2$ and E_{mp} measured directly during deposition is −850 mV versus SCE. Therefore, comparison between the values of the mixed potential and the rate of deposition via direct determination with those derived from the mixed potential theory is very good. Thus, the mixed potential theory was verified for this case of electroless Ni deposition. In 2013, Xie et al. [28] investigated the polarization behavior of electroless Ni–P plating on magnesium alloys. The bath and operation condition are: $NiSO_4$ 15–35 g/L, NaH_2PO_2 15–35 g/L, $C_6H_8O_7$ 5 g/L, HF(40%) 12 cm^3/L, NH_4HF_2 10 g/L, $CS(NH_2)_2$ 1 g/L, NH3(25%) ~30 cm^3/L, pH 4–6, temperature 78–90°C, plating time 60 min. They compared the deposition rates of EN alloys by gravimetrical measurement and electrochemical measurement. The influences of concentration of nickel ions, hypophosphite, and hydrogen ion on the polarization curves and EN deposition in a complete bath were also studied and discussed. From these variations in the deposition potential and current density, a kinetic expression employing the Butler–Volmer equation and mixed potential theory was set up and was in good agreement with the experimental findings. Therefore, these authors believed that the EN deposition in their research is a mixed controlled electrochemical system.

Besides EP Cu and EN, the MPT model was also applied to a number of other electroless metal deposition systems, such as Co–W–P [28,29], Ni–Zn–P [30], Co–W–B [31], Fe–P [32], and Ni–Fe–P [33]. The Shacham–Diamand group compared the average plating rate and the mixed potential determined during the deposition process with those that were calculated from the electrochemical measurements more in detail in 1997 and 2002 [28,29]. The bath solution they

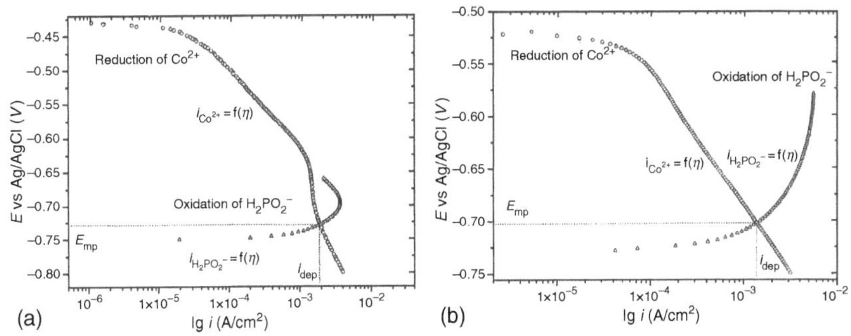

FIGURE 10.5 ***I–E* curves for reduction of Co^{2+} ions and for oxidation of hypophosphite combined into one graph (Evans diagram) for the electroless deposition system of a, Co–P, b, Co–W–P (Na_2WO_4 0.03 M)** [29].

used was $CoSO_4$ 0.082 mol/L, $Na_3C_5H_5O_7$ 0.492 mol/L, H_3BO_3 0.502 mol/L, NaH_2PO_2 0.169 mol/L, Na_2WO_4 0.03 mol/L (for Co–W–P), $H_3[P(W_3O_{10})_4]$ 0.0061 mol/L (for Co–W–P), KOH to adjust pH to 8.8–9.0, and temperature 90 ± 5°C. Figure 10.5 shows the Evans diagrams of mixed potential theory for the EP Co–P and Co–W–P alloys. So, from the figure, the mixed potentials and deposition rates for different Co alloys were determined from the MPT. The group measured the average rates of electroless deposition of Co alloys gravimetrically (by weighting before and after deposition), at 90°C, after 1 h plating. Table 10.1 lists the theoretical and experimental results for the plating rate and mixed potential of EP Co alloys. From the table, it can clearly be seen that the mixed potentials that were determined during the plating are in good agreement with those obtained from electrochemical measurements for both Co–W–P and Co–P alloys. For the EP Co–W–P alloy, the average deposition rate calculated by the mixed potential theory showed good agreement with that obtained from electrochemical measurements. However, this was different for the EP Co–P alloy where the calculated magnitude (2.09 mg/h·cm^2) was significantly higher

TABLE 10.1 Comparison of Average Deposition Rates of EP Co Alloys and Mixed Potential [29]

	Average deposition rate (mg/h·cm^2)		Average mixed potential, V versus Ag/AgCl	
Alloy	Experimental	Calculated from MPT	Experimental	Obtained from MPT
Co–P	1.12	2.09	−0.75	−0.728
Co–W–P	1.34	1.51	−0.702	−0.702

than the average measured deposition rate (1.12 mg/h·cm^2). The authors' explanation for this fact may result from that the cathodic and anodic processes that occur simultaneously are interdependent. Another possible explanation is that the autocatalytic process of the Co–P reduction is suppressed due to the formation of the cobalt phosphides and poisoning of the catalytic surface with byproducts of hypophosphite oxidation. Such processes can be inhibited in the presence of tungstate ions in the solution. The authors' conclusion is that the mixed potential theory was verified for the autocatalytic chemical reduction reaction of Co–P and Co–W–P alloys with hypophosphite.

As for the Co–W–B alloy, the same group in 2005 [31] obtained the similar results. Their conclusion is that the electroless deposition process of Co–W–B alloys has an electrochemical nature that may be described using the mixed potential theory. This was confirmed by a comparison of Faraday's law of deposition rate with experimental observations (weight method). Furthermore, E_{MP} values obtained from Tafel's plots actually coincide with values measured *in situ*.

In 2004, Veeraraghavan et al. [30] used the rotating disc electrode (RDE) studies and surface characterization techniques to research the Ni–Zn–P electroless process and to determine how the alloy composition varies with the electrolysis conditions. They also developed a mathematical model based on the mixed potential theory, and used this to optimize a nonanomalous Ni–Zn–P electroless deposition process. Their conclusion is that a theoretical kinetic model based on the mixed potential theory was sufficient to explain the processes occurring during the electroless deposition of the Ni–Zn–P alloy from alkaline electrolytes. The model simulations were compared to the experimental data obtained using RDE experiments. In other words, the MPT fits for the EP Ni–Zn–P alloy.

In 2006, Huang [32] tested the mixed potential theory on Fe–P electroless deposition. They measured values of the mixed potential during deposition and the rate of deposition, and compared them with those derived theoretically from I–E curves for partial reactions. Table 10.2 gives the plating rate and mixed potential of electroless Fe–P alloys obtained by gravimetrical measurements and by derivation from the MPT theory. It can be seen from this table that the mixed potentials determined during the plating process are in good agreement with those obtained from electrochemical measurements. The plating rates determined by two methods show similar trends, that is, the plating rates increase with pH and reach a maximum value at pH equal to about 11. However, the instantaneous plating rates calculated from electrochemical measurement are significantly higher than the average plating rates determined from gravimetrical measurements. These authors conclude that the possible reason for the plating rate determined by electrochemical measurement is higher than the average plating rate determined from gravimetrical measurements is due to the hydrogen evolution. Moreover, they still believe that these results indicate the electroless Fe–P deposition, obeys the mixed potential theory.

In 2009, Huang et al. [33] investigated the influence of $CeCl_3$ and deposition parameters on electroless Ni–Fe–P deposition. They found that the deposition

TABLE 10.2 Plating Rate and Mixed Potential of Electroless Fe–P Alloys Obtained by Gravimetrical and Electrochemical Methods [32]

pH value	Average deposition rate (mg/h·cm^2)			Average mixed potential, V versus Ag/AgCl	
	Experimental	Electrochemical method			
		Deposition current density (A/dm^2)	Calculated from MPT	Experimental	Obtained from MPT
9.0	0.972	0.240	2.509	−1.05	−1.05
9.5	1.255	0.295	3.079	−1.08	−1.11
10.0	1.518	0.381	3.976	−1.10	−1.17
10.5	1.634	0.416	4.343	−1.16	−1.2
11.0	1.814	0.415	4.333	−1.21	−1.23
11.5	1.714	0.369	3.853	−1.21	−1.26

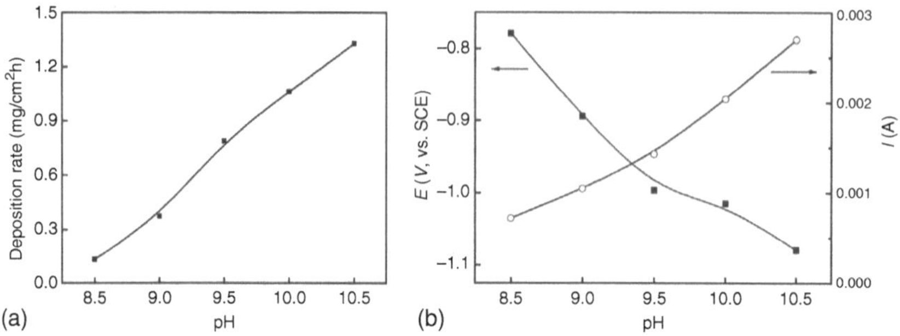

FIGURE 10.6 Effect of bath pH on the deposition rate determined by gravimetrical measurements (a), the dependence of the mixed potential and deposition current density on the pH, as derived from the polarization curves (b) [33].

rate deduced from polarization experiments shows a similar trend to that determined by gravimetrical measurements, as shown in Fig. 10.6. Moreover, the mixed potential shifts cathodically with the pH, the same trend as was observed for electroless Co–W–P and Co–P deposition [29]. So, these authors believed that the electroless Ni–Fe–P deposition obeys the mixed potential theory. It should be said that it is unfortunate that the plating rate data were not supplied or the mixed potential. If they gave these data, as they had in their previous paper on EP Fe–P [32], then the description would be clearer.

From the above discussions, it is evident that the MPT can be used to describe the EP process in many systems, such as Cu, Ni–P, Co–P, Fe–P, Co–W–P, Co–W–B, Ni–Zn–P, and Ni–Fe–P. It should be said that the MPT has had considerable success. However, this EP mechanism still has defects and shortcomings, which will be discussed in Section 10.2 together with other mechanisms.

10.1.3 The Uniform Electrochemical Mechanism

The so-called uniform electrochemical mechanism (universal mechanism) was proposed by Van den Meerakker in 1981 [34]. As for why Van den Meerakker proposed this mechanism and seemed not to have special reasons for favoring this method, she just said, "We propose that for all electroless plating solutions one mechanism must be valid." However, O'Sullivan in 2001 [35] observed: "The failure of the preceding mechanisms, ... Observations such as coevolution of H_2 gas known to be derived from the reductant in electroless deposition, ... prompted Van den Meerakker to postulate a general mechanism that involves reductant adsorption followed by dissociation."

The uniform mechanism may be outlined as follows:

In alkaline media

Anodic:

$$RH \xrightarrow{M} R + H \tag{10.23}$$

$$R + OH^- \rightarrow ROH + e \tag{10.24}$$

$$\text{Recombination}: H + H \rightarrow H_2\uparrow \tag{10.25}$$

$$\text{Oxidation}: H + OH^- \rightarrow H_2O + e \tag{10.26}$$

Cathodic:

$$\text{Metal deposition}: M^{n+} + ne \rightarrow M\downarrow \tag{10.27}$$

$$\text{Hydrogen evolution}: 2H_2O + 2e \rightarrow H_2\uparrow + OH^- \tag{10.28}$$

In acid media, reactions (10.26) and (10.28) become

$$\text{Oxidation}: H + H^+ \rightarrow H_2\downarrow \tag{10.26'}$$

$$\text{Hydrogen evolution}: 2H^+ + 2e \rightarrow H_2\downarrow \tag{10.28'}$$

where RH denotes the reducing agent, which yields the adsorbed radical species R and atomic H upon adsorption and dissociation. More could be said to explain it clearly (especially for the case in alkaline media). In this mechanism, the first step of the oxidation reaction is a dehydrogenation (Eq. (10.23)), producing an R radical and an atomic hydrogen. This first step determines the catalytic nature of the electroless deposition processes. In Eq. (10.24), an electron is produced from the oxidation of the R radical. Depending on the metal species, which catalyze the reaction, the hydrogen atom gives either H_2 gas (Eq. (10.25)) or H_2O (Eq. (10.26)). Therefore, the reaction can be divided into three steps, namely, adsorption, dehydrogenation, and oxidation.

Using this uniform mechanism, Van den Meerakker claims that this mechanism explains the validity of the mixed potential theory and the catalytic nature and the high activation energy of electroless processes. It also accounts for the influence of pH on the deposition rate and it also explains the fact that in electroless copper, nickel–boron, and nickel–phosphorus solutions a reducing agent efficiency of at most 50% is found, whereas this efficiency can exceed 50% when metals like platinum, rhenium, and tungsten are codeposited. In a word, she believed that the universe mechanism is very successful for explaining the EP processes.

10.2 SHORTCOMINGS AND DEFICIENCIES OF EXISTING REACTION MECHANISMS OF EP DEPOSITS

From the above discussions, we can clearly see that all these EP mechanisms can explain most characteristics of the electroless solutions for which they were originally proposed. The MPT mechanism can also explain other experimental facts obtained by later research groups such as in EP Cu, Ni–P, Co–P, Fe–P, Co–W–P, Co–W–B, Ni–Zn–P, and Ni–Fe–P systems. Then, can the existing mechanisms be used to give a universal description of electroless processes in general? No! There are shortcomings and deficiencies of existing reaction mechanisms of EP deposits. We will describe these briefly in the next few paragraphs.

The atomic hydrogen mechanism cannot explain why the stoichiometric utilization of hypophosphite is always less than 50%, and does not account for the simultaneous reduction of nickel and hydrogen [3]. In addition, according to this mechanism, adsorbed atomic H serves to both reduce Ni ions and generate H_2 through a recombination reaction. Thus, one could argue that experimental conditions could be found in which the amount of H_2 evolved is strongly dependent on experimental conditions, for example, solution pH. However, there seems to be no clear evidence that adsorbed H atoms participate to a significant extent in direct reduction of metal ions [35].

According to the hydride transfer mechanism, hydrogen evolution should be at least equivalent to metal deposition; however, it was contrary to some experimental observations [33]. Enyo in his review of the H_2 electrode reaction in 1983 [36] concluded that involvement of H^- intermediates in H_2 electrode reactions had not gained sufficient experimental evidence to warrant serious consideration. Therefore, the hydride transfer mechanism proposes H^- as an intermediate reducing agent has not been affirmed yet [35].

The electrochemical mechanism predicts that the nickel ion concentration must have significant effect on the rate of nickel deposition; however, the results show just the converse [3]. O'Sullivan in 2001 [35] pointed two drawbacks of this mechanism. One is that, in its most basic form as expressed in Eqs. (10.10)–(10.13), it does not take into account the nature of the surface, either the catalytic activity of the seed, or initial catalyst, or that of the electroless deposit. Another one is that in this mechanism evolved H_2 is derived from the electrochemical reduction of protons; this is again a case of a key reaction species, here electrons, participating in more than one critical reaction. However, O'Sullivan did not give experimental data to support his arguments.

In addition, the metal hydroxide mechanism cannot explain the experiments in which the anodic and cathodic reactions take place in a cell with two compartments separated by a glass frit [3]. O'Sullivan pointed out that this mechanism involves NiOH in four reactions from Eqs (10.14) to (10.17), and $Ni(OH)_2$ as the active Ni reactant in solution [35]. Since increasing the concentration of the complexant(s) in solution will decrease the concentration of both unhydrolyzed and hydrolyzed metal ions, arguments of complexation cannot be readily employed to either support or discount this mechanism.

In the preceding section, we said that many scientists used the mixed potential mechanism to explain their experimental data or to modify the theory [27–33]. Is this mechanism perfect? No, many groups have pointed out its shortcomings and deficiencies. We will introduce them briefly later.

In 1987, Bindra and Roldan [37] investigated the mixed potential theory and the interdependence of partial reactions in electroless copper plating. These authors distinguish two kinds of reactions. The first is when the rate of reaction is controlled by the rate of diffusion of reactants to the plating surface. This type of reaction is said to be under diffusion control. The second takes place when the rate is limited by some slow electrochemical step; this type of reaction is said to be under electrochemical control. In the case of electroless copper plating, experimental results have shown that the formaldehyde partial reaction is electrochemically controlled while the metal deposition partial reaction is diffusion controlled. In other words, the two partial reactions are not independent of each other. Therefore, in the authors' opinion the mixed potential theory is not suitable for EP Cu [37].

In 1992, Gafin and Orchard [38] used the quartz crystal microbalance electrode (QCME) to study two similar electroless nickel baths, using hypophosphite and dimethylamine borane as respective reducing agents. The data (Fig. 10.7) showed that the rate of nickel deposition is substantially higher in the presence of hypophosphite than it is with either DMAB or no reductant at

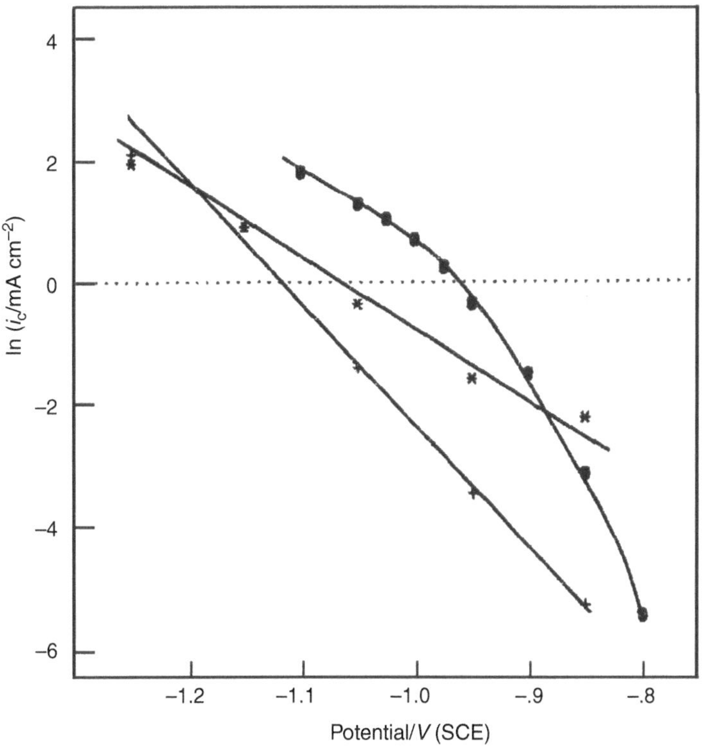

FIGURE 10.7 **Plots of ln i_c against E for data, as well as for a solution containing no reducing agent. All data obtained at 20°C (●) $H_2PO_2^-$ bath, (*) DMAB bath, (+) no reducing agent [38].**

all. For example, the relative cathodic rates (hypophosphite:DMAB:no reductant) are in the approximate ratio 22:7:1 when the potential equals −0.95 V. This clear indication of substantial mechanistic differences between the systems is further emphasized by the differing slopes and curvature of the three plots. In both baths, the presence of the anodic half-reaction leads to a substantial increase in the rate of nickel deposition, indicating the existence of important interactions between the half reactions at the mixed potential. Thereby, they believe that their results cannot be explained by the MPT mechanism. As for why there are significant interactions between the half reactions in both baths, these authors honestly pointed out that "the mechanism of interaction remains to be determined."

In 1992, Bittner et al. [39] investigated the microstructure of copper deposits by impedance measurements, by surface-enhanced Raman scattering, and by STM. The simultaneous determination of the copper deposition rate with a quartz microbalance and the mixed potential under electroless plating conditions allows the separation of the two partial processes copper deposition and formaldehyde oxidation. The measured results show that a microscopically rough surface was formed on deposition, leading the authors to conclude that the rough surface contained specific copper sites that acted as formaldehyde oxidation catalysts; that is, the morphology evolved because of the strong interdependence between the cathodic and anodic processes. Thereby, these authors believe that the mixed potential approach was not valid for copper deposition with formaldehyde.

In 1994, Abrantes and Correia [40] studied the mechanism of EP Ni–P. They paid particular attention to the initial stages of the electroless process using sodium hypophosphite as the reducing agent to understand the interactions and reactions that lead to the deposition of Ni–P. They found that the reduced metal mainly results from the direct reaction among the nickel ions and the reductant radicals rather than from the consumption of the electrons produced by the anodic oxidation of the reducing agent. This process involves a current that is not measurable in the electrochemical oxidation of hypophosphite ion that, in their experimental conditions, represents less than 1% of the real charge-transfer taking place in the electroless experiment. These authors therefore conclude that this is the main reason for the failure of the mixed potential theory. They proposed their own mechanism but no name was given to it.

In 1996, Tzeng [41] reported the effect of halide ions ((F^-, Cl^-, Br^-, I^-)) on electroless nickel plating. He used five baths. Three components and compositions are the same for the five baths: $NiSO_4$ 0.1 mol/L, NaH_2PO_2 0.25 mol/L, CH_3COONH_4 0.13 mol/L. But the halide is different: bath A without any halide, bath B with 0.1 mol/L NaF, bath C with 0.1 mol/L NaCl, bath D with 0.1 mol/L NaBr, and bath E with 0.1 mol/L NaI, respectively. His experimental results show the plating potentials of baths A, B, C, and D are nearly equal and are about 0.15 V versus SCE. The order of the plating current in the electroless nickel plating is: bath A (no halide ion, 10^{-4} A) ≅ bath B (F^-, 10^{-4}A) > bath C (Cl^-, 5×10^{-5} A) > bath D (Br^-, 1.5×10^{-5}A) > bath E (I^-, 0 A). According to the mixed potential theory, the order of plating rate of electroless nickel deposit should be consistent with the sequence of plating currents, and it is interesting to find that, except for bath B, the sequence was fully consistent with the experimental observation. Under the same plating conditions, the experimental measured order of plating rate for various baths follows the sequence: bath A (no halide ion)> bath C (Cl^-) > bath D (Br^-) > bath B (F^-) > bath E (I^-). So, Tzeng concluded that evidently, the mixed potential theory cannot fully explain the electroless nickel plating mechanism and it is possible that other factors affect the deposition rate.

In 2012, using dimethylamine borane complex as a reducing agent Kulyk et al. [42] studied electrochemically the EP Cu at low pH (7–9) using linear sweep voltammetry, cyclic voltammetry, and chronopotentiometry in full electrolytes. Figure 10.8 shows their measured polarization curves of EP Cu at different conditions. From Fig. 10.8a and b, it can be seen that the anodic current due to DMAB oxidation strongly depends on the pH value and decreases monotonically as pH is decreased. But, the cathodic current for the Cu–EDTA complex reduction only slightly depends on the solution pH value. Figure 10.8c indicates that the anodic and inverted cathodic polarization curves only intersected when the electrolyte pH was 7.5. For all other pH levels applied, the anodic and cathodic curves never intersected within the measured range (the original plots of anodic and uninverted cathodic curves are shown in Fig. 10.8a and b, respectively; the corresponding plots other than for pH = 7.5 in Fig. 10.8c are omitted for clarity). However, according to MPT, EP should occur at potentials where the anodic and inverted cathodic polarization curves intersect. Thereby, these authors show that the partial reactions in this system are not independent of each other. Moreover, both cathodic and anodic reactions are catalyzed by each other, and the real deposition rate must be higher than that predicted by cathodic EP Cu. In other words, both the working potential and

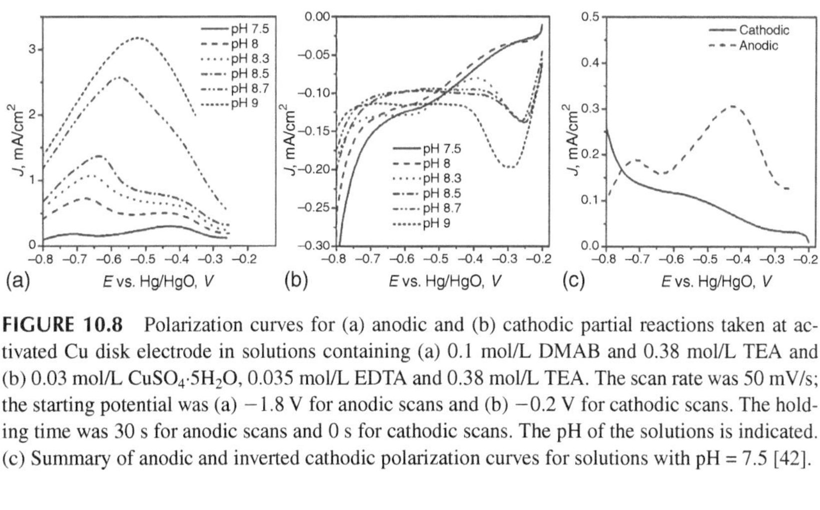

FIGURE 10.8 Polarization curves for (a) anodic and (b) cathodic partial reactions taken at activated Cu disk electrode in solutions containing (a) 0.1 mol/L DMAB and 0.38 mol/L TEA and (b) 0.03 mol/L $CuSO_4{\cdot}5H_2O$, 0.035 mol/L EDTA and 0.38 mol/L TEA. The scan rate was 50 mV/s; the starting potential was (a) −1.8 V for anodic scans and (b) −0.2 V for cathodic scans. The holding time was 30 s for anodic scans and 0 s for cathodic scans. The pH of the solutions is indicated. (c) Summary of anodic and inverted cathodic polarization curves for solutions with pH = 7.5 [42].

the rate-controlling mechanism are dependent on the pH of the solution. Thus, these authors conclude that the mixed potential theory is not applicable to the described EP Cu system.

In 2013, Yu et al. [43] studied the interactions between the cathodic and anodic processes during electroless Cu deposition using the glyoxylic acid reducing agent. Their measured polarization curves (Fig. 10.9) demonstrate that simplistic application of the mixed potential theory to this system leads to erroneous predictions of the surface mixed potential and the electroless deposition rate. The predicted mixed potential (−0.38 V) is ~120 mV, more anodic than the experimentally measured mixed potential (−0.5 V). The deposition rate predicted from the mixed potential analysis of the partial polarization curves is one-sixth of the value obtained by deposit weight measurement in a complete electroless bath. This suggests an interdependence between the cathodic and anodic half reactions. Moreover, these authors used chronoamperometry under various conditions of bath composition, electrode surface preparation, and applied potential, and further confirmed this interdependence. Then, they proposed an autocatalytic mechanism in glyoxylic acid-based electroless Cu deposition.

As for the so-called uniform mechanism, because it is just the summary of reactions in existing mechanisms, and does not essentially improve the theoretical situation of electroless plating, so it is rare for reports of research into drawbacks to be published. Moreover, just as we pointed out in the very beginning of this chapter, EP is a very complex process, including several reactions such

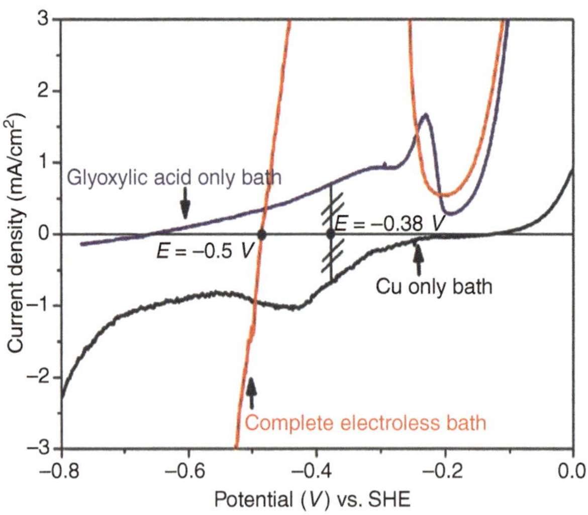

FIGURE 10.9 Polarization curves obtained on a Cu foil from a "glyoxylic acid only" bath (blue curve), a "Cu only" bath (black curve), and a "complete electroless" bath (red curve). The graph clearly shows the predicted mixed potential (−0.38V) is ~120 mV more anodic than the experimentally measured mixed potential (−0.5 V) [43].

as oxidation reaction of reducing agent, deposition reaction of metal ion and catalytic reaction, and is affected by various factors such as metal salt(s), complexing agent, stability agent, additives, pH value, and temperature, it is very difficult to develop a universe mechanism explaining all kinds of experimental facts. Therefore, based on such consideration, some scientists think that finding a generalized correct mechanism may not be a clever idea [44].

In the rather lengthy discussion above, scientists reported the shortcomings and deficiencies of existing reaction mechanisms of EP deposits from the very beginning of discovery of EN until the writing of this book (2014–2015). Thus, these studies make people understand the EP mechanisms in a more deep and comprehensive way.

From the contents in the above two sections, the following conclusions can be drawn.

First, the existing reaction mechanisms of EP deposits make us understand the nature of EP process from various respects; especially the MPT theory has been confirmed to describe the EP process in many systems, such as Cu, Ni–P, Co–P, Fe–P, Co–W–P, Co–W–B, Ni–Zn–P, and Ni–Fe–P. It should be said that the research achievement of the electroless plating mechanism is worthy of recognition.

Second, the shortcomings and deficiencies of existing reaction mechanisms of EP deposits are also obvious. In 2013, scientists confirmed that the MPT is invalid for the EP Cu deposition.

Third, none of the existing reaction mechanisms of EP can explain all the experimental phenomena observed in the EP process completely. Generally speaking, the nature of the electroless deposition process is still fairly understood, so the chemical plating mechanism is still in need of further research.

Fourth, sometimes the reports from various groups are distinctly different, one is that the MPT is valid; another is that the MPT is invalid. Such conflicting phenomena appear on many occasions in the EP Cu and EN systems, but no one has compared them, discussed them, or pointed out the reasons why this is the case. Maybe, this point should be considered in future research.

10.3 KINETICS AND RECENT PROGRESS

The above problems revealed though the investigation of EP mechanisms has caused scientists to try other ways of understanding the nature of the EP process broadly and deeply. That is the research into kinetics, especially the real-time monitoring of the initial electroless plating. *In situ* measurement of electroless plating rates, morphologies, compositions, and so on in the initial deposition stage and subsequent deposition processes using various experimental methods can help researchers see the phenomena under scrutiny more immediately and directly.

In the above chapters, for example, in the chapters of affecting parameters and deposition rate, composition, and microstructure, all of the discussions are concentrated on the whole EP process, virtually no attention has been paid to the initial deposition stage, especially they have not used any real-time monitoring methods. Therefore, outcomes such as the plating rate and microstructure may be different from those obtained in the previous chapters.

There are many methods that can be used for *in situ* monitoring of electroless deposition, including optical absorption measurements [45,46], electric resistance [47], Rutherford backscattering spectrometry (RSB) [48], various electron microscopes [49–54], and quartz crystal microbalance (QCM) method [55–68]. Usually, scientists did not only use a single method, but also used combining them, for example, QCM methods with microscopy in their *in situ* monitoring process.

QCM was first introduced by Sauerbrey in 1959 [69]. It constitutes of a thin quartz crystal blade sandwiched between two metal electrodes, which produces an alternating electric field across the crystal, causing vibrational motion of the crystal at its resonant frequency [1]. This resonant frequency is sensitive to mass changes of the crystal and its deposited electrodes. So, a very small weight change can be measured accurately as a change in oscillation frequency [69]. Sauerbrey found that the shift in frequency due to deposition of a film of the same acoustic impedance as quartz is proportional to the deposited mass per unit area of the film, $\Delta m/A$; thereby he obtained a relationship [70,71]:

$$\Delta f = -C\Delta m/A \qquad (10.29)$$

where Δf is the change in the resonant frequency of a piezoelectric crystal, Δm is the deposited mass, A is the area, and C is the Sauerbrey constant, which is a material-specific parameter. For an AT-cut 5 MHz crystal at room temperature, $C = 56.6$ Hz/(μg/cm^2). Because it is easy to measure frequencies to ±0.01Hz, changes in mass per unit area of <1 ng/cm^2 are measurable with QCM [70]. Usually, the deposition rate is estimated from the weight gain after a given immersion time. The gravity balances we use today in our laboratories, for example, beam, cantilever, torsion, or spring balances can detect 10^{-10} kg (0.1 μg) [72]. However, the QCM can detect up to 10^{-16} kg (10^{-7} μg). In other words, the measurement accuracy of mass for QCM is higher by at least a factor of 10^6 than that of conventional analytical microbalances. Why is QCM more sensitive than conventional analytical microbalances? According to Mecea [72], it is because that the product of the minimum detectable mass and the acceleration acting on that mass is a constant for both QCM and beam balances. The interested readers can refer to the original paper for more detail.

Since then, QCM has become a widely used instrument for small mass measurements. QCM can act as an *in situ* technique, monitoring in real time the thickness of the film during the changes that occur in any surface, giving access to the rate of deposition. That is why QCM has become a very useful instrument in the research of initial electroless deposition in recent years. Of course, QCM does not give any information on microstructure. The various electron microscopes are very effective equipment for this task.

10.3.1 Real-Time Monitoring of Initial EP

In 1983, Svendsen et al. [48] investigated the first stages of EP Ni–P and Ni–P-W deposits by means of Rutherford backscattering spectrometry (RBS). Measurements of the kinetics during the first 60 s of precipitation indicate that the deposition rates are 37% higher than the rates found in the previous study [73,74]. According to these authors, this discrepancy may be explained by two different types of deposition rates being measured. The gravimetric method yields the deposition rate integrated over 1 h of precipitation, that is, it measures the long-term deposition rate, whereas the RBS method gives the short-term rate within the first 60 s. The large difference in deposition rates for different catalytic systems may be seen only in the short-term rate. In addition, investigations into the stoichiometry indicate that the codeposition of phosphorus, on average, decreases from 10.0 to 3.1% when tungsten is present in the layer. For brief deposition periods (1 and 3 s), the phosphorus content is found to be higher than the average. The phosphorus content in the first 50 nm of a layer seems to be increased compared with the layer average. These authors believe that the increase in phosphorus content for the very thin layers deposited for only 1 and 3 s has not been reported previously. The larger content of phosphorus in the initial plating stage may be caused by the presence of decomposition reaction of hypophosphite due to the direct contact of a palladium catalyst.

In 1990, Honma and Noguchi [46] investigated the initial deposition process of EN measured *in situ* by laser beam. They concluded that initial deposition comprises a number of characteristic stages. An induction period was found before the plating reaction. The induction time was largely dependent on catalyst adsorption and dissolved oxygen concentration. After this induction time, the initial deposition reaction took place very rapidly at the activated catalytic sites, and the reaction then stagnated and eventually reached a steady state. TEM images show that initially the nickel grew laterally followed by steady state growth, and the plating reaction proceeded three dimensionally.

Except for an induction period, two peaks in the plating rate versus time curve obtained from monitoring in real time using QCM have been reported in the chemical bath deposition of Zn(Se,O) thin films [62]. Figure 10.10 shows the growth rate of Zn(Se,O) thin films on three different substrates: Au-polished, Au-unpolished, and Au covered with isolating SiO_2 ($R_d = 1\ k\Omega$). Similar trends can be observed on the three substrates, consisting of an induction period, an initial peak, a second broader peak, and a final decay. The induction time shows little dependence on substrate type, whereas rate and time of appearance of the peaks are substrate dependent. The original authors believed that the mechanisms are different in these three parts. First, the induction period ends up with the formation of first ZnSe particles. These particles activate an electroless reaction induced by a reducer, hydrazine, or hydroxide, which gives rise to deposition of a film with ZnO and ZnSe composition. Second, the electroless mechanism is decelerated after deposition of some nanometers of film; hence, a second growth

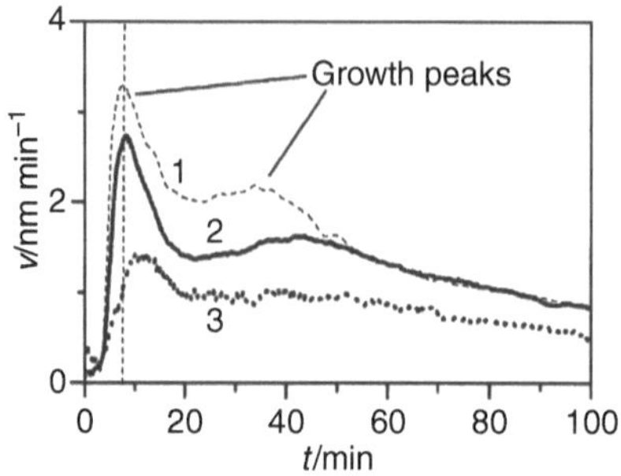

FIGURE 10.10 Plating rate curves corresponding to chemical deposition of ZnSe thin films on three different substrates, Au-unpolished (1), Au-polished (2), Au/SiO_2 (3). The base composition was used at 50°C. The vertical line indicates the time at which a precipitate is distinguishable in solution [43].

mechanism starts to predominate consisting of the chemical reaction of soluble selenide with hydroxide and Zn^{2+} cations on the substrate surface.

The layers resulting from this second mechanism have higher proportion of ZnSe than the previously electroless generated layers. Third, at longer times, the predominant growth mechanism is the deposition of ZnSe clusters formed in the solution, which produces less compact and poorly adherent top layers. This sequence of mechanisms gives rise to films with heterogeneous morphology and composition in depth.

In 2002, Matsubara et al. [63] observed the initial deposition process of EN by the QCM method and microscopy. They found an anomalous deposition rate in the initial deposition and four stages were noted: an induction period (ca. 20 s) before the initiation of deposition, an acceleration period with an increase in the deposition rate, a deceleration period with a decrease in the deposition rate, and a stationary period at a constant deposition rate, as shown in Fig. 10.11. The same authors further reported the rate of initial deposition using two kinds of EN films plated from hypophosphite baths in 2006 [65]. Anomalous initial deposition rate and four stages were found as well. Moreover, they found that in both films, the transformation of the island structure deposits into a continuous structure clearly occurred near the maximum deposition rate. As for the microstructures of the initial stages, they will be discussed in Section 10.3.2.

EN initial deposition rates from a glycine bath were studied by the QCM method in 2005 [64]. Ashassi-Sorkhabi et al. used two different methods for activation of the surface of the substrate: the preplate method and the palladium activation method. In the first method, the substrate was coated with 8 nm coats from the same electroless plating bath at desired temperature using a preplate rate of 0.05 μm/min. In the second method, a sensitizer–activator catalization process was used (one cycle). Their results show different plating rates in the initial deposition. The deposition rate in the initial deposition has four stages

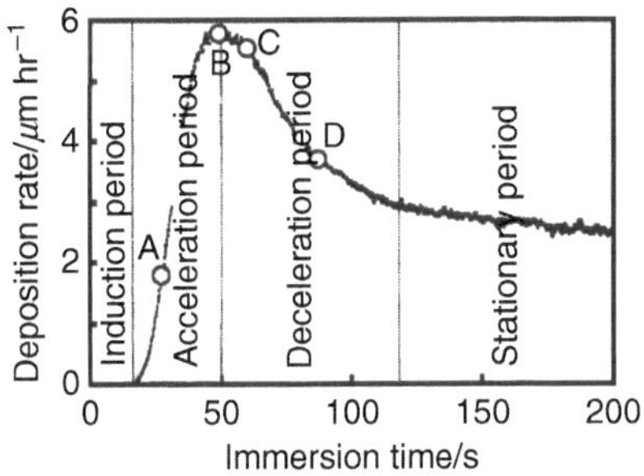

FIGURE 10.11 **Deposition rate observed in EN initial deposition** [63].

(introduction, acceleration, deceleration, and stationary periods) for the second method (Fig. 6b in [64]), which is just as reported by Matsubara et al. [63], but no introduction period (Fig. 3b in [64]) in the first method. These authors conclude that the palladium activation is preferable. However, they did not give other explanations.

A Chinese group in 2011 [68] reported similar results (Fig. 9B in [68]) when they studied the real-time monitor electroless plating of Ni–P film. The deposition starts after a short induction period. Four stages in Refs [63–65] were also observed. It is noteworthy that they explained the deposition rate curves (original Fig. 9 in [68]) more clearly. After the addition of a reducing agent, the palladium nuclei on the substrate surface grow through the reduction of NaH_2PO_2. The deposition of Ni–P then is initiated by the catalytic action of palladium nuclei. Meanwhile, the Ni–P nuclei themselves have the catalytic action for the sequent deposition. Hence, the deposition rate speeds up as the area of catalyst increases. As the deposition proceeds, catalytic nuclei grow and merge into a larger size. The deposition rate reaches a maximum when the substrate surface is fully covered. Then two factors cause a decrease in the deposition rate. One is that the catalytic activity of Ni–P nuclei is less than that of palladium nuclei. The deposition rate will be reduced as the palladium nuclei are covered by Ni–P nuclei. Another factor is that the rough substrate surface is smoothed by the deposition layer. The decrease in the effective catalytic area decreases the deposition rate. Finally, the catalytic area and catalytic activity of the deposition layer approach stable levels; thereby, a stationary period at a constant deposition rate was observed.

A group from UK in 2010 [66] reported a novel one-step photocatalytically initiated electroless deposition (PIED) process, which is used to generate layers of various metals including Ag and Pd on mesoporous TiO_2 (m-TiO_2) coated quartz glass slides and Polyvinylidene fluoride (PVDF) membranes. Rates of metal deposition were measured *in situ* using QCM. They found that deposition occurs only onto areas of the substrate both sensitized with TiO_2 and irradiated with ultraband gap energy light. Visual inspection and SEM analysis show that, even after 2 h immersion in Ag and Pd electroless plating solutions, no deposition occurs on m-TiO_2-sensitized surfaces in the dark. No deposition occurred on samples that had not undergone illumination. However, any period of illumination >30 s initiated a PIED process that produced coherent layers after a sufficient period of dark growth. The results of plating rates versus plating time for Ag (A) and Pd (B) deposits are shown in Fig. 10.12. $\Delta f/\Delta t$ is directly proportional to the metal deposition rate. It can be seen that not only the shapes between A and B are different, but also those between (a), (b), and (c) in A and B are not the same. Only the form of the QCM traces of Pd layer (Fig. 10.12b(a)) is described as follows to make the meaning clear. Figure 10.12b(a) indicates that there is first an induction period of ~40 s during which conventional Pd-on-Au nucleation is initiated. Then, a period of deposition rate acceleration follows, reaching a peak at ~200 s, as nucleation sites are formed and nucleate growth begins. Finally, at $t > 200$ s, the rate decreases as nuclei coalesce through lateral growth and precursor metal ion depletion occurs into solution. Once the surface is covered, deposition reaches a stable rate determined by the balance between the mixed potential (E_{mp}) controlled electrochemical kinetics and mass transfer. Comparison of Fig. 10.12A and B indicates that peak deposition rates obtained from the Pd/hydrazine system at m-TiO_2-coated piezoelectrodes are at least 4.5× and up to 16× greater than those for the Ag/tartrate system. Specifically, peak rates for the Ag system are 9.9 Hz/s (from Fig. 10.12A(b)) and 6.7 Hz/s (Fig. 10.12a(c)) in the dark and light respectively, while peak rates for the Pd system are 49 (Fig. 10.12B(b)) and 56.8 Hz/s (Fig. 10.12B(c)) in the dark and light, respectively. Steady state rates at $t > 6000$ s show similar trends where, from the above, rates for Ag deposition are 3.56 Hz/s or 0.9 nm/min on m-TiO_2 in the dark and 2.4 Hz/s or 0.6 nm/min under illumination. Analogous rates for the Pd system are 37.4 Hz/s or 8.7 nm/min in the dark and 40.9 Hz/s or 9.5 nm/min under illumination. As for the reasons of the

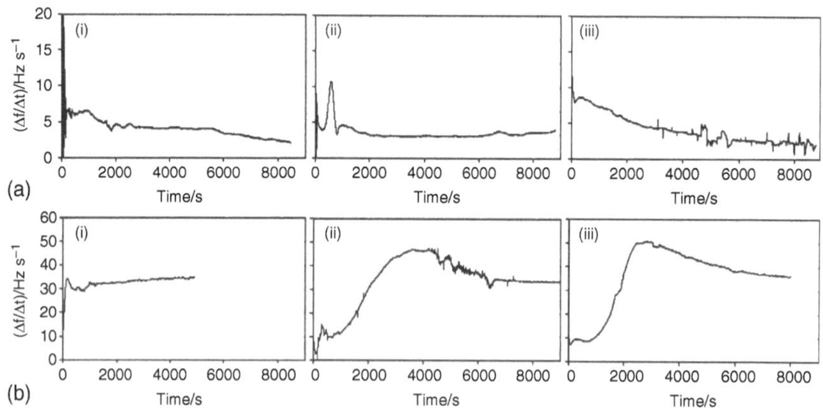

FIGURE 10.12 QCM measurements showing the rate of change of frequency (rate of change of crystal mass) as a function of time: (a) for PIED of Ag/tartrate on (i) bare gold in the dark, (ii) m-TiO_2-coated gold in dark and (iii) m-TiO_2-coated gold in light; (b) for PIED of Pd on (i) bare gold in the dark, (ii) m-TiO_2-coated gold in dark, and (iii) m-TiO_2-coated gold in light [66].

enhancement in Pd plating rate with respect to Ag, these authors believe at least in part derived from the difference in E_{mp} values observed, these in turn being determined by the reducing power of the reductant employed. That a difference in the steady-state deposition rate can be seen on variation of thermodynamic driving force for metal ion reduction indicates that, at the very least, Ag deposition in the Ag/tartrate system is under electrochemical control.

From above 30 years' research (from the very beginning 1983 to today) in the real time monitoring the initial EP, several remarks can be drawn as follows.

The real time monitoring the initial EP in several systems (Ag, Pd, EN, Ni–W–P, and ZnSe) has been investigated. An induction period was found in nearly all of the systems, but the incubation time was different in various systems. In 2011, a Korean group [67] reported that the incubation time for electroless Ag decreases from 100 to 60 s in the ultrasonic-assisted activation step when they investigated the formation of electroless Ag coating. The *in situ* monitoring of electroless Ag deposition at the initial stage was conducted using QCM measurement.

The initial deposition comprises a number of characteristic stages. But the number of stages is different in various systems. Especially, the shapes of the plating rate versus plating time curves are distinctly different as shown in Figs. 10.10–10.12. Even though the four stages are found by several groups, the shapes are not the same.

There are two peaks in the plating rate versus plating time curves for the Ag, Pd, and ZnSe systems (Figs 10.10 –10.12). No explanation has been given.

It is obvious that further work is needed in the future to understand the real-time monitoring of the initial EP more clearly.

10.3.2 Microstructure in Initial Stage of EP

Here, the description is concentrated on the microstructure in initial stage of EP, which is observed by using various *in situ* and *ex situ* instruments. Many papers have studied this issue, and most of them were published recently [49–54,76–86]. Early in 1968, Marton and Schlesinger [49] reported the nucleation, early growth, and structure of Ni–P on $SnCl_2$–$PdCl_2$ activated dielectric substrates using the transmission electron microscope as the probe. They found that the activation is visualized to produce small catalytic sites dispersed on the surface, serving as the nuclei for Ni–P deposition. The sites are estimated to be less than 1 nm in diameter. The growth of Ni–P on these sites is isotropic, resulting in a spherical island-type film geometry. The circular geometry is shown for islands ranging in size from 20 to 200 nm. In 1972, Chow et al. [50] investigated the initial as-deposited structures of Ni–P, Co–P, and Ni–Co–P depositions using a transmission electron microscope. From the results, they conclude that nickel, when deposited from strongly alkaline solutions, assumes a fcc structure. Evidence is presented that the structure of cobalt deposits can be either fcc or hcp depending on the chemistry of the metalizing solution. Mixed cobalt–nickel films appear to form a fcc structure like nickel. In 1972, Pai et al. [47] reported the average crystal size in the initial as-deposited samples to be in the range of 1–2 nm.

In 1991, Homma et al. [51] investigated electroless Ni–P and Cu films at initial deposition stage comparatively by *ex situ* TEM observation. First, they used the different substrates (50 μm thick polyimide (PI) and acetylcellulose (AC) films) and different catalyzing procedures to compare the features of the TEM bright field images of the electroless Ni–P films and Cu at the initial deposition stage on these substrates. The plating time is from 1 to 10 s for EN, and less than 60 s for EP Cu. They found that the difference in the initial deposition conditions between Ni–P and Cu films was clearly observed. The Ni–P deposits consist of very fine component crystallites that grow isotropically to form uniform and spherical grains. On the other hand, the Cu deposits are composed of larger component crystallites. Second, they monitored the initial growth process of the films. Figure 10.13 shows the TEM bright field images and the corresponding transmission high energy electron diffraction (THEED) patterns of electroless-plated Ni–P and Cu films in the initial deposition stages, deposited by AC substrates. In the figure, three stages were found. In stage I of the nucleation stage, the amount of the nucleated particles shows an apparent difference between the Ni–P and the Cu deposits, and the density of nucleated particles is of the order of 10^3 particles/μm^2 for the Ni–P deposits and of 10^2 particles/μm^2 for the Cu deposits, respectively. Stages II and III are called the intermediate stage between stages I and III and the stage of forming completely continuous films, respectively. The thicknesses of the film in stages I, II, and III were approximately 10, 20–30, and 50 nm, respectively. In the growth process of grains of II and III stages, their growth modes and features are shown quite different as well. The grains of Ni–P grow spherically, with uniformity in size and shape, and the diffraction patterns of the Ni–P broaden in all stages, suggesting that the deposits consist of very fine crystals. These results confirm [47,75] that the initial deposits consist of component crystallites whose average size is estimated to be 1–2 nm. Homma et al. thus concluded that the numerous nucleations of such fine component crystallites from the whole surface of each Ni–P grain result in their seemingly isotropic growth, forming spherical grains. However, the Cu deposits show angular and irregular conditions, and each grain consists of a single crystallite or a few larger component crystallites. The Cu grains may grow in the direction of the crystal axes of each component crystallite that orients randomly, resulting in an irregular growth condition of the initial deposits. Moreover, some small particles seem to cease growing and remain smaller in size; thereby irregular conditions appear in the films. The results with the PI substrate show that the general conditions of nucleation and the growth of both deposits are similar to those with the AC substrate described above (see Fig. 5 in [51]).

In 1992, the same group [52] also investigated EN processes *in situ* by STM. A highly ordered pyloritic graphite (HOPG) was used as the substrate electrode.

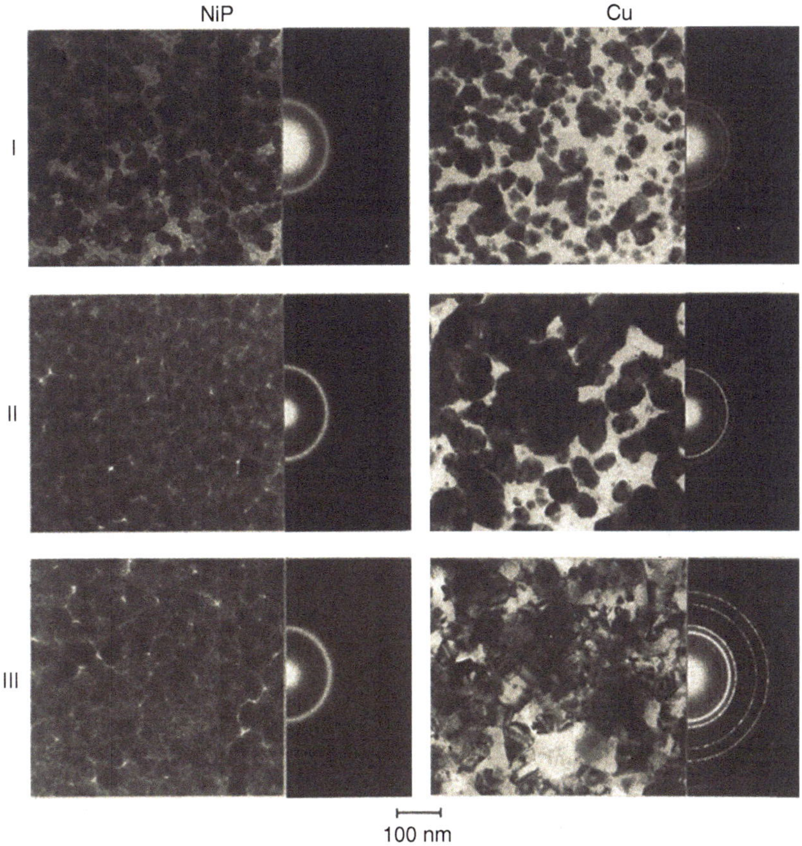

FIGURE 10.13 TEM bright field images and corresponding THEED patterns of electroless-plated Ni–P and Cu films in the initial deposition stages, deposited by AC substrates. Plating time of stages I, II, and III is 1, 3, and 5 s for NiP and 30, 45, and 60 s for Cu, respectively [51].

The features of the HOPG from the TEM and STM images are cluster-like structures, which consist of fine particles about 10 nm in diameter and which are thought to be the Pd-active nuclei. Observation of the initial deposition process (3 min deposition) shows representative STM images of EN film at the nucleation stage, in which particles 20–30 nm in diameter nucleate along a step in the substrate where the catalytic nuclei are thought to be. As film growth proceeds, small particles gather, form larger ones and then a continuous film. However, STM observation of the sequential deposition process indicates that even though the deposition time is up to 70–80 min, and the growth of large grains more than 100 nm in diameter is observed (the original Fig. 5 in [51]), these large grains consist of smaller particles 20–30 nm in diameter, as shown by measurement of a smaller area. The size of these particles is almost constant, regardless of the size of the growth of the larger grains appearing in the STM images. These results suggest that the electroless deposition process applied in this study is mainly the result of the continuous nucleation of 20–30 nm diameter grains and their three-dimensional growth.

On the basis of the above studies, in 1997, Homma et al. [53] quantitatively investigated the growth process of EN films on nonconducting substrate (40-μm-thick polyimide films) using tapping mode atomic force microscopy (TMAFM) with focus on the nucleation density on the substrate surface. Three kinds of nucleation densities of 900, 750, and 550 nuclei/μm^2 were obtained and used. The TMAFM observation showed that film growth proceeds mainly through successive nucleation of fine "particles" with several nanometers in diameter and their three-dimensional growth, and that the nucleation density influences the state of aggregation of the particles and hence the resulting surface topography. The lower nucleation density developed the rougher surface with larger "grains" with a diameter of ca. 80 nm. However, the effect of nucleation density decayed with film growth and eventually vanished at a thickness of ca. 2000 nm due to successive nucleation. A higher nucleation density of 3000 nuclei/μm^2 was also examined. This process resulted in a smooth surface without the formation of aggregates at the initial stage.

In 2004, Tsai and Chao [78] reported the growth morphology and crystallinity of EN deposited on Si substrate in early deposition stage (2, 4, 6, 15, and 60 s) pretreated by $SnCl_2$/HCl and $PdCl_2$/HCl solutions using TEM, field emission scanning electron microscope (FESEM), and EDS. Their observations show that the reaction mechanism of EN on Si is an electrochemical mechanism and the deposits are composed of a columnar structure grown along the vertical direction of the substrate surface. The peak of Ni presents together with the peak of Pd in the EDS analysis of the deposits for 2 s, which implies that the initial deposition of Ni–P is related to the presence of Pd. Observing the cross-sectional images, the initially deposited Ni–P particles were sparsely dispersed on the substrate. However, from the EDS spectrum detected in some places of the deposits for 4 s, it reveals that these places consist mainly of Ni and have no Pd appearing. Therefore, these authors infer that deposition of Ni–P is nucleated not only on the Pd particles but also between them. The result is different from the above observations of Marton and Schlesinger [49]. These two authors concluded that the growth of Ni–P starts only on Pd particles based on the chemical reduction mechanism. However, in 2006, Liu et al. [79] obtained the similar results to observations in Ref. [78] because they found that at initial Ni–P deposition stage, the Ni–P particles can be deposited on Si everywhere and concluded that the reaction mechanism appears to be an electrochemical mechanism. The size of the Ni–P particles increases but their number decreases as the plating time increases, and when the Ni–P particles have touched each other in the direction parallel to the substrate surface, they grow only in the direction vertical to the substrate surface and develop into a columnar structure. Khan et al. in 2007 [81] also obtained similar results when they studied the surface characterization of zincated aluminum and selected alloys at the early stage (at immersion time intervals of 5, 10, 20, and 40 s, 1, 2, and 4 min) of the autocatalytic EN immersion process. They found that during the initial stages of electroless immersion, a large quantity of nickel is deposited on catalytic centers. However, there is still a fraction of the surface area consisting of voids or gaps between nickel islands. In addition, they found that the average size of the crystallites within 4 min of electroless Ni–P deposition was around 0.25 nm. In 2011, Qin et al. [83] investigated the morphology variation of Ni–P coating on AZ31 Mg alloy in initial stage by OCP curve (open circuit potential), SEM, and EDS. They found rapid formation of spherical nodules on the exposed Mg substrate in the initial stage. The formation mechanism of the spherical nodules in the initial deposition stage had not been reported. These authors believe that the spherical nodules are the product of the autocatalytic reaction with sodium hypophosphite participating. The process of electroless Ni–P plating contains the coating formation stage and the coating growth stage. The formation stage includes three procedures, that is, the nucleation and growth of Ni crystallites, the extension of the coating in the two-dimensional (2D) direction, and the coalescence of the coating along the three-dimensional (3D) direction.

There are very few papers on the microstructure of initial stages in EP Ni–B [84–86]. We mentioned in Chapter 9 that Vitry et al. [85] reported the surface morphologies of such kinds of EP deposits for the samples plated at plating times from 5 s to 1 h by SEM observation. Figure 9.60 shows the SEM images after an immersion of 5 s to 1 h (f) in the Ni–B deposition bath. In order to compare conveniently the microstructures of initial stages in other EP alloy deposits, we still repeat briefly the main SEM features of Ni–B deposits as enhancing the immersion time in the bath. In the period up to 15 s, there was no evidence of nickel deposition detectable by SEM observation. This period can be considered an induction period. After 15 s, some nickel is already detectable by EDX analysis and by SEM. The first nodules of electroless deposit are observed after an immersion of 30 s and are preferentially concentrated on the scratches and defects of the surface, but do not form a continuous layer yet. Their size is in the range of 0.1–0.2 μm. After 60 s, the nodules are bigger and the nickel layer appears nearly continuous. The size of the nodules is bigger after an immersion of 90 s to 4 min, and a leveling of the surface is observed. A refining of the size of the top nodules is observed after 10 min. After 1 h, the intercolumnar spaces appear to be nearly completely filled out and the apparent size of the nodules is small. The typical cauliflower-like surface texture is then observed clearly. As for the mechanisms for the beginning of the deposition process for EP Ni–B on an un-catalyzed steel substrate, these authors affirm that catalytic oxidation of the borohydride ion by the substrate is the main factor for deposit initiation.

In 2001, Inberg et al. [76] investigated the film evolution at the initial stage of deposition and the structure of EP silver films that contain tungsten Ag(W) on single crystal silicon (100) using STM, HSSEM, and TEM. Figure 10.14a–c shows the microstructures after 1, 10, and 15 s deposition, respectively. They observed some atomic-sized spots at the A and B areas in Fig. 10.14 (a). The

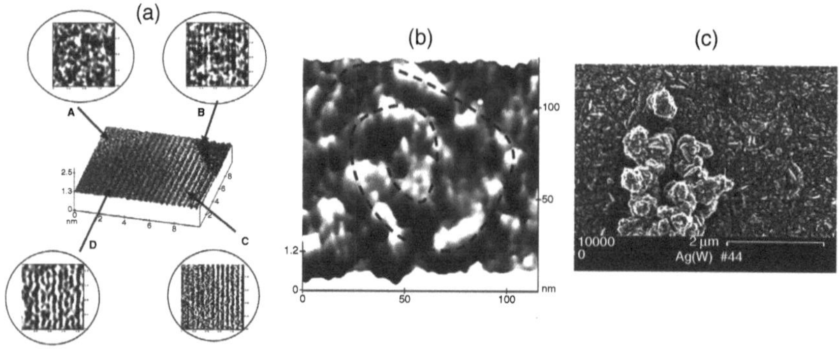

FIGURE 10.14 (a) A representative STM image of Ag(W) surface (10 × 10 nm) after 1 s deposition. (b) Spiral growth presentation of Ag(W) film. (c) Spiral growth presentation of Ag(W) film by HRSEM [76].

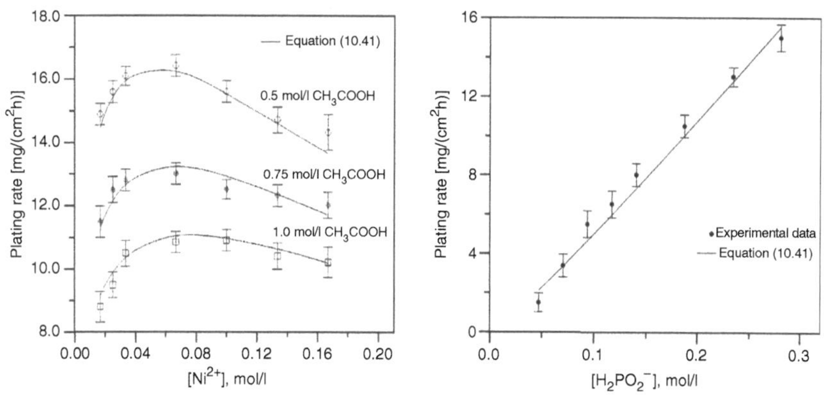

FIGURE 10.15 (Left) The plating rate of electroless Ni–P deposition as a function of nickel ion concentration in baths having various acetic acid concentrations ($[H_2PO_2^-]$ = 0.28 mol/l, pH 4.8, temperature 360 K). (Right) The plating rate of electroless Ni–P deposition as a function of hypophosphite ion concentration ($[Ni^{2+}]$ = 0.1 mol/l, $[CH_3COOH]$ = 0.5 mol/l, pH 4.8, temperature 360 K) [2].

spot profiles demonstrate a height difference between the bottom and top in the range of 0.1–0.14 nm and a width extent of 0.16–0.28 nm, respectively. The atomic spacing as measured by STM is almost equal to that of Ag and much different from what is observed for bare Si substrate. Moreover, chronovoltametry measurements indicate that silver is deposited on the silicon surface from the very beginning. Therefore, the image in Fig. 10.14a is most likely that of silver. In region C in Fig. 10.14a some parallel rows can be observed. And in part D in Fig. 10.14a, the Ag row lines became blurred and the Ag adatoms filled the grooves of rows. After 10 s of deposition the surface of the film exhibits a spiral growth (Fig. 10.14b). The collapsed regions caused by voids become the source of screw dislocations that yield the spiral growth of Ag. These authors believed that one possible explanation is that the voids are produced by H_2 evolution during electroless deposition. Figure 10.14c shows the spiral growth after 15 s deposition.

Although there are many studies on the morphology variation of the coating in the initial deposition stage as discussed earlier, an obvious randomness for the timing for observing the morphology of the coating is clearly presented. No one compares them together and explains the reasons for such obvious randomness. Such a situation also has been pointed out in the literature, for example, in Ref. [83]. Further works are needed in the future to understand such issues deeply.

10.3.3 Kinetics and Empirical Modeling of EP

The kinetics of EP plating describes the variation of deposition process with plating time. Much kinetic experimental data for EP plating is described and discussed in Sections 10.3.1 and 10.3.2. In order to understand the kinetics of EP deeply, it is necessary to model it using equations.

10.3.3.1 Mathematical Model for EP Process

Some authors have built up the mathematical model for the EP process [87–95], although little has been done on developing mathematical models for the electroless deposition processes. In 1978, Paunovic [87] developed a computer model for studying the effect of pH on the deposition rate. In 1980, Donahue [88] developed a mathematical study of electroless copper deposition by assuming diffusion to be the major mode of transfer of ions to the electrode surface. He also took into account equations for microconvection resulting from bubble formation and evolution at the substrate surface. In 1996, Kim and Sohn [89] developed a mathematical model for the electroless plating of nickel on a rotating disk electrode under steady state. They used the concepts of mixed potential theory and applied Butler–Volmer kinetics for determining the plating rate as a function of the various concentrations in solution at steady state. The model was complicated and its agreement with experimental data remained unclear.

In 1999, Ramasubramanian et al. [90] proposed a mathematical model for the electroless deposition of copper on a planar electrode. This was used to make time-dependent predictions of the various quantities in the system. The model takes into account mass transport by diffusion and migration, Butler-Volmer kinetics at the electrode surface, and mixed potential theory. The resultant model was used to predict the concentration profiles, potential response, and plating rate as a function of time and concentration of various reactive components. In 2003, Touhami et al. [91] proposed a mathematical model to reproduce the main features observed in the polarization curves and impedance spectra during the EN deposition in the various ammoniacal solutions, with various nickel sulfate and hypophosphite concentrations. In 2003, Yin et al. [92] proposed a one-dimensional diffusion model to explain the dependence of the bath stability on the plating time under different bath loadings. This mathematical model contains V_d, defined as the decomposition volume, a measure to judge the chemical stability of a plating solution. In 2005, Kantola [93] proposed a mathematical model for the electroless nickel plating process. Based on the model an online monitoring and a new control strategy are proposed. The methods are evaluated by simulations and it is shown that the most important deposition parameters can be controlled and the effect of the process perturbation can be eliminated with the proposed methods.

In 2006, Pfisterer and Hyötyniemi [94] using simple mathematics combined with multilinear tools lead to linear models that can be used for monitoring and control such as electroless nickel plating. The linear model provides a very accurate estimation of the alloy thickness and an accurate estimation of its phosphorus content. Since this model is a linear combination of measured data, it is also possible to see the importance of each measured variable to the value of each output variable. In 2012, Beygi et al. [95] performed simulation of the process using artificial neural network (ANN) media. On the basis of the presented model they consider that it is possible to design a high-efficiency EP bath, while minimum received materials are used and maximum plating rate is obtained.

It must be pointed out that although some mathematical models for the EP process have been proposed since the 1970s, and the proposers of the models have claimed that their models can model and simulate the EP process well, as mentioned above, the proposers of the models did not compare the advantages and disadvantages of other models with their own. In particular, you also cannot see other author groups to cite these models describing their experimental data. In other words, it is difficult to say that these models showed their superiority and success. Future work is needed on this issue.

10.3.3.2 Plating Rate Empirical Equation for EP Process

In Chapter 6, we described and discussed the parameters that influenced plating rate in EP as comprehensively and in as much detail as possible. Moreover, we have paid particular attention to the explanations of experimental data. However, the plating rate for the EP process and the activation energy of EP plating did not refer to basics. Here, this issue will be discussed in the following subsections.

It is well known that the plating rate in the EP process is very important, as has been stressed in the preceding chapters. Researchers not only want to explain the plating rate qualitatively, but also want to describe it quantitatively to understand it deeply and broadly. Describing it quantitatively with mathematical equations is very difficult because this issue is very complicated. From the 1980s [96], scientists used "formal kinetic rate law" to start describing this issue, to the present when many research groups to use the equation modeled on their experimental data [2,55,59,60,96–110]. All the proposed kinetic rate laws never got further than the "empirical nature", that is, they simulated the experimental data. Therefore, the contents of only one empirical rate law, which was proposed by Mallory [3], will be discussed in detail later.

In 1990, Mallory [3] derived an empirical rate equation for Ni–P plating, based on the variables that affect the plating rate:

$$\text{Rate} = \frac{\text{dNi}^0}{\text{d}t} = k[\text{H}_2\text{PO}_2^-]^{\alpha}[\text{Ni}^{+2}]^{\beta}[\text{L}]^{\delta}[\text{H}^+]^{\gamma}[\text{H}_2\text{PO}_3^-]^{\varepsilon}\exp\left[-\frac{E_\text{a}}{RT}\right] \quad (10.30)$$

where α, β, γ, δ, and ε are the reaction orders. k is the frequency factor that is also called order constant, E_a the activation energy, T absolute temperature, and R the gas constant. If all the other factors are kept constant, then from Eq. (10.30), the activation energy, E_a, can be computed from the slope of the straight lines obtained in log rate versus $1/T$ plots, that is,[1]:

$$\frac{\partial \log \text{rate}}{\partial 1/T} = -E_\text{a}/2.3R = -\text{slope} \quad (10.31)$$

$$E_\text{a} = 2.3(1.98\,\text{cal/deg/mole})(\text{slope}) \quad (10.32)$$

Using these equations, Mallory calculated the activation energies for several electroless nickel baths with different complexing agents (see Table 2.6 in [3]). However, he did not simulate the plating rate curve of EN deposition.

Nearly all the plating rate empirical equations for EP processes in various plating systems such as in Co–P [97], Ni–P [2,59,96,98–102], pure Ni [103], Pd and Pd–Ag [60,104], Ni-Cu–P [105], Ni–W–P [106], Co–Fe–B [107], pure Co [108], Ni–P–B [109], and Ni–Co–P [110] are derived from Mallory's formal kinetic rate law. The following is an example. In 2008 Dadvand et al. [107] reported the EP Co–Fe–B depositions. In order to quantify the influence of each parameter more precisely, they follow Mallory, writing the general equation of deposition rate as a function of experimental parameters as follows:

$$R = \frac{\text{d}M}{\text{d}t} = k[\text{DMAB}]^{\alpha}[\text{Co}^{+2}]^{\beta}[\text{Fe}^{2+}]^{\gamma}[\text{H}^+]^{\delta}[\text{A}]^{\sigma}[\text{L}]^{\varepsilon}\exp\left[-\frac{E_\text{a}}{RT}\right] \quad (10.33)$$

where R is the deposition rate, α, β, γ, δ, σ and ε the reaction orders, T the temperature, E_a the activation energy, "L" the complexing agent and "A" the accelerator. Since the concentrations of the complexing agents were kept constant for all experiments, the corresponding reaction order (ε) is equal to zero. To determine the reaction orders (α, β, γ, δ, and σ), the logarithms of deposition

1. Because ln 10 = 2.3, so there is 2.3 in Eq. (10.31). We just cited the original usage from Mallory, therefore 2.3 appears in that equation. In fact, if using ln data not log data, then 2.3 will not be appeared.

rates were plotted versus the logarithm of DMAB, cobalt, iron, and hydrogen ion concentrations, respectively:

$$\log R = \log k + \alpha \log[\mathrm{DMAB}] + \beta \log[\mathrm{Co}^{2+}] + \gamma \log[\mathrm{Fe}^{2+}] + \delta \log[\mathrm{H}^+] + \sigma \mathrm{Log}[\mathrm{A}] - E_a/2.3RT \tag{10.34}$$

$$\alpha = \left(\frac{\mathrm{d}\log R}{\mathrm{d}\log[\mathrm{DMAB}]}\right)_\mathrm{a} \tag{10.35}$$

$$\beta = \left(\frac{\mathrm{d}\log R}{\mathrm{d}\log\left[\mathrm{Co}^{2+}\right]}\right)_\mathrm{b} \tag{10.36}$$

$$\gamma = \left(\frac{\mathrm{d}\log R}{\mathrm{d}\log\left[\mathrm{Fe}^{2+}\right]}\right)_\mathrm{c} \tag{10.37}$$

$$\delta = \left(\frac{\mathrm{d}\log R}{\mathrm{d}\log\left[\mathrm{H}^{+}\right]}\right)_\mathrm{d} \tag{10.38}$$

$$\sigma = \left(\frac{\mathrm{dlog}R}{\mathrm{dlog}[\mathrm{A}]}\right)_\mathrm{e} \tag{10.39}$$

It must be noted that only the linear parts of the plots were considered to calculate the reaction orders. The subscripts a, b, c, d, and e indicate each set of variables, which were held constant for the corresponding partial derivative. The reaction orders were $\alpha = 1.01$, $\beta = 0.77$, $\gamma = -1.31$, $\delta = -0.40$, and $\sigma = 0.29$. Substituting the values into Eq. (10.33), the overall reduction reaction rate can be obtained as

$$R = k[\mathrm{DMAB}]^{1.012}[\mathrm{Co}^{+2}]^{0.773}[\mathrm{Fe}^{2+}]^{-1.307}[\mathrm{H}^+]^{-0.395}[\mathrm{A}]^{0.287}\exp\left[-\frac{E_\mathrm{a}}{RT}\right] \tag{10.40}$$

From Eq. (10.40), it is easy to see that the concentrations of cobalt sulfate, DMAB and accelerator have positive effects on the deposition rate whereas iron sulfate and hydrogen ion concentrations have negative effects. As well, it can be seen that the reaction order for iron sulfate concentration is larger than those of other constituents. Therefore, a slight increase in iron ions concentration in plating solution can decrease the deposition rate remarkably. It should be said that Dadvand et al. still did not compare the data calculated from Eq. (10.40) with the experimental data of the plating rate, that is, did not simulate the plating rate curve of EN deposition. This was true of most cited papers [2,55,59–60,96–109].

Using a similar method to Mallory, in 2000 Małecki and Micek-Ilnicka [2] obtained the following empirical rate equation for EN baths under isothermal conditions.

$$R_{Ni-P} = R_{Ni} + R_P = 0.603\left(\frac{10.9\left[Ni^{2+}\right]}{F(c_i)}\right)^{0.32}\left(\frac{0.61\left[H_2PO_2^-\right]}{F(c_i)}\right)^{1.13} + 0.053\left(\frac{0.61\left[H_2PO_2^-\right]}{F(c_i)}\right)^{1.35} \quad (10.41)$$

$$F(c_i) = 1 + 10.9\,[Ni^{2+}] + [0.61H_2PO_2^-] + 5.72[CH_3COOH]_{undiss} + 10474[H^+ \quad (10.42)$$

Figure 10.15 shows that the experimental data of the plating rate are in agreement with the data calculated from Eq. (10.41). This is not strange because the theoretical equation is an empirical rate equation for EN baths. Li and An in 2008 [100] using nearly the same method as Małecki and Micek-Ilnicka [2] derived a similar equation to Eq. (10.41) for EN deposition. They also simulated several plating curves (R_{Ni-P} versus Ni^{2+}, $H_2PO_2^-$, and pH). However, they did not cite the paper of Małecki and Micek-Ilnicka, which is somewhat strange because their paper was published 8 years after that of Małecki and Micek-Ilnicka.

10.3.3.3 The Activation Energy, E_a, of Electroless Deposition

The activation energy, E_a, of electroless deposition is the necessary energy for EP to commence. Comparing the magnitude of the activation energy, the degree of difficulty for EP can be evaluated. The activation energy is an important kinetic parameter. Therefore, many scientists measured the activation energy of EP in their experiments. It must be remembered that the activation energy for the EP process is different with the crystallization activation energy discussed in Chapter 8. Here the activation energy of EP deposition is the energy that promotes the EP process, but the crystallization activation energy is the energy for producing crystallization transformation of amorphous or metastable EP alloy films. By measuring the plating rate of EP deposits at different deposition temperatures, the activation energy, E_a, can be easily obtained from Eq. (10.32).

In order to understand the kinetic nature of the EP process deeply, the activation energies in various EP systems are listed in Table 10.3. The following conclusions can be drawn from the data in the table.

Activation energies have been measured in many EP systems including pure metals, binary alloys, and ternary alloys. The measurements concentrated on EN, for which the measurements have been made in the time period 1959 to 2013.

TABLE 10.3 Activation Energies of EP Process

No	EP deposits	Reductant or complexing agent	Activation energy (kJ/mol)	References
1	Cu	Formaldehyde	48.98	Shippey FL, Donahue FM, Plating 1973;60:43.
2	Cu	Formaldehyde	60.9	Schumacher R, et al, J Phys Chem 1985;89:4338.
3	Cu on Si	Formaldehyde	52.97	Wu et al, J University Sci Technol Beijing 2007;14(1):1.
4	Ag	Glucose method	113.46	Mayaux P, Descureux S, Silicates Ind 1968;33:381.
5	Ag	Rochelle salt method	92.1	Mayaux P, Descureux S, Silicates Ind 1968;33:381.
6	Au	Borohydride	79.55	Okinaka Y, Ptating 1970; 57:914.
7	Pd	Hydrazine	52.34	Cheng YS, Yeung KL, J Memb Sci 1999;159:107.
8	Ni	Hydrazine	91.65	Cheng SL, Peng HC, J Electrochem Soc 2010;157(2):D81.
9	Co	Hydrazine	66.57	Levy DJ, Proc Amer Electroplating Soc, 1963; 50:29.
10	Co on Si	Hydrazine	32.6	Cheng SL, et al, Appl Surf Sci, 2013;264:732.
11	Ni–P	Hypophosphite	74.1	Gutzeit G, Plating 1959;46:1158.
12	Ni–P	Hydroxyacetate Lactate Aminoacetate p-Alanine Malate Citrate	96.3 67 71.18 54.43 71.18 67	Mallory GO, Lloyd VA, Plat Surf Finish, 1985;72(9):52.
13	Ni–P	Hypophosphite	113.8	Chen L, et al, Electroplat Finish 1997;16(2):9.

TABLE 10.3 Activation Energies of EP Process *(cont.)*

No	EP deposits	Reductant or complexing agent	Activation energy (kJ/mol)	References
14	Ni–P on ZrO_2 powders	Hydrazine No surfactants I* II NaF I II Sodium dodecyl sulfate I II	 217.08 90.21 156.73 52.75 70.98 30.00	Shao Z, et al, J Inorg Mater 1999;14(3):397.
15	Ni–P	Hypophosphite	69.5	Ban C, et al, J Shenyang Uni Technol, 2000;22(2):119.
16	Ni–P on Si	pH 5.2, 4.8, and 4.2	68.8, 69.8, and 79.4	Liu WL, et al, Thin Solid Films 2006;510:102.
17	Ni–P on Fe/Si	Hypophosphite pH 4.2, 4.6, and 5.0	99.58, 87.4, and 41.55	Liu WL, et al, J Electrochem Soc, 2008;155:D192.
18	Ni–P on Fe powder	Hypophosphite	50.775	Yan C, et al, Chem Eng (China) 2011;39(1):73.
19	Ni–P	Hypophosphite No accelerator MeTU N,N′-EtTU TU	 59.16 79.56 78.66 67.39	Karthikeyan S, et al, Intern J ChemTech Res, 2013;5(1):246.
20	Ni–B	DMAB	44.38–64.1	Schmeckenbecher AF, Plating 1971;58:905.
21	Ni–B on Cu	DMAB	128 ±10	Wehner S, et al, J Appl Electrochem, 2003;33:457.
22	Co–P on Cu	Hypophosphite	56.8	Wang D, et al, Plat Finish, 1998;20(6):27.
23	Co–P on Si	Hypophosphite	65.61	Liu WL, et al, J Electrochem Soc, 2004;151(10):C680.
24	Co–P on Cu	Hypophosphite	92.6	Rudnik E, Skrzyniarz P, Trans IMF 2007;85:82.

(Continued)

TABLE 10.3 Activation Energies of EP Process *(cont.)*

No	EP deposits	Reductant or complexing agent	Activation energy (kJ/mol)	References
25	Ni–Co–P on Si	Hypophosphite Co/(Co + Ni) in bath		Liu WL, et al, Thin Solid Films 2011;519(15):4749
		0.1	49.66	
		0.3	52.54	
		0.5	59.54	
		0.7	68.83	
		0.9	87.63	
26	Co–Fe–B on Cu, stainless steel	DMAB	64.2	Dadvand N, et al, J Mater Sci, 2008;19:51.

**The deposition experiments are classified into three groups according to the different surfactants (no surfactants, NaF, and sodium dodecyl sulfate). Experiments indicate that reaction rate versus time curve can be divided into two stages, the reaction rate is slow in the first stage (I), and that is increased significantly in the second stage (II).*

The substrate influences the activation energy enormously. For example, No. 23 for Co–P on Si substrate, the activation energy is 65.61 kJ/mol, but No. 24 for Co–P on Cu substrate the activation energy is 92.6 kJ/mol, which means that EP Co–P on Si is easier than that on Cu.

It is found that even for the same deposits the activation energies measured from different groups are very different, for example, looking at Nos. 1 and 2 for Cu, the first is 48.98 kJ/mol, another is 60.9 kJ/mol. Also comparing Nos. 22 and 24 for Co–P deposits, they all are obtained on Cu substrate and use sodium hypophosphite as reductant, but their activation energies are 56.8 and 92.6 kJ/mol, respectively.

The pH value greatly affects the activation energy. No. 17 shows that the activation energies in EN hypophosphite bath on Fe/Si substrate are 99.58, 87.4, and 41.55 kJ/mol when the pH values are 4.2, 4.6, and 5.0, respectively. The activation energy of electroless Ni–P deposition on Fe/Si increases as pH decreases. No. 16 also shows a similar variation. The authors did not explain this phenomenon.

No. 25 shows that the effect of the atomic ratio of Co to Ni + Co in baths on the activation energy, thereby on the growth behavior of the electroless Ni–Co–P films, was significant. The activation energies are 49.66, 52.54, 59.54, 68.83, and 87.63 kJ/mol as Co/(Co + Ni) in bath 0.1, 0.3, 0.5, 0.7, and 0.9, respectively. In other words, the activation energy increases with enhancement of the content of the Co^{2+} ion in the bath solution, which means that increasing the Co^{2+} ion in the bath, the plating reaction rate reduces. Real experimental results indicate that the plating rate of the electroless Ni–Co–P films is decreased with atomic ratio of Co to Ni + Co in baths.[2] The original authors try to explain such phenomenon using the phase diagram of the Ni–Co system [111], but they found that such an explanation is contrary to thermodynamics. Therefore, they conclude that the clear reason for this contrariety is unknown at the present. The interested reader is referred to [111] to read their detailed explanation.

No. 12 shows the activation energies for several electroless nickel baths with six different complexing agents obtained by Mallory and Lloyd. The effect of complexing agent on the activation energy is very obvious. For example, the activation energy is 96.3 kJ/mol using hydroxyacetate as a complexing agent, but that is only 54.43 kJ/mol with p-alanine.

The effect of surfactant on the activation energy also is significant. No. 14 shows that for EN plated on the ZrO_2 powders, the activation energies in the initial stage are 217.08 kJ/mol, 156.73 kJ/mol and 70.98 kJ/mol when the added surfactants is either none, NaF, and sodium dodecyl sulfate, respectively. There is approximately a three-fold difference between highest and lowest activation energy using these surfactants. The original authors [98] explained this phenomenon as follows: The Ni atoms produced from the redox reaction diffuse to the surrounding area of the surface of ZrO_2 micro particles. The surfactants NaF and sodium dodecyl sulfate are all anionic surfactants, and adsorb cations easily, thereby the Ni^{2+} ions trend segregating onto the surface of ZrO_2 particles. Moreover, the surfactant can reduce the surface energy of the ZrO_2 particles, so the electroless reaction energy of the Ni on the surface of the particles decrease, promoting such reaction easily performed.

In 2013 Karthikeyan et al. [102] studied the influence of accelerators on the activation energies in EN plating process. From the results in No. 19, it is evident that all the accelerators in trace quantities are able to bring down the activation energy of EN process compared to that in their absence. The order of performance with all the compounds is well reflected in their activation energy values.

It is obvious that many conflicting results appear in the existing data as pointed out in the preceding description. In order to clarify these contradictions and to understand their nature deeply, further studies are needed in the future.

10.4 SUMMARY

In this chapter, we concentrated on describing the existing EP mechanism, kinetics, and recent progress in the EP process. Now, the following remarks can be made.

---

2. In the conclusion of the original paper [111], the authors said that "The activation energy is decreased with increase in atomic ratio of Co to Co + Ni in bath for the baths with 0.3, 0.5, and 0.7 atomic ratio of Co to Co + Ni," which is just opposite with their data in Table 3 and statement in abstract.

First, the discussion of the known reaction mechanisms of EP deposits enable us to understand the nature of EP process from various respects. In particular MPT theory has been used to describe the EP process in many systems, such as Cu, Ni–P, Co–P, Fe–P, Co–W–P, Co–W–B, Ni–Zn–P, and Ni–Fe–P. It should be said that the research achievement of electroless plating mechanism is worthy of recognition.

Second, the shortcomings and deficiencies in the existing reaction mechanisms of EP deposits are also obvious, and have been described in more detail here. In addition, in 2012 and 2013, scientists confirmed that MPT is invalid for EP Cu deposition [42,43].

Third, generally speaking, the nature of the electroless deposition process is fairly well understood. However, none of the existed reaction mechanisms of EP can explain completely all the experimental phenomena observed in the EP process, so further research into the mechanisms of chemical plating mechanism is still needed.

Fourth, the problems revealed by the existing EP mechanisms have encouraged scientists to look for other ways to study the nature of the EP process. Various equipment and methods have been tried, such as electron microscopes, QCM to measure and investigate the kinetics, especially the real-time monitoring of the initial electroless plating. Such *in situ* measurement of electroless plating rates, morphologies, compositions, etc. in initial deposition stage and subsequent deposition process is different from the research on the whole EP process that has been described in all the preceding chapters. In the past few years, scientists have performed real-time monitoring of the initial EP for several systems (Ag, Pd, EN, Ni-W–P, and ZnSe). Several stages including induction period from the very beginning in EP process have been found. In addition, many recently published papers studied the microstructure in the initial stages of EP. So, we can say that this is new progress in EP, and shows promise as a technique.

Fifth, sometimes the reports from various groups are distinctly different, for example, one is that the MPT is valid; another is that the MPT is invalid. Such conflicting phenomena appear on many occasions in the EP Cu and EN systems, but no one gives any comparison of them, discusses them, or points out the reasons why the difference arose. Maybe, this ought to be considered in further research. Similarly, results from research into the real-time monitoring of the initial phases of electroless plating appear many conflict phenomena as pointed out in the text. In order to clarify these contradictions and to understand their nature deeply, further studies are needed in the future.

Lastly, the measured activation energy of the electroless deposition in various pure metals (Cu, Ag, Au, Pd. Ni and Co), binary alloys (Ni–P, Ni–B, Co–P) and two ternary alloys (Ni–Co–P and Co–Fe–B) have been collected and analyzed. Comparing the magnitude of the activation energy of EP process, the degree of difficulty for EP can be evaluated. Because of the dispersion of the measured data of the activation energy in EP, especially those in many other alloy systems have not been measured, it is necessary to study this subject further.

REFERENCES

[1] Brenner A, Riddell G. J Res Natl Bur Stand 1946;37:31. Proc Amer Electropl Soc 1946;33:16.
[2] Małecki A, Micek-Ilnicka A. Surf Coat Technol 2000;123:72.
[3] Mallory GO. The electroless nickel plating bath: effect of variables on the process BT Electroless plating: fundamentals and applications. Orlando, FL: American Electroplaters and Surface Finishers Society; 1990. p. 57–99.
[4] Hersch P. Trans Inst Metal Finish 1955–56;33:417.
[5] Jusys Z, Liaukonis J, Vaškelis A. J Electroanal Chem 1992;325(34):247.
[6] de Minjer CH. Electrodepos Surf Treat 1975;3:261.
[7] Cavallotti P, Salvago G. Electrochim Metall 1968;3:239.
[8] Gutzeit G. Plating 1959;46:1158. ibid., 47 (1960) 63.
[9] Brenner A. J Franklin Inst 1947;244(6):479.
[10] Hari Krishnan K, John S, Srinivasan KN, Praveen J, Ganesan M, Kavimani PM. Metall Mater Trans A 2006;37A:1917.
[11] Lukes RM. Plating 1964;51:969.
[12] Gutzeit G. Chemical reactions. In: Symposium on electroless nickel plating. Philadelphia, PA: ASTM STP 11265; 1959. p. 3–12.
[13] Puttlitz K. J IEEE Trans Comp Hybrids Manuf Technol 1990;13(4):467.
[14] Brenner A, Riddell G. J Res Natl Bur Stand 1947;39:385.
[15] Machu W, El-Gendi S. Metalloberfliiche 1959;13:97.
[16] Hwang BJ, Lin SH. J Electrochem Soc 1995;142(11):3749.
[17] Randin JP, Hintermann HE. J Electrochem Soc 1970;117:160.
[18] Wagner C, Traud W. Z Electrochem 1938;44:391.
[19] Saito M. J Met Finish Soc Jpn 1966;17:14.
[20] Paunovic M. Plating 1968;55:1161.
[21] Paunovic M. Electroless deposition of metals and alloys. In: Paunovic M, Ohno I, editors. Pennington, NJ: The Electrochemical Society, Inc.; 1988.
[22] Paunovic M, Schlesinger M. Fundamentals of electrochemical deposition. 2nd ed Hoboken, NJ: Wiley; 2006.
[23] Donahue FM. Oberflaeche-Surf 1972;13(12):301.
[24] Molenaar A, Holdrinet MF, van Beek LKH. Plating 1974;61:238.
[25] El-Raghy SM, Abo-Salama AA. J Electrochem Soc 1979;126:171.
[26] Hung A, Ohno I. J Electrochem Soc 1990;137(3):918.
[27] Paunovic M. P&SF 1983;70:62.
[28] Xie Z, Yu G, Hu B, Zhang X, Li L, Wang W, Zhang D. Int J Electrochem Sci 2013;8:6664.
[28] Kim YS, Lopatin S, Shacham-Diamand Y. In: Proceedings of the IEEE Cornell conference on advanced concepts in high speed semiconductor devices and circuits. 1997. p. 192–200.
[29] Petrov N, Sverdlov Y, Shacham-Diamand Y. J Electrochem Soc 2002;149(4):C187.
[30] Veeraraghavan B, Kim H, Popov B. Electrochim Acta 2004;49:3143.
[31] Sverdlov Y, Bogush V, Einati H, Shacham-Diamand Y. J Electrochem Soc 2005;152(9):C631.
[32] Huang G-F, Huang W-Q, Wang L-L, Meng Y, Xie Z, Zou BS. Electrochim Acta 2006;51:4471.
[33] Huang W-Q, Deng J-Q, Huang Guo-F. Int J Electrochem Sci 2009;4:377.
[34] Van den Meerakker JEAM. J Appl Electrochem 1981;11:395.
[35] O'Sullivan EJ. Fundamental and practical aspects of the electroless deposition reaction. In: Alkire RC, Kolb DM, editors. Advances in electrochemical science and engineering, vol. 7, Wiley-VCH, Verlag GmbH; 2001. p. 225–73.

[36] Enyo M. In: Conway BE, Bockris JO, Yeager E, Khan SUM, White RE, editors. Comprehensive treatise of electrochemistry. New York: Plenum Press; 1983. p. 241.

[37] Bindra P, Roldan J. J Appl Electrochem 1987;17:1254.

[38] Gafin AH, Orchard SW. J Appl Electrochem 1992;22:830.

[39] Bittner A, Wanner M, Weil KG. Ber Bunsenges Phys Chem 1992;96:647.

[40] Abrantes LM, Correia JP. J Electrochem Soc 1994;141(9):2356.

[41] Tzeng GS. J Appl Electrochem 1996;26:969.

[42] Kulyk N, Cherevko S, Chung C-H. Electrochim Acta 2012;59:179.

[43] Yu L, Guo L, Preisser R, Akolkar R. J Electrochem Soc 2013;160:D3004.

[44] Djokic SS. Electroless deposition of metals and alloys. In: Conway BE, White RE, editors. Modem aspects of electrochemistry, vol. 35. New York: Kluwer Academic/Plenum; 2002. p. 51–133.

[45] Dumesic J, Koutsky JA, Chapman TW. J Electrochem Soc 1974;121:1405.

[46] Honma H, Noguchi M. Jpn J Surf Finish 1990;41:164.

[47] Pai ST, Marton JP, Brown J. J Appl Phys 1972;43:282.

[48] Svendsen LG, Osaka T, Koiwa I, Sawai H. J Electrochem Soc 1983;130(11):2255.

[49] Marton JP, Schlesinger M. J Electrochem Soc 1968;115(1):16.

[50] Chow SL, Hedgecock NE, Schlesinger M. J Electrochem Soc 1972;119(12):1614.

[51] Homma T, Naito K, Takai M, Osaka T, Yamazaki Y, Namikawa T. J Electrochem Soc 1991;138:1269.

[52] Homma T, Yamazaki T, Osaka T. J Electrochem Soc 1992;139:732.

[53] Homma T, Tanabe M, Itakura K, Osaka T. J Electrochem Soc 1997;144(2):4123.

[54] Plana D, Campbell AI, Patole SN, Shul G, Dryfe RAW. Langmuir 2010;26:10334.

[55] Schumacher R, Pesek JJ, Melroy OR. J Phys Chem 1985;89:4338.

[56] Kanazawa KK, Doss SK. P&SF 1987;74:52.

[57] Wiese H, Weil KG. Ber Bunsenges Phys Chem 1987;91:619.

[58] Ariizumi N, Shibata M, Furuya N. Electrochemistry 1999;67:445.

[59] Yeung KL, Christiansen SC, Varma A. J Memb Sci 1999;159:107.

[60] Cheng YS, Yeung KL. J Memb Sci 1999;158:127.

[61] Kobayakawa K, Morita M, Miyauchi K, Sato Y, Fujimoto E. P&SF 2000;87:77.

[62] Chaparro AM, Gutiéerrez MT, Herrero J. Electrochim Acta 2001;47:977.

[63] Matsubara H, Yonekawa T, Ishino Y, Nishiyama H, Saito N, Inoue Y. Electrochim Acta 2002;47:4011.

[64] Ashassi-Sorkhabi H, Mirmohseni A, Harrafi H. Electrochim Acta 2005;50:5526.

[65] Matsubara H, Yonekawa T, Ishino Y, Saito N, Nishiyama H, Inoue Y. Electrochim Acta 2006;52:402.

[66] Bromley MA, Boxall C, Galea S, Goodall PS, Woodbury S. J Photochem Photobiol A Chem 2010;216:228.

[67] Lee C-M, Lee M-H, Hur J-Y, Lee H-N, Lee H-K. World Acad Sci Eng Technol 2011;59:1079.

[68] Kang Q, Sheng R, Li Y, Zhang Z, Wu S, Shen D. Sens Actuators B 2011;157:533.

[69] Sauerbrey G. Z. Phys. 1959;155:206.

[70] Smith A. The quartz crystal microbalance. In: Brown ME, Gallagher PK, editors. The handbook of thermal analysis and calorimetry, vol. 5. Amsterdam, The Netherlands: Elsevier; 2008. p. 133–69. [chapter 5].

[71] Kurtz O, Barthelmes J, Rüther R, Wünsche M, Donner C. J Electrochem Plat Technol 2010;5:1.

[72] Mecea VM. Anal Lett 2005;38:753.

[73] Osaka T, Sawai H, Kasai N. J Met Fin Soc Jpn 1981;32:615.

[74] Osaka T, Sawai H, Otoi F, Nihei K. Metal Fin 1982;80:31.
[75] Cortijo RO, Schlesinger M. J Electrochem Soc 1983;130:2341.
[76] Inberg A, Zhu L, Hirschberg G, Gladkikh A, Croitoru N, Shacham-Diamand Y, Gileadi E. J Electrochem Soc 2001;148:C784.
[77] Hong SW, Lee YS, Park K-C, Park J-W. J Electrochem Soc 2003;150(1):C16.
[78] Tsai T-K, Chao C-G. Appl Surf Sci 2004;233:180.
[79] Liu WL, Hsieh SH, Tsai TK, Chen WJ, Wu SS. Thin Solid Films 2006;510:102.
[80] Zhang X, Ren F, Goorsky MS, Tu KN. Surf Coat Technol 2006;201:2724.
[81] Khan E, Oduoza CF, Pearson T. J Appl Electrochem 2007;37:1375.
[82] Khan E, Oduoza CF, Pearson T. Surf Coat Technol 2007;201:5416.
[83] Qin T, Ma L, Yao Y, Ni C, Zhao X, Ding Y. Trans Nonferrous Met Soc China 2011;21:2790.
[84] Vitry V, Kanta A-F, Delaunois F. Mater Sci Eng B 2010;175:266.
[85] Vitry V, Sens A, Kanta A-F, Delaunois F. Appl Surf Sci 2012;263:640.
[86] Vitry V, Sens A, Kanta A-F, Delaunois F. Initial deposition of electroless nickel from borohydride-reduced bath on various ferrous alloys. In: 21st international conference on metallurgy and materials (Metal 2012), Brno, Czech Republic, May 23–25, 2012.
[87] Paunovic M. J Electrochem Soc 1978;132:173.
[88] Donahue FM. J Electrochem Soc 1980;127:51.
[89] Kim Y-S, Sohn H-J. J Electrochem Soc 1996;143(2):505.
[90] Ramasubramanian M, Popov BN, White RE, Chen KS. J Electrochem Soc 1999;146(1):111.
[91] Touhami ME, Chassaing E, Cherkaoui M. Electrochim Acta 2003;48:3651.
[92] Yin X, Hong L, Chen B-H, Ko T-M. J Coll Interf Sci 2003;262:89.
[93] Kantola K. Modelling, monitoring and controlling electroless nickel plating process of plated through hole boards. In: Proceedings 19th European conference on modelling and simulation, Merkuryev Y, Zobel R, Kerckhoffs E, ECMS, 2005.
[94] Pfisterer H-C, Hyötyniemi H. In: Hyvönen E, Kauppinen T, Kortela J, Laukkanen M, Raiko T, Viljanen K, editors. Proceedings of the 12th Finnish artificial intelligence conference STeP 2006. Espoo, Finland: Helsinki University of Technology; October 26–27, 2006. p. 97–102.
[95] Beygi H, Vafaeenezhad H, Sajjadi SA. Appl Surf Sci 2012;258:7744.
[96] Chen L, Wang H, Zhao Z, Teng X. Electroplat Finish 1997;16(2):9.
[97] Wang D, Cheng H, Li Yq, Chen Z, Che R, Yang F, Hu W, Liu Z. Plat Finish 1998;20(6):27.
[98] Shao Z-C, Tian Y-W, Zhai Y-C, Li B-S. J. Inorg, Mater. 1999;14(3):397.
[99] Ban C-Y, Zhang L-T, Zhang L-R, Chen L-J. J Shenyang Uni Technol 2000;22(2):119.
[100] Li L. An Maozhong, Rare Metal Mater Eng 2008;37(11):191.
[101] Yan C, Dai Y-L, Xu H, Hou F, Yao Z-y. Chem Eng (China) 2011;39(1):73.
[102] Karthikeyan S, Narayanan S, Jeeva PA, Srinivasan KN. Intern J ChemTech Res 2013;5(1):246.
[103] Cheng SL, Peng HC. J Electrochem Soc 2010;157(2):D81.
[104] Engin AM, Ma YH. J Membr Sci 2009;330:233.
[105] Chen Z, Chen X. Basic Sci J Textile Uni 2001;14(2):102.
[106] Hu G, Wu H, Yang F, Wang S. Acta Phys Chem Sin 2004;20(3):327.
[107] Dadvand N, Jarjoura G, Kipouros GJ. J Mater Sci Mater Electron 2008;19:51.
[108] Cheng SL, Hsu TL, Lee T, Lee SW, Hu JC, Chen LT. Appl Surf Sci 2013;264:732.
[109] Jiang B, Xiao L, Hu S, Peng J, Zhang H, Wang M. Opt Mater 2009;31:1532.
[110] Liu WL, Hsieh SH, Chen WJ. Thin Solid Films 2011;519(15):4749.
[111] Brandes EA. Smithells metals reference book. 6th ed London, UK: Butterworths & Co; 1983.

Chapter 11

Formation Theory and Formation Range of Electroless Amorphous Alloys

Chapter Outline

Amorphous and Nano Alloys Electroless Depositions. http://dx.doi.org/10.1016/B978-0-12-802685-4.00011-X

After describing and discussing the structure of EP alloys, it is obvious that their structure is mostly amorphous or a mixture of amorphous and microcrystalline and/or nanocrystalline states. The key problem is why EP alloy deposits can form an amorphous state. If an alloy system can form an amorphous phase in the EP process, how large is the formation range of the electroless amorphous alloy? Can we calculate the formation range of the electroless amorphous alloy? All of these topics will be discussed in this chapter.

11.1 GENERAL DESCRIPTION OF FORMATION OF ELECTROLESS AMORPHOUS ALLOYS

Why do the EP alloy deposits form amorphous states? How do we understand the formation of amorphous structure? In particular, can we describe the formation of EP amorphous alloys theoretically and quantitatively? It must be pointed out that these problems have not been solved yet. Only a very few authors refer to this issue. For example, Agarwala and Agarwala in 2003 [1] pointed out that during the deposition of electroless nickel films, an increase in phosphorus content produces amorphous deposition due to the increase in lattice distortion caused by phosphorus atoms situated in the interstitial position of the nickel. It has been found that a number of crystalline nonequilibrium phases are present due to the compositional inhomogeneities prevailing in the Ni–P as-deposited layer. It is obvious that there is a theoretical gap in our understanding of the formation of EP amorphous alloy systems that needs to be filled.

The author studied a similar issue more in detail in the 1980s [2–7] when his group investigated the theory of formation for amorphous alloy systems produced by melt quenching (MQ), ion beam mixing (IBM), and direct ion implantation (II) methods. He believes that such considerations can be basically applied to the formation of amorphous alloy deposits in the EP process. Of course, the details will differ.

At the very beginning of a redox reaction a large quantity of the metal and metalloid (P or B etc.) atoms reduced from metal salt and reductant are present in the bath solution and are deposited onto the substrate. These metal and metalloid atoms tend to be closer to each other because the interaction between the unlike atoms would be stronger than one between like atoms. Usually, the number of the reduced metal atoms is much larger than that of the metalloid atoms in any of the EP alloy systems. So the metalloid atom is surrounded by the metal atoms, forming a group of dissimilar atoms. The number of such atomic groups is large. The interaction between the metal and metalloid atoms consists of both the electron factor as the author called it [2] and the size factor. The electron factor is determined by the bond effect, the difference in electronegativity, and the alloying effect of an element. They represent the rather complex effects of the atomic nuclei on the outer shell, valence electrons, as well as the complex interactions between valence electrons and between the ions themselves. They are generally connected with the behavior of electrons; therefore, the author calls them the electron factor. It is obvious that if the difference between the electron factors of two kinds of atoms which is denoted by y [2,3] is large, the short-range interaction between the two dissimilar atoms then combines closely, and there is not enough motive force and enough time for the atoms in the group to move any great distance, or to rearrange cooperatively in order to produce a regular atomic configuration suitable for forming crystalline structure in longer distance. This means that crystallization is hindered, thereby promoting the easy formation of amorphous alloys.

Besides y, there is another factor that has also been proved to play a role and cannot be ignored in the formation of amorphous alloys and this is the size factor [2,4,5]. The size factor is denoted by x, and indicates the difference of atomic size between two kinds of atoms, which is represented with half of the empirical interatomic distance of elements. If x is large, the difference between the two atomic sizes is large, and the two dissimilar atoms combined more tightly together and the atomic group is more stable. It is, therefore, more difficult for the atoms in the group to move any large distance or to rearrange cooperatively. The formation of crystalline structure over longer distances is therefore impeded, thereby leading to amorphous alloys being formed easily. Considering together these two kinds of factors, that is, y and x, when they are large, the short-range interaction for the two dissimilar atoms is strong, and groups of dissimilar atoms then combine tightly. Therefore, it is more difficult for the atoms in the group to move any great distance, or to rearrange cooperatively in order to produce a regular atomic configuration suitable for forming crystalline structure in longer distance. That is to say, the formation of a crystalline structure is hindered, thereby leading to the easy formation of amorphous alloys.*

---

* Many experimental determinations on amorphous alloys containing metalloids (P, B) such as Ni–P, Ni–B, Co–P, Co–B, Fe–P, and Fe–B alloys prepared by various methods actually found that a very strong interaction existed between metal and metalloid elements as well as short-range ordering in these amorphous alloys. In addition, metalloid elements donated electrons to metal elements or contrary metalloids accepted electrons from metal elements. For example, O'Handley et al. in 1976 (R. C. O'Handley, R. Hasegawa, R. Ray and C.-P. Chou, Appl. Phys. Lett., 29 (1976) 330.) found that the different electron-donor effects of boron and phosphorus are clearly seen for the first time; each atom of these metalloids gives ~1.6 and 2.4 electrons, respectively, to the TM (iron, cobalt, and nickel) d bands. The determination of the magnetic properties in Fe–B metallic glasses from Kemeny et al. in 1979 (T. Kemeny, I. Vincze, and B. Fogarassy, and S. Arajs, Phys. Rev. B 20 (2) (1979) 476.) also found that 1.6 electron/(B atom) transferred from boron to iron. However, ▶

It must be pointed out that for EP the formation of amorphous alloys in the bath solution, the operation temperature is not too high–always less than 100°C, and the diffusion action caused by the temperature should be rather weak. This benefits the stability of the group of dissimilar atoms. Once the atomic group consisting of the metalloid element surrounded with metal atoms is formed through the oxidation–reduction reaction, the operation temperature cannot promote the atoms moving long enough distance to create the atomic regular arrangement suitable for a crystalline structure.

It is clear that although the affected factors (electron and size factors) for the formation of amorphous alloys in EP technology are basically the same as those in conventional MQ [2–4], or in IBM [5], but the formation processes and types in all these techniques are different. The formation of amorphous alloys in MQ is due to temperature quenching under splat cooling. That in IBM results from the "quenching" of energy state as we called it [5]. There is also a rapid "quenching" process in EP, which is due to electroless reduction, and we may call it redox "quenching." As described above, the redox "quenching" promotes the formation of EP amorphous alloys.

If the metalloid atoms formed in the redox process are rather few, the number of the atomic groups is not much greater. Many of the metal atoms cannot surround the metalloid elements, and they may move rather longer distance to form some types of atomic regular arrangement due to the repulsive action between like atoms. Then the microcrystalline or nanocrystalline structure will appear beside the amorphous structure. That is why a mixture of microcrystalline or nanocrystalline and amorphous phases occur in the low P or low B content alloys. If the P or B content is very low in the deposits, then their structure is only crystalline phase with no amorphous phase. Such experimental facts have been discussed in Chapter 8.

From the above formation theory of electroless amorphous alloys, the problem discussed in Chapter 8, that is, how the EP Ni–B coating structure transforms from completely amorphous to a mixture of amorphous and crystalline

---

▶ A group from Japan (J. Tamaki, and T. Imanaka, Chem. Lett., 15 (5) (1986) 679) found that in the Ni–P amorphous alloys prepared by an RF glow discharge sputtering technique that below 25 atom %P phosphorus donated electrons to nickel but above 25 atom% P phosphorus accepted electrons from nickel. There were other experimental results, which may be different from these data. That is to say that though there existed a very strong interaction between metal and metalloid elements, the electron-donor effects between the metalloids and transition metals were not usually the same but different from various experimental determination data. This problem is still being investigated experimentally today . For example, (Hou Yongjiang, Wang Yaquan, He Fei, Han Sen, Mi Zhentao, Wu Wei, and Min Enze, Mater. Lett., 58 (2004) 1267.) It has determined the Ni–B amorphous alloy prepared by chemical reduction method and found that B in Ni–B amorphous alloys donated electrons to Ni and thus became electron-deficient. Other groups are trying more theoretical approaches. For example, a group in China (Fang Zhigang, Fan Kangnianb, and Jiang Hualiang, Chin. J. Chem. Phys., 17 (6) (2004) 684) calculated the Co_nP and Co_nP_2 cluster models with the DFT method and showed that P accepted electrons from Co in single-P cluster models.

as depth increases [8] can be better understood. If the substrate is mild steel, the structure is basically bcc. The structure of the Ni crystalline state is fcc. At the very beginning of the deposition, a number of atomic groups consisting of metalloid atom B surrounded by metal atoms Ni are deposited onto the steel substrate. The structure of these atomic groups is not crystalline fcc structure, and there is a large incoherence between these atomic groups and the crystalline mild steel substrate. A high distortion energy is produced by the large incoherence, causing these atomic groups to be relatively unstable. Opposite to the electron factor y and size factor x, which are attractive and promote the atomic groups to be closely combined and stable, the distortion energy is positive and causes them to bind loosely. Only the outer atoms in the groups move any distance. If some of them ever move a rather longer distance, producing a regular atomic configuration suitable for forming crystalline structure, the distortion energy then can be released gradually. The total system including the substrate and the coating then can be trended to stable. Because the plating rate increases rapidly and the atoms reduced increase rapidly, a lot of microcrystalline nuclei grow faster in the direction perpendicular to the substrate. It should not be forgotten that besides the microcrystalline phase, the basic phase still is the amorphous state. Why does the microcrystalline phase not grow in the lateral direction? A lot of atomic groups formed at the beginning in the deposition spread over almost the whole substrate, the atoms coming from the lateral direction are very few, but a lot of atoms come from the direction perpendicular to the substrate where the whole liquid interface is. As the coating grows, the release of the distortion energy gradually decreases; finally, it comes to zero at a certain place. In that EP Ni–B experiment (see Fig. 8.22, left), the place is located at ca. 3 μm from the substrate. After the end of the microcrystalline phase, the formation of atomic groups onto the top of the microcrystalline phase is just like at the beginning of deposition. Therefore, the amorphous phase forms and grows in the perpendicular and lateral directions for the above-mentioned reasons. Thereby, the amorphous layer has a pronounced granular morphology, but the mixture of microcrystalline and amorphous phases shows a columnar morphology.

11.2 FORMATION THEORY OF ELECTROLESS AMORPHOUS ALLOYS

If the above general description of the formation of electroless amorphous alloys is the qualitative description of the formation theory of electroless amorphous alloys, then we will describe and discuss the quantitative theory of EP amorphous alloys in this section. The formation of EP amorphous alloys can be understood more deeply only when a quantitative description can be given. Once the quantitative theory is developed, the formation of electroless amorphous alloys then can be very clear, needing not so many words to describe it. Its importance is very obvious.

11.2.1 A Brief Retrospect of the Quantitative Theory of Metallic Glass Formation

Ordinary amorphous alloys are prepared by the MQ method, which was first discovered by Duwez et al. in 1960 [9,10]. Using the so-called splat quenching equipment [9], this group quenched a 25 at% Si–Au alloy from ~1300°C to room temperature to obtain an alloy with an amorphous structure [10]. Since then, not only the fabrication, structure, property, and application of amorphous alloys developed rapidly, promoting the amorphous alloy materials as a new field of materials science and engineering, but also the formation theory of amorphous alloys appeared and developed.

The quantitative theory of metallic glass formation can be subdivided into two classes. One is a theory based on the features and parameters of phase diagram of the alloy system. These theories are included: (1) theory of the reduced glass temperature [11,12], (2) theory of departure of the alloy liquidus temperature from the idea solution liquidus temperature [13,14], (3) theory of departure of the alloy liquidus temperature from the rule of mixtures liquidus temperature [14], (4) theory of the liquid alloy relative stability index and changing trend of liquidus parameters [15], and (5) theory of the ratio of $T_x/(T_g + T_l)$ (T_x is the onset crystallization temperature, T_g the glass transition temperature, and T_l the liquidus temperature) [16].

The other kind of theory is based on the characteristic parameters of constituents of an alloy system. These theories consist of (1) theory of heat of mixing and difference of liquidus temperatures diagram [17], (2) theory of pure bond parametric diagram [18], (3) theory of bond parameters and size factor diagram [2], (4) theory of formation heat of liquid alloy and ratio of atomic radium plot ($\Delta H - r/R$ plot) [19,20], (5) theory of pure Miedema coordinates diagram [3], (6) theory of chemical coordinates (Miedema coordinates and size factor) diagram [4], and (7) theory of free electron concentration [21,22]. All of these approaches are two-dimensional maps showing promise as a means of associating range of glass formation (RGF) with two system parameters. In addition, these theoretical methods were proposed in the 1970s to the 1980s. Since then, although most research activity concentrated on the bulk metallic glasses (BMG), that is, bulk amorphous alloys, many theoretical approaches [23–29] dealt with the alloying effect (in multicomponent systems), size factor, and negative heat of mixing to discuss and predict the formation of BMG systems. No new two-dimensional diagram methods have been proposed. It is interesting to note that among these theoretical approaches, Egami in his two papers claimed that his method was the "universal criterion for metallic glass formation" [23,24]. However, the scientific community seems not to agree with his method as a universal criterion for metallic glass formation because he did not describe the formation of amorphous alloys prepared by any other methods except MQ. Moreover, only a limited number of amorphous alloys can be described using his method. Therefore, this method, which is just used to obtain the size factor to explain the formation of amorphous systems, does not really qualify as a "universal criterion."

As for the theoretical methods based on phase diagrams, the basic condition is knowledge of the phase diagram of the alloy system. This depends entirely on the characteristics of the alloy constituents, which in turn depend on electron and size factors. Therefore, the electron and size factors are more basic parameters, and they reflect more deeply the nature of the alloying elements. In fact, compared to the theoretical methods based on the phase diagram, the existing theoretical approaches based on the characteristic parameters of constituents of an alloy system are in better agreement with the experimental data.

Although these two kinds of theoretical methods for the formation of amorphous alloys obtained by MQ have existed for more than 30 years, there are no reports describing the formation of electroless amorphous alloy systems. Only the author's group has published any papers on such a subject. We used the pure Miedema coordinates diagram and chemical coordinates diagram to describe the formation of electroless amorphous Ni–P and Ni–Sn–P, and Ni–Sn–Cu–P alloys [30,31]. Here, this approach will be applied to describe the formation of more electroless amorphous alloy systems systematically.

The author used the pure Miedema coordinates diagram and chemical coordinates diagram to successfully describe the formation of amorphous alloys prepared by the MQ method [3,4], by the IBM method [5], and by the ion implantation (II) method [6]. The author's group used these two 2D diagram methods to describe the formation of the binary amorphous alloys of transition metals by mechanical alloying (MA) [32,33] and the formation of the binary amorphous alloys by laser quenching (LQ) [34]. The number of alloy systems described by the MQ method was 136 (formed and nonformed binary amorphous alloys were 82 and 54, respectively), which by the IBM method was 69 (formed and nonformed binary amorphous alloys were 55 and 14 respectively), which by the II method was 25 (formed and nonformed binary amorphous alloys were 15 and 10), which by the MA method was 51 (formed and nonformed binary amorphous alloys were 37 and 14 respectively), and which by the LQ method was 65 (formed and nonformed binary amorphous alloys were 51 and 14). The regularities (empirical criteria) of formation of amorphous alloys in all cases are a linear and hyperbola curve for the pure Miedema coordinates and chemical coordinates approaches, respectively. The total accuracy in all cases ranges from 80% to 90%.

11.2.2 Formation Theory of Electroless Amorphous Alloy Systems

Since 1973, Miedema et al. published three papers [35–37], and proposed the cellular model (also called "macroscopic atoms" model) to quantitatively predict the heat of formation of binary alloy systems. In 1988, they published the monograph "Cohesion in metals—Transition metal alloys" [38]. A systematical theory for calculating the heat of formation of binary transition alloy systems therefore has been established. If you input "Miedema model" into Google, about 1,200,000 results can be found. And if you input "Miedema model" into Elsevier, 7411 articles can be found. Later, some groups extended the original Miedema model from binary alloys to ternary and quaternary alloy systems [39–43]. Today, the application of the Miedema theory is applied so widely and deeply that its absence cannot be imagined.

When Miedema et al. built the cellular model, they believed that the atomic cells of metallic atoms in binary alloys are in first order the same as in the pure metallic elements. When an alloy is formed by two elements, the alloying effect is considered as resulting from the changes in boundary conditions comparing the alloy with the pure elements. First, a negative term in the heat of formation is caused by the difference in the chemical potential for electrons at the boundary between dissimilar cells in the alloy. The chemical potential of pure metals, in general, is different. With alloying this difference leads to some transfer of charge, and thus to lowering of the energy of the alloy. On the basis of the fact that there is approximately linear dependence between work function Φ and electronegativity and between chemical potential for electrons and Pauling electronegativity, respectively, Miedema et al. concluded that there must also be a linear dependence between chemical potential and work function Φ. As a consequence, the negative term of the heat of formation is described in terms of the difference in work function $\Delta\Phi = \Phi_1 - \Phi_2$ (1 and 2 are the two elements in binary alloy). Second, another term in heat of formation which is positive is caused by the requirement that the electron density must be equalized at the boundary between dissimilar cells in the alloys. The requirement implies that in order to smooth the discontinuity in the electron density originally existing at the cellular boundary energy is needed, which is a "kinetic energy" type of consideration. So this provides a positive term in the heat of formation. Miedema et al. described the energy of contribution by $\Delta n^{-1/3} = n_1^{-1/3} - n_2^{-1/3}$; they have found that n varies systematically with bulk modulus B, so the n values can be estimated from B. Thus, it is evident that the two Miedema's coordinates belong to the class of electron factors as we called them [2,3].

Originally, when ZBW proposed the theoretical schemes to describe the formation of binary amorphous alloys, he found that two kinds of deciding factors (election and size factors) were similar as in the formation of the ordinary alloy phases [2,3]. So, he used the two Miedema parameters $\Delta n^{-1/3}$ and $\Delta\Phi$ as ordinate and abscissa to construct a two-dimensional diagram. Putting all the values of coordinate points of the formed and nonformed amorphous alloys onto the graph, an empirical criterion can be worked out; then the accuracy comparing between the experimental and theoretical results can be calculated. He called this method the pure Miedema coordinate scheme.

As mentioned above, the two Miedema's coordinates belong to the electron factor. Another factor, that is, the size factor affects the formation of amorphous alloys. A theoretical approach must include such a factor; otherwise, some errors will be produced. The author therefore combined the two Miedema coordinates properly as one factor Y (electron factor), and used the size factor X

of alloys as abscissa, a similar two-dimensional map can also be constructed. The law for describing the formation of amorphous alloys and the accuracy for the criterion can be worked out. This scheme is called the chemical coordinate method. It is easy to imagine that the accuracy would be usually improved for the chemical coordinate scheme compared with the pure Miedema coordinate approach, which will be seen in the following.

Before describing the theoretical schemes and their results for the formation of EP various amorphous alloys systematically, we just analyze the formation of EP Ni–P, Ni–Sn–P, and Ni–Sn–Cu–P amorphous alloys using the chemical approach for describing the formation of EP binary alloys.

11.2.2.1 Analyzing the Formation of EP Ni–P, Ni–Sn–P, and Ni–Sn–Cu–P Amorphous Alloys With the Chemical Approach for Binary Alloys [30,31]

There are six binary alloys in EP Ni–P, Ni–Sn–P, and Ni–Sn–Cu–P alloys: Cu–Ni, Ni–Sn, Ni–P, Cu–Sn, Sn–P, and Cu–P. Figure 11.1 shows the amorphous formation plot using the chemical coordinate scheme. The size and electron factors are

$$X = \left|(R_1 - R_2)/R_2\right| \tag{11.1}$$

$$Y = \left|\left|10\Delta\Phi\right| - 39\left|\Delta n^{-1/3}\right| - 1\right| \tag{11.2}$$

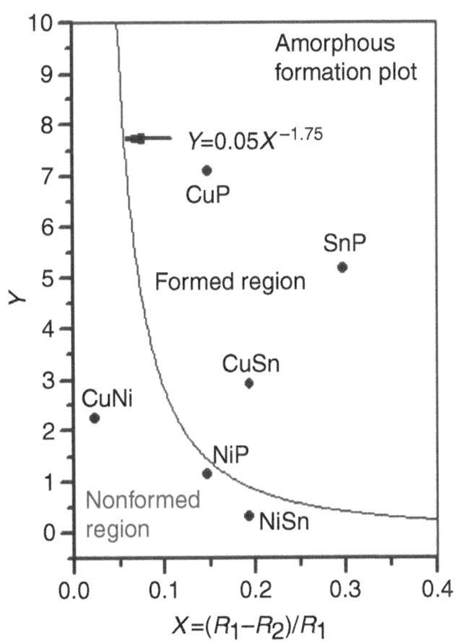

FIGURE 11.1 Amorphous formation plot of the relevant binary alloys for the Ni–P, Ni–Sn–P, and Ni–Sn–Cu–P alloy systems in chemical coordinates [30,31].

where R_1 and R_2 are the atomic radii of elements A and B in binary alloy, the values are quoted from [44]. The values of Φ and n are quoted from [38]. As mentioned above, when both the factors X and Y are large, the strong interaction of dissimilar kinds of atoms causes crystallization in the redox "quenching" process of an alloy to be impeded, and amorphous formation becomes easy: otherwise, it becomes difficult. The equation for separating the amorphous formed and nonformed areas is

$$Y = 0.05X^{-1.75} \tag{11.3}$$

From the location in the plot constructed by the two coordinates for an alloy, its amorphous-forming ability (AFA) could be found. The theoretical results for relevant binary alloys can be found in Fig. 11.1. The AFA of the binary alloys increases according to the following sequence: Cu–Ni, Ni–Sn, Ni–P, Cu–Sn, Sn–P, and Cu–P. Of the six binary alloys, Cu–Ni and Ni–Sn are located in the nonamorphous-forming region, and there are no amorphous alloys formed in melt quenching, also in electroless plating [45,46]. Though the model predicts amorphous formation for the binary Cu–P system, the experimental data showed no amorphous phase formed by MQ [47] and EP [48]. The reason could be that the interaction between Cu and P is very strong and the intermediate compound Cu_3P is formed. The simple structure of the Cu_3P compound phase means the atoms need not move a long distance in order to suit the atomic rearrangement of crystalline structure, and the small elastic strain energy of the a-Cu(P) solid solution implies rather small energy is needed to form a-Cu(P). These facts are favorable for the formation of the crystalline phases Cu_3P and a-Cu(P), but not amorphous phase for the binary Cu–P system.

For the binary Ni–P alloys, there are two kinds of atomic bonds, Ni–P and Ni–Ni. Ni–P is an amorphous forming alloy, so the bond Ni–P is favorable for glass formation. However, Ni–P is located in the boundary area between the amorphous-forming and nonforming regions in the plot, indicating its AFA is not strong. It is well known that pure Ni forms no amorphous phase under the usual splat quenching: the bond Ni–Ni therefore is not favorable for glass formation. The higher the P content, the more the atomic bonds of Ni–P and the less the bonds of Ni–Ni are present in the binary Ni–P system. Therefore, only when the P content exceeds a critical value in the system can the amorphous phase form; otherwise the crystalline phase is formed. That is why, P is a favorable element for amorphous phase formation. After adding Sn to the Ni–P system, two new kinds of atomic bonds, Ni–Sn and Sn–P, can be formed. The location of Sn–P in the amorphous-forming region of the plot is not deeper than that of Cu–P, indicating a tendency to form intermediate compound that is less than that of Cu–P. Therefore its AFA is strong. Though Ni–Sn is located in the nonforming region, its ability to promote the formation of the crystalline phase is less than that of Cu–Ni, causing the ability to hinder the formation of amorphous phase to be decreased. As a whole, adding Sn into Ni–P improves its AFA. Therefore, Sn is also a favorable element for amorphous phase formation.

TABLE 11.1 Composition (At%) and Structure of Electroless Ni–P Based Alloy Deposits [31]

Alloys	No.	Ni	Sn	Cu	P	Structure
Ni–P	1	95.4			4.6	Crystalline
	2	92			8	Crystalline
	3	87.5			12.5	Amorphous
Ni–Sn–P	4	94.8	0.7		4.5	Crystalline
	5	93.3	1.7		5	Amorphous
	6	95.7	0.8		3.5	Crystalline
	7	85.1	10.4		4.5	Amorphous
Ni–Cu–P	8	90		1.8	8.2	Crystalline
Ni–Sn–Cu–P	9	87.1	0.33	2.97	9.6	Crystalline
	10	85.45	1.64	6.98	5.93	Amorphous
	11	63.83	1.17	20.82	14.18	Crystalline
	12	66.05	1.22	17.47	15.26	Amorphous

According to the author's experimental data shown in Table 11.1 [31], the P content in 2 samples of No.5 $Ni_{93.3}Sn_{1.7}P_5$ and No.4 $Ni_{94.8}Sn_{0.7}P_{4.5}$ is nearly the same (5 at% vs. 4.5 at%), but the former is present as the amorphous phase and the latter as the crystalline phase. The reason is that the former contains 1.7 Sn at% but the latter only 0.7 Sn at%.

Adding Cu into the Ni–P system causes two new kinds of atomic bonds, Cu–Ni and Cu–P, to form as well as the Ni–P bond. The AFA of Cu–Ni is the worst, and therefore it impedes the amorphous phase formation. In the binary Cu–Ni system there is no amorphous formation under quenching splat and in EP [45]. As mentioned above, the atomic interaction of Cu–P is too strong, and it too easily forms the crystalline phase. Therefore, Cu is a nonfavorable element for the formation of the amorphous phase.

Just because of the favorable and nonfavorable effects of Sn and Cu elements, respectively, for the formation of amorphous phase, adding lowest P contents into Ni–P deposits to form ternary amorphous Ni–Sn–P and Ni–Cu–P deposits should be different. Indeed, this analysis is confirmed by experimental data. From Table 11.1 [31] the lowest P contents for forming EP amorphous Ni–Sn–P and Ni–P alloys are 4.5 and 12.5 at%, respectively. But that measured by Yu et al. [49] for the formation of amorphous phase in Ni–Cu–P alloys is 12.6 wt% (21 at%).

Now we consider the formation of EP quaternary amorphous Ni–Sn–Cu–P alloys, and the left problem in Section 8.1.4.2 will be analyzed more in detail. According to the reasons described above, in this system, P and Sn are favorable for, but Cu hampers, amorphous phase formation. They are competing factors in amorphous phase formation. Their relevant content in an alloy decides the structure of the alloy. For example, the content of 9.6 P at% in sample 9 of Table 11.1 is up to the critical value for amorphous formation of the Ni–P alloy: it also contains 0.33 Sn at%; therefore its structure should be amorphous. However, it is crystalline. This results from the 2.97 Cu at% content in it. Samples 11 and 12 are other interesting examples. The content of 17.47 Cu at% in sample 12 is much higher than that of 2.97 Cu at% in sample 9, but its structure is amorphous not crystalline. The reason is that its contents of 15.26 P and 1.22 Sn at% are much higher than those (9.6 P and 0.33 Sn at%) in sample 9. However, the structure of sample 11 is crystalline because its content of 20.82 Cu at% is too high and its contents of 14.18 P and 1.17 Sn at% are somewhat less than those in sample 12. In a word, the experimental data could be understood well with this model and analysis.

11.2.2.2 Systematical Description of Formation of Binary EP Amorphous Alloys

Searching the literature carefully, one can find that 7 formed and 24 nonformed EP binary amorphous alloy systems have been determined experimentally up to date, as shown in Table 11.2 [50].

For the pure Miedema's coordinate approach, the results are shown in Fig. 11.2. The values of Miedema's coordinates are quoted from [38]. From Fig. 11.2, the regions of formed and nonformed amorphous alloys can be

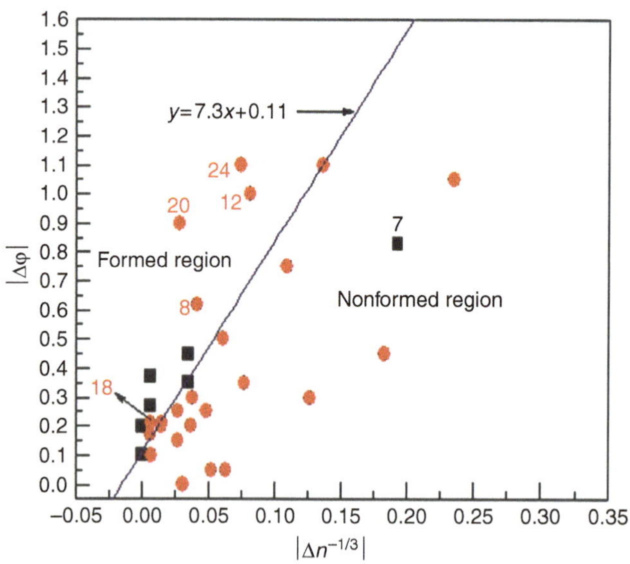

FIGURE 11.2 **Amorphous formation plot of all the EP binary alloys in pure Miedema's coordinates. ■ formed amorphous alloys, ● nonformed amorphous alloys [50].**

TABLE 11.2 EP Formed and Nonformed Binary Amorphous Alloy Systems [50]

Formed amorphous alloys		Exp.	Theor.		Ref.	Nonformed amorphous alloys		Exp.	Theor.		Ref.
			Method 1*	Method 2*					Method 1*	Method 2*	
1	Ni–P	a	a	a	51–53	8	Fe–P	c	a	a	67
2	Ni–B	a	a	a	54–56	9	Pd–Ag	c	c	c	68, 69
3	Co–P	a	a	a	57	10	Pd–Au	c	c	c	70
4	Co–B	a	a	a	58–60	11	Pd–Ni	c	c	c	71
5	Fe–B	a	a	a	61–63	12	Pd–Cu	c	a	c	72
6	Fe–Ni	a	a	c	64	13	Pd–Ru	c	c	c	73
7	Fe–Zn	a	c	a	65,66	14	Pd–P	c	c	c	74
						15	Pd–B	c	c	c	75
						16	Pd–Pt	c	c	c	76
						17	Pt–Rh	c	c	c	77
						18	Pt–Ru	c	a	c	78
						19	Co–Fe	c	c	c	79
						20	Cr–P	c	a	a	80
						21	Ni–Sn	c	c	a	46
						22	Cu–Zn	c	c	c	81
						23	Cu–Sn	c	c	c	82, 83

(Continued)

TABLE 11.2 EP Formed and Nonformed Binary Amorphous Alloy Systems [50] (*cont.*)

Formed amorphous alloys	Exp.	Theor.		Ref.	Nonformed amorphous alloys	Exp.	Theor.		Ref.
		Method 1*	Method 2*				Method 1*	Method 2*	
					24 Cu–P	c	a	a	48
					25 Cu–Ni	c	c	c	45
					26 Ag–W	c	c	c	84
					27 Ni–Re	c	c	c	85
					28 Sn–Pb	c	c	c	86
					29 In–Sb	c	c	c	87
					30 In–Sn	c	c	c	88
					31 In–Pb	c	c	c	88

**Method 1 is the pure Miedema coordinates approach, Method 2 is the chemical coordinates approach. a and c represent amorphous and crystal states, respectively.*

separated quite well by an empirical straight line. The equation of the straight line is

$$y = 7.3x + 0.11 \quad (11.4)$$

Of the seven EP amorphous alloys only one alloy (Fe–Zn) is located in the nonformed area, and of the 24 nonformed EP binary amorphous alloy systems five alloys (Fe–P, Pd–Cu, Pt–Ru, Cr–P, and Cu–P) are located in the formed area. The accuracy of predicting formed amorphous alloys is 85.7%, and for nonformed amorphous alloys is 79.2%, respectively. Therefore, the criterion of pure Miedema's coordinate approach is successful in predicting the formation of EP amorphous alloys is 80.6%.

This result is compatible to the respective total accuracies for predicting the formation of amorphous alloys prepared by other methods using the same pure Miedema coordinate approach, which are 77.7 % (MQ method) [3], and 84.2% (IBM method) [5], respectively. The regularities (empirical criteria) of formation of amorphous alloys in all these cases are a linear line for the pure Miedema coordinates approach. So, it can be concluded that the theoretical method is successful for predicting the formation of amorphous alloys prepared by EP, and by all above-mentioned techniques.

When using the chemical coordinate approach to the formation of binary amorphous alloys prepared by electroless plating, the results are shown in Fig. 11.3. The values of R are quoted from [44]. Figure 11.3 also clearly shows

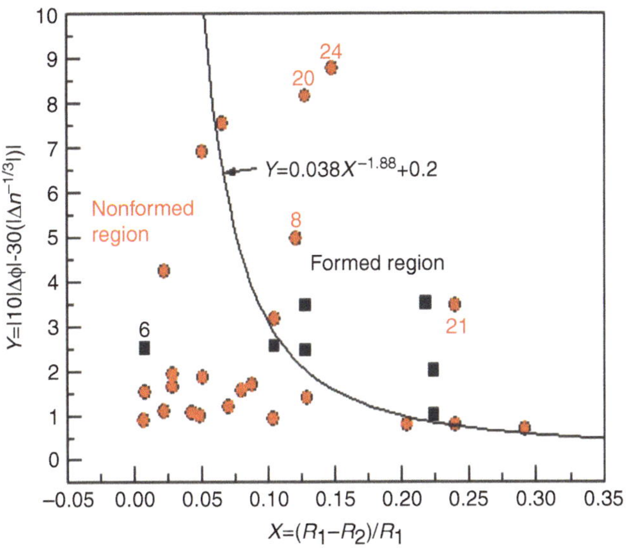

FIGURE 11.3 Amorphous formation plot of all the EP binary alloys in chemical coordinates. ■ formed amorphous alloys, ● nonformed amorphous alloys [50].

that the regions of formed and nonformed amorphous alloys are now separated by a hyperbolic curve, which is given by

$$Y = 0.038X^{-1.88} + 0.2 \quad (11.5)$$

where $X = |(R_1 - R_2)/R_1|$ and $Y = \left|10|\Delta\Phi| - 30\,|\Delta n^{-1/3}|\right|$. From Fig. 11.3, of the seven EP amorphous alloys one alloy (Fe–Ni) is located in the nonformed area, and of the 24 nonformed EP binary amorphous alloy systems four alloys (Fe–P, Cr–P, Ni–Sn, and Cu–P) are located in the formed area. Thus, the accuracy of predicting formed amorphous alloys is 85.7 %, and for nonformed amorphous alloys is 83.3%, respectively. Therefore, the criterion of chemical coordinate approach is successful in predicting the formation of EP amorphous alloys in 83.9%. The result represents that the accuracy for the formation criterion of amorphous alloys is obviously improved when the size factor is added to the pure Miedema coordinate criterion.

11.2.2.3 Systematical Description of Formation of Ternary EP Amorphous Alloys

As mentioned above, the original Miedema model is applicable only to binary alloys. Some other groups [39–43] extended the Miedema model to describe the *n*-component ($n \geq 3$) alloy systems. For a ternary alloy, Zhang and Jesser [40] only considered the binary interaction between atoms, and the ternary atomic interactions between the constituents were neglected when the Miedema model extends to the ternary alloys. Here, we adapt the similar consideration, then, in the two pure Miedema's coordinate method, the ordinate and abscissa can be represented as

$$\mathrm{x1} = \left|\left(\Delta n^{-\frac{1}{3}}\right)1\right| = \left|\left|n_1^{-1/3} - n_2^{-1/3}\right| + \left|n_1^{-1/3} - n_3^{-1/3}\right| + \left|n_2^{-1/3} - n_3^{-1/3}\right|\right| \quad (11.6)$$

$$\mathrm{y1} = |(\Delta\Phi)1| = \left||\Phi_1 - \Phi_2| + |\Phi_1 - \Phi_3| + |\Phi_2 - \Phi_3|\right| \quad (11.7)$$

where 1, 2, and 3 indicate the three constituents in ternary alloy. $n^{-1/3}$ and Φ are the two Miedema's coordinates for the element in the alloy. Using these two coordinates, a two-dimensional map can be constructed. Putting the Miedema's coordinate points of all ternary EP formed and nonformed amorphous alloys onto the two-dimensional map, a regularity (quantitative equation) can be worked out, and the agreement between the calculations and experimental data can be obtained easily.

Except the electron factor, the size factor also affects the formation of amorphous alloys as in the formation of solid solution. We used the half of the empirical interatomic distance of elements, *R*, to represent the size factor for an element. For ternary alloys, it is not as easy to represent the size factor as in a binary alloy. Experiments indicate that there are one or two metalloid (P or B) elements present in nearly all ternary EP amorphous alloys. We can consider approximately such ternary alloys as a pseudo-binary system. Therefore, the size factor in a ternary alloy can be indicated as

$$X1 = \left| \left(\frac{R_1 + R_2}{2} - R_3 \right) / \left(\frac{R_1 + R_2}{2} \right) \right| \tag{11.8}$$

or

$$= \left| \left[R_1 - \left(\frac{R_1 + R_2}{2} \right) \right] / R_1 \right| \tag{11.9}$$

In Eq. (11.8), the equality corresponds to the fact that there are two metal elements in a ternary alloy such as Ni–Fe–P, and the equality (11.9) corresponds to the fact that two metalloids exist in a ternary alloy such as Ni–P–B. R_1, R_2, and R_3 are the half of the empirical interatomic distance of three elements in a ternary alloy.

Now the two Miedema coordinates should be combined properly into one electron factor in order to construct a two-dimensional diagram in the chemical coordinate approach, which is given by

$$Y1 = \left| \left| (\Delta\Phi)1 \right| - 5 \left| \left(\Delta n^{-\frac{1}{3}} \right) 1 \right| \right| \tag{11.10}$$

$(\Delta\Phi)1$ and $\left(\Delta n^{-\frac{1}{3}} \right) 1$ are shown in Eqs. (11.6) and (11.7), respectively.

Similarly as in the two pure Miedema's coordinate methods, putting the values of the two coordinate points for all ternary EP formed and nonformed amorphous alloys onto the two-dimensional map for the chemical coordinate approach, a quantitative equation separating the formed and nonformed amorphous alloy regions of binary alloy systems can be obtained. Then, the accuracy between the calculations and experimental data can be calculated easily.

Up today, 32 formed and 8 nonformed EP ternary amorphous alloy systems have been determined experimentally, as shown in Table 11.3 [89].

Figure 11.4 shows the results for the pure Miedema's coordinate approach. The values of Miedema's coordinates are quoted from [38]. From Fig. 11.4, the regions of formed and nonformed amorphous alloys can be separated quite well by an empirical straight line. The equation of the straight line is

$$y1 = 51.9 \times 1 - 5.6 \tag{11.11}$$

Of the 32 EP ternary amorphous alloys 10 alloys (6 Ni–Cr–P, 16 Ni–Cu–P, 17 Ni–Cu–B, 18 Ni–Sn–P, 19 Ni–Sb–P, 20 Ni–Tl–P, 27 Co–Zn–P, 30 Fe–Sn–B, 31 Ni–Pb–B, 32 Ni–Tl–B) are located in the nonformed area, and of the

TABLE 11.3 EP Formed and Nonformed Ternary Amorphous Alloy Systems [89]

Formed amorphous alloys		Exp.	Theor.		Ref.	Nonformed amorphous alloys		Exp.	Theor.		Ref.
			Method 1*	Method 2*					Method 1*	Method 2*	
1	Ni-Fe–P	a	a	c	90, 91	33	Ni–Pt–P	c	a	c	139
2	Ni–Fe–B	a	a	a	92, 93	34	Ni–Ce–P	c	c	a	140
3	Ni–Co–P	a	a	c	94, 95	35	Ni–Zn–P	c	c	a	141, 142
4	Ni–Co–B	a	a	a	96, 97	36	Ni–Sn–B	c	c	c	143, 144
5	Ni–P–B	a	a	a	98–100	37	Co–Re–P	c	c	c	145
6	Ni–Cr–P	a	c	a	101, 102	38	Co–Ir–P	c	c	c	146
7	Ni–Cr–B	a	a	a	103	39	Co–Cu–P	c	c	c	147
8	Ni–Mo–P	a	a	a	104, 105	40	Fe–W–B	c	c	a	148
9	Ni–Mo–B	a	a	a	106, 107						
10	Ni–W–P	a	a	a	108, 109						
11	Ni–W–B	a	a	a	110, 111						
12	Ni–Re–P	a	a	c	112, 113						
13	Ni–Re–B	a	a	a	114						
14	Ni–Ru–P	a	a	c	115						
15	Ni–Pd–P	a	a	a	116, 117						
16	Ni–Cu–P	a	c	a	49, 118						
17	Ni–Cu–B	a	c	a	119						

18	Ni–Sn–P	a	c	a	120, 121					
19	Ni–Sb–P	a	c	c	122					
20	Ni–Tl–P	a	c	a	123					
21	Co–Fe–P	a	a	c	124					
22	Co–Fe–B	a	a	a	125					
23	Co–Mo–P	a	a	a	126					
24	Co–Mo–B	a	a	a	127					
25	Co–W–P	a	a	a	128, 129					
26	Co–W–B	a	a	a	130, 131					
27	Co–Zn–P	a	c	a	132					
28	Co–P–B	a	a	a	133					
29	Fe–Mo–B	a	a	a	134, 135					
30	Fe–Sn–B	a	c	a	136					
31	Ni–Pb–B	a	c	a	137					
32	Ni–Tl–B	a	c	a	138					

**Method 1 is the pure Miedema coordinates approach, Method 2 is the chemical coordinates approach. a and c represent amorphous and crystal states, respectively.*

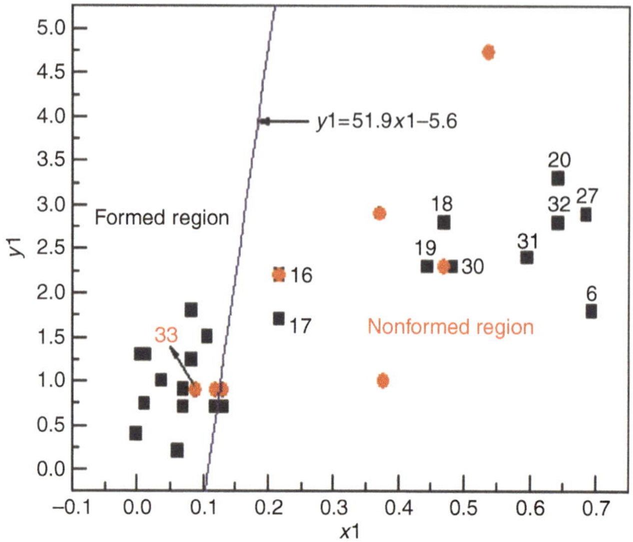

FIGURE 11.4 Description of the pure Miedema's coordinates for the ternary amorphous alloys formed by electroless plating. ■ formed amorphous alloys, ● nonformed amorphous alloys [89].

eight nonformed EP ternary amorphous alloy systems only one alloy (33 Ni–Pt–P) is located in the formed area. The accuracy of predicting all the formed and nonformed amorphous alloys is 72.5%. Though the accuracy for the criterion of pure Miedema's coordinate approach in predicting the formation of EP ternary amorphous alloys is not so high, but still it is quite good and can be accepted.

When using the chemical coordinate approach to the formation of ternary amorphous alloys prepared by electroless plating, the results are shown in Fig. 11.5. The values of R are quoted from [44]. It can be seen evidently from Fig. 11.5 that the regions of formed and nonformed amorphous alloys now are separated properly by a hyperbolic curve, which is given by

$$\mathrm{Y1} = 0.0158\mathrm{X1}^{-2.2} - 0.265 \tag{11.12}$$

From Fig. 11.5, it can be seen that of the 32 EP ternary amorphous alloys just six (1 Ni–Fe–P, 3 Ni–Co–P, 12 Ni–Re–P, 14 Ni–Ru–P, 19 Ni–Sb–P, 21 Co–Fe–P) are located in the nonformed area, and of the eight nonformed EP binary amorphous alloy systems three (34 Ni–Ce–P, 35 Ni–Zn–P, 40 Fe–W–B) are located in the formed area. Therefore, the criterion of chemical coordinate approach is quite good in predicting the formation of EP amorphous alloys in 77.5%. The result represents that the accuracy for the formation criterion of amorphous alloys is obvious improved when the size factor is added to the pure Miedema's coordinate criterion.

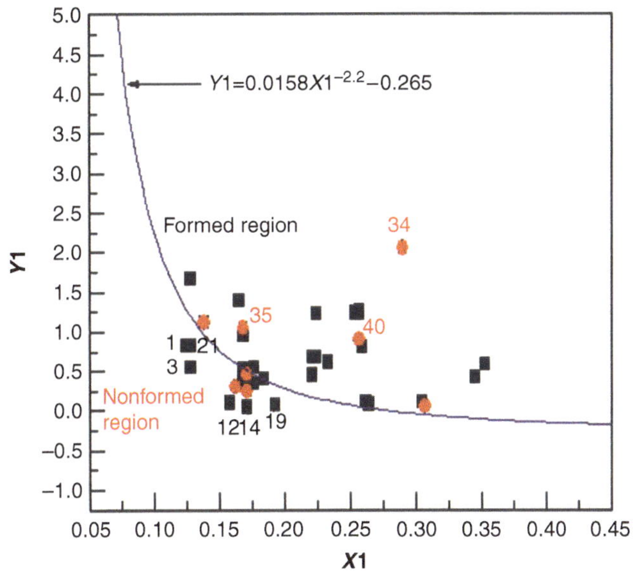

FIGURE 11.5 **Description of the chemical coordinates for the ternary amorphous alloys formed by electroless plating. ■ formed amorphous alloys, ● nonformed amorphous alloys [89].**

11.2.2.4 Systematical Description of Formation of Quaternary EP Amorphous Alloys

To obtain a theoretical scheme for describing the formation of quaternary amorphous alloys, the treatment of ternary alloy systems is simply extended.

The ordinate and abscissa in the pure Miedema coordinate approach can be represented as

$$x2=\left|\left(\Delta n^{-\frac{1}{3}}\right)2\right|=\left\|n_1^{-\frac{1}{3}}-n_2^{-\frac{1}{3}}\right|+\left|n_1^{-\frac{1}{3}}-n_3^{-\frac{1}{3}}\right|+\left|n_1^{-\frac{1}{3}}-n_4^{-\frac{1}{3}}\right| \\ +\left|n_2^{-1/3}-n_3^{-1/3}\right|+\left|n_2^{-1/3}-n_4^{-1/3}\right|+\left|n_3^{-1/3}-n_4^{-1/3}\right\| \tag{11.13}$$

$$\begin{aligned} y2&=|(\Delta\Phi)2| \\ &=\|\Phi_1-\Phi_2|+|\Phi_1-\Phi_3|+|\Phi_1-\Phi_4|+|\Phi_2-\Phi_3|+|\Phi_2-\Phi_3|+|\Phi_3-\Phi_4\| \end{aligned} \tag{11.14}$$

It is obvious that only the binary interactions between atoms have been considered, but the ternary and quaternary atomic interactions between the constituents are all neglected.

The electron factor in the chemical approach for quaternary alloys is combined properly with $\left(\Delta n^{-\frac{1}{3}}\right)2$ and $(\Delta\Phi)2$ as follows:

$$Y2=\left|\left|(\Delta\Phi)2\right|-8\left|\left(\Delta n^{-\frac{1}{3}}\right)2\right|\right| \tag{11.15}$$

$\left(\Delta n^{-\frac{1}{3}}\right)2$ and $(\Delta\Phi)2$ in Eq. (11.15) are the sum of $\Delta\Phi$ and $\Delta n^{-\frac{1}{3}}$ for the six binary alloys in a quaternary alloy, respectively, which are given in Eqs. (11.13) and (11.14), respectively.

As for the calculation of the size factor X2 in a quaternary alloy system, it is still just considered approximately as a pseudo-binary system [89]. Therefore, the size factor in a quaternary alloy containing one or two metalloid elements is

$$X2=\left|\left[(R_1+R_2+R_3)/3-R_4\right]/(R_1+R_2+R_3)/3\right| \tag{11.16}$$

or

$$=\left|\left(\frac{R_1+R_2}{2}-\frac{R_3+R_4}{2}\right)/\left(\frac{R_1+R_2}{2}\right)\right| \tag{11.17}$$

where R_1, R_2, R_3, and R_4 are the half of the empirical interatomic distance of the four elements in a quaternary alloy.

Putting the values of the coordinate points (x2, y2) and (X2, Y2) for all quaternary EP formed and nonformed amorphous alloys onto the two-dimensional map for the pure Miedema coordinate and chemical coordinate approaches, respectively, a quantitative equation separating the formed and nonformed amorphous alloy regions of quaternary alloy systems can be drawn out. Then, the agreement between the calculated and experimental data can be calculated easily.

For the formation of quaternary amorphous alloy systems, 26 alloys have been determined experimentally up to date, in which 15 are formed and 11 are nonformed amorphous alloys, respectively, as shown in Table 11.4 [89]. Using the two coordinates x2 and y2, that is, Eqs (11.13) and (11.14), a two-dimensional graph Fig. 11.6 can be constructed by putting the values of coordinate points for all of the 26 alloys onto the map. The values of Miedema's coordinates are quoted from [38].

From Fig. 11.6, the regions of formed and nonformed amorphous alloys can be separated quite well by an empirical straight line. The equation of the straight line is

$$y2=15.1\times 2-1.6 \tag{11.18}$$

TABLE 11.4 EP Formed and Nonformed Quaternary Amorphous Alloy Systems [89]

Formed amorphous alloys	Exp.	Theor.		Ref.	Nonformed amorphous alloys	Exp.	Theor.		Ref.
		Method 1*	Method 2*				Method 1*	Method 2*	
1 Ni–Fe–Co–P	a	a	a	149, 150	16 Ni–Co–Re–P	c	c	c	145
2 Ni–Fe–Ru–P	a	a	a	151	17 Ni–W–Cu–P	c	c	c	163
3 Ni–Fe–W–P	a	a	a	152	18 Ni–W–Sn–P	c	c	a	164
4 Ni–Fe–P–B	a	a	a	89	19 Ni–Co–Fe–B	c	a	a	165
5 Ni–Co–Cu–P	a	c	c	153	20 Co–Pt–W–P	c	a	a	166
6 Ni–Co–Zn–P	a	c	c	154	21 Co–Ni–Mn–P	c	a	c	167
7 Ni–W–Cr–P	a	a	a	155	22 Co–Ni–Ce–P	c	c	c	168
8 Ni–W–Re–P	a	a	a	112	23 Co–Ni–Y–B	c	c	c	169
9 Ni–Re–B–C	a	a	a	156	24 Co–Ni–Ce–B	c	c	c	170
10 Ni–Sn–Cu–P	a	a	a	120	25 Co–Fe–La–B	c	c	c	171
11 Co–Fe–Mn–P	a	a	a	157	26 Co–Fe–Ce–B	c	c	c	172
12 Ni–Co–Tl–B	a	c	a	158					
13 Co–W–P–B	a	a	a	159, 160					
14 Fe–Mo–W–B	a	a	a	161					
15 Ni–Co– W–P	a	c	c	162					

**Method 1 is the pure Miedema coordinates approach, Method 2 is the chemical coordinates approach. a and c represent amorphous and crystal states, respectively.*

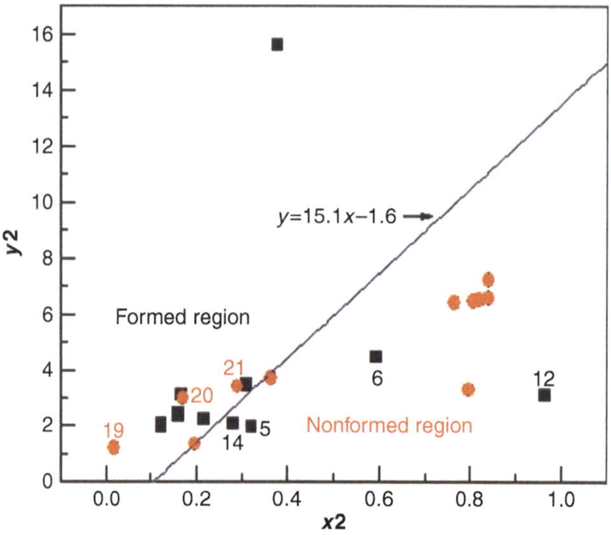

FIGURE 11.6 Description of the pure Miedema's coordinates for the quaternary amorphous alloys formed by electroless plating. ■ formed amorphous alloys, ● nonformed amorphous alloys [89].

Of the 15 EP quaternary amorphous alloys four (5 Ni–Co–Cu–P, 6 Ni–Co–Zn–P, 12 Ni–Co–Tl–B, 14 Fe–Mo–W–B) are located in the nonformed area, and of the 11 nonformed EP quaternary amorphous alloy systems three (19 Ni–Co–Fe–B, 20 Co–Pt–W–P, 21 Co–Ni–Mn–P) are located in the formed area. The accuracy of predicting all the formed and for nonformed amorphous alloys is 73.1%. The accuracy for the criterion of pure Miedema's coordinate approach in predicting the formation of EP quaternary amorphous alloys is nearly the same as that for the above predicting the formation of EP ternary amorphous alloys, but is somewhat lower than those for predicting the formation of MQ (82.4%) [3] or IBM (84.2%) [5] binary amorphous alloys.

When using the chemical coordinate approach to the formation of quaternary amorphous alloys prepared by electroless plating, the results are shown in Fig. 11.7. The values of R are quoted from [44]. It can be seen evidently from Fig. 11.7 that the regions of formed and nonformed amorphous alloys now are separated properly by a hyperbolic curve, which is given by

$$Y2 = 0.006X2^{-2.2} + 0.225 \quad (11.19)$$

From Fig. 11.7, it is evident that of the 15 EP quaternary formed amorphous alloys three alloys (5 Ni–Co–Cu–P, 6 Ni–Co–Zn–P, 14 Fe–Mo–W–B) are located in the nonformed area, and of the 11 nonformed EP quaternary amorphous alloy systems three alloys (18 Ni–W–Sn–P, 19 Ni–Co–Fe–B, 20 Co–Pt–W–P) are located in the formed area. Therefore, the criterion of chemical coordinate approach is quite good at predicting the formation of EP amorphous alloys,

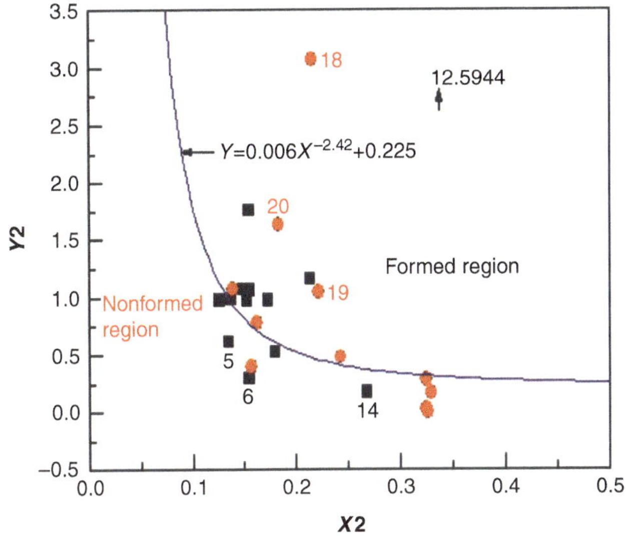

FIGURE 11.7 Description of the chemical coordinates for the quaternary amorphous alloys formed by electroless plating. ■ formed amorphous alloys, ● nonformed amorphous alloys [89].

at 76.9%. The accuracy for the criterion of chemical coordinate approach in predicting the formation of EP quaternary amorphous alloys is near the same as that for the above predicting the formation of EP ternary amorphous alloys, but is somewhat lower than those for predicting the formation of MQ (86%) [4] or IBM (89.5%) [5] binary amorphous alloys. In addition, the result represents that the accuracy for the formation criterion of quaternary amorphous alloys is obviously improved when the size factor is added to the pure Miedema's coordinate criterion.

11.2.2.5 Analysis of Some Problems from the Above Theoretical Schemes and Results for the Formation of Various EP Amorphous Alloys

Table 11.5 summarizes the criteria and accuracies for various amorphous alloys prepared by different techniques. From the comparison in Table 11.5, the results indicate that either using the pure Miedema's coordinate scheme or the chemical coordinate approach, not only the empirical criteria but the accuracies for the formation of EP binary, ternary, and quaternary amorphous alloys are similar to or comparable with those from the binary formation of amorphous alloys prepared by MQ [3,4] and IBM [5] techniques. How do we understand these similarities? ZBW [5] discussed in detail the nature of the formation of amorphous alloys by IBM, which is basically similar as for that by MQ. From Section 11.1, it is also indicated that the basic characteristics are similar for the EP and MQ methods. There are all existed rapidly "quenching" process in these two

TABLE 11.5 Comparison of the Laws and Accuracies for Various Amorphous Alloys Formed by Different Techniques

		EP			MQ	IBM
Prepared technique		Binary [49]	Ternary [88]	Quaternary [88]	Binary [3,4]	Binary [5]
Pure Miedema coordinate criterion	Coordinates	$x = \|\Delta n^{-1/3}\|$ $y = \|\Delta\Phi\|$	$x1 = \|\Delta n1^{-1/3}\|$ $y1 = \|\Delta\Phi1\|$	$x2 = \|\Delta n2^{-1/3}\|$ $y2 = \|\Delta\Phi2\|$	$x = \|\Delta n^{-1/3}\|$ $y = \|\Delta\Phi\|$	$x = \|\Delta n^{-1/3}\|$ $y = \|\Delta\Phi\|$
	Alloy number					
	Formed a*	7	32	15	82	29
	Nonformed a*	31	8	11	54	9
	Wrong number	6	11	7	24	6
	Empirical law	$y = 7.3x + 0.11$	$y1 = 51.9x1 - 5.6$	$y2 = 15.1x2 - 1.6$	$10y = 39x + 1$	$y = 5x$
	Accuracy,%	80.6	72.5	73.1	82.4	84.2
Chemical coordinate criterion	Coordinates	$X = \|(R_1 - R_2)/R_1\|$ $Y = \|10\|\Delta\Phi\| - 30\|\Delta n^{-1/3}\|\|$	X1 Y1	X2 Y2	$X = \|\Delta R/R_1\|$ $Y = \|10\|\Delta\Phi\| - 30\|\Delta n^{-1/3}\| - 1\|$	$X = \|\Delta n^{-1/3}\|$ $Y = \|10\|\Delta\Phi\| - 35\|\Delta n^{-1/3}\|\|$
	Alloy number					
	Formed a*	7	32	15	82	29
	Nonformed a*	24	8	11	54	9
	Wrong number	5	9	6	19	4
	Empirical law	$Y = 0.038X^{-1.88} + 0.2$	$Y1 = 0.0158X1^{-2.2} - 0.265$	$Y2 = 0.006X2^{-2.2} + 0.225$	$Y = 0.05X^{-1.75}$	$Y = 0.25X^{-1}$
	Accuracy, %	83.9	77.5	76.9	86	88.9

a*- amorphous alloys. In Table 11.5, x1, y1, X1, Y1, x2, y2, X2 and Y2 in columns 4 and 5 are indicated as follows:

$$x1 = \left|\left(\Delta n^{-\frac{1}{3}}\right)1\right| = y1\left|\left|n_1^{-1/3} - n_2^{-1/3}\right| + \left|n_1^{-1/3} - n_3^{-1/3}\right| + \left|n_2^{-1/3} - n_3^{-1/3}\right|\right|$$

$$y1 = \left|(\Delta\Phi)1\right| = X1\left|\left|\Phi_1 - \Phi_2\right| + \left|\Phi_1 - \Phi_3\right| + \left|\Phi_2 - \Phi_3\right|\right|$$

$$X1=\left|\frac{\frac{R_1+R_2}{2}-R_3}{\frac{R_1+R_2}{2}}\right| \quad \text{or} \quad =\left|\left[R_1-\left(\frac{R_1+R_2}{2}\right)\right]/R_1\right|$$

$$Y1=\left|\left|(\Delta\Phi)1\right| - 5\left|\left(\Delta n^{-\frac{1}{3}}\right)1\right|\right|$$

$$x2=\left|\left(\Delta n^{-\frac{1}{3}}\right)2\right| = y2\left|\left|n_1^{-\frac{1}{3}}-n_2^{-\frac{1}{3}}\right| + \left|n_1^{-\frac{1}{3}}-n_3^{-\frac{1}{3}}\right| + \left|n_1^{-\frac{1}{3}}-n_4^{-\frac{1}{3}}\right| + \left|n_2^{-1/3}-n_3^{-1/3}\right| + \left|n_2^{-1/3}-n_4^{-1/3}\right| + \left|n_3^{-1/3}-n_4^{-1/3}\right|\right|$$

$$y2=\left|(\Delta\Phi)2\right| = x2\left|\left|\Phi_1-\Phi_2\right| + \left|\Phi_1-\Phi_3\right| + \left|\Phi_1-\Phi_4\right| + \left|\Phi_2-\Phi_3\right| + \left|\Phi_2-\Phi_3\right| + \left|\Phi_3-\Phi_4\right|\right|$$

$$x2=\left|\left[(R_1+R_2+R_3)/3-R_4\right]/(R_1+R_2+R_3)/3\right| \quad \text{or} \quad =\left|\left(\frac{R_1+R_2}{2}-\frac{R_3+R_4}{2}\right)/\left(\frac{R_1+R_2}{2}\right)\right|$$

$$Y2=\left|\left|(\Delta\Phi)2\right| - 8\left|\left(\Delta n^{-\frac{1}{3}}\right)2\right|\right|$$

prepared techniques, one is splat cooling during MQ, another is redox "quenching" due to electroless reduction. So, it is obvious that though the methods are different, the substance is the same. They are all "quenching" processes that promote the formation of amorphous alloys, thus the similarities in the results.

In a binary alloy, the $\Delta\Phi = \Phi_1 - \Phi_2$ represents the negative contribution to the alloying formation, which will decrease the energy of the system, and thereby will benefit the formation of amorphous phase. Contrarily, the $\Delta n^{-1/3} = n_1^{-1/3} - n_2^{-1/3}$ indicates the positive contribution to the alloying formation, which will increase the energy in the system, which hinders the formation of amorphous phase. An alloy either or not forms amorphous phase, depending on the competitive result of these two factors. Therefore, when using these two Miedema's coordinates to predict the formation of amorphous alloys, the left part in the two-dimensional graph must be the region for the formation of amorphous phases, and the right part is the region for the nonformation of amorphous phases. The separating line just represents the balance between these two factors. Figures 11.2, 11.4, and 11.6 for predicting results from the pure Miedema's scheme for EP binary, ternary, and quaternary, respectively, indicate such behaviors. In fact, the similar results exist for the pure Miedema's coordinate approach for describing the binary alloy by MQ [3,4] and IBM [5].

As for what are the reasons that large Y and X in Figs 11.3, 11.5, and 11.7 lead to easy formation of EP amorphous alloys? One can also easily understand from the description in Section 11.1. When Y and X are large, the short-range interaction for the two dissimilar atoms is strong, the group of the dissimilar atoms then combine tightly. Therefore, it is more difficult for the atoms in the group to move any great distance, or to rearrange cooperatively in order to produce a regular atomic configuration suitable for forming crystalline structure in longer distance. That is to say, the formation of crystalline structure is hindered, thereby leading to the easy formation of amorphous alloys. Also, one can easily understand from these discussions that the accuracies for the formation criterion of EP ternary and quaternary amorphous alloys, respectively, are obviously improved when the size factor is added to the pure Miedema's coordinate criterion.

The accuracies from the pure Miedema coordinate and chemical coordinate schemes for predicting the formation of EP binary amorphous alloys are 81 to 84%, which are comparable to those for predicting the heats of formation of binary alloy systems by the original Miedema model because the errors are 10 to 20% for such predictions [38,173]. Although the accuracies for predicting those for EP ternary and quaternary alloys are somewhat lower, they are still 73 to 78%. It should be said that the prediction for the formation of EP amorphous alloys using these two theoretical schemes is good. Why are the accuracies from the pure Miedema coordinate and chemical coordinate schemes for the formation of ternary or quaternary amorphous alloys all somewhat less than those for the formation of binary amorphous alloys? There are several reasons.

First, only the binary interaction between atoms is considered, and the ternary atomic interactions between the constituents were neglected in the formation of amorphous alloys for the ternary alloy systems. For the formation of amorphous alloys in the quaternary alloy systems, not only ternary but also quaternary atomic interactions between the constituents are all neglected. These ternary and quaternary interactions between atoms although in general these are much fewer than the binary interaction between atoms but the accuracies must lead to some errors when neglecting these much weak interactions.

Second, the selection of the electron factors Y1 (11.10) in ternary alloy systems and Y2 (11.15) in quaternary alloy systems was not derived theoretically, but was just empirically determined. Some errors would be result from such empirical selection.

Third, the size factors used will cause some errors because we just considered the ternary and quaternary as a pseudo-binary alloy system. These approximate representations ((11.8), (11.9), (11.16), and (11.17) equalities) were not rigorously theoretically proven. In fact, we do not know how to represent the real size factor in a ternary and quaternary alloy system right now. So, there is a problem to be solved.

Finally, if the formed metalloid atoms in the redox process are rather few, fewer atomic groups will form. Many of the metal atoms cannot surround the metalloid elements, and may move rather longer distances to form some types of regular atomic arrangement due to the repulsive action between like atoms. Then the microcrystalline or nanocrystalline structure will appear beside the amorphous structure. That is why a mixture of microcrystalline or nanocrystalline and amorphous phases occurs in the low P or low B content alloys. If the P or B content is very low in the deposits, then their structure becomes crystalline with no amorphous phase. Such experimental facts can be seen in some papers about EP ternary or quaternary alloys. For example, in 2010 Manna and Bandyopadhyay [91] prepared the EP Ni–Fe–P alloys (P 11–13.5 wt%), and all were amorphous structures, but those prepared by Schmeckenbecher in 1966 [174] were crystal (P 05–1 wt%). Our group in 1999 [120] and Zou et al. in 2010 [121] both measured the structures of EP Ni–Sn–P alloys, and found that all are amorphous [the former of P 14.7–20.3 at%, the latter of P 6–9 wt%], but Shimauchi et al. in 1994 [175] found those are crystalline for the same EP alloy (P less than 10 at%). Zou et al. in 2004 [162] reported the structure of an EP Ni–Co–W–P alloy coating to be amorphous with content of P 7.66 wt%. But in 1981 Osaka and Kasai [176] reported the structure of the same alloy coating as crystalline with content of P 1.4 wt%. In 2008, Cheng et al. [177] also prepared crystalline EP Ni–Co–W–P films with around P 5.5 at%. Although we classified all of these EP alloy systems (Ni–Fe–P, Ni–Sn–P, Ni–Co–W–P) into amorphous but not into crystal states in Tables 11.3 and 11.4, but it is obvious that such a classification would have some errors.

11.2.2.6 Summary

The formation of EP binary, ternary, and quaternary amorphous alloys has been described systematically with the pure Miedema's coordinate and chemical coordinate schemes proposed originally by the author.

The quantitative equations for predicting the formation of EP binary, ternary, and quaternary amorphous alloys in these two schemes are all linear and hyperbolic curves, respectively, which are similar to those for predicting the formation of EP binary amorphous alloys prepared by MQ and IBM. The accuracies of predicting the formation of EP amorphous alloys for these two theoretical schemes are 81 to 84% (EP binary alloys), but are 73 to 78% (EP ternary and quaternary alloys), which are similar to or somewhat less than those predicting the formation of EP binary amorphous alloys by MQ and IBM using the same two theoretical schemes. The results indicate that the two theoretical approaches can also be used to describe the formation of EP amorphous alloys successfully.

The main affected factors (electron and size factors) for the formation of amorphous alloys in the EP technology are basically the same as those in the MQ technique, and the nature of formation processes is also similar except the types of preparation are different. The formation of amorphous alloys in MQ is due to the temperature quenching under the splat cooling. And that in EP stems from the redox "quenching" process in the solution bath. That is why using either the pure Miedema's coordinate scheme or the chemical coordinate approach, not only the empirical criteria but the accuracies for the formation of EP binary, ternary, and quaternary alloys are similar as or comparable to those from the formation of binary amorphous alloys prepared by MQ [3,4] and IBM [5] techniques. That is to say, there exists obvious similarity between the EP technique and other techniques such as MQ, IBM, etc., in the formation of amorphous alloys.

The reasons that the accuracies are rather low (somewhat less than 80%) for predicting the formation of EP ternary and quaternary amorphous alloy films are because of neglecting some factors such as the ternary and quaternary atomic interactions between the constituents, and treating the ternary and quaternary alloy films as pseudo-binary systems, etc.

11.3 FORMATION RANGE OF ELECTROLESS AMORPHOUS ALLOYS: EXPERIMENTAL FACTS AND THEORETICAL CALCULATIONS

After describing the formation theory for the EP amorphous alloys, another important problem, that is, the range of amorphous formation (RAF) must be considered either from fundamental or from technological importance.

11.3.1 Experimental Data of RAF of EP Amorphous Alloys

To date, many papers on the RAF of EP amorphous alloys have been published, as shown in Table 11.6. From this table, several conclusions can be drawn.

The first is that most of the papers on the RAF of EP amorphous alloys concentrate on the EP binary and ternary alloy deposits, those on EP quaternary alloy deposits are rather few, and no reports can be found on RAF of EP quinary amorphous alloys.

TABLE 11.6 Composition, Structure, and Formation Range of Amorphous Phases for Various EP Alloy Deposits

No.	Alloys	Parameters	Range of parameters							Range of amorphous phases	Refs
1	Ni–P	Bath chemical	Bath	Bath+Pb ions	Bath		Bath without H_3BO_3	Bath without $NiCl_2$	Bath+ Pb ions and allylthiourea	26.5–30.8 at% P	178
		Composition	P, at%	30.8	26.5		15.8	14.2	10.8		
		Structure[1]		A	A		C	C	C		
2	Ni–P	Bath chemical	pH	5	9	10				20.8 at% P	179
		Composition	P, at%	20.8	13.3	6.4					
		Structure[1]		A	A+C	C					
3	Ni–P	Bath chemical	Bath name	Bath A	Bath B	Bath C				22.5 at% P	180
		Composition	P, at%	22.5	12	6.2					
		Structure[1]		A	Micro C	C					
4	Ni–P	Bath chemical	pH	6					3.6	20 at% P	181
		Composition	P, at%	12					20		
		Structure[1]		fcc Ni crystallites in an amorphous matrix					A		
5	Ni–P	Bath chemical	NaH_2PO_2, g/L	0.1[2]	02[3]	0.2[4]				13.8 at% P	182
		Composition	P, at%	4.0	6.9	13.8					
		Structure[1]		C	C	A					
6	Ni–P	Bath chemical	$NiSO_4$, g/L	120	226	340					183
		Composition	P, at%	21.5	18.4	12.3					
		Structure[1]	Nano C, nm	5.2	7.8	17					
7	Ni–P	Bath chemical	$NiCl_2$, g/L	25	35	45	55	65	75	12.53–28.71 at% P	30
		Composition	P, at%	4.60	8.03	12.53	23.09	27.42	28.71		
		Structure[1]		C	C	A	A	A	A		
8	Ni–Sn–P	Bath chemical	NaH_2PO_2, g/L	6	12	24	30	36		5 –26 at% P 1.2–1.7 at% Sn	120
		Composition	Sn, at%	0.8	1.7	1.2		1.6			
			P, at%	3.5	5	12.6		26			
		Structure[1]		C	A	A	A	A			

(Continued)

TABLE 11.6 Composition, Structure, and Formation Range of Amorphous Phases for Various EP Alloy Deposits (*cont.*)

No.	Alloys	Parameters	Range of parameters								Range of amorphous phases	Refs
9	Ni–Sn–P	Bath chemical Composition Structure[1]	 Sn, at% P, at%	 0.7 4.5 C	 0.8 3.5 C	 1.7 5.0 A	 10.4 4.5 A				4.5–5.0 at% P 1.7 –10.4 at% Sn	184
10	Ni–Sn–P	Bath chemical Composition Structure[1]	 Sn, at% P, at%	 1.09 14.50 A	 1.39 14.43 A	 0.35 20.51 A					14.43–20.51 at% P 0.35 –1.39 at% Sn	31
11	Ni–Cu–P	Bath chemical Composition Structure[1]	$CuSO_4$, g/L Cu, wt% P, wt%	0.1 3.0 15.0 A	0.5 9.5 11.5 A	1.0 18.0 10.0 A	1.5 28.0 9.0 A				9–15 wt% P 3–28 wt% Cu	185
12	Ni–Cu–P	Bath chemical Composition Structure[1]	$CuSO_4$, g/L Cu, wt% P, wt%	0.2 3.4 12.6 A	1.0 18.0 9.3 C	3.0 40.2 6.4 C	5.0 55.8 3.9 C	10.0 81.5 1.1 C		x	12.6 wt% P 3.4 wt% Cu	49
13	Ni–Cu–P	Bath chemical Composition Structure[1]	$Na_3C_6H_5O_7$, g/L Cu, wt% P, wt%	0.04 15.3 12 A	0.07 12.7 12.1 A	0.09 10.5 14.1 A	0.11 5.7 12.3 A				12–14.1 wt% P 5.7–15.3 wt% Cu	186
14	Ni–Cu–P	Bath chemical Composition Structure[1]	 Cu, wt% P, wt%	 11 12 A							12 wt% P 11 wt% Cu	118

15	Ni–Sn–Cu–P	Bath chemical	$NiCl_2$, g/L	20	30	55	65	70	80		14.50–21.86 at% P	31
		Composition	Sn, at%	11.40	4.65	1.25	1.09	0.32	0		1.09–11.40 at% Sn	
			Cu, at%	15.10	5.14	2.45	0	2.97	2.84		0–15.10 at% Cu	
			P, at%	16.10	21.86	15.32	14.50	9.63	6.43			
		Structure[1]		A	A	A	A	C	C			
	Ni–Sn–Cu–P	Bath chemical	$CuCl_2$, g/L	0.1	0.3	0.5	1.0	2.0	3.0		11.11–20.86 at% P	31
		Composition	Sn, at%	2.14	1.38	1.70	1.26	1.22	1.17		1.22–2.14 at% Sn	
			Cu, at%	2.70	4.23	7.71	8.61	17.47	20.82		2.70–17.47 at% Cu	
			P, at%	15.61	11.11	20.86	18.35	15.26	14.19			
		Structure[1]		A	A	A	A	A	C			
16	Ni–B	Bath chemical	Stabilizer	$TlNO_3$	None	$PbCl_2$+2–mereaptobenzotbiasole					24.7–27.1 at% B	187
		Composition	B, at%	19.6	24.7	27.1						
		Structure[1]		C	A	A						
17	Ni–B	Bath chemical	$NaBH_4$, mg/L	200	400	800					41 at% B	188
		Composition	B, at%	31	41	42						
		Structure[1]		C	A	A						
18	Ni–B	Bath chemical	DMAB, mol/m^3	5.1	12.7	50.9		84.9			17.1 at% B	189
		Composition	B, at%	5.5	7.9	12.9		17.1				
		Structure[1]		C	Micro C	Micro C		A				
19	Ni–B	Bath chemical									25 at% B	59
		Composition	B, at%	2.26	21	25						
		Structure[1]		C	C	A						
20	Ni–B	Bath chemical		Alkaline cleaning+EP			Acid pickling+EP			EP	26.6 at% B	56
		Composition	B, at%	14.4			24.1			26.6		
		Structure[1]		C			Micro C+A			A		
21	Co–P	Bath chemical									about 12 at% P	190
		Composition	P, at%	~10	10–12	>12						
		Structure[1]		C	Part A	Total A						

(Continued)

TABLE 11.6 Composition, Structure, and Formation Range of Amorphous Phases for Various EP Alloy Deposits (*cont.*)

No.	Alloys	Parameters	Range of parameters								Range of amorphous phases	Refs
22	Co–P	Bath chemical	$NaH_2PO_2/CoSO_4$ (mole ratio)	1.42	2.485	3.55	4.615	5.68			16.5–17.1 at% P	191
		Composition	P, at%	14.6	15.1	15.9	16.5	17.1				
		Structure[1]		C	C	C	A	A				
23	Co–B	Bath chemical									25.8 at% B	59
		Composition	B, at%	2.26	22	25.8						
		Structure[1]		C	C	A						
24	Co–B	Bath chemical									5.1 at% B	192
		Composition	B, at%	5.1								
		Structure[1]		A								
25	Fe–P	Till today, no reports for amorphous structure of EP Fe–P can be found.										
26	Fe–B	Bath chemical	KBH_4, g/L	1	3	8	15	20			18–27.4 at% B	61
		Composition	B, at%	10.6	18.0	24.0	26.1	27.4				
		Structure[1]		C	A	A	A	A				
27	Fe–B	Bath chemical	KBH_4, g/L	0	4	8	12	16	20		23 at% B	62
		Composition	B, at%	0	7	11.5	14.5	19.5	23			
		Structure[1]		C	C	C	C	C	A			
28	Fe–W–B	Bath chemical	KBH_4, g/L	4	8	12	16	20			None	148
		Composition	W, at%	3.51	4.32	2.50	3.61	2.51				
			B, at%	6.33	2.01	11.91	4.28	9.29				
		Structure[1]		C	C	C	C	C				
29	Fe–Mo–B	Bath chemical	KBH_4, g/L	2	6	10	15	20			10.1–30 at% B 11.3–16.3 at% Mo	135
		Composition	Mo, at%	14.0	13.3	11.3	16.3	18.9				
			B, at%	14.9	23.3	30.0	10.1	6.4				
		Structure[1]		A	A	A	A	C				

30	Fe–Sn–B	Bath chemical	KBH_4, g/L	6	8	10	15	20	24		12.1–27 at% B	136
		Composition	Sn, at%	18.4	17.4	23.0	13.3	26.9	13.1		13.3–26.9 at% Sn	
			B, at%	10.9	12.1	14.9	19.9	27.0	30.0			
		Structure[1]		C	A	A	A	A	C			
31	Fe–W–Mo–B	Bath chemical	KBH_4, g/L	2	6	8	10	12	15	20	9.8–27.3 at% B	161
		Composition	Mo, at%	16.7	17.7	13.5	12.1	18.2	10.0	13.5	10–18.2 at% Mo	
			W, at%	0.6	0.6	0.7	1.3	1.3	trace	trace	Trace–1.3 at% W	
			B, at%	8.8	9.8	12.5	22.9	21.6	27.3	28.3		
		Structure[1]		C	A	A	A	A	A	C		
32	Ni–W–P	Bath chemical	Na_2WO_4, M	0.05	0.1	0.2	0.3	0.4			5.6–32.5 at% P	108
		Composition	W, at%	10.5	7.0	1.6	1.7	1.8			1.7–10.5 at% W	193
			P, at%	32.5	15.9	7.2	5.6	4.0				
		Structure[1]		A	A	A	A	C				
33	Ni–W–P	Bath chemical	Na_2WO_4, M	0.03	0.05	0.07	0.09				10.5–12 at% P	194
		Composition	W, at%	11	5	8	9				8–9 at% W	
			P, at%	13	11	10.5	12					
		Structure[1]		C	C	A	A					
34	Ni–W–P	Bath chemical	NaH_2PO_2, M	0.086					0.114		18 at% P	180
		Composition	W, at%	3.6					2.7		2.7 at% W	
			P, at%	14					18			
		Structure[1]		Ni–W crystallites in amorphous matrix					A			
35	Ni–W–P	Chemicals	Substrate	Brass	Brass	Brass	Brass	Mild steel	Mild steel		6.0–6.5 at% P	195
		Compositions	W, wt%	8.7	9.5	23.4	21.5	7.0	7.4		21.5–23.4 at% W	
			P, wt%	2.5	2.5	6.0	6.5	3.0	3.8			
		Structure[1]		C	C	A	A	C	C			
36	Ni–Mo–P	Chemicals	Na_2MoSO_4 mol/L	0.0005	0.0015	0.002	0.0030	0.010	0.020		18.4–15.5 at% P	196
		Compositions	Mo, at%	1.1	3.8	4.2	7.9	12.8	14.9		1.1–3.8 at% Mo	
			P, at%	18.4	15.5	4.1	2.3	1.3	1.5			
		Structure[1]		A	A	C	C	C	C			

(*Continued*)

TABLE 11.6 Composition, Structure, and Formation Range of Amorphous Phases for Various EP Alloy Deposits (*cont.*)

No.	Alloys	Parameters	Range of parameters						Range of amorphous phases	Refs
37	Ni–Mo–P	Chemicals	Na_2MoSO_4 mol/L	0.0015	0.002	0.005	0.01		8.1–8.9 wt% P	197
		Compositions	Mo, wt%	3.5	5.4	14.1	19.4		3.5–5.4 at% Mo	
			P, wt%	8.9	8.1	2.1	1.8			
		Structure[1]		A	A	C	C			
38	Ni–Mo–P	Bath chemical	$MoSO_4^{-2}$, mol/L	0.002	0.003	0.0035			11.2 at% P	198
		Composition	Mo, wt%	3.07	8.81	12.42			3.07 at% Mo	
			P, wt%	11.2	6.68	1.0				
		Structure[1]		A	Mixing crystal	Crystal				
39	Ni–Mo–P	Chemicals							8.2–14.2 at% P	105
		Compositions	Mo, wt%	4.7	5.5	5.7	5.9	6.7	4.7–6.7 at% Mo	
			P, wt%	14.2	1.3	8.2	13.9	11.3		
		Structure[1]		A	C	A	A	A		
40	Ni–Co–P	Bath chemical	$\frac{CoSO_4}{CoSO_4+NiSO_4}$	0.3	0.5	0.7			8.1–9 wt% P	95
		Composition	Co, wt%	31	40	52			31–52 wt% Co	
			P, wt%	9	8.6	8.1				
		Structure[1]		A	A	A				
41	Ni–Co–P	Bath chemical	$\frac{CoSO_4}{CoSO_4+NiSO_4}$	0.315	0.514	0.714			6.5–9.35 wt% P	199
		Composition	Co, wt%	11.01	36.87	45.06			11–45 wt% Co	
			P, wt%	9.346	8.20	6.49				
		Structure[1]		A	A	A				

42	Ni–Fe–P	Bath chemical	$\frac{FeSO_4}{FeSO_4+NiSO_4}$	0.095	0.289	0.486					18.3–19 at% P	200
		Composition	Fe, at%	9.6	10.7	16					9.6–16 at% Fe	
			P, at%	19	20.3	18.3						
		Structure[1]		A	A	A						
43	Ni–Fe–P	Film depth profiles		Outer coating layer			Substrate coating interface				11–13.5 wt% P 1.5–5 wt% Fe	91
		Composition	Fe, wt%	1.5			5					
			P, wt%	13.5			11					
		Structure[1]		A			A					
44	Co–Fe–P	Bath chemical	$\frac{FeSO_4}{FeSO_4+CoSO_4}$	0.01	0.05	0.1	0.15				5.02 wt% P	201
		Composition	Fe, wt%	0.54	2.75	7.75	14.56				0.54 wt% Fe	
			P, wt%	5.02	3.78	3.18	2.37					
		Structure[1]		A	C	C	C					
45	Ni–Co–Fe–P	Bath chemical									4.13 wt% P 7.98 wt% Co 9.75 wt% Fe	202
		Composition	Co, wt%	7.98								
			Fe, wt%	9.75								
			P, wt%	4.13								
		Structure[1]		A								
46	Ni–Re–P	Bath chemical	ReO_4^{-1}, wt%	7.5×10^{-4}	2.44×10^{-3}	3.75×10^{-3}	5.15×10^{-3}		7.5×10^{-3}		13–14.7 at% P 0.3–3.5 at% Re	112
		Composition	Re, at%	0.3	1	1.2	2.6		3.5			
			P, at%	14.7	13.5	13.8	13.5		13			
		Structure[1]		A	A	A	A		A			
47	Ni–Re–P	Bath chemical										113
		Composition	Re, wt%	9.2								
			P, wt%	13.0								
		Structure[1]		A								

(Continued)

TABLE 11.6 Composition, Structure, and Formation Range of Amorphous Phases for Various EP Alloy Deposits (*cont.*)

No.	Alloys	Parameters	Range of parameters								Range of amorphous phases	Refs
48	Ni–W–Re–P	Bath chemical Composition Structure[1]	Film thickness, μ Re, wt% W, wt% P, wt%	0.7 3.1 4.0 12.2 A	2.1 3.2 2.6 13.3 A						12.2–13.3 wt% P 3.1–3.2 wt% Re 2.6–4.0 wt% W	113
49	Co–W–P	Bath chemical Composition Structure[1]	 W, wt% P, wt%	 2.9–3.4 2.7–2.4 C	 3.6–3.7 7.4–8.3 A						3.6–3.7 wt% P 7.4–8.3 wt% W	203
50	Co–W–P	Bath chemical Composition Structure[1]	pH W, at% P, at%	7.8 2.34 6.65 C	8.0 2.29 6.73 C+μC	8.2 2.21 7.32 A+nC	8.4 1.63 8.09 A+nC	8.6 1.27 9.06 A	8.8 1.07 10.13 A	9.0 0.89 12.94 A	9.06–12.94 wt% P 1.27–0.89 wt% W	128
51	Co–W–P	Bath chemical Composition Structure[1]	$CoSO_4$, M W, at% P, at%	0.01 6 12 A	0.03 3.3 8 A+nC	0.05 2.6 5.6 C	0.07 2 4.4 C	0.09 1 2 C			12 at% P 6 at% W	129
52	Co–W–B	Bath chemical Composition Structure[1]	W^{2+}/Co^{2+} W, at% B, at% O, at%	0.03 8.73 5.06 1.15 A	0.075 6.5 1.39 0.54 C	0.15 6.99 0.98 0.32 C	0.3 13.71 0.28 0.47 C				5.06 at% B 8.73 at% W	130
53	Co–W–B	Bath chemical Composition Structure[1]	 W, at% B, at%	 1 34 A							34 at% B 1 at% W	131

54	Co–W–P–B	Bath chemical Composition Structure[1]	W, at% P, at% B, at%	0.3 2.3 27.1 A							2.3 at% P 27.1 at% B 0.3 at% W	159
55	Co–Ni–B	Bath chemical Composition Structure[1]	$\frac{NiSO_4}{NiSO_4 + CoSO_4}$ Ni, wt% B, wt%	0.3 24.2 2.6 A	0.5 14.4 2.1 A	0.7 8.1 1.0 A					1.0–2.6 wt% B 8.1–24.2 wt% W	204
56	Co–Fe–B	Bath chemical Composition Structure[1]	pH Fe, at% B, at%	6.5 6.25 3.7 A	7.5 10 5.5 C	8 11.9 6.5 C	9 10.2 7 μC+A	9.5 12 6.3 μC+A			3.7 at% B 6.25 at% Fe	205
57	Ni–P–B	Bath chemical Composition Structure[1]	DMAB, M P, wt% B, wt%	0.01 7.82 0.28 A	0.025 6.80 0.70 A						7.82–6.8 wt% P 0.28–0.7 wt% B	98
58	Ni–P–B	Bath chemical Composition Structure[1]	P, at% B, at%	15.2 4.1 A							15.2 at% P 4.1 at% B	99
59	Ni–P–B	Bath chemical Composition Structure[1]	$NaBH_4$, g/L P, wt% B, wt%	0.2 7.78 1.26 A	0.4 6.83 3.37 A	0.8 4.04 5.89 A					7.78–4.04 wt% P 1.26–5.89 wt% B	206
60	Ni–P–B	Bath chemical Composition Structure[1]	NaH_2PO_2, g/L P, at% B, at%	10 2.3 5 A	18 3.4 4.9 A	25 3.7 4.3 A	35 4,7 4.2 A				2.3–4.7 wt% P 5–4.2 wt% B	100

1 A, amorphous, C, crystal, μC, micro crystal, nC, nano crystal.
2 pH=5.5; 3 pH=5.5; 4 pH=4.8.

Second is that the data of the RAF of EP amorphous alloys measured from various author groups are not the same, but usually are different either for the EP binary and ternary or for the quaternary alloy deposits. For example, the lower limitation of P for EN in Nos. 5 and 7 in Table 11.6 is 13 to 14 at%, but that in No. 1 is 26.5 at%; those of B for Ni–B in Nos. 18 and 17 are 17.1 and 41 at%, and those of B for Co–B in Nos. 24 and 23 are 5.1 and 25.8 at%, respectively. The reasons are that the preparation techniques of samples are difficult, not the same; moreover, the measurement error is difficult to avoid from different authors.

The third is that there is only a lower limit, but no upper limit for the RAF of EP binary amorphous alloys. Nearly all the measured data for the RAF of EP binary amorphous alloys contain one TM (Ni, Co, or Fe) and one metalloid (P or B) element. P and B help to form amorphous phases, and the higher the content of P or B in the deposits, the easier it is to form an amorphous alloy. Once the deposits became fully amorphous, adding more P or B in the deposits cannot turn them into crystal but still maintains their amorphous phase. Therefore, no upper limitation of metalloid in such binary amorphous deposits can be reached. In fact, the EP binary amorphous deposits, for example, EN with very high content of P (>80 at%) have not been reported up today because of the experimental difficulty.

The fourth is about the effect of addition of the metal(s) to the EP binary TM-M alloy deposits. In Section 11.2.2.1, we analyzed the actions of Sn and Cu when added to the EP Ni–P films, and found that Sn is a favorable element and Cu is a nonfavorable element for the formation of amorphous alloys. The experimental facts in Table 11.6 readily confirm the correctness of these conclusions. For example, in No. 8 the lower limitation of P for the RAF of EP Ni–Sn–P amorphous alloys decreases to 5 at%, but that for the RAF of EP Ni–P amorphous alloys is 13 to 14 at%. However, when adding Cu to EP Ni–P deposits, the lower limitation of P for the RAF of Ni–Cu–P amorphous alloys (Nos. 13 to 14) is about 12 wt% (=20.7 at% according to the calculation of the weight percentage of $Ni_{77}Cu_{11}P_{12}$), which indicates that the lower limitation of P for the RAF in EP Ni–Cu–P amorphous alloys increases and is larger than that for the EP Ni–P amorphous alloys.

The last point is that though many experimental data for the RAF of EP amorphous alloys have been reported as shown in Table 11.6, the full data for the RAF of EP amorphous alloy films in many alloy systems are still absent. In addition, the explanations to the experimental data near cannot be found in the literature. Therefore, broad and deep research in such fields is needed and welcomed.

11.3.2 Theoretical Calculation of RAF of EP Amorphous Alloys

Theoretical calculation of RAF of EP amorphous alloys is very few. The following works come from the author' group [30].

11.3.2.1 Binary Alloys

Using the scheme to calculate the RAF with the Miedema model for the heat of mixing of binary alloys proposed by Miedema and co-workers [207], our group in 2002 [30] calculated the RAF of EP Ni–P alloy deposits. The key of the scheme for estimating the RAF is to compare the free energies of the crystalline and the amorphous phases. Because the entropy contribution to the free energy is very small at the room temperature of 300 K and so the driving force for amorphization comes mainly from the enthalpy contribution, we therefore just calculated the enthalpy-composition plots instead of the free energy-composition plots in our study. The enthalpies of solid solution can be represented as [208]

$$\Delta H_{\mathrm{AB}} = \Delta H_{\mathrm{AB}}^{\mathrm{c}} + \Delta H_{\mathrm{AB}}^{\mathrm{e}} + \Delta H_{\mathrm{AB}}^{\mathrm{s}} \tag{11.20}$$

where the superscripts c, e, and s correspond to the chemical, elastic, and structure contributions to the enthalpy of solid solution, respectively. It is obvious that crystalline pure elements were chosen to be the standard state and their enthalpies were assigned to be zero in the calculations, hence providing the representation of Eq. (11.20). The equation of the chemical interaction contribution to the enthalpy of a solid solution derived and used by Miedema belongs to the regular model (Equation (2.25) in [38]). However, we used the subregular model for calculating the mixing enthalpies [209]:

$$\Delta H_{\mathrm{AB}}^{\mathrm{c}} = x_{\mathrm{A}} x_{\mathrm{B}} \left(x_{\mathrm{A}} \Delta H_{\mathrm{BinA}}^{0} + x_{\mathrm{B}} \Delta H_{\mathrm{AinB}}^{0} \right) \tag{11.21}$$

where x_{A} and x_{B} are the contents of compositions A and B, respectively. $\Delta H_{\mathrm{BinA}}^{0}$ is the enthalpy of solution of one element A in other element B at an infinite dilution for a binary alloy system, which can be calculated from the Miedema model [38].

The elastic contribution to the enthalpy arises from the difference in atomic volume between the solute and the solvent. Similarly, it can be presented as

$$\Delta H_{\mathrm{AB}}^{\mathrm{e}} = x_{\mathrm{A}} x_{\mathrm{B}} \left(x_{\mathrm{A}} \Delta H_{\mathrm{BinA}}^{\mathrm{e}} + x_{\mathrm{B}} \Delta H_{\mathrm{AinB}}^{\mathrm{e}} \right) \tag{11.22}$$

where $\Delta H_{\mathrm{AinB}}^{\mathrm{e}}$ is the size-mismatch contribution to the enthalpy of solution in a binary system. The Friedel formalism [210] is used in the study, given by

$$\Delta H_{\mathrm{AB}}^{\mathrm{e}} = \frac{24\pi B_{\mathrm{A}} \mu_{\mathrm{B}} R_{\mathrm{A}} R_{\mathrm{B}} (R_{\mathrm{A}} - R_{\mathrm{B}})^2}{3B_{\mathrm{A}} R_{\mathrm{B}} + 4\mu_{\mathrm{B}} R_{\mathrm{A}}} \tag{11.23}$$

where B_{A} is the bulk modulus of the solute, and μ_{A} the shear modulus of the solvent, for which their values have been tabulated by Gschneidner [211]. R_{A} and R_{B} are the radii of the solute and solvent, respectively.

$\Delta H_{\mathrm{AB}}^{\mathrm{s}}$ in Eq. (11.20) is a structural contribution to the enthalpy of the alloys taking into account the difference between the valences and the crystal structure of the solute and solvent, and the latter is expected to have only a minor effect when compared with the elastic energy contribution [212]. Further, it is difficult to calculate the structure contribution. Therefore, this term will not be considered in the calculation, as a first approximation.

As for the enthalpy of amorphous alloy, there are no contributions from the elastic and structural terms, so that its enthalpy of formation can be written as

$$\Delta H_{\mathrm{AB}}^{\mathrm{a}} = \Delta H_{\mathrm{AB}}^{\mathrm{a,c}} + \Delta H_{\mathrm{A}}^{\mathrm{a}} + \Delta H_{\mathrm{B}}^{\mathrm{a}} \tag{11.24}$$

where $\Delta H_{\mathrm{A}}^{\mathrm{i}}$ is the enthalpy of the amorphous pure element i. According to van der Kolk et al. [213], it is given by

$$\Delta H_{\mathrm{A}}^{\mathrm{i}} = \alpha T_{\mathrm{m,i}} \tag{11.25}$$

where $\alpha = 3.5/\mathrm{mol} \cdot \mathrm{K}$, and $T_{\mathrm{m,i}}$ is the melting temperature of element i. $\Delta H_{\mathrm{AB}}^{\mathrm{a,c}}$ is the chemical contribution to the enthalpy of the amorphous alloy. Considering the short-range order observed in the amorphous phases [214], it is given by

$$\Delta H_{\mathrm{AB}}^{\mathrm{a,c}} = \Delta H_{\mathrm{AB}}^{\mathrm{c}} \left[1 + 5\left(x_{\mathrm{A}}^{\mathrm{s}} x_{\mathrm{B}}^{\mathrm{s}} \right)^2 \right] \tag{11.26}$$

where

$$x_{\mathrm{A}}^{\mathrm{s}} = \frac{x_{\mathrm{A}} V_{\mathrm{A}}^{2/3}}{x_{\mathrm{A}} V_{\mathrm{A}}^{2/3} + x_{\mathrm{B}} V_{\mathrm{B}}^{2/3}} \tag{11.27}$$

$$x_{\mathrm{B}}^{\mathrm{s}} = 1 - x_{\mathrm{A}}^{\mathrm{s}} \tag{11.28}$$

in which $x_{\mathrm{i}}^{\mathrm{s}}$ is the surface concentration of element i, where V_{i} is the volume of the element i.

The results of these calculations for the Ni_{1-x}–P_x alloy system have been plotted as enthalpy versus content of P in Fig. 11.8. It can be seen that the predicted lower limitation of RAF from the theoretical calculation is P 16 at%. It is well known that the enthalpies calculated from the Miedema model cause some errors, as Miedema himself accepted [38]. On the other hand, the elastic term of enthalpy is calculated using the Friedel formalism obtained from the theory of elasticity. It is expected that this term will cause some errors in the calculations of enthalpy. Furthermore, the neglect of the structure term in the enthalpy and the effect of temperature on the enthalpy would also produce some errors in the calculations. All of these factors will make the calculated enthalpies somewhat different from the actual values for the alloys. In addition, considering the dispersion of experimental data by various groups as shown in Table 11.6, we can say that the theoretically predicted RAF is in pretty good agreement with the experimental results. Theoretical calculation (Fig. 11.8) also indicates that there is no upper limitation of RAF for Ni–P alloys. This result is also in good agreement with the various experiments in Table 11.6. In Section 11.3.1, we discussed qualitatively the reasons for this point.

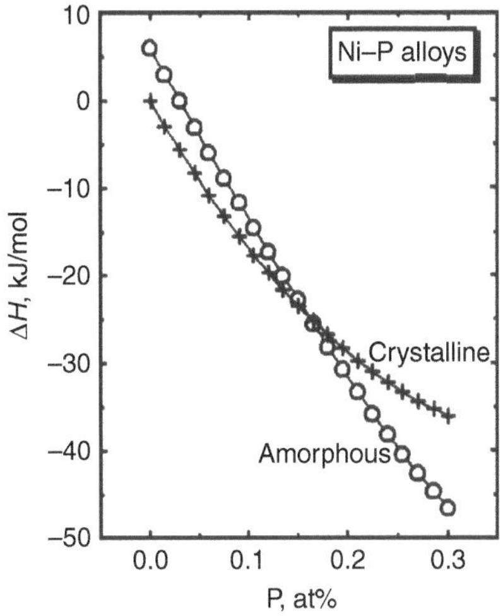

FIGURE 11.8 Enthalpy vs. content of phosphorus for the Ni–P alloys: (+) crystalline; (O) amorphous [30].

11.3.2.2 Ternary Alloys

As pointed out in Section 11.2.2.3, for a ternary alloy, Zhang and Jesser [40] only considered the binary interaction between atoms, and the ternary atomic interactions between the constituents were neglected when the Miedema model extends to the ternary alloys. Therefore, we only need to extend the above (11.21), (11.22), (11.24), and (11.26) equations as follows, then the RAF for the ternary amorphous alloys can be easily calculated [215].

$$\begin{aligned}\Delta H^{\mathrm{c}}_{\mathrm{ABC}} &= x_{\mathrm{A}}x_{\mathrm{B}}\left(x_{\mathrm{A}}\Delta H^{0}_{\mathrm{BinA}} + x_{\mathrm{B}}\Delta H^{0}_{\mathrm{AinB}}\right)\\ &+ x_{\mathrm{A}}x_{\mathrm{C}}\left(x_{\mathrm{A}}\Delta H^{0}_{\mathrm{CinA}} + x_{\mathrm{C}}\Delta H^{0}_{\mathrm{AinC}}\right)\\ &+ x_{\mathrm{B}}x_{\mathrm{C}}\left(x_{\mathrm{B}}\Delta H^{0}_{\mathrm{CinB}} + x_{\mathrm{C}}\Delta H^{0}_{\mathrm{BinC}}\right)\end{aligned} \tag{11.29}$$

$$\begin{aligned}\Delta H^{\mathrm{e}}_{\mathrm{ABC}} &= x_{\mathrm{A}}x_{\mathrm{B}}\left(x_{\mathrm{A}}\Delta H^{\mathrm{e}}_{\mathrm{BinA}} + x_{\mathrm{B}}\Delta H^{\mathrm{e}}_{\mathrm{AinB}}\right)\\ &+ x_{\mathrm{A}}x_{\mathrm{C}}\left(x_{\mathrm{A}}\Delta H^{\mathrm{e}}_{\mathrm{CinA}} + x_{\mathrm{C}}\Delta H^{\mathrm{e}}_{\mathrm{AinC}}\right)\\ &+ x_{\mathrm{B}}x_{\mathrm{C}}\left(x_{\mathrm{B}}\Delta H^{\mathrm{e}}_{\mathrm{CinB}} + x_{\mathrm{C}}\Delta H^{\mathrm{e}}_{\mathrm{BinC}}\right)\end{aligned} \tag{11.30}$$

$$\Delta H^{\mathrm{a}}_{\mathrm{ABC}} = \Delta H^{\mathrm{a,c}}_{\mathrm{ABC}} + \Delta H^{\mathrm{a}}_{\mathrm{A}} + \Delta H^{\mathrm{a}}_{\mathrm{B}} + \Delta H^{\mathrm{a}}_{\mathrm{C}} \tag{11.31}$$

$$\begin{aligned}\Delta H^{\mathrm{a,c}}_{\mathrm{ABC}} = \Delta H^{\mathrm{c}}_{\mathrm{AB}}\left[1+5\left(x^{\mathrm{s}}_{\mathrm{A}}x^{\mathrm{s}}_{\mathrm{B}}\right)^{2}\right] + \Delta H^{\mathrm{c}}_{\mathrm{AC}}\left[1+5\left(x^{\mathrm{s}}_{\mathrm{A}}x^{\mathrm{s}}_{\mathrm{C}}\right)^{2}\right]\\ + \Delta H^{\mathrm{c}}_{\mathrm{BC}}\left[1+5\left(x^{\mathrm{s}}_{\mathrm{B}}x^{\mathrm{s}}_{\mathrm{C}}\right)^{2}\right]\end{aligned} \tag{11.32}$$

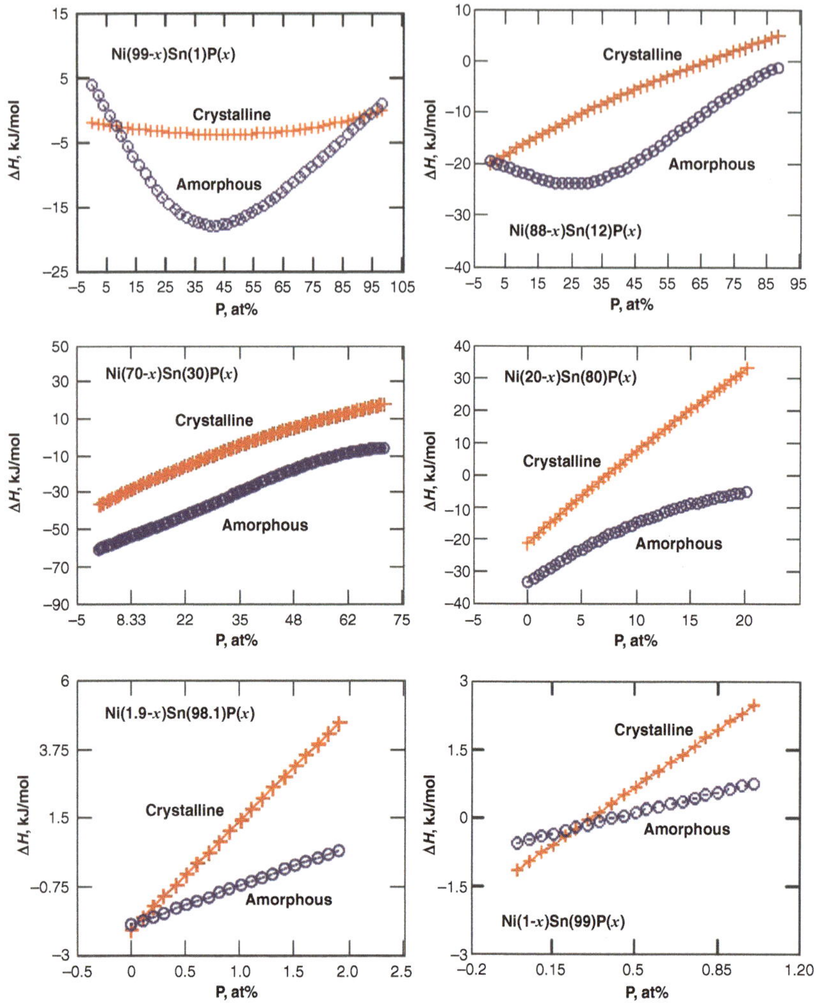

FIGURE 11.9 Calculated plots of amorphous formation of Ni–Sn–P alloys. Upper row: $Ni_{99-x}Sn_1P_x$, and $Ni_{88-x}Sn_{12}P_x$; middle row: $Ni_{70-x}Sn_{30}P_x$, and $Ni_{20-x}Sn_{80}P_x$; lower row: $Ni_{1.9-x}Sn_{98.1}P_x$, and $Ni_{1-x}Sn_{99}P_x$ [215].

First, the FRA plots of the amorphous alloy systems for FP Ni–Sn–P alloy systems were calculated. Figure 11.9 shows the calculated typical samples for the alloy systems that change the P contents [215]. Three cases can be clearly seen from the figure.

1. When Sn is less than 12 at%, there are one or two non-FRAs besides one FRA in every alloy system, as shown in Table 11.7 [215]. For example, the FRA and non-FRA in $Ni_{95-x}Sn_5P_x$ alloy system are $Ni_{87-0}Sn_5P_{8-95}$ and $Ni_{95-87}Sn_5P_{0-8}$, respectively. But there are two non-FRAs of $Ni_{99-90}Sn_1P_{0-9}$

TABLE 11.7 Calculated FRA, Non-FRA and Enthalpies for EP Ni–Sn–P Ternary Alloy Systems [215]

Theoretical calculations (present)							
Alloys Ni–Sn–P		**FRA, at%**	**Non-FRA, at%**	**Content of P, at%**	**ΔH, kJ/mol**		**Reference**
					Amorphous	**Crystalline**	
P_x	$Ni_{99-x}Sn_1P_x$	$Ni_{90\sim5}Sn_1P_{9\sim94}$	$Ni_{99\sim90}Sn_1P_{0\sim9}$ $Ni_{5\sim0}Sn_1P_{94\sim99}$	P =9 at% P =94 at%	−2.818 −0.523	−2.48 −0. 527	Present
	$Ni_{95-x}Sn_5Px$	$Ni_{87\sim0}Sn_5P_{8\sim95}$	$Ni_{95\sim87}Sn_5P_{0\sim8}$	P =8.0 at%	−8.888	−8.348	
	⋮	⋮					
	$Ni_{88-x}Sn_{12}P_x$	$Ni_{87\sim0}Sn_{12}P_{1\sim88}$ A critical case (not exactly)	$Ni_{88\sim87}Sn_{12}P_{0\sim1}$	P =0 at%	−19.372	−19.723	
	$Ni_{70-x}Sn_{30}P_x$	$Ni_{70\sim0}Sn_{30}P_{0\sim70}$	No crystalline alloys are formed in all alloy systems				
	⋮	⋮					
	$Ni_{20-x}Sn_{80}P_x$	$Ni_{80\sim0}Sn_{0\sim80}P_{20}$					
	⋮	⋮					
	$Ni_{1.9-x}Sn_{98.1}P_x$	$Ni_{1.9\sim0}Sn_{98.2}P_{0\sim1.9}$ A critical case		P =0 at%	−2.019	−2.144	
	$Ni_{1.8-x}Sn_{98.2}P_x$	$Ni_{1.7\sim0}Sn_{98.2}P_{0.1\sim1.8}$	$Ni_{1.8\sim1.7}Sn_{98.2}P_{0\sim0.1}$	P =0.1 at%	−1.721	−1.67	
	⋮	⋮					
	$Ni_{1-x}Sn_{99}P_x$	$Ni_{0.74\sim0}Sn_{99}P_{0.26\sim1.0}$	$Ni_{1.0\sim0.74}Sn_{99}P_{0\sim0.26}$	P =0.26 at%	−0.201	−0.19	

(Continued)

TABLE 11.7 Calculated FRA, Non-FRA and Enthalpies for EP Ni–Sn–P Ternary Alloy Systems [215] (*cont.*)

Sn_x	$Ni_{99-x}Sn_xP_1$	$Ni_{90.5\sim4}Sn_{8.5\sim95}P_1$	$Ni_{96\sim90.5}Sn_{0\sim8.5}P_1$ $Ni_{4\sim0}Sn_{95\sim99}P_1$	Sn =8.5 at% Sn =96 at%	−2.818 −0.523	−2.48 −0.527	Present
	⋮	⋮					
	$Ni_{87-x}Sn_xP_{13}$	$Ni_{87\sim0}Sn_{0\sim87}P_{17}$ A critical case	No crystalline alloys are formed in all alloy systems	Sn =0 at%	−21.675	−21.03	
	$Ni_{85-x}Sn_xP_{15}$	$Ni_{85\sim0}Sn_{0\sim85}P_{15}$					
	⋮	⋮					
	$Ni_{5-x}Sn_xP_{95}$	$Ni_{5\sim0}Sn_{0\sim5}P_{95}$					
	⋮	⋮					
	$Ni_{2.1-x}Sn_xP_{97.9}$	$Ni_{2.1\sim0}Sn_{0\sim2}P_{97.9}$ A critical case		Sn =0 at%	−2.235	−2.369	
	⋮	⋮					
	$Ni_{1-x}Sn_xP_{99}$	$Ni_{0\sim0.7}Sn_{1\sim0.3}P_{99}$	$Ni_{1\sim0.7}Sn_{0\sim0.3}P_{99}$	Sn =0.3	−0.149	−0.046	

Experimental data

Alloys	FRA, at%	Non-FRA, at%	Reference
Ni–Sn–P	$Ni_{93.3}Sn_{1.7}P_5$, $Ni_{86.2}Sn_{1.2}P_{12.6}$, $Ni_{72.4}Sn_{1.6}P_{26}$	$Ni_{95.7}Sn_{0.8}P_{3.5}$	[120]
	$Ni_{84.18}Sn_{1.39}P_{14.43}$, $Ni_{79.4}Sn_{0.35}P_{20.51}$, $Ni_{84.49}Sn_{1.09}P_{14.5}$		[31]
	$Ni_{93.3}Sn_{1.7}P_5$, $Ni_{85.1}Sn_{10.4}P_{4.5}$	$Ni_{94.8}Sn_{0.7}P_{4.5}$, $Ni_{95.7}Sn_{0.8}P_{3.5}$	[184]

and $Ni_{5-0}Sn_1P_{94-99}$ and one FRA of $Ni_{90-5}Sn_1P_{9-94}$, in the $Ni_{99-x}Sn_1P_x$ alloy system. At Sn = 12 at%, ΔH^a (−19.372 kJ/mol) ≅ ΔH^c (−19.723 kJ/mol) at P = 0 at%, we may call this case a critical state. Then, $\Delta H^a < \Delta H^c$ in all composition ranges of the alloy system, that is, all of the alloys from P = 0 at% to 87 at% are amorphous structures.

2. In the range from Sn = 12 at% to 98.1 at%, all the curves of ΔH^c locate above those of ΔH^a, which mean that no any crystalline phases can form. The two middle figures ($Ni_{70-x}Sn_{30}P_x$ and $Ni_{20-x}Sn_{80}P_x$) show these phenomena clearly. When Sn approaches 98.1 at%, another critical case appears, because ΔH_a(−2.019 kJ/mol) ≅ ΔH_c(−2.144 kJ/mol) at P = 0 at%.
3. When Sn > 98.1 at%, an FRA and a non-FRA occur for every alloy system. The last figure for $Ni_{1-x}Sn_{99}P_x$ alloy system in Fig. 11.9 is a sample, indicating these phenomena evidently. Table 11.7 shows the FRA and non-FRA of this alloy system are $Ni_{0.74-0}Sn_{99}P_{0.26-1.0}$ and $Ni_{1.0-0.74}Sn_{99}P_{0-0.26}$, respectively.

When considering the alloy systems with varying Sn content, the calculated typical ranges of amorphous formation of Ni–Sn–P alloys are similar to Fig. 11.9 [215], so it is not shown here. Six typical plots of the range of amorphous formation correspond to the contents of Sn variable and P = 1, 13, 15, 95, 97.9, and 99 at%, respectively. When the contents of Sn are 13 and 97.9 at%, the ΔH^a (−21.675 kJ/mol) ≅ ΔH^c (−21.03 kJ/mol) or ΔH^a(−2.235 kJ/mol) ≅ ΔH^c(−2.369 kJ/mol)] at Sn = 0 at% of the alloys, respectively and $\Delta H^a < \Delta H^c$ in the whole range of composition, thereby the AFR is the whole range of composition in these two alloy systems. These two cases are the critical cases. Between these two points of compositions of P, all of the ΔH^c curve is wholly located over above the ΔH^a curve, in a similar way to the two middle figures of Fig. 11.9, the AFR for all the alloys in the range certainly is the whole range of composition in these alloy systems. Outside the range, that is, P<13 or >97.9 at% there exist an AFR and a non-AFR in those alloy systems, and two examples are as shown in the first and last figures. The details of the AFR and non-AFR for the alloy systems when varying the Sn contents in the alloys are shown in Table 11.7 as well.

Besides the calculated typical range of amorphous formation of Ni–Sn–P alloys either changing P or Sn contents in the alloy systems listed in Table 11.7, the existing experimental data are also indicated in the table. From the table, it can be seen clearly that the measured FRAs or non-FRAs in Ni–Sn–P alloy systems are all included in the calculated results. In addition, the calculated FRA tends to be greater than that experimentally determined value for all of the systems. Why are the calculated FRAs larger than those measured? The reasons could be as follows. First, as pointed out in the binary Ni–P alloys, the model used is rather simple. In addition, the ternary atomic interactions between the constituents in the ternary alloy systems are neglected, and no kinetic factors for the formation of the amorphous phase in the present model are taken into account. The final reason is that all of the compositions have not been studied systematically using the EP method, leading to limitation of the number of experimental results by EP. It is interesting to point out that this matter was also found by Takeuchi and Inoue in 2001 [216] when they calculated the amorphous-forming composition range for ternary alloy systems manufactured by the MQ method using a model similar to Miedema's model.

Then, we calculated the FRA of the EP Ni–Cu–P amorphous alloy deposits. The results are shown in Fig. 11.10 and Table 11.8. In the calculations, two series of alloys are calculated, one is that the Cu is changed but P is unchanged (Fig. 11.10a–d), another is that the P is changed but Cu is unchanged (Fig. 11.10e–h). From this figure, the FRA, and non-FRA in EP Ni–Cu–P can be obtained, as shown in Table 11.8.

From Fig. 11.10, one can see that the features are different for these two series of alloys. First, let us consider the feature in the series of Fig. 11.10a–d. The enthalpy $\Delta H^a \cong \Delta H^c$ for nearly every alloy in Fig. 11.10c, that is, $Ni_{83-x}Cu_xP_{17}$ alloy system. So, this is the critical case in the Ni–Cu–P system with changing Cu but unchanging P. When the P content is less than 17 at%, the ΔH^c curve wholly locates below the ΔH^a curve, no amorphous alloys can be formed for whole composition range in the alloy system. Contrarily when the P content is larger than 17 at%, then the ΔH^c curve wholly locates above the ΔH^a curve, and amorphous alloys can be formed for whole composition range. For the series of Fig. 11.10e–h, that is, the P is changed but Cu is unchanged, the feature of the figures is somewhat like in Fig. 11.9 (EP Ni–Sn–P). When the content of Cu = 83 at% in the alloy, ΔH^a $(-12.932) \cong \Delta H^c$ (-12.793) only at P = 17 at%, otherwise $\Delta H^a > \Delta H^c$ in whole range of composition, no amorphous alloys are formed. This is the critical case in the series of Fig. 11.10e–h. Since then, the crossover phenomenon between the ΔH^a curve and ΔH^c curve occurs as decreasing the content of Cu. Thereby, an RFA and a non-RFA are formed in the alloy systems, as indicated in Fig. 11.10g and h.

Several groups have measured the FRAs for EP Ni–Cu–P alloy films [31,49,118,184–186], and two groups measured the range of nonamorphous formation in this alloy system [31,49,184]. The measured FRA and non-FRA are also listed in Table 11.8.

It can be seen that the experimental data for the FRA and non-FRA of EP Ni–Cu–P alloy depositions are included in the ranges of our calculations, which means that the calculated results are in agreement with the measured data. However, the calculated FRA tends to be greater than that experimentally determined for all of the systems. Such a result is similar to that occurred in the Ni–Sn–P alloy systems. The reasons for such a result are also similar as mentioned above; therefore there is no need to repeat again.

11.3.2.3 Quaternary Alloys

In the calculations of RAF for the EP quaternary amorphous alloy systems, we omitted the ternary and quaternary atomic interactions between the constituents and only considered the binary interaction between atoms, which is similar as in the above calculations of RAF of ternary amorphous alloys. Therefore, we only need to extend the above (11.21), (11.22), (11.24) and (11.26), respectively equations as fellows, then the RAF for the quaternary amorphous alloys can be easily calculated.

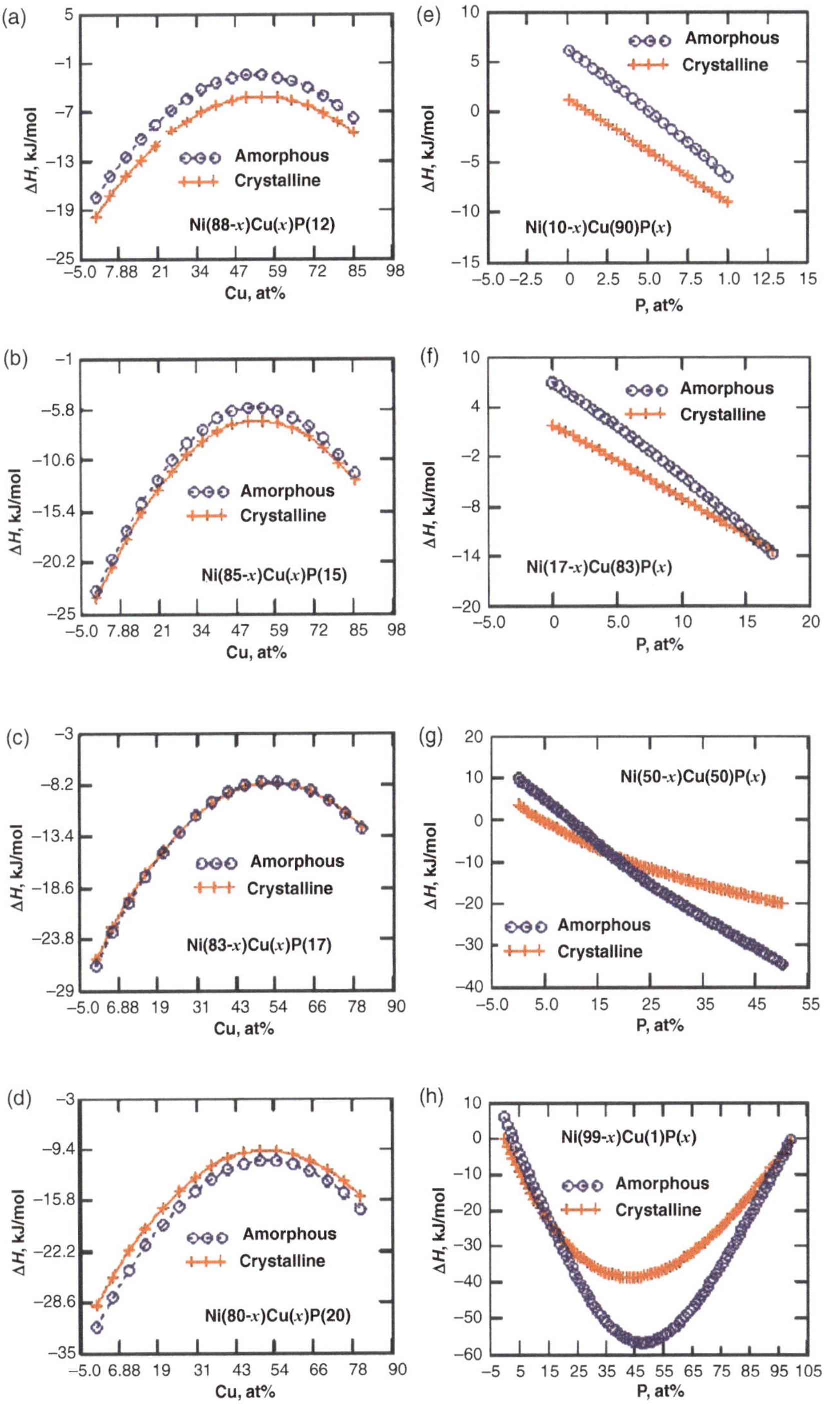

FIGURE 11.10 Calculated plots of amorphous formation of Ni–Cu–P alloys: (a) $Ni_{88-x}Cu_xP_{12}$, (b) $Ni_{85-x}Cu_xP_{15}$, (c) $Ni_{83-x}Cu_xP_{17}$, (d) $Ni_{80-x}Cu_xP_{20}$, (e) $Ni_{10-x}Cu_{90}P_x$, (f) $Ni_{17-x}Cu_{83}P_x$, (g) $Ni_{50-x}Cu_{50}P_x$, and (h) $Ni_{99-x}Cu_1P_x$ [215].

TABLE 11.8 Calculated FRA, Non-FRA, and Enthalpies for EP Ni–Cu–P Ternary Alloy Systems [215]

Theoretical calculations (present)						
Alloys Ni–Cu–P		FRA, at%	Non-FRA, at%	Content of P, at%	ΔH, kJ/mol	
					Amorphous	Crystalline
Cu_x	$Ni_{88-x}Cu_xP_{12}$	No amorphous phases are formed for all these alloy systems.	$Ni_{88-0}Cu_{0-88}P_{12}$			
	⋮		⋮			
	$Ni_{85-x}Cu_xP_{15}$		$Ni_{85-0}Cu_{0-85}P_{15}$			
	$Ni_{83-x}Cu_xP_{17}$	No. All enthalpies of amorphous and crystalline states are just nearly equal in the whole range of components. A critical case.	No			
	$Ni_{80-x}Cu_xP_{20}$	All are amorphous alloys. $Ni_{80-0}Cu_{0-80}P_{20}$	No			
P_x	$Ni_{10-x}Cu_{90}P_x$	No amorphous phases are formed in these alloy systems	$Ni_{10-0}Cu_{90}P_{0-10}$			
	⋮		⋮			
	$Ni_{17-x}Cu_{83}P_x$	$Ni_{0-0}Cu_{83}P_{17-17}$. A critical case.	$Ni_{17-0}Cu_{83}P_{0-17}$	17.0	−12.932	−12.793
	⋮		⋮			
	$Ni_{50-x}Cu_{50}P_x$	$Ni_{32-0}Cu_{50}P_{18-50}$	$Ni_{50-32}Cu_{50}P_{0-18}$	18.0	−8.258	−8.214
	$Ni_{80-x}Cu_{20}P_x$	$Ni_{62.5-0}Cu_{20}P_{17.5-80}$	$Ni_{80-62.5}Cu_{20}P_{0-17.5}$	17.5	−15.012	−14.803
	$Ni_{85-x}Cu_{15}P_x$	$Ni_{68-0}Cu_{15}P_{17-85}$	$Ni_{85-68}Cu_{15}P_{0-17}$	17.0	−16.69	−16.672
	$Ni_{90-x}Cu_{10}P_x$	$Ni_{73-0}Cu_{10}P_{17-90}$	$Ni_{90-73}Cu_{10}P_{0-17}$	17.0	−19.309	−19.205
	$Ni_{95-x}Cu_5P_x$	$Ni_{78-0}Cu_5P_{17-95}$	$Ni_{95-78}Cu_5P_{0-17}$	17.0	−22.241	−22.053
	$Ni_{99-x}Cu_1P_x$	$Ni_{80.5-0}Cu_1P_{17.5-99}$	$Ni_{99-80.5}Cu_1P_{0-17.5}$	17.5	−25.716	−25.096

TABLE 11.8 Calculated FRA, Non-FRA, and Enthalpies for EP Ni–Cu–P Ternary Alloy Systems [215] *(cont.)*

Experimental data

Alloys	FRA, wt%	Non-FRA, wt%	Reference
Ni–Cu–P		$Ni_{90.73}Cu_{2.84}P_{6.43}$, $Ni_{90.02}Cu_{1.78}P_{8.2}$	[31]
		$Ni_{90}Cu_{1.8}P_{8.2}$	[184]
	$Ni_{63-82}Cu_{28-3}P_{9-15}$		[185]
	$Ni_{84}Cu_{3.4}P_{12.6}$	$Ni_{17.4-72.7}Cu_{81.5-18}P_{1.1-9.3}$	[49]
	$Ni_{82.3-70.6}Cu_{5.7-15.3}P_{12-14.1}$		[186]
	$Ni_{77}Cu_{11}P_{12}$		[118]

$$\Delta H^{c}_{ABCD} = x_A x_B\left(x_A \Delta H^{0}_{BinA} + x_B \Delta H^{0}_{AinB}\right) + x_A x_C\left(x_A \Delta H^{0}_{CinA} + x_C \Delta H^{0}_{AinC}\right) + x_A x_D\left(x_A \Delta H^{0}_{DinA} + x_D \Delta H^{0}_{AinD}\right) + x_B x_C\left(x_B \Delta H^{0}_{CinB} + x_C \Delta H^{0}_{BinC}\right) + x_B x_D\left(x_B \Delta H^{0}_{DinB} + x_D \Delta H^{0}_{BinD}\right) + x_C x_D\left(x_C \Delta H^{0}_{DinC} + x_D \Delta H^{0}_{CinD}\right) \tag{10.33}$$

$$\Delta H^{e}_{ABCD} = x_A x_B\left(x_A \Delta H^{e}_{BinA} + x_B \Delta H^{e}_{AinB}\right) + x_A x_C\left(x_A \Delta H^{e}_{CinA} + x_C \Delta H^{e}_{AinC}\right) + x_A x_D\left(x_A \Delta H^{e}_{DinA} + x_D \Delta H^{e}_{AinD}\right) + x_B x_C\left(x_B \Delta H^{e}_{CinB} + x_C \Delta H^{e}_{BinC}\right) + x_B x_D\left(x_B \Delta H^{e}_{DinB} + x_D \Delta H^{e}_{BinD}\right) + x_C x_D\left(x_C \Delta H^{e}_{DinC} + x_D \Delta H^{e}_{CinD}\right) \tag{10.34}$$

$$\Delta H^{a}_{ABCD} = \Delta H^{a,c}_{ABCD} + \Delta H^{a}_{A} + \Delta H^{a}_{B} + \Delta H^{a}_{C} + \Delta H^{a}_{D} \tag{11.35}$$

$$\Delta H^{a,c}_{ABCD} = \Delta H^{c}_{AB}\left[1+5\left(x^{s}_{A}x^{s}_{B}\right)^{2}\right] + \Delta H^{c}_{AC}\left[1+5\left(x^{s}_{A}x^{s}_{C}\right)^{2}\right] + \Delta H^{c}_{AD}\left[1+5\left(x^{s}_{A}x^{s}_{D}\right)^{2}\right] + \Delta H^{c}_{BC}\left[1+5\left(x^{s}_{B}x^{s}_{C}\right)^{2}\right] + \Delta H^{c}_{BD}\left[1+5\left(x^{s}_{B}x^{s}_{D}\right)^{2}\right] + \Delta H^{c}_{CD}\left[1+5\left(x^{s}_{C}x^{s}_{D}\right)^{2}\right] \tag{11.36}$$

We calculated the RAF of three class systems in Ni–Sn–Cu–P quaternary alloys. Three cases are changing the contents of Cu, Sn, and P, respectively, in the alloys. The results are shown in Figs 11.11–11.13 and Table 11.9 [215].

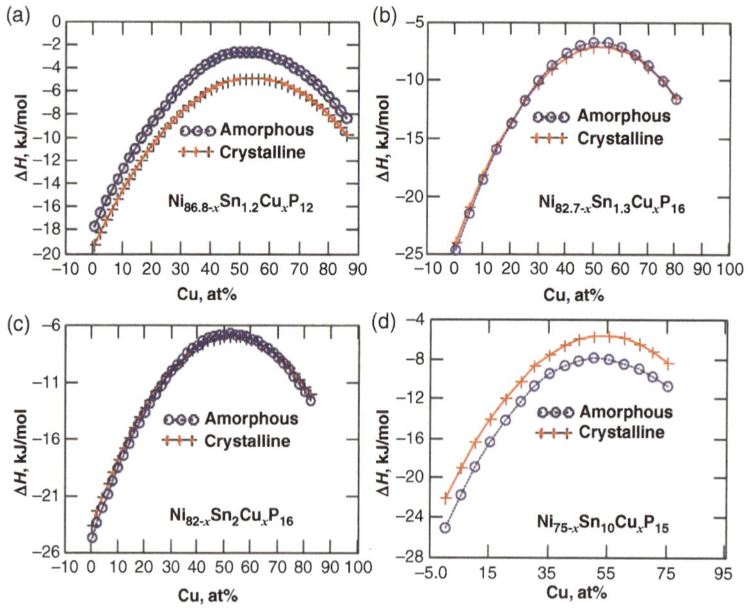

FIGURE 11.11 **Calculated plots of amorphous formation of Ni–Sn–Cu–P alloys: (a) $Ni_{86.8-x}Sn_{1.2}Cu_xP_{12}$, (b) $Ni_{82.7-x}Sn_{1.3}Cu_xP_{16}$, (c) $Ni_{82-x}Sn_2Cu_xP_{16}$, and (d) $Ni_{75-x}Sn_{10}Cu_xP_{15}$ [215].**

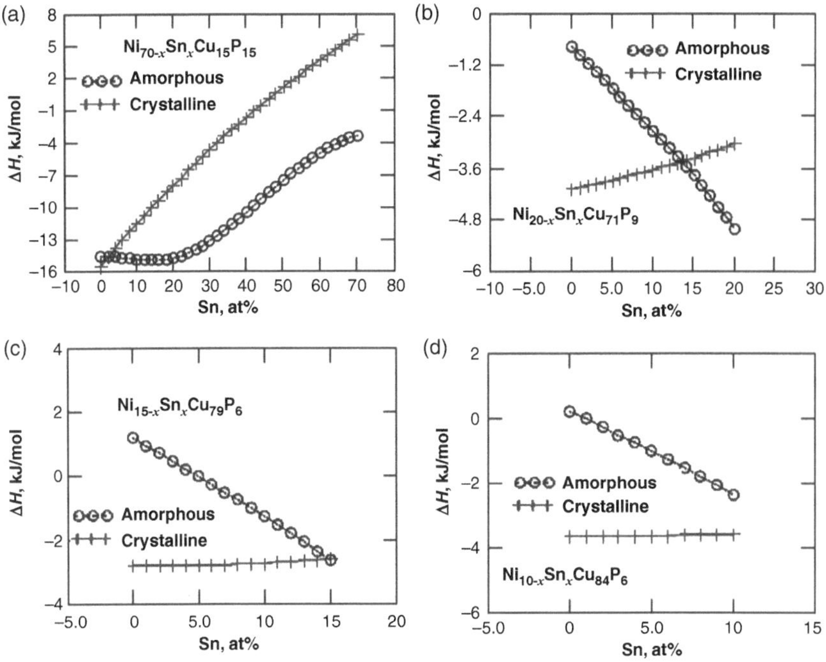

FIGURE 11.12 Calculated plots of amorphous formation of Ni–Sn–Cu–P alloys: (a) $Ni_{65-x}Sn_xCu_{15}P_{20}$, (b) $Ni_{20-x}Sn_xCu_{71}P_9$, (c) $Ni_{15-x}Sn_xCu_{79}P_6$, and (d) $Ni_{10-x}Sn_xCu_{84}P_6$ [215].

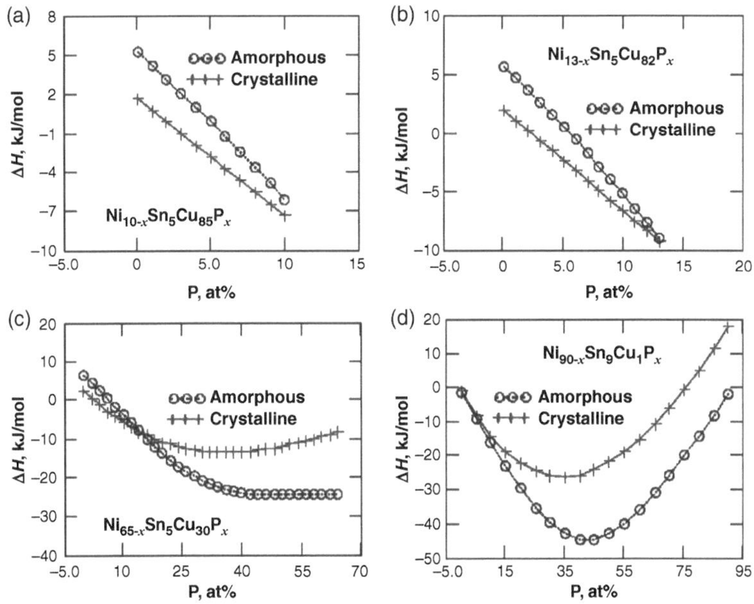

FIGURE 11.13 Calculated plots of amorphous formation of Ni–Sn–Cu–P alloys: (a) $Ni_{10-x}Sn_5Cu_{85}P_x$, (b) $Ni_{13-x}Sn_5Cu_{82}P_x$, (c) $Ni_{65-x}Sn_5Cu_{30}P_x$, and (d) $Ni_{90-x}Sn_9Cu_1P_x$ [215].

TABLE 11.9 Range of Amorphous Formation of EP Ni–Sn–Cu–P Alloy Deposits [215]

	Alloy systems		FRA, at%	Range of non-amorphous formation, at%	Reference
Theo.	Cu_x	$Ni_{86.8-x}Sn_{1.2}Cu_xP_{12}$	no FRA	$Ni_{86.8\text{–}0}Sn_{1.2}Cu_{0\text{–}86.8}P_{12}$ All are crystalline alloys	Present
		$Ni_{82.7-x}Sn_{1.3}Cu_xP_{16}$	$Ni_{82.7\text{–}60.7}Sn_{1.3}Cu_{0\text{–}22}P_{16}$	$Ni_{60.7\text{–}0}Sn_{1.3}Cu_{22\text{–}82.7}P_{16}$	
		$Ni_{82-x}Sn_2Cu_xP_{16}$	no FRA A critical case $\Delta H^a \cong \Delta H^c$ for every alloy	no non-FRA	
		$Ni_{75-x}Sn_{10}Cu_xP_{15}$	$Ni_{75\text{–}0}Sn_{10}Cu_{0\text{–}75}P_{15}$ All are amorphous alloys	no non-FRA	
	Sn_x	$Ni_{70-x}Sn_xCu_{15}P_{15}$	$Ni_{67\text{–}0}Sn_{3\text{–}70}Cu_{15}P_{15}$	$Ni_{70\text{–}67}Sn_{0\text{–}3}Cu_{15}P_{15}$	
		$Ni_{20-x}Sn_xCu_{71}P_9$	$Ni_{6\text{–}0}Sn_{14\text{–}20}Cu_{71}P_9$	$Ni_{20\text{–}6}Sn_{0\text{–}14}Cu_{71}P_9$	
		$Ni_{15-x}Sn_xCu_{79}P_6$	Only 15 at% Sn is amorphous alloy	$Ni_{15\text{–}0}Sn_{0\text{–}15}Cu_{79}P_6$ All are crystalline alloys	
		$Ni_{10-x}Sn_xCu_{84}P_6$	no AFR	$Ni_{10\text{–}0}Sn_{0\text{–}10}Cu_{84}P_6$ All are crystalline alloys	
	P_x	$Ni_{10-x}Sn_5Cu_{85}P_x$	no AFR	$Ni_{10\text{–}0}Sn_5Cu_{85}P_{0\text{–}10}$ All are crystalline alloys	
		$Ni_{13-x}Sn_5Cu_{82}P_x$	no AFR A critical case $\Delta H^a \cong \Delta H^c$ at P=13 at%	$Ni_{13\text{–}0}Sn_5Cu_{82}P_{0\text{–}13}$ All are crystalline alloys	
		$Ni_{65-x}Sn_5Cu_{30}P_x$	$Ni_{49\text{–}0}Sn_5Cu_{30}P_{16\text{–}65}$	$Ni_{65\text{–}49}Sn_5Cu_{30}P_{0\text{–}16}$	
		$Ni_{90-x}Sn_9Cu_1P_x$	$Ni_{90\text{–}0}Sn_9Cu_1P_{0\text{–}90}$ All are amorphous alloys, but $\Delta H^a \cong \Delta H^c$ at P=0 at%	no non-FRA	

TABLE 11.9 Range of Amorphous Formation of EP Ni–Sn–Cu–P Alloy Deposits [215] (*cont.*)

	Alloy systems	FRA, at%	Range of non-amorphous formation, at%	Reference
Exp.	Ni–Sn–Cu–P	14.50–21.86 at% P 1.09–11.40 at% Sn 0–15.10 at% Cu i.e. $Ni_{84.41-51.64}Sn_{1.09-11.40}Cu_{0-15.10}P_{14.50-21.86}$	6.43–9.63 at% P 0–0.32 at% Sn 2.84–2.92 at% Cu i.e. $Ni_{90.73-87.13}Sn_{0-0.32}Cu_{2.84-2.92}P_{6.43-9.63}$	31
		11.11~20.86 at% P 1.22~2.14 at% Sn 2.70~17.47 at% Cu	14.19 at% P 1.17 at% Sn 20.82 at% Cu	
		i.e. $Ni_{84.97\sim59.53}Sn_{1.22\sim2.14}Cu_{2.70\sim17.47}P_{11.11\sim20.86}$	i.e. $Ni_{63.82}Sn_{1.17}Cu_{20.82}P_{14.19}$	
		11.01–22.47 at% P 0.49–3.66 at% Sn 1.85–5.09 at% Cu i.e. $Ni_{86.65-68.78}Sn_{0.49-3.66}Cu_{1.85-5.09}P_{11.01-22.47}$		
		1.66_17.79 at% P 1.28–3.75 at% Sn 2.33–12.3 at% Cu i.e. $Ni_{94.73-66.16}Sn_{1.28-3.75}Cu_{2.33-12.3}P_{1.66-17.79}$		

The features of the calculated FRA plots for three series alloys are different. For the alloy series that change the content of Cu (Fig. 11.11), there is a watershed at the alloy system of $Ni_{82-x}Sn_2Cu_xP_{16}$. At this watershed, ΔH^a is just around equal to the ΔH^c for every alloy in the whole range of composition. There is no any FRA or non-FRA formed. A critical case appears. If Ni > 82 at%, and Sn < 2 at%, the ΔH^a curve will be higher than the ΔH^c curve gradually, and finally the ΔH^a curve is wholly above the ΔH_c curve, as shown in Fig. 11.11a ($Ni_{86.8-x}Sn_{1.2}Cu_xP_{12}$ alloy system). No amorphous structure but all of crystalline phases are formed in the whole range of alloy system. Otherwise, the ΔH^a curve will be lower than the ΔH^c curve gradually, and finally the ΔH^a curve is wholly below the ΔH^c curve, as shown in Fig. 11.11d ($Ni_{75-x}Sn_{10}Cu_xP_{15}$ alloy system). Now, all the alloys in the system are amorphous phases. This feature for changing the content of Cu in the Ni–Sn–Cu–P alloy system is similar to that in the Ni–Cu–P alloy system when changing the content of Cu (Fig. 11.10a–d).

Two other features of FRA plots are changing the contents of Sn and P in the Ni–Sn–Cu–P quaternary alloys. These elements are something like the plots for changing content of P or Sn in the Ni–Sn–P and Ni–Cu–P alloy systems. Figure 11.12 shows the features in the case of changing the content of Sn (Fig. 11.12) in the Ni–Cu–Sn–P alloy system. For the system of $Ni_{15-x}Sn_xCu_{79}P_6$ (Fig. 11.12c), ΔH^a (–2.646 kJ/mol) ≅ ΔH^c (–2.577 kJ/mol) at Sn = 15 at%, but $\Delta H^a > \Delta H^c$ when Sn = 0–15 at%, which means no FRA is formed, but all alloys are crystalline structure in this alloy system. So, a critical case appears in the $Ni_{15-x}Sn_xCu_{79}P_6$ alloy system. When Ni > 15 at%, and P > 6 at% (Fig. 11.12a and b as samples), the crossover phenomenon between the ΔH^a curve and ΔH^c curve occurs, and an FRA, and a non-FRA are formed. Otherwise (Fig. 11.12d as a sample), no amorphous alloys are formed in whole range of composition. Figure 11.13 shows the feature of FRA plots for changing the content of P in the Ni–Sn–Cu–P quaternary alloys. Now the critical case appears in the $Ni_{13-x}Sn_5Cu_{82}P_x$ alloy system (Fig. 11.13b), because at P = 13 at%, ΔH^a (−8.9996 kJ/mol) ≅ ΔH^c (−9.163 kJ/mol), but $\Delta H^a > \Delta H^c$ when P = 0–13 at%, which means no FRA is formed, but all alloys are crystalline structure in this alloy system. When Ni<13 at% and Cu>82 at%, the ΔH^a curve is wholly above the ΔH^c curve, no amorphous alloys are formed in whole range of composition (Fig. 11.13a as a sample). Otherwise, the crossover phenomenon between the ΔH^a curve and ΔH^c curve occurs, and an FRA and a non-FRA are formed (Fig. 11.13c and d as samples).

From Figs 11.11–11.13 and Table 11.9, it can be seen that the calculated results match the experimental data of the RAF in the quaternary Ni–Sn–Cu–P alloy systems. For example, the calculated amorphous formation alloys of $Ni_{67-0}Sn_{3-70}Cu_{15}P_{15}$ and $Ni_{82.7-60.7}Sn_{1.3}Cu_{0-22}P_{16}$ are located in the AFR of $Ni_{84.41-51.54}Sn_{1.09-11.40}Cu_{0-15.10}P_{14.50-21.86}$ and $Ni_{84.97-59.53}Sn_{1.22-2.14}Cu_{2.70-17.47}P_{11.11-20.86}$ alloys measured by the author's group, respectively [31]. The calculated nonamorphous formation alloys of $Ni_{90-8}Sn_2Cu_{0-82}P_8$ and $Ni_{70-67}Sn_{0-3}Cu_{15}P_{15}$ are located in the non-AFR of $Ni_{90.73-87.13}Sn_{0-0.32}Cu_{2.84-2.92}P_{6.43-9.63}$

and $Ni_{63.82}Sn_{1.17}Cu_{20.82}P_{14.19}$ alloys measured by the author's group, respectively [31]. In other words, these semiempirical calculations for the RAF of Ni–Sn–Cu–P alloy systems can basically predict the amorphous formation of these alloys. It is a shame that there were only one set of experimental results for the RAF of Ni–Sn–Cu–P alloy deposits available, otherwise we can make more comparisons to see if our calculations can describe more measurement data.

The calculated results also show that the calculated FRAs are greater than those experimentally determined for the quaternary Ni–Sn–Cu–P system. The reasons can be thought as similar as in the ternary alloy systems. Moreover, besides the ternary atomic interactions between the constituents, the quaternary atomic interactions between the constituents in the quaternary alloy systems are not considered in the present model. Such neglects of course result in some error as well.

11.4 SUMMARY

Early in 1987 [217], Liou and Chien pointed out that "the formation and the stability of the amorphous state are questions of fundamental and technological importance. … In particular, what is the concentration range or ranges in which the amorphous state exists at a finite temperature (e.g., 400K) ?" They referred to the conventional amorphous alloys, but it should be suitable for the EP amorphous alloys. From Section 11.2 and Section 11.3, such a statement is true.

In this section, not only the experimental data on formation range of electroless amorphous alloys were summarized, but also the theoretical calculations on the EP RAF were described more in detail.

Nearly all published data on the RAF of EP amorphous alloys currently available have been summarized. Five features for the measured data were analyzed and discussed. Because the existing experimental data were not completed and no one has yet given interpretation to the existing experimental data, broad and deep studies in this field are welcomed.

The RAF on EP Ni–P, Ni–Sn–P, Ni–Cu–P, and Ni–Sn–Cu–P alloy systems have been calculated theoretically. The features of the calculated RAF plots of EP amorphous alloys are described in detail. The calculated results can predict the experimental data. However, the calculated FRAs either in ternary Ni–Sn–P and Ni–Cu–P alloy systems or in quaternary Ni–Sn–Cu–P alloy system are greater than those experimentally determined in the responding alloy systems. The reasons of such phenomena have been analyzed.

The calculated results show that the proposed theoretical approaches for describing the RAF of EP amorphous alloys are effective.

REFERENCES

[1] Agarwala RC, Agarwala V. Sadhana 2003;28:475.
[2] Zhang B. Acta Metall Sin 1981;17:285.
[3] Zhang B. Physica 1983;121B:405.

[4] Zhang B. Physica 1985;132B:319.
[5] Zhang B. Phys Stat Sol (a) 1987;102:199.
[6] Zhang B. J Hunan Uni (Natual Sci) 1986;13(4):262.
[7] Zhang B, Tan Z. J Mater Sci Lett 1988;7:681.
[8] Rao Q, Bi G, Lu Q, Wang H, Fan X. Appl Surf Sci 2005;240:28.
[9] Duwez P, Willens RH, Kelement W. J Appl Phys 1960;31:1136.
[10] Klement W, Willens RH, Duwez P. Nature 1960;187:869.
[11] Davis HA, Lewis BG. Scr Met 1975;9:1107.
[12] Davis HA, Lewis BG. RQM II, Section I. In: Grant NJ and Giessen BC, editors. Cambridge, MA: MIT Press; 1976. p. 259–266.
[13] Marcus M, Turnbull D. Mater Sci Eng 1976;23:211.
[14] Donald IW, Davis HA. J Non-Cryst Solids 1978;30:77.
[15] Ning Y, Zhou X. Acta Metall Sin 1981;17(3):273.
[16] Lu ZP, Liu CT. Phys Rev Lett 2003;91(11):115505–11.
[17] Zeilinski PG, Ostatek J, Kijek M, Matyja H. RQM III. Brighton, UK: University of Sussex; 1978. p. 337–46.
[18] Shi T, Zheng L, Chen N. Acta Metall Sin 1979;15(1):94.
[19] Giesen BC, Whang SH. J Phys (Paris) 1980;41:c8–c95.
[20] Giesen BC, Whang SH. Mater Res Soc Symp Proc 1983;19:289.
[21] Nagel SR, Tauc J. Phys Rev Lett 1975;35:380.
[22] Mizutani U, Hartwig KT, Massalski TB, Hopper RN. Phys Rev Lett 1978;41:661.
[23] Egami T. J Non-cryst Solids 1996;205–207:575.
[24] Egami T. Mater Sci Eng A 1997;226–228:261.
[25] Inoue A, Zhang T, Masumoto T. Non-cryst Solids 1993;156–158(2):473.
[26] Takeuchi A, Inoue A. Intermetallics 2010;18:1779.
[27] Jonson WL. MRS Bull 1999;24(10):42.
[28] Xu D, Duan G, Johnson WL. Phys Rev Lett 2004;92(24). 245504-1.
[29] Kim D, Lee B-J, Kim NJ. Scr Mater 2005;52:969.
[30] Xie H, Zhang B. J Mater Proc Technol 2002;124:8.
[31] Zhang B, Xie H, Xu X. Metal Finish 1999;97(10):35.
[32] Zhang H, Zhang B, Wu L, Wan Z. Acta Phys Sin 1994;43(10):1638.
[33] Zhang H, Zhang B. Physica B 1995;205:263.
[34] Yuan X, Li S, Zhang B, Wang L, Li Y, Guo R. Funct Mater (Sup) 2001;10:1638.
[35] Miedema AR. Philips Tech Rev 1973;33:149.
[36] Miedema AR. J Less Common Met 1973;32:117.
[37] Miedema AR, de Boer FR, de Chatel PF. J Phys F: Met Phys 1973;3:1558.
[38] de Boer FR, Boom R, Mattens WCM, Miedema AR, Niessen AR. Cohesion in metals, transition metal alloys. Amsterdam: North-Holland; 1988.
[39] Goncalves AP, Almeida M. Physica B 1996;228:289.
[40] Zhang B, Jesser WA. Physica B 2002;315:123.
[41] Takeuchi A, Inoue A. Mater Trans JIM 2000;41:1372.
[42] Ouyang YF, Zhong XP, Du Y, Feng YP, He YH. J Alloys Compd 2006;420:175.
[43] Ray PK, Akinc M, Kramer MJ. J Alloys Compd 2010;489:357.
[44] Smithells CJ. Metals reference book. London: Butterworths; 1976.
[45] Nawafune H, Uegaki T, Mizumoto S, Ishikawa M, Nakamura T. Jpn J Surf Finish 1998;9(7):797.
[46] Kong Y, Shao J, Wang W, Liu Q, Chen Z. J Alloys Comp 2009;477:328.
[47] Zhang H, Zhang B. Physica 1993;B 192/193:247.

[48] Faraji S, Rahim AA, Mohamed N, Sipaut CS, Raja B. Mater Chem Phys 2011;129:1063.

[49] Yu H, Luo S, Wang Y. Surf Coat Technol 2001;148:143.

[50] Shu X, Zhang B, Liao S, Xie H. To be published.

[51] Cargill GS III. Appl Phys 1970;41(1):12.

[52] Srinivasan KN, John S. J Alloys Comp 2009;486:447.

[53] Palaniappa M, Seshadri SK. Mater Sci Eng A 2007;460–461:638.

[54] Watanabe T, Tanabe Y. J Japan Inst Metals 1975;39(8):831.

[55] Krishnaveni K, Sankara Narayanan TSN, Seshadri SK. Surf Coat Technol 2005;190:115.

[56] Wang ZC, Jia F, Yu L, Qi ZB, Tang Y, Song G-L. Surf Coat Technol 2012;206:3676.

[57] Fisher RD, Koopman DE. J Electrochem Soc 1964;111(2):263.

[58] Chang YH, Lin CC, Hung MP, Chin TS. J Electrochem Soc 1986;133(5):985.

[59] Saito T, Sato E, Matsuoka M, Iwakura C. J Appl Electrochem 1998;28(5):559.

[60] Jiang D, Zhang C, Liao S, Zhang B. Electroplat Finish 2007;26(2):5.

[61] Hu W, Zhang B. Physica B 1991;175:396.

[62] Fujita N, Tanaka A, Makino E, Patrick T, Squire PT, Lim PB, Inoue M, Fujii T. Appl Surf Sci 1997;113/114:61.

[63] Qadri SB, Dinderman MA, Dressick WJ, Schoen PE, Lubitz P, He JH, Tonucci RJ, Cross J. Appl Phys A 2007;89:493.

[64] Chang YS, Heieh IJ. P&SF 1990;77(1):52.

[65] Chen H, Zhang J, Yan Y, Zhong X. Guang Xi Sci 2009;16(2):161.

[66] Wang X, Zhang Y, Hu X, Wu J, Lu H. Electroplat Finish 2011;30(5):17.

[67] Yasumasa A, Yoshio N, Masao S, Taizo U. J Metal Finish Soc Jpn 1974;25(12):666.

[68] Tanaka DAP, Tanco MAL, Niwa S, Wakui Y, Mizukami F, Namba T, Suzuki TM. J Memb Sci 2005;247:21.

[69] Ayturk ME, Ma Y. J Membr Sci 2009;330:233.

[70] Gade SK, E. Payzant A, Park HJ, Thoen PM, Way JD. J Memb Sci 2009;340:227.

[71] Keuler JN, Lorenzen L, Sanderson RD, Prozesky V, Przybylowicz WJ. Nucl Instrum Meth Phys Res, B 1999;158:678.

[72] Roa F, Way JD, McCormick RL, Paglieri SN. Chem Eng J 2003;93:11.

[73] Sabina KG, Keeling MK, Davidson AP, Hatlevik O, Way JD. Inter J Hydrogen Energ 2009;34:6484.

[74] Byeon JH, Kim Y-W. ACS Appl Mater Interf 2011;3:2912.

[75] Nawafune H, Kamiura Y, Mizumoto S, Haga M, Uchida E. J Jpn Inst Interconn Pack Electr Circ 1996;11(4):262.

[76] Wang C, Peng B, Xie H-N, Zhang H-X, Shi F-F, Cai W-B. J Phys Chem C 2009;113:13841.

[77] Kozlov AS, Palanisamy T, Narasimhan D. US Patent 6,706,420, Electroless platinum-rhodium alloy plating, March16, 2004.

[78] Mustain WE, Kim H, Narayanan V, Osborn T, Kohl PA. J Fuel Cell Sci Technol 2010;7:041013.

[79] Balci S, Hahn K, Kopold P, Kadri A, Wege C, Kern K, Bittner AM. Nanotechnology 2012;23:045603.

[80] Tharamani CN, Begum NS, Mayanna SM. Mater Chem Phys 2004;83:278.

[81] Domínguez-Ríos C, Moreno MV, Torres-Sánchez R, Antúnez W, Aguilar-Elguézabal A, González-Hernández J. Surf Coat Technol, 2008;202:4848.

[82] Duncan GT, Banter JC. P&SF 1989;76(9):54.

[83] Peng C, Chen A, Wang W. Welding (China) 2006;(4):2.

[84] Bogush V, Inberg A, Croitoru N, Dubin V, Shacham-Diamand Y. Microelectr Eng 2003;70:489.

[85] Duhin A, Inberg A, Eliaz N, Gileadi E. Electrochim Acta 2011;56:9637.
[86] Cheng J, Hu B, Shen J. Electroplat Pollu Cont 2001;21(3):18.
[87] Senda A, Takano Y, Kanbayashi Y, Morita K. Jpn J Surf Finish 1994;45(9):923.
[88] Senda A, Takano Y, Nakagawa T. Jpn J Surf Finish 1992;43(7):694.
[89] Zhang B, Liao S, Shu X, Xie H. J Mechatro. Accepted.
[90] Wang L, Zhao L, Huang G, Yuan X, Zhang B, Zhang J. Surf Coat Technol 2000;126:272.
[91] Manna M, Bandyopadhyay N. ISIJ Intern 2010;50(4):556.
[92] Aoki K, Kim D-H, Takano O. Jpn J Surf Finish 1994;45(12):1287.
[93] Nie M, Zou YC, Huang YM, Wang JQ. Intern J Hydro Ener 2012;37:1568.
[94] Takano O, Ishibashi S. J Metal Finish Soc Jpn 1969;20(7):348.
[95] Sankara Narayanan TSN, Selvakumar S, Stephen A. Surf Coat Technol 2003;172:298.
[96] Zhang L, Xuan T, Huang Q. Chin Rare Earths 2003;24(3):20.
[97] Wang S. Thin Solid Films 2007;515:8419.
[98] Mital CK, Shrìvastava PB. Metal Finish 1986;84(10):67.
[99] Zhang H, Zhang B, Tan Z, Lei Q. Mater Mech Eng 1995;19(1):19.
[100] Stremsdoerfer G, Omidvar H, Roux P, Meas Y, Ortega-Borges R. J Alloys Comp 2008;466:391.
[101] Zhang X, Wu J. Electroplt Pollu Cont 1990;16(5):14.
[102] Wu M, Lou B. Adv Mater Res 2011;284–286:2187.
[103] Abdel Hamid Z, Abdel Mouez F, Morsy FA, Khalifa NA. Inter J Eng Pract Res 2013; 2(3):112.
[104] Koiwa I, Nishikawa M, Yamda K, Osaka T. Bull Chem Soc Jpn 1986;59:133.
[105] Liu D, Yang Z, Zhang C. Mater Sci Eng B 2010;166:67.
[106] Osaka T, Arai K, Masubuchi N. J Metal Finish Soc Jpn 1988;39(12):778.
[107] Wang W, Yang Y, Bao J, Luo H. Catal Commun 2009;11:100.
[108] Zhang B, Hu W, Zhang Q, Qu X. Mater Charac 1996;37:119.
[109] Balaraju JN, Kalavati, Manikandanath NT, William Grips VK. Surf Coat Technol 2012;206:2682.
[110] Osaka T, Takano N, Kurokawa T, Kaneko T, Ueno K. Surf Coat Technol 2003;169–170:124.
[111] Drovosekov AB, Ivanov MV, Krutskikh VM, Lubnin EN, Polukarov YM. Prot Metals 2005;41(1):55. Translated from Zashchita Metallov 2005;41(1):61.
[112] Mencer D. J Alloys Comp 2000;306:158.
[113] Valova E, Armianov S, Franquet A, Hubin A, Steenhaut O, Delplancke JL, Vereecken J. J Appl Electrochem 2001;31:1367.
[114] Kim M, Yokoshima T, Osaka T. J Electrochem Soc 2001;148(11):C753.
[115] Liu BS, Zhang WD, Dai WL, Deng JF. J Memb Sci 2004;244:243.
[116] Mizumoto S, Nawafune H, Okada T, Haga M. Jpn J Surf Finish 1989;40(5):680.
[117] Vrublevskaya ON, Vorobyova TN. Thin Solid Films 2012;524:197.
[118] Cissé M, Abouchane M, Anik T, Himm K, Belakhmima RA, Touhami ME, Touir R, Amiar A. Intern J Corr 2010:Article ID 246908, 9 pages.
[119] Hu W, Wu L, Wang L, Zhang B, Guan H. Physica B 1995;212:195.
[120] Xie H, Zhang B, Yang Q. Trans IMF 1999;77(3):99.
[121] Zou Y, Cheng Y, Cheng L, Liu W. Mater Trans 2010;51(2):277.
[122] Krasteva N, Armyanov S, Georgieva J, Avramova N, Fotty V. J Electr Mater 1995;24(8):941.
[123] Kwang RI, Korean J. Sanitation 2003;18(3):35.
[124] Yuan XY, Wu GS, Xie T, Geng BY, Lin Y, Meng GW, Zhang LD. Solid State Sci 2004; 6:735.
[125] Wang S, Hong L. J Alloys Comp 2007;429:99.

[126] Shacham-Diamand Y, Zylberman A, Petrov N, Sverdlov Y. Proceedings of 2001 6th international conference on solid-state and integrated circuit technology, IEEE, Shanghai, China, 2001. p. 410–5.
[127] Dai H, Gao L, Liang Y, Kang X, Wang P. J Power Sour 2010;195:307.
[128] Wu WC, Hsieh T-E, Pan H-C. J Electrochem Soc 2008;155(5):D369.
[129] Dulal SMSI, Kim TH, Rhee H, Sung JY, Kim C-K. J Alloys Comp 2009;467:370.
[130] Einati H, Bogush V, Sverdlov Y, Rosenberg Y, Shacham-Diamand Y. Microelectr Eng 2005;82:623.
[131] Dai HB, Liang Y, Wang P, Yao XD, Rufford T, Lu M, Cheng HM. Int J Hydro Ener 2008;33:4405.
[132] Yuan X, Du C, Sun G, Pan N. Appl Surf Sci 2007;253:4546.
[133] Patel N, Fernandes R, Bazzanella N, Miotello A. Thin Solid Films 2010;518:4779.
[134] Wang L, Zhang B, Yi G, OuYang Y, Hu W. J Alloys Comp 1997;255:231.
[135] Yi G, Zhang B, Wang L, Ouyang Y, Xie H. J Electrochem Soc 2013;160(9):D403.
[136] Wang L, Zhao L, Zhang B, Hu W, Shu X, Sheng X, Fang Z. J Alloys Comp 1999;287:234.
[137] Delaunois F, Lienard P. Surf Coat Technol 2002;160:239.
[138] Sankara Narayanan TSN, Seshadri SK. J Alloys Comp 2004;365:197.
[139] Aoki K, Takano O, Kim D-H. Jpn J Surf Finish 1995;46(1):79.
[140] Balaraju JN, Chembath M. Appl Surf Sci 2012;258:9692.
[141] Valova E, Georgiev I, Armyanov S, Delplancke J-L, Tachev D, Tsacheva Ts, Dille J. J Electrochem Soc 2001;148(4):C266.
[142] Ranganatha S, Venkatesha TV, Vathsala K. Appl Surf Sci 2010;256:7377.
[143] Koji A, Osamu T, Yuki S. Jpn J Surf Finish 1993;44(10):821.
[144] Mallory GO, Horhn TR. P&SF 1979;66(4):40.
[145] Valova E, Armyanov S, Dille J, Van Ingelgem Y, Hubin A, Steenhaut O. J Electrochem Soc 2008;155(6):D449.
[146] Liao W-K. Master Thesis, Study on cobalt alloy thin films as cu/si diffusion barrier, Dep Appl Chem, Chaoyang University of Technology, Taiwan, China, June, 2007.
[147] Zhang Y, Wang X, Li X. Mater Prot 2003;36(7):34.
[148] Yi G, Wang L, Ouyang Y, Mao L, Guo Y, Zhang B, Xie H. Metal Finish 1996;94(11):31.
[149] Simpson AW, Clements WG. IEEE Trans Mag 1975;11(5):1338.
[150] Wang S, Zhan J. The Chin J Nonferrous Metals 2008;18(6):1105.
[151] Guo H, Liu H, Li W, Xie G. Electroplat Finish 2010;29(7):20.
[152] Huang Y, Zhao L, Shi K, Wang L, Zhao W. Rare Metal Mater Eng 2006;35(11):1725.
[153] Yu G, Li Y, Zhang Z, Chen J. Ordnance Mater Sci Eng 1994;17(4):21.
[154] Yang X. Mater Protec 1996;19(5):18.
[155] Zhang L, Jin Y, Peng B, Zhang Y, Wang X, Yang Q, Yu J. Appl Surf Sci 2008;255:1686.
[156] Kim M, Yokoshima T, Kubomiya T, Osaka Y. Jpn J Surf Finish 2000;51(6):617.
[157] Wang S, Zhang Q. Rare Metal Mater Eng 2007;36(12):2224.
[158] Edward C. US Patent 5,019,163, Corrosion/wear-resistant metal alloy coating compositions, May 28, 1991.
[159] Ofek Almog R, Sverdlov Y, Goldfarb I, Shacham-Diamand Y. Microelectr Eng 2007; 84:2450.
[160] Yang J, Cheng F, Liang J, Chen J. Inter J Hydro Ener 2011;36:1411.
[161] Wang L, Zhao L, Zhang B, Liao S, OuYang Y, Hu W. Z Metallkd 1997;88:945.
[162] Zou Y, Fei X, Xiao Z, Li W, Long G. Mater Prot 2004;37(9):17.
[163] Balaraju JN, Ezhil Selvi V, William Grips VK, Rajam KS. Electrochim Acta 2006;52:1064.
[164] Balaraju JN, William Grips VK, Jain A, Rajam KS. J Alloys Comp 2007;436:319.

[165] Azizi A, Mohammadi M, Sadrnezhaad SK. International conference on materials heat treatment (ICMH 2010), Islamic Azad University, Majlesi Branch, Isfahan, Iran, May 11–14, 2010.

[166] Wei G, Jin Y, Yu Y, Ge H, Wang X, Jiang L, Sun L. Thin Solid Films 2010;519:1259.

[167] Liu Z, Chen H, Yang C, He Y. J Phys Condens Matter 1991;3:9007.

[168] Xuan T, Jia W. Plat Finish 2009;31(7):4.

[169] Zhang L, Xuan T, Huang Q. Chin Rare Earths 2003;24(3):20.

[170] Xuan T, Zhang L, Huang Q. Trans Nonferrous Met Soc China 2006;16:363.

[171] Xuan T, Zhang L, Huang Q. Trans Mater Heat Treat 2003;24(4):66.

[172] Xuan T, Zhang L, Huang Q. Nonferr Metals 2004;56(2):12.

[173] Zhang B. Shanghai Met 1993;15(5):23.

[174] Schmeckenbecher AF. J Electrochem Soc 1966;113(8):778.

[175] Shimauchi H, Ozawa S, Tamura K, Osaka T. J Electrochem Soc 1994;141(6):1471.

[176] Osaka T, Kasai N. J Metal Finish Soc Jpn 1981;32(6):309.

[177] Cheng H, Xu S, Zhao W, Zhao S, Zhai P, Sun Z, Zhang Q. J Mater Sci Technol 2008; 24(5):712.

[178] Mimani T, Mayanna SM. Surf Coat Technol 1996;79:246.

[179] Gao Y, Zheng ZJ, Zhu M, Luo CP. Mater Sci Eng A 2004;381:98.

[180] Antonelli SB, Allen TL, Johnson DC, Dubin VM. J Electrochem Soc 2006;153(6):J46.

[181] Sankara Narayanana TSN, Baskaranb I, Krishnavenia K, Parthiban S. Surf Coat Technol 2006;200:3438.

[182] Fundo AM, Abrantes LM. J Electroanal Chem 2007;600:63.

[183] Cheng YH, Zou Y, Cheng L, Liu W. Surf Coat Technol 2009;203(12):1559.

[184] Zhang B, Xie H. Mater Sci Eng 2000;A281:286.

[185] Wu Y, Zhao Y, Yao M, Hu X. Mater Prot 1995;28(9):6.

[186] Hsu J, Lin K. J Electrochem Soc 2003;150(9):C653.

[187] Gorbunova KM, Ivanov MV, Moiseev VP. J Electrochem Soc 1973;120(5):613.

[188] Watanabe T, Tanabe Y. Mater Sci Eng 1976;23:97.

[189] Watanabe T, Tanabe Y. Trans Jap Inst Metals 1983;24(6):396.

[190] Chekanova LA, Denisova EA. IEEE Trans Mag 1997;33(5):3730.

[191] Zeng P, Zeng GZ, Wang D, Huang N, Hu S, Wnag G. Mater Protec 1999;32(12):4.

[192] Xuan T, Zhang L, Huang Q. J Rare Earths 2002;20(5):512.

[193] Zhang B, Hu W, Qu X, Zhang Q, Zhang H, Tan Z. Trans IMF 1996;74(2):69.

[194] Stepanova LI, Bodrykh TI, Sviridov VV. Metal Finish 2001;99(1):50.

[195] AlZahrani A, Alhamed Y, Petrov L, Armyanov S, Valova E, Georgieva J, Dille J. J Solid State Electrochem 2014;18.

[196] Koiwa I, Usuda M, Yamada K, Osaka T. J Electrochem Soc 1988;135(20):718.

[197] Lee S, Liang H. P&SF 1991;78(9):82.

[198] Lu Z, Gao Y, Zheng Z. China Surf Eng 2005;(2):44.

[199] Kumar A, Singh A, Kumar M, Kumar D, Barthwal S. J Mater Sci Mater Electron 2011;22:1495.

[200] Wang S. Surf Coat Technol 2004;186:372.

[201] Ma J, Wu Y, Hang Y, Wang W, Qiao Y, Zheng Y, Liu Y, Huang X, Li G, Zhang L. Metall Funct Mater 2004;11(1):4.

[202] Yang W, Fu Y, Xia A, Zhang K, Wu Z. J Alloys Comp 2012;518:6.

[203] Armyanov S, Valova E, Franquet A, Dille J, Delplancke J-L, Hubin A, Steenhaut O, Kovacheva D, Tatchev D, Vassilev Ts. J Electrochem Soc 2005;152(9):C612.

[204] Sankara Narayanana TSN, Stephanb A, Guruskanthan S. Surf Coat Technol 2004;179:5696.

[205] Dadvand N, Jarjoura G, Kipouros GJ. J Mater Sci Mater Electron 2008;19:51.
[206] Jiang T, Wu H. Electrochem 2000;6(1):17–24.
[207] Loeff PI, Weeber AW, Miedema AR. J Less-common Metals 1988;140:299.
[208] Niessen AK, Miedema AR. Ber Bunsenges Phys Chem 1983;87:717.
[209] Zhang B, Liao S, Xie H, Yuan X, Shu X. EPL 2010;89:56002.
[210] Friedel J. Adv Phys 1954;3:446.
[211] Gschneidner KA Jr. Solid State Phys 1964;16:275.
[212] Murty BS, Ranganathan SR, Rao MM. Mater Sci Eng A 1992;149:231.
[213] Van der Kolk GJ, Miedema AR, Niessen AK. J Less-common Metals 1988;145:1.
[214] Weeber AW. J Phys F 1987;17:809.
[215] Liao S, Zhang B, Shu X, Xie H. Afri Rev Phys. Accepted.
[216] Takeuchi A, Inoue A. Mater Trans 2001;42(7):1435.
[217] Liou SH, Chien CL. Phys Rev B 1987;35(5):2443.

Chapter 12

Microscopic Theory of Electroless Plating

Chapter Outline

Amorphous and Nano Alloys Electroless Depositions. http://dx.doi.org/10.1016/B978-0-12-802685-4.00012-1

In Chapter 1, we pointed out that after 2000, electroless plating got into its fourth stage, in-depth development (deep investigation into the mechanism and theory of electroless plating) and the nanoelectroless plating stage. Nanoelectroless plating has been analyzed and described in Chapter 4 in detail, so our discussion here will concentrate on the in-depth development of the mechanism and theory in electroless plating.

There are three main features in the in-depth investigation of the EP mechanism and theory. Using methods such as the QCM method etc. these are: to in situ monitor the deposition to understand the mechanism; proposing mathematical models to study the mechanism; and starting to investigate the mechanism of electroless plating by using microscope theory such as the ab initio molecular orbital approach, density function theory, etc.

The former two features have been described in more detail in Chapter 10 and 11; therefore we only need to concentrate on the microscopic theory of electroless plating in this chapter.

12.1 WHY USE THE MICROSCOPIC *AB INITIO* THEORY TO INVESTIGATE THE EP PROCESS?

After studying EP for more than 60 years, we have a thorough in-depth understanding of the EP process from nearly all aspects. In Chapter 10, we described the mechanism and kinetics of the EP process from both experimental and nonmicroscopic theory results in detail. However, the mechanisms for the EP process have not been fully understood as yet. The main reasons are that the EP process is very complicated, the reaction process is composed of oxidation reaction of reducing agent, deposition reaction of metal ion, and catalytic reaction, and many factors such as complexing agent, additives, pH, temperature, and reaction field affect the process. In addition, these reactions and factors intertwine each other, and thereby it is difficult to elucidate the whole mechanism using the experimental measurements.

The formation theory and formation region of EP amorphous alloy deposits have been also described in detail in Chapter 11. Although it took a significant step, the theory has not built at atomic level, but belonged to the semiempirical theoretical approach.

Ab initio methods are based on the laws of quantum mechanics; they use no experimental parameters in the calculations, but only the basic physical constants, such as the velocity of light, masses, and charges on electrons and nuclei, and Planck's constant. Once these methods are applied to the EP reaction process, it has the capability to estimate the mechanisms quantitatively and systematically at atomic level. It is expected that this theoretical approach can achieve breakthrough for the electroless deposition process. In 1999, Homma's group used the *ab initio* molecular orbital (MO) method for the first time to investigate the electroless deposition reaction [1]. As is well known, an electroless deposition reaction takes place at the solid/liquid interface; thus the reaction is influenced by a number of factors, such as catalytic activities of the deposited metals and a solvation effect. Therefore, Homma's group using *ab initio* methods and relative models concentrates their study on the most important factors of the electroless deposition, that is, the oxidation of the reductants and the catalytic activities of the deposited metals. Then the group and other groups step by step published their calculation results on the EP process with *ab initio* methods more deeply.

It should be pointed out that although the microscope theory has a very profound physical meaning, some approximations must be introduced in the calculations because an accurate solution of the multivariable Schrödinger equation is very difficult [2,3]. Therefore, the calculated results are not definitely in good agreement with the experimental data. In contrast, good results can usually be obtained, in agreement with experimental data by using the empirical and semiempirical theories. In addition, it is difficult for *ab initio* methods to be used to carry out systematic research and calculation. For example, such as the systematic calculations for the formation and RAF of many EP amorphous alloy systems discussed in Chapter 11, *ab initio* methods are nearly almost impossible to carry out at least until now. Perhaps these are the shortcomings and weaknesses of *ab initio* methods.

12.2 *AB INITIO* COMPUTATIONAL METHODS

As mentioned above, exact solutions of the Schrödinger equation are computationally unrealistic; therefore, various mathematical approximations must be introduced in the *ab initio* methods.

12.2.1 Calculation Methods and Program Package

Until now, nearly all *ab initio* theoretical calculations applied to the EP process were carried out with a Gaussian software package [4–7]. The B3LYP-level density functional theory was usually used in the packages. B3LYP is the Becke-style three-parameter density functional theory, which is a combination of the Becke's three-parameter hybrid exchange functional (B3) [8] and the Lee–Yang–Parr correlation functional (LYP) [9]. And the electronic exchange and correlation energy were estimated with the generalized gradient approximation in the B3LYP functional. Theoretical researches indicate that the Gaussian is a powerful tool for investigating many chemical subjects such as reaction mechanisms and potential energy surfaces. The reasons are that the Gaussian package is capable of predicting many properties of molecules and reactions, for example, molecular energies and structures, energies and structures of transition states, chemical bond and reaction energies, reaction pathways, molecular orbitals, atomic charges, electron affinities, ionization potentials, etc.

12.2.2 Selection of the Basis Sets

In the calculations, selection of the basis sets is necessary in order to obtain reliable results. What is a basis set? A basis set is a mathematical representation of the molecular orbital within a molecule. The basis set can be interpreted as each electron of a particular region of space. Or we can say that basis sets assign a group of basis functions to each atom within a molecule to approximate its orbitals.

Right now, the 6–31 ++G** basis set was assigned to all atoms [10]. The name can be understood as follows. "3G" indicates three Gaussian primitives per basis function. A basis set can be made larger by increasing the number of basis functions per atom. Split valence basis sets, such as 3-21G and 6-31G, have two (or more) sizes of basis function for each valence orbital. Similarly, triple split valence basis sets, such as 6-311G, use three sizes of contracted functions for each orbital type.

How do we indicate polarized basis sets? This is done by adding orbitals with angular momentum beyond what is required for the ground state to description of each atom. For example, polarized basis sets add d functions to carbon atoms and f functions to transition metals, and some of them add p functions to hydrogen atoms. Therefore, the polarized basis set 6-31G (d), that is, 6-31G* indicates that it is the 6-31G basis set with d functions added to heavy atoms. The polarized basis set 6-31G (d, p), that is, 6-31G**, indicates adding p functions to hydrogen atoms in addition to the d functions on heavy atoms. These two basis sets are very common for calculations involving up to medium-sized systems.

The 6-31+G* basis set is the 6-31G* basis set with diffuse functions added to heavy atoms. The double plus version, 6-31++G*, adds diffuse functions to the hydrogen atoms as well.

In 2001, Homma's group [11] compared their calculated results of the MP2 (the second-order Møller–Plesset perturbation method) calculation for the electron affinity of OH^- using various basis sets, which is shown in Table 12.1. It is clear from the table that values reasonably close to the experimental one (176.56 kJ/mol) can be obtained by using the 6-31+G** and AUG-cc-pVDZ basis sets in which diffuse functions are argumented. But, the calculated values obtained by using the 6-31G** and cc-pVDZ basis sets are negative.

TABLE 12.1 Experimental and Calculated Values for the Electron Affinity of OH^- [11]

Basis sets	Electron affinity of OH^- (kJ/mol)
(a) 6-31G**	−15.77
(b) 6-31+G**	160.33
(c) cc-pVDZ	−45.14
(d) AUG-cc-pVDZ	183.55
Experimental value	176.56

Basis sets (b) and (d) consist of basically the same structure as those of (a) and (c), respectively, while diffuse s- and p-type functions are augmented to the (b) and (d) basis sets, respectively.

This calculation evidently confirmed that selection of the basis sets is significant in order to obtain reliable results. Especially, those for the anions such as OH^- require careful consideration.

How do we consider the effect of the core electrons for heavier elements such as Pd, Ni, and Cu metal atoms in the basis sets? Usually, the basis sets with an effective core potential (ECP) energy function developed by Hay and Wadt [12–14] were used.

12.2.3 Selection of Oxidation Pathway

In order to investigate the oxidation mechanism of a reducing agent for the electroless deposition process, the oxidation pathway, and adequate calculation models such as solvation model have to be selected.

Reaction pathways for the oxidation process of a reductant represent which intermediate products have been produced in the oxidation–reduction reaction. Early in 1981, van den Meerakker proposed a general reaction pathway for the various reductant species [15]. According to her mechanism, the reaction was divided into three steps, namely, adsorption, dehydrogenation, and oxidation. We have discussed it in more detail in Section 10.1.3, please refer to it there. Homma's group believed that this reaction proceeds via three-coordinate intermediate species, to distinguish another pathway via five-coordinate intermediates proposed by themselves. We will say more about this in Section 12.3.

Comparing the calculating energy diagrams for the possible pathways in the elementary reaction process of a reductant, the most likely pathway can be worked out.

12.2.4 Solvation Effect

Because the electroless deposition reaction takes place at the solid/liquid interface in the bath solution, the reaction has to take the solvation effect and catalytic activity of the deposited metal surface into account.

The solvation effect makes the reaction endothermic, indicating that the reaction preferably proceeds at the solid/liquid interface, that is, the surface of the deposits, rather than in the solution bulk.

How do we evaluate the effect of solvation? The solvation effect was usually examined using the self-consistent reaction field method with an isodensity surface polarized continuum model (SCRF-IPCM) [16]. IPCM expresses the effect of solvation using the dielectric constant, namely, IPCM at $\varepsilon = 1.0$ corresponds to the gas phase (nonpolar solvent) and the value for bulk pure water is $\varepsilon = 78.3$. The dielectric constants of the inner and outer Helmholtz layers are estimated by the Bockris–Devanathan–Muller (BDM) model [17]. So the solvation effect can be quantitatively examined in terms of the dielectric constant of the solution. In addition, its effect on the heat of reaction is evaluated.

12.2.5 Catalytic Activity of Metal Surfaces

The catalytic activity of a metal surface for the oxidation reaction of the reducing agent has been evaluated by the electrochemical methods. However, the origin of the catalytic activity for the electroless deposition processes has not been elucidated. In order to estimate theoretically the catalytic activity of the metal surface, the first task is to model the metal surface, using appropriate-sized clusters of atoms. The second task is to design possible geometries for adsorption of reductant ions onto the cluster surface, and then, calculating the stabilization energy of each geometry to determine the catalytic activity of metal surface. We cited some data published by Homma's group in 2001 [11] to illustrate this more clearly later.

They used four to seven metal atom cluster with fixed geometry (Pd–Pd distance of 0.275 nm) as a model of a palladium (111) surface (see the original Fig. 5 in [11]). Four kinds of adsorbed geometries and the stabilization energy of hypophosphite ion on the Pd_4 cluster have been calculated. From their calculated stabilization energy for each geometry, it has been found that the most stable geometry is the adsorption of the oxygen side of the hypophosphite ion at the hollow site of the Pd surface (original Fig. 5(d)) because its stabilization energy is −149.4 kJ/mol, and those for other three kinds are −66.1, −80.3, and −136.8 kJ/mol, respectively. By calculating the effect of Pd cluster size on the instability of adsorbed $H_3PO_3^{2-}$ species on the Pd cluster with respect to the H_2PO^{2-} species on the Pd cluster (Fig. 12.1a), Homma et al. found that instability becomes smaller with an increase in the cluster size due to the stabilization effect of the Pd surface. They believed that this could be one of the origins of the "catalytic activity." Then, they tried to evaluate the "catalytic activity" of various metal surfaces by calculating the energy diagrams for the oxidation reaction process of hypophosphite ion via 5-coordinate intermediates on various metal surfaces (Fig. 12.1b), comparing the plots with the dotted line to those

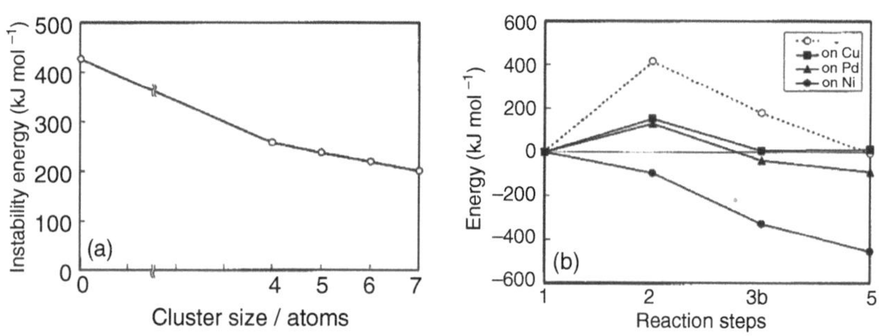

FIGURE 12.1 (a) Effect of the size of Pd cluster on instability energy of adsorbed $H_2PO_2(OH)^{2-}$ with respect to adsorbed H_2PO_2; (b) energy diagram for the oxidation reaction process of hypophosphite ion via 5-coordinate intermediates on various metal surfaces; without the metal surface (○), on copper surface (■), on palladium surface (▲), and on nickel surface (●) [11].

with the solid lines. It shows that the reaction pathway with the nickel cluster shows the lowest energy, indicating that the reaction is catalytically enhanced at the Ni surface. The total reaction is also expected to be exothermic on the Pd surface. However, the reaction with the Cu cluster is not stabilized at step 5, suggesting that the copper surface is not catalytically active for the reaction. In fact, Ohno et al. in 1985 [18] proved experimentally that Pd surfaces display strong catalytic activity on hypophosphite reactions, but Cu surfaces show only weak activity. Thereby, the Homma group confirmed these experimental facts from their calculations of microscopic theory. That is why, these authors said that "this is the significant contribution for understanding the electroless deposition processes" [11].

In order to analyze the origin of the catalytic activity, it is necessary to use the energy density analysis (EDA). What is EDA? EDA was proposed by Nakai in 2002 [19]. It has a capability to analyze the local energies of each atom comprising the molecule, while conventional calculation methods can estimate the energy just as "total" value. Thereby, local interactions between molecules or atoms can be directly evaluated and various reaction mechanisms have been elucidated using EDA [20,21]. In 2004, Nakai et al. [22] investigated the adsorption of H_2 and hydrocarbon onto the Si (100) surface. The electronic energy densities in the Kohn–Sham formulation [23] are given by

$$E_{ELC}[\rho] = T_S[\rho] + E_{Ne}[\rho] + E_{CLB}[\rho] + E_{XC}[\rho] \tag{12.1}$$

where $T_S[\rho]$ is the noninteracting kinetic energy, $E_{Ne}[\rho]$ is the nucleus–electron attraction energy, $E_{CLB}[\rho]$ is the classical coulomb interaction, and $E_{XC}[\rho]$ is the exchange-correlation energy. The EDA resolves each term into atomic contribution.

12.3 THEORETICAL RESULTS OBTAINED BY *AB INITIO* METHODS

After more than 10 years in the new century, a number of promising results have been obtained by using the *ab initio* theoretical methods to investigate the EP process from scratch. We will introduce them from three aspects.

12.3.1 Reaction Mechanisms of EP Processes for Various Reductants

The redox reaction mechanism has been investigated with the microscopic theory for nearly all the usually used reducing agents on metal surfaces such as hypophosphite ions [11,24–26], DMAB [1,27], titanium trichloride [28], aldehydes [29], and hydrazine [30]. We will describe and discuss each of these in the following sections.

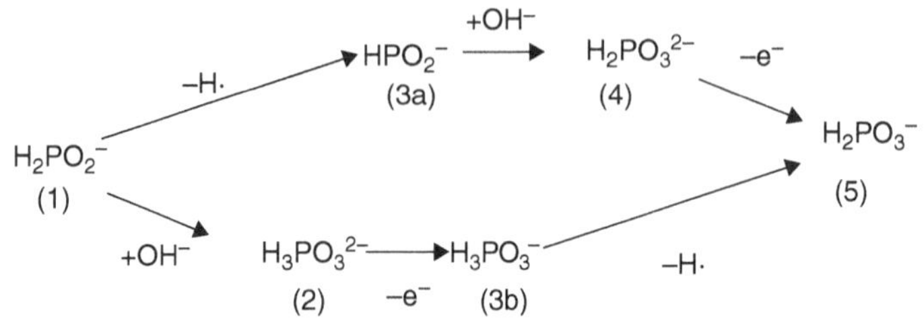

FIGURE 12.2 Reaction pathways for the oxidation process of hypophosphite ion via 3-coordinate intermediates (3a) and 5-coordinate intermediates (3b) [11].

12.3.1.1 Reaction Mechanisms of EP Processes of Hypophosphite Ions

In 2001, Homma et al. [11] using an *ab initio* molecular orbital method studied the possible two pathways (Fig. 12.2) and the corresponding energy diagrams for the oxidation reaction process of hypophosphite ion.

From the energy calculations (Fig. 12.3), it can be seen that the pathway via 5-coordinate intermediates is favorable for the reaction processes because the energy level of the 5-coordinate intermediate $H_2PO_2(OH)^-$ (3b) is ca. 200 kJ/mol lower than that of the 3-coordinate HPO_2^- (3a). In other words, their conclusion is that the oxidation reaction occurs via five-coordinate compounds by primary addition of OH^- (the electron is emitted after the coordination of OH^-), not the pathway via three-coordinate compound obtained by primary dehydrogenation. These authors of another paper [24] also obtained the same conclusion for hypophosphite ion. Moreover, they believed that this reaction pathway via five-coordinate intermediate could be a general mechanism to the reductant species for the EP process. In Fig. 12.3, the energy levels of dianion-type intermediate species, $H_3PO_3^{2-}$ (2) and $H_2PO_3^{2-}$ (4), are also shown, which are formed by the addition of OH^- to $H_2PO_2^-$ (1) and HPO_2^- (3a), respectively.

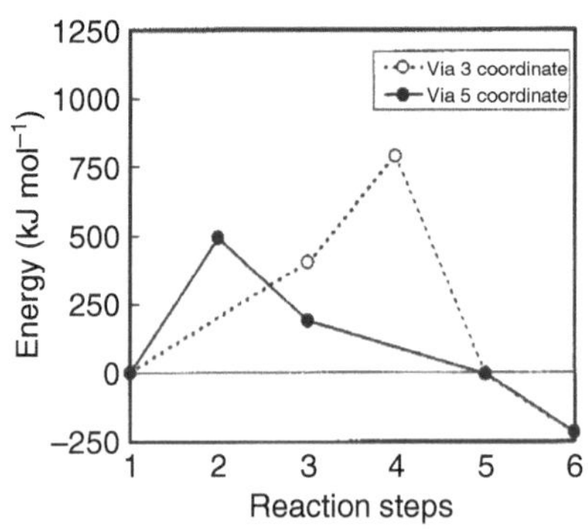

FIGURE 12.3 Energy diagram for the oxidation reaction process of hypophosphite ion via 3-coordinate intermediates (○) and 5-coordinate intermediates (●) [11].

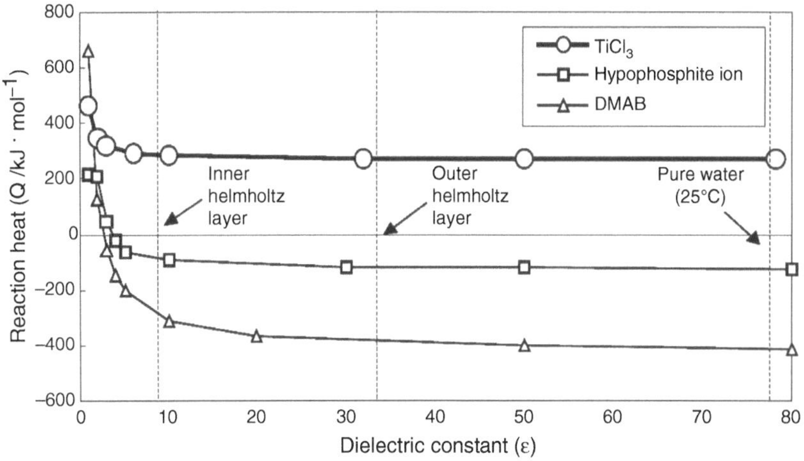

FIGURE 12.4 Effect of dielectric constant of the solution on the heat of reaction of the oxidation reaction of the hypophosphite ion [11].

It was shown that such species are quite unstable; thus immediate emission of one electron should occur to transform them to mono anionic species.

Through the analysis of solvation effect on the reactions they believed that the reactions preferably proceed at the interfacial region, that is, the surface of the metal, rather than in the "bulk" solution region. The heat of reaction, Q, is evaluated using the following reaction:

$$H_2PO_2^- + OH^- \rightarrow H_2PO_3^- + 1/2H_2 + e^-, Q\,(\text{kJ/mol}) \qquad (12.2)$$

If the Q value is positive, the reaction is exothermic, while it is negative, that is endothermic. Figure 12.4 shows representative results of the Q value versus the dielectric constant. From the figure, one can see that the reaction is originally exothermic with a Q value of 218 kJ/mol*, but it turns to endothermic with a Q value of -126 kJ/mol by taking the effect of solvation, that is, dielectric constant of 78.3, into account. The value of Q decreases exponentially as the increase of dielectric constant.

The dielectric constants of the outer and inner Helmholtz layers for the BDM model [17] are estimated to be ca. 32 and ca. 6, respectively. In addition, the Q value sharply shifts to the exothermic direction in the lower dielectric constant region. The results for the reductant hypophosphite ion show that only the reaction in the inner layer is exothermic. Therefore, their oxidation reaction is expected to proceed preferentially at the solid/liquid interfacial region, that is, the surface of the metal, rather than at the "bulk" solution region.

* There may be some confusion of the concept of exothermic and endothermic in the original papers [11,24,27,28]. Especially, Fig. 4 in [11] and Fig. 3 in [24] are the same figure, the contents are completely the same, without any difference, but the curves are just contrary, one is upward (Fig. 4), another is downward (Fig. 3). There seems some error that has been corrected.

In addition, they also evaluated the "catalytic activity" of various metal surfaces by calculating the energy diagrams for the oxidation reaction process of hypophosphite ion via 5-coordinate intermediates on various metal surfaces. We discussed the results in Section 12.2.5.

In 2005, a group from China [25] studied the oxidation mechanism of hypophosphite anion ($H_2PO_2^-$) in acidic and alkaline media in the presence of Ni(II) species by using electrochemical impedance spectroscopy (EIS) and density functional theory (DFT). They took this route because they believed that though Homma et al.'s theoretical contributions to the oxidation of hypophosphite ion showed that the primary pathway of the addition of OH^- is energetically favorable in the reaction [11,24], however, the pathway model cannot be used to explore the charge transfer process and the effect of pH value involved in the mechanism. In addition, their experimental works show that the electron transfer process and pH value were found to be two key factors in the process [26]. Therefore, it is necessary to further understand the effect of pH on the reaction mechanism. After emulating two reactions, these authors found that path (I) passes through a primary dehydrogenation, and path (II) includes a primary addition of OH^- on the hypophosphite anion. By comparison of the energy levels of all species, they concluded that path (II) is energetically favorable under both acidic and alkaline conditions. Even in both acidic and alkaline conditions, they obtained the same conclusion for the reaction pathway of hypophosphite ion as that from Homma et al. [11, 24]. These authors used the following steps to connect their measurements of ELS and the calculation results of DFT. The steps are: ELS measurement → DFT simulation → connecting ELS and DFT. How do they connect both the results? They said: "These products can provide evidence for the formation of a capacitive loop at the low-frequency domain in EIS." It should be pointed out that their connection and description are not so clear, so we did not cite more contents here.

12.3.1.2 Reaction Mechanisms of EP Processes of DMAB

The first paper of *ab initio* molecular orbital to study the oxidation mechanism for reductant DMAB in the EP process was published in 1999 by Homma's group [1]. They examined two types of reaction pathways (Fig. 12.5), via either three-coordinate borane molecules obtained by primary dehydrogenation reactions, or five-coordinate ones by primary additions of OH^- for the oxidation of the DMAB (Fig. 12.5). The first one corresponds to Meerakker's mechanism in the gas phase [15].

The calculations of the energy diagram for the oxidation reactions of DMAB are shown in Fig. 12.6. The reaction proceeds from left to right, and the energy is shown in kcal/mol relative to that of DMAB. The energy level of a produced electron is defined as zero. The change in energy level indicates that the stepwise oxidation reactions of DMAB proceed exothermically via four-coordinate borane molecules $BH_{4-n}(OH)^{n-}$ (n = 1–4). However, the processes from these four-coordinate compounds to the three- or five-coordinate ones are

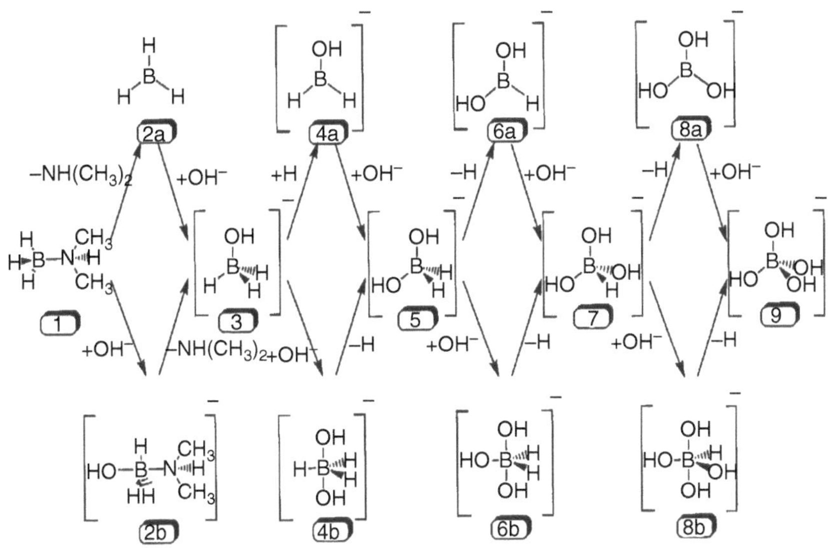

FIGURE 12.5 The two reaction pathways from DMAB to $B(OH)^{4-}$ via three-coordinate and via five-coordinate [1].

endothermic. On the whole, exothermic reactions proceed naturally. From the energy diagram, it is evident that the reaction pathway is via the five-coordinate borane molecules, not via the three-coordinate borane molecules. In addition, the energy difference between three- and five-coordinate compounds is gradually increased: 21, 200, 272, and 309 kJ/mol from 1→3 (The first step from

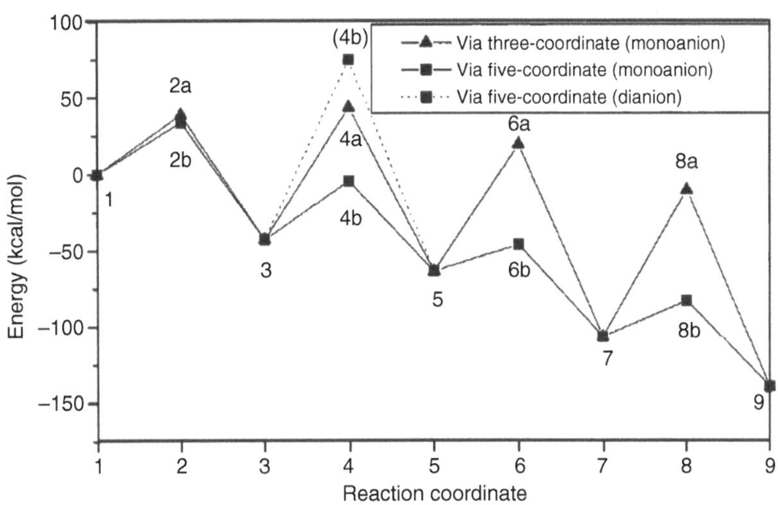

FIGURE 12.6 The energy diagram for the reaction of DMAB handled by the *ab initio* Hartree–Fock molecular orbital method [1].

DMAB to $BH_3(OH)^-$ is a ligand replacement reaction from dimethylamine to OH^-), 3→5 (the second step is from $BH_3(OH)^-$ to $BH_2(OH)^{2-}$), 5 → 7 (the third step is from $BH_2(OH)^{2-}$ to $BH(OH)^{3-}$) to 7 → 9 (the fourth step is from $BH(OH)^{3-}$ to $B(OH)^{4-}$), respectively.

The existences of the five-coordinate borane molecules result from the calculations by microscopic theory, which are never found by experimental, macroscopic, or semiempirical theories. Therefore, Homma et al. in their paper [1] said that these are first clarified theoretically.

In 2003, Homma et al. [27] further studied the oxidation mechanism of DMAB using more accurate basis sets, and the effects of solvation and interaction with metal surface sites on the oxidation mechanism. They proposed a modified pathway that includes the dianion states, shown as 4b, 5a, 7b, 8a, 10b, and 11a in Fig. 1 of [27]. Figure 12.7 shows energy profiles for the oxidation process, which indicated that the total reaction is stabilized and the energy pathway becomes exothermic. It indicates that the result is the same as that of the above result, that is, the reaction pathway via the five-coordinate intermediates (4b, 7b, and 10b) is more favorable than that via the three-coordinate intermediate (5a, 8a, and 11a), due to the stabilization effect caused by the three-center three- or four-electron bondings of boron. They found that the heat of oxidation was transferred to an exothermic reaction with decreasing dielectric constant by taking the solvation effect into account using the self-consistent reaction field method with an isodensity polarized continuum model (SCRF-IPCM). This indicated that the reaction preferably proceeds at the solid/liquid interface. In addition, combined with Cu(111) and Pd(111) neutral cluster models as metal surface sites, they found that the oxidation reaction proceeds preferentially at the metal surface sites. They also believed that the catalytic activity of the deposited metal is caused by its electron acceptivity.

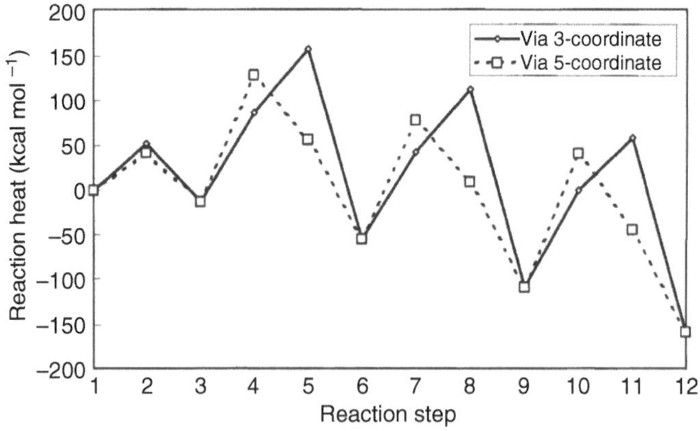

FIGURE 12.7 The energy diagram for the reaction of DMAB handled by the *ab initio* Hartree–Fock molecular orbital method [27].

These authors also examined the solvation effect for this reaction using the self-consistent reaction field method with an isodensity surface polarized continuum model (SCRF-IPCM). The reaction heat is defined as follows:

$$\mathrm{DMAB} + 4\mathrm{OH}^- \rightarrow \mathrm{B(OH)_4}^- + 3/2\mathrm{H_2} + 3\mathrm{e}^-, Q(\mathrm{kJ/mol}) \qquad (12.3)$$

The calculated results of Q versus ε are also shown in Fig. 12.4. It can be seen from Fig. 12.4 that two regions, corresponding to the dielectric constants of the inner and outer Helmholtz layers, are indicated, which are estimated by the BDM model [17]. The Q value decreases exponentially with an increase in the dielectric constant. The variation of characteristics corresponds to the calculated result for hypophosphite ion [11]. Thereby, the conclusion is similar to that for hypophosphite ion, that is, the oxidation reaction of the DMAB is expected to take place at the nearest-neighbor surface region of the deposited metals. In other words, it is expected that the overall reaction is enhanced at the surface region of the metals.

12.3.1.3 Reaction Mechanisms of EP Processes of $TiCl_3$

In 2001 [11] and 2004 [28], the Homma group investigated the oxidation mechanism of $TiCl_3$ as a reductant for an electroless deposition process by the *ab initio* molecular orbital method. Figure 12.8 is the reaction pathways for the oxidation process of $TiCl_3$ via 5-coordinate intermediates they used, in which the reaction proceeds with the substitution of Cl to OH^-. It should be noted that now the electron is emitted by oxidation of titanium from trivalent Ti(III) to tetravalent Ti(IV), but not supplied from the OH^- since the chloride leaves as Cl^-. This is just contrary to the above two cases (hypophosphite ion and DMAB); there the electron emission is due to the substitution of hydrogen to OH^-. Therefore, the timing of the electron emission is a significant step to understand the overall reaction of $TiCl_3$.

The calculated energy diagrams for the oxidation reaction process of $TiCl_3$, either in trivalent Ti(III) or in tetravalent Ti(IV) state, are shown in Fig. 12.9a, in which the energy values are plotted with respect to those of the initial states, $TiCl_3$. One can see that the trivalent species are more stable than the tetravalent species up to step 6, whereas the latter turns out to be more stable at the final step in which the $Ti(OH)_4$ is formed. The total heat of reaction is estimated to be

$$\underset{(1)}{\mathrm{TiCl_3}} \xrightarrow{+\mathrm{OH^-}} \underset{(2)}{\mathrm{TiCl_3(OH)^-}} \xrightarrow{+\mathrm{OH^-}} \underset{(3)}{\mathrm{TiCl_3(OH)_2}^{2-}} \xrightarrow{-\mathrm{Cl^-}} \underset{(4)}{\mathrm{TiCl_2(OH)_2^-}} \xrightarrow{+\mathrm{OH^-}}$$

$$\underset{(5)}{\mathrm{TiCl_2(OH)_3}^{2-}} \xrightarrow{-\mathrm{Cl^-}} \underset{(6)}{\mathrm{TiCl(OH)_3^-}} \xrightarrow{+\mathrm{OH^-}} \underset{(7)}{\mathrm{TiCl(OH)_4}^{2-}} \xrightarrow{-\mathrm{Cl^-}} \underset{(8)}{\mathrm{Ti(OH)_4^-}}$$

FIGURE 12.8 Reaction pathways for the oxidation process of $TiCl_3$ via 5-coordinate intermediates [11].

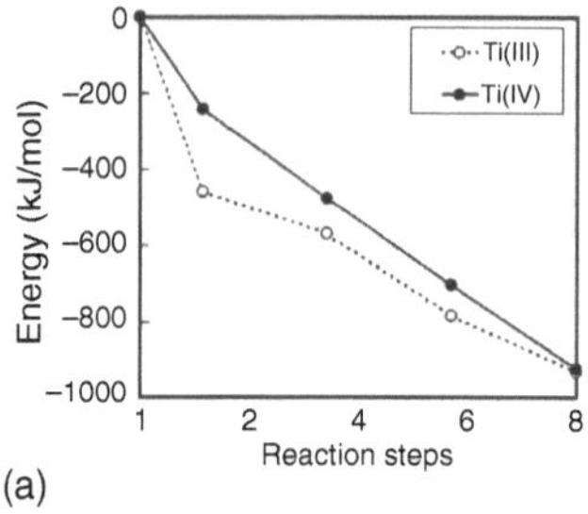

(a)

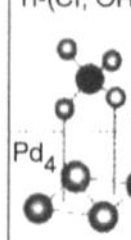

Reaction steps		1	2	4	6	8
Species		$TiCl_3$	$TiCl_3OH$	$TiCl_2(OH)_2$	$TiCl(OH)_3$	$Ti(OH)_4$
Ti-(Cl, OH)$_4$	Net charge		−0.668	−0.841	−0.231	−0.089
	Spin density	1.000	1.000	1.000	0.189	0.016
Pd_4	Net charge		−0.332	−0.159	−0.769	−0.911
	Spin density	0.000	0.000	0.000	0.811	0.984

(b) Change in net charge and spin density in the reaction of $TiCl_3$ at the palladium cluster surface.

FIGURE 12.9 (a) Energy diagram for the oxidation reaction process of $TiCl_3$ intermediates, assuming trivalent Ti (○) and tetravalent Ti (●). (b) Change in net charge and spin density in the reaction of $TiCl_3$ at the palladium cluster surface [11].

920 kJ/mol till this step, indicating high reducibility of this reductant. However, the energy difference between the trivalent and tetravalent species is quite small. After taking the effect of metal surface into account, the representative results of the analysis of the net charges and spin densities for TiX_4 (X = Cl or OH) species and palladium cluster are shown in Fig. 12.9b. We can see from the figure, at the initial stage (steps 1, 2, and 4), that the charge and spin remains at the TiX_4 part, not changed. However, they shift to the metal cluster part at step 6 and finally they are transferred to the metal cluster at step 8. These results evidently show that the electron emission, that is, the oxidation of titanium from trivalent to the tetravalent state takes place when all the Cl is replaced by OH^- to form $Ti(OH)_4$ species.

These authors have also taken the effect of solvation into account in terms of the dielectric field constant. The calculated results are shown in Fig. 12.4 too. Though the value of Q decreases exponentially as the dielectric constant increases, from 465.6 kJ/mol at $\varepsilon = 1.0$ to 269.1 kJ/mol at $\varepsilon = 78.3$, but the Q value are all positive, that is, the heat of $TiCl_3$ is still exothermic till the dielectric constant is near the pure water, indicating that $TiCl_3$ is reactive even in pure water. This nature is different from those for the reductants of hypophosphite ion and DMAB. Just because $TiCl_3$ is reactive even in pure water, Homma et al. believed that appropriate stabilization such as formation of complex is required for the application of the $TiCl_3$ to the electroless deposition process.

It is worth noting that in 2001 Homma et al. [11] defined the "reducibility" of the reductant species in EP from their results of theoretical calculations. The results of their theoretical analysis indicate that the reaction via 5-coordinate intermediates is favorable for several reductant species. Then they used the heat of reaction for each process, that is the thermodynamic energy of the reaction, to indicate qualitatively the "reducibility" of the reductant species in the EP process. Because their calculated values of the heat of reaction for each process for three reductant species (DMAB, hypophosphite ion, and formaldehyde) are 669, 218, and 142 kJ mol/L, respectively, they believed that the orders of the quantitative "reducibility" of these three reductant species are successively dimethylamine borane > hypophosphite ion > formaldehyde. However, Homma's group did not compare the reducibility from the *Q* value of $TiCl_3$ with those of DMAB and hypophosphite ion in their papers. In addition, it is difficult to understand why after calculating the *Q* values of other reductants, for example, aldehydes including formaldehyde, acetaldehyde, and glyoxylic acid (see Section 12.3.1.4), they did not mention the qualitative evaluation of the "reducibility" of the reductant species again, or make any comparison of the "reducibility" of these three reductants (Fig. 12.12) with the other three reductants in Fig. 12.4.

12.3.1.4 Reaction Mechanisms of EP Processes of Aldehydes

Aldehydes, including formaldehyde, acetaldehyde, and glyoxylic acid, are also usually used as reductants in EP processes. In 2005, Homma's group [29] reported a DFT study on the oxidation mechanisms of aldehydes as reductants for electroless Cu deposition process. Figure 12.10 shows the total reaction pathways of aldehydes proposed by these authors in their paper. The reaction's intermediate species include three-coordinate and five-coordinate species.

The calculated energy diagram of all intermediate species for the oxidation process of formaldehyde in the isolated system is shown in Fig. 12.11. It is evident that the reaction pathway via five-coordinate intermediates is more favorable than that via three-coordinate ones, since a dianion state formed via five-coordinate intermediate (2 5) is more stable than those formed via three-coordinate intermediate (3 4). But the most favorable pathway is the one via

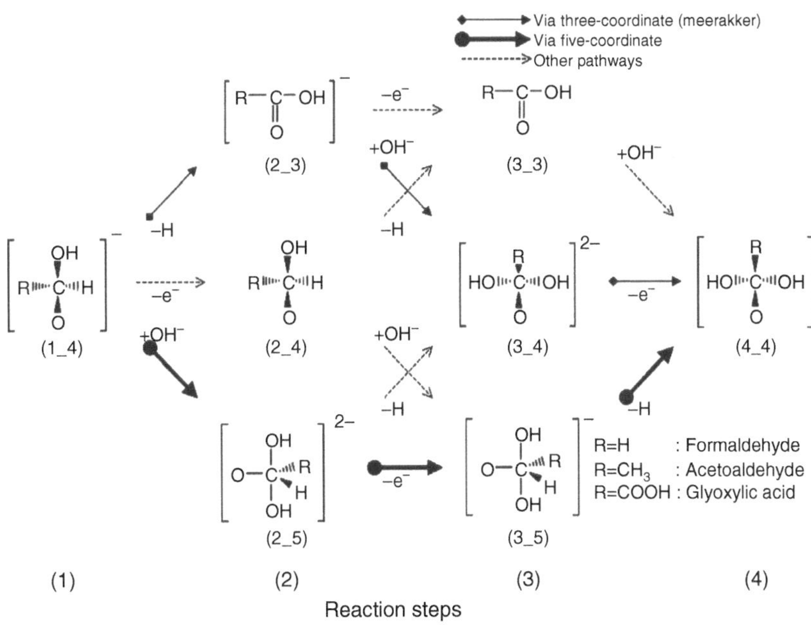

FIGURE 12.10 Reaction pathways for the oxidation process of aldehyde via all intermediates. (*n m*): *n*: reaction steps; *m*: coordinate number [29].

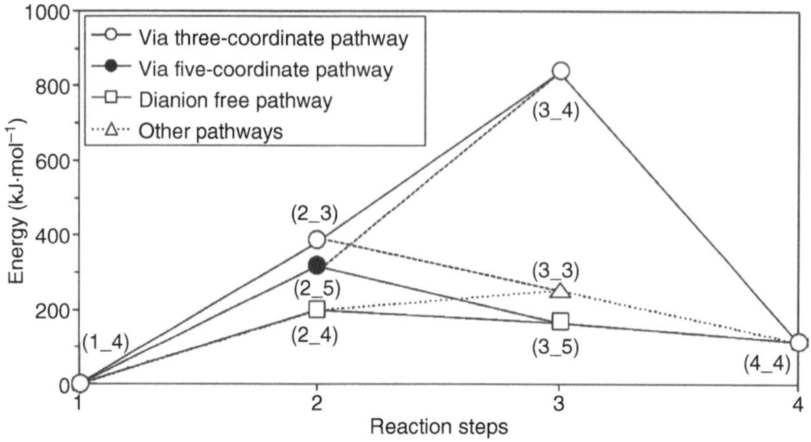

FIGURE 12.11 Energy diagram for the oxidation reaction process of formaldehyde via three-coordinate intermediates (○), via five-coordinate intermediates (●), via dianion free intermediates (□), and other pathways [29].

1 4, 2 4, 3 5, and 4 4, for only this pathway can proceed through dianion-free state. This pathway will be denoted as the "dianion-free pathway." The energy profiles for the oxidation process of acetaldehyde in the isolated system (original Fig. 3 in [29]) are the same as that for reductant formaldehyde, that is, Fig. 12.11. So, acetaldehyde and formaldehyde possess the same oxidation mechanism including the energy level of each intermediate, which means that methyl group of acetaldehyde hardly affects the oxidation reaction in the isolated system. For glyoxylic acid, the energy profiles for the oxidation process in the isolated system (original Fig. 4 in [29]) also indicate that the reaction pathway via the dianion-free intermediate is the most favorable. However, the difference between the energy levels of the pathway via five-coordinate and dianion-free intermediates at reaction step 2 is smaller than those of the other aldehydes. In addition, the energy levels of all intermediates are higher, indicating that this pathway is less stable than the other aldehydes. This indicates that carboxyl group, which is the functional group of glyoxylic acid, considerably affects the oxidation reaction. Why is the diagram of acetaldehyde similar to that of formaldehyde, whereas that of glyoxylic acid differs from that of the others? The reason is the substituent effect derived from the polarity of the functional groups. The effect of acetaldehyde (functional groups: CH_3) is weaker than that of glyoxylic acid (functional groups: COOH, electron-withdrawing).

What is the reaction mechanism of aldehydes at the Cu surface? These authors investigated the oxidation pathways of aldehydes at the Cu_{22} cluster as a model surface. It was found that in general, the energy levels of the intermediate species are stabilized at the Cu surface compared with the results for the isolated system. They believed the reason is that the charge transfer from the adsorbate to the surface leads to stabilization, which promotes the oxidation reaction of aldehydes to take place preferentially at the surface. Calculations indicate that

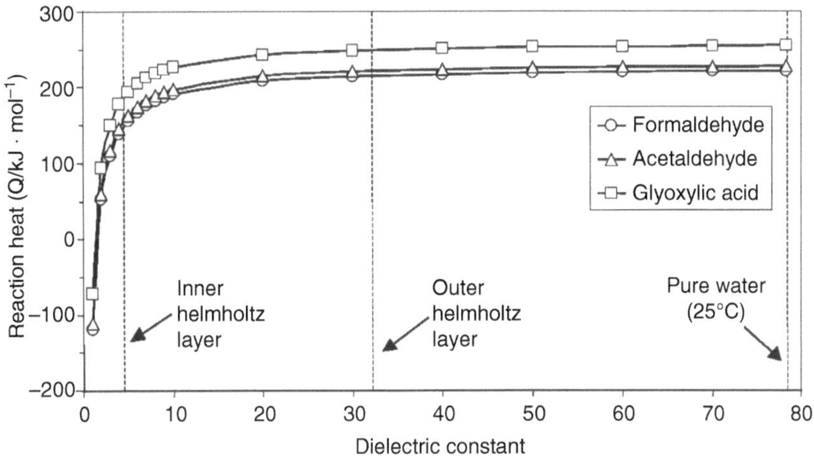

FIGURE 12.12 Effect of dielectric constant of the solution on the reaction heat of the oxidation reaction of the three aldehydes [29].

the pathway via five-coordinate intermediates is the most stable at the Cu surface, whereas the reaction pathway via dianion-free intermediates is the most stable pathway in the isolated system. In other words, the oxidation reaction of aldehydes preferentially proceeds at the Cu surface than in the isolated system. However, comparing the energy profiles (original Fig. 7 in [29]) for the oxidation process of three aldehydes via five-coordinate intermediates at the Cu surface, it is evident that the energy of formaldehyde and acetaldehyde shows the same level, while that of glyoxylic acid indicates a more unstable state. That is to say that though the oxidation reaction mechanisms of aldehydes at the Cu surface are all the same, since the behavior of energy profiles is the same, but glyoxylic acid is strongly influenced by the interaction between the surface and the functional groups in terms of energy.

When the solvation effect is taken into account, the calculations indicate that the oxidation reactions of the three aldehydes preferably proceed at the solid/liquid interface. Figure 12.12 shows the calculations of the change in reaction heat for three aldehydes by taking the solvation effect into account. The reaction heat (Q) is defined by

$$RCHO(OH)^- + OH^- \rightarrow RCO(OH)_2^- + 12H_2 + e^-, Q(\text{kJ/mol}) \quad (12.4)$$

From Fig. 12.12, Shimada et al. [29] said that the Q value increases exponentially as the increase of dielectric constant**. It also shows that only the

** It is obvious that the representation of this figure is just opposite to that in Fig. 12.4, that is, the effect of dielectric constant of the solution on the heat of reaction of the oxidation reaction for the hypophosphite ion, DMAB, and $TiCl_3$. There may also be somewhat confusion of the concept of exothermic and endothermic in the original papers [11,29]. Here, we used the original statements by Homma's group, but readers have to consider carefully according to what we said in this footnote.

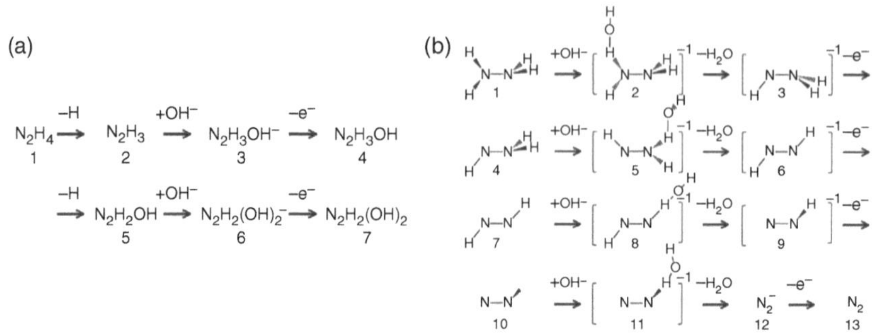

FIGURE 12.13 (a) The oxidation reaction scheme of hydrazine proposed by Meerakker [15]. (b) The oxidation reaction scheme of hydrazine proposed by Harrison [31].

reaction in the inner layer is endothermic and those on the outer layer and in the bulk solution are exothermic. These results correspond to the value of Q for formaldehyde, acetaldehyde, and glyoxylic acid at the Cu surface, which are calculated to be 79.27, 73.11, and 45.73 kJ/mol, respectively. Thus, these authors concluded that the oxidation reactions of the aldehydes are expected to take place near the metal surface. As the Q value of glyoxylic acid is more exothermic than the others, Homma's group believed that the solvation is caused by the functional groups.

12.3.1.5 Reaction Mechanisms of EP Processes of Hydrazine and Hydroxylamine

In 2007, Homma's group [30] using MO calculations studied the oxidation mechanism of hydrazine and hydroxylamine. They investigated two pathways as shown in Fig. 12.13. One pathway is proposed by van den Meerakker in 1981 [15], in which hydrogen is eliminated from the reducing agent, followed by coordination of hydroxyl group to the central nitrogen atom accompanied by electron release. Another pathway was proposed by Harrison and Khan in 1970 [31], in which the hydroxyl group did not coordinate to nitrogen atom but to az hydrogen atom, to form H_2O, which is eliminated later, followed by electron emission.

The calculated energy profiles for these two pathways are indicated in Fig. 12.14. It can be seen from Fig. 12.14a that the energy barriers in the pathway of $N_2H_4 \rightarrow N_2H_3OH^-$ (1→3) and $N_2H_3OH \rightarrow N_2H_2(OH)^{2-}$ (4→6) decrease by the stability of hydrogen gas generation. But energy barriers remain high and the heat of reaction for the whole reaction (1→7) is ca. 200 kJ/mol. Therefore, it is difficult for the reaction to take place via the pathway proposed by Meerakker. The result implies instability derived from the geometry of hydrazine where the hydroxyl group directly coordinates to the center atom. However, the results in Fig. 12.14b show that the energy barriers were lower than those shown in Fig. 12.14a, and the calculated value of the heat of reaction (1→13)

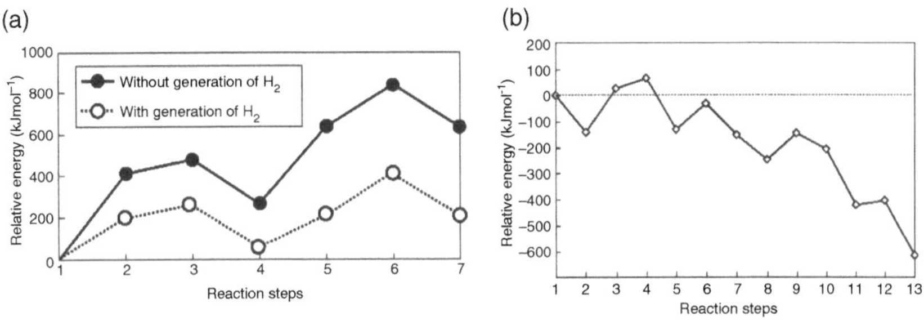

FIGURE 12.14 (a) Energy diagram for the oxidation reaction process of hydrazine proposed by Meerakker [30]. (b) Energy diagram for the oxidation reaction process of hydrazine proposed by Harrison [30].

was sufficiently low, thereby the oxidation reaction of hydrazine takes place via elimination reaction of H_2O accompanied with direct coordination of hydroxyl group to hydrogen atom, rather than H–OH^- substitution reaction accompanied with direct coordination to nitrogen atom. As for the reaction mechanisms of hydroxylamine, these authors' calculated energy profiles also lead to the same conclusions as for hydrazine above, that is, the oxidation pathways of this agent in an isolated system did not proceed via the substitution reaction of $H \leftrightarrow OH^-$ but the coordination of OH to H to form H_2O, which is eliminated later.

What are the reaction mechanisms of hydrazine and hydroxylamine on the Cu surface? From the energy profile for the oxidation of hydrazine at the Cu surface (original Figure 5 in [30]), Homma et al. found that the initial oxidation reaction from N_2H_4 to N_2H_3 takes place through the adsorption to the Cu surface, following reaction to form N_2H_2 from N_2H_3 is unstabilized through the adsorption. But no stabilization effect of the Cu surface is observed during the reaction from N_2H_2 to N_2. So, the Cu surface catalyzes only the initial step of the hydrazine oxidation, and following reactions do not take place on the Cu surface. That is to say, the Cu surface will receive only one electron from the first step of the hydrazine oxidation reaction, which triggers the following reaction explosively over the deposited Cu surface/liquid interface. As for the oxidation of hydroxylamine at the Cu surface, the conclusion obtained by these authors is that although the oxidation of hydroxylamine also proceeded via the elimination reaction of H_2O, but after the oxidation, OH radical, generated from N_2O as a product, decomposed hydroxylamine.

12.3.2 Detailed Investigation of Atomic Interaction Between Reductants and Metal Surfaces

In Section 12.3.1, the catalytic activity of the metal surfaces (mainly Cu and Pd) for reductants (hypophosphite ion, DMAB, $TiCl_3$, aldehydes, and hydrazine and hydroxylamine) in the electroless deposition process was investigated. However, the origin of the catalytic activity for the metal surfaces has not yet been elucidated because only the total energy of the system has been calculated there. In such a way, the local energies of each atom composing the molecule cannot be known. EDA has the capability to analyze the local energies of each atom in the molecule. Using EDA methods, local interactions such as the "catalytic activity" can be investigated, since the local energies of molecules or atoms can be evaluated directly. Thereby, the origin of the catalytic activity for the metal surfaces can be elucidated further. Since 2011, several papers by Homma et al. have just focused on this subject, and we will introduce them in the next section.

12.3.2.1 Origin of Catalytic Activity by Using EDA for Reductants on Ag Surface

In order to verify the origin of the catalytic activity, Homma et al in 2007 [32] investigated catalytic activity of an Ag surface for the oxidation reaction of reductants, such as DMAB, formaldehyde, and hypophosphite ion, by using an EDA approach, focusing upon the local interaction of the species. First, they investigated the oxidation reaction mechanisms of these three reductants for the EP Ag process using DFT calculations. From their calculations (original Fig. 2 in [32]), it was found that the energy levels of every intermediate species on the Ag surface are all less than those in the isolated system, indicating that the oxidation processes of the reductants on the Ag surface are initiated by the addition of OH^- and proceed via a five-coordinate intermediate species. These authors believed that the catalytic activity of OH^- itself is less than that of the Ag surface because the energy level of the intermediate decreases on the Ag surface rather than in the isolated system. However, it is known that the Ag surface does not act as a catalyst for the oxidation reaction of hypophosphite ion, whereas it does act for that of formaldehyde. Therefore, the origin of the catalytic activity could not be explained simply by the total energy change of the reductants and the Ag surface. Thus, in order to investigate the reaction from the local viewpoint, the EDA method is employed.

For the adsorption state of reductants on the Ag surface, the calculated results of EDA show that the energy level of the reductant part is high, that is, unstable (DMAB: 336.8 kJ/mol, hypophosphite ion: 609.4 kJ/mol, and formaldehyde: 641.9 kJ/mol), whereas that of the Ag_{22} cluster part is low (DMAB: −463.8 kJ/mol, hypophosphite ion: −716.2 kJ/mol, and formaldehyde: − 753.8 kJ/mol). The total energy levels are the sum of these two terms, which also indicates the stable condition (DMAB: − 127.0 kJ/mol, hypophosphite ion: −106.8 kJ/mol, and formaldehyde: − 111.8 kJ/mol). In addition, EDA calculations indicate directly that total stabilization by adsorption is mainly contributed to greater stabilization of the Ag cluster through the adsorption site, rather than the destabilization of particular atoms in the reductant. So, the conclusion is that the adsorption of the reductants onto the Ag surface is driven by destabilization of the reductant and stabilization of the Ag surface.

EDA calculations for the oxidation reaction of hypophosphite ion and formaldehyde on the Ag surface (original Figs 5 and 6 in [32]) indicate that the energy changes of Ag atoms during the reaction of hypophosphite ion are small, whereas those of formaldehyde are large, that is, the Ag surface does not activate the oxidation reaction of hypophosphite ion, but it does activate the oxidation reaction of formaldehyde.

Further, they calculated the energy changes of reductant and OH^- parts by the oxidation reactions in isolated system and on the Ag surface for hypophosphite ion and formaldehyde. The results (Table 2 in [32]) indicate that the energy change of hypophosphite ion on the Ag surface at step 2 is 254.1 kJ/mol, which is smaller than that in the isolated system, and the change at step 3 is 215.1 kJ/mol, which is about 100 kJ/mol larger than that in the isolated system. However, the energy changes of formaldehyde on the Ag surface at steps 2 and 3 are larger than those in the isolated system, for example, at step 3 is larger than about 200 kJ/mol. It is obvious that the oxidation reaction of formaldehyde is strongly activated on the Ag surface, whereas that of hypophosphite ion is rather weakly activated on the Ag surface. In other words, for the reaction of formaldehyde, in which the Ag surface acts as a catalyst, the intermediates were considerably activated at the Ag surface, whereas such an effect was not clearly seen with the hypophosphite ion, for which the Ag surface is not catalytic. Therefore, destabilization of the five-coordinate intermediate species at the Ag surface is a key factor of the "catalytic activity" for the oxidation reaction. Or, one can say that the factors for the catalytic activity of the oxidation reaction originate from the activation of reductants at the five-coordinate intermediate state. These authors believed that such a difference is caused by the coordination behavior of OH^-. Because OH^- coordinated formaldehyde was more destabilized in the presence of the Ag surface, whereas the coordinated hypophosphite ion at the Ag surface was stabilized, thereby the behavior of adsorbed OH^- at the Ag surface plays a significant role in the overall reaction of the reductant species.

It should be pointed out that only using the EDA approach, the above fine calculations for the local energies between molecules or atoms can be obtained, the local interactions such as the "catalytic activity" could be analyzed, whereas the calculations of total energy cannot obtain such results.

12.3.2.2 Detailed Investigation of Reaction Mechanism of Hypophosphite Ions on Metal Surfaces

In 2006, a group from China [33] using DFT at B3LYP/6-31++g(d, p) level investigated surface structures and electronic properties of hypophosphite, $H_2PO_2^-$, molecularly adsorbed on Ni (111) and Cu (111) surfaces†. They used a four-metal-atom cluster as the simplified model for the surface. Calculations

† These authors published another article in 2005 (Y. Zeng, J. Yi, H. Wang, G. Zhou, and S. Liu, J. Mol. Struct.: THEOCHEM, 724 (2005) 81), in which the contents are substantially similar to those in this article. However, they did not cite that paper in this paper; of course they did not mention what is the difference between these two papers. It seems that their treatment is not in agreement with the common usage in academic community.

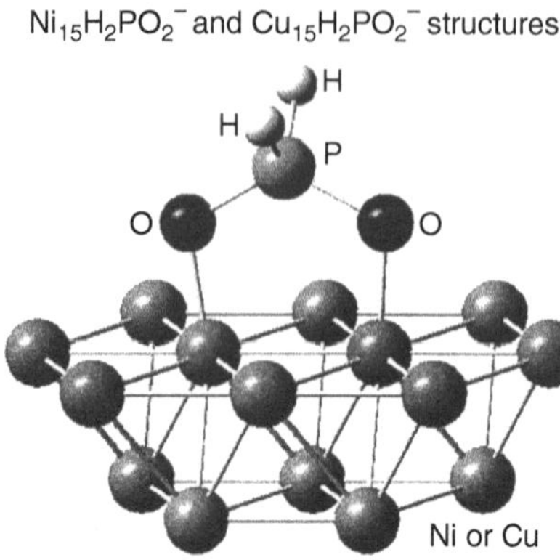

FIGURE 12.15 **$Ni_{15}H_2PO^{2-}$ and $Ni_{15}H_2PO^2$ – structures for the single-point calculations [33].**

indicate that six stable orientations have been discovered on both Ni (111) and Cu (111) surfaces. And the most stable orientation of $H_2PO_2^-$ was found to have its two oxygen atoms interacting with the surface with two P–O bonds pointing downward. Then, they used an extended model of a 15 metal-atom cluster (Fig. 12.15) for all stable patterns of $M_{15}H_2PO_2^-$ by keeping the optimized geometries obtained for $M_4H_2PO_2^-$ to calculate total energies, E_{tot}, and adsorption energies, E_a for each pattern. The results are shown in Fig. 12.16. It can be seen from the figure that patterns A and E were the most stable structures for $H_2PO_2^-$ absorption on Ni and Cu substrates. So, two oxygen atoms, instead of two hydrogens, should play the anchoring role, which is illustrated in Fig. 12.15. This result is contrary to the proposal by Abrantes et al. [34,35], who claimed that the two hydrogen atoms play the anchoring role in the adsorption process. Considering cases such as metal–olefin complexes the bonding nature of electron feedback donation has widely been used to explain electron density shift from filled d-orbitals of a metal atom to any empty orbitals of a ligand; these authors anticipate that there exists a similar situation in their systems. Therefore, they calculated Mulliken for all stable structures of both $M_4H_2PO_2^-$ (M = Ni and Cu) to estimate the amounts of the forward and backward electron donations between hypophosphite and substrate atoms in the systems. It has been found

	$Ni_{15}H_2PO_2^-$			$Cu_{15}H_2PO_2^-$		
	E_{tot} (a.u.)	E_{ad} (eV)	Δ^a (eV)	E_{tot} (a.u.)	E_{ad} (eV)	Δ^a (eV)
Pattern A	−23115.8553	−2.03	0.00	−25098.8417	−1.03	0.06
Pattern B	−23115.8339	−1.45	0.58	−25098.8217	−0.48	0.61
Pattern C	−23115.8270	−1.26	0.77	−25098.8197	−0.43	0.57
Pattern D	−23115.7956	−0.48	1.55	−25098.8133	−0.26	0.83
Pattern E	−23115.8525	−1.96	0.07	−25098.8438	−1.09	0.00
Pattern F	−23115.8269	−1.26	0.77	−25098.8395	−0.67	0.42

[a] The difference from the lowest energy of $M_{15}H_2PO_2^-$.

FIGURE 12.16 Total energies, E_{tot}, adsorption energies, E_{ad}, in various stable adsorption patterns for $M_{15}H_2PO_2^-$, M = Ni and Cu [33].

from the calculated Mulliken population that the back-donation from 3d orbitals of Ni and Cu substrates to the unfilled 3d orbital of the P atom in $H_2PO_2^-$ and 4s orbital's acceptance of electron donation from one lone pair of the O atom in $H_2PO_2^-$ plays very important roles in the $H_2PO_2^-$ adsorption on the transition metals. The averaged electron configuration of Ni in the Ni_4 cluster is $4s^{0.63}4p^{0.02}3d^{9.35}$ and that of Cu in the Cu_4 cluster is $4s^{1.00}4p^{0.03}3d^{9.97}$. This difference in bonding features has the following consequences: (1) for the first step of the electroless deposition, $H_2PO_2^-$ adsorption is more stable on a Ni substrate than on a Cu surface; (2) in the second step of the electroless deposition, the Mulliken charge of the P atom becomes more positive when $H_2PO_2^-$ is adsorbed on a metal substrate, and this charge is larger on Ni than on Cu, making the P atom in $H_2PO_2^-$ on the Ni substrate more likely to be attacked by nucleophiles such as OH^-; (3) in the key step of the electroless depositions, there are more electrons transferred from $H_2PO_2^-$ onto the Ni_4 moiety of $Ni_4H_2PO_2^-$ than onto the Cu_4 moiety of $Cu_4H_2PO_2^-$, indicating that oxidation of $H_2PO_2^-$ can take place more favorably on the Ni substrate than on the Cu substrate. In conclusion, these authors believed that these characteristics can be served as the reasons why $H_2PO_2^-$ is a good reductant in the Ni-electroless deposition but cannot be used in the Cu-electroless deposition.

After 2011, Homma's group carried out further investigations into the reaction mechanism of hypophosphite ions on various metal surfaces [36–38].

First, they examined the elementary steps (adsorption, dehydrogenation, and oxidation) of the reactions of hypophosphite ions on Cu, Ni, and Pd surfaces using the DFT method [36]. For the adsorption step, the calculated adsorption energies of hypophosphite ion on Cu, Ni, and Pd metal surfaces are −49.36, −244.50, and −84.09 kJ/mol, respectively. Thus, Cu (111) should show a weak interaction with hypophosphite ion, while Pd (111) shows a stronger interaction, Ni (111) a very strong interaction, that is, hypophosphite ions adsorb onto each surface spontaneously with stabilities in the order of Ni (111) > Pd (111) > Cu (111). These authors believed that differences of their valence electronic structures for these metals (Cu: $d^{10}s^1$, Ni and Pd: d^8s^2) determine the difference in stabilities of the adsorbate on their respective surfaces. In addition, it is well known that Ni has a high magnetic moment originating from its high exchange interaction energy, so these authors conclude that such high exchange interaction energy caused the high adsorption energy on the Ni surface. For the dehydrogenation step, they calculated the energy change in the system for hypophosphite ions at each metal surface by using a simulation of a P–H bond elongation. The calculated energy changes during dehydrogenation are shown in Fig. 12.17a for the gas phase. It can be seen that the energy levels for dehydrogenation of hypophosphite ion on Cu (111) Ni (111) and Pd (111) surfaces are around 100, 50, and 0 kJ/mol, respectively. Since the adsorption and dehydrogenation steps are reactions in series, the destabilization in the dehydrogenation can be compensated for by adsorption energy. As mentioned above, the calculated adsorption energies of hypophosphite ion on Cu, Ni, and

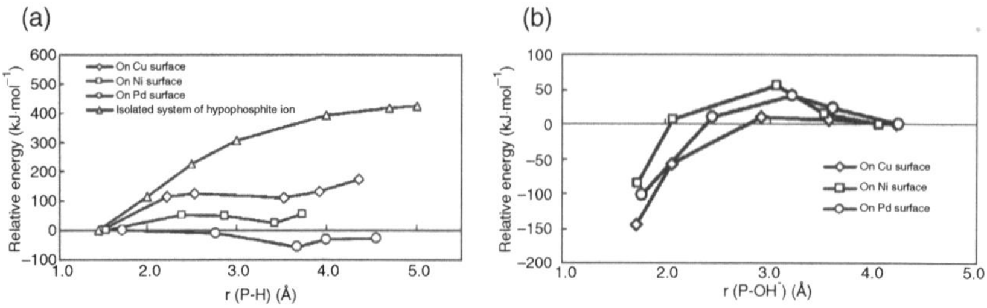

FIGURE 12.17 Energy profiles for dehydrogenation (a) and oxidation (b) of hypophosphite ion on each metal surface [36].

Pd metal surfaces are −49.36, −244.50, and −84.09 kJ/mol, respectively. So, the adsorption energy of hypophosphite ion on Cu (111) cannot compensate for the destabilization of dehydrogenated hypophosphite ion on Cu. The adsorption energy on Ni (111) can compensate completely for the destabilization of dehydrogenated hypophosphite ion on Ni. As for Pd, there is no need to compensate for any destabilization, because there is only negative energy change when dehydrogenating on Pd (111). Thus, when the results from the calculation of the dehydrogenation step are compared with that of the adsorption step, the destabilization of Cu (111) is significant for reaction progress, while those of Ni (111) and Pd (111) are insignificant. According to these authors, these differences are still to arise from the electronic structure of each metal. They believed that the large destabilization of Cu (111) should link to the large energy barrier on this dehydrogenation step. Such a large energy barrier should be one of the origins of the low catalytic activity of Cu (111) on hypophosphite ion reactions. In such a way, Homma et al.'s microscopic calculations found the origins of the low catalytic activity of Cu (111) on hypophosphite ion reactions, which were not known before.

Then these authors analyzed the oxidation step, and calculated energy profiles for each metal surface. Figure 12.17b is the calculated energy profiles for the oxidation step. It can be seen that the barriers in this step are not so significant, and the adsorption energy on each surface can compensate for the barriers in this step. The reason is that in this oxidation step, dehydrogenated anions on each metal surface react spontaneously with the hydroxyl base. Finally, consideration of all elementary steps indicates that a reaction barrier should exist for the dehydrogenation step on Cu (111), causing its low catalytic activity toward the hypophosphite ion reaction. However, on Ni (111) and Pd (111), the hypophosphite ion oxidizes spontaneously; thereby these metals exhibit high catalytic activity to the same reaction. Microscopic theory calculations finally explain the corresponding experimental facts mentioned in the former chapters: that is, Ni and Pd surfaces display strong catalytic activity toward hypophosphite reactions, whereas Cu surfaces show only weak activity. These calculation results show clearly that dehydrogenation should be a rate-determining step for reducing agent reaction and one of the most significant factors.

In order to explain the reactivity difference of hypophosphite ions for electroless deposition on Pd and Cu metal surfaces, Homma et al. in 2011 [37] further analyzed electronic structures of the activation states of hypophosphite ion oxidation on these surfaces by using the DFT method. They focused on the dehydrogenation reaction, that is, a rate-determining step in the elementary reaction steps in the calculations. Because smaller values of the Mulliken population indicate strong bond cleavage tendency between P and H, promoted by metal surfaces, it is necessary to calculate populations. Their calculated Mulliken population values between P and H for gas phase and on Pd and Cu are 0.5553, 0.3301, and 0.4910 respectively, that is, Mulliken population values between P and H on Pd are much smaller than those on Cu and those in the gas phase. From calculations they found a particular orbital interaction between the hypophosphite ion and the metal surface, that is,

$$\text{Pd}(111)\text{: s-orbital of H (hypophosphite ion)} \overset{\text{interacts singly}}{\Leftrightarrow} \text{d- or p-orbital (Pd)} \quad (12.5)$$

$$\begin{aligned}&\text{Cu}(111)\text{: s-orbital of H}\\ &\quad + \text{s-orbital of P (hypophosphite ion)} \overset{\text{interacts simultaneously}}{\Leftrightarrow} \text{p-orbital (Cu)}\end{aligned} \quad (12.6)$$

Interaction (12.5) induces an anti-bonding interaction between H and P in the hypophosphite ion, which is responsible for P–H cleavage. But interaction (12.6) barely induces an anti-bonding interaction between H and P in the hypophosphite ion. These authors believe that such a difference in orbital interaction structures should be related to P–H cleavage activity and the reactivity difference of hypophosphite ion on each metal surface. So, based on atomic orbital interaction, they further studied the catalytic activity of metal surfaces in each elementary step of the hypophosphite ion reaction.

However, the origin of such orbital interaction has not yet been clarified; therefore these authors in 2012 [38] once again investigated in detail the orbital interaction between adsorbed hypophosphite ions and metal surfaces as a basic factor, in order to find which one of the factors can determine the catalytic activity of metal surfaces on P–H cleavage. Pd (111) and Cu (111) were chosen as the surfaces with high and low catalytic activity, respectively. They calculated the Mulliken population and projected density of states (PDOS) profiles of the frontier orbitals in the adsorption states in order to analyze the electronic states of adsorption systems and the interaction between these orbitals and the metal surface. The results showed that on the Pd surface, the d-band lies near the Fermi level (an area from -5 to 3 eV) and contains a vacancy. So, the d-band promotes electron donation from the highest occupied molecular orbital (HOMO) of distorted hypophosphite to the d-band, which reduces repulsion between distorted hypophosphite and the surface. Moreover, the d-band also promotes back-donation from the d-band to the lowest unoccupied molecular orbital (LUMO), which enhances the stabilization of distorted hypophosphite on the surface and promotes P–H cleavage. On the basis of such electron transfer effects, the Pd surface exhibits high catalytic activity on P–H bond cleavage of hypophosphite ions. Therefore, other metal surfaces with high catalytic activity on dehydrogenation of hypophosphite should have similar mechanisms to the Pd surface. However, on the Cu surface, its d-band lies in a deep energy region (an area from −6 eV to −1 eV) and contains no vacancy, which allows very little electron donation or back-donation to occur, unlike in the case of the Pd surface. Thereby, the Cu surface does not possess a mechanism for reducing the repulsion with hypophosphite and cannot promote P–H cleavage. Consequently, hypophosphite ions adsorb weakly on Cu to avoid repulsion and the Cu surface exhibits little catalytic activity on P–H bond cleavage. The conclusion is obviously that the position of the d-band plays a key role in determining the catalytic activity on P–H bond cleavage of hypophosphite.

In all the above studies, including Section 12.3.1 and this Section 12.3.2, the models of the metal surfaces were simplified as being flat (111) surfaces in order to study the most fundamental characteristics of metals. However, the actual surface has many defects such as steps, kinks, pits, and adatoms, which have not yet been included in the analysis model. On the one hand, reports indicate that such surface defects often play key roles in surface reactions [39–41]. On the other hand, there is limited research to show the effects of defects on the reaction of chemical agents used in the EP process, which is necessary to completely understand the different chemical phenomena involved in this process. That is why, Homma et al. in 2013 [42] theoretically compared the reaction on a stepped surface with that on a flat surface, and proposed a mechanism by which a step influences the reaction of a reducing agent. They chose the hypophosphite ion as reductant to analyze theoretically the reaction behavior of hypophosphite ion around the defect using DFT calculations. Pd and Cu surfaces were chosen as the metal surfaces. Still they focused on the dehydrogenation reaction, because it is the rate-determining step in the oxidation pathway. Figure 12.18 shows the calculated reaction energy profiles of the dehydrogenation of hypophosphite ion on Pd and Cu metal surfaces by Homma et al. [42]. From the figure, two factors are apparent: catalytic activity and the effect of the step can be determined. The first is that the dehydrogenation on Pd is strongly stabilized, whereas that on Cu is unstable. Thus on Pd, dehydrogenation reactivity of H_2PO^{2-} is enhanced, but on the Cu surface, such a behavior is not present. They found that donation and back-donation on a Pd surface with defects promote P–H bond cleavage. This clear difference in the promotion effect on the P–H bond cleavage between the two metals should result in the difference in the dehydrogenation behavior of H_2PO^{2-} on each surface. The second is that comparing the reaction on the stepped surface and that on the flat surface reveals that the final state of dehydrogenation should be more stabilized on the stepped surface than on the flat surface in both the Pd and Cu systems, which indicates that the defect enhances the dehydrogenation of H_2PO^{2-}, and accelerates the oxidation of the

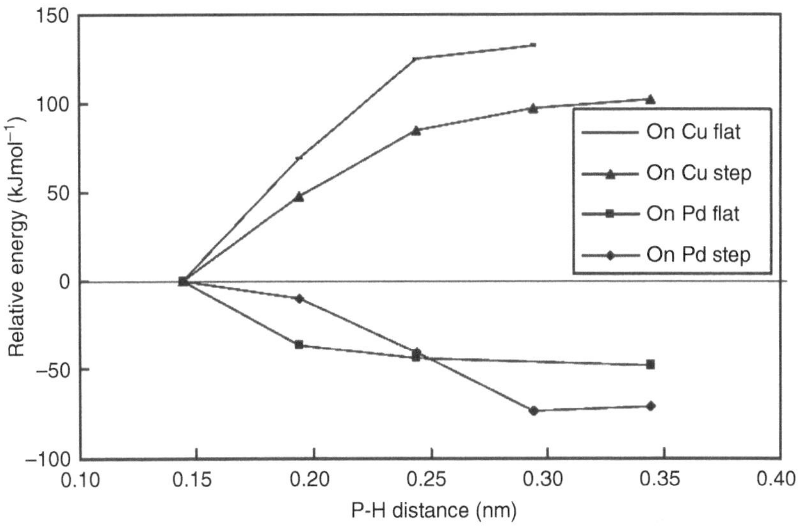

FIGURE 12.18 Reaction energy profiles of the dehydrogenation of hypophosphite ion on each metal surface [42].

hypophosphite ion. From this evidence, these authors proposed a mechanism for the effect of a surface defect on the oxidation of the hypophosphite ion. That is, the hydrogen atom dissociated from H_2PO^{2-} migrates along the slope of the stepped surface and is highly stabilized on the hollow site of the slope, which has a high electron density and is stimulated by atoms in the downstairs layer to interact with some atoms. Such interaction provides a highly stabilized final state of dehydrogenation of H_2PO^{2-}, which in turn accelerates oxidation on the stepped surface when compared to the same reaction on the flat surface.

We can summarize the step-by-step, in-depth study for the catalytic activity of metal surfaces carried out by Homma et al. as follows:

Examining the elementary steps (adsorption, dehydrogenation, and oxidation), the catalytic activities of metals on the hypophosphite ion reaction are primarily dependent on the dehydrogenation step → orbital interaction between hypophosphite and the metal surface in the adsorption step plays a critical role in initiating the dehydrogenation → the position of the d-band of the metal surface is one of the determining factors for how well the catalytic reaction proceeds → comparing the reaction on a stepped surface with that on a flat surface, and proposed a mechanism by which a step defect influences the reaction of a reducing agent.

12.3.3 The Role of Stabilizer and Plating Rate in EP

The effect of a stabilizer such as Pb or thiourea (Tu) is very important, as they are essential to the EP process. We described the experimental data for them in detail in Chapter 6. The plating rate is the key subject in the EP process, and its experimental facts were described in that chapter. Using microscopic theory to investigate them can allow us to understand their nature deeply. Until now, only a very few papers on this subject have been published [43–45]. We will introduce them in the following sections.

12.3.3.1 Effect of Pb^{2+} Stabilizer and Calculation of Plating Rate in EN Process

In 2004, a group from Singapore [43] reported the effect of Pb^{2+} stabilizer and calculation of plating rate in the EN process. The composition of the electroless plating bath that they used is: $NiSO_4 \cdot 6H_2O$ 26 g/L, $NaH_2PO_2 \cdot H_2O$ 28 g/L, NaAc 10 g/L, DL-malic acid 15 g/L, lactic acid (90%) 15 mL/L, borax 6 g/L, $Pb(Ac)_2$ variable, pH adjuster NaOH (0.2M) (pH 5), and the bath temperature was controlled within 90 ± 1°C. Their measurement results indicated that the concentration of $Pb(Ac)_2$ influences the deposition rate of (a) phosphorus and (b) of nickel on brass substrate, as shown in Fig. 12.19. The figure indicates that the P deposition rate slips very slightly in the low-concentration range (0–10 ppm)

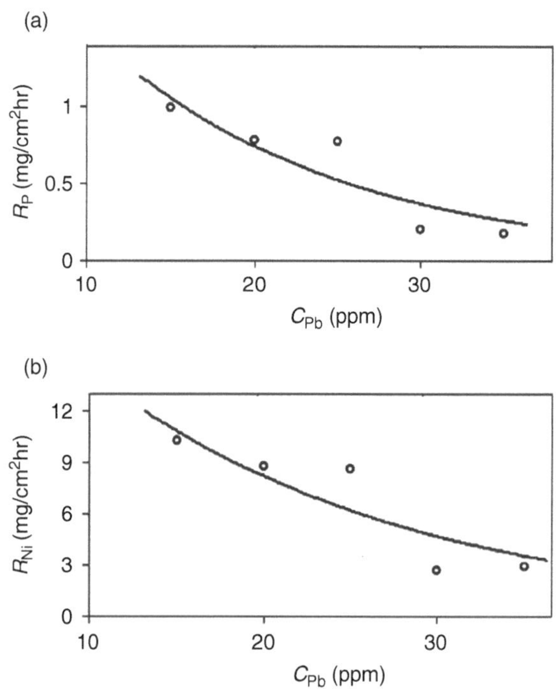

FIGURE 12.19 Influences of the $Pb(Ac)_2$ concentration on the deposition rate of (a) phosphorus and (b) of nickel on brass substrate [43].

but decreases in the higher concentration range. In the low stabilizer concentration range, both the nickel deposition rate (R_{Ni}^0) of 12.23 mg/cm^2 · h and phosphorus deposition rate (R_P^0) of 1.24 mg/cm^2 · h are nearly constant (see the original Figure 7 in [4]). Using the electronic tunneling theory of the quantum mechanics and mathematical models to explain the EN plating mechanism and through a complex and very long derivation process, they finally derive two formulas for the plating rates of Ni and P:

$$R_{Ni} = \frac{R_{Ni}^0}{\sqrt{E_f^0}}\sqrt{E_f^0 + 4\pi pa\left(C_B^{Pb} - C_B^{Pb-C}\right)}$$
$$\exp\left[\left(16\times 4\pi pa - \frac{4\pi k_G pa}{kT}\right)\left(C_B^{Pb} - C_B^{Pb-C}\right)\right] \tag{12.7}$$

$$R_P = R_P^0 \exp\left[-\frac{4\pi k_G pa}{kT}\left(C_B^{Pb} - C_B^{Pb-C}\right)\right] \tag{12.8}$$

where R_{Ni}^0 is the nickel deposition rate without interference from the Pb^{2+} stabilizer. R_P^0 is the phosphorus deposition rate in the absence of Pb^{2+} at the OHP (outer Helmholtz plane) of the desired plating surface. E_f is the Fermi energy, and the subscript or superscript "0" represents under the condition without the codeposited stabilizer Pb. k is the Boltzmann constant, T stands for the absolute temperature. C_B^{Pb} is the Pb^{2+} concentration in the bulk. C_B^{Pb-C} is the critical bulk concentration of lead ion, over which the lead ion concentration influences both the plating rate and phosphorus content at the desired plating surface. k_G is a proportional constant describing the influence of work function on the adsorption tendency of $H_2PO_2^-$. a is a constant signifying the fraction of those Ni surface sites on an individual Ni particle at which the displacement of Ni with Pb^{2+} ions can carry out. p is the dipole moment of adsorbed species. The values of four physical terms in Eqs. (12.7) and (12.8) need to be determined: $\frac{4\pi k_G pa}{kT}$, $4\pi pa$, E_f^0 and C_B^{Pb-C}. After determining these values and substituting the simulation parameters into Eqs. (12.7) and (12.8), these authors obtained the equations of the total plating rate, R, and the phosphorus content of the deposited films, P% as follows:

$$R = 14.33\sqrt{3 + 0.0009C_B^{Pb}}\exp\left(-0.0556C_B^{Pb}\right) + 3.02\exp\left(-0.07C_B^{Pb}\right) \tag{12.9}$$

$$P\% = \frac{100\%}{1 + 4.75\sqrt{3 + 0.0009C_B^{Pb}}\exp\left(0.0144C_B^{Pb}\right)} \tag{12.10}$$

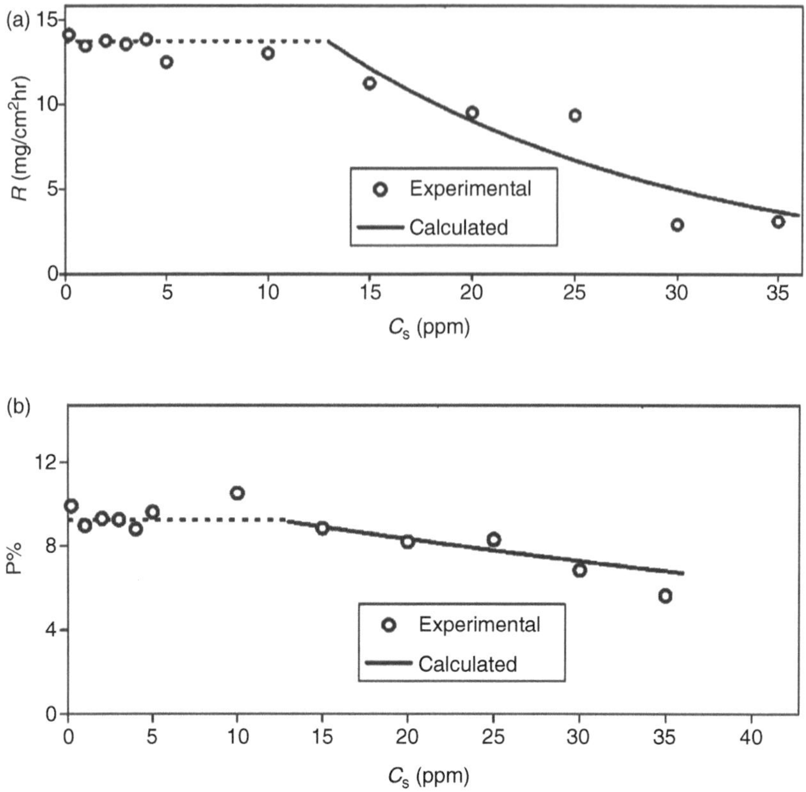

FIGURE 12.20 (a) Change of the total plating rate (R) with the lead ion concentration (C_S). (b) Change of phosphorus content (P%) of the Ni–P films on the lead ion concentration (C_S) [43].

Figure 12.20 compares the experimental data and simulation results. The calculations are in good agreement with the experimental data. Moreover, one can see that both the total plating rate (R) and phosphorus content (P%) of the Ni–P films stay constant till the lead ion concentration (C_S) is around 13 ppm, and then decreases as C_s increases. However, the decrease of R is faster than that of the P content. Through analysis, these authors found that the stabilizing effect of the lead ion is attributed to its participation as the neutral atom in the lattice of the Ni–P film at the plating frontier, which results in the rise of Fermi energy. Thus, the oxidation of hypophosphite at the plating frontier is retarded, which in turn slows down the deposition of both Ni and P as these need electrons. Beyond this, it seems that these authors did not say anything further to explain the facts in Fig. 12.20. In addition, they said nothing about other experimental determination of the plating rate versus concentration of Pb^{2+}. Anyway, these authors derived a representation of the plating rate by using the microscopic theory and it was a good attempt.

12.3.3.2 The Effect of Thiourea (Tu) Stabilizer

In Section 6.1.4.1 Thiourea, the experimental results obtained to date were described in more detail, and some qualitative interpretation was reported. In order to elucidate the mechanism of the effect of thiourea, Homma et al. [44,45] analyzed the influence of Tu on the deposition process using both experimental and DFT theoretical procedures. First, they designed and constructed a bath system in which the anodic and cathodic reactions are physically separated [44]. No Ni ions were added to the anodic reaction bath, and none to the cathodic reaction bath. These two separated baths were connected by a salt bridge. From the experimental determination, it was found that the current flowing between the two separated baths increased when thiourea was added only to the anodic reaction bath. This indicated that the acceleration effect of thiourea was mainly on the anodic reaction of the hypophosphite ions.

Homma's group then used the DFT approach to calculate and analyze the effect of thiourea on the reaction of the hypophosphite ion by applying a coadsorption model consisting of thiourea, hypophosphite, and a Ni metal surface. As mentioned before, there are three steps (adsorption, dehydrogenation, and oxidation with hydroxyl group attack) in the reductant reaction; thereby, the influence of thiourea on each reaction step of the hypophosphite ion for each of these three steps needs to be determined.

In the adsorption step, the calculated adsorption energies of hypophosphite ion with and without Tu are −147.2 and −93.44 kJ/mol, respectively. This indicates that the adsorption of hypophosphite ion should be promoted by the presence of thiourea. Moreover, the difference between these adsorption energies is evaluated to be ~50 kJ/mol. In the second step of dehydrogenation, the calculated energy profiles of dehydrogenation of hypophosphite, both with and without thiourea, are shown in Fig. 12.21a. The results illustrate suppression effects of thiourea on the oxidation of hypophosphite ions on a Ni surface because the energy with Tu is higher than that without Tu. Or we can say that thiourea destabilizes the reaction system by ~20 kJ/mol. Figure 12.21b indicates the

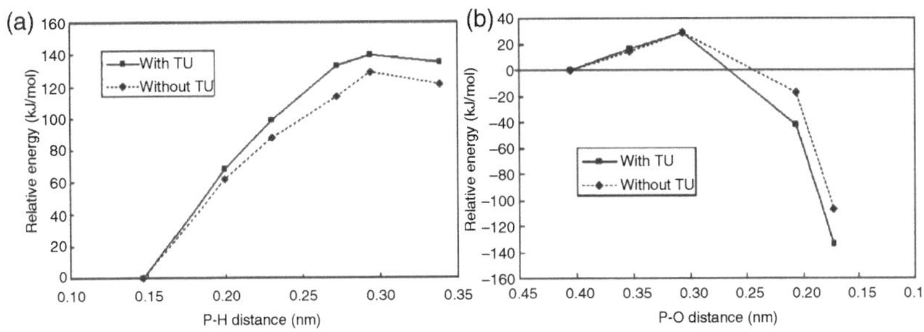

FIGURE 12.21 Energy profiles of dehydrogenation of hypophosphite (a) and of oxidation of hypophosphite attacked by hydroxyl group (b), both with and without thiourea [44].

calculated energy profiles of oxidation of hypophosphite attacked by hydroxyl group, both with and without thiourea. The energy behavior in the oxidation step just balances to that in the dehydrogenation step, that is, thiourea should stabilize the reaction system by ~20 kJ/mol. The energy contributions of the destabilization (step 2) and the stabilization (step 3) by thiourea are both ~20 kJ/mol, which indicates that the destabilization and stabilization effects of thiourea compensate for each other to make these effects insignificant. Throughout these three steps, therefore, the energy difference in the adsorption step, ~50 kJ/mol, is thermodynamically important in discussing the effect of thiourea on the entire hypophosphite oxidation reaction, which is thought to be the origin of the acceleration effect of thiourea. Why does thiourea primarily promote the adsorption step in a hypophosphite reaction? These authors further analyzed the reasons, which originate in two factors. One was the molecular interaction between thiourea and the hypophosphite ion. The other was the restructuring of the electronic states of the Ni surface owing to the adsorption of thiourea, which enhanced the interaction between hypophosphite and its adsorption site on the Ni surface.

In 2013, Homma et al. [45] further investigated the effect of thiourea on the oxidation of hypophosphite ions on the Ni surface by Raman spectroscopy and DFT calculation. They mainly attempted to illustrate in their paper that while thiourea works as an accelerating factor in the electroless Ni deposition in an acidic bath, it suppresses the deposition in an alkaline bath. The acceleration and suppression mechanisms of thiourea in acidic/alkaline electroless deposition baths are not yet fully understood. First, these authors characterized the adsorption of thiourea and the hypophosphite ions on a Ni surface experimentally using surface-enhanced Raman spectroscopy. Then, by comparing the results with those obtained from the theoretical calculation, coadsorption of thiourea and hypophosphite ions on the Ni surface was confirmed with the DFT-calculated results. In addition, they analyzed the acceleration or suppression effects of thiourea on the oxidation of hypophosphite ions on a Ni surface in acidic/alkaline bath from the experimental and computational perspectives. In such a way, they proposed an explanation in terms of its own fundamental characteristics that unifies both the acceleration and suppression mechanisms of thiourea. It should be said that the use of the calculated data to illustrate the problem is not so clear and direct in their paper as in [44] or other papers cited above from Homma et al. themselves. However, these authors promised in the end of their paper that "In our future work, thiourea and hypophosphite ions on various metal surfaces and at various pH values will be characterized from both experimental and theoretical perspectives."

12.4 SUMMARY AND PROSPECTIVE

A very new field, microscopic theory of electroless plating is introduced in this chapter. We will summarize and consider future developments in the rest of this chapter.

Many reports on this subject have been published to date. The main studies were from Homma's group in Japan; other groups were from Singapore, China, and elsewhere. It should be said that there were not many groups participating in the study. In order to understand the nature of the electroless deposition process deeply and to achieve more precise control of that process for fabricating further advanced devices, it is necessary to elucidate the reaction mechanism of the process at molecular and atomic levels. Therefore, extensive research in this field is welcomed.

Some significant results for the EP process have been achieved by using microscopic theory calculations. These achievements include reaction mechanisms of EP processes for various reductants (hypophosphite ions, DMAB, titanium trichloride, aldehydes, and hydrazine); reaction pathways via 5-coordinate intermediates; the catalytic activity of the metal surfaces and its origin; and the role of stabilizer (Pb^{2+}, and Tu). Especially, a representation of the plating rate in EN deposition has been derived.

Too many aspects in the EP process have not been investigated. With the exception of the reductant and two stabilizers, all other impact parameters such as metal salt, complexing agent, accelerating agent surfactants, pH value, plating temperature, plating time, and bath loading have all not been studied with an *ab initio* approach. As for the subjects described in Chapter 11, that is, formation theory and formation range of electroless amorphous alloys, no one studied them at all until now.

A complete theory not only can explain all, at least most of the existing experimental facts, but should also predict new phenomena or help design new experimental schemes or new products. There is still a long way to go for the microscopic theory of the EP process to achieve this objective. Fortunately, the ice was broken, and some achievements have been made. The microscopic theory of the EP process has no doubt got bright prospects.

REFERENCES

[1] Homma T, Nakai H, Onishi M, Osaka T. J Phys Chem B 1999;103:1774.

[2] Pettifor DG, Cottrell AH. Electron theory in alloy design. London: Institute of Materials; 1994.

[3] Zhang B, Hu W, Xu X. Theory of embedded atom method and its application to materials science-atomic scale materials design theory. Changsha: Hunan University Press; 2003.

[4] Frisch MJ, Gordon MH, Trucks GW, Foresman JB, Schlegel HB, Raghavachari K, Robb MA, Binkly JS, Gonzalez C, Defrees DJ, Fox DJ, Whiteside RA, Seeger R, Melius CF, Baker J, Martin RL, Stewart JJP, Topol S, Pople JA. Gaussian 94. Pittsburgh: Gaussian Inc.; 1994.

[5] Frisch MJ, Trucks GW, Schlegel HB, Scuseria GE, Robb MA, Cheeseman JR, Zakrzewski VG, Montgomery JA, Stratmann RE, Burant JC, Dapprich S, Millam JM, Daniels AD, Kudin KN, Strain KN, Farkas O, Tomasi J, Barone V, Cossi M, Cammi R, Mennucci B, Pomelli C, Adamo C, Clifford S, Ochterski J, Petersson GA, Ayala PY, Cui Q, Morokuma K, Malick DK, Rabuck AD, Raghavachari K, Foresman JB, Cioslowski J, Ortiz JV, Stefanov BB, Liu G, Liashenko A, Piskorz P, Komaromi I, Gomperts R, Martin RL, Fox DJ, Keith T, Al-Laham

MA, Peng CY, Nanayakkara A, Gonzalez C, Challacombe M, Gill PMW, Johnson BG, Chen W, Wong MW, Andres JL, Head-Gordon M, Replogle ES, Pople JA. Gaussian 98. Pittsburgh: Gaussian Inc.; 1998.

[6] Frisch MJ, Trucks GW, Schlegel HB, Scuseria GE, Robb MA, Cheeseman JR, Montgomery JA, Vreven T Jr, Kudin KN, Burant JC, Millam JM, Iyengar SS, Tomasi J, Barone V, Mennucci B, Cossi M, Scalmani G, Rega N, Petersson GA, Nakatsuji H, Hada M, Ehara M, Toyota K, Fukuda R, Hasegawa J, Ishida M, Nakajima T, Honda Y, Kitao O, Nakai H, Klene M, Li X, Knox JE, Hratchian HP, Cross JB, Bakken V, Adamo C, Jaramillo J, Gomperts R, Stratmann RE, Yazyev O, Austin AJ, Cammi R, Pomelli C, Ochterski JW, Ayala PY, Morokuma K, Voth GA, Salvador P, Dannenberg JJ, Zakrzewski VG, Dapprich S, Daniels AD, Strain MC, Farkas O, Malick DK, Rabuck AD, Raghavachari K, Foresman JB, Ortiz JV, Cui Q, Baboul AG, Clifford S, Cioslowski J, Stefanov BB, Liu G, Liashenko A, Piskorz P, Komaromi I, Martin RL, Fox DJ, Keith T, Al-Laham MA, Peng CY, Nanayakkara A, Challacombe M, Gill PMW, Johnson B, Chen W, Wong MW, Gonzalez C, Pople JA. Gaussian 03, Revision C.02. Wallingford, CT: Gaussian Inc.; 2004.

[7] Frisch MJ, Trucks GW, Schlegel HB, Scuseria GE, Robb MA, Cheeseman JR, Scalmani G, Barone V, Mennucci B, Petersson GA, Nakatsuji H, Caricato M, Li X, Hratchian HP, Izmaylov AF, Bloino J, Zheng G, Sonnenberg JL, Hada M, Ehara M, Toyota K, Fukuda R, Hasegawa J, Ishida M, Nakajima T, Honda Y, Kitao O, Nakai H, Vreven T, Montgomery, Jr. JA, Peralta JE, Ogliaro F, Bearpark M, Heyd JJ, Brothers E, Kudin KN, Staroverov VN, Kobayashi R, Normand J, Raghavachari K, Rendell A, Burant JC, Iyengar SS, Tomasi J, Cossi M, Rega N, Millam JM, Klene M, Knox JE, Cross JB, Bakken V, Adamo C, Jaramillo J, Gomperts R, Stratmann RE, Yazyev O, Austin AJ, Cammi R, Pomelli C, Ochterski JW, Martin RL, Morokuma K, Zakrzewski VG, Voth GA, Salvador P, Dannenberg JJ, Dapprich S, Daniels AD, Farkas Ö, Foresman JB, Ortiz JV, Cioslowski J, Fox DJ. Wallingford, CT: Gaussian Inc.; 2009.

[8] Becke AD. J Chem Phys 1993;98:5648.

[9] Lee C, Yang W, Parr RG. Phys Rev B 1988;37:785.

[10] Hehre WJ, Random L, Schleyer PVR, Pople JA. *Ab initio* molecular orbital theory. New York: Willey; 1986. p. 1–548.

[11] Homma T, Komatsu I, Tamaki A, Nakai H, Osaka T. Electrochim Acta 2001;47:47.

[12] Hay PJ, Wadt WR. J Chem Phys 1985;82:270.

[13] Wadt WR, Hay PJ. J Chem Phys 1985;82:284.

[14] Hay PJ, Wadt WR. J Chem Phys 1985;82:299.

[15] van den Meerakker JEAM. J Appl Electrochem 1981;11:395.

[16] Foresman JB, Keith TA, Wiberg KB, Snoonian JM, Frisch MJ. J Phys Chem 1996;100:16098.

[17] Bockris JO, Devanathan MAV, Muller K. Proc Roy Soc London Ser A Math Phys Sci 1963;274:55.

[18] Ohno I, Wakabayashi O, Haruyama S. J Electrochem Soc 1985;132:2323.

[19] Nakai H. Chem Phys Lett 2002;363:73.

[20] Nakai H, Sodeyama K. Chem Phys Lett 2002;365:203.

[21] Kawamura Y, Nakai H. Chem Phys Lett 2003;368:673.

[22] Nakai H, Katouda M, Kawamura Y. J Chem Phys 2004;121:4893.

[23] Kohn W, Sham LJ. Phys Rev 1965;140:1133.

[24] Nakai H, Homma T, Komatsu I, Osaka T. J Phys Chem B 2001;105:1701.

[25] Cui G, Liu H, Wu G, Zhao J, Song S, Shen PK. J Phys Chem C 2008;112:4601.

[26] Cui G, Li N, Li D, Chi M. J Electrochem Soc 2005;152(10):C669.

[27] Homma T, Ta maki A, Nakai H, Osaka T. J Electroanal Chem 2003;559:131.

[28] Shimada T, Komatsu I, Homma T, Nakai H, Osaka T. Electrochemistry 2004;72:462.

[29] Shimada T, Sakata K, Homma T, Nakai H, Osaka T. Electrochim Acta 2005;51:906.
[30] Shimada T, Tamaki A, Nakai H, Homma T. Electrochemistry 2007;75:45.
[31] Harrison JA, Khan ZA. J Electroanal Chem 1970;26:1.
[32] Shimada T, Nakai H, Homma T. J Electrochem Soc 2007;154(4):D273.
[33] Zeng Y, Liu S, Ou L, Yi J, Yu S, Wang H, Xiao X. Appl Surf Sci 2006;252:2692.
[34] Abrantes LM, Bewick A, Kalaji M, Oliveira MC. J Chem Soc Faraday Trans 1996;92:4663.
[35] Abrantes LM, Oliveira MC, Correia JP, Bewick A, Kalaji M. J Chem Soc Faraday Trans 1997;93:1119.
[36] Kunimoto M, Shimada T, Odagiri S, Nakai H, Homma T. J Electrochem Soc 2011;158:D585.
[37] Kunimoto M, Nakai H, Homma T. J Electrochem Soc 2011;158:D626.
[38] Kunimoto M, Nakai H, Homma T. Electrochemistry 2012;80(3):126.
[39] Liu Z-P, Hu P. J Am Chem Soc 2003;125:1958.
[40] Ge Q, Neurock M. J Phys Chem B 2006;110:15368.
[41] Huo C-F, Li Y-W, Wang J, Jiao H. J Phys Chem C 2008;112:14108.
[42] Kunimoto M, Otomo A, Takahashi N, Nakai H, Homma T. Electrochim Acta 2013;113:785.
[43] Yin X, Hong L, Chen B-H. J Phys Chem B 2004;108:10919.
[44] Kunimoto M, Endo K, Nakai H, Homma T. Electrochim Acta 2013;100:311.
[45] Jiang Bin, Kunimoto M, Yanagisawa M, Homma T. J Electrochem Soc 2013;160(9):D366.

Index

A

D

E

H

I

K

L

M

N

O

P

Q

R

S

T

U

V

W

X

Z

Kolibri 360 GmbH
Ferdinand-Jühlke-Straße 7
99095 Erfurt, Deutschland
produktsicherheit@kolibri360.de